TRAFFIC AND GRANULAR FLOW '07

C. Appert-Rolland F. Chevoir P. Gondret
S. Lassarre J.-P. Lebacque M. Schreckenberg
Editors

Traffic and Granular Flow '07

With 397 Figures, 197 in Colour, and 29 Tables

Springer

Cécile Appert-Rolland
Université Paris-Sud XI
Laboratoire de Physique Théorique
Batiment 210
91405 Orsay
France
Cecile.Appert-Rolland@th.u-psud.fr

François Chevoir
Université Paris Est
Institut Navier
UMR LCPC-ENPC-CNRS
2 allée Kepler
77420 Champs-sur-Marne
France
chevoir@lcpc.fr

Philippe Gondret
Université Paris-Sud 11
Laboratoire FAST
Batiment 502
91405 Orsay
France
gondret@fast.u-psud.fr

Sylvain Lassarre
GARIG/INRETS
23 rue Alfred Nobel
77420 Champs-sur-Marne
France
sylvain.lassarre@ inrets.fr

Jean-Patrick Lebacque
INRETS
Le Descartes 2 - 2 rue de la Butte Verte
F93166 Noisy-Le-Grand Cedex
France
lebacque@inrets.fr

Michael Schreckenberg
Universität Duisburg-Essen
Physik von Transport und Verkehr
Lotharstr. 1
47048 Duisburg
Germany
schreckenberg@ptt.uni-due.de

ISBN 978-3-540-77073-2 e-ISBN 978-3-540-77074-9

DOI 10.1007/978-3-540-77074-9

Library of Congress Control Number: 2008944305

Mathematics Subject Classification (2000): 90B20, 90B06, 90B15, 90B18, 60K30, 76T25, 65C20, 65C35, 68U20, 92C05, 92D50, 60K20

Cover design: WMXDesign GmbH, Heidelberg

Printed on acid-free paper

9 8 7 6 5 4 3 2 1

springer.com

Preface

The "Traffic and Granular Flow '07" conference was the seventh of a series of international conferences that started in 1995 in Jülich (Germany). Since then, the conference took place in Duisburg (1997), Stuttgart (1999), Nagoya (2001), Delft (2003) and Berlin (2005).

The aim of TGF conferences is to facilitate the exchanges between various fields dealing with transport. When the conference was created, the fields that were represented were road traffic and granular flow – hence the name of the series. Since then, the scope of the conference has been enlarged to include in particular collective motion in biology (molecular motors), a subject which turns out to have many connections with the two original ones.

Transversal themes have emerged. For TGF07, a session was specifically devoted to the subject of networks. An important theme is also the one of self-propelled particles. It ranges from granular flows with anisotropic grains, to collective motion of animals, and to pedestrian traffic.

We were very happy to organize the 2007 occurrence of TGF in Orsay (France), at the University Paris-Sud. The conference was organized mainly by the Laboratory of Theoretical Physics (LPT), with the help of the Laboratory FAST (Fluides, Automatique et Systèmes Thermiques) – these two laboratories are both associated to the CNRS (Centre National pour la Recherche Scientifique) – and of the GARIG Group at INRETS.

With more than 2000 researchers or teaching researchers, University Paris-Sud represents 4% of the french public research. More than 25000 students are studying at this university. It is known for its high scientific level. Recently, the Fields medal (2006) honored Wendelin Werner, professor at the mathematical departement at the Orsay campus. In 2007, the physics Nobel price was given to A. Fert, professor at University Paris-Sud.

Besides, the Orsay campus is located in a scientifically very active area, very near for example Ecole Polytechnique and CEA.

In France, research on road traffic is mainly performed in specialized public research centers, the main two being INRETS (Institut National de Recherche sur les Transports et leur Sécurité) and LCPC (Laboratoire Central des Ponts et Chaussées). Both of them have important research centers in region Ile-de-France. Besides, for the sake of building roads, LCPC also has an active research activity in the granular field.

The TGF07 conference was the opportunity to gather national actors as well as international researchers. It took place from the 20th to 22nd of June 2007. 127 participants from 24 nationalities were present. The 8 plenary talks and the 70 posters allowed exchanges between the various communities. Parallel sessions (48 oral presentations) allowed more specialized and thorough discussions within each field.

We would like to thank our sponsors. We are especially grateful to Region Ile-de-France, first because it was our main sponsor, but also because it was the only one who gave us an answer enough in advance, so that we could decide to go on. Without Region Ile-de-France, this conference would not have taken place. The conference was supported also by the CEA (Commissariat à l'énergie atomique), the ministry of research, the RTRA "triangle de la physique", CNRS, INRETS, DGA, University Paris-Sud, the French embassy in India, and the European Physical Journal.

We would like to thank Olivier Dauchot and Cécile Sykes for their valuable help on scientific and pratical issues, and Henk Hilhorst, the director of the Laboratory of Theoretical Physics, for his support during the preparation of the conference. Odile Heckenauer and Mireille Calvet deserve a special thank-you for all the administrative and organizational work they did before the conference, and for their presence during the whole conference at the welcome desk. Thank-you also to Gérard Hoffeurt, Manuel Ramos, Antoine Seguin and Yann Bertho for their help before and during the conference. We are grateful to Mrs. Dahm-Courths for the great work she did on the proceedings.

Paris, Duisburg
October 2008

Cécile Appert-Rolland
François Chevoir
Birgit Dahm-Courths
Philippe Gondret
Sylvain Lassarre
Jean-Patrick Lebacque
Michael Schreckenberg

Avant-Propos

La conférence internationale "Traffic and Granular Flow '07" (TGF07) était la 7ème d'une série qui a débuté en 1995 à Jülich (Allemagne). Depuis, des conférences TGF ont eu lieu à Duisburg (1997), Stuttgart (1999), Nagoya (2001), Delft (2003) et Berlin (2005).

L'objectif des conférences TGF est de faciliter les interactions entre divers domaines de recherche touchant au transport. Lors de la création de cette série de conférences, les domaines représentés étaient le trafic routier et les écoulements granulaires – d'où le nom de la série. Depuis, les thèmes abordés se sont multipliés, en particulier pour inclure les mouvements collectifs en biologie (moteurs moléculaires), un sujet qui s'est avéré avoir des problématiques communes avec les thèmes d'origine de la conférence.

Des thèmes plus transversaux émergent peu à peu. Lors de TGF07, une session a été spécialement consacrée aux réseaux. Un autre thème important est celui des particules auto-propulsées, qui va des écoulements granulaires à grains anisotropes aux mouvements collectifs d'animaux, et au trafic piéton.

Nous étions très heureux de pouvoir organiser TGF07 à Orsay (France), à l'Université Paris-Sud. La conférence a été essentiellement organisée par le Laboratoire de Physique Théorique (LPT), avec l'aide du Laboratoire FAST (Fluides, Automatique et Systèmes Thermiques) – ces deux laboratoires sont tous deux associés au CNRS (Centre National pour la Recherche Scientifique) – et du groupe GARIG à l'INRETS.

Avec plus de 2000 chercheurs ou enseignants-chercheurs, l'Université Paris-Sud représente 4% de la recherche publique française. Plus de 25000 étudiants y étudient. L'Université Paris-Sud est reconnue pour son haut niveau scientifique. Récemment, la médaille Fields (2006) a été remise à Wendelin Werner, professeur au Département de Mathématiques du campus d'Orsay. En 2007, le prix Nobel de physique a été décerné à A. Fert, professeur à l'Université Paris-Sud.

Le campus d'Orsay est de plus situé dans un environnement scientifique très actif, proche par exemple de l' Ecole Polytechnique et du CEA.

En France, la recherche sur le trafic routier est principalement effectuée dans des centres de recherche publique spécialisés, les deux principaux étant l'INRETS (Institut National de Recherche sur les Transports et leur Sécurité) et le LCPC (Laboratoire Central des Ponts et Chaussées). Tous les deux ont d'importants centres de recherche en région Ile-de-France. De plus, le LCPC a aussi une grosse activité de recherche sur les milieux granulaires, liée à la fabrication et à l'entretien des infrastructures (routes...).

La conférence TGF07 a été l'occasion de réunir acteurs nationaux et chercheurs étrangers. Elle a eu lieu du 20 au 22 Juin 2007. 127 participants de 24 nationalités différentes étaient présents. Les 8 sessions plénières et les 70 posters ont permis de nombreux échanges entre les diverses communautés, tandis que les sessions parallèles (communications orales) ont été le lieu de discussions plus spécialisées et plus approfondies au sein de chaque sujet.

Nous souhaitons maintenant remercier nos sponsors. Nous sommes tout spécialement reconnaissants envers la région Ile-de-France, d'une part parce que c'était notre sponsor principal, mais aussi parce que c'était le seul qui nous ait donné une réponse suffisamment à l'avance pour que nous puissions décider de nous lancer dans l'organisation de cette conférence. Sans la région Ile-de-France, cette conférence n'aurait pas eu lieu. La conférence TGF07 a aussi été soutenue financièrement par le CEA (Commissariat à l'énergie atomique), le ministère de la recherche, le RTRA "triangle de la physique", le CNRS, l'INRETS, la DGA, l'Université Paris-Sud, l'ambassade de France en Inde, et le European Physical Journal.

Nous aimerions remercier ici Olivier Dauchot et Cécile Sykes pour leur aide précieuse tant sur le plan scientifique que pratique, et Henk Hilhorst, directeur du Laboratoire de Physique Théorique, pour son soutien tout au long de la préparation de la conférence. Nous devons des remerciements tout particuliers à Odile Heckenauer et Mireille Calvet pour tout leur travail administratif et d'organisation avant la conférence, et pour leur présence pendant toute la conférence à la table d'enregistrement. Merci aussi à Gérard Hoffeurt, Manuel Ramos, Antoine Seguin et Yann Bertho pour leur aide avant et pendant la conférence. Nous sommes reconnaissants envers Mme Dahm-Courths pour le travail effectué pour les Actes de la conférence.

Paris, Duisburg
Octobre 2008

Cécile Appert-Rolland
François Chevoir
Birgit Dahm-Courths
Philippe Gondret
Sylvain Lassarre
Jean-Patrick Lebacque
Michael Schreckenberg

Contents

Part I Traffic

Modelling of Traffic Flow from an Engineer's Perspective
Partha Chakroborty and Akhilesh Kumar Maurya 3

Cellular Automaton Approach to Highway Traffic: What do we Know?
Andreas Schadschneider 19

Replications in Stochastic Traffic Flow Models: Incremental Method to Determine Sufficient Number of Runs
Nicolas Chiabaut and Christine Buisson 35

Capacity and Capacity Drop of a Revolving Door
Winnie Daamen, Serge P. Hoogendoorn, and Henk van Wijngaarden ... 45

Risk Index Modeling for Real-Time Motorway Traffic Crash Prediction
Habib Haj-Salem and Jean-Patrick Lebacque 55

Colliding Particles: Beyond Accident-Free Car Following Models
Samer H. Hamdar and Hani S. Mahmassani 65

Dynamic First-Order Modeling of Phase-Transition Probabilities
Serge P. Hoogendoorn, Hans van Lint, and Victor Knoop 85

Traffic Dynamics on a Rotary
Ding-wei Huang 93

Effect of Adaptive Cruise Control Vehicles on Phase Transition in a Mixture with Manual Vehicles
Rui Jiang, Mao-Bin Hu, Bin Jia, Ruili Wang, and Qing-Song Wu 105

Calibration of Car-Following Models Using Floating Car Data
Arne Kesting and Martin Treiber 117

A Queueing Theory Approach for a Multi-Speed Exclusion Process
Cyril Furtlehner and Jean-Marc Lasgouttes 129

A Stochastic Macroscopic Traffic Model Devoid of Diffusion
Megan M. Khoshyaran and Jean-Patrick Lebacque 139

A Multiclass Car-Following Rule Based on the LWR Model
Ludovic Leclercq and Jorge A. Laval 151

A Cross Entropy Based Multi-Agent Approach to Traffic Assignment Problems
Tai-Yu Ma and Jean-Patrick Lebacque 161

Handling of Contacts in Crowd Motion Simulations
Bertrand Maury and Juliette Venel 171

Optimal Velocity Model Via Considering Multi-Velocity Difference Ahead
Mo Yeliu, Yu Xue, Shi Wei, and Liu Zhang 181

Empirical Data for Pedestrian Flow Through Bottlenecks
Armin Seyfried, Bernhard Steffen, Andreas Winkens, Tobias Rupprecht, Maik Boltes, and Wolfram Klingsch 189

Modeling Synchronized Flow at Highway Bottlenecks
Florian Siebel, Wolfram Mauser, Salissou Moutari, and Michel Rascle .. 201

Modeling Lane-Changing Decisions with MOBIL
Martin Treiber and Arne Kesting 211

Kinetic Derivation for a Traffic Flow Model
Alma R. Méndez and Rosa M. Velasco 223

Qualitative Change of Car-Following Behavior Observed in Real Traffic
Yasushi Yokoya, Yoichi Asano, and Nobuyuki Uchida 233

Phase Transition Induced by Homogeneous Initial Distribution of Cars
Xue Yu, Yanfang Wei, and Shi Wei 245

Similarity Between Temporal and Spatial Structures in Pattern Formation of Dissipative Non-Equilibrium System
Rintarou Akiyama, Hiroshi Watanabe, and Yūki Sugiyama 253

Pattern Formation in Traffic Microscopic Model
Tomas Apeltauer, Petr Holcner, and Jiri Macur 259

Stability of Flow on a Ring with Three Links
Alexander P. Buslaev, Alexander G. Tatashev, and Marina V. Yashina . . 265

Improving Pedestrian Micro-Simulations with Event Steps
Mario C. Campanella, Serge P. Hoogendoorn, and Winnie Daamen 273

Road Traffic Models Using Petri Nets and Minplus Algebra
Nadir Farhi, Maurice Goursat, and Jean-Pierre Quadrat 281

Vehicular Traffic Flow at a Non-Signalised Intersection
M. Ebrahim Fouladvand and Somayyeh Belbasi 287

Traffic Anticipation Effect in the Lattice Hydrodynamic Model
Hong-xia Ge 293

Is the Nearest Neighbor Distribution Enough to Describe the Statistical Behavior of a Domain System?
Diego Luis González and Gabriel Téllez 301

Statistical Properties of Disordered Driven Lattice Gases with Open Boundaries
Philip Greulich and Andreas Schadschneider 307

Control of Traffic Congestion in the Modified Coupled Map Car-Following Model Based on Intelligent Transportation System
Han Xianglin, Ge Hongxia, Jiang Changyuan, Li Xingli, and Dai Shiqiang 315

Generic Driving Behavior Modeling by Differential Game Theory
Serge P. Hoogendoorn and Piet Bovy 321

A Model for City Traffic
Ding-wei Huang and Wei-neng Huang 333

Traffic Behaviors of Mixed Bicycle System in the Multi-Value Cellular Automata Model
Bin Jia, Xin-Gang Li, Rui Jiang, and Zi-You Gao 339

Empirical Differences Between Time Mean Speed and Space Mean Speed
Victor Knoop, Serge P. Hoogendoorn, and Henk van Zuylen 351

Anisotropy and Stabilization of Traffic Flow in Extended Cooperative Driving Lattice Hydrodynamic Models Based on Backward-Looking Effect
Li Xingli, Li Zhipeng, Han Xianglin, and Dai Shiqiang 357

Complex Fundamental Diagram of Traffic Flow in the Deep Lefortovo Tunnel (Moscow)
Ihor Lubashevsky, Cyril Garnisov, and Boris Lifshits 365

Three-Frame Algorithm of Car Path Reconstruction from Airborne Traffic Data
Ihor Lubashevsky, Namik Gusein-zade, Dmitry Klochkov, and Sergey Zuev 373

Vehicular Motion and Traffic Breakdown: Evaluation of Energy Balance
Christof Liebe, Reinhard Mahnke, Jevgenijs Kaupužs, and Hans Weber . 381

Detailed Data of Traffic Jam Experiment
Akihiro Nakayama, Minoru Fukui, Katsuya Hasebe, Macoto Kikuchi, Katsuhiro Nishinari, Yuki Sugiyama, Shin-ichi Tadaki, and Satoshi Yukawa 389

Prediction Accuracy of Evacuation Times for High-Rise Buildings and Simple Geometries by Using Different Software-Tools
Christian Rogsch, Wolfram Klingsch, Armin Seyfried, and Henning Weigel 395

Statistical Approach to Traffic Flow
Anton Šurda 401

A Public Conveyance Model and Analysis on Clustering of Vehicles
Akiyasu Tomoeda, Katsuhiro Nishinari, Debashish Chowdhury, and Andreas Schadschneider 407

Modelling and Simulating Several Time-Delay Mechanisms in Human and Automated Driving
Martin Treiber and Arne Kesting 413

Multi-Class First Order Traffic Flow Modeling
Hans van Lint, Serge P. Hoogendoorn, and Marco Schreuder 421

Dynamics and Thermodynamics of Traffic Flow
Hans Weber, Reinhard Mahnke, Christof Liebe, and Jevgenijs Kaupužs . 427

Phase Transitions and Even/Odd Effects in Asymmetric Exclusion Models
Marko Woelki and Michael Schreckenberg 435

Simulation and Mean Field Analysis on Pedestrian Dynamics Around an Exit
Daichi Yanagisawa and Katsuhiro Nishinari 441

A Density-Dependent NaSch Model for Traffic Flow Controlled by a Traffic Light
Hui-bing Zhu, Hong-xia Ge, and Shi-qiang Dai 447

Part II Granular Matter

Large-Scale Collective Behavior of Swimming Microorganisms at High Concentrations
Andrey Sokolov, Igor S. Aranson, John O. Kessler, and Raymond E. Goldstein 455

Flow and Jamming of Granular Matter Through an Orifice
Angel Garcimartín, Cristian Mankoc, Alvaro Janda, Roberto Arévalo, J. Martín Pastor, Iker Zuriguel, and Diego Maza 471

A Fluctuating Energy Model for Dense Granular Flows
Riccardo Artoni, Andrea Santomaso, and Paolo Canu 487

Influence of Particle Shape on Shear Stress in Granular Media
Emilien Azéma, Farhang Radjaï, Robert Peyroux, and Gilles Saussine . . 497

Motion of an Intruder Through a Two Dimensional Disordered Granular Medium
Pierre Cixous, Evelyne Kolb, Jean-Claude Charmet, and Chay Goldenberg 507

Silo Collapse: An Experimental Study
Gustavo Gutiérrez, Philippe Boltenhagen, José Lanuza, and Eric Clément 517

Rough Cylindrical Object Immersed in a Granular Stream of Hard Disks
Payman Jalali and Pertti Sarkomaa 525

Stability and Jamming Transition in Hard Granular Materials: Algebraic Graph Theory
Nicolas Rivier ... 535

Vertical Granular Transport in a Vibrated U-Tube
Ivan Sánchez, José Ramón Darias, Ricardo Paredes, Christopher J. Lobb, and Gustavo Gutiérrez ... 545

Dynamics of Dissipative System with Asymmetric Interaction and N-Body Problem for the Emergence of Moving Cluster
Yūki Sugiyama, Katsutoshi Masuoka, and Takahiro Ishida ... 555

Leidenfrost Granular Flows
Nicolas Taberlet, Patrick Richard, and Renaud Delannay ... 565

Granular Rayleigh-Taylor Instability
Jan Ludvig Vinningland, Øistein Johnsen, Eirik G. Flekkøy, Renaud Toussaint, and Knut Jørgen Måløy ... 577

Morphology of Polydisperse Granular Media
Charles Voivret, Jean-Yves Delenne, Moulay Saïd El Youssoufi, and Farhang Radjaï ... 587

Kinetic Energy Fluctuations and Diffusivity in a 2D Vibrated Granular Packing
Rim Harich, Geoffroy Lumay, and Eric Clément ... 597

Modelling and Simulation of a Maze-Forming Process in Granular-Fluid Systems
Henning Arendt Knudsen, Bjørnar Sandnes, Eirik Grude Flekkøy, and Knut Jørgen Måløy ... 603

Description of Wire-Reinforced Geomaterials by Numerical Discrete Experiments
Romain Laniel, Pierre Alart, and Stéphane Pagano ... 609

Granular Flow Under the Action of Centrifugal Force: A Critical Dynamic Friction Coefficient
Aurélie Le Quiniou, François Rioual, and Yuri Lapusta ... 615

Interfacial Instability of a Confined Suspension Under Oscillating Shear
Maniya Maleki, Héctor Pacheco, Carlos Ruiz Suárez, and Eric Clément . 621

Pattern Formation in Slowly Drained Granular-Fluid Systems
Bjørnar Sandnes, Henning Arendt Knudsen, Knut Jørgen Måløy, and Eirik Grude Flekkøy ... 629

Emerging Stripe Patterns in Drying Suspension Droplets
Bjørnar Sandnes and David Molenaar 635

Behaviour of Railway Ballast Under Cycling Loading
Gilles Saussine and Pierre Etienne Gautier 641

Penetration of a Projectile by Impact into a Granular Medium
Antoine Seguin, Yann Bertho, and Philippe Gondret 647

Rheological Transition in Granular Media
Zahra Shojaaee, Lothar Brendel, and Dietrich E. Wolf 653

Stress Transmission in a Multi-Phase Granular Packing
Vincent Topin, Jean-Yves Delenne, Farhang Radjaï, and Frédéric Mabille 659

Part III Molecular Motors and Motion in Biology

Non-Equilibrium Collective Transport on Molecular Highways
Andrea Parmeggiani 667

Traffic of Molecular Motors: From Theory to Experiments
Paolo Pierobon 679

Numerical Investigations on Coupling of Asymmetric Exclusion Process with Zero Range Process
Rui Jiang, Bin Jia, Mao-Bin Hu, Ruili Wang, and Qing-Song Wu 689

Traffic by Small Teams of Molecular Motors
Melanie J.I. Müller, Janina Beeg, Rumiana Dimova, Stefan Klumpp, and Reinhard Lipowsky 695

Traffic Flow on Ant Trails: Empirical Results vs. Theoretical Predictions
Alexander John, Andreas Schadschneider, Katsuhiro Nishinari, and Debashish Chowdhury 701

Part IV Networks

Modeling Metropolis Public Transport
Christian von Ferber, Taras Holovatch, Yurij Holovatch, and Vasyl Palchykov 709

Attack Vulnerability of Public Transport Networks
Christian von Ferber, Taras Holovatch, and Yurij Holovatch 721

Traffic Dynamics Based on Local Routing Strategy in a Weighted Scale-Free Network
Mao-Bin Hu, Yong-Hong Wu, Rui Jiang, Qing-Song Wu, and Wen-Xu Wang 733

Performance Evaluation of VANET Under Realistic Vehicular Traffic Assumption
Mabiala Muriel, Anthony Busson, and Véronique Véque 739

Packet Flow and its Temporal Properties in the Internet
Shin-ichi Tadaki 745

List of Contributors 751

Part I

Traffic

Modelling of Traffic Flow from an Engineer's Perspective

Partha Chakroborty and Akhilesh Kumar Maurya

Department of Civil Engineering, Indian Institute of Technology, Kanpur, Kanpur 208016, India partha@iitk.ac.in, akmaurya@iitk.ac.in

1 Introduction

The purpose of this short paper is threefold. First, it highlights the requirements of a traffic engineer from a model of traffic flow; that is, it discusses what properties of a flow model will make it attractive and useful to an engineer. Second, the paper points out the basic features of the existing models of traffic flow. Third, the paper presents some of the work done by the authors to develop models which attempt to meet some of the needs of traffic engineers.

The paper is divided into five sections of which this is the first. The second, third and fourth sections are devoted to the three points mentioned in the previous paragraph. The last section summarizes this paper.

2 An Engineer's Requirements from Traffic Flow Models

A traffic engineer is entrusted with the duty of designing and efficiently operating facilities which aid in the mobility of goods and vehicles. This task is the most challenging when it comes to road traffic for a variety of reasons: (i) there are a large number of vehicles on roads, (ii) these vehicles belong to different classes with widely varying operating characteristics, and (iii) each vehicle is driven by a human driver whose nature and characteristics vary. In physical terms, traffic engineers dealing with road traffic essentially deal with a system which has a large number of particles of various shapes and sizes, whose response to different situations are different and are "motivated." In addition to introducing the issue of "motive," the involvement of human drivers in the system cause roadway and traffic features to have psychological (in addition to physical) impact on the stream behaviour.

Roadway features include various aspects of the road like road width, shoulder width, grade (or slope), curves — their frequency and curvature, surface conditions, etc. Traffic features include, the vehicle mix in the stream, the driver mix in the stream, traffic rules (like speed limit) and other control measures (like signals, signs, etc.). Based on how these different aspects are present in a situation (or facility), a traffic engineer often divides the flow into two broad categories: uninterrupted flow and interrupted flow. The former relates to flow seen on freeways and expressways while the latter relates to flow seen on arterials and other urban streets. In the former type, the traffic stream primarily gets impacted by the roadway features and some traffic rules but unlike in the interrupted stream case does not go through intersections with conflicting movements. For instance, the Highway Capacity Manual defines uninterrupted flow as flow (or stream) that "results from the interactions among vehicles in the stream and between vehicles and the geometric and environmental characteristics of the roadway;" while interrupted flow is the flow on facilities which have elements (like traffic signal, stop signs) which stop traffic periodically irrespective of the flow or traffic that exists [1].

In this paper the discussion concentrates on uninterrupted traffic flow and what are required of models of such flows. The reason for concentrating on uninterrupted traffic flow is that a large part of a nation's road network (for example the expressway system) as well as some sections of arterials (with large inter-intersection separation) and rural roads carry traffic which are largely uninterrupted. To begin the discussion, first, the way in which an engineer views a transport facility catering to uninterrupted traffic flow is described. For an engineer, a facility catering to uninterrupted flow is a system characterized by a set of parameters which the engineer can control; for example, the lane width and the number of lanes, the shoulder width, the grades on the road, the types and frequency of horizontal curves on the road, etc. What is of paramount importance to the engineer is to know what impact changes in these parameters will have on the ability of the facility to provide efficient transportation. More specifically, one is interested in knowing, for a given combination of the various design parameters, what will be the average speed, density of the stream at various flow levels (or loosely speaking, demand levels), what is the maximum value of flow (or capacity) that can be handled by the facility, what kind of level of service the road will provide to its users, how sensitive will travel time on the road be to minor changes in demand levels, and many more such questions.

Stated differently, the speed-flow, (or flow-density or speed-density) relation of a road and how this relation depends on the various parameters outlined above is of interest to a traffic engineer. These relations, in a way, give the engineer a sense of how a design will respond to various demand levels. Thus the engineer needs a model of traffic flow (or uninterrupted traffic flow) which will be able to provide, at the very basic, the relationship between speed and flow for heterogeneous streams (both with respect to vehicles and drivers) going through roads of varying characteristics (geometry). In addition to the points mentioned here, an engineer would also expect the uninterrupted traffic flow models to (i) predict the impact of incidents on the traffic flow, (ii) understand the perturbations that occur when streams merge or diverge (like at an on or off ramp); and (iii) incorporate lane changing whenever lane discipline is present. It must be mentioned that there exists streams where lane discipline is absent; for example traffic on Indian expressways hardly ever move along lanes. One needs to model traffic for such roads also for the same reasons as above.

Enhancing the need for such models is the fact that a traffic engineer does not have a laboratory where he/she can perform controlled experiments or what-if studies; for example, one will not be able to "see" what might happen if an extra curve is introduced on the road or if an obstacle is placed on the road. This increases a traffic engineer's reliance on models of traffic flow which can simulate traffic realistically under a variety of conditions.

3 Uninterrupted Traffic Flow and its Models

In this section, uninterrupted flow and certain parameters which help quantify the flow are discussed. Next, the existing classes of models for such uninterrupted flow are briefly described. This description is excerpted from a previous paper by the author [2].

3.1 Uninterrupted Traffic Streams

In the previous section a small description was provided on what is meant by uninterrupted flow and why it is important to study such flow. In this section uninterrupted flow and its characteristics are looked at in slightly greater detail.

An uninterrupted traffic stream is an outcome of the responses of individual drivers to the immediate driving scenario. The responses of drivers are in terms steering and speed control. These controls are achieved through a sequence of steering angle and acceleration rate choices. Three different interactions cause a particular flow to occur; these are driver-vehicle interactions, vehicle-vehicle interactions, vehicle-road interactions. These are briefly described in the following paragraphs.

A vehicle has certain operating characteristics and these characteristics impact how a driver interacts with the vehicle he/she is driving which in turn shows up in the behaviour of a driver-vehicle pair. For example, drivers driving two-wheelers often react differently to a situation than say a driver driving an eighteen-wheeler because the former has greater acceleration capabilities, respond faster to driver actions, and has higher maneuverability. This variation has implications on stream behaviour when a stream has a mix of vehicles with widely varying operating characteristics.

When lane discipline is present, vehicles typically interact with the vehicle ahead in the same lane while driving. Observations tend to suggest that the following vehicle (FV) reacts to the relative speed and distance headway between itself and the leading vehicle. The reaction is often impacted by the speed at which the FV is travelling. This phenomenon is called car-following and has been studied extensively. The next section briefly describes some of the existing models of car-following. Vehicles also interact with other vehicles in the vicinity when passing or changing lanes. In cultures where lane discipline is weak the vehicle-vehicle interaction is far more complex than when lane discipline is present. In such situations vehicles interact with a host of vehicles in its vicinity and a clear lead vehicle does not exist. That is, there is both longitudinal and lateral interaction between vehicles. Recent work (see Gunay [3]) also suggests that even when lane discipline is present vehicles other than the one directly ahead have an impact on how a vehicle moves. From a theoretical perspective, the two situations are qualitatively different; when lateral interactions are absent the flow is an outcome of unidimensional (or linear) interaction between vehicles, whereas when lateral interactions are present the flow is an outcome of two-dimensional interaction between vehicles. The next section presents some models which attempt to incorporate the two-dimensional interaction between vehicles.

Static features of the road like road (or lane) width, horizontal curves, surface deformities, etc. also impact the vehicle (driver) behaviour. Observations suggest that the nature of the speed-flow relations (or alternatively speed-density relations) change with various geometric features of the road [1]. The impact of surface deformities (like large potholes) or obstacles (like an out-of-order vehicle) can cause large disruptions to traffic flow. For example, as per HCM 2000 [4] estimates, if an obstacle causes a lane blockage on a three lane expressway, then capacity of the road falls by 50% (notice that it is much greater than 33%). The interesting aspect of a vehicle-road interaction is that often (or for large parts) the interaction is due to lateral "forces" since there is generally no binding on ones forward motion (except when obstacles occupy the carriageway); for example drivers drive more carefully (i.e., maintain larger distances at lower speeds than they would normally do) on narrow roads because they feel constrained by the closeness of the road edges from the lateral directions. The reason for bringing this up here is to highlight the fact that if one attempts to develop models which incorporate vehicle-road interactions then such models must be able to handle lateral interactions.

Uninterrupted flow has certain defining characteristics; some are microscopic in nature while others are macroscopic. In the following some of the important microscopic and macroscopic characteristics of uninterrupted traffic streams are mentioned and the reader is directed to certain sources which present these characteristics in greater detail.

Microscopic Features

1. Drivers in uninterrupted streams at reasonable flow levels, for the most parts, follow the vehicle ahead and is impacted by the actions of that vehicle. The behaviour exhibited by drivers in such situations is referred to as car-following behaviour and has been studies extensively. Car-following behaviour has certain characteristics properties like, local and asymptotic

stability, closing-in and shying-away, etc. One may refer to May [5] or Chakroborty and Kikuchi [6] for a detailed description of the properties of car-following behaviour.

2. Acceleration noise is the root mean square deviation of acceleration of a vehicle over a period of time (see Herman et al. [7], Jones and Potts [8]) and is a parameter which can characterize the behaviour of a driver in a given driving environment. Winzer [9] did a detailed study on acceleration noise observed in real traffic streams at various flow conditions. Among other things, he found that average acceleration noise of vehicles tend to rise with density before reducing again; typically acceleration noise is below 0.6 m/s^2, etc.
3. The time headway distributions of uninterrupted streams have certain distinguishing features. At low flows, the distribution closely resembles a shifted exponential distribution (i.e., the vehicle arrivals are Poisson like); as flow increases the distribution gets right skewed and can be described reasonably well through Gamma distributions. The mode of the distribution is generally less than the median which is less than the mean; typically the mean is around the 67-percentile value. Further, the ratio of standard deviation of time headway to its mean generally approaches 1 from below as flow increases. The interested reader may refer to May [5] for more details on time headway distributions.
4. Speed distributions are typically symmetric with the mean value reducing with flow in the free flow region. Another interesting feature of such streams is that average speed varies transversely across the stream; it is highest on the median lane (left most lane on roads with keep-right policy) and lowest on the shoulder lane. One may refer to May [5] or Kang and Chang [10] for more details.

Macroscopic Features

1. Relationships exist between the three basic parameters of traffic streams, namely, flow, speed and density (or occupancy). Given the fundamental relation of traffic flow (see Equation 1 provided later), issues of data collection, and the engineering requirements, often speed-flow relationships are the only relations that are looked at. A large amount of empirical work has been done to study these relations (for example, see Hall [11], Hall et al. [12], Banks [13]) and the basic points that emerge are: (i) there are three distinct regimes, namely, uncongested, queue discharge, and congested regimes, (ii) the general shape can be approximated as that shown in Figure 1, (iii) the speed remains more or less constant till about 75% of the maximum flow, (iv) there seems to be a drop in the maximum flow value along the different arms of the relationship (see Figure 1); this value, though difficult to observe, has been estimated to be around 3 to 5% of the pre-congestion maximum flow (see Banks [14, 15] and Agyemang-Duah and Hall [16]), (v) the location of the arm corresponding to the

congested regime is often very difficult to fix and some even doubt the credibility of having such a precise relation in this region (see Ross [17]). Before leaving this point, it may be mentioned that, often streams do not operate at "equilibrium" and hence the observations on speed and flow at a given time may not indicate the equilibrium state of the system.

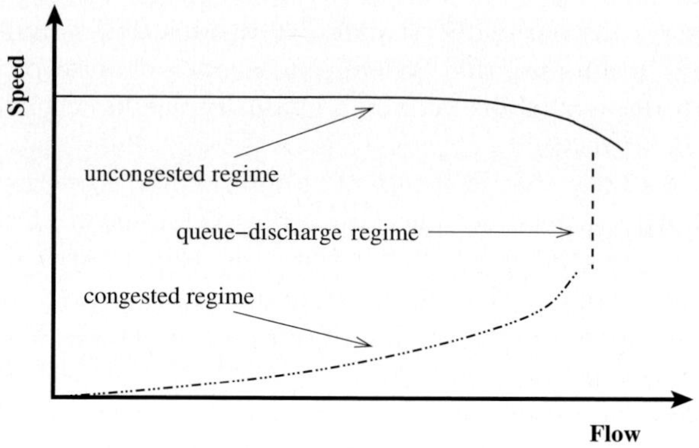

Fig. 1. Schematic showing the understanding of the shape of speed-flow relationships.

2. Geometry affects the dynamics of a traffic stream because of the impact it has on drivers. The shape and nature of the relationships between stream parameters change with changing geometry. Geometric features like lane (or road) width, lateral clearance (or shoulder width), grades, curves, etc. affect the speed-flow (and all other) relations. One may refer to HCM [1], or any other text book on transportation for more discussion on this.
3. Vehicles of different types have different impact on the stream behaviour due to differences in operating characteristics. Often these impacts are large; for example, the maximum flow rate (in vehicles per hour) on a plain terrain road may be reduced by a third if all the vehicles are trucks and not passenger cars. As can be imagined, different types of drivers have different impacts on the stream behaviour also. The reader may refer to HCM [1] for an idea as to how these impacts are taken into account in engineering studies with traffic streams.

In this section, some microscopic and macroscopic characteristics of an uninterrupted traffic stream are highlighted. It is felt that realistic models of traffic flow should strive to achieve these properties. A microscopic model of traffic flow — that is, one which models traffic flow by modelling individual vehicle motion, must show the microscopic as well as macroscopic properties of traffic streams (some of which are highlighted here).

3.2 Existing Models

The aim of this section is to look at some of the models of traffic flow, in general, and to point out, from an engineer's standpoint, those areas which need improvement. Consequently, this section is not to be treated as an exhaustive review of the various models developed in traffic flow. The models are presented under two classes, namely macroscopic models and microscopic models. The macroscopic models describe traffic flow in terms of average stream parameters while the microscopic models try to capture stream behaviour by describing individual vehicle motion. The microscopic models, at least in theory, can be used to study macroscopic properties of traffic streams.

Macroscopic Models

Macroscopic models of uninterrupted traffic streams typically include the fundamental relation (see Equation 1), the continuity equation (see Equation 2) and some relationship reflecting driver behaviour. Over the years, different models have been proposed for this relationship. In a latter paragraph brief descriptions of these models are provided. First, the fundamental and continuity equations are presented; in the equations, q is the flow, k is the density, u is the speed of the stream, x is the distance and t is time.

$$q = uk \tag{1}$$

$$\frac{\partial q}{\partial x} + \frac{\partial k}{\partial t} = 0 \tag{2}$$

Driver behaviour is often defined through a relation between u and k. Although, other pair wise relations (like $u - q$, $q - k$ relations) are often reported and studied, it is felt that, while describing driver behaviour, $u - k$ relations are the most fundamental as they are a direct outcome of the driving process (note that it is difficult to imagine that individual drivers have any notion of q while driving).

Over the years various models on u-k relations have been suggested. In the thirties, Greenshields [18] proposed a linear relation; later Greenberg [19] proposed a logarithmic relation based on fluid flow analogies of traffic stream movement. One of the most general descriptions of the u-k relation is the generalized polynomial model derived from microscopic models of driver behaviour (see May [5] for more details). This model is given in Equation 3; where, u_f and k_j are free flow speed and jam density, respectively and m and ℓ are calibration constants.

$$u^{1-m} = u_f^{1-m} \left[1 - \frac{k}{k_j} \right]^{\ell - 1} \tag{3}$$

Many researchers have raised objections to the use of a single function to describe the u-k relation over the entire density range (the so-called single regime models) on the grounds that humans do not behave according to the same rules over the entire range of density values. There is some merit to these objections. Many multi-regime models were proposed with Edie's [20] model being one of the first. Yet other researchers argued that it is better to look at the relationship among all the three parameters at once (for example see Navin [21], and Persaud and Hall [22]).

Unfortunately, there has been very little work on developing a model which describes the impact of geometry and other road features on driver behaviour. The Highway Capacity Manual [1], however, makes an attempt to relate geometric and other conditions to the speed-flow relation through a procedure which uses various empirically derived look-up tables to determine what the free flow speed will be in a given situation and then uses this free speed to choose an appropriate speed-flow curve from a template. In spite of such attempts, the fact remains that one does not have a model which given the geometric conditions will be able to determine the nature of the driver behaviour.

It must be pointed out here, at the cost of being repetitive, that a lot of research is going on in trying to replicate the observed macroscopic relations from simple driving rules; however, little or no research is currently on to relate roadway features and traffic features (like vehicle mix, driver mix, etc.) to the flow behaviour. Hence, reliance of traffic engineers on ad hoc procedures, to relate these aspects, continues. Despite advances in computation abilities and theoretical insight this reliance has not changed in the last half-a-century. Surely, this needs to change and the author feels researchers must now channelize their energy to evolve models which will reduce such reliance.

Microscopic Models

As opposed to the macroscopic models, microscopic models attempt to define the behaviour of a traffic stream by describing the behaviour of individual drivers in different driving situations. In general, drivers have two basic tasks, (i) controlling the vehicle's position along the direction of motion, and (ii) controlling the vehicle's position along the width of the road or lane. The first task is referred to as longitudinal control and is achieved by controlling the vehicle's speed (through acceleration / deceleration). The second task of lateral control is achieved through proper choice of steering angles. In reality both these activities are inter-dependent and goes on concurrently.

However, in order to simplify the understanding of driving behaviour, often it is assumed that the primary task of a driver is the longitudinal control of the vehicle. This assumption is largely true where the road characteristics are reasonably same for long distances, vehicles have well demarcated travel paths (like lanes) and vehicles do not generally cross these demarcations; even when they do it is a discrete event (like lane changing). Under these assumptions, the vehicle is assumed to be only under the influence of vehicles traveling in the same path (or lane); that is, only longitudinal interactions are taken into account. In the following, some of the properties of longitudinal control behaviour and its models are described.

The driver's behaviour in situations where the driver primarily performs longitudinal control can be broadly divided into three regimes: (i) free flow behaviour, (ii) car-following behaviour, and (iii) stop-and-go behaviour.

In free flow behaviour the driver is not encumbered by other drivers. The driver can choose his speed and maintain it purely at will. No models, except ones which can determine the choice of speeds (or free speeds) of vehicles given the road conditions are necessary. These are not within the purview of longitudinal control models, rather such choice of speeds are affected by the lateral impact of roads edges and other such static obstacles on the driver's mind. Further, free flow behaviour occurs when density is very low and the average distance headway between vehicles is much larger than what can be reasonably assumed to be a value at which leading vehicles (LVs) can hinder the following vehicle's (FV's) motion.

As densities increase vehicles start traveling closer to one another. In such situations the actions of a vehicle are affected by the state (or actions) of the LV; speeds of vehicles fall below their desired speed and there is constant tug-o-war between two conflicting motivators — the need to reach the destination as quickly as possible (i.e., urgency) and the concern for ones safety. Further, it is human nature to feel threatened if distance headway is small at high speeds; hence as distances reduce so does speed (this can be seen on a macroscopic scale from any data on u and k). So what happens is that the following vehicle constantly tries to increase the speed (effect of urgency) but in so doing closes in; this increases the threat to safety and the person reduces the speed. This behaviour is referred to as car-following behaviour. This is the prevalent form of driving, meaning this is the mode in which drivers are for the largest range of densities (may be from 8 to 10 vehicles per km per lane to about 60 vehicles per km per lane). In the latter parts of this section more is discussed about this important driving behaviour. One may also refer to Chakroborty and Kikuchi [6] for a better exposition.

At the other end of the density scale, where densities are large, vehicles move with frequent halts or near halts. This kind of traffic is referred to as stop-and-go traffic. Driver behaviour in this region is impacted by the vehicles ahead. However, the strength of the relationship is not as strong as in the car-following case; often it is seen that vehicles keep longer than safe distances, vehicles do not immediately respond to spacing increments; etc. It seems that the primary motivator in these cases is only safety and urgency plays a lesser role. Similar observations have been made by others (for example, see Minderhoud and Zurbier [23]). It is felt that very little empirical research has been reported on stop-and-go traffic and more needs to be done to understand the behaviour better. In the rest of the section, the discussion is on models of car-following behaviour.

Over the years various models of car-following have been proposed. An overview of the models can be found in Brackstone and McDonald [24]. Here two of these models are mentioned. The models described here are the GHR models (one of the first set of models in car-following) and the fuzzy rule-based models (one of the most recent developments in this area).

The GHR models [7, 25–27] proposed the following stimulus-response car-following rule:

$$\ddot{x}_{FV}(t+\delta t) = \alpha_{\ell,m} \frac{\{x_{FV}(t+\delta t)\}^m}{\{x_{LV}(t) - x_{FV}(t)\}^\ell} \{\dot{x}_{LV}(t) - \dot{x}_{FV}(t)\} \tag{4}$$

where $x_i(t)$ is the position of vehicle i at time t measured from an upstream point, ℓ and m are calibration constants and are the same as those used in Equation 3. This rule with a proper choice of the exponents yields actions which give rise to stability. However, its reliance on only relative speed as the stimulus (the others simply modify the response to the stimulus) gives rise to certain problems; for example, this model cannot replicate the closing-in and shying-away behaviour. For a detailed discussion on this model and its shortcomings one may refer to Chakroborty and Kikuchi [6]; which it should not be. These drawbacks, notwithstanding, what this model showed for the first time was that expressions on the FV's actions could be derived which gave rise to stable behaviour. The authors believe that stability is an important car-following property which every microscopic model of traffic flow must exhibit.

The fuzzy rule-based model of car-following was initially proposed by Kikuchi and Chakroborty [28]. These models form the "latest distinct stage in their development, as it represents the next logical stage in attempting to accurately describe driver behaviour" (Brackstone and McDonald [24]). The fuzzy rule-based model [6, 28, 29] simply models driver behaviour by specifying a set of linguistic rules on what to do under different circumstances. For example, a rule could be: *IF (at time t) the Distance headway is very large AND Relative speed moderately negative AND Acceleration of LV is negative THEN (at time $t + \delta t$) FV should accelerate mildly.*

The results from the model show that all the properties of car-following behaviour are satisfied (see Chakroborty and Kikuchi [6]). The author believes that this model illustrated that simple rules-of-thumb can explain in all details such complicated behaviour as car-following. In this sense, it is felt that rule based structures like the ones that can be employed in cellular automata models can be successful in realistically representing driver behaviour and hence the macroscopic behaviour of the traffic stream.

Before leaving this section it must be pointed out that all models which are in essence microscopic models of traffic flow (like the cellular automata based models) must be subjected to tests which determine whether these models possess the microscopic properties mentioned earlier. Certain car-following properties, like stability, independence of stable conditions from initial conditions and perturbations, etc. are essential for any microscopic model. If a large number of the microscopic properties are absent, then the model's predictions at a macroscopic level also become suspect.

Another point which must be reiterated is that the existing microscopic models of traffic flow ignore lateral interactions. When lane discipline is not maintained or in situations of extensive weaving (merging or diverging of traffic streams like at roundabouts or near on or off ramps) there is considerable lateral interactions between vehicles. In such cases studying the process of longitudinal control of vehicles will not suffice; one has to look at the process of longitudinal and lateral control of vehicles in a comprehensive manner. In fact, models which can account for both lateral and longitudinal interactions between vehicles can in general be used to study the interactions of vehicles with other features of the road like road edges, geometry, static obstacles like parked vehicles, etc. It is felt that such models will provide a basis for relating capacity of roads (or more generally flow behaviour) to engineering features of the road like width, radius of curves, lateral clearance, and the like. Hence, the ultimate goal of modeling traffic flow must be to evolve a model of driver behaviour which can handle both longitudinal and lateral interactions and is simple enough to be used to simulate a large number of vehicles at a time so that macroscopic properties of the road can be studied. In the next section, some recent ideas on modelling both lateral and longitudinal interactions are discussed.

4 Some New Directions to Modelling Uninterrupted Traffic

The final goal in microscopic modelling is to be able to devise a comprehensive model of driver behaviour; a model which under one framework can explain a driver's choice of steering angle and acceleration values under various different driving situations. That is, such models should be able to describe the path of every vehicle over space and time.

Two types of comprehensive models have been developed in the recent past. The first are force field models and the second is a model which combines the concepts of utility-based discrete choice models and longitudinal control models. The force-field based models were developed at IIT Kanpur (for example, see Gupta et al. [30] and Chakroborty et al. [31]). The force field idea was also used by Helbing and Tilch [32] to model traffic dynamics; but the study was limited to only longitudinal control and hence did not contribute towards the development of a comprehensive model as envisaged here. The force-field based comprehensive model [30, 31] of traffic flow relate the steering angle and acceleration values to force (or potential) field in a driver's vicinity and is based on the following simple ideas:

(i) every goal (or "local" destinations like "ahead of the previous vehicle") emanates attractive (or negative) potentials and every other feature on the road (like road edges, parked vehicles, moving vehicles, etc.) are considered as obstacles which emanate repulsive (or positive) potentials,

(ii) the potential at a point on the road is assumed to be the algebraic sum of all the potentials from the various obstacles and goals,

(iii) the potential at a point is perceived as a threat to a driver's safety; the threat increases with speed; hence, it is assumed that the sustainable speed (a speed at which a driver feels comfortable) at a point is inversely related to the potential at that point,

(iv) given that a driver wishes to reach his destination quickly, he chooses the path which minimizes the potential (and hence maximizes the speed),

(v) the acceleration value is related to the potential values and their gradient along the chosen path and in some sense embody the notions of longitudinal control models.

These models are computationally intensive and therefore limited in their application for simulating large streams. Some work is on going at IIT Kanpur to develop a comprehensive model which is computationally efficient and yet reasonably realistic in its representation of a stream both microscopically and macroscopically.

A preliminary version of such a model has been developed and named CUTSiM (Comprehensive, Unidirectional, Uninterrupted Traffic Simulation Model). This model assumes that drivers choose a path based on various properties of that path like, headway available, closeness to obstacles, difficulty of re-orienting the vehicle to that path, etc. Once this choice is made, drivers drive along that path using principles similar to those used for longitudinal control model described earlier. CUTSiM incorporates the impact of geometry (so far only in terms of road width) and vehicle mix while simulating a traffic stream; further CUTSiM does not assume lane discipline. Even so, it has been shown that under certain conditions of "choice parameters" vehicles can be made to follow "lanes [33]." It is shown that CUTSiM satisfies most of the microscopic and macroscopic properties of traffic streams described earlier and is also computationally efficient. For example, CUTSiM can simulate traffic streams with upward of 4000 vehicles for an hour in less than 30 minutes on a desktop computer with dual core 2.2 GHz processor and 32 GB RAM.

Some macroscopic results from CUTSiM are presented here to illustrate its ability to incorporate the impact of road width and vehicle-mix on stream behaviour and also to show that it can model streams without lane discipline reasonably well. The model and its detailed analysis will be presented in a forthcoming paper based on Maurya [33]; details of a primitive version of CUTSiM can be found in Maurya and Chakroborty [34]. Figure 2 shows three speed-flow scatter plots obtained from streams simulated using CUTSiM; on comparing (a) and (b) parts of the figure one can see that as the lane width reduces the speed-flow relations obtained from streams simulated using CUTSiM also changes and the maximum flow reduces. On comparing (a) and (c) one can see that as the percentage of trucks is increased the stream behaviour also changes and the maximum flow in vehicles per hour (vph) falls; another interesting feature is that at low flow values the effect of trucks on stream behaviour is much less pronounced than at high flows as is expected. Figure 3 shows a comparison of the speed-flow data collected from Delhi-Gurgaon Highway in India (road width = 14.6 m; no lane discipline) with those obtained

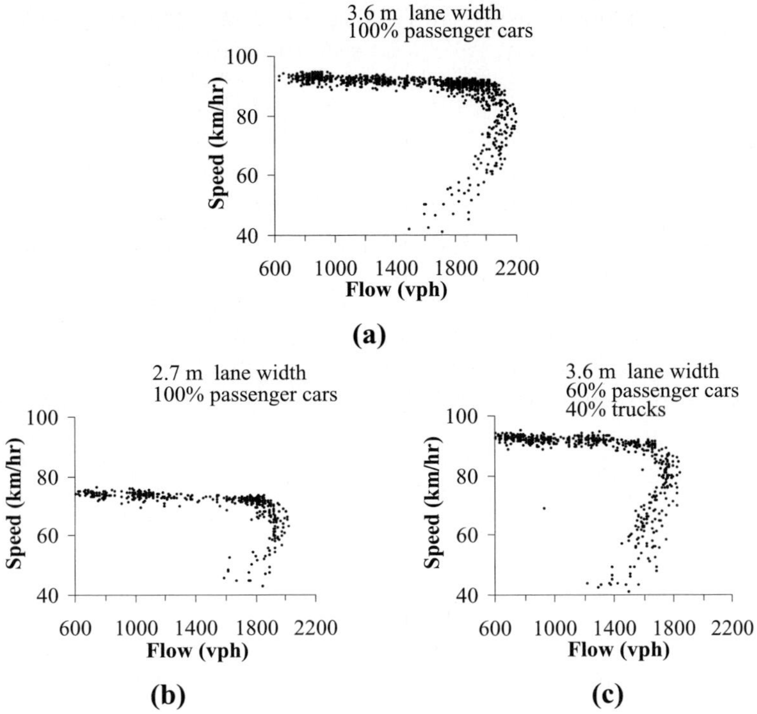

Fig. 2. Speed-flow data obtained from streams simulated using CUTSiM for different lane widths and vehicle-mix.

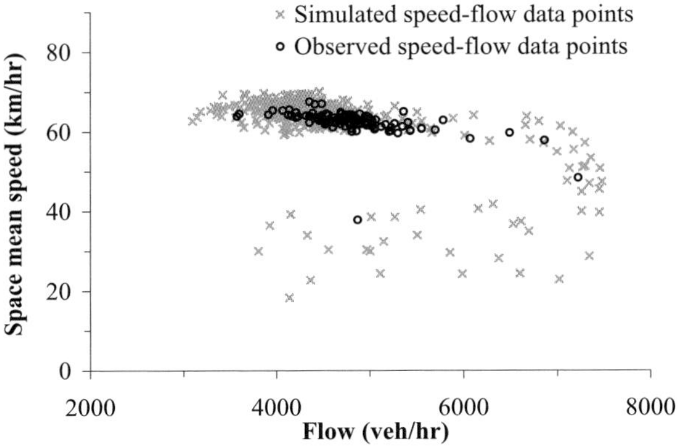

Fig. 3. Comparison of speed-flow data obtained from Delhi-Gurgaon Highway (road width = 14.6 m, no lane discipline) with those obtained from a CUTSiM simulated stream.

from a stream simulated using CUTSiM for the same conditions as those on the Delhi-Gurgaon Highway. As can be seen the match is good.

5 Summary

The purpose of the paper was to highlight (i) the importance of flow models to traffic engineers who often cannot conduct experiments to see the impact of certain design alternatives on traffic behaviour, and (ii) the ingredients needed in a model of traffic flow for it to be really useful for traffic engineering purposes. It was mentioned that developing models which can incorporate both lateral and longitudinal interactions from both roadway and traffic features should be the ultimate goal of the exercise of traffic flow modelling. Some new models which attempt to achieve this goal were also briefly discussed.

References

1. Transportation Research Board (1998) Highway Capacity Manual.
2. Chakroborty, P (2006) Phys A 372:151–161.
3. Gunay, B (2007) Transp Res 41B:722–735.
4. Transportation Research Board (2000) Highway Capacity Manual.
5. May, A (1990) Traffic Flow Fundamentals. Prentice Hall, Englewood Cliffs, New Jersey.
6. Chakroborty, P, Kikuchi, S (1999) Transp Res 7C:209–235.
7. Herman, R, Montroll, E W, Potts, R B, Rothery, R W (1959) Oper Res 7:86–106.
8. Jones, T R, Potts, R B (1962) Oper Res 10:745–763.
9. Winzer, T (1981) Transp Res 15C:437–443.
10. Kang, K P, Chang, G L (2004) Proc IEEE Intel Transp Sys Conf, Washington, USA.
11. Hall, F L (1999). In: Gartner, N, Messer, C J, Rathi, A K (eds) Traffic Flow Theory — A State-of-the-Art Report. Transportation Research Board, Washington, D.C.
12. Hall, F L, Hurdle, V F, Banks, J H (1992) Transp Res Rec 1365:12–18.
13. Banks, J H (1998) Introduction to transportation engineering. McGraw Hill, Boston.
14. Banks, J H (1991) Transp Res Rec 1320:83–90.
15. Banks, J H (1991) Transp Res Rec 1320:234–241.
16. Agyemang-Duah, K, Hall, F L (1991) In: Brannolte, U (ed) Proc Int Symp on Highway Cap, Karlsruhe. Balkema.
17. Ross, P (1987) Public Roads 51(3):90–96.
18. Greenshields, B D (1935) Proc High Res Board 14:448–477.
19. Greenberg, H (1959) Oper Res 7:78–85.
20. Edie, L (1961) Oper Res 9:66–76.
21. Navin, F (1986) Transp Plann Tech 11:19–25.
22. Persaud, B N, Hall, F L (1989) Trans Res 23A:103–113.

23. Minderhoud, M M, Zurbier, F (2004) Proc IEEE Intel Veh Symp, Parma:676–681.
24. Brackstone, M, McDonald, M (1999) Trans Res 2F:181–196.
25. Gazis, D, Herman, R, Rothery, R W (1961) Oper Res 9:545–567.
26. Chandler, R E, Herman, R, Montroll, E W (1958) Oper Res 6:165–184.
27. Gazis, D C, Herman, R, Potts, R B (1959) Oper Res 7:499–505.
28. Kikuchi, S, Chakroborty, P (1992) Transp Res Rec 1365:82–91.
29. Chakroborty, P, Kikuchi, S (2003) Transp Res 11C:91–119.
30. Gupta, S, Chakroborty, P, Mukerjee, A (1998) Proc Int Symp Ind Rob Sys 98, Bangalore.
31. Chakroborty, P, Agarwal, S, Vasishtha, K (2004) J Transp Engg 130:438–451.
32. Helbing, D, Tilch, B (1998) Phys Rev 58E:133–139.
33. Maurya, A K (2007) Unpublished Ph.D. Thesis, IIT Kanpur.
34. Maurya, A K, Chakroborty, P (2007) Proc 11th WCTR, Berkeley.

Cellular Automaton Approach to Highway Traffic: What do we Know?

Andreas Schadschneider[1,2]

[1] Institut für Theoretische Physik, Universität zu Köln, 50937 Köln, Germany
as@thp.uni-koeln.de
[2] Interdisziplinäres Zentrum für komplexe Systeme, 53117 Bonn, Germany

Summary. Cellular automata for the simulation of highway traffic have become increasingly popular in recent years. But how realistic are these models? We review the current state of this approach, focussing on their ability to reproduce the macroscopic structure of traffic flow as well as the single-vehicle data obtained empirically. Furthermore the role of accidents is discussed. Although of general relevance for traffic modelling this aspect has not been attracted much attention so far.

1 Introduction

Modelling of traffic flow has a long history in various fields ranging from engineering to physics and applied mathematics [1–4]. The model classes used for their description can be classified in various ways.

Microscopic models represent each car separately. This allows to introduce different vehicle (or drivers) types with individual properties as well as issues like route choice. In *macroscopic models* the state of the system is described by densities, e.g. a mass density derived from the positions of the cars.

Another characterisation is based on the properties of the variables used to describe a system of vehicles, namely space, time and state variable (e.g. velocity). Each of these can either be *discrete* (i.e. an integer number) or *continuous* (i.e. a real number). Discrete time is usually realized through a *parallel* or *synchronous update* where all particles or sites are moved at the same time. This introduces a timescale that can be used for calibration. Models that are discrete in all variables are usually called *cellular automata (CA)*.

Interactions between cars can be implemented in at least two different ways: In a *rule-based approach* drivers make "decisions" based on their current situation and that in their neighbourhood as well as their goals etc. These rules are therefore often motivated by psychology. In contrast, *force-based models* specify interactions directly on the level of equations of motion, similar to classical mechanics although the forces are not necessarily "physical" forces. Typical examples are *car-following* and *hydrodynamic models* [3, 5, 6].

The dynamics of vehicles can either be *deterministic* or *stochastic*. In the first case the behaviour at future times is completely determined by the past. In stochastic models, the behaviour is controlled by probabilities and agents can react differently in the same situation. As known from other examples, the introduction of stochasticity into rather simple systems can generate rather complex behaviour. On the other hand, the stochasticity reflects our lack of knowledge of the underlying physical processes that e.g. determine the decision-making of the drivers. Stochastic behavioural rules then often lead to a rather realistic representation of the actual behaviour. This "intrinsic" stochasticity should be distinguished from *external noise* terms which are sometimes added to *macroscopic* observables, like position or velocity. Often the main effect of these terms is to avoid certain special configurations which are considered to be unrealistic. Otherwise the behaviour is very similar to the deterministic case. For true stochasticity, on the other hand, the deterministic limit usually has very different properties from the generic case.

Finally, one can distinguish models according to their *fidelity* which here refers to the *apparent* realism of the modelling approach. High fidelity models try to capture the complexity of decision making, actions etc. that constitute vehicular motion in a realistic way. In contrast, in the simplest models vehicles (drivers) are represented by particles without any intelligence. Roughly speaking, the number of parameters in a model is a good measure for fidelity in the sense introduced here. But note that higher fidelity does not necessarily mean that empirical observations are reproduced better! One should therefore distinguish between *modelling*, which implies the identification of the fundamental mechanisms responsible for the observed phenomena, and *imitation*, which is just their reproduction without any additional insight.

2 Empirical Results

The quality of any model has to be decided by comparison with empirical observations on a qualitative or quantitative level. *Qualitative* results are usually related to the occurrence of spatio-temporal structures among which *jams* are the most prominent (and annoying!). "Jam" in the following will refer to a sequence of cars standing with small intervehicular distances (see Sec. 2.1). On the other hand, more *qualitative results* exist which in principle allow for a calibration of models (Sec. 2.2).

2.1 Spontaneous Jam Formation

In principle two types of jams can be distinguished. The first type is created by a *bottleneck*, i.e. locations of reduced capacity, if the inflow than this capacity. Apart from these *bottleneck-induced jams*, *spontaneous jams* or *phantom jams* exist for which this is not true, at least not in an obvious way.

Fig. 1(left) shows empirical observations [7] indicating that growing instabilities can lead to spontaneous formation of jams even in the absence of bottlenecks. This is confirmed by controlled experiments (Fig. 1(right) and [8]) in which drivers were advised to drive as fast as possible around a circular course. At intermediate densities after a while (typically of the order of 10 minutes) free-flow breaks down and a jam is created. Its origin is the imperfect driving of human drivers. Typically if a driver approaches the preceding car too fast and is forced to brake to avoid an accident, this braking maneuver will be stronger than actually necessary to avoid the accident. At sufficiently large densities this *overreaction* will start a chain reaction which forces other drivers to brake (and thus overreact) until finally a car has to stop, marking the beginning of the jam. In the beginning of the experiment drivers were still focussed and could avoid such behaviour, but they quickly lost concentration. This shows the importance of psychological effects in traffic.

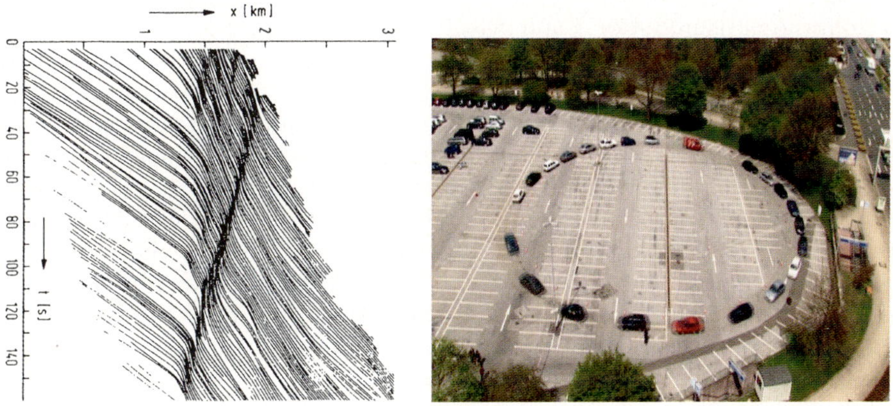

Fig. 1. Left: Empirical data for the trajectories of single cars showing spontaneous jam formation (from [7]). Right: Experiment for German TV station WDR demonstrating the spontaneous formation of jams.

Daganzo and collaborators [9], on the other hand, have argued that all jams are created by bottlenecks which are just sometimes not easy to identify. Often jams occur at the same location every day, especially close to road inhomogeneities like ramps, sharp bends etc. However, these jams are not necessarily bottleneck-induced and they might occur even though the local capacity has not yet been reached. Probably both mechanisms play a role here, e.g. in the sense that inhomogeneities increase the probability of spontaneous jamming. In addition, the experiments clearly show that the underlying instability mechanism has to be taken into account in any modelling approach since it reflects an essential part of the vehicle-vehicle interactions.

2.2 Fundamental Diagram and Traffic Phases

At many highway locations, empirical data are nowadays collected automatically by stationary inductive loops. For each vehicle the time of passing is recorded from which several other quantities can be derived. The *flow* J is given by the number of cars $N(T)$ passing the detector per time interval T, i.e. $J = \frac{1}{T}N(T)$. The *velocity* v can be derived from the time interval needed to pass two loops at close separation. Other quantities that can be obtained are the length of the vehicle and the headways, i.e. the time interval between the signals of consecutive cars (*temporal headway*) and the spatial distance between them (*spatial headway*). However, the determination of the *density* ρ is rather problematic. This has two reasons, a fundamental and a technical one: 1) density is a "spatial" quantity that is difficult to measure locally, especially since the objects are extended, and 2) many detectors are event-driven and therefore only moving cars can be detected. Problem 1) is often avoided by using the hydrodynamic relation $J = \rho v$ to calculate the density ρ from the measured flow J and velocity v. The average velocity v is usually overestimated (if, due to problem 2, standing or slow cars are not detected) and thus the density ρ is underestimated.

The most important quantitative characterization of traffic flow is the *fundamental diagram*. In physics it usually means the density-dependence of the flow, $J(\rho)$. Due to the hydrodynamic relation this is equivalent to the relations $v(\rho)$ or $v(J)$. The latter avoids the problems in the density measurement and is frequently used in traffic engineering.

Usually data are averaged over 1 to 5 minute intervals to obtain a time-series of data points. To understand the microscopic structure one needs, however, data which are not averaged and allow to distinguish between the cars (*single-vehicle data*).

Nowadays three different phases of traffic flow can be distinguished in the fundamental diagram, supported by spatio-temporal observations [11], although some points remain controversial [12]:

- **Free flow:** In this phase interactions between vehicles are rare. Every car moves with its desired velocity corresponding e.g. to a speed limit. Therefore the flow increases linearly with the density of cars. The free flow branch F can clearly be seen in Fig. 2(left). The part of the branch with flows larger than J_{out} is called *metastable branch.* It corresponds to a region where the flow is not uniquely determined by the density.
  All states not of free flow type are called *congested states.* They are characterized by an average velocity that is smaller than the "desired" velocity of the drivers. Two congested phases can be distinguished.
- **Wide moving jams:** Jams can form spontaneously, i.e. without any obvious external reason (see Sec. 2.1). Wide (moving) jams are regions of very high density and negligible average velocity and flow. Their width is much larger than the fronts at both ends where the speed of vehicles changes

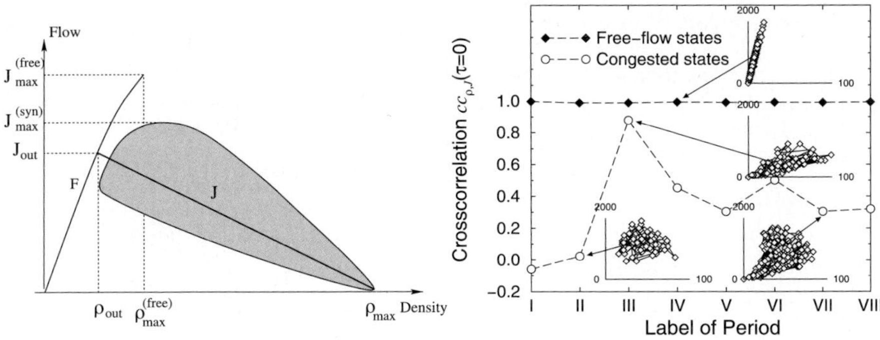

Fig. 2. Left: Schematic form of the fundamental diagram. F denotes the free flow branch and the jam line J is determined by the properties of wide moving jams. Right: Empirical cross-correlation function obtained on a German highway [10]. Different periods of free-flow and congested traffic are labeled by I through VIII.

sharply. The jam front moves upstream (i.e. opposite to the driving direction) at a typical velocity $v_{\mathrm{Jam}} \approx 15$ km/h [13]. Other characteristic properties of wide jams are its density and the outflow J_{out}, which does not depend on the inflow into the jam [11]. The velocity v_{Jam} and the corresponding flow rate is only determined by the density inside a wide jam and the delay-time between two vehicles leaving the jam [13].

- **Synchronized flow:** In synchronized flow [11] the average velocity is significantly lower than in free flow, but the flow can be much larger than in wide moving jams. The main characteristic is the apparent absence of a functional flow-density form, i.e. the corresponding data points are spread irregularly over a large two-dimensional area (see Fig. 2(left)). In time-series of flow-density measurements, the flow can increase or decrease with increasing density, in sharp contrast to the free-flow (jammed) phase where flow always increases (decreases). This "irregularity" of the time-series can be quantified by using the cross-correlation function $cc_{\rho,J}(\tau) \propto \langle \rho(t)J(t+\tau)\rangle - \langle \rho(t)\rangle\langle J(t+\tau)\rangle$ between density ρ and flow J [10]. In free flow $cc_{\rho,J}(\tau)$ is very close to 1, but in synchronized flow it is almost 0 [10]. Here flow and density are independent of each other. This quantifies the earlier statement that the data points are "spread irregularly". Furthermore, on a multilane highway, the time-series of measurements on different lanes are highly correlated, i.e. synchronized. This was the motivation for denoting this traffic state as "synchronized traffic".

Above we have only given the most important properties of the phases. A more detailed characterization and discussion of the related spatio-temporal organization of traffic states can be found e.g. in [11] (and references therein).

2.3 Microscopic Structure

Modern detectors measure single-vehicle data that can provide important information about the microscopic structure of traffic flow which, of course, determines the macroscopic properties as well.

The distribution of time-headways, i.e. the time intervals between two consecutive cars passing the detector, shows a surprisingly large fraction of headways which are much shorter than allowed by (German) legal regulations (1.8 sec) [10, 14]. The distributions in the three phases differ, e.g. it becomes much broader in the synchronized phase (Fig. 3) than in free-flow. In the outflow region of wide jams a typical headway of approx. 2 sec is observed which is consistent with a jam velocity of $v_{\mathrm{Jam}} \approx 15$ km/h.

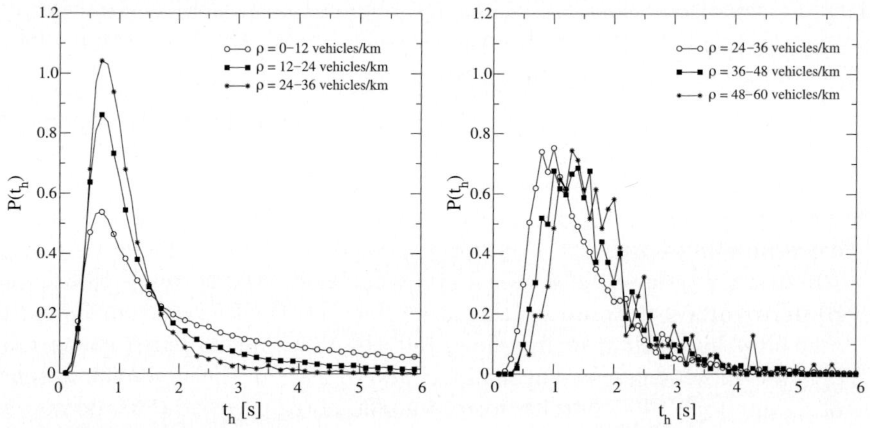

Fig. 3. Distribution of time-headways obtained empirically on German highways. Left: Free flow regime; Right: Synchronized regime.

Other interesting quantities like velocity distributions and optimal-velocity curves, i.e. the dependence of the velocity on the distance-headway, are discussed in more detail in [10, 14].

3 Cellular Automaton Models

In recent years, modern approaches adopted from statistical physics have become quite popular for modelling interdisciplinary problems due to the powerful methods that have been developed to deal with interacting many-particle systems [6]. Among those, a cellular automaton-based description has become most fruitful due to its the relative simplicity and flexibility. Especially the fact that the dynamics can be implemented in the form of intuitive "rules" has allowed to include rather complex aspects (e.g. psychology) in a rather simple way.

3.1 General Aspects

In a cellular automaton (CA), space, time and state variables are discrete which makes them ideally suited for high-performance computer simulations. For continuum models, discretization is necessary for their numerical treatment and one then has to perform the limit Δx, $\Delta t \to 0$ in order to obtain accurate results. In CA the discretizations Δx and Δt are finite and accurate results can be obtained since the discreteness is already taken into account in the definition of the model and its dynamics.

Due to the existence of more and more powerful computers, the higher speed of simulations based on CA will become less important in the future. However, a point that will become more relevant is the fact that CA models are rule-based (Sec. 1). This is a great advantage for modelling of interdisciplinary problems where the interactions between "agents" are not based on physical forces. Instead rule-based models allow to take into account e.g. psychological aspects in a natural and efficient way.

3.2 Nagel-Schreckenberg Model

The Nagel-Schreckenberg (NaSch) model [15] is a probabilistic CA able to reproduce many of the basic features of traffic flow. The state of each car n ($n = 1, 2, \ldots, N$) is characterized by its velocity v_n which can take one of the $v_{\max} + 1$ integer values $v_n = 0, 1, ..., v_{\max}$ (Fig. 4). The position of the

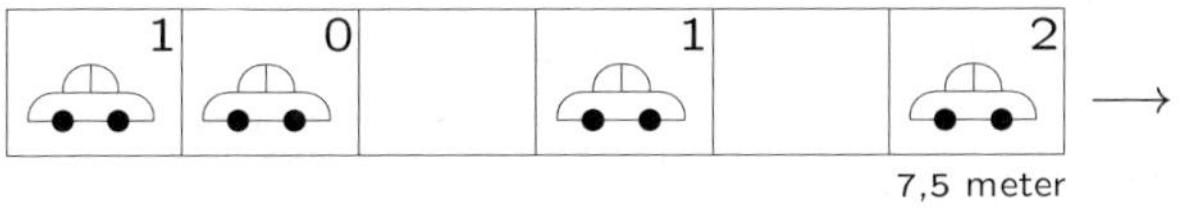

Fig. 4. A typical configuration in the NaSch model. Cars are moving to the right. The number in the upper right corner is the speed v_n of the vehicle.

n-th vehicle is denoted by x_n. Then $d_n = x_{n+1} - x_n - 1$ is its headway, i.e. the number of empty cells in front of it. At each time step $t \to t + 1$, the arrangement of the N cars on a finite lattice of length L (i.e. for a global density $\rho = N/L$) is updated *in parallel* according to the following "rules":

Step 1: Acceleration.
If $v_n < v_{max}$, velocity is increased by 1, i.e. $v_n^{(1)} = \min(v_n + 1, v_{\max})$.

Step 2: Deceleration (due to other cars).
If $d_n < v_n^{(1)}$, velocity is reduced to d_n, i.e. $v_n^{(2)} = \min(v_n^{(1)}, d_n)$.

Step 3: Randomization.
If $v_n^{(2)} > 0$, velocity is decreased randomly by 1 with probability p, i.e.

$$v_n^{(3)} = \begin{cases} \max(v_n^{(2)} - 1, 0) & \text{with probability } p, \\ v_n^{(2)} & \text{with probability } 1-p. \end{cases}$$

Step 4: Vehicle movement.
Each car is moved forward according to its new velocity $v_n = v_n^{(3)}$ determined in Steps 1–3, i.e. $x_n \to x_n + v_n$.

All rules have a simple interpretation. *Step 1* expresses the desire of the drivers to move as fast as possible (or allowed). Often v_{max} corresponds to a speed limit which is the same for all the cars. *Step 2* reflects the interactions between vehicles and guarantees the absence of collisions in the model. Here the velocity of the preceding car is not taken into account. This is already sufficient to reproduce the basic properties of real traffic. In order to obtain good agreement with the 'fine-structure' of empirical data a more sophisticated braking rule is necessary (see Sec. 4). *Step 3* incorporates many effects that play an important role, e.g. it is responsible for spontaneous jam formation. Acceleration is delayed since a car will accelerate only with probability $1-p$ if possible. On the other hand, deceleration processes are enhanced: If a car has to brake due to another car ahead (*Step 2*) with probability p it will even brake further in *Step 3*. Such overreactions are the origin of spontaneous jam formation. Another effect incorporated in *Step 3* are natural fluctuations. Even on a free road a driver will not keep a strictly constant velocity which will then show small fluctuations. Finally, in *Step 4* all cars will move with their new velocity as determined in the first three steps.

This set of rules is minimal in the sense that every subset will no longer produce realistic behaviour. E.g. by leaving out *Step 3* – which is equivalent to the deterministic limit $p = 0$ – one no longer finds spontaneous jam formation [16]. Also the order of the rules is important. Applying *Step 2* before *Step 1* would no longer guarantee the absence of crashes. By interchanging *Step 2* and *Step 3*, on the other hand, no spontaneous jams would occur.

An important point is the timescale corresponding to one update step. It should not be too small to allow efficient computer simulations, but also not too large. It can be estimated in different ways [15], e.g. by identifying the average 'free' velocity $v_{\text{max}} - p$ with a typical speed limit on highways, e.g. 120 km/h. Taking into account that the length of a cell corresponds to 7.5 m in reality one then finds that for $v_{\text{max}} = 5$ and $p = 0.5$ one timestep corresponds to approximately 1 sec in real time. Since this time is of the same order of magnitude as the smallest relevant timescale in real traffic, the reaction time of the drivers, these parameter values are a reasonable choice.

The fundamental diagram of the NaSch model consists of two branches, a free flow and a congested branch, corresponding to the lines F and J in Fig. 2(left). However, the free flow branch F ends at flow J_{out}, i.e. the NaSch model does not reproduce the metastable states with flow $J_{\text{out}} < J < J_{\text{max}}^{(\text{free})}$. In Sec. 3.3 we will show that already a small modification of the NaSch rules will remedy this problem. Finally, the NaSch model also does not exhibit a

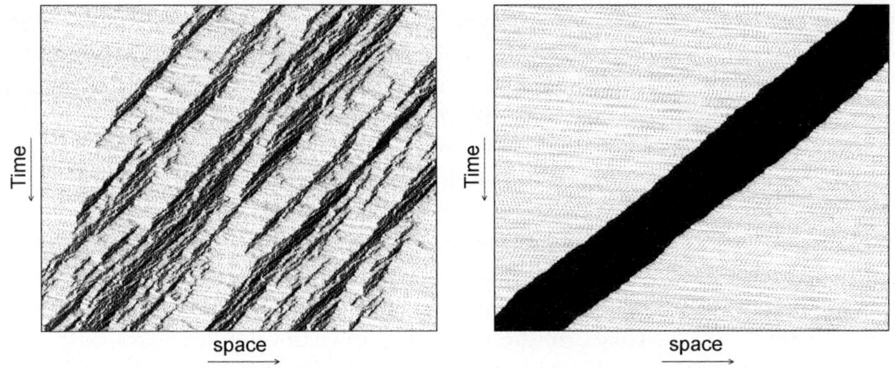

Fig. 5. Typical space-time diagrams of the NaSch model (left) and the VDR model for $p \ll p_0$ (right). One can clearly see the different structure of the jams.

synchronized phase. Here larger modifications of the rules are necessary. In Sec. 4 we will show that the human desire for comfortable driving plays an important role here. This is not incorporated into the basic NaSch rules which only try to avoid accidents.

The NaSch model is also able to reproduce the spontaneous formation of jams (see Fig. 5 and [17]) which have their origin in the randomization step 3 that leads exactly to the overreaction mechanism described in Sec. 2.1. Fig. 5(left) shows a typical space-time diagrams obtained from simulations of the NaSch model [17]. For $p > 0$ one can clearly see various jams (compare with the empirical result in Fig. 1(right)) with different lifetime and interesting dynamics of these jams, e.g. branching. In the deterministic case $p = 0$ overreactions are not possible and therefore no spontaneous jams occur [16].

3.3 VDR Model

In order to reproduce the metastable states of high flow corresponding to flows $J > J_{\text{out}}$ on the free flow branch F in Fig. 2 only a simple modification of the NaSch model is necessary. In contrast to the original NaSch model, in the so-called *Velocity-Dependent-Randomization (VDR) model* [18] the randomization parameter depends on the velocity of the car, $p = p(v)$. The rules of Sec. 3.2 are supplemented by a new rule:

Step 0: Determination of the randomization parameter.
The randomization parameter for the n-th car is given by $p = p(v_n(t))$.

This new step has to be carried out before the acceleration *Step 1.* The randomization parameter used in *Step 3* depends on the velocity $v_n(t)$ of the n−th car after the previous timestep. Metastable states occur for so-called *slow-to-start rules* where one chooses [18]

$$p(v) = \begin{cases} p_0 & \text{for } v = 0, \\ p & \text{for } v > 0, \end{cases} \tag{1}$$

with $p_0 > p$. This means that cars which have been standing in the previous timestep have a higher probability p_0 of braking in the randomization step than moving cars, explaining the name 'slow-to-start' rule. Other slow-to-start rules have been suggested, see [2, 19] and references therein.

Fundamental diagrams obtained from simulations of the VDR model consist of the two branches F and J in Fig. 2, including the states with $J > J_{\text{out}}$. However, no synchronized traffic is found in this simple modification.

The macroscopic structure of the congested state in the VDR model is very different from that of the NaSch model [17, 18]. It exhibits phase separation into a free flow region and a large jam which is almost compact for $p \ll 1$ (see Fig. 5(right)). In contrast, in the original NaSch model stop-and-go waves are found (Fig. 5(left)). On the other hand, the structure of the free flow branch is very similar to that of the NaSch model. However, for $J > J_{\text{out}}$ the homogeneous free flow states are not stable, but can decay to a congested state through fluctuations or small perturbations.

The fundamental diagram of the VDR model can be understood from heuristic arguments [18]. For small densities every car moves with free-flow velocity $v_f = v_{\max} - p$ and the flux is given by $J_{\text{hom}}(\rho) = \rho(v_{\max} - p)$ which is identical to the NaSch model with randomization p. For densities close to $\rho = 1$, only velocities $v = 0$ or $v = 1$ occur and therefore the random braking is controlled by p_0. A simple waiting time argument [18] shows that the flow in the phase-separated regime is given by $J_{\text{sep}}(\rho) = (1 - p_0)(1 - \rho)$ as in the jammed branch of a NaSch model with randomization p_0.

If the difference between the parameters p and p_0 becomes smaller, the behaviour will be more similar to that of the NaSch model. Jams are no longer compact, i.e. small holes appear. Furthermore small jams can be formed in the outflow region of the jam. For $p = p_0$ the NaSch model is recovered.

The behaviour found in the VDR model is generic for models with slow-to-start rules. The outflow from a large jam is smaller than the maximal possible flow, i.e. the maximum of the fundamental diagram. Note that the large jam is only stable if the outflow J_{out} equals the inflow J_{in}. For $J_{\text{out}} = J_{\max}^{(\text{free})}$ it is very unlikely that this condition can be satisfied. Therefore the density far downstream is rather small and the vehicles propagate almost freely in the low-density region. The spontaneous formation of jams is suppressed if p is not too large. This is the basic mechanism which leads to the formation of the phase-separated state and hysteresis.

4 CA Modeling of Synchronized Flow

Although the VDR model shows metastable states and related phenomena like hysteresis it is not able to reproduce the characteristic properties of synchronized traffic. Furthermore the agreement with empirical data on a microscopic level is not very satisfactory [20]. This indicates that an important ingredient is still missing. So far the dynamics was mainly based on the avoidance of accidents. This appears to be insufficient. Therefore it has been suggested that the desire of the drivers for smooth and comfortable driving is responsible for the occurrence of synchronized traffic [21]. The avoidance of accidents implies that only interactions with the next car ahead are important and any information about the velocity or the velocity difference to the preceding car is not taken into account. As we know from our own experience usually one tries to "anticipate" the behaviour of the predecessor. E.g. if its headway is rather large than we know that it is very unlikely that (s)he will brake abruptly. Therefore we are willing to accept a much smaller headway than in denser traffic. Thus anticipation allows for a much smoother driving.

Thus the three observed traffic phases correspond to different driving strategies. In free flow, drivers try to drive as fast as possible and interactions are rare. In the jammed phase, the avoidance of accidents determines the behaviour and in synchronized traffic it is the desire to drive in a smooth and comfortable way. Drivers try to avoid abrupt velocity changes which requires to observe the behaviour of the surrounding traffic in a more detailed way than in the other two phases.

The following aspects appear to be important for realistic models, but are not included in the NaSch or VDR models:
(i) Velocity anticipation: The empirically observed very short temporal headways (Sec. 2.3) can only be explained by strong anticipation effects. Furthermore, at inhomogeneities, e.g. created by ramps, the anticipation of the leaders velocity avoids abrupt braking of the traffic behind and therefore reduces the probability to form jams.
(ii) Retarded acceleration: Comfortable driving also implies that cars do not accelerate immediately in case of a larger gap ahead if they observe slow downstream traffic. This leads to a sub-optimal gap usage because the velocity is smaller than the headway allows. On the other hand, larger gaps in a dense region reduce the car-car interactions and cut-off the chain reactions of braking overreactions which are responsible for spontaneous jam formation.
(iii) Timely braking: Finally timely braking suppresses another mechanism of jam formation: When the velocity adjustment is only based on the distance to the next car ahead, jams often emerge in the layer between free-flow and synchronized traffic. In these models the jam formation arises from cars approaching a slow-moving cluster with high speed which leads to a compactified region. This artificial mechanism of jam formation can be avoided by allowing the drivers to adjust their speed to the vehicles ahead.

An important feature for realizing *(i)*–*(iii)* is *anticipation* of the actions of other drivers in the next timestep. Thus more information about the next car $n+1$ ahead is needed, not just its distance d_n as in the NaSch model. From its velocity $v_{n+1}(t)$ and the headway $d_{n+1}(t)$, an anticipated velocity $v_{\text{anti}} = \min(d_{n+1}(t), v_{n+1}(t)) - 1$ can be estimated such that $v_{n+1}(t+1) \leq v_{\text{anti}}$. This allows to determine a safe velocity $v_n(t+1)$ not leading to a collision.

The *brake light model* [22] incorporates these aspects by extending the VDR model. Velocity anticipation is realized through brake lights which indicate (within an interaction horizon) velocity changes of the preceding car. Another feature is a reduced cell length of 1.5 m instead of 7.5 m as in the standard NaSch model. Therefore the vehicles now have a length of 5 cells. This allows for a smoother acceleration and deceleration.

These refinements of the model lead to the following characteristics of the drivers' behaviour:
(*i*) For large headways cars move, apart from fluctuations, with their desired velocity $v_{\max}$.
(*ii*) For intermediate headways drivers react to velocity changes of the next vehicle downstream. This is implemented by introducing brake lights.
(*iii*) For small headways drivers adjust their velocity such that safe driving is possible.
(*iv*) The acceleration is delayed for standing vehicles (slow-to-start) and also directly after braking events.

The model is able to reproduce all three phases (free-flow, synchronized, and wide jams) observed in real traffic. Furthermore it shows good agreement with detailed empirical single-vehicle data in all phases [20]. The model passes also a rather sensitive test, namely it allows to reproduced the empirically observed coexistence of phases [23], especially the upstream propagation of wide jams through both free flow and synchronized traffic with constant velocity and without disturbing these states.

Similar ideas have been used in the model of Kerner, Klenov and Wolf [24]. Here drivers change their behaviour within a certain synchronization distance to the preceding vehicle. In this situation they try to move at the same velocity as the preceding car instead of maximizing their speed.

Lee et al. [25] have emphasized the conflict between human overreaction and limited mechanical capabilities as possible origin of congested traffic states. Their model takes the limited acceleration and deceleration capabilities of the vehicles into account. In all other models discussed so far deceleration is not limited in order to avoid accidents. Furthermore different driving strategies are distinguished, depending on the local traffic situation. *Optimistic driving* controls the behavior in free flow where drivers accept "unsafe" gaps which are too small to react to an emergency braking of the leading vehicle. *Pessimistic driving* occurs at high densities where interactions between the cars are strong and braking is likely so that drivers have to remain aloof.

The core of the model is an inequality that defines a velocity $c_n(t+1)$ which is considered to be safe by the driver:

$$x_n(t)+\Delta+\sum_{i=0}^{\tau_f(c_n(t+1))}(c_n(t+1)-Di) \le x_{n+1}(t)+\sum_{i=1}^{\tau_l(v_{n+1}(t))}(v_{n+1}(t)-Di)\,. \quad (2)$$

Δ represents the minimum gap between the vehicles and is at least the length l of the leading vehicle. Each summation in (2) denotes a deceleration cascade with maximum braking capability D. As long as both $\tau_f(v)$ and $\tau_l(v)$ are set to v/D and $\Delta = l$, the deceleration would end in a bumper-to-bumper configuration. But this is weakened if the human factor is introduced. In [25] functions $\tau_f(v)$, $\tau_l(v)$ and D have been proposed that lead to realistic behaviour. In general, they are different for optimistic and pessimistic behaviour. For a detailed description of the update rules, we refer to [25].

It turns out the model is not intrinsically accident-free, although accidents are rare [26]. Usually collisions are avoided by introducing a strict hardcore repulsion between the individual cars. However, this typically leads to processes with a very large deceleration. Therefore it is rather difficult to define a model that at the same time captures the finite deceleration capabilities of vehicles and is accident-free.

In the presence of limited deceleration capabilities crashes have to be avoided by choosing the dynamics appropriately. Therefore in [26] a slight modification of the model has been suggested that appears to be accident-free and, at the same time, keeps the realistic behaviour of the original model. A key ingredient are again brake lights which provide a way to communicate the presence of a hindrance and therefore a possible change of the driving behaviour (from optimistic to pessimistic) to the following cars.

5 Accidents

An important aspect of real traffic that is usually neglected in modelling approaches is the occurrence of accidents although these are responsible for a considerable fraction of jams, either directly or indirectly (e.g. through passing drivers that slow down).

With regard to accidents, traffic models can be classsified into two classes: (i) models that are *intrinsically accident-free*, and (ii) models which are not. "Intrinsically" means that the dynamics is defined in such a way that accidents are strictly avoided (for every initial condition). A typical example are CA like the NaSch model where Step 2 in the dynamics enforces the absence of accidents. However, this can lead to unrealistically large decelerations. Extreme braking maneuvres might have a strong effect on the dynamics and are a possible origin of jams.

Other models are not intrinsically accident-free, like the model of Lee et al. (Sec. 4). Some, like the optimal-velocity model [27], are believed to be accident-free in the stationary state, although accidents can occur for certain initial conditions during relaxation. However, since here cars are point-like objects the problem is less important and accidents can be interpreted as overtaking maneuvres. The problem is not restricted to microscopic models. Also in hydrodynamic models accident-like behaviour can be observed, e.g. cars moving backwards [28].

In principle, two types of accidents in real traffic can be distinguished, namely those due to careless drivers and those related to technical failures (tyre blowout etc.). The latter are difficult to implement explicitly in simple models and could be treated as random events. Careless and aggressive drivers, on the other hand, can be taken into account, e.g. by modification of the dynamical rules in CA models.

One general problem with models that are not intrinsically accident-free is the need to specify how to deal with these accidents if they occur in the simulations. This is an important issue since it might have a strong influence on the dynamics, especially for microscopic quantities. Surprisingly this is not discussed in most publications, although this means that the dynamics of the model has not been fully specified! Sometimes authors seem even not to be aware of the accidents occurring in their models and then it depends on the specific implementation how these accidents are treated. Obviously such situations can even lead to artefacts in the dynamics. However, it appears that in most investigations accidents are only occurring virtually and the simulated dynamics is always accident-free. Therefore it is better to speak about *dangerous situations* instead of *accidents*[3].

There are only a few qualitative studies for CA models that try to elucidate the importance of accidents. In these approaches the braking rule of the NaSch model (or other models) is modified such that with a certain probability q drivers do not respect the safety distance in Step 2, i.e. with probability q the velocity after this step is $v_n^{(2)} = d_n + 1$. This leads to an accident, if the car in front of the careless driver will not move in the same timestep. This has first been studied for the deterministic limit of the NaSch model in [29, 30] and then been generalized to other models [31–35].

Since accidents do not occur during the simulations but are replaced by a "safe" dynamics, quantities like the *accident probability* are just special correlation functions of the original model. In fact it is closely related to the spatial headway distribution.

In the original definition, accidents or dangerous situations only occur with stopped cars and thus only at high densities. Therefore modifications of the "accident criteria" have been suggested, e.g. in [36]. In reality, accidents due to careless driving do not only occur with standing cars, but more generally involving vehicles with large velocity differences. Another source of dangerous situations are abrupt velocity changes and small safety gaps.

This poses the question how relevant these investigations are for real traffic accidents. It appears that the probability of dangerous situations is systematically overestimated due to the correlations induced by the fact that accidents are not explicitly modelled. If an "accident" has occurred it is rather likely that another one involving the same vehicles will occur shortly after that because the conditions are usually still fulfilled.

This short discussion shows that additional work is required to understand the conditions better under which accidents are likely to occur. In addition

[3] This is by now the standard terminology due to my insistence in several referee reports.

this might help to further improve the dynamical properties of the models since accident avoidance is a major factor in the driving behaviour. A special challenge is the combination of realistic car-following behaviour with limited deceleration capabilities.

A first step in the direction of a more realistic investigation of traffic accidents has been in [37]. A model is proposed that describes the dynamics of a platoon of vehicles undergoing emergency braking, but taking into account reaction times and braking capabilities. The model can be applied to real platoons obtained from empirical studies which takes into the effects of short headways (Sec. 2.3) and allows to investigate the impact of legal regulations (speed limits or minium headways).

6 Conclusions and Outlook

We have seen that cellular automata are able to reproduce the empirically observed properties of highway traffic. On a macroscopic level, the basic observed spatio-temporal structures like spontaneous jams can already be found in the NaSch model. More sophisticated models even show the microscopic characteristics obtained from single-vehicle data and related structures like the occurrence of a synchronized phase. However, these models still need several parameters that have to be adjusted. Since this also applies to other model classes, we need a better understanding of the underlying mechanisms to reduce the number of necessary parameters. Maybe a more detailed look at the dynamical processes leading to accidents can be useful in this respect.

Nowadays, cellular automata are so realistic that they form the basis for reliable traffic forecasts. These extend the idea of *online simulation* [38] which is an (iterative) algorithm to derive the state of a traffic network from incomplete information obtained from detectors at certain points in the network. This is possible due to efficiency and accuracy of the underlying dynamical models. The information is made available on the internet (see e.g. [39]) and can be used to plan a trip. Nowadays even travel times can be predicted. Drivers may then decide to change the travel route or travel time or even try to avoid the trip or use public transport.

References

1. Prigogine I, Herman R (1971) Kinetic Theory of Vehicular Traffic. American Elsevier, New York.
2. Chowdhury D, Santen L, Schadschneider A (2000) Phys. Rep. 329:199.
3. Helbing D (2001) Rev. Mod. Phys. 73:1067.
4. Schadschneider A, Pöschel T, Kühne R, Schreckenberg M, Wolf DE (eds) (2007) Traffic and Granular Flow '05. Springer.
5. Rothery RW (1992) Car following models. In: Gartner N, Messer CJ, Rathi AK (eds) Traffic Flow Theory. Transportation Research Board, Special Report 165.

6. Chowdhury D, Nishinari K, Santen L, Schadschneider A (2008) Stochastic Transport in Complex Systems: From Molecules to Vehicles. Elsevier.
7. Treiterer J (1975) Ohio State Technical Report No. PB 246 094.
8. Sugiyama Y, Nakayama A, Fukui M, Hasebe K, Kikuchi M, Nishinari K, Tadaki S, Yukawa S (2005) In Hoogendoorn S, Luding S, Bovy PHL, Schreckenberg M, Wolf DE (eds) Traffic and Granular Flow '03. Springer.
9. Daganzo CF, Cassidy MJ, Bertini RL (1999) Transp. Res. A 33:365.
10. Neubert L, Santen L, Schadschneider A, Schreckenberg M (1999) Phys. Rev. E 60:6480.
11. Kerner BS (2004) The Physics of Traffic. Springer.
12. Helbing D, Treiber M (2002) Cooperative Transportation Dynamics 1:2.1.
13. Kerner BS, Rehborn H (1996) Phys. Rev. E 52:R1297.
14. Knospe W, Santen L, Schadschneider A, Schreckenberg M (2002) Phys. Rev. E 65:056133.
15. Nagel K, Schreckenberg M (1992) J. Physique I 2:2221.
16. Schreckenberg M, Schadschneider A, Nagel K, Ito N (1995) Phys. Rev. E 51:2339.
17. A Java applet for the simulation of the NaSch and VDR model can be found at `http://www.thp.uni-koeln.de/~as`.
18. Barlovic R, Santen L, Schadschneider A, Schreckenberg M (1998) Eur. Phys. J. B 5:793 (1998).
19. Maerivot S and De Moor B (2005) Phys. Rep. 419:1.
20. Knospe W, Santen L, Schadschneider A, Schreckenberg M (2004) Phys. Rev. E 70:016115.
21. Knospe W, Santen L, Schadschneider A, Schreckenberg M (2002) Phys. Rev. E 65:015101(R).
22. Knospe W, Santen L, Schadschneider A, Schreckenberg M (2000) J. Phys. A 33:L477.
23. Kerner BS (2001) Networks and Spatial Economics 1:35.
24. Kerner BS, Klenov SL, and Wolf DE (2002) J. Phys. A 35:9971.
25. Lee HK, Barlovic R, Schreckenberg M, Kim D (2004) Phys. Rev. Lett. 92:23702.
26. Pottmeier A, Thiemann C, Schadschneider A, Schreckenberg M (2007) In [4], p. 503.
27. Bando M, Hasebe K, Nakayama A, Shibata A, Sugiyama Y (1995) Phys. Rev. E 51:1035.
28. Daganzo CF (1995) Transp. Res. 29B:277.
29. Boccara N, Fuks H, Zeng Q (1997) J. Phys. A 30:3329.
30. Huang D (1998) J. Phys. A 31:6167.
31. Yang X, Ma Y (2002) J. Phys. A 35:10539.
32. Jiang R, Wang XL, Wu QS (2003) J. Phys. A 36:4763.
33. Yang X, Zhang W, Qiu K, Xu W, Tang G, Ren L (2007) Physica A 384:589.
34. Jiang R, Jia B, Wang XL, Wu QS (2007) Physica A 377:633.
35. Moussa N (2005) Int. J. Mod. Phys. C 16:1133.
36. Moussa N (2004) Eur. Phys. J. B 41:421.
37. Appert-Rolland C, Santen L (2007) In [4], p. 623.
38. Wahle J, Neubert L, Esser J, Schreckenberg M (2001) Parallel Comp. 27:719.
39. `http://www.autobahn.nrw.de`.

Replications in Stochastic Traffic Flow Models: Incremental Method to Determine Sufficient Number of Runs

Nicolas Chiabaut and Christine Buisson

Laboratoire Ingénierie Circulation Transport - INRETS / ENTPE - Université de Lyon, Rue Maurice Audin, 69518 Vaulx En Velin CEDEX, France
chiabaut@entpe.fr, christine.buisson@entpe.fr

Summary. This paper tackles the issues of the minimal and sufficient number of replication needed to evaluate correctly the mean value of a stochastic simulation results but also the shape of the results' distribution. Indeed, stochasticity is more and more widespread in traffic flow models.

On one hand, microscopic models try to reproduce inter-vehicular deviation through stochastic algorithm. Distributions are sources of randomness. Even if many articles discuss the need for a certain number of simulations to obtain reliable results, they seldom if ever suggest a way to determine this number. Different simulations runs can produce various results due to a randomly assignment of desired speed of each car for example.

On other hand macroscopic models have no individual parameter. This can prevent them from representing some traffic phenomena as roundabout insertion; lane-changing; various desired speed... Stochasticity can overcome those weaknesses. Recently [1] presents a microscopic Lagrangian solution of LWR model allowing individual fundamental diagram.

Thereby, numerous runs have to be computed to estimate the mean value of a measure of effectiveness (MOE) but also to test if the results come from a particular stochastic process. The knowledge of the whole distribution allows us to determine every percentile needed (for example the 5% worse situations).

The aim of this paper is to propose a way to identify such distributions and to estimate the minimal number of replications that one should make to obtain a given confidence level.

We will focus on car-following component of models while lane changing, insertion (ramps, roundabout), and assignment will not be considered. However the proposed methodology can be applied to any components of traffic models.

1 Review of Distribution in Microscopic Traffic Simulation Models

1.1 Car Following Behavior Experimental Distributions

As a human process, traffic flow is full of individual parameters. These parameters hardly measurable can have a high dispersion. It will be quite difficult to provide an exhaustive overview, so note the commonly behavioral parameters used in both microscopic and macroscopic traffic flow models:

- Reaction time: Historical measurement can be found in [2]. Values are comprised into 0.7 and 2 second.
- Safe headway,
- Desired speed,
- Acceleration and deceleration ability of vehicles,
- Vehicles weight and size,
- Drivers aggressiveness ...

1.2 Simulation Tools Input Parameters Distribution

Most of the microscopic traffic simulations models have the tremendous ambition to reproduce each individual driver behavior. Actually, they are reproducing only a part of the above-mentioned distribution of individual vehicle or driver parameters. For instance, reaction times are often chosen identically for any vehicle. Even if this major assumption permits to reduce computational difficulties, it neglects effects of inter-individual dispersion of behaviors.

For other individual parameters (desired speed or safe headways), the distributions are usually embedded into microscopic models. Default values for parameters' mean and variance are usually proposed to microscopic simulation users.

1.3 Output Analysis

The measure of effectiveness (MOE) has a strong impact in the analysis of the simulations outputs. Two families of MOE can be easily distinguished:

- Aggregated measures as the queue length: theses variables result from the global behavior of the traffic flow and their values are somehow the sum of individual dispersion. Thereby, these MOEs are less sensible to the number of replications than some other MOEs.
- Individual measures as the travel time: these outputs must be studied not only through mean values but also through extremal values.

Furthermore, the final use of the MOE plays a role in the needed replication number. It is addressing at least two main questions:

- In the most frequent case, general results of the simulated system are searched. In this case, the minimal number of replication must be chosen to ensure that the complete distribution will appear within the replication set and that mean and variance are reflecting with fidelity the traffic conditions that will occur if the simulation is realistic.
- One may want to be sure that the system will never or in very rare cases be in situation resulting in MOE higher or lower than a given threshold. For instance, this can be the maximum waiting time in the on-ramp signal: if more than 5% of the incoming flow suffers a waiting time of 5 minutes then we will exclude this policy, even if mean waiting time value is correct. We will then speak of an "exclusion simulation study".

The total number of replications must be determined carefully, in order to avoid, on the one hand the waste of computation time, on the other hand the lack of been able to predict an extreme case of simulation results. We will here present an *a posteriori* method that can ensure that the minimal and necessary number of replications is attained. We will take into account various aspects:

- Is the MOE a result of a distribution (mean or extreme value)?
- Is it a global and aggregated value?
- Is simulation study an exclusion or a comprehensive one?

2 Illustrative Example on Particular Stochastic Model

This section presents the test bed that we used to develop our method. The well-known LWR macroscopic model is resolved at a microscopic scale [1]. It allow us to capture multiple behaviors in our model whereas the robustness and readiness of a first order macroscopic model are kept.

2.1 The Lagrangian Godunov Resolution of the LWR Model

We will present briefly the numerical scheme used for the Lagrangian resolution of the LWR model. Readers could refer to [3] for the whole theory.

Let us define $Q(k, x, t)$ as the fundamental diagram depending on k the density, x space coordinate, and t the time. In the remainder of the paper, the fundamental diagram will be triangular. k_c will define the critical density, k_m the jam density, v_m the free flow speed, i.e. $\partial_k Q(k, x, t)$ for $k \leq k_c$, and w the wave speed i.e. $-\partial_k Q(k, x, t)$ for $k \geq k_c$.

Recently, [3] proposes to solve the LWR model in Lagrangian coordinates inspired by the variational principle explained in [4]. The LWR solution is $X(n, t)$ the solution of a Hamilton-Jacobi equation (1).

$$\partial_t X = V^*(-\partial_n X, n, t) \tag{1}$$

where V^* is speed-spacing relationship depending on spacing s, i.e. $-\partial_n X = s = 1/k$, n the vehicle number and t the time. $v(n,t) = \partial_t X$ is the speed of the vehicle n.

The conservation equation associated is the usual LWR hyperbolic equation express in s and $v = V^*(s,n,t)$ variables:

$$\partial_t s(n,t) + \partial_n V^*(s,n,t) = 0 \tag{2}$$

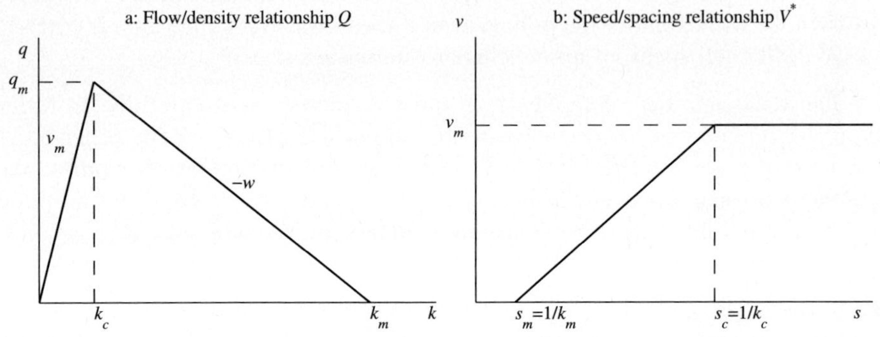

Fig. 1. (a) Fundamental diagram, (b) speed-spacing relationship.

Numerically one can compute the model on a GODUNOV scheme with parameters Δt and Δn. The discrete solution becomes [3] for a triangular fundamental diagram Q:

$$X(n,t+\Delta t) = \min\Big(X(n,t) + v_m \Delta t, (1-\alpha)X(n,t) + \alpha X(n-\Delta n,t) - w\Delta t\Big) \tag{3}$$

where:

- v_m is the free speed of vehicle n,
- $\alpha = wk_m \Delta t/\Delta n$ where w (respectively v_m) is the wave speed (respectively the free flow speed) of the vehicle $n - \Delta n$.

As in the usual GODUNOV resolution of the LWR model, a CFL condition is needed to avoid numerical diffusion.

$$\Delta n \geq \max_{s,n,t} \left|\partial_s\big(V(s,n,t)\big)\right| \Delta t \tag{4}$$

2.2 Distribution of Fundamental Diagrams

The α parameter is particular to each vehicle n and linked to the FD. Assuming that time-step Δt and volume-step Δn stay constant during the simulation and common to all the vehicles, wave speed w and maximal density k_m could be distributed. Formally each probability distribution could be used. We are focusing here on normal distribution, Poisson and bimodal processes. Physical meaning of the bimodal distribution is a flow composed only by heavy trucks and light vehicles. Both of the parameters have the same process centered on a mean value.

3 General Framework

The aim of this paper is to propose a method to determine the sufficient number of replication needed to well evaluate the mean value of the results but also the shape of the result's distribution. The tests we proposed are done *a posteriori* and are based on a statistical test of adequacy.

3.1 Goals of the Proposed Method

The goal of the paper is to propose a method to evaluate adequacy of the result distribution with theoretical laws. The following method is built on existing statistical test.

The first stage of the proposed method is to determine if the number of runs made is sufficient to estimate the mean value of the results according to a given confidence level. For practitioners, it is often sufficient to have an estimation of the mean scenario. The paper will illustrate a statistical test easy to compute which ensure by an incremental method that enough simulations have been done.

The second stage consists on a adequacy test to determine the shape of the result distribution. Once again, we use existing statistical tests. This incremental process accept or reject particular shape of the distribution.

These two steps allow to completely determine the distribution of the results with a given confidence level. Analysis of the MOE can be followed like percentile studies, standard deviation estimation ...

3.2 Statistical Tests

We used three different statistical tests for evaluating mean value and adequacy with a theoretical process. To the author's knowledge publications about replications propose only formula to obtain a good point estimate of mean value. Therefore our method is build on two step. First incremental approach taken from [5] and [6] allows us to obtain an estimate of the mean. Then the second stage comprises of statistical tests of the null hypothesis that data come from a specified process.

Student Test

First of all, real mean value μ must be well estimated by $\bar{X}$. [6] propose a method based on Student test (t-test) to assure that enough replications have been ran. The algorithm could be found in the report [6]. It is based on the construction of the confidence interval for the mean value. This interval can be determined using the STUDENT distribution:

$$\bar{X}(n) \pm t_{n-1,1-\alpha/2}\sqrt{\frac{S^2(n)}{n}} \tag{5}$$

where:

- $\bar{X}(n)$ is the estimate of μ the real mean value from n simulation runs.
- $S^2(n)$ is the estimate of σ the standard deviation from n simulation runs.
- n is the number of replication.
- α is the level of significance.
- $t_{n-1,1-\alpha/2}$ is the critical value of the t-test for $n-1$ degrees of freedom and significance α.

The STUDENT test only concerns samples identically independently distributed. This assumption will be confirmed with tests of adequacy.

Jarque-Bera Test for Normal Distribution

The JARQUE-BERA test (JB-test) is a measure of goodness-of-fit with a normal distribution. It is based on the result's sample kurtosis and skewness. Kurtosis can be interpreted as the measure of the "peakedness" of the probability process of a real-valued random variable. Similarly skewness is the measure of the asymmetry. A distribution has positive skew if the right tail is longer or fatter and negative skew if the left tail is longer or fatter.

Normal distribution has some particular values for skewness and kurtosis. The JB-statistic can be used to test the hypothesis that the results come from a normal process with unknown parameters.

$$JB = \frac{n}{6}\left(S^2 + \frac{(K-3)^2}{4}\right) \tag{6}$$

where:

- n is the number of observations (i.e. runs),
- S is the skewness of the data $S = \frac{\frac{1}{n}\sum_{i=1}^{n}(x-\bar{x})^3}{\left(\frac{1}{n}\sum_{i=1}^{n}(x-\bar{x})^2\right)^{3/2}}$
- K is the kurtosis of the data $K = \frac{\frac{1}{n}\sum_{i=1}^{n}(x-\bar{x})^4}{\left(\frac{1}{n}\sum_{i=1}^{n}(x-\bar{x})^2\right)^{2}}$

JB-statistic calculated is compared with a critical value to accept or reject hypothesis.

The JB-test is more recommended than the χ_2-test to test the null hypothesis that the data come from a normal distribution. The χ_2-test is a discrete test and do not use all the characteristics of the Gaussian probability process on the contrary of JB-test.

Kolmogorov-Smirnov Test for any Distribution

The one sample KOLMOGOROV-SMIRNOV (KS-test) compares the empirical distribution function with the cumulative distribution function specified. In theory any distribution function can be used. KS-test is more sensitive at point near the median of the distribution than at its tails. To well evaluate the 5% percentile of worth situation when we expect a Gaussian distribution process, JB-test is much more powerful. We will only use the KS-test to test the hypothesis that the results come from a Poisson distribution function. The KS-statistic is calculated and compared with the critical value of KS-test.

$$KS = \max_x |F_n(x) - F(x)| \tag{7}$$

where:

- F_n is the empirical distribution function for n observation xi defined as:

$$F_n(x) = \frac{1}{n} \sum_{i=1}^{n} \begin{cases} 1 & \text{if } y_i \leq x \\ 0 & \text{otherwise.} \end{cases}$$

- F is the theoretical repartition function of the process (Poisson distribution).

4 Results

4.1 Experimental Test Bed

Two theoretical scenarios have been defined: (i) a lane restriction scenario and (ii) a traffic signal on a single lane. The aim of these simulations is to illustrate the method and to stress out differences between estimating the mean value of a MOE and estimating the whole distribution.

Minimal replication number found may have no physical meaning. However, it still gives a glance at how proposed statistical tests can run.

For both simulation cases, parameters such as jam density and desired speed have been distributed through (i) a normal process, (ii) a Poissonian law and (iii) a bimodal process whereas the in-flow is constant.

Finally, we determine the minimal replication number to obtain both estimated mean value and distribution's shape with a level-confidence of 95% for maximal travel time and mean travel time.

4.2 Results

Lane Restriction

As explained before, first we test our method with the lane restriction scenario for parameters normally distributed. The mean value of the results distribution is well estimated with 9 replications of the simulation (figure 2). The

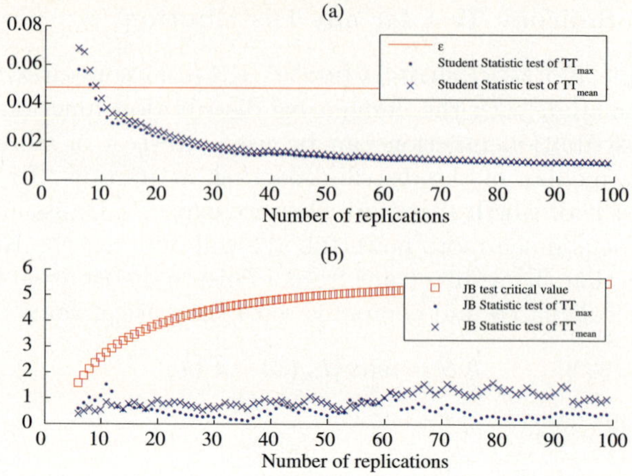

Fig. 2. Test of adequacy for a normal distribution: (a) Student test, (b) JB test.

hypothesis of adequacy with a normal distribution of the results is always accepted according to the JB-TEST. We can conclude that results are normally distributed and their distribution is totally determined by only 9 runs.

At a glance, the results are much more complicated when parameters are distributed through a bimodal process. Thus, this time 13 replications are needed to well evaluate the mean value of the maximum travel time and more than 16 runs for the mean travel time. Despite that the JB-TEST is true for a few number of replications, once the mean value is well estimated according to the student test, the JB-TEST rejected the hypothesis that the results are normally distributed. A KS-TEST has to be done to consider that MOEs are coming from a particular process. However, figure 3 shows that we can not conclude results distribution fitted a theoretical process.

Results with parameters distributed through a Poissonian process are not displayed here. We find that only 5 replications are needed to estimate the mean value of the both MOEs but more than 50 to obtain a positive results of the test of adequacy of the results with a normal distribution.

Traffic Signal

Figure 4 shows results of statistical tests for the traffic signal scenario normally distributed parameters. The mean value of both MOEs are well estimated with only 5 runs. However, according to the JB-TEST more than 18 replications are needed to ensure that results come from a normal process.

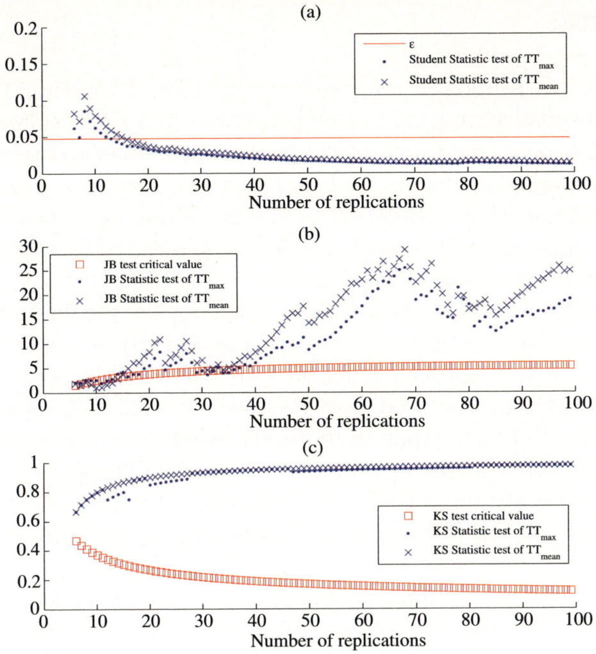

Fig. 3. Test of adequacy for a bimodal distribution: (a) Student test, (b) JB test, (c) KS test.

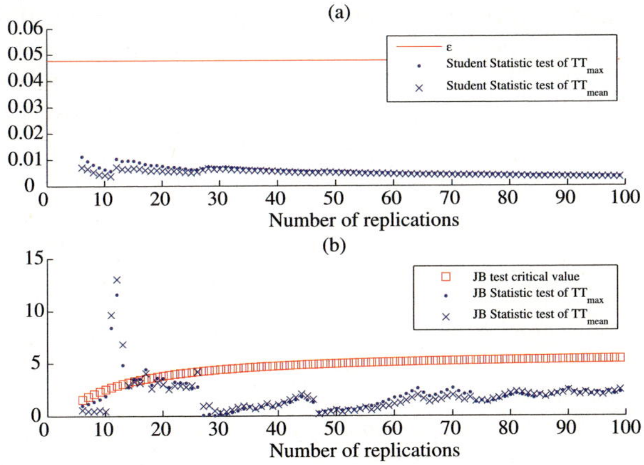

Fig. 4. Test of adequacy for a normal distribution: (a) Student test, (b) JB test.

We do not display results of Poissonian process and bi-modal law here. Note that parameters distributed through a bi-model law leads to normally distributed results where results of parameters distributed through a Poissonian have not a particular shape.

5 Discussion

In this paper, we proposed a operational method to ensure that (i) mean value of the result distribution is well estimated and (ii) result distribution is adequate with a particular theoretical process. This method relies on existing statistical tests.

We illustrate it with a two theoretical scenarios based on a microscopic resolution of the well-known LWR model. It allows to distribute parameters through vehicles according stochastic laws.

The obtained results stress out the efficiency of the method. They also give a first idea of the number of replication needed to well-estimate the MOE distribution for a very simple model on a very simple scenario.

Though these results depend strongly on the model and scenario we choose. The main insight of the paper is more the method we proposed. It can be used with every microscopic or macroscopic model. Those statistical test can be implemented aside of the model. Even if the method is *a posteriori*, we strongly recommend to add it when calibration and validation of stochastic model are made.

Finally, some phenomenon are smoothing stochasticity in traffic flow such as traffic signal. Authors are now studying these phenomenon.

References

1. Ludovic Leclercq. Hybrid approaches to the solutions of the "LIGHTHILL-WITHAM-RICHARDS" model. *Transportation Part B*, 41:701–706, 2007.
2. M. Brackstone and M. Mc Donald. Car-following: a historical review. *Transportation Research Part F: Psychology and Behaviour*, (2):181–196, 1999.
3. Ludovic Leclercq, Jorge Andres Laval, and Estelle Chevallier. The Lagrangian coordinates and what it means for first order traffic flow models. In Richard E. Allsop, Michael G.H. Bell, and Benjamin G. Heydecker, editors, *Transportation and Traffic Theory 2007 (ISTTT17)*, pages 735–754, London, U.K., 2007. Elsevier.
4. C. F. Daganzo. A variational formulation of kinematic waves: basic theory and complex boundary conditions. *Transportation Research Part B*, 39:187–196, 2005.
5. A. M. Law and W. D. Kelton. *Simulation, Modeling and Analysis.* McGraw-Hill, Boston, 3rd edition, 2000.
6. W. Burghout. A note on the number of replication runs in stochastic traffic simulation models. Technical Report CTR2004:01, Center for Traffic Research, 2004.

Capacity and Capacity Drop of a Revolving Door

Winnie Daamen[1], Serge P. Hoogendoorn[1], and Henk van Wijngaarden[2]

[1] Delft University of Technology, Stevinweg 1, 2628 CN Delft, The Netherlands
w.daamen@tudelft.nl, s.p.hoogendoorn@tudelft.nl
[2] Boon Edam Group Holding B.V., Ambachtstraat 4, 1135 GG Edam, The Netherlands hwn@boon.nl

1 Introduction

Designing safe and efficient pedestrian infrastructure is complicated [1–7]. The main reason for this is the complex interaction between the infrastructure users (two directional pedestrian flows, crossing flows, waiting pedestrians, etc.), and the infrastructure. These interactions make it nearly impossible to predict the infrastructure functioning beforehand, without using proper tools, such as computer simulation.

Usually, entrance points (such as doors, sliding doors, revolving doors and turnstiles) are the active bottlenecks in the design, both in normal and in exceptional conditions (e.g. evacuations). It is therefore remarkable that available computer software does generally not model entrance points explicitly and accurately [8–13]. Moreover, new entrance concepts and technologies are developed, of which the functioning can only be correctly predicted using accurate simulation tools correctly considering pedestrian behavior at the entrance point.

Before simulation tools can be extended with pedestrian behavior in and around entrance points (specifically revolving doors), we need to further our insight into the behavior of pedestrians and pedestrian flows near these points. This contribution describes laboratory experiments involving a revolving door and in particular focusses on the capacity drop suggested by the first results obtained from the data.

2 Laboratory Experiments Involving a Revolving Door

To derive a model describing pedestrian behavior in and around revolving door accurately, detailed insights into this behavior are required. These data can be obtained by collecting real-life data or by performing laboratory experiments. The chosen data collection method should fulfill the following requirements:

1. The resulting data set should contain microscopic pedestrian behavior, e.g. individual walking speeds, acceleration, interaction behavior and individual pedestrian characteristics.
2. The resulting data set should contain macroscopic flow behavior, e.g. flow, density and (average) speed.
3. Observation conditions should be stable to get the best possible results.
4. Flow conditions should be stable to have multiple observations in similar flows.
5. The observed behavior should be realistic.
6. The revolving door should perform in conditions varying between low flows and saturated flows in which congestion is likely to occur.
7. The population should correspond to the population in different environments, e.g. transfer stations and shopping centers.

Apart from the aforementioned requirements, some practical requirements came up with respect to the observation equipment, the locations available to attach the equipment at the observed site and the time needed to perform the observations. During real-life observations, flow conditions may vary only slightly, which makes congestion hard if not impossible to observe. To observe different populations (requirement 7) different areas will have to be observed. Advantages are that the observed behavior is natural (if people do not recognize the observation equipment) and that a random population is observed. In a controlled environment the researcher is in control of all conditions, concerning not only external factors, but also flow composition and flow size, variation of the flow over time and revolving door characteristics (e.g. revolving speed and safety sensor adjustment). Disadvantages are the cost to build up a revolving door in the laboratory and the fact that the observations might be biased since participants do not behave naturally. The latter can be overcome by keeping the experiments on a skill-based level, that is that pedestrians will behave in a subconscious way. Since pedestrians are familiar with revolving doors, they will not expose 'new' behavior, thus behaving as in real-life. These considerations as well as our experiences in performing experiments to observe pedestrian behavior [14, 15] lead to the decision to perform laboratory experiments.

2.1 Experimental Set Up

Two types of variables are distinguished for the experimental set up, namely *experimental variables* and *context variables*. Experimental variables are systematically changed during the experiments to isolate and to describe their influence within the system. Context variables are variables that cannot be influenced by the researcher, but do influence the system performance.

The experimental variables are free door rotation speed, population and pedestrian flow composition, flow and walking pattern.

The initial rotation speed of the revolving door is the speed with which the tips of the door wing travel over the circumference of the entrance. At low rotation speed, pedestrians have much time to fill a segment, but their travel time through the door will be higher. At high rotation speed, the opposite occurs: pedestrians have less time to fill the segment, but they are able to pass the door faster.

Populations are combinations of different types of pedestrians. Multiple variables are brought together in this experimental variable, such as pedestrian type (gender, trip motive), free speed indication and the degree in which pedestrians want to maintain their free or desired speed (if someone has to catch a train, he will put more effort into keeping a high speed than when he is strolling along a shopping street). The size of the pedestrian flow determines the load upon the revolving door. The aim is to gradually load the entrance to and beyond its capacity and then gradually decrease the load again.

Walking patterns indicate predefined routes that pedestrians follow during the experiments. Changing those routes will lead to loads from different directions, thus distinguishing between unidirectional flows, bidirectional flows and crossing flows. When the door is only loaded from one direction, pedestrians have a longer time to fill the segment than when an opposite flow is present that first has to empty the segment. In case of opposite flows, strict separation of the two flows (e.g. by using a fence) might decrease the hinder and thus optimize the throughput of the revolving door.

As indicated before, the above described experimental variables are systematically changed during the experiments to identify their influence on the system performance and pedestrian behavior. In an ideal situation, all combinations of variables should be tested. Due to the time restrictions, a selection has been made of the values of the observed variables. The performed scenarios are shown in Table 1.

2.2 Performing the Experiments

The experiments have been performed in the Stevin Laboratory of the Delft University of Technology. The walking area was 10 meters wide and 20 meters long (see Fig. 1(a)). Two video cameras were mounted perpendicular above the pedestrian flow at a height of 10 meters. The revolving door was located in the middle of the walking area. A digital camera observed the walking area upstream of the revolving door (see Fig. 1(b)), while an analogue camera observed the revolving door (see Fig. 1(c)). A second analogue camera next to the door was used to register the characteristics of pedestrians entering the door (see Fig. 1(d)).

Eighty persons participated in the experiment, whose age varied between 14 and 80 years. Each pedestrian was given a white T-shirt and a colored cap, the color of which depended on the role of the participant (normal walking pedestrian, fast walking pedestrian, walking as a couple or walking with a shopping trolley). The participants were subdivided into 8 groups of 10 people. Acquaintances were separated as much as possible to stimulate individual behavior. 'Real' couples were grouped together to form realistic couples in the

Table 1. Overview of experimental scenarios

	Pedestrian flow		Revolving door			Environment				
	Population		Rotation speed			Obstacle		Walking pattern		
Scenarios	50%N,50%F	50% N, 40% C, 10% T	0.50 m/s	0.75 m/s	1.00 m/s	Free	Wicket	Unidirectional flow	Opposite flows	Crossing flows
1	X		X			X		X		
2	X			X		X		X		
3	X				X	X		X		
4	X			X			X		X	
5		X	X			X			X	
6		X		X		X			X	
7		X			X	X			X	
8		X		X		X				X

scenarios five to eight. All group compositions were heterogeneous, indicating that groups consisted of men and women of different age.

2.3 Data Analyses

Before the gathered data can be analyzed, the raw video images need to be processed. For this, dedicated software has been developed at the department of Transport & Planning to identify and track pedestrians. The software results in a data set containing for each pedestrian his or her location over time (each 0.04 second), thus allowing to draw trajectories and perform detailed traffic analyses.

To obtain data on door capacity, we use cumulative curves (see Fig. 2). A cumulative curve describes the time moments that passengers pass a given cross-section. We have determined two cross-sections: one at the start of the walking area and a second one downstream of the revolving door (see Fig. 2(a)). Since the revolving door is considered to be the bottleneck, the departure curve can be used to determine the capacity of the revolving door.

The use of cumulative curves implies a first-in-first-out order of pedestrians. Since the system is empty at the start of each scenario, the travel time w of the Nth pedestrian corresponds to the horizontal distance between the departure curve $D^{-1}(N)$ and the arrival curve $A^{-1}(N)$ (see Fig. 2(c)) and can be calculated by

$$w\left(N\right) = A^{-1}\left(N\right) - D^{-1}\left(N\right) \tag{1}$$

(a) Laboratory overview

(b) Top view of area upstream of door

(c) Top view of revolving door

(d) Image from the side

Fig. 1. Overview of the laboratory and images from the three video cameras.

To calculate the delay of an individual pedestrian, a 'free' departure curve V(t) has been drawn, which is similar to the arrival time, but shifted over the free walking time to the right. This free walking time is obtained by calculating the travel time at low flows, since the delay due to the door is still included, while a pedestrian does not have waiting time due to congestion. When the realized departure curve is on top of the 'free' departure curve, pedestrians encounter no delays. The horizontal distance between the realized departure curve and the 'free' departure curve indicates the delay of each pedestrian. The total delay is then the cumulative delay of all pedestrians that arrived at X2 during the considered time period, while the average delay is the total delay divided by the number of pedestrians. While the arrival curve is rather smooth, the departure curve has a clear step-wise character. The varying step sizes show the varying capacity of the door over time. During oversaturation (queuing) full segments empty quickly, but because of the door breakdowns it takes time before the next segment opens again. Hence, the higher the load upon the door, the more distinctive the steps become. The height of the steps indicates the number of pedestrians inside a segment, while the duration of a

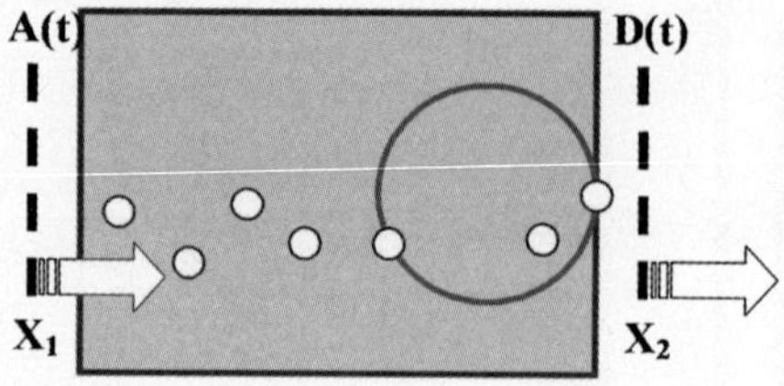

(a) Arrival and departure cross-section

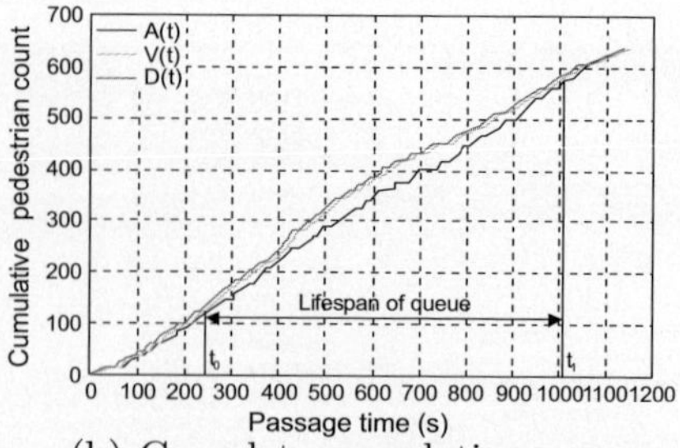

(b) Complete cumulative curve

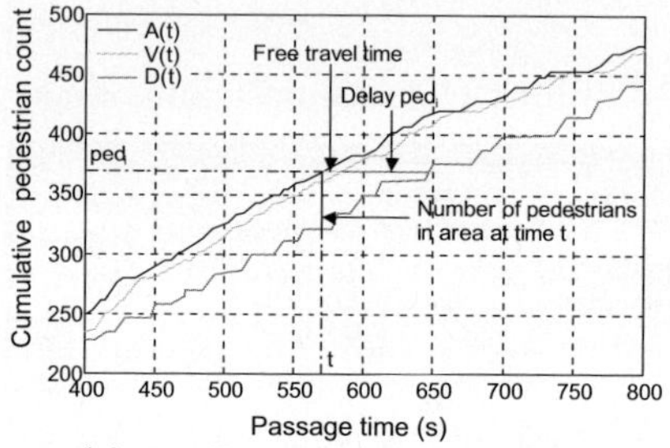

(c) Traffic variables in cum. curves

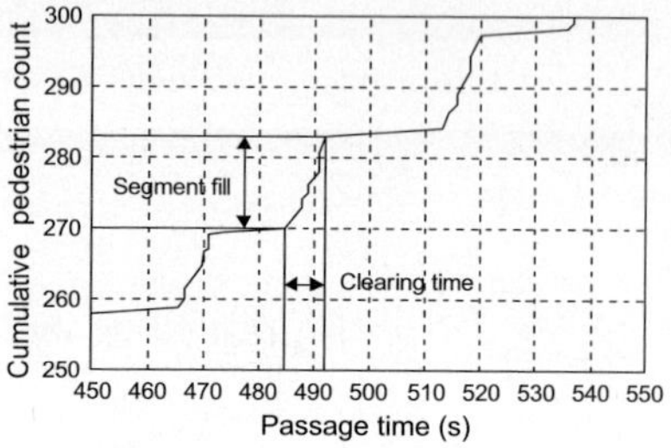

(d) Departure curve characteristics

Fig. 2. Cumulative curves.

step (so-called *clearing time*) indicates the time it takes to empty the segment (see Fig. 2(d)).

3 Capacity of the Revolving Door

Capacity is defined as the maximum number of passengers passing a cross-section during a given time period. Capacity is stochastic, with a certain mean and standard deviation. We can distinguish between two capacities: the *pre-queue capacity* (outflow from bottleneck just before congestion sets in) and the *queue discharge rate* (outflow from bottleneck during congestion).

The nature of a revolving door (stepwise release of pedestrians each time a segment opens) limits the number of useful observations to determine the capacity. To overcome this problem, we will apply slanted cumulative curves. The slope of a cumulative curve (in this case the departure curve) describes the throughput and capacity of the revolving door. However, it is hard to determine variations in the slope directly for the cumulative curves. Therefore, we apply slanted cumulative curves, which are cumulative curves in which the variations are shown with respect to a specific flow q_0 (see Fig. 3):

$$D_{slanted}(t) = D(t) - q_0(t - t_0) \tag{2}$$

where $D_{slanted}(t)$ is the slanted cumulative curve, $D(t)$ is the departure curve, q_0 is a first crude estimate for the global average capacity and t_0 is the time the observations started.

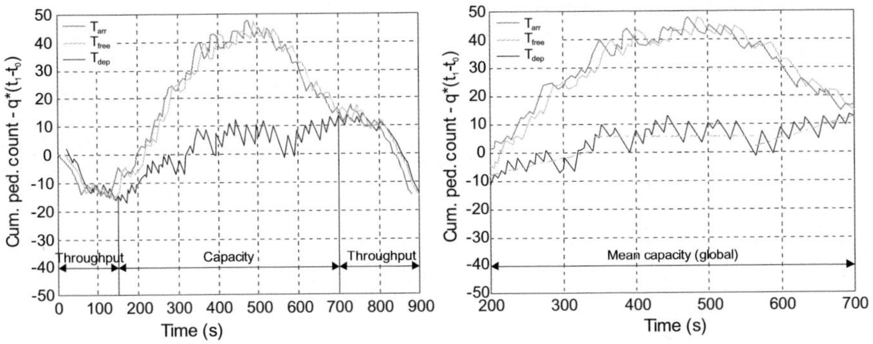

Fig. 3. Slanted curves of sc. 1 to determine capacity of revolving door.

The interaction between pedestrians and revolving door results in a varying system performance. The fluctuating capacity is indicated by the trends in the slanted departure curve during queuing time (see the figure on the righthand side in figure 3). The total capacity C is here determined as the total number of pedestrians passing the door during oversaturation:

$$C = \frac{D_{slanted}(t_{qe}) - D_{slanted}(t_{qb})}{t_{qe} - t_{qb}} + q_0 \tag{3}$$

where t_{qb} indicates the start of the considered period and t_{qe} indicates the end of this time period.

Table 2 shows an overview of the capacity trends and the global capacity of each scenario. For each scenario at least three trends can be distinguished. The first and the last trend are usually highest, while the trends in between are lower. This suggests that the capacity depends on the length on the queue: the shorter the queue, the higher the capacity.

The observed capacity of the revolving door varies between 25.3 P/min (scenario 5) and 38.9 P/min (scenario 3). For both commuter and shopping population, an increase in door rotation speed results in a higher mean capacity (up to 19%). The opposite flow in scenario 4 clearly shows a reduction in the capacity (about 15%). Also the population has a large effect on the door capacity: with commuters the capacity is 10.1%–26.2% higher than with a shopping population. Shoppers are strolling and less aggressive than commuters, which benefits the use of the entrance. However, couples and individuals with a trolley require more space and do not feel the urge of completely filling a segment. The occupancy inside the door is therefore far from optimal, which has a negative influence on the capacity.

Table 2. Capacity trends and global capacity of each scenario

Capacity [P/min]	Scenario 1 $v_r = 0.5m/s$	Scenario 2 $v_r = 0.75m/s$	Scenario 3 $v_r = 1.0m/s$	Scenario 4 $v_r = 0.75m/s$
Trend 1	33.0	44.2	48.5	35.2
Trend 2	30.6	35.3	37.8	16.7
Trend 3	34.1	25.1	39.5	27.2
Trend 4	–	36.8	–	33.8
Trend 5	–	–	–	26.7
Global mean	32.7	36.3	38.9	30.6
	Scenario 5 $v_r = 0.5m/s$	Scenario 6 $v_r = 0.75m/s$	Scenario 7 $v_r = 1.0m/s$	Scenario 8 $v_r = 0.75m/s$
Trend 1	25.9	25.2	15.0	25.6
Trend 2	25.6	25.0	32.2	30.9
Trend 3	–	30.0	24.5	27.5
Trend 4	–	45.8	–	25.3
Trend 5	–	25.6	–	–
Global mean	25.3	26.8	28.8	27.5

4 Capacity Drop of the Revolving Door

When the load upon the door increases, the door suffers severe breakdowns caused by pedestrians constantly activating the safety sensors. The breakdown frequencies, and subsequently the observed capacity, are due to the interaction between pedestrians and revolving door. The idea is that only the pedestrians at the head of the queue influence the functioning of the entrance, since they are the ones interacting with the mechanism. However, their behavior might be influenced by the amount of pedestrians that are in the queue. The more pedestrians in the queue, the higher the pressure on the pedestrians in front, the more pushy their behavior and consequently the more often a sensor gets activated. This might lead to a throughput that is higher before congestion sets in than during congestion. To illustrate this, Fig. 4 shows for each scenario the observed throughput (during a complete experiment, indicated by crosses) and the throughput during congestion (capacity, indicated by circles).

For all scenarios, it appears that the highest observed throughput is higher than the highest observed capacity. The difference varies between $2.5P/min$ in scenario 7 and $12.0P/min$ in scenario 3. On average, the difference is more than 17% of the observed capacity.

A similar phenomenon can be observed in car traffic. Here, the pre queue capacity occurs just before congestion sets in. This capacity is higher than the queue discharge rate, which is the capacity during congestion. However, the difference between the pre queue capacity and the queue discharge rate is lower than the difference observed here for pedestrian in revolving doors. Also the cause is somewhat different: where in car traffic the pre queue capacity is higher due to more efficient driver behavior, in pedestrian traffic around

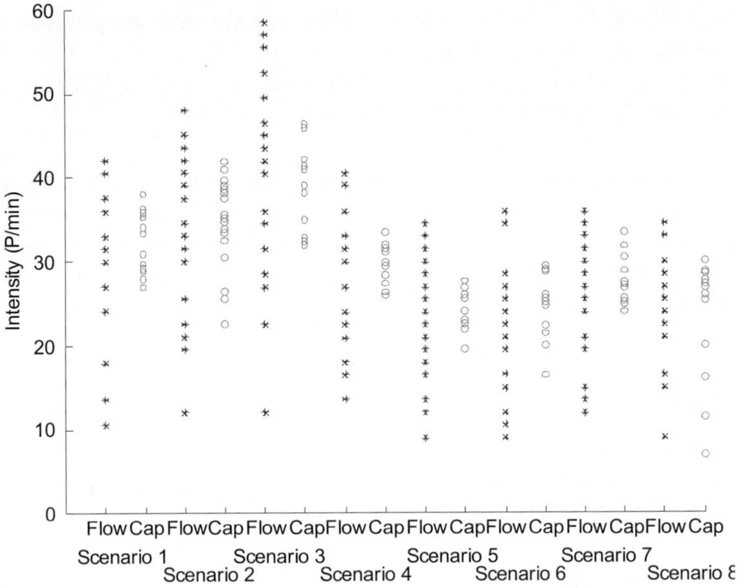

Fig. 4. Observed flows and capacities for all scenarios.

revolving doors, the difference is caused by the activation of the safety sensors in the door.

5 Conclusions and Future Work

Revolving doors are a means to provide access to buildings, while maintaining the interior climate. Although they are frequently used in designs of pedestrian facilities, their behavior is generally not included sufficiently accurate in the pedestrian simulation tools used to assess these designs. Many issues appear to be unresolved, such as the capacity of a revolving (given its characteristics and the characteristics of the pedestrian population), and the queuing processes in front of the door. This contribution has focussed on the capacities observed in a laboratory experiments with a revolving door.

To this end, after automatically tracking the pedestrians on the collected video images, we derived pedestrian trajectories, cumulative curves and slanted cumulative curves. These are used to asses the capacity as well as a possible capacity drop of revolving doors.

For both commuter and shopping population, an increase in door rotation speed results in a higher average capacity (up to 19%). Opposite flows cause a reduction in the capacity (about 15%). When the door is used by commuters, the door capacity is 10.1%–26.2% higher than with a shopping population. This is mainly due to the larger space requirements and the lack of urge to completely fill a door segment. For the different experiments, a capacity drop of 17% on average was observed.

Future work will consist of further elaboration of data analyses and theory and model formulation for pedestrian behavior in and around revolving doors. However, we will start with further analyses of the available data, e.g. with respect to self-organized phenomena. Based on the results of the data analyses we will derive theories describing this pedestrian behavior. These theories will then be translated into models, each describing different aspects of the pedestrian behavior. Our final aim is to include these models in our pedestrian simulation tool Nomad [9].

References

1. Daamen W (2002) A quantitative assessment on the design of a railway station. In: Allen J, Hill RJ, Brebbia CA, Sciutto G, Sone S (eds) Computers in railways VIII. WIT Press, Southampton:191–200.
2. Fruin JJ (1971) Pedestrian Planning and Design. Metropolitan Association of Urban Designers and Environmental Planners Inc., New York.
3. Hoogendoorn SP, Daamen W (2004) Design assessment of Lisbon transfer stations using microscopic pedestrian simulation. In: Allen J, Hill RJ, Brebbia CA, Sciutto G, Sone S (eds) Computers in railways VIII. WIT Press, Southampton:135–147.
4. Helbing D, Buzna L, Johansson A, Werner T (2005) Trans Sci 39(1):1–24.
5. De Neufville R, Grillot M (1982) J of Trans Eng 11(TEI):87–102.
6. TRB (2000) Highway Capacity Manual, Special Report 209. National Academy of Sciences, Washington.
7. Willis A, Gjersoe N, Havard C, Kerridge J, Kukla R (2004) Environment and Planning B: Planning and Design 31(6):805–828.
8. Helbing D, Molnar P (1995) Phys Rev E 51(5):4282–4286.
9. Hoogendoorn SP, Bovy PHL (2003) Optimal Control Applications and Methods 24:153–172.
10. Klüpfel HL (2003) A cellular automaton Model for Crowd Movement and Egress Simulation. PhD thesis, University Duisburg-Essen, Germany.
11. Still GK (2000) Crowd Dynamics. PhD thesis, University of Warwick, United Kingdom.
12. Teknomo K (2002) Microscopic Pedestrian Flow Characteristics: Development of an Image Processing Data Collection and Simulation Model. PhD thesis, Tohoku University, Japan.
13. Willis A, Kukla R, Kerridge J, Hine J (2001) Laying the foundations: The use of video footage to explore pedestrian dynamics in PEDFLOW. In: Schreckenberg M, Sharma S (eds) Pedestrian and Evacuation Dynamics. Springer, Berlin:181–186.
14. Daamen W, Hoogendoorn SP (2003) EJTIR 3:39–59.
15. Daamen W, De Boer E, De Kloe R (2007) The gap between vehicle and platform as a barrier for the disabled; an effort to empirically relate the gap size to the difficulty of bridging it. In: Proceedings of the Transed conference 2007, Montreal, Canada.

Risk Index Modeling for Real-Time Motorway Traffic Crash Prediction

Habib Haj-Salem and Jean-Patrick Lebacque

INRETS/GRETIA, 2 av du General Malleret-Joinville, F-94114 Arcueil, France
haj-salem@inrets.fr, lebacque@inrets.fr

1 Introduction

Control measures introduced to improve traffic performance in motorway traffic include speed limit control, ramp metering, user information aiming at homogenizing the practical speed along the motorway sections and at minimizing the number and the severity of accidents and consequently increasing safety [1]. On the other hand, introduction of electronics and computer systems in the vehicle technologies have significantly contributed to safety and comfort. However, the prediction of the crash in real time is still in investigation phase and some research efforts are dedicated in this area [2]. During the last five years, there is an increasing focus on the development of real time ("potential crash") prediction algorithm on urban motorway traffic [3–5].

With respect to the safety evaluation analysis, the classical approaches consist in collecting incidents/accidents traffic data during the experimented scenarios (traffic control strategies, modification of the infrastructure etc.), and in proceeding to traffic impact and statistical safety analysis of the number of accidents before and after the implementation of these scenarios. However, the constitution of the accident database must includes a minimum number of accident in order to guaranty the statistical significance before undertaking the evaluation procedure. This means that the field data collection period must have a long time comprising between 5 to 10 years, which is the "price to pay" for having significant safety evaluation results.

This paper is focused on the development of risk index modeling using real-data measurements. The risk model can be used either off-line as a safety evaluation index leading to the dramatical reduction of the field test periods, or in real-time like safety monitoring tool (e.g. real time safety user warning system). Another important issue concerns the optimal traffic control strategies such as coordinated ramp metering, speed limit control, route guidance, etc. The risk index can be introduced in the global criterion which can be considered as a multi-criterion function to be optimized in real time (safety index combined with traffic index).

2 Data Base Characteristics

The ring way of Paris is fully equipped with loop detectors located at around every 500 meters apart for traffic measurements of traffic volume, occupancy rate and speed. The developed model index is based on the collection of traffic measurements synchronized with incidents/accidents data on the ring way of Paris. The incidents/accidents data characteristics are collected from the real time Automatic Incident Detection system (AID). These characteristics include: starting time, end time, location, weather conditions, severity. The first step was the collection of the incidents/accidents on the overall ring-way of Paris. The second step was the extraction from the traffic database of the associated traffic measurements. For each incident/accident the considered measurement stations are depicted in the figure 1 and includes two upstream and two downstream measurement stations.

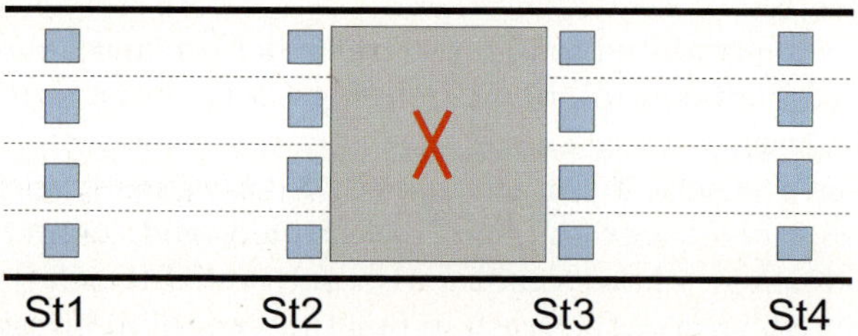

Fig. 1. Topology of the considered stretch measurements for each crash.

The collected traffic data covers two hours (one before and one after the crash) of the measurement stations of the considered stretch for each accident. The time interval of the measurements is equal to one minute.

The final constituted database includes the overall accidents occurred and traffic data during 3 years (2002–2004). The total number of accidents collected is around 900.

After measurements data cleaning, among the 900 accidents, only 85 accidents are considered, corresponding to sunny days. The accidents occurring during the night periods are excluded.

3 Methodology

The applied methodologies are based on statistical analysis of the traffic conditions before the accident [5, 6]. The main objective of this step is to analyze the traffic conditions before the occurrence of the accident and to extract the most important traffic variables to be considered for the risk modelling. A series of multivariate statistical methods is used, with the aim of finding the relationship between the accident and the traffic measurements. Two well-known statistical methods are used: clustering analysis and the most common form of factors analysis [7]. In particular, the principal components analysis is applied to find the non-correlated variables to be used for the risk model building. According to the selected stretch by accident, the total number of variables characterizing the traffic database is equal to 4(stations) $*$ 3(volume, occupancy rate and speed) $*$ 4(number of lanes) which corresponding to 48 variables. For the clustering analysis, several possibilities are investigated:

- Clustering by upstream occupancy rates/lane
- Clustering by downstream occupancy rates/lane
- Clustering by all occupancy rates/lane
- Clustering by upstream occupancy rates/station
- Clustering by downstream occupancy rates/station
- Clustering by all occupancy rates/station

The principal components analysis results demonstrate that the traffic volume, speed and occupancy rates of consecutive measurement stations are very correlated. Consequently, only two variables are considered for the clustering studies and in particular the traffic volumes and the occupancy rates (the speeds variables are omitted).

Lastly, based on the clustering output results, linear regression and non linear logistic modelling approaches are applied using two variables (Oc, Q) for the risk index model building.

In the following section, two cluster analysis cases are reported:

- Clustering by all occupancy rates/lane
- Clustering by all occupancy rates/station

4 Clustering Results

In order to exhibit the particular cluster of traffic conditions which prevail at the time just before the accident, the hierarchical ascending clustering using the statistical tool named "SAS" is performed.

4.1 Clustering Results by all Occupancy Rates per Lane

In this case, five main representative clusters are found. The first cluster is characterized by a homogeneous average occupancy (MOcc) on the 16 measurement locations. The occupancy rates (Occ) are between 9% and 15%, and characterize a low occupancy rate which corresponds to light traffic conditions. This cluster contains 2191 observations and represents 42.96% of all measurements. The cluster 2 gathers the observations with higher and inhomogeneous average occupancy. Indeed, the MOcc are lower on the fast lane; their values vary around the critical occupancy (18% to 23%). The two central lanes have higher occupancy rates and correspond to unstable traffic states. All lanes of the last station (St4) are congested. The occupancy rates range from 24% to 27%. This cluster represents 27.37% of the samples. As regards cluster 3 (representing 8.8% of the data), a clear transition is observed between the MOcc of the upstream stations, which are very high (36% to 52%), and the low MOcc of the downstream stations (6% to 11%). The MOcc of cluster 4 are homogeneous on the 16 measurement locations, with high values, ranging from 30% to 40%. This cluster represents 19.39% of the population. These states correspond to a high level of congestion. Lastly, cluster 5 is characterized by average upstream MOcc (16% to 21%), particularly on the first two lanes, and very congested downstream (52% to 68%). Moreover, we observe that station St_4 is more fluid than station St_3. This cluster is less representative (1.15% of the data). Screening the time evolution of the clusters (one hour before the crash) of all accident records (85 in total), 41 accidents indicate a change of cluster during the last six minutes, i.e. 48% of the cases. If only the last observation before the accident is observed, among the total number of accidents, 39 accidents (46%) are moved to cluster 3. Cluster 3 represents upstream congestion and downstream fluid conditions. The risk modelling is based on the traffic state of this cluster.

4.2 Logistic Regression

The database is split into two parts. The first half is dedicated to the calibration of the linear regression and the second half is used for the validation of the risk model. During the validation process, the same parameters found during the calibration step are used.

The logistic regression is performed by considering this cluster (cluster 3) which presenting the highest level of crash. In this case, the risk model value is set to 1, otherwise it set to 0. Hence, the logistic regression model found is the following:

$$Risk = \frac{1}{1+\exp-(\alpha+f(St_1Oc_i)+G(St_2Oc_i,Q_j)+H(St_3Oc_i,Q_j)+\Phi(St_4Oc_i,Q_j))} \quad (1)$$

where: $f(St_1,Oc) = \beta St_1Oc_1 + \nu St_1Oc_3$
$G(St_2,Oc,Q) = \gamma St_2Oc_1 + \delta St_2Oc_3 + \theta St_2Oc_4 + \xi St_2Q_2$
$H(St_3,Oc,Q) = \vartheta St3Oc_4 + \mu St_4Oc_4 + \sigma St_3Q_2$
$\Phi(St_4,Oc,Q) = \pi St_4Oc_2 + \chi St_4Oc_4 + \varphi St_4Q_1$
and:
$\alpha = -5.7335; \beta = 0.01107; \nu = -0.0827;$
$\gamma = 0.02601; \delta = 0.1102; \theta = 0.1886; \xi = -0.00407;$
$\vartheta = 0.3851; \mu = -0.5809; \sigma = 0.00663;$
$\pi = -0.4483; \chi = -0.5809; \varphi = 0.00449.$

N.B: StiOcj variable corresponds to the occupancy measurements of the station(i) and lane(j).

Using all measurements of traffic volume (Q) and occupancy rate (Oc), the Risk model found includes 13 parameters and 12 variables. The obtained results during the validation step are not satisfactory. As a matter of fact, applying the Risk model on the second half of the database generates a large oscillation of the Risk value between 0 and 1. This behavior is probably du to the high number of parameters and variables.

In order to reduce the number of parameters and variables, the same approach is applied on the aggregate variables by measurement stations aiming at the reduction of the number of parameters.

4.3 Model Based on the Clustering by Station

Using the measurements by station, the number of used variable used is dramatically reduced leading to the simplification of the risk model and in particular its interpretation. During the clustering step, the same number of cluster (5 clusters) is found as already reported in the previous section (by lane).

Cluster 1 is the most dense one (more than 36% of all observations cases). It is characterized by quite homogeneous Occ and average flow over the 4 stations, (Occ of 11% to 12% and a flow of 1450 to 1500 vehicles per hour and per lane) characterizing fluid traffic conditions. Cluster 2 presents a very high Occ on the (upstream) stations St_1 and St_2 and rather average downstream (14% to 18%). As for the flow, it is rather stable and low compared to other clusters. This cluster contains 20% of the overall observations. Cluster 3 presents high occupancy rates over all stations. The flow is higher upstream. Cluster 4 has average Occ close to the usual 20% critical value, increasing from upstream to downstream (26.7% at station St_4). The flows are higher than the other clusters, up to 1774 veh/h/lane at station St_2. Cluster 5 has a high average Occ (around 37%) at all stations and a lower flow (around 1230 veh/h/lane). When we consider the accidents and attribute to each time step the cluster number to which it belongs, we observe that 43 accidents out of 85 studied (50.58%) present cluster change during the last six minutes. For 60 accidents (i.e. more than 70% of them), the last time step belongs to cluster 2, characterized by a rarefaction shock wave (congested upstream and fluid downstream).

The same approach as the one previously described is applied. The Risk model is set to 1 for the observations belonging to cluster 2 and 0 elsewhere. The calibration of the Risk model is based on 70% of the full observations. The logistic regression model output is the following:

$$Risk = \frac{1}{1 + \exp -(\alpha + \beta Oc_1 + \gamma Oc_2 + \delta Oc_3 + \theta Oc_4 + \Phi Q_1)} \qquad (2)$$

where:
$\alpha = -7.1677; \beta = 0.0.1383; \gamma = -0.1061; \delta = -0.2052;$
$\theta = 0.1886; \phi = 0.00038;$
and the Oc_i variable is the occupancy measurements by station.

In this model, we can observe that the number of parameters is reduced. Compared to the first one (equation 1), only 6 parameters must be calibrated instead of 13.

The risk model validation has been proceed on the rest of the database which represents (30%) of the total observations. The same model parameters obtained are applied on 1000 observations which are not used for the calibration. The output results of this model indicate that the risk index value is maximal just before the occurrence of the accident. The risk index time evolution results are depicted in the figures 2 and 3.

In the figure 3, we can observe that before the occurrence of the real accident, the risk index is very high without accident. However, around 30

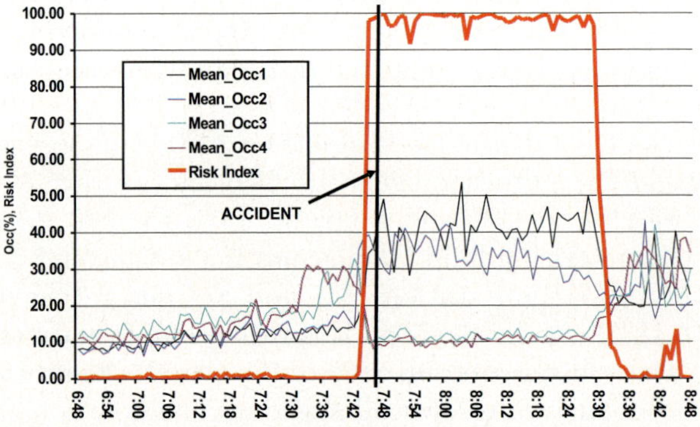

Fig. 2. Acc-1: Time evolution of the cumulative risk index.

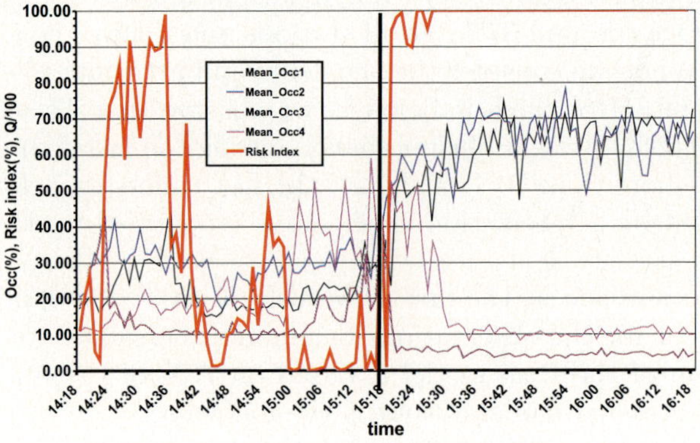

Fig. 3. Acc-2: Time evolution of the cumulative risk index.

minutes later, the accident is occurred. This means that the traffic flow is completely instable leading the generation of the accident.

5 Risk Model Application

In frame of the European project "EURAMP", field trials have been conducted aiming at traffic impact evaluation of several ramp metering strategies. Four control strategies has been tested:

- no control: reference case
- ALINEA: traffic responsive strategy
- Variable cycle ALINEA
- coordinated strategy

The test site is located in the south of the Ile de France Motorway network (figure 4). The total length of the experimental area is approximately 20 kilometers (only the direction towards Paris is considered). This part of the motorway includes 5 consecutive on-ramps, which are fully equipped with loop detectors and traffic signals. The carriageway is equipped with detector stations (each 500 m) for traffic volume, occupancy and speed measurements.

The overall period of these field trials are limited to 6 months. During the evaluation process, the risk index was applied for the safety evaluation. However, before using the safety index model, it is necessary to undertake the model validation process. The used data corresponds to the collected accidents synchronized with measurement traffic data. The total number of collected accidents is equal to 60. After data cleaning and accidents selection criteria (sunny day, night excluded), only 20 accidents are selected for the analysis

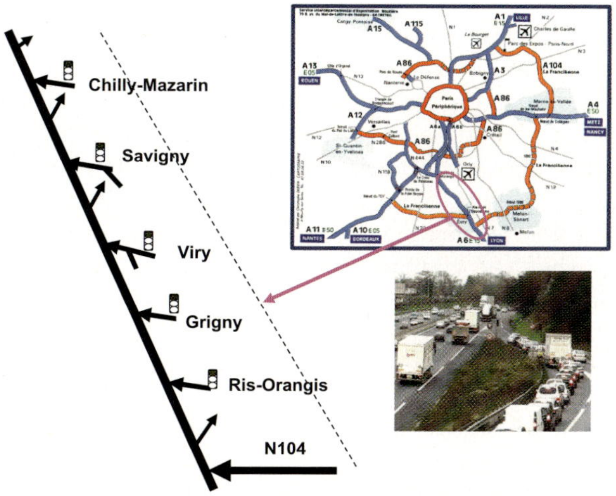

Fig. 4. Field trial test site.

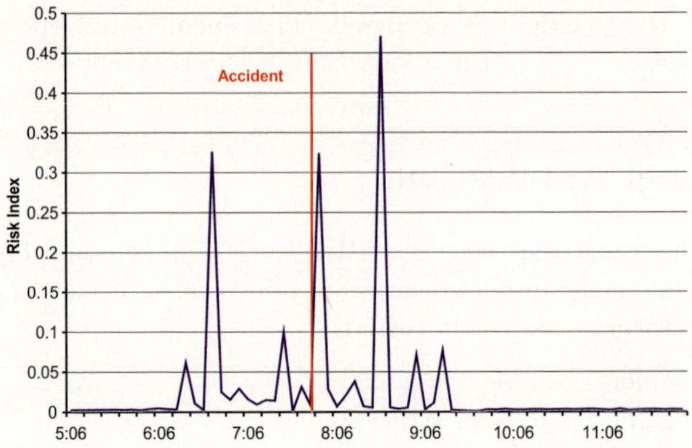

Fig. 5. Risk index time evolution: case acc 1.

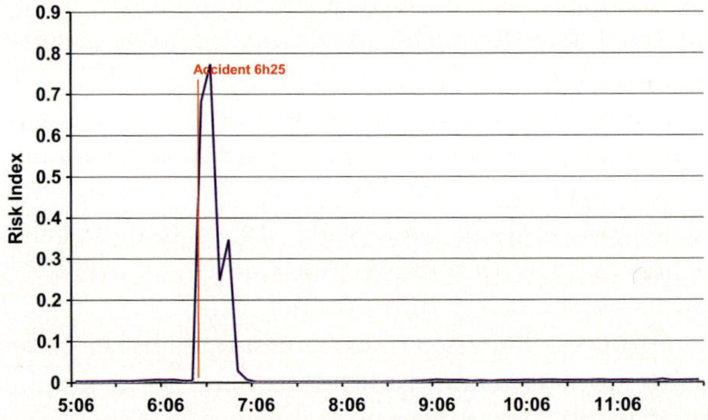

Fig. 6. Risk index time evolution: case acc 2.

and in particular the risk model validation. The same risk model parameters found on the ring way of Paris are conducted on A6W motorway. Figures 5 and 6 depict the risk index time evolution of two selected accidents.

We can underline that: using the same parameters of the build risk model by station on the ring way of Paris and applying to the A6W motorway, the obtained results of the risk index is very promising. Without any calibration, the risk index value is maximal before the accident occurs. Consequently, we can assume that the computation of risk index can be considered as a robust safety index and can be compared between the tested ramp metering strategies.

Figure 7 depicts the time evolution of the risk index of the four tested strategies on the A6W motorway. The risk index is computed at every station for 6 minute time interval, using the traffic measurements of the two upstream and downstream stations; all local indices are cumulated, resulting in a Cumulative Risk Index pertaining to the whole motorway section.

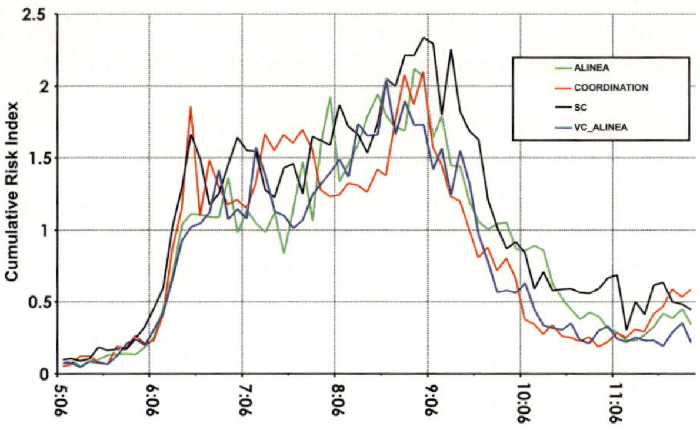

Fig. 7. Risk index time evolution of the four ramp metering strategies.

The cumulative risk index by strategy is computed on the overall motorway sections (19 sections) and on the overall time period ($6h$–$12h$). The obtained results indicate that the implementation of the ramp metering strategies improve the safety aspect (reduction of the number of accidents) by 20%. In particular, the safety benefit is more important in case of the coordination (29%). These results are very similar than the other safety evaluation impact of the ramp metering based on the accidents/incident collection data collection during several years. Extensive results can be found in [6].

6 Conclusion and Next Steps

As indicated, the obtained results are very promising. The number of parameters of the risk model was limited to 5 which can minimize the effort of the calibration process. The behavior of the risk index is satisfactory as regards the real occurrence of accidents. In particular, with the same topology of the used motorway axis, the risk index model can be applied without any calibration which is an important results for the real time application. Investigations on other motorway sites are necessarily for the validation results. However, more investigations are needed in order to take into account other parameter conditions such as the weather conditions, luminosity and in particular the night.

References

1. J. Dilmore (2005) Implementation strategies for Real-Time traffic safety Improvement on Urban Freeways. MA Thesis, University of Central Florida, Orlando.
2. M. Abdel-Aty, A. Pande (2005) Identifying crash propensity using specific traffic speed conditions. Journal of safety Research 36:97–108.

3. F. Thomas Glob, Wilfred W. Recker (2002) Method for relating type of crash to traffic characteristics on urban freeway. Transportation Research Part A 38:55–80.
4. C. Lee, B. Hellinga, F. Saccomanno (2006) Evaluation of variable speed limits to improve traffic safety. Transportation Research Part C 14:213–228.
5. H. Haj-Salem, C. Ramananjoana, J.P. Lebacque (2006) Risk Index Model for Real-Time Prediction of Potential traffic Crash on Urban Motorway. Proceedings of the EWGT2006, Advances in Traffic and transportation Systems Analysis, BARI, Italy.
6. EURAMP Project (2006). Deliverable D3.5, Safety Critical Issue For Ramp metering, European commission, Brussels.
7. B. Everitt (1974) Cluster Analysis. Heinemann Educ Books, London.

Colliding Particles: Beyond Accident-Free Car Following Models

Samer H. Hamdar and Hani S. Mahmassani

Northwestern University, The Transportation Center, Chambers Hall, 600 Foster Street, Evanston, Illinois 60208, USA hamdar@northwestern.edu, masmah@northwestern.edu

Summary. This paper explores specifications of microscopic traffic models that could capture congestion dynamics and model accident-prone behaviors on a highway section in greater realism than models currently used in practice. A comparative assessment of several major acceleration models is conducted, especially in regards to congestion formation and incident modeling. Based on this assessment, alternative specifications for a car-following/lane changing model are developed and implemented in a microscopic simulation framework. The models are calibrated and compared in terms of resulting vehicle trajectories and macroscopic flow-density relationships. Experiments are conducted with the models under different degrees of relaxation of the safety constraints typically applied in conjunction with simulation codes used in practice. The ability of the proposed specifications to capture traffic behavior in extreme situations is examined. The results suggest that these specifications offer an improved basis for microscopic traffic simulation for situations that do not require an accident free environment. As such, the same basic behavior model structure could accommodate both extreme situations (evacuation scenarios, over-saturated networks) as well as "normal" daily traffic conditions.

1 Introduction

Acceleration models are at the core of operational traffic behaviors. Fundamentally, car-following models aim at describing the trajectory of the nth vehicle in a traffic lane given the trajectory of the $(n-1)^{\text{th}}$ vehicle in the same lane. Accordingly, the main assumption in these models is that a relationship exists between a leader and a follower traveling on the same lane when inter-vehicle spacing is within a given range, typically between 0 to 125 meters [1]. According to Boer [2], more recent acceleration models are structured to account for several factors such as i) task scheduling and attention management; ii) the use of perception rather than Newtonian variables; and iii) satisfying a performance evaluation strategy, rather than an optimal one.

Existing car following models are generally limited in their ability to capture congestion dynamics [3], especially the onset of traffic breakdown and hysteresis phenomena. Moreover, existing car-following models are embedded in a "crash-free" environment, reflecting important gaps in understanding of driver behavior under extreme conditions and during incidents.

This paper explores specifications of microscopic traffic models that could capture congestion dynamics and model accident-prone behaviors on a highway section in greater realism than models currently used in practice (commercial software). A brief review of car-following models is first presented in Section 2; the focus will be on the following five models:

- Gazis, Herman and Potts' (GHP) model [4]
- Gipps' model [5]
- Cellular automaton Model (CA) [6]
- IDM or Intelligent Driver Model [3]
- IDMM or Intelligent Driver Model with Memory [7]

After implementing the above models and estimating the corresponding parameters, Section 3 presents the obtained results in terms of macroscopic flow-density relationship and the ability to model accidents when relaxing the safety constraints. Based on this analysis, a modified car-following model with a simplified lane changing framework is constructed in Section 4. This model with its new specification is able to capture some traffic characteristics during breakdown. Moreover, the results in Section 5 suggest that these specifications offer an improved basis for microscopic traffic simulation for situations that do not require an accident free environment.

2 Review of Acceleration Models

In previous continuous-time single-lane car-following models, the main response to a given stimulus is through acceleration or deceleration. The stimulus consists of the velocity of the driver, the relative velocity between a vehicle and the front vehicle and finally, the corresponding space-gap [4]. The main limitation of these models is that they are not applicable for very low traffic densities. Moreover, in dense traffic, small gaps will not induce braking reactions if the front vehicle is traveling at the same velocity (zero relative velocity).

Newell [8] addressed this concern by introducing the concept of the velocity depending adiabatically on the gap. As all previous car-following models, the Newell model is collision free. Moreover, since there is an immediate dependence of the velocity on the density (gap), very high and unrealistic accelerations can be produced.

To overcome this limitation, Bando et al. [9] suggested a modified Newell model, called the Optimal Velocity Model (OVM), in which the change in velocity is controlled by a relaxation time. Unfortunately, when the relaxation time is less than 0.9 seconds, the accelerations produced were still unrealistic.

The generalized force model [10] offered a generalized optimal velocity function that incorporates reaction to velocity differences and different rules for acceleration and braking. However, even though this model was able to produce time-dependent gaps and velocities, unrealistic small accelerations and decelerations were produced. All the above models (Newell, OVM and Generalized Force Model) offer important insights to the car-following existing logic but will not be considered here due to the known problems they face. The remaining models are numerically assessed in greater detail in the following section.

3 Numerical Assessment

To be able to compare the performance of the 5 models mentioned in Section 1, their respective logic was implemented using Visual C++ language. These models were then calibrated as explained in the following subsection.

3.1 Model Calibration

The data used for calibration are provided through the US FHWA's Next Generation Simulation (NGSIM) project. The data set includes trajectory data for 4733 vehicles over one-half hour (2:35 p.m.–3:05 p.m.) on December 3, 2003 [11]. The data are collected on Interstate 80 in Emeryville, California, USA by researchers at the University of California, Berkeley. The study area is a straight 2950 feet freeway section consisting of six lanes (lane 1 through lane 6) with an on-ramp (lane 7) at the beginning of the section and an off-ramp (lane 8) at the end. The x and y coordinate location is captured every $1/15^{\text{th}}$ second. These data are also processed so aggregate traffic measures such as flows and space-mean speeds are calculated over the time period of the study. As for the implemented models, the network considered is a single-lane straight freeway section of length $L = 10$ kilometers. The vehicles are generated and injected into the network following an exponential inter-arrival time. 3600 vehicles are generated in a period of two hours before ending the simulation.

This calibration is performed so acceptable parameters values are used for each model in the comparative analysis in the next subsections. The parameter values obtained for the observed traffic data are presented in Tables 1 through 4. Average flows, densities, speeds and headways for the five models are collected by placing "virtual" detectors at the end of each one-kilometer section of the freeway (10 detectors), and collecting average data for each 30 minutes, resulting in 40 data points in each simulation run. The results are posted in Table 5. The average flow and average speed values are found to be similar to the results obtained in the NGSIM data. To increase the confidence in the comparison, three additional simulation runs were used to

Table 1. Calibrated parameter values for the GHP model

Parameter	Calibrated value
Constant parameter c	12.1 m/s (40.2 mi/hr)
Reaction time (Lag) T	1 s

Table 2. Calibrated parameter values for the Gipps model

Parameter	Calibrated value
Desired velocity V_n	29 m/s
Mean reaction time τ_n	0.66 s
Mean acceleration a_n	0.73 m/s^2
Deceleration b_n	$-(2^*$ an) m/s^2
Mean vehicle length s_n	6.5 m

Table 3. Calibrated parameter values for the continuous CA model

Parameter	Calibrated value
Maximum velocity $v_{\max}$	28 m/s
Maximum acceleration $a_{\max}$	1.37 m/s^2
Maximum deceleration σ	1.73 m/s^2
Vehicle length	4 m

Table 4. Calibrated parameter values for the IDM/IDMM model

Parameter	Calibrated value (IDM)	Calibrated value (IDMM)
Desired velocity v_0	31 m/s	31 m/s
Safe time headway T	1.6 s	–
Maximum acceleration a	0.73 m/s^2	0.8 m/s^2
Desired deceleration b	1.67 m/s^2	–
Acceleration exponent δ	4	
Jam distance s_0	2 m	1.6 m
Jam distance s_1	0 m	–
Vehicle length $l = \frac{1}{\rho_{\max}}$	5 m	6 m
Netto time gap T_0	–	0.85 s
Comfortable deceleration b	–	1.8 m/s^2
Adaptation factor $\beta_T = T_{jam}/T_0$	–	1.8 s
Adaptation time τ	–	600 s

test the effect of randomness on the obtained results. The same range of output values as well as the same patterns of behavior were observed. Moreover, common random number (CRN) method is used when generating the vehicle characteristics and their inter-arrival times.

3.2 Data Analysis

To compare the five models in a consistent manner, the parameters found suitable in the calibration study are kept the same during the data analysis. The comparison is performed in terms of flow-density relationships and incident formation when relaxing the safety constraints.

3.2.1 Fundamental Diagrams

In this section, the mean inter-arrival time was modified so that the flow-density data points can cover most of the fundamental diagram space. Five fundamental diagrams were obtained for the 5 models using the parameters calibrated in Section 3.1. These fundamental diagrams are illustrated in Figures 1 through 5.

It can be seen that the GHP model (Figure 1) and the original Gipps model (Figure 2) did not capture either the metastable congested state or the instability encountered during or at the beginning of the traffic breakdown. This problem was already mentioned in several references in the literature in the case of the GHP model [4] and the Gipps model [5].

In its turn, the CA model showed the beginning of a traffic breakdown when reaching a flow capacity of 1800 veh/hr (Figure 3). However, as men-

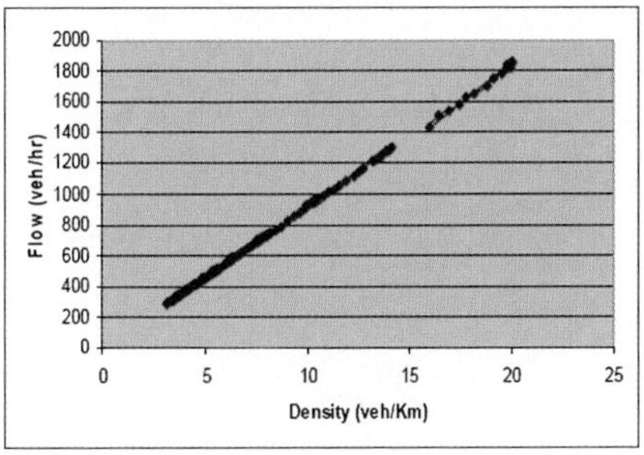

Fig. 1. GHP flow-density.

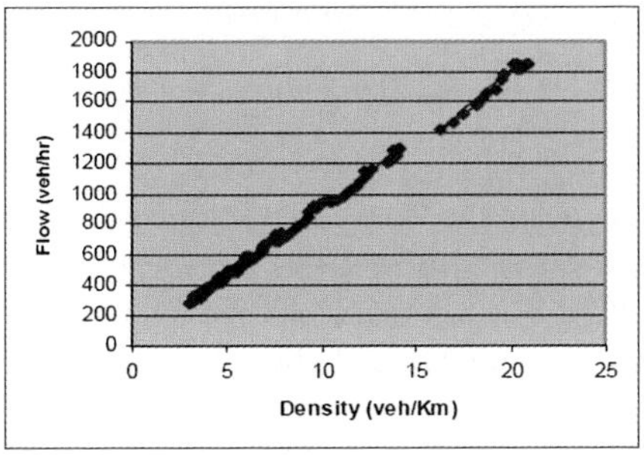

Fig. 2. Gipps flow-density.

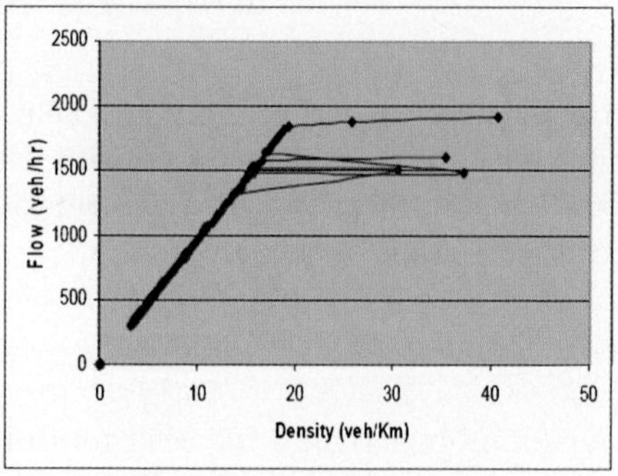

Fig. 3. CA flow-density.

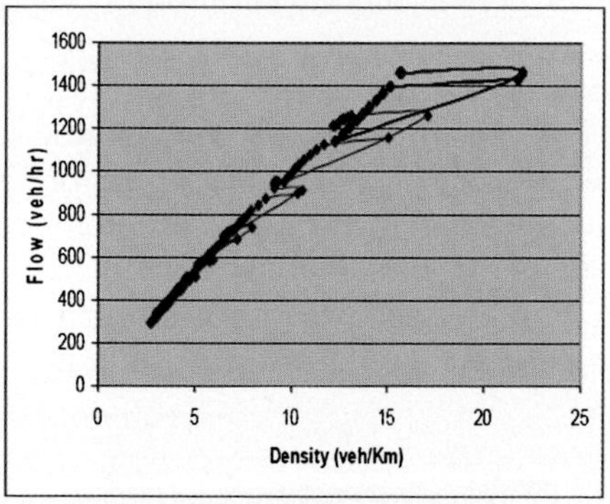

Fig. 4. IDM flow-density.

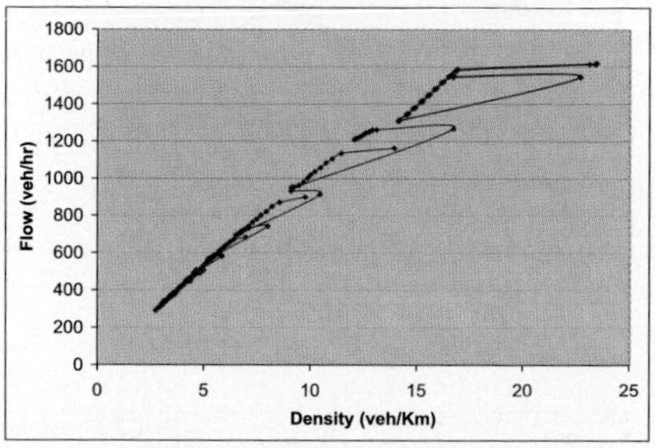

Fig. 5. IDMM flow-density.

tioned earlier, the CA lacks the cognitive logic behind it, making the model "mechanical" and sometimes, unrealistic. This is due to the fact that the model is controlled heavily by the constant deceleration rate attributed to the drivers. Finally, both IDM and IDMM showed realistic fundamental diagrams with a stable region and an unstable region (Figures 4 and 5 respectively).

3.2.2 Accident Modeling

In this section, the safety constraints imposed in each model will be discussed briefly. These safety constraints will be then relaxed when the vehicles traverse the first kilometer of the highway stretch. The parameters presented in Tables 1 through 4 are kept the same. The corresponding results are discussed afterward. Moreover, when an accident is generated (space separation between two consecutive vehicles is less than 0), the vehicles involved is the accident decelerates at a maximum deceleration rate of 6 m/s^2 until they come to a complete stop. The separation between the two consecutive vehicles is then set to zero.

3.2.2.1 GHP

The main factors that allow an accident-free environment in the GHP model is the form of the sensitivity term λ and the exact relationship between acceleration and relative speed. It can be seen that having the spacing s in the denominator $\lambda = \frac{c}{s}$ will reduce the acceleration response tremendously for smaller headways. Moreover, assuming that the driver will be able to observe and measure exactly the relative speed term, the vehicle will travel at the same speed of the leader.

To relax this safety condition, the λ term is treated as a random variable with a normal distribution, a mean of $\lambda_{mean} = \frac{c}{s}$, a standard deviation of $\lambda_{std} = 0.1$ and a range of $\lambda_{mean} - 0.1 \leq \lambda \leq \lambda_{mean} + 0.1$. This modification did not cause the formation of any accidents. The accidents were created when treating the relative speed stimulus as a normally distributed random variable with a standard deviation of 0.5.

Even though the relaxation is in the order of 0.1, a complete breakdown with 561 chain-type accidents is immediately produced. Such unrealistic scenario is due to the little elasticity the GHP Model offers between the particles. The vehicles are not even able to stop once an accident is generated. Once traffic is disrupted (accident, shockwave), the GHP model is not feasible.

3.2.2.2 Gipps Model

The safety constraint is the Gipps Model is presented in the expression: $x'_{n-1} - -s_{n-1} > x'_n$. It indicates that when a driver starts decelerating so his vehicle will stop at a given location x'_{n-1}, the following vehicle will decelerate and come to rest at x'_n before hitting the rear end of the preceding vehicle. To relax this condition, the risk term D_n is subtracted from $x'_{n-1} - s_{n-1}$. In this case, even if $x'_{n-1} - s_{n-1} - D_n > x'_n$, the distance between two vehicles can be negative and an accident may be generated.

If the safety conditions are kept, the following relations are obtained:

$$x'_{n-1} = x_{n-1}(t) - \frac{v_{n-1}(t)^2}{2b_{n-1}} \quad (1)$$

$$x'_n = x_n(t) + [v_n(t) + v_n(t+\tau_n)]\,\frac{\tau_n}{2} - \frac{v_n(t+\tau_n)^2}{2b_n} \tag{2}$$

and

$$x'_{n-1} - s_{n-1} \geq x'_n \tag{3}$$

After introducing $\mathbf{D_n}$, Equation 3 will be transformed to:

$$\begin{aligned} x_{n-1}(t) - \frac{v_{n-1}(t)^2}{2b_{n-1}} - s_{n-1} - D_n \geq x_n(t) + [v_n(t) + v_n(t+\tau_n)]\frac{\tau_n}{2} \\ + v_n(t+\tau_n)\theta - \frac{v_n(t+\tau_n)^2}{2b_n} \end{aligned} \tag{4}$$

The final expression for the velocity of Vehicle n at Time $t + \tau_n$ is:

$$\begin{aligned} &v_n(t+\tau_n) \\ &= \min \begin{cases} v_n(t) + 2.5a_n\tau_n\left(1 - \dfrac{v_n(t)}{V_n}\right)\left(0.025 + \dfrac{v_n(t)}{V_n}\right)^{1/2}; \\ b_n\tau_n + \sqrt{b_n^2\tau_n^2 - b_n\left[2(x_{n-1}(t) - s_{n-1} - x_n(t)) - v_n(t)\tau_n - \dfrac{v_n(t)^2}{b_{n-1}} + D_n\right]} \end{cases} \end{aligned} \tag{5}$$

Treating the risk factor $\mathbf{D_n}$ as a normally distributed random variable with a mean of 0.1, a standard deviation of 0.1 and a range of $0 < \mathbf{D_n} < 0.2$, 47 accidents were created. Even though much less accidents were obtained than in the GHP Model (561 accidents), chain type accidents can still be seen with a relatively high number compared to real life situations. The logic is still not suited for incident scenarios.

3.2.2.3 CA Model

The safety condition is imposed by setting the maximum velocity equal to the space gap $s_{gap}(t)$ between two successive vehicles irrespectively of the required deceleration needs to be changed. By allowing the velocity to be equal to $s_{gap}(t) + 0.1$ m (increase of 10 cm), 29 accidents are produced. This relatively low number of accidents compared to the GHP and Gipps models is due to the absence of constraints on the deceleration (braking) rate that can be applied: unrealistic behavior can still be seen.

3.2.2.4 IDM and IDMM Models

The IDM and the IDMM models assume that the acceleration is a continuous function of the velocity v_α, the gap s_α and the velocity difference Δv_α:

$$\dot{v}_\alpha = a^{(\alpha)}\left[1 - \left(\frac{v_\alpha}{v_0^{(\alpha)}}\right)^\delta - \left(\frac{s^*(v_\alpha, \Delta v_\alpha)}{s_\alpha}\right)^2\right] \tag{6}$$

The desired gap s^* is set by the following equation:

$$s^*(v, \Delta v) = s_0^{(\alpha)} + s_1^{(\alpha)} \sqrt{\frac{v}{v_0^{(\alpha)}}} + T^{(\alpha)} v + \frac{v \Delta v}{2\sqrt{a^{(\alpha)} b^{(\alpha)}}} \tag{7}$$

In these two models, the last term of Equation 7 is responsible of preventing crashes even though the safe time headway $T^{(\alpha)}$ is already included. Accordingly, accidents are obtained by removing this safety buffer. A complete traffic breakdown with 1211 accidents for IDM and 674 accidents for IDMM is observed. It should be noted that trying to decrease the desired minimum gap $s^*(v, \Delta v)$ by a value up to 1 did not produce any accidents.

4 Improved Model Formulation

The advantage of the Gipps model is its ability to model driving behavior following some plausible cognitive thinking that may be adopted by the driver. This led to explicitly incorporating the reaction time τ_n and an asymmetric application of acceleration versus deceleration. Moreover, the Gipps model showed an acceptable degree of stability (relatively low number of accidents) when relaxing its safety constraints. Motivated by these properties, the objective is to follow the same logic applied in the Gipps model while modifying it so that interactions between drivers during high density situations can be captured. In the speed-density relationship, this will be indicated by observing a metastable state and a sort of traffic break down in the fundamental diagram.

Moreover, further specifications and a simplified lane changing logic are added so an acceptable incident modeling framework can be offered. The conditions where incidents are created are called "extreme conditions" for the rest of the paper.

4.1 Acceleration Model

The model aims to capture driver behavior under "incident" or extreme conditions and congested situations. The modification that can be made on the variables included in the simulation model (the rest already described in the original Gipps Model) as well as the description of the initial risk factor $\mathbf{D_{0n}}$ are listed below:

1. $\mathbf{a_n}$ (m/s^2): under extreme conditions, drivers typically can be willing to apply higher acceleration rates than under normal conditions, causing irregularities and possible instabilities in traffic flow patterns [12]. This variable is drawn from a truncated Gaussian-shaped (Normal) distribution with a given mean and variance. The truncation is performed through a range variable and it is based on the value of the mean chosen during the sensitivity analysis.
2. $\mathbf{b_n}$ (m/s^2): the value of $\mathbf{b_n}$ can increase in absolute value. The hypothesis is that under extreme conditions, drivers tend to have higher braking rates

or increased use of emergency braking [12]. This value is also drawn from a truncated normal distribution with a given mean, variance and range at the beginning of the simulation.

3. $\mathbf{V_n}$ (m/s) is the speed at which the driver of vehicle n wishes to travel. In extreme conditions, the value can be randomly chosen from a probabilistic mixture of two normal distributions. For the first distribution, the mean is higher than the suggested mean in the Gipps Model. For the second distribution, the mean is lower than the suggested Gipps mean. This choice is consistent with an illustration by [13] of the disruptions and the irregularities in traffic flow resulting from velocity differentials (idealized as two classes of drivers, so-called "slugs" versus "rabbits").
4. $\mathbf{D_{0n}}$ (m): initial risk factor: it represents the distance a driver is willing to travel beyond the safety threshold. The safety threshold indicates the distance between the driver and the leading vehicle at which the driver would start decelerating so that his vehicle can come to a complete stop before hitting the preceding vehicle. This value is added to the model to allow potential accidents to be generated (See Subsection 3.2.2.2). It reflects the willingness of a driver to take a risk. The value of D_n for each vehicle n is initially drawn from a truncated normal distribution. When this value is positive, the driver is willing to take risk and this may increase the probability of causing an accident. If this value is negative, the driver prefers to stay within the safety margin so he/she can come to a stop without hitting the vehicle in front.
5. $\mathbf{s_n}$ (m) is size of vehicle n instead of its effective size as suggested by the original Gipps Model.
6. $\boldsymbol{\tau_n}$ (s) is the reaction time corresponding to the driver of Vehicle n.

Using the same logic adopted in Subsection 3.2.2.2 to relax the safety constraint of the Gipps Model, the relationship of $x'_{n-1} - s_{n-1} - D_n > x'_n$ will allow the distance between two vehicles to be negative and an accident may be generated. Moreover, another safety factor illustrated by the safety margin $v_n(t+\tau)\theta$ in Equation 4 (θ is normally equal to $\tau/2$) is removed. After introducing $\mathbf{D_n}$ (function of D_{0n}), Equation 4 will be transformed to:

$$x_{n-1}(t) - \frac{v_{n-1}(t)^2}{2b_{n-1}} - s_{n-1} - D_n \geq x_n(t) + [v_n(t) + v_n(t+\tau_n)]\frac{\tau_n}{2} - \frac{v_n(t+\tau_n)^2}{2b_n} \tag{8}$$

The new relationship dominating the Gipps logic is:

$$v_n(t+\tau_n) = \min \begin{cases} v_n(t) + 2.5a_n\tau_n\left(1 - \frac{v_n(t)}{V_n}\right)\left(0.025 + \frac{v_n(t)}{V_n}\right)^{1/2}; \\ b_n\left(\frac{\tau_n}{2}\right) + \sqrt{\frac{b_n^2\tau_n^2}{4} - b_n\left[2(x_{n-1}(t) - s_{n-1} - x_n(t)) - v_n(t)\tau_n - \frac{v_n(t)^2}{b_{n-1}} + D_n\right]} \end{cases} \tag{9}$$

4.2 Lane Changing Decision Logic

Lane-changing decisions are strongly related to the desirable speed at which a driver wishes to travel. A driver traveling at a speed less than his desirable speed will seek to increase his speed in the same lane. If another vehicle is in the way (space headway between the two vehicles is less than 5 meters, which is the average length of a car), the following driver will consider changing lanes. However, the driver must check first if this maneuver is possible with the gaps offered in the adjacent lane. Checking these gaps is a procedure to be specified as part of the lane changing model.

On the other hand, it was found that the average lead or lag times for all traffic conditions are almost equal [14]. Accordingly, it may be suggested that neither the lead nor the lag dominates the gap-acceptance decision in lane-changing. Therefore, both the leading and the lagging vehicles in an adjacent lane are objects of interest in this study.

The theoretical estimate of the minimum safe lead value based an assumed desirable deceleration rate and an average braking perception/reaction time is given by the following equation:

$$L_1 = v_n(t)\tau_n + \frac{v_n(t+\tau_n)^2}{|2b_n|} - \frac{v_m(t+\tau_n)^2}{|2b_m|} \tag{10}$$

where:
m = subscript for a leading vehicle in destination lane
L_1 = safe "lead" distance for lane changing (m)
$v_n(t+\tau_n)$ = speed of lane-changing vehicle n (m/sec)
$v_m(t+\tau_m)$ = speed of leading vehicle m in destination lane (m/sec)
b_n = deceleration rate vehicle n can sustain (m/sec^2)
b_m = deceleration rate vehicle m can sustain (m/sec^2)
τ_n = apparent reaction time for vehicle n (braking perception/reaction time) (sec)
τ_m = apparent reaction time for vehicle m (braking perception/reaction time) (sec).

With the same logic, the theoretical estimate of the safe lag value is:

$$L_2 = v_{m+1}(t)\tau_{m+1} + \frac{v_{m+1}(t+\tau_{m+1})^2}{|2b_{m+1}|} - \frac{v_n(t+\tau_{m+1})^2}{|2b_n|} \tag{11}$$

where:
$m+1$ = subscript for a lagging vehicle in destination lane
L_2 = safe "lag" distance for lane changing (m)
$V_n(t+\tau_n)$ = speed of lane-changing vehicle n (m/sec)
$v_{m+1}(t+\tau_{m+1})$ = Speed of lagging vehicle $m+1$ in destination lane (m/sec)
b_n = deceleration rate vehicle n can sustain (m/sec^2)
b_{m+1} = deceleration rate vehicle m can sustain (m/sec^2)

τ_n = apparent reaction time for vehicle n (braking perception/reaction time) (sec)
τ_{m+1} = apparent reaction time for vehicle $m+1$ (braking perception/reaction time) (sec).

However, it is suggested that both lag and lead distances are over-estimated [14]. The use of different parameters during extreme conditions will help deal with this subject. First, the higher deceleration rates in absolute value will decrease the safe leads and lags to be accepted. This is expected during panic behavior especially on the part of aggressive drivers, since their patience is limited and they tend to accept shorter gaps.

4.3 Incident Modeling

The above model still faces several limitations in its ability to represent car following behavior during incidents, especially with regard to uncontrollable chain type accidents. On the other hand, it should be noted that every time the headway between two vehicles is less than zero, the speed of both vehicles will decrease at a rate of -6 m/s^2 that is the maximum deceleration a vehicle can sustain. The vehicles will come to rest at the end. The headway of the two vehicles will be reset to zero governed by the location of the front vehicle.

Accidents will be still possible in this lane changing model due to the duration required for the lane-changing maneuver. The general idea is that if the lane-changing maneuver is seen as possible, the respective locations of vehicles n, m, and $m+1$ are computed in the target lane after a given lane changing time. This time is also drawn from a normal distribution varying from one driver to another. Moreover, to capture sudden lane changing [12] to escape an incident, lane changing time is reduced to have a mean of 2 seconds: a value that is found feasible by sensitivity analysis. If $x_m(t) - x_{m+1}(t) - s_m$ is less than or equal to s_n, the respective velocities of the three vehicles are set to be equal to zero by the same manner it was done for car-following. This indicates the occurrence of an accident in that lane. It may be suggested that accidents due to lane changing may block both lanes of travel.

To prevent traffic breakdown in a fast manner, it is suggested that once an accident occurs, another vehicle-specific variable enters into the equation. This variable is called "alert distance" (R_n). If the vehicle is within an alert distance from the incident location, the risk of this vehicle will be reset to zero. In other words, drivers will tend to be more attentive and may slow down when they see an accident in their sight and for a while after they pass it [12]: even if on the side of the road, accidents have an effect on the traffic conditions.

4.4 Parameter Change Over Time

The change in some driver's characteristics over time is performed through linking them to dynamic traffic properties in his or her immediate surrounding:

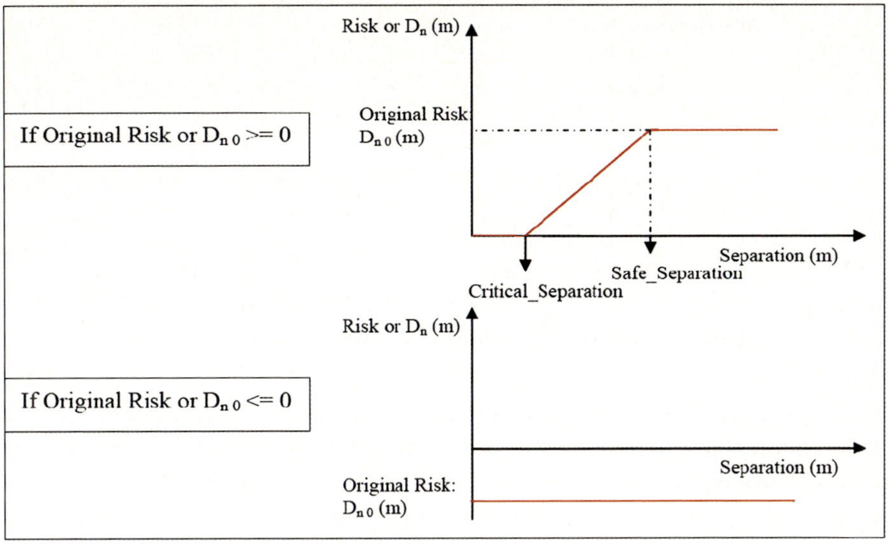

Fig. 6. Parameters' change over time: $D_n = f_1$ (separation).

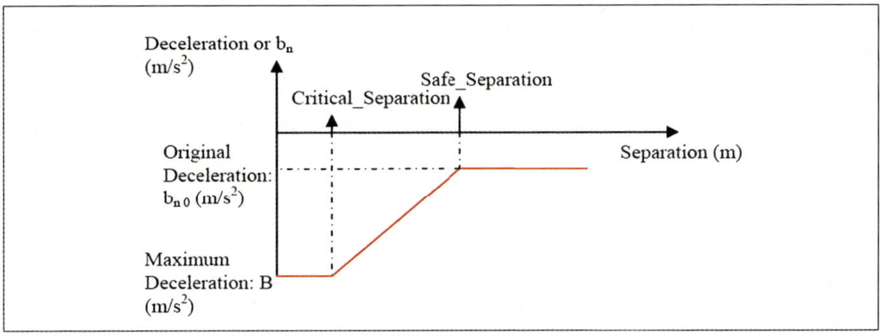

Fig. 7. Parameters' change over time: $b_n = f_2$ (separation).

"deceleration (braking) rate b_n" and "risk factor D_n" are considered functions of the vehicle's separation (distance separating the front end of a lagging vehicle to the rear end of the leading one). Moreover, the "alert distance R_n" depends on the "initial risk factor D_{0n}". Since literature offers general and not detailed description of how these characteristics may change, the functional relations are assumed to be piece-wise linear; they are presented in Figures 6, 7 and 8.

The Critical_Separation will be considered equal to the average vehicle length that is 4 meters. It is the space separation below which drivers will tend to react in an exaggerated manner to any stimulus [12]. This includes the usage of the maximum deceleration rate a vehicle can support ($b_n = -6$ m/s^2). Below this separation, drivers are not willing to take any risk

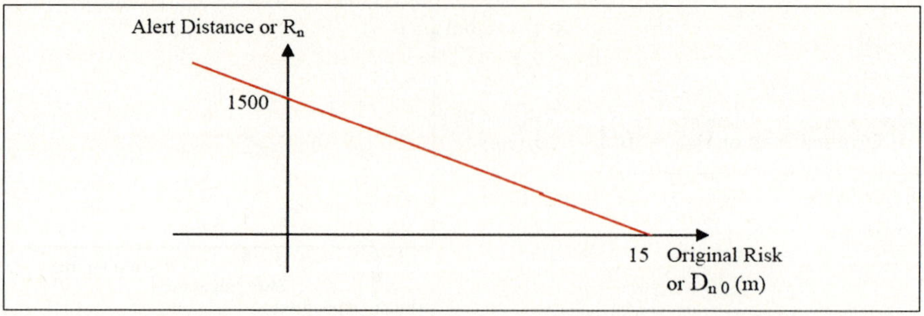

Fig. 8. Parameters' change over time $R_n = f_5\ (aD_{n0})$.

($D_n = 0$ if the original risk factor D_{0n} was greater than zero) but this may be too late for a collision. However, if the original risk factor was greater than zero ($D_{0n} < 0$: drivers already conservative), the risk factor will be always equal to the original risk factor and will not depend on the space separation. As for the Safe-Separation, it is chosen to be equal 10 average vehicle lengths (40 meters) and it is seen as the distance-separation above which drivers do not tend to take extra precautions for preventing collisions: the initial deceleration rate b_n and the initial risk factor D_{0n} already drawn for normal distributions are used.

As for Figure 8, the alert distance (alertness of the drivers) is seen negatively correlated to the initial risk factor (aggressiveness). The functional relationship is also assumed to be linear.

In Conclusion, the above model is rich in behavioral driving parameters offering flexibility and model dynamics to imitate real-life cognitive drivers' behaviors. Once implemented, the next section will offer the results obtained in:

1- Incident-Free Congested Situations
2- Incident Conditions

5 Analysis of Results

This section is divided into two parts. Calibrating the modified Gipps Model using the NGSIM data will allow a better comparison with the 5 analyzed models. The flow density relationship of the modified Gipps Model is presented in Subsection 5.1. In Subsection 5.2, different incident scenarios are produced and the resulting behavior will be analyzed.

5.1 Incident Free Environment

The NGSIM data described in Section 3.1 is used to calibrate the acceleration model presented in section 4.1 and 4.4. For comparison purposes, the same

Table 5. Results obtained by each model after calibration

Model	Avg. travel time (s)	Avg. speed (m/s)	Avg. flowrate (veh/hr)	Avg. density (veh/km)
GHP-Calibrated	29.3	25.6	1533.1	16.7
Gipps-Calibrated	40.0	25.0	1532.0	17.1
IDMM-Calibrated	40.0	25.2	1525.5	17.0
IDM-Calibrated	39.9	25.3	1423.1	15.9
CA-Calibrated	40.8	25.6	1503.0	17.1
Modified Gipps-Calibrated	34.3	24.1	1457.6	17.9
NGSIM Data	–	25.6	1578	–

Table 6. Calibrated parameter values for the modified Gipps model

Parameter	Calibrated value
Desired velocity V_n	33 m/s (one distribution adopted)
Mean reaction time τ_n	0.66 s
Mean acceleration a_n	1.7 m/s^2
Mean deceleration b_n	-3.4 m/s^2
Critical space separation s_0	2 m
Safe space separation s_s	50 m
Mean vehicle length s_n	5 m
Original risk factor D_{0n}	0 m

simulation environment created in Section 3 is created here. Calibrated parameters can be seen in Table 6. The macroscopic results of this calibration are presented in Table 5.

Flow-Density Relationship

The fundamental diagram corresponding to the modified Gipps model is presented in Figure 9. Congestion patterns were captured when the flows reach a value of 1600 veh/hr. At that time, two states are seen: the first state is a slow traffic movement state (metastable state) until hitting the flow capacity bar of 1900 veh/hr. The other state is the traffic breakdown state and it is located under the slow-moving vehicle line. This kind of shape has some common aspects compared to the hysterisis loop.

5.2 Extreme Conditions

Since there are no field data to calibrate or validate this model with the presence of incidents, the model will rely on sensitivity analysis to examine the feasibility of the logic used. The initial input parameters and the different scenarios performed are presented in Tables 7 and 8.

At the beginning, the trajectories of 50 sample vehicles were dressed for one simulation run using Scenario 1 and Scenario 3. Figures 10 and 11 show these trajectories in the form of "Time-Space Diagrams".

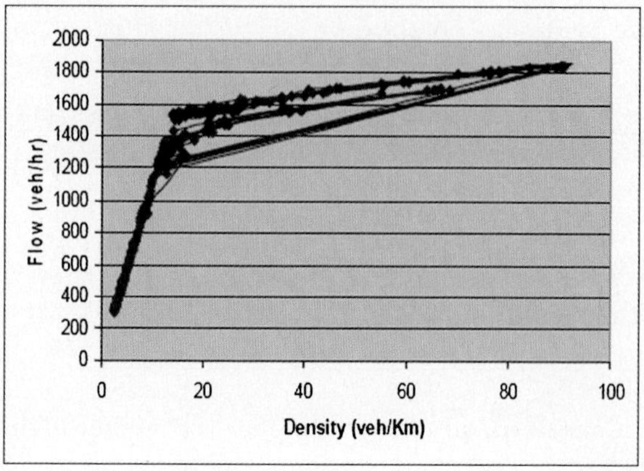

Fig. 9. Flow-density relationship for the modified Gipps model.

Table 7. Main input parameters used in the simulation of extreme events

Input variable	Value
Time T (s)	7200
Road length L (m)	10000
Mean reaction time (s)	1
Mean Vd1 (m/s): mean desired velocity for slugs	20
Mean Vd2 (m/s): mean desired velocity for rabbits	35
Percent Vd1(%): percent of slug drivers	40
Mean LCT (s): mean of the Lane Changing Time	2.5
Mean acceleration (m/s^2)	2
Mean deceleration (m/s^2)	−3

Table 8. Different scenarios adopted in the sensitivity analysis

Scenario description	Scenario #	# of vehicles	Mean interrarrival time (s)	Risk (m)
Free-Flow Risk Free	1	1200	6	0
Free-Flow Minor Risk	2	1200	6	1
Free-Flow Major Risk	3	1200	6	5
Free-Flow Break Down	4	1200	6	10
Congested Risk Free	5	3600	2	0
Congested Minor Risk	6	3600	2	1
Congested Major Risk	7	3600	2	5
Congested Break Down	8	3600	2	10

In the first set of trajectories, we can observe two families of drivers; those with a steeper velocity lines possess higher velocities and thus are the "rabbits" we mentioned in Section 4.1. The others are the slower slug drivers. Moreover, in Figure 11, the risk factor is not zero anymore. Accordingly, the horizontal lines the last until the end of the simulation indicates that an accident has

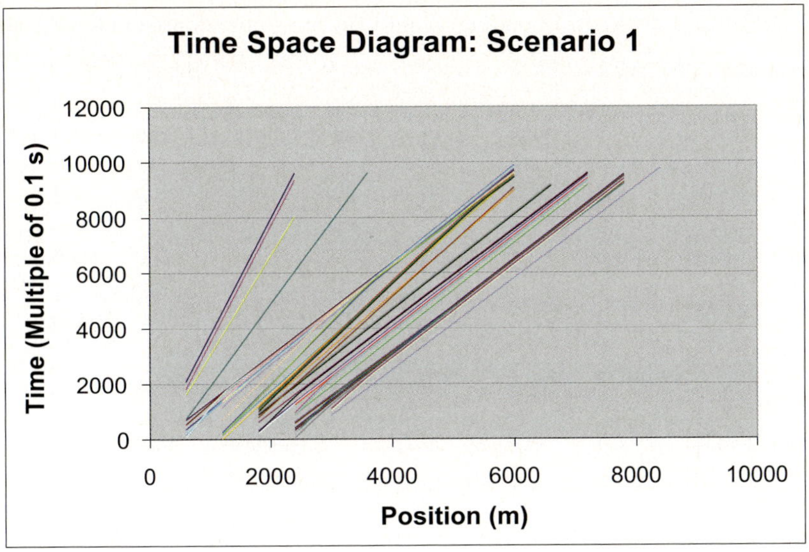

Fig. 10. Time-space diagram for one run for scenario 1.

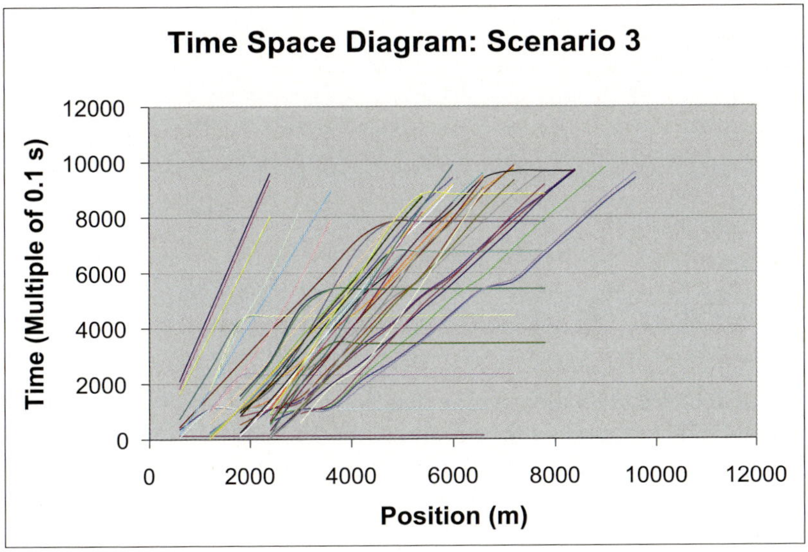

Fig. 11. Time-space diagram for one run for scenario 3.

occurred. Some of the vehicles will be stuck behind these accidents for a while before being able to perform a lane change. Accordingly, chain type accidents can be avoided by just waiting behind the vehicles involved in the crash.

After talking about the microscopic trajectory data, the macroscopic data of average travel times, average speeds, total number of lane changes, total

Table 9. Output data corresponding to different scenarios adopted in the sensitivity

Performance measure	Sc. 1	Sc. 2	Sc. 3	Sc. 4	Sc. 5	Sc. 6	Sc. 7	Sc. 8
Average travel time (s)	54.9	46.7	48.9	50.9	60.1	59.9	74.0	49.2
Average speed (m/s)	18.3	27.4	25.4	22.8	16.7	27.8	23.7	22.4
Number of cars crashed	0.0	14.0	24.0	161.0	0.0	14.0	39.0	108.0
Number of lane changes	74.0	8733.0	11753.0	3247.0	268.0	37162.0	50542.0	995.0
Average flow rate (veh/hr)	291.3	290.0	287.4	43.7	874.9	822.8	842.4	15.9
Average density (veh/(km.lane))	4.5	4.7	5.2	8.1	14.6	15.2	19.6	6.1

Table 10. Avoidance of chain-type accidents: escape by lane changing twice on the same road segment

Road segment index (1 km)	Number of vehicles crashed		Number of lane changes
	Lane 1	Lane 2	
1	0	2	1023
2	2	2	1575
3	0	2	1086
4	2	0	1135
5	2	0	1022
6	0	2	1193
7	0	2	1036
8	2	2	1363
9	2	0	1128
10	0	2	1192

number of vehicles crashed, average headway, average flows and average densities (over 10 simulation runs) are shown in Table 9.

It was found that the variance of the speed will increase with the risk factor; this is expected since accidents will bring an increasing number of traffic fluctuations. In terms of the means, with higher risks, accidents, travel times and lane changes will increase. On the other hand, speeds and flows decrease. This special rule is violated sometimes in the traffic breakdown situation. This is due to the fact that most vehicles are trapped at the beginning of the road length due to the accidents. Few vehicles will escape and can thus travel at higher speeds and with higher headways. This "fewer number" of vehicles is reflected in the low density value for scenario 8 compared to scenario 7.

Table 10 and the associated visual representation focuses on Scenario 3 and how vehicles escape and travel between accidents avoiding the unrealistic chain effect observed in and discussed in Subsection 3.2.2.

6 Conclusion

This paper analyzes existing car-following models with an assessment of their qualities and limitations. The focus was on 5 models:

- GHP model
- Gipps model
- Cellular automaton Model (CA)
- IDM model or Intelligent Driver Model
- IDMM model or Intelligent Driver Model with Memory

The above models were implemented, calibrated and tested in terms flow-density relationships and ability to model driver behavior during incident situations.

This kind of review has enabled the formulation of an improved car-following model that shows instability during congestion without the necessity of having an accident-free environment. The modified model could capture some congested flow dynamics (hysteresis effect, beginning of traffic break-down). Moreover, when incorporating the space risk factor, drivers showed an acceptable degree of maneuverability by either standing behind accidents without being involved in a rear-end collision or by changing lanes. Long and unrealistic chain type accidents causing a total traffic breakdown could be avoided.

The results suggest that the new specifications presented in this paper offer an improved basis for microscopic traffic simulation for situations that do not require an accident free environment. As such, the same basic behavior model structure could accommodate both extreme situations (evacuation scenarios, over-saturated networks) as well as "normal" daily traffic conditions.

Acknowledgements

This study is based in part on research funded by the National Science Foundation's Human and Social Dynamic Systems under Award Number 0527573. Of course, the authors remain solely responsible for the contents of this paper.

References

1. R. W. Rothery, 1999. Traffic Flow Theory: A State-of-the-Art Report-Revised Monograph on Traffic Flow Theory. Transportation Research Board, National Research Council, Washington, D. C.

2. E. R. Boer, 1999. Car Following from the Driver's Perspective. Transportation Research, Part F, No. 4, pp. 201-206.
3. M. Treiber, K. Hennecke, and D. Helbing, 2000. Congested Traffic States in Empirical Observations and Microscopic Simulations. Physical Review E, Volume 2, No. 2, pp. 1805-1824.
4. D. C. Gaziz, R. Herman, and R. Potts, 1959. Car-Following Theory of Steady State Traffic Flow. Operations Research, Issue 7, pp. 499-505.
5. P. G. Gipps, 1981. A Behavioral Car-Following Model for Computer Simulation. Transportation Research 15B, pp. 101-115.
6. S. Krauss, P. Wagner, and C. Gawron, 1996. Continuous Limit of Nagel-Shreckenberg Model. Physical Review E, Vol. 54, No. 4, pp. 3707-3712.
7. M. Treiber and D. Helbing, 2003. Memory Effect of Microscopic Traffic Models and Wide Scattering in Flow-Density Data. Physical Review E68, 046119.
8. G. Newel, 1961. Nonlinear Effects in the Dynamics of Car-Following. Operations Research 9, pp. 209-229.
9. M. Bando, K. Hasebe, A. Nakayama, A. Shibata, and Y. Sugiyama, 1995. Dynamical Model of Traffic Congestion and Numerical Simulation. Physical Review E 51, pp. 1035-1042.
10. B. Tilch and D. Helbing, 1998. Generalized Force Model of Traffic Dynamics. Physical Review E, Vol. 58, Issue 133.
11. FHWA, 2004. NGSIM Task E. 1-1: Core Algorithms Assessment, Final Report. Cambridge Systematic, Inc., Massachusetts.
12. A. Querejeta-Iraola and U. Reiter, 1991. Calibration, Validation and Tesing of Multi-Lane Simulation Model. Deliverable of EC DRIVE Project ICARUS (V-1052), Brussels.
13. C. F. Daganzo, 1999. A Behavioral Theory of Multi-Lane Taffic Flow, Part I: Long Homogeneous Freeway Sections. Institute of Transportation Studies, University of California, Berkeley.
14. FHWA, 1969. Lane Changing on Multi-Lane Highways, Final Report. U. S. DOT, Department of Civil and Environmental Engineering, Northwestern University, Evanston, Illinois.

Dynamic First-Order Modeling of Phase-Transition Probabilities

Serge P. Hoogendoorn, Hans van Lint, and Victor Knoop

Delft University of Technology, Stevinweg 1, 2628 CN Delft, The Netherlands
s.p.hoogendoorn@tudelft.nl, j.w.c.vanlint@tudelft.nl,
v.knoop@tudelft.nl

Summary. Modeling breakdown probabilities or phase transition probabilities is an important issue when assessing and predicting the reliability of traffic flow operations. Looking at empirical spatio-temporal patterns, these probabilities clearly are not only a function of the local prevailing traffic conditions (density, speed), but also of time and space. For instance, the probability that start-stop wave occurs generally increases when moving upstream away from the bottleneck location.

The dynamics of the breakdown probabilities are the topic of this paper. We propose a simple partial differential equation that can be used to model the dynamics of breakdown probabilities, in conjunction with a first-order model. The main assumption is that the breakdown probability dynamics satisfy the way information propagates in a traffic flow, i.e. they move along with the characteristics.

The main result is that we can reproduce the main characteristics of the breakdown probabilities, such as observed by Kerner. This is illustrated by means of two examples: free flow to synchronized flow (F-S transition) and synchronized to jam (S-J transition). We show that the probability of an F-S transition increases away from the on-ramp in the direction of the flow; the probability of an S-J transition increases as we move upstream in the synchronized flow area. Note that all the examples shown in the paper are deterministic.

1 Introduction

The research and claims of Kerner [1] has resulted in quite a stir in the traffic flow theory community. Amongst the issues raised by Kerner are the fact that there are three phases (free flow, synchronized flow and jams), rather than two (free flow and congestion), the fact that the breakdown phenomenon is a stochastic process stemming from the fact that small or large disturbances can trigger phase transitions with a certain probability, and the fact that the fundamental diagram does not exist since the congested branch is a 2D area, rather than a straight line.

Furthermore, Kerner claims that none of the current microscopic or macroscopic traffic flow models captures correctly the different flow characteristics that are observed from empirical analyzes.

This paper focuses on the breakdown phenomenon. More specially, the main contribution of the paper is that we show a first-order macroscopic modeling framework that allows us to model the dynamics of the breakdown or phase-transition probabilities in an intuitive and simple manner.

Different researchers have considered the dynamic modeling of breakdown probabilities (see [2–4]), commonly using (stochastic) queuing analysis. In this contribution, we propose using coupled set of partial differential equations describing both the traffic dynamics (using a simple first-order model) and the dynamics of the phase-transition probabilities. In other words, the proposed modeling framework can be considered as a relatively straightforward generalization of the kinematic wave theory.

Note that we focus on the dynamic modeling of the phase-transition probabilities, and the implications this has for the properties of the first-order model. Other issues discussed by Kerner (such as the 2D area depicting stable states in synchronized flow) are not considered.

2 Mathematical Model of Breakdown Probability

This contribution describes dynamic modeling of the breakdown (or rather phase transition) probability, which is denoted by $P = P(t, x)$. Note that the probability is a function of time t and space x, and is thus not only determined by the prevailing traffic conditions such as the density.

The macroscopic dynamic model consists of the following set of equations:

$$\frac{\partial \rho}{\partial t} + c(\rho)\frac{\partial \rho}{\partial x} = 0 \tag{1}$$

$$\frac{\partial P}{\partial t} + c(\rho)\frac{\partial P}{\partial x} = \pi(\rho, P) \tag{2}$$

In Eqs. (1) and (2), $c(\rho) = \frac{dQ}{d\rho}$ denotes the kinematic wave speed, describing the speed (and direction!) at which (small) perturbations propagate through the traffic flow. The kinematic wave speed is equal to the derivative of the fundamental diagram $Q = Q(\rho)$. This follows directly from the shockwave equation, stating that the speed of a shockwave S separating regions (ρ_1, q_1) and (ρ_2, q_2) is given by:

$$\omega = \frac{q_2 - q_1}{\rho_2 - \rho_1} = \frac{Q(\rho_2) - Q(\rho_1)}{\rho_2 - \rho_1} \tag{3}$$

yielding:

$$\lim_{\rho_2 \to \rho_1} \frac{Q(\rho_2) - Q(\rho_1)}{\rho_2 - \rho_1} = \frac{dQ}{d\rho}(\rho_1) \tag{4}$$

In Eq. (2), $\pi = \pi(\rho, P)$ denotes the rate of change in the breakdown probabilities P, which are assumed to be a function of the density $\rho = \rho(t, x)$ and the probability P itself. Also note that P can describe both an F-S transition ($P = P_{F-S}$) or a S-J transitions ($P = P_{S-J}$). Both examples will be shown in the ensuing of the contribution.

2.1 Model Justification

The concept behind the mathematical model is the assumption that the phase-transition probability P changes along the *characteristic curves* (just as the density). This means that if we consider a perturbation in the flow, the phase-transition probability P will change along with this perturbation.

To understand this property, let us consider a platoon of vehicles. Suppose that the platoon leader will brake briefly. This perturbation will move from the one vehicle to the next, possibly changing in amplitude while moving upstream. The speed at which the perturbation moves is equal to the characteristic speed. If the perturbation becomes sufficiently large, it may induce a phase transition. Alternatively, the perturbation may damp out implying that the probability of a phase transition will reduce along the perturbation.

Let us now take a closer look at the characteristic curves. These curves are parameterized curves C that are defined by the path:

$$C = \{t(s), x(s)\} \tag{5}$$

where $t = t(s)$ and $x = x(s)$ are defined by the following differential equations:

$$\frac{dt}{ds} = 1 \text{ and } \frac{dx}{ds} = c(\rho) \tag{6}$$

Now, let $\rho(s) = \rho(t(s), x(s))$ denote the (parameterized) density along the characteristic curve. We have:

$$\frac{d\rho}{ds} = \frac{d\rho}{dt}\frac{dt}{ds} + \frac{d\rho}{dx}\frac{dx}{ds} = \frac{\partial \rho}{\partial t} + c(\rho)\frac{\partial \rho}{\partial x} = 0 \tag{7}$$

We thus see that the density ρ is conserved along the characteristic C (i.e. $\rho(s) = \rho(0)$). Since the characteristic speed $c(\rho)$ depends on ρ, the speed is constant as well, and thus the characteristic C is a straight line.

For (2), we can use the same characteristic curves. Let $P = P(s)$ denote the breakdown probability along C. We can thus show that:

$$\frac{dP}{ds} = \frac{dP}{dt}\frac{dt}{ds} + \frac{dP}{dx}\frac{dx}{ds} = \frac{\partial P}{\partial t} + c(\rho)\frac{\partial P}{\partial x} = \pi(\rho, P) \tag{8}$$

Since $dP/ds = \pi(\rho, P)$, $\pi(\rho, P)$ can be interpreted as the rate at which the breakdown probability changes over time along the characteristic.

Please note inside a congested region, $c(\rho) \approx -15$ km/h, implying that the breakdown probability increases as we move upstream away from the point at which the congestion originated. If we consider $P = P_{S-J}$ (transition from synchronized to free flow), we can thus model the fact that the probability of a transition form synchronized flow to wide moving jams increases when moving away from the head of the queue (in the upstream direction).

Outside congestion, we have $c(\rho) \approx 85$ km/h. If we now consider $P = P_{F-S}$ (probability that we have a transition form free flow to synchronized flow), we can thus model the observed increases in this probability as we proceed downstream from the bottleneck, e.g. the fact that congestion sets in downstream of an on-ramp rather than at the location of the on-ramp itself.

2.2 Discretization of the Equations

To numerically solve the problem, we propose using the standard Godunov scheme for the conservation of vehicle equation [5]. For the transition probability, basically any discretization scheme will work. We propose the following standard scheme:

$$\begin{aligned} P_{i,j+1} &= P_{i,j} + \Delta t \cdot \pi(\rho_{i,j}, P_{i,j}) \\ &\quad - \Delta t \cdot \left(c^{+}(\rho_{i,j}) \cdot \frac{P_{i+1,j} - P_{i,j}}{\Delta x} + c^{-}(\rho_{i,j}) \cdot \frac{P_{i,j} - P_{i-1,j}}{\Delta x} \right) \end{aligned} \tag{9}$$

where

$$c^{+}(\rho_{i,j}) = \max(0, c(\rho_{i,j})) \text{ and } c^{-}(\rho_{i,j}) = \min(0, c(\rho_{i,j})) \tag{10}$$

3 Example Application of Theory

Let us now show some results of applying the model. In this section, we will consider both the F-S transitions ($P = P_{F-S}$) and the S-J transitions ($P = P_{S-J}$). Before showing these examples, we will present the specification of the transition probability rates π used in the remainder of the contribution.

3.1 Specification of the Fundamental Diagram

In this contribution, we use a simple linear fundamental diagram:

$$Q(\rho) = \begin{cases} C_{free} \cdot \frac{\rho}{\rho_{crit}} & \rho \leq \rho_{crit} \\ C_{queue} \cdot \frac{\rho_{jam} - \rho}{\rho_{jam} - \rho_{crit}} & \rho > \rho_{crit} \end{cases} \tag{11}$$

In Eq. (11), C_{free} denotes the free flow capacity (typically 2250 veh/h/lane), while C_{queue} denotes the queue discharge rate (between 1800 and 2000 veh/h/lane); ρ_{crit} denotes the critical density (25 veh/km/lane), and finally ρ_{jam} denotes the jam density. Note that for $\rho \leq \rho_{crit}$ we have $c(\rho) = C_{free}/\rho_{crit}$. For $\rho > \rho_{crit}$ we have $c(\rho) = C_{queue}\rho_{jam}\ (\rho_{jam} - \rho_{crit})$. This shows clearly how the parameters of the fundamental diagram determine the way perturbations propagate through the flow.

3.2 Specification of the Transition Probability Rate

We will use the following linear expression for the rate $\pi(\rho, P)$ (both for the F-S transitions, and the S-J transitions, be it with different parameter values):

$$\pi(\rho, P) = \begin{cases} (\pi_0 + \pi_1 P) \cdot \frac{\rho - \rho_0}{\rho_1 - \rho_0} & \text{for } \rho_0 \leq \rho \leq \rho_1 \\ 0 & \text{elsewhere} \end{cases} \tag{12}$$

Additionally, we will assume that $P = 0$, if the density is less than ρ_0. Furthermore, P will be limited to values between 0 and 1. After some straightforward computations, it follows that along the characteristic curves $C = \{t(s), x(s)\}$, the transition probability equals:

$$P(s) = \min\left\{\frac{\pi_0}{\pi_1}\left(e^{\pi_1\left(\frac{\rho-\rho_0}{\rho_1-\rho_0}\right)s} - 1\right), 1\right\} \tag{13}$$

3.3 F-S Transition Probability Behavior

Let us consider two-lane 10 km road with an on-ramp at $x = 6$ km. For the piecewise linear fundamental diagram, we assume $C_{free} = 4500$ veh/h, $C_{queue} = 4000$ veh/h, $\rho_{crit} = 50$ veh/km and $\rho_{jam} = 250$ veh/km. For the scenario at hand, we choose $Q_{main} = 3500$ veh/h and $Q_{on-ramp} = 1250$ veh/h. After playing around with different parameter values, we have chosen $\pi_0 = 1$ and $\pi_1 = 100$ (for illustration purposes); $\rho_0 = 40$ veh/km and $\rho_1 = \rho_{crit}$.

Fig. 1 below shows the results of the numerical experiment. It shows the density profile, the location of the points on the fundamental diagram, and the transition probabilities. The F-S breakdown probability increases non-linearly after the on-ramp at $x = 6$ km. In other words, the occurrence of a breakdown becomes more likely further downstream of the on-ramp.

If we, for the sake of argument, assume that we would have an F-S transition (in this case, modeled by temporarily assume that the capacity is reduced from C_{free} to C_{queue}) when $P_{F-S} > 0.5$, the simulation shows that at a certain time instant, the transition occurs (downstream of the bottleneck), moves upstream, and passing the on-ramp location. There it leads to the on-set of congestion (because the capacity is reduced); see Fig. 2.

3.4 S-J Transition Probability Behaviour

For the S-J transitions, we find similar behavior. In this case, we have again used $\pi_0 = 1$ and $\pi_1 = 100$ (for illustration purposes) to describe the S-J transition; $\rho_0 = \rho_{crit}$ and $\rho_1 = 200$ veh/km. For this scenario, we assume that $Q_{main} = 4000$ veh/h and $Q_{on-ramp} = 1500$ veh/h, implying that the bottleneck is oversaturated.

The result is indeed similar to the result we found for the F-S transition (see Fig. 3): the probability on an S-J transition is zero at the on-ramp (where

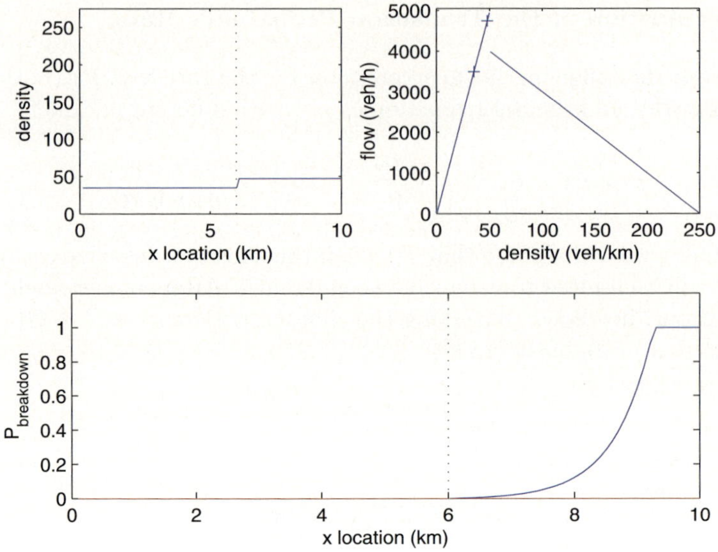

Fig. 1. On-ramp scenario showing increase in the F-S transition probability.

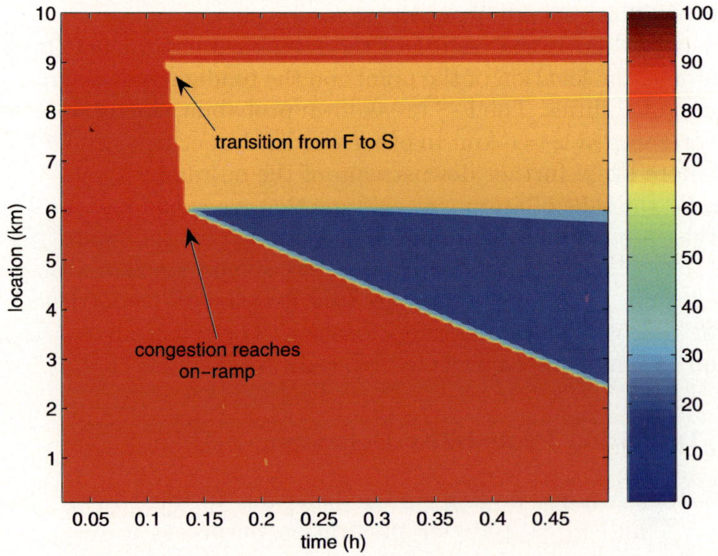

Fig. 2. Speed contours showing F-S transition. Figure shows how congestion sets in downstream of on-ramp and moves upstream.

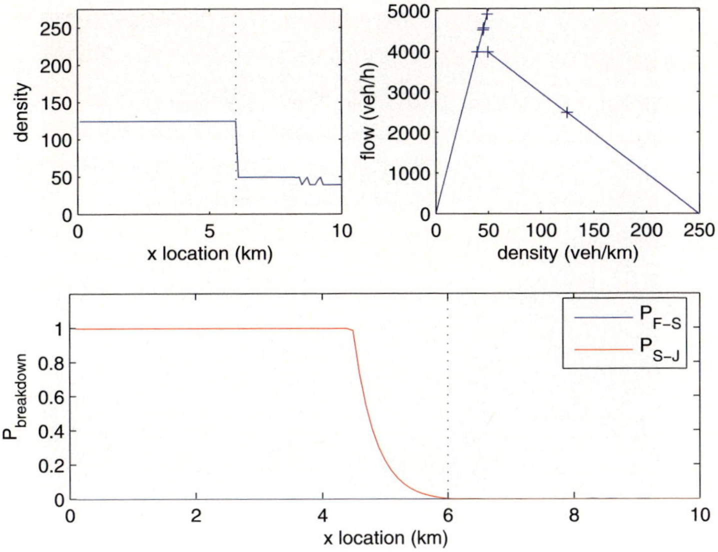

Fig. 3. Probability of an S-J breakdown, increasing as we move from the on-ramp in the upstream direction.

this model assumes that the head of the queue is located), and increases non-linearly as we move upstream away from the bottleneck.

As a final example, let us assume that an S-J transition occurs when $P_{S-J} > 0.5$ (i.e. it is in a way deterministic). Fig. 4 shows the results of this analysis. Clearly, the precise values are not realistic, but the general picture appears to be correct. It is also interesting to note the chaotic-like patterns that emerge even when this simple example is used.

4 Conclusions and Future Work

In this paper we have proposed a relatively simple extension of the first-order model pertaining to the inclusion of breakdown probabilities. The breakdown probability is modeled using a partial differential equation. The main assumption is that information regarding the breakdown probability moves along the characteristic curves.

The workings of the model are illustrated by means of an example featuring flow breakdown due to an on-ramp. Using this example, it is shown that the model can capture the main features of the different phase transitions (free flow to synchronized flow, synchronized flow to jam).

Future research is aimed at modeling the phase-transition itself. In the examples provided in this paper, this was achieved using a simple threshold value for the transition probabilities. A stochastic approach is however more realistic. Clearly, this would yield a stochastic first-order macroscopic model.

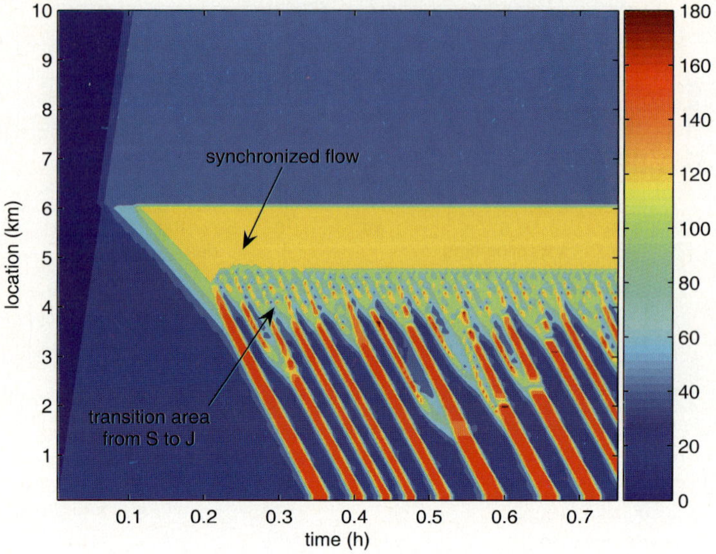

Fig. 4. Speed contours showing transition from S to J upstream of the on-ramp.

Another extension of the theory is to use a multi-class traffic flow model, distinguishing between person-cars and trucks. In doing so, the dynamics of the phase-transitions can be made dependent on the traffic composition, since clearly this has a strong effect on the breakdown probability dynamics.

References

1. Kerner B S (2004) The Physics of Traffic: Empirical Freeway Pattern Features. Springer, Berlin.
2. Heidemann D (2002) Mathematical Analysis of non-stationary queues and waiting times in traffic flow with particular consideration of the coordinate transformation technique. In: Transportation and Traffic Theory in the 21st Century. M P Taylor (ed.). Pergamon, 675–696.
3. Brilon W, J Geistefeldt, M Regler (2005) Reliability of freeway traffic flow: a stochastic concept of capacity In: Flow, Dynamics And Human Interaction. H Mahmassani (ed.) Pergamon, 125–144.
4. Kuehne R, R Mahnke (2005) Controlling traffic breakdowns In: Flow, Dynamics And Human Interaction. H Mahmassani (ed.) Pergamon, 229–244.
5. J-P Lebacque (1996) The Godunov scheme and what it means for first order traffic flow models. In: Proceedings of the 13th International Symposium on Transportation and Traffic Theory. J P Lesort (ed.) Pergamon, 647–677.

Traffic Dynamics on a Rotary

Ding-wei Huang

Department of Physics, Chung Yuan Christian University, Chung-li, Taiwan
dwhuang@phys.cycu.edu.tw

Summary. We present a simple cellular automaton model to study the traffic dynamics on a 4-ramp rotary. Vehicles can move in and out of the rotary through on-ramp and off-ramp, respectively. On the rotary, vehicles move deterministically; while the ramps operate stochastically. We show that, both numerically and analytically, the traffic states on the rotary are completely determined by the ramps. The ramps provide a means to stabilize the density difference on the rotary and to support the maximum flow as a distinct phase. We are able to obtain exact solutions in the full parameter space. The complete phase diagram can be derived. We compare the results between a closed system and an open system. We show that the bulk density is not a good parameter. The traffic states are better characterized by the stochastic rates in the boundaries.

1 Introduction

Traffic related problems are all important to a modern society. Some of the traffic phenomena can only be resorted to the particularity of a few specific vehicles, for example, the accidents caused by mechanical failure or drunk driving. However, there are still prominent phenomena could be understood without attributing to such particularity. The emergence of traffic jam as vehicular density increases is just one of the example. The complicated phenomena can be understand in terms of simple effective interactions among identical vehicles [1–4]. Some basic ingredients are: (a) in the steady state, we still expect vehicles to move asymptotically; (b) the movement of vehicles should be confined by the roadway; (c) collisions should be avoided since safety must be a high priority. Along this line, traffic dynamics can be characterized as non-equilibrium, one-dimensional, and hard-core repulsive.

The asymmetric simple exclusion process (ASEP) can be taken as the most basic model of traffic dynamics [5]. The roadway is divided into discrete cells, i.e., a lattice. Vehicles are identical particles hopping on the lattice. Each cell can only accommodate one particle. Accumulation is forbidden. For each cell, there are only two possible states, i.e., either empty or occupied by a particle. Time is also discretized. And the motion is uni-directional. At each time step, a particle will hop forward to the next cell if that cell is empty in the previous time step; otherwise, the particle will stay where it is. Particles are not allowed to hop backward nor to pass each other. When the periodic boundary condition is imposed, the simple roadway becomes a loop and the density ρ is a conserved quantity. It is easy to deduce that there are two different phases as ρ varies. In the low density regime, $\rho < \frac{1}{2}$, all particles can move freely. In this dimensionless model, the traffic flow is equal to the density ρ. Even if one starts with a situation where all particles were jammed pack together, they will self-arrange to separate from each other. When the asymptotic steady state is reached, there are at least one empty cell in between any two particles. In the high density regime, $\rho > \frac{1}{2}$, the free flow becomes inaccessible. It is impossible to have all the particles moving together. The system will then self-arrange to have all the empty cells kept away from each other, i.e., separated by at least one particle. Only those particles located right behind the empty cells will be able to move in the next time step. We have a traffic flow at $(1 - \rho)$. It is easy to observe that these two phases are related by the time reversal symmetry or the so-called particle-hole symmetry.

When the closed boundary is replaced by the open boundary, we have to introduce two parameters (α, β) to specify the two open ends. Particles will be injected from one end, hopping through the lattice, and then removed from the other end. The hopping dynamics is deterministic; while the boundary conditions are stochastic. If the first cell is empty, a new particle will be injected with a finite probability α; if the last cell is occupied, that particle will be removed with a finite probability β. Although the parameter space becomes two-dimensional, there are still only two phases: when $\alpha < \beta$, we have the free flow; when $\alpha > \beta$, we have the congestion. In practice, these two phases are exactly the same as in the case of a closed loop. With open boundaries, the density is not a conserved quantity. Yet in the asymptotic steady states, uniform density profiles can still be expected. In the low-density free flow, the bulk properties are controlled by the injection boundary. The self-consistent condition is $\alpha(1 - \rho) = \rho$, which can be solved to give $\rho = \alpha/(1 + \alpha)$. The bulk properties are independent of the removal boundary. When β varies, the effects can only be observed in a boundary layer near the last cell. On the other hand, in the high-density congestion, the system is controlled by the removal boundary. The self-consistent condition is $\beta\rho = (1 - \rho)$, which leads to $\rho = 1/(1 + \beta)$. When α varies, only the boundary layer near the first cell will be affected. As regards to the distinct traffic states, the closed loop and the open roadway seem to provide the same results. However, such results are invalided when one goes beyond the homogeneous roadway. With junctions, the results of the open system and the closed system can be very different.

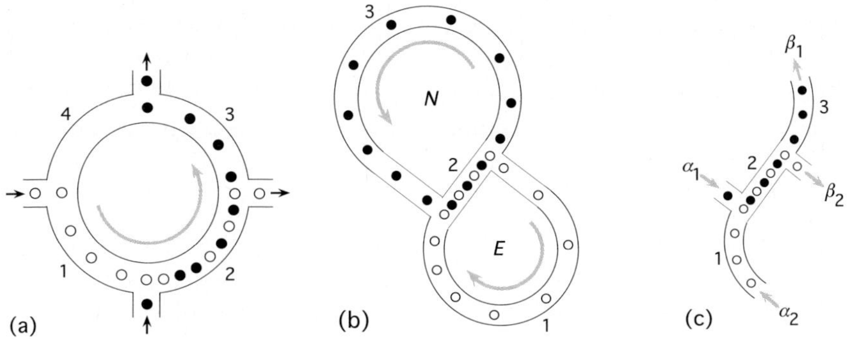

Fig. 1. **(a)** Configuration of a 4-ramp rotary. North-bound traffic and east-bound traffic are represented by different symbols. The four segments on the rotary are labeled by number. **(b)** A 2-loop configuration by imposing periodic boundary conditions. The east-bound traffic consists of segments 1 and 2; the north-bound traffic consists of segments 2 and 3. **(c)** Configuration of an open rotary. The east-bound traffic is controlled by (α_2, β_2); the north-bound traffic by (α_1, β_1).

2 Closed Rotary

We consider a four-ramp rotary to regulate the traffic from two different directions as shown in Fig. 1(a). The north-bound traffic enters from the bottom side and leaves from the top side; the east-bound traffic enters from the left side and leaves from the right side. These two types of traffic are totally distinguishable and represented by different symbols. In the rotary, all vehicles have to follow each other orderly to move forward counter-clockwise [6]. At the bottom entrance, newly arrived vehicle has to yield to those vehicles in the rotary, i.e., vehicles entering from the left side get the right-of-way. At the right exit, if the traffic leaving the rotary is not smooth, the vehicles behind have to wait patiently. Passing is not allowed in the rotary. Together, the priority is given to the east-bound traffic, which may block the north-bound traffic both at the bottom entrance and at the right exit. In this configuration, the traffic mixes in segment 2. In the upstream (segment 1), only the east-bound traffic presents; in the downstream (segment 3), only the north-bound traffic presents. Segment 4 is irrelevant.

When the periodic boundary conditions are imposed separately on the north-bound traffic and on the east-bound traffic, the rotary is equivalent to a system consists of two closed loops sharing a finite roadway, see Fig. 1(b). There are three length scales in the model: L_i denotes the length of segment i, where $i = 1, 2, 3$. As expected, the dynamics can be characterized by two parameters ρ_E and ρ_N, which denote the densities of east-bound vehicles (in loop E) and north-bound vehicles (in loop N), respectively. When both ρ_E and ρ_N are small, the traffic is smooth. In practice, the free flow can be expected on all roadways in the following regime,

$$\rho_E + \rho_N < \frac{1}{2} \ , \tag{1}$$

where the congestion will not emerge.

Since the two loops share a finite roadway (segment 2 with a length L_2), the two densities are upper bounded as

$$\rho_E(L_1 + L_2) + \rho_N(L_2 + L_3) < (L_1 + L_2 + L_3) \ . \tag{2}$$

When ρ_E and/or ρ_N are large, gridlock emerges inevitably. For the gridlock attributed to the high density in loop E, the required conditions are

$$\rho_E(L_1 + L_2) + \rho_N(L_2 + L_3) > (L_1 + L_2) \ , \tag{3}$$

$$\rho_E(L_1 + L_2) > (L_1 + \frac{1}{2}L_2) \ . \tag{4}$$

The first condition implies that the loop E along is not enough to accommodate all the vehicles; the second condition implies that the shared roadway would still be congested even if all the vehicles try to stay in other place (segment 1 and segment 3) to make room in segment 2. Similarly, the gridlock can also be attributed to the high density in loop N.

In between the free flow and gridlock, various kinds of congestion manifest in this simple model. The complete phase diagram is shown in Fig. 2(a). First, we consider the homogeneous phases where vehicles distribute homogeneously on each segment. Of course, different densities can be expected for different segments. To clarify the notations, ρ_E and ρ_N denote the global densities on the roadways (L_1, L_2) and (L_2, L_3), respectively. For each single segment, the density is denoted by ρ_i, where $i = 1, 2, 3$. There are two homogeneous phases: F-J-J and J-J-F. The notation refers to the segment index 1-2-3 in Fig. 1(b). On each segment, the traffic conditions can be either free-flowing (F) or jamming (J). When ρ_E is small and ρ_N is large, the traffic in loop N is congested. Yet free flow can still be maintained in loop E except for the shared roadway. Such a traffic state is denoted as F-J-J. The regime of this phase can be obtained as

$$\rho_E(L_1 + L_2) + \rho_N(L_2 + L_3) < (L_2 + L_3) \ , \tag{5}$$

$$\rho_E(L_1 + L_2) + \rho_N(L_1 - L_2) < L_1 - \frac{L_2}{2}\frac{(L_1 + L_3)}{(L_2 + L_3)} \ , \tag{6}$$

$$\rho_E(L_2 - L_3) + \rho_N(L_2 + L_3) > \frac{1}{2}(L_2 + L_3) \ . \tag{7}$$

The first condition marks the boundary to gridlock, i.e., $\rho_3 < 1$. The second condition ensures that free flow can still be maintained in segment 1, i.e., $\rho_1 < (1 - \rho_3)$. The third condition implies that the loop N is congested, i.e., $\rho_2 > \frac{1}{2}$. Similarly, the regime for J-J-F phase can also be derived.

Besides the homogeneous phases, the phenomena of phase separation can also be observed in this simple model. On a homogeneous roadway, congestion

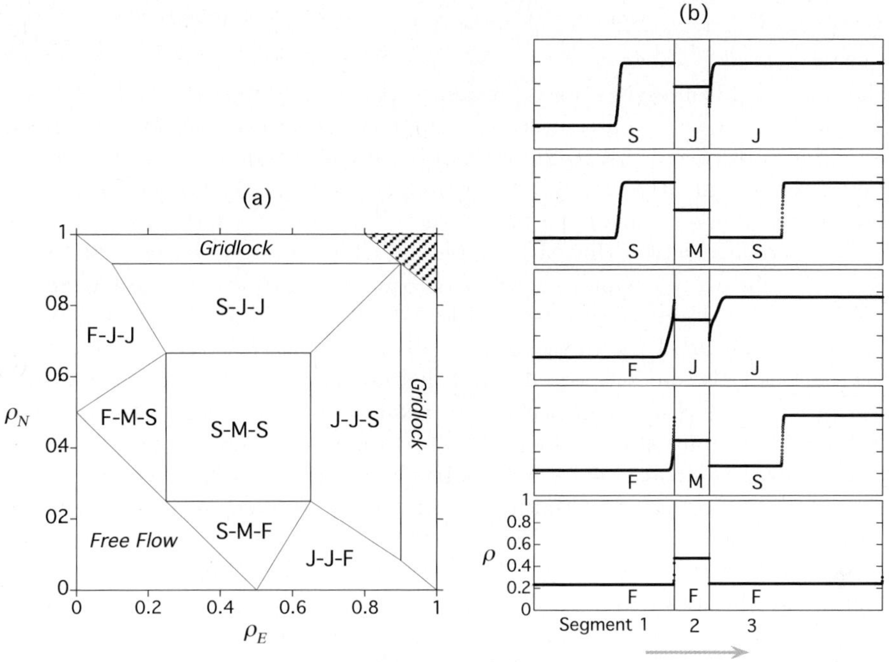

Fig. 2. (a) Typical phase diagram for a closed rotary shown in Fig. 1(b) with length scales $L_1 : L_2 : L_3 = 4 : 1 : 5$. The forbidden regime is shown by the shade. **(b)** Typical density profiles of various phases: F-F-F, F-M-S, F-J-J, S-M-S, S-J-J. The traffic direction is shown by the gray arrow. The length scales are $L_1 = 1200$, $L_2 = 300$, and $L_3 = 1500$.

only emerges near one end (downstream) and free flow can still be maintained near the other end (upstream). Somewhere in between these two ends, a stationary shock front can be sustained. Start with the F-J-J phase, where ρ_N is large and ρ_E is small. If ρ_E is further increased, the homogeneous free flow in segment 1 will not be able to sustain. Yet a homogeneous congestion will not result when the homogeneous free flow broken down. In the resultant S-J-J phase, the phase separation (S) is observed in segment 1 while the loop N is still congested. In such a case, though the average density ρ_1 can still be defined mathematically, it is no longer a good parameter to characterize the dynamics. In fact, the traffic flow in segment 1 is locked into the traffic flow in segment 3. Thus the dynamics in segment 1 is better characterized by ρ_3, instead of ρ_1. The regime of S-J-J phase can be obtained by the constraints: $(1 - \rho_3) < \rho_1 < \rho_3$ and $\rho_2 > \frac{1}{2}$. After some algebraic manipulations, we have

$$\rho_E(L_1 + L_2) + \rho_N(L_1 - L_2) > L_1 - \frac{L_2}{2}\frac{(L_1 + L_3)}{(L_2 + L_3)} , \tag{8}$$

$$(\rho_E - \rho_N) < \frac{L_2(L_1 - L_3)}{2(L_1 + L_2)(L_2 + L_3)} , \tag{9}$$

$$\frac{1}{4} + \frac{L_3}{2(L_2 + L_3)} < \rho_N < \frac{1}{2} + \frac{L_3}{2(L_2 + L_3)} \quad . \tag{10}$$

The first condition gives a lower bound to the density in segment 1, i.e., $\rho_1 > (1 - \rho_3)$; the second condition gives an upper bound to the density in segment 2, i.e., $\rho_1 < \rho_3$. The third condition can be retrospected to $\frac{1}{2} < \rho_2 < 1$, where the upper bound of ρ_2 marks the boundary to gridlock. For another phase separation known as the J-J-S phase, the regime can be derived similarly.

Now we have shown that this simple model can provide a rich structure of various traffic states, which include free flow (F), gridlock (G), homogeneous jamming (J), and phase separation (S). Yet these phases have not exhausted the two-dimensional parameter space (ρ_E, ρ_N). We find that the maximum flow (M) can also be supported as an distinct phase on the shared roadway. Start with the F-J-J phase, where ρ_E is small and ρ_N is large. If ρ_N is now decreased, the congestion in loop N will be resolved. However, before the free flow is restored, we observed a distinct phase denoted as F-M-S, where segment 2 is saturated to the maximum flow and the phase separation appears in segment 3. The break down of homogeneous congestion in segment 3 provides a mean to stabilize the maximum flow in segment 2. The traffic flow in segment 3 locks into the traffic flow in segment 1 indirectly. The dynamics in segment 3 is better characterized by ρ_1, instead of ρ_3. On the shared roadway, the north-bound vehicles are still the majority. The phase regime can be determined by the constraints: $\rho_1 < \frac{1}{4}$ and $(\frac{1}{2} - \rho_1) < \rho_3 < (\frac{1}{2} + \rho_1)$. In terms of ρ_E and ρ_N, the conditions for F-M-S phase become

$$\rho_E < \frac{1}{4} \quad , \tag{11}$$

$$\frac{1}{2} < (\rho_E + \rho_N) < \frac{1}{2} + \rho_E \, \frac{2L_3}{L_2 + L_3} \quad . \tag{12}$$

The conditions for S-M-F phase can be obtained similarly. For the rest area of the parameter space (ρ_E, ρ_N), phase separation can be observed in both segments 1 and 3, which is then denoted as S-M-S phase. In summary, we have derived all the distinct traffic phases in this model. The complete phase diagram is obtained, see Fig. 2(a). In between the free flow and gridlock, there are seven distinct phases. Some of the typical density profiles are shown in Fig. 2(b). For each segment, the traffic states can be free flow (F), jamming (J), phase separation (S), or maximum flow (M). However, not all the combinations are possible. The maximum flow only presents in segment 2, i.e., the shared roadway. Yet free flow and phase separation never appear in segment 2. In this simple model, the above mean-field considerations yield the exact results.

It is interesting to notice that F-J-F phase is absent. With naive intuition, it is plausible to have a situation where vehicles are congested on the shared roadway; yet they can move freely on other parts of the roadways, i.e., segment 2 becomes a bottleneck to the traffic. However, such an intuition is flawed. Similarly, the J-J-J phase is also unstable, which can only exist at the boundary between S-J-J and J-J-S phases.

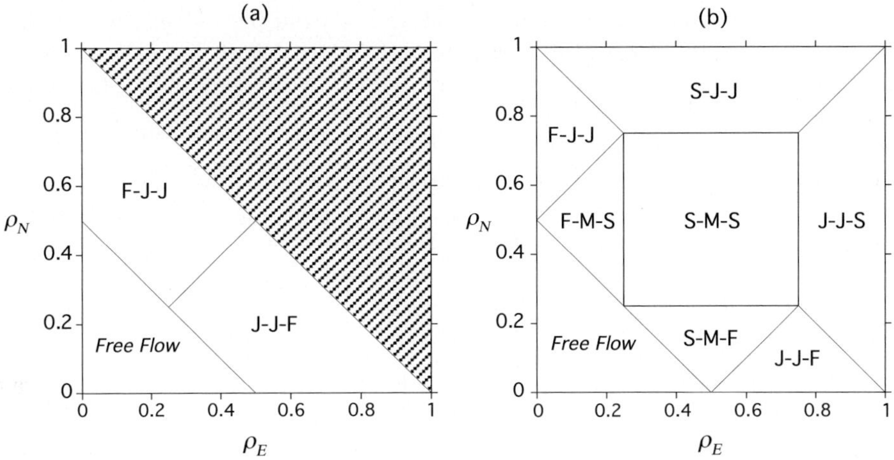

Fig. 3. Phase diagrams in the extreme limits of **(a)** a dominant $L_2 \gg L_1, L_3$; **(b)** a negligible $L_2 \ll L_1, L_3$.

Except for the free flow regime, the phase diagram depends on the length scales (L_1, L_2, L_3). When the ratio of the shared roadway L_2 increases, the forbidden regime in the parameter space (ρ_E, ρ_N) increases. As the maximum flow only appears on the shared roadway, naive intuition would expect an extended M-phases (F-M-S, S-M-S, S-M-F) as L_2 increases. However, the results are quite in the contrary, where the area of M-phases shrinks as the ratio of L_2 increases. In the extreme limit when L_2 becomes dominant, i.e., $L_2 \gg L_1, L_3$, all the M-phases and S-phases disappear completely. The phase diagram is greatly simplified as shown in Fig. 3(a). In the other extreme limit when L_2 is negligible, i.e., $L_2 \ll L_1, L_3$, the phase diagram is shown in Fig. 3(b). In this case, the entire parameter space becomes accessible. The gridlock disappears. It is interesting to note that, in these two extreme limits of L_2, the phase diagram becomes independent of the remaining length scales (L_1, L_3).

3 Open Rotary

Now we discuss the case of an open rotary [7]. The periodic boundary conditions are replaced by open ramps through which vehicles can move in and out of the rotary. The system configuration is shown in Fig. 1(c). Instead of the two-dimensional parameter space (ρ_E, ρ_N), now we have a four-dimensional parameter space $(\alpha_1, \beta_1, \alpha_2, \beta_2)$. The in and out of the north-bound vehicles are controlled by α_1 and β_1, respectively; the east-bound vehicles by α_2 and β_2. The phase regime for each distinct phase can be deduced by the mean-field theory. As a ramp sits in between two segments, the effects can be replaced by two effective rates: one to remove vehicles from the segment before and the other to add vehicles to the segment after. For example, α_1 in the junction between segments 1 and 2 can be replaced by two rates (a_1, b_1), where a_1 controls the end point of segment 1 and b_1 controls the start point of segment 2. Similarly, β_2 can be replaced by (a_2, b_2), where b_2 controls the end point of segment 2 and a_2 controls the start point of segment 3. These four effective rates (a_1, b_1, a_2, b_2) can be related to the four physical rates $(\alpha_1, \beta_1, \alpha_2, \beta_2)$ by balancing the flow across the ramp and by imposing the self-consistence from the mean-field theory. Then each segment in the rotary can be treated as a homogeneous roadway, where the traffic conditions are determined by the two boundary. The traffic conditions in segment 1 are determined by (α_2, b_1); segment 2 by (a_1, b_2); segment 3 by (a_2, β_1).

It is interesting to observe that F-J-F phase becomes sustainable in the open rotary. In contrast, such a phase cannot be supported in the closed rotary discussed in last section. Across the on-ramp (α_1), i.e., the entrance of north-bound vehicles, the flow balance can be written as

$$\frac{\alpha_2}{1+\alpha_2} + (1-\rho_1^*)\alpha_1 = \frac{b_2}{1+b_2} \quad , \tag{13}$$

where ρ_1^* denotes the local density at the on-ramp. The free flow in segment 1 has a traffic flow controlled by the injection: $\alpha_2/(1+\alpha_2)$; the congestion in segment 2 has a traffic flow controlled by the removal: $b_2/(1+b_2)$. The other term represents the inflow from the ramp. The self-consistence at the end point of segment 1 gives $\rho_1^*\, b_1 = \alpha_2/(1+\alpha_2)$. The self-consistence at the start point of segment 2 gives $(1-\rho_1^*)a_1 = b_2/(1+b_2)$. Across the off-ramp (β_2),i.e., the exit of east-bound vehicles, the flow balance becomes

$$\frac{b_2}{1+b_2} = \frac{a_2}{1+a_2} + \rho_{2E}^*\, \beta_2 \quad , \tag{14}$$

where ρ_{2E}^* denotes the local density of east-bound vehicles at the off-ramp. The free flow in segment 3 has a traffic flow $a_2/(1+a_2)$ controlled by the injection. For steady states, the outflow at the off-ramp should equal to the traffic flow in segment 1; the inflow at the on-ramp should equal to the traffic flow in segment 3, i.e., $\rho_{2E}^*\, \beta_2 = \alpha_2/(1+\alpha_2)$ and $(1-\rho_1^*)\alpha_1 = a_2/(1+a_2)$. Thus the flow balance at the off-ramp, Eq. (14), is satisfied automatically. The free flow in segment 3 supports a local density at the off-ramp as $\rho_{2N}^* = a_2/(1+a_2)$, where ρ_{2N}^* denotes the local density of north-bound vehicles at the off-ramp. With the effective removal rate b_2 in the congested segment 2, the local density at the off-ramp can also be written as $\rho_2^* = 1/(1+b_2)$, where ρ_2^* denotes the effective local density at the off-ramp, which includes two kinds of vehicles. The vehicle number conservation implies $\rho_2^* = \rho_{2E}^* + \rho_{2N}^*$ which can be written as

$$\frac{1}{1+b_2} = \frac{\alpha_2}{\beta_2(1+\alpha_2)} + \frac{a_2}{1+a_2} \quad . \tag{15}$$

With these equations, the effective rates (a_1, b_1, a_2, b_2) and the local densities $(\rho_1^*, \rho_{2E}^*, \rho_{2N}^*)$ can all be solved explicitly. The regime of F-J-F phase can then be marked self-consistently as $\alpha_2 < b_1$, $a_1 > b_2$, and $a_2 < \beta_1$. In terms of $(\alpha_1, \beta_1, \alpha_2, \beta_2)$, these constraints becomes

$$\frac{\beta_2 - \alpha_2}{\alpha_2 + \beta_2} < \alpha_1 < \frac{\beta_2 - \alpha_2}{2\alpha_2\beta_2} \quad , \tag{16}$$

$$\frac{\beta_2 - \alpha_2}{\alpha_2 + \beta_2 + 2\alpha_2\beta_2} < \beta_1 \quad . \tag{17}$$

In practice, F-J-F phase can only be supported in a regime less than as shown in above. The upper bound of the effective injection rate $a_1 < 1$ places a much stringent constraint than $\alpha_2 < b_1$ (the free flow condition in segment 1) does. Thus the upper bound of α_1 in Eq. (16) should be replaced by

$$\alpha_1 < \frac{\beta_2 - \alpha_2}{\beta_2 - \alpha_2 + 2\alpha_2\beta_2} \quad . \tag{18}$$

The regime for other distinct phases can also be obtained similarly. It is interesting to note that the phase diagram becomes independent of the length scales and can be completely specified by the boundary rates. For the entire four-dimensional parameter space $(\alpha_1, \beta_1, \alpha_2, \beta_2)$, there are only five distinct phases: F-F-F, F-J-F, F-J-J, J-J-F, and J-J-J. When the closed boundary conditions are replaced by the open boundary conditions, the parameter space extends hugely; yet the number of distinct phases reduces. The gridlock, phase separation, and the maximum flow are all disappeared. Only the homogeneous phases can be supported. With naive combinations of free flow (F) and jam (J), there could be eight homogeneous phases. But three of them cannot be realized. When there is only one segment congested, it must be the shared roadway, i.e., F-F-J and J-F-F- are both impossible. When the free flow can only be supported in one segment, it cannot be the shared roadway, i.e., J-F-J is also impossible.

The phase diagram in four-dimensional parameter space is hard to visualize. With two of the parameters fixed, a conventional phase diagram in two dimension is shown in Fig. 4(a). The typical density profiles of various phases are shown in Fig. 4(b). The numerical simulations can be exactly reproduced by the analytical formulas obtained from mean-filed theory.

4 Conclusion

We propose a simple cellular automaton model to study the traffic dynamics in a rotary. The bulk properties can be successfully analyzed by the mean-field theory. The numerical simulations can be reproduced by analytical results. The complete phase diagram can be constructed to display all the distinct traffic states in the full parameter space. Although the vehicular interactions are simple and the configuration of a rotary is just a small step to go beyond a homogeneous roadway, the model does provide a rich structure of various

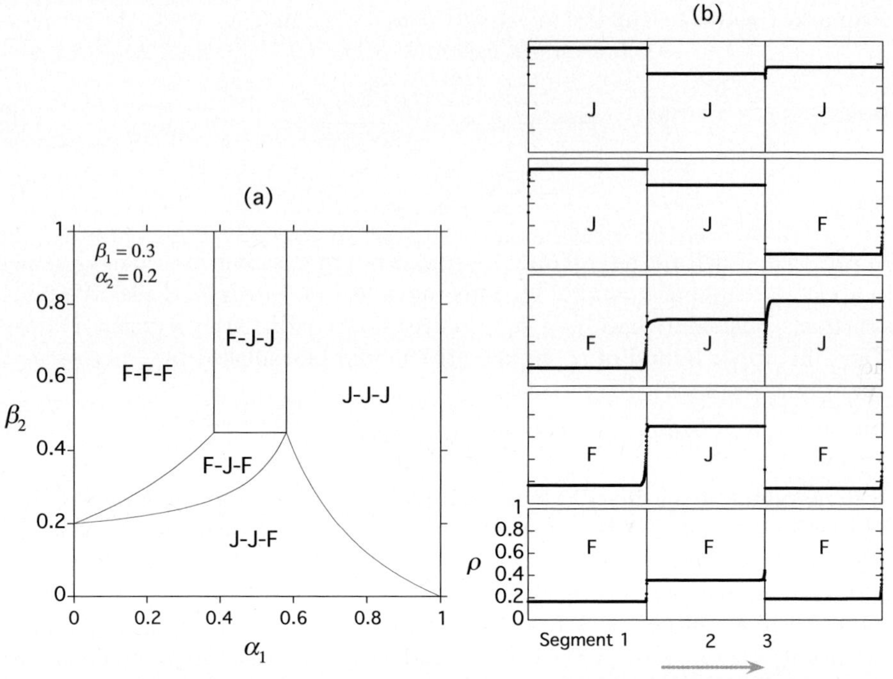

Fig. 4. (a) Phase diagram in (α_1, β_2) for an open rotary. The other two parameters are fixed at $(\beta_1, \alpha_2) = (0.3, 0.2)$. **(b)** Typical density profiles of various phases: F-F-F, F-J-F, F-J-J, J-J-F, J-J-J. The traffic direction is shown by the gray arrow. The length scales are $L_1 = L_2 = L_3 = 1000$.

traffic states. A rotary can be taken as a system consists of homogeneous segments separated by the ramps. For each segment, the traffic states can be free flow, jamming, maximum flow, phase separation, or gridlock. Yet not all the combinations can be realized. The traffic dynamics in the rotary is a system driven far away from equilibrium, which imposes very strong constraints on the possible combinations.

In this model, the dynamics in bulk is deterministic; in boundary, the ramps operate stochastically. Although the ramps only constitute a negligible part of the roadway, the phase diagram is highly sensitive to the operation of ramps. When the ramps are connected to form a closed rotary, the global density is a conserved quantity. The phase diagram is two-dimensional. The phase separation is prominent. The characteristics of a phase diagram depends on the length scale. With an open rotary, the phase diagram becomes four-dimensional. The phase separation disappears. The homogeneous phases become dominant. The phase diagram is independent of the length scale.

By conventional wisdom, congestion is often attributed to too many vehicles on road. When the vehicular density is high, the homogeneous jamming becomes inevitable. In that sense, the emergence of phase separation can be taken as a signature to the inefficient usage of roadway. In the steady state on a homogeneous roadway, traffic flow is not a local property, but a global one. In the phase separation, the free flow in the upstream is limited by the congestion in the downstream. If the vehicles can be redistributed uniformly on the roadway, the traffic flow can be significantly enhanced. When the dynamics is characterized by the density, phase separation dominates the phase diagram shown in Fig. 2(a); on the other hand, when the dynamics is characterized by the ramps, phase separation disappears completely. Thus we have a simple conclusion that, instead of density, the ramp should be monitored and/or controlled to have a smooth traffic.

To further justify the insufficiency of density, we note that the homogeneous jamming (J-J-J) is absent in the phase diagram shown in Fig. 2(a). When the density is large, the gridlock emerges instead. With naive intuition, such a severe congestion is often blamed to a few special vehicles, which might involve in either accidents, mechanical broken down, or violating the traffic regulations intentionally. However, all these particularities are not addressed in this simple model, where all the vehicles are exactly the same. And all the vehicles follow the same rules perfectly. Yet the gridlock still appears, which can only be attributed to the insufficiency of the regulations. When the dynamics is characterized by the ramps, the gridlock also disappears completely.

References

1. B. Kerner, *The Physics of Traffic* (Springer-Verlag, Berlin 2004).
2. S. Maerivoet and B. De Moor, *Phys. Rep.* **419**, 1 (2005).
3. D. Chowdhury, L. Santen, and A. Schadschneider, *Phys. Rep.* **329**, 199 (2000).
4. D. Helbing, *Rev. Mod. Phys.* **73**, 1067 (2001).
5. B. Derrida, *Phys. Rep.* **301**, 65 (1998).
6. FHWA, *Roundabouts: an informational guide*, Publication No. FHWA-RD-00-067, US Department of Transportation, Federal Highway Administration, Washington, D.C. 2000.
7. D. W. Huang, *Physica* **A383**, 603 (2007).

Effect of Adaptive Cruise Control Vehicles on Phase Transition in a Mixture with Manual Vehicles

Rui Jiang[1], Mao-Bin Hu[1], Bin Jia[2], Ruili Wang[3], and Qing-Song Wu[1]

[1] School of Engineering Science, University of Science and Technology of China, Hefei 230026, China
[2] School of Traffic and Transportation, Beijing Jiaotong University, Beijing 100044, China
[3] Institute of Information Sciences and Technology, Massey University, New Zealand

Summary. In this paper, we have investigated the effects of adaptive cruise control (ACC) vehicles in a mixture with manually-controlled (manual) vehicles. The manual vehicles are simulated by using the modified comfortable driving model, which can describe synchronized traffic flow. The phase transition probabilities from free flow to synchronized flow and from synchronized flow to jams are studied. The dependence of microscopic properties of traffic flow, including the spatiotemporal patterns and the velocity distribution, is explored. Our results are expected to be useful for developing ACC systems.

1 Introduction

Recently, research on vehicles equipped with adaptive cruise control (ACC) systems has attracted the interest from both physicists and engineers [1–10]. ACC is a driver assistance system designed to provide more convenience and comfort to a driver. An ACC-equipped vehicle can detect the presence of a preceding vehicle and measures the distance (range) as well as the relative speed (range rate) by using a forward-looking sensor. It automatically adjusts the vehicle speed to keep a proper range when a preceding vehicle is detected. Obviously, ACC vehicles will have some impact on the characteristics of traffic flow, including highway safety, efficiency and capacity, because of their different behavior compared with human drivers. Therefore, before ACC vehicles are deployed on a large scale, their effects on the traffic flow characteristics need to be carefully investigated.

Previous studies on ACC vehicles focus on the range policy, effect of ACC vehicles on traffic flow stability, road capacity, traffic safety, and environmental benefits. However, to our knowledge, the effect of ACC vehicles on the phase transition behavior in a mixture of ACC vehicles and manual vehicles has not been studied. In this paper, we investigate the mixture of ACC vehicles with manual vehicles described by a CA model, which can describe first order phase transition from free flow to synchronized flow.

In the next section, the models for simulating ACC vehicles and manual vehicles are briefly reviewed. The simulation results are presented and analyzed in section 3. The conclusions are given in section 4.

2 CA Models for ACC Vehicles and Manual Vehicles

In this section, we briefly review the CA models for modelling ACC vehicles and manual vehicles. In CA models, a road is divided into cells. Each cell is either empty or occupied by a vehicle. The vehicles move with an integer velocity $1, 2, \ldots, v_{max}$ with v_{max} the maximum velocity of vehicles.

The model for ACC vehicles is recently presented in Ref. [10], in which a constant time headway (CTH) policy is adopted. The parallel updating rules of the model are as follows.

(i) Acceleration or deceleration: $v_n(t+1) = \min(v_{max}, \lceil d_n(t)/T \rceil)$;
(ii) Randomization: $v_n(t+1) = \max(0, v_n(t+1) - 1)$ with probability $p = \lceil d_n(t)/T \rceil - d_n(t)/T$ for $d_n(t)/T < v_{max}$ and $p = 0$ for $d_n(t)/T \geq v_{max}$. Note $v_n(t+1)$ on the right hand side of the equation refers to that determined in the previous step in this and following equations.
(iii) Motion of a vehicle: $x_n(t+1) = x_n(t) + v_n(t+1)$.

Here v_n is the velocity of vehicle n, $d_n = x_{n+1} - x_n - L$ is the inter-vehicle distance, L is the vehicle length, x_n is the position of vehicle n, and vehicle $n+1$ precedes vehicle n. T is the time headway preferred by ACC vehicles. $\lceil x \rceil$ denotes the minimum integer that is not smaller than x.

The model for manual vehicles is presented in Refs. [11–13]. The parallel updating rules of the model are as follows.

1. Determination of the randomization parameter $p_n(t+1)$:
   $p_n(t+1) = p(v_n(t), b_{n+1}(t), t_{h,n}, t_{s,n})$
2. Acceleration:
   if $((b_{n+1}(t) = 0$ or $t_{h,n} \geq t_{s,n})$ and :
   $(v_n(t) > 0))$ then : $v_n(t+1) = \min(v_n(t) + 2, v_{max})$
   else if $(v_n(t) = 0)$ then : $v_n(t+1) = \min(v_n(t) + 1, v_{max})$
   else : $v_n(t+1) = v_n(t)$
3. Braking rule:
   $v_n(t+1) = \min(d_n^{eff}, v_n(t+1))$
4. Randomization and braking:
   if $(rand() < p_n(t+1))$ then: $v_n(t+1) = \max(v_n(t+1) - 1, 0)$
5. The determination of $b_n(t+1)$:
   if $((v_n(t+1) > v_n(t))$ or $(v_n(t+1) \geq v_c$ and $t_{f,n} > t_{c1}))$ then: $b_n(t+1) = 0$
   else if $(v_n(t+1) < v_n(t))$ then: $b_n(t+1) = 1$

else ($v_n(t+1) = v_n(t)$) then: $b_n(t+1) = b_n(t)$

6. The determination of $t_{st,n}$:
   if $v_n(t+1) = 0$ then: $t_{st,n} = t_{st,n} + 1$
   if $v_n(t+1) > 0$ then: $t_{st,n} = 0$
7. The determination of $t_{f,n}$:
   if $v_n(t+1) \geq v_c$ then: $t_{f,n} = t_{f,n} + 1$
   if $v_n(t+1) < v_c$ then: $t_{f,n} = 0$
8. Motion of a vehicle:
   $x_n(t+1) = x_n(t) + v_n(t+1)$.

Here b_n is the status of the brake light (on(off)→ $b_n = 1(0)$). The two times $t_{h,n} = d_n/v_n(t)$ and $t_{s,n} = \min(v_n(t), h)$, where h determines the range of interaction with the brake light, are introduced to compare the time $t_{h,n}$ needed to reach the position of the leading vehicle with a velocity dependent interaction horizon $t_{s,n}$. The effective distance is $d_n^{eff} = d_n + \max(v_{anti} - gap_{safety}, 0)$, where $v_{anti} = \min(d_{n+1}, v_{n+1})$ is the expected velocity of the preceding vehicle in the next time step and gap_{safety} controls the effectiveness of the anticipation. $rand()$ is a random number between 0 and 1, $t_{st,n}$ denotes the time that the car n stops, $t_{f,n}$ denotes the time that car n is in the state $v_n \geq v_c$. The randomization parameter p is defined:

$$p(v_n(t), b_{n+1}(t), t_{h,n}, t_{s,n}) = \begin{cases} p_b & : \text{ if } b_{n+1} = 1 \text{ and } t_{h,n} < t_{s,n} \\ p_0 & : \text{ if } v_n = 0 \text{ and } t_{st,n} \geq t_c \\ p_d & : \text{ in all other cases} \end{cases} .$$

Here v_c, t_{c1}, and t_c are parameters.

3 Simulation Results

In this section, the simulation results are presented and analyzed. In the simulations, the parameter values are $t_c = 9$, $t_{c1} = 30$, $v_c = 18$, $v_{max} = 20$, $p_d = 0.25$, $p_b = 0.94$, $p_0 = 0.5$, $h = 6$ and $gap_{safety} = 7$. Each cell corresponds to 1.5m and a vehicle has a length of five cells. One time step corresponds to 1s. Periodic boundary conditions are adopted in the simulations and the road length is set to $L_r = 10000$.

Figure 1 shows the fundamental diagram of 100% ACC vehicles (i.e., $R = 1$) at different values of T. Here R denotes the ratio of ACC vehicles. A triangular curve of flow rate versus density is reproduced as the CTH policy is adopted. With the increase of parameter T, both the maximum flow rate and the corresponding density decrease.

The fundamental diagram of 100% manual vehicles ($R = 0$) is also shown in Fig. 1. Initially, the vehicles are homogeneously distributed on the road. We focus on the probability of the phase transition (see Ref. [14] for details of phase transition probability).

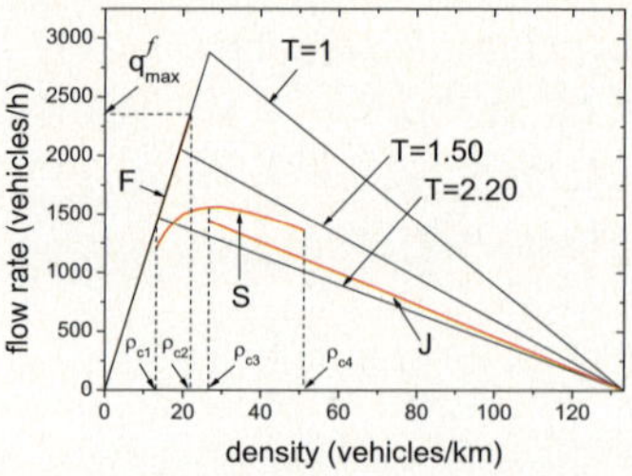

Fig. 1. Fundamental diagrams of 100% ACC vehicles at different values of T (black lines) and of 100% manual vehicles (red lines). F, S and J mean the free flow branch, synchronized flow branch, and jam branch of 100% manual vehicles, respectively.

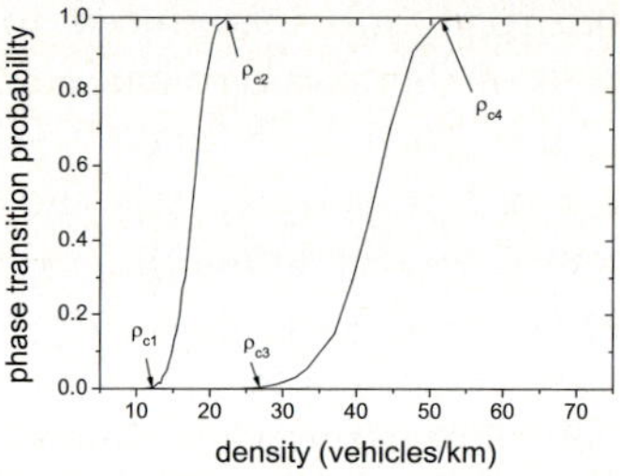

Fig. 2. Phase transition probabilities from free flow to synchronized flow (left curve) and from synchronized flow to jams (right curve) for 100% manual vehicles.

The simulations show there are four critical densities. When the density is smaller than ρ_{c1}, the probability of the phase transition from free flow to synchronized flow is zero; when the density is larger than ρ_{c2}, the probability of the phase transition from free flow to synchronized flow is one; when $\rho_{c1} < \rho < \rho_{c2}$, the phase transition probability is shown in Fig. 2. We denote the maximum flow rate corresponding to ρ_{c2} as $q^f_{\max}$. Similarly, when the density is smaller than ρ_{c3}, the probability of the phase transition from synchronized flow to jam is zero; when the density is larger than ρ_{c4}, the probability of the phase transition from synchronized flow to jam is one; When the density $\rho_{c3} < \rho < \rho_{c4}$, the phase transition probability is also shown in Fig. 2.

Comparing the fundamental diagrams of 100% ACC vehicles and 100% manual vehicles, one can see that the maximum flow rate corresponding to $T = 1.0$ is larger than $q^f_{\max}$, and the maximum flow rate corresponding to $T = 1.5$ is smaller than $q^f_{\max}$. In addition, the congested branches corresponding to $T = 1.0$ and $T = 1.5$ are above the congested branch of manual vehicles, and the congested branch of ACC vehicles is below that of manual vehicles for $T = 2.20$. Next we study the mixture of manual vehicles and ACC vehicles with $T = 1.0, 1.5$ and 2.2 respectively.

3.1 Effects of ACC Vehicles on $F \rightarrow S$

In this subsection, we focus on the effects of ACC vehicles on the phase transition from free flow to synchronized flow in a mixture of ACC vehicles and manual vehicles. Figure 3 shows the phase transition probability curve under different values of T and R. It can be seen that:

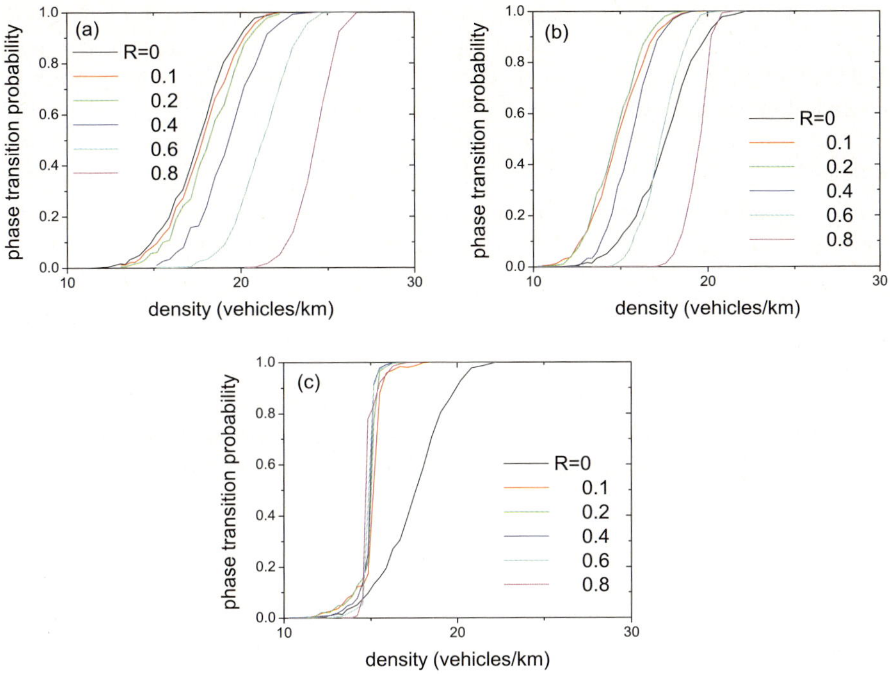

Fig. 3. Phase transition probability from free flow to synchronized flow for a mixture of ACC vehicles and manual vehicles. (a) $T = 1.0$; (b) $T = 1.5$; (c) $T = 2.20$.

(i) For $T = 1.0$, one can see that with the introduction of ACC vehicles, phase transition probability curve moves right. This means that phase transition occurs more difficult, i.e., ACC vehicles could enhance the stability of free flow.

(ii) For $T = 1.50$, the simulations show that when R is small, the phase transition probability curve moves left compared with the curve of $R = 0$. This means that the phase transition occurs more easily. When the ratio of ACC vehicles is large (e.g., $R > 0.2$), the probability curve moves towards the right with the increase of R.

Comparing Figs. 3(a) and (b), one can see that the curves of $T = 1.50$ are on the left of that corresponding to $T = 1.0$ at any ACC vehicle ratio. This means that the phase transition occurs more easily if T increases. Furthermore, one can see that with the increase of T and/or R, the probability curve becomes steeper, which means that the metastable density range corresponding to the transition from free flow to synchronized flow shrinks.

(iii) For $T = 2.20$, the results are quite different. One can see that the probability curves are very steep. Moreover, they only slightly depend on R. These curves are on the left of that corresponding to $R = 0$, which means that a phase transition occurs more easily.

The dependence of the phase transition curve on the parameters R and T might be relevant to the following effects. (1) In the free flow, the headway of ACC vehicles is $v_{max}T$ because ACC vehicles move with the maximum velocity. The total headway of ACC vehicle is, therefore, $NRv_{max}T$. Here NR is the number of ACC vehicles. This leads to the effective headway for manual vehicles $h_{\text{eff}} = \frac{L_r - NRv_{max}T}{N(1-R)}$, which may be different from the average headway $h = \frac{L_r}{N}$. As a result, the stability is enhanced when $h_{\text{eff}} > h$ and is weakened when $h_{\text{eff}} < h$. (2) With the increase of ACC vehicle ratio, the chance that a large platoon of manual vehicles could exist decreases. This will enhance the stability. (3) An ACC vehicle will change the moving behavior of the manual vehicle that follows it, because the velocity of the ACC vehicle $v_{\text{ACC}} = d_{n+1}/T$ might be much smaller than the expected velocity $v_{anti} = \min(d_{n+1}, v_{n+1})$ of the manual vehicle, especially for a large T. Therefore, the collision will happen if the expected velocity of a manual vehicle, which follows an ACC vehicle, remains unchanged. In our simulations, the expected velocity is changed by using $v_{anti} = \min(d_{n+1}/T, v_{n+1})$. Obviously, this will weaken the stability if $T > 1$.

3.2 Effect of ACC Vehicles on $S \rightarrow J$

In this subsection, we focus on the effects of ACC vehicles on the phase transition from synchronized flow to jam. Figure 4 shows the phase transition probability curve under different values of T and R. It can be seen that

(i) For $T = 1.0$: With the increase of R, the probability curve shifts right. This means that with the introduction of ACC vehicles, the phase transition from synchronized flow to jams more unlikely occurs. Furthermore, with the increase of R, the probability curve becomes a little flatter, which means that the metastable density range corresponding to the transition from synchronized flow to jams expands.

(ii) For $T = 1.50$ and $T = 2.20$, similar results are observed. When $T = 1.50$ and $R > 0.8$, a jam will not spontaneously appear from synchronized flow. This is because the propagating speed of the downstream front of a jam is so large that the jam will soon dissolve even if it forms due to reasons such as car accidents. When $T = 2.20$, a jam will not spontaneously appear from synchronized flow provided $R > 0.4$.

We have compared the probability curves of $T = 1.0$, $T = 1.50$ and $T = 2.20$, and found that the curves of $T = 2.20$ are always rightmost and the curves of $T = 1.0$ are always leftmost at any ACC vehicle ratio (not shown).

This means that the phase transition more unlikely occurs if T increases. This is opposite to the effects of ACC vehicles on $F \rightarrow S$ discussed in the previous subsection.

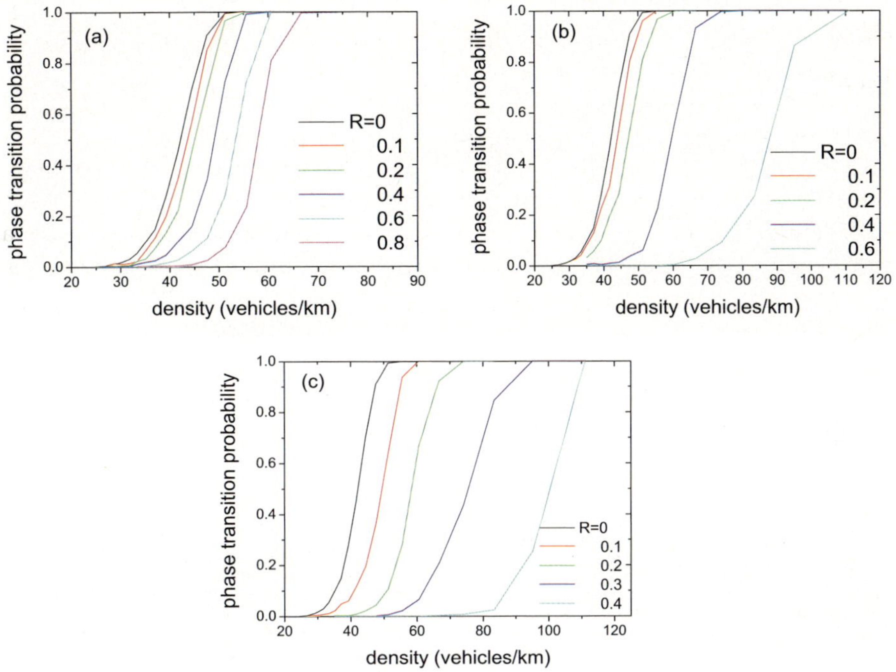

Fig. 4. Phase transition probability from synchronized flow to jams for a mixture of ACC vehicles and manual vehicles. (a) $T = 1.0$; (b) $T = 1.5$; (c) $T = 2.20$.

3.3 Effects of ACC Vehicles on Microscopic Properties

In this subsection, we study the effects of ACC vehicles on spatiotemporal patterns and velocity distribution in synchronized flow. Figure 5 shows the typical spatiotemporal patterns in synchronized flow for different values of R, with the parameter $T = 1.0$ and the density $\rho = 37.04$ vehicles/km (which corresponds to headway 27 m). One can see that with the increase of R, the traffic flow gradually evolves into a mixture of free flow and synchronized flow. Even at $R = 0.99$, the traffic pattern is still characterized by the mixture. However, when $R = 1$, the traffic flow is quite homogeneous (not shown).

Figure 6 shows the spatiotemporal patterns in synchronized flow with the parameter $T = 1.50$ and the density $\rho = 37.04$ vehicles/km. One can see that there also appears a mixture of free flow and synchronized flow. However, when R is large, the mixture is suppressed and the traffic flow gradually becomes homogeneous (see Figs. 7(c) and (d)).

Fig. 5. Typical spatiotemporal patterns in synchronized flow for different values of R, with the parameter $T = 1.0$ and the density $\rho = 37.04$ vehicles/km. The vehicles move from left to right and the time is increasing in up direction. (a) $R = 0$; (b) $R = 0.4$; (c) $R = 0.8$; (d) $R = 0.99$.

Fig. 6. Typical spatiotemporal patterns in synchronized flow at different values of R, with the parameter $T = 1.5$ and the density $\rho = 37.04$ vehicles/km. The vehicles move from left to right and the time is increasing in up direction. (a) $R = 0.4$; (b) $R = 0.8$.

Figure 8 shows the snapshots of velocity in synchronized flow when $T = 2.20$. It is found that with the increase of R, the traffic flow gradually becomes homogeneous and there is no mixture of free flow and synchronized flow.

Figure 9 shows the velocity distribution in the synchronized flow. One can see that when R is given, the distribution becomes narrower with the increase of T. This further demonstrates that if R is fixed, the traffic flow becomes more homogeneous with the increase of T.

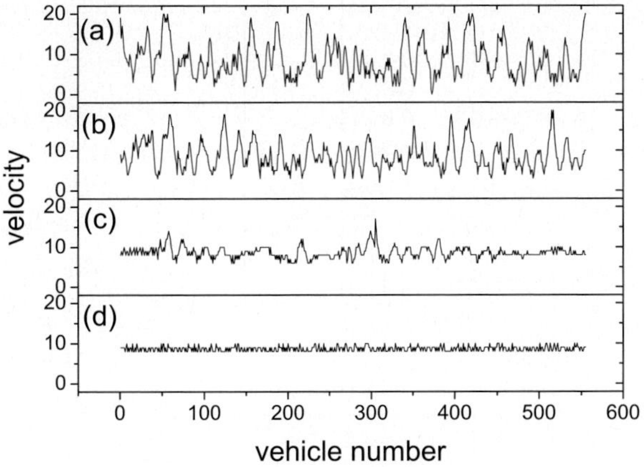

Fig. 7. Snapshots of velocity corresponding to the traffic flow in Fig. 6. (a) $R = 0.4$; (b) $R = 0.8$; (c) $R = 0.99$; (d) $R = 1$.

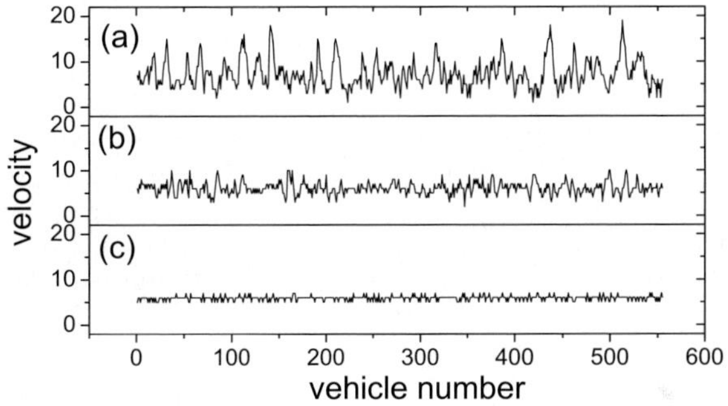

Fig. 8. Snapshots of velocity in synchronized flow for different values of R, with the parameter $T = 2.2$ and the density $\rho = 37.04$ vehicles/km. (a) $R = 0.4$; (b) $R = 0.8$; (c) $R = 1$.

4 Conclusions

Previous research on ACC vehicles has studied the impact of ACC vehicles on traffic flow stability, safety and exhaust emissions. In this paper, we mainly concentrate on the effects of ACC vehicles on the phase transition in traffic flow of a mixture of ACC and manual vehicles. Our simulations show that when the preferred time headway of ACC vehicles is small (e.g., $T = 1$), the introduction of ACC vehicles will enhance the free flow stability. However, when T is large, the introduction of ACC vehicles will reduce the phase transition threshold ρ_{c2}. Different from the phase transition from free flow to synchronized flow, the introduction of ACC vehicles will generally increase the threshold from synchronized flow to jams.

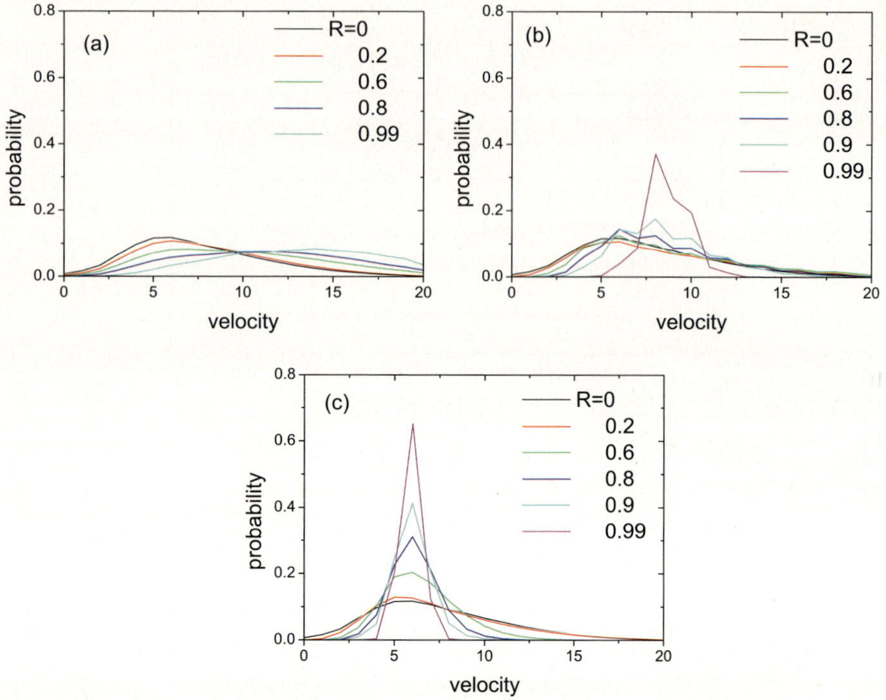

Fig. 9. Velocity distribution in the synchronized flow for different values of R, with the parameter (a) $T = 1.0$ (b) $T = 1.50$ (c) $T = 2.20$, and the density $\rho = 37.04$ vehicles/km.

Furthermore, the spatiotemporal patterns and velocity distribution of mixed ACC vehicles and manual vehicles are also studied. It is interesting to report that when T is small and the traffic is in synchronized flow, a mixture of free flow and synchronized flow will appear, even though $R = 0.99$. In other words, only several manual vehicles will seriously destroy the homogeneity of traffic flow. This is undesired and we need to find a way to suppress this phenomenon.

In our future work, this research needs to be extended to multi-lane traffic and we also need to consider the impact of various kinds of bottlenecks.

Acknowledgements

We acknowledge the support of National Basic Research Program of China (No. 2006CB705500), the NNSFC under Project No. 10532060, 70601026, 10672160. We are grateful to M. Gillingham for proofreading this manuscript.

References

1. Van Arem B, van Driel CJG, Visser R (2006) IEEE Trans. ITS **7**, 429.
2. Zhou J, Peng H (2005) IEEE Trans. ITS **6**, 229.
3. Yi JG, Horowitz R (2006) Transportation Research, Part C, **14**, 81.
4. Ioannou PA, Stefanovic M (2005) IEEE Trans. ITS **6**, 79.
5. Zhang JL, Ioannou PA (2006) IEEE Trans. ITS **7**, 92.
6. Kerner BS, e-print cond-mat/0309017.
7. Treiber M and Helbing D, e-print cond-mat/0210096; Kesting A, Treiber M, Schönhof M, et al., physics/0601096, also in Proc. Traffic and Granular Flow '05.
8. Davis LC (2004) Phys. Rev. E **69**, 066110.
9. Davis LC (2007) Physica A **379**, 274.
10. Jiang R, Wu QS (2006) Phys. Lett. A **359**, 99.
11. Jiang R, Wu QS (2005) Euro. Phys. J. B **46**, 581.
12. Jiang R, Wu QS (2004) J. Phys. A **37**, 8197.
13. Jiang R, Wu QS (2003) J. Phys. A **36**, 381.
14. Kerner BS, Klenov SL, Wolf DE (2002) J. Phys. A **35**, 9971.

Calibration of Car-Following Models Using Floating Car Data

Arne Kesting and Martin Treiber

Technische Universität Dresden, Institute for Transport & Economics,
Andreas-Schubert-Straße 23, D-01062 Dresden, Germany

Summary. We study the car-following behavior of individual drivers in real city traffic on the basis of publicly available floating car datasets. By means of a nonlinear optimization procedure based on a genetic algorithm, we calibrate the "Intelligent Driver Model" and the "Optimal Velocity Model" by minimizing the deviations between the observed driving dynamics and the simulated trajectory when following the same leading vehicle. The reliability and robustness of the nonlinear fits can be assessed by applying different optimization criteria, i.e., different measures for the deviations between two trajectories. We also investigate the sensitivity of the model parameters. Furthermore, the parameter sets calibrated to a certain trajectory are applied to the other trajectories allowing for model validation. We found that the calibration errors of the Intelligent Driver Model are between 11% and 28%, while the validation errors are between 22% and 30%. The calibration of the Optimal Velocity Model led to larger calibration and validation errors, and stronger parameter variations regarding different objective measures. The results indicate that "intra-driver variability" rather than "inter-driver variability" accounts for a large part of the fit errors.

1 Introduction

As microscopic traffic flow models are mainly used to describe collective phenomena such as traffic breakdowns, traffic instabilities, and the propagation of stop-and-go waves, these models are traditionally calibrated with respect to macroscopic traffic data. Nowadays, as microscopic traffic data have become more and more available, the problem of analyzing and comparing microscopic traffic flow models with real microscopic data has raised some interest in the literature [1–3].

We will consider three empirical trajectories of different drivers that are publicly available and that have been provided by the Robert Bosch GmbH [4]. The datasets have been recorded in 1995 during an afternoon peak hour on a fairly straight one-lane road in Stuttgart, Germany. A car equipped with a radar sensor in front provides the relative speed and distance to the car ahead.

The duration of the measurements are 250 s, 400 s and 300 s, respectively. All datasets show complex situations of daily city traffic with several acceleration and deceleration periods including standstills due to traffic lights. Here, we will apply the "Intelligent Driver Model" [5] and the "Optimal Velocity Model" [6] to these empirical trajectories. By means of a nonlinear optimization, we will determine the "optimal" model parameters which fit the given data best. Since the fit errors alone do not provide a good basis for an understanding or benchmarking of the applied car-following models, we will address further investigations on the role of the objective functions, the structure of the parameter space and the robustness concerning validation.

In the following section, we introduce the car-following models under investigation, the simulation set-up, the objective functions and the nonlinear optimization algorithm. In Sec. 3, the calibration and validation results will be presented. Finally, the factors influencing the calibration errors will be discussed in Sec. 4.

2 Calibration Methodology

2.1 Intelligent Driver Model

Car-following models are formulated as ordinary differential equations and, consequently, space and time are treated as continuous variables. This model class is typically characterized by an acceleration function that depends on the actual velocity $v(t)$, the gap $s(t)$ and the velocity difference $\Delta v(t)$ to the leading vehicle. Note that Δv is defined as approaching rate, i.e., positive if the following vehicle is faster than the leading vehicle. The Intelligent Driver Model (IDM) [5] is defined by the acceleration function

$$\dot{v}_{\mathrm{IDM}}(s, v, \Delta v) = \frac{\mathrm{d}v}{\mathrm{d}t} = a\left[1 - \left(\frac{v}{v_0}\right)^4 - \left(\frac{s^*(v, \Delta v)}{s}\right)^2\right]. \tag{1}$$

This expression combines the acceleration strategy $\dot{v}_{\mathrm{free}}(v) = a[1 - (v/v_0)^4]$ towards a *desired velocity* v_0 on a free road with a *maximum acceleration* a with a braking strategy $\dot{v}_{\mathrm{brake}}(s, v, \Delta v) = -a(s^*/s)^2$ which is dominant if the current gap $s(t)$ to the preceding vehicle becomes smaller than the "desired minimum gap"

$$s^*(v, \Delta v) = s_0 + vT + \frac{v\Delta v}{2\sqrt{ab}}. \tag{2}$$

The *minimum distance* s_0 in congested traffic is significant for low velocities only. The dominating term of Eq. (2) in stationary traffic is vT which corresponds to following the leading vehicle with a *desired time gap* T. The last term is only active in non-stationary traffic and implements an "intelligent" driving behavior including a braking strategy that, in nearly all situations, limits braking decelerations to the *comfortable deceleration* b. Note, however, that the IDM brakes stronger than b if the gap becomes too small. This braking strategy makes the IDM collision-free [5].

2.2 Optimal Velocity Model

The Optimal Velocity Model (OVM) [6] is defined by the acceleration function

$$\dot{v}_{\mathrm{OVM}}(s,v) = \frac{v_{\mathrm{opt}}(s) - v}{\tau}. \tag{3}$$

The *relaxation time* τ describes the adaptation to an "optimal velocity" v_{opt} due to changes in s and v, but not in velocity differences. The properties of the OVM are defined by the function for $v_{\mathrm{opt}}(s)$. For example, Bando et al. [6] suggested a function of the following form:

$$v_{\mathrm{opt}}(s) = \frac{v_0}{2}\left[\tanh\left(\frac{s}{l_{\mathrm{int}}} - \beta\right) - \tanh(-\beta)\right]. \tag{4}$$

The parameter v_0 defines the *desired velocity* under free traffic conditions. The *"interaction length"* l_{int} determines the transition regime for the s-shaped function (4) going from $v_{\mathrm{opt}}(s=0) = 0$ to $v_{\mathrm{opt}} \to v_0$ when the distance to the leading vehicles becomes large. Finally, the *"form factor"* β defines (together with l_{int}) the shape of the equilibrium flow-density relation, i.e., the fundamental diagram (see Sec. 3.3 below). In contrast to the IDM, the OVM exhibits collisions for some regimes of the parameter space.

2.3 Simulation Setup

The Bosch trajectory data contains velocities of both the leading and the following (measuring) vehicle. These data allow for a direct comparison between the measured driving behavior and trajectories simulated by a car-following model with the leading vehicle serving as externally controlled input. Initialized with the empirically given distance and velocity differences, $v^{\mathrm{sim}}(t=0) = v^{\mathrm{data}}(0)$ and $s^{\mathrm{sim}}(t=0) = s^{\mathrm{data}}(0)$, the model is used to compute the acceleration and, from this, the trajectory of the following car (see Fig. 1 below). The gap to the leading vehicle is then given by the difference between the simulated trajectory $x^{\mathrm{sim}}(t)$ (front bumper) and the given position of the rear bumper of the leading vehicle $x^{\mathrm{data}}_{\mathrm{lead}}(t)$:

$$s^{\mathrm{sim}}(t) = x^{\mathrm{data}}_{\mathrm{lead}}(t) - x^{\mathrm{sim}}(t). \tag{5}$$

This can be directly compared to the empirical gap $s^{\mathrm{data}}(t)$. In addition, the distance $s^{\mathrm{sim}}(t)$ has to be reset to the value in the dataset when the leading object changes as a result of a lane change of one of the considered vehicles. For example, the leading vehicle of the dataset 3 (cf. Fig. 1) turns into another street at $t \approx 144\,\mathrm{s}$, which leads to a jump in the gap of the considered follower.

2.4 Optimization Criteria

The calibration process aims at minimizing the difference between the measured driving behavior and the driving behavior simulated by the car-following model under consideration. Basically, any quantity can be used as error measure that is not fixed in the simulation, such as the velocity, the velocity difference, or the gap. In the following, we will use the error in the gap $s(t)$ for conceptual reasons: When optimizing with respect to s, the average velocity errors are automatically reduced as well. This does not hold the other way round, as the error in the distance may incrementally grow when optimizing with respect to differences in the velocities $v^{\text{sim}}(t)$ and $v^{\text{data}}_{\text{follow}}(t)$.

For the parameter optimization, we need an objective function as a quantitative measure of the error between the simulated and observed trajectories. As the objective function has a direct impact on the calibration result, we consider the following three different error functionals $\mathcal{F}[\cdot]$ of the empirical and simulated time series:

$$\mathcal{F}_{\text{rel}}[s^{\text{sim}}] = \sqrt{\left\langle [(s^{\text{sim}} - s^{\text{data}})/s^{\text{data}}]^2 \right\rangle}\,, \tag{6}$$

$$\mathcal{F}_{\text{abs}}[s^{\text{sim}}] = \sqrt{\langle (s^{\text{sim}} - s^{\text{data}})^2 \rangle \,/\, \langle s^{\text{data}} \rangle^2}\,, \tag{7}$$

$$\mathcal{F}_{\text{mix}}[s^{\text{sim}}] = \sqrt{\langle (s^{\text{sim}} - s^{\text{data}})^2 / |s^{\text{data}}| \rangle \,/\, \langle |s^{\text{data}}| \rangle}\,. \tag{8}$$

Here, the expression $\langle \cdot \rangle$ means the temporal average of a time series of duration ΔT, i.e.,

$$\langle z \rangle := \frac{1}{\Delta T} \int_0^{\Delta T} z(t)\,\mathrm{d}t. \tag{9}$$

Since the relative error $\mathcal{F}_{\text{rel}}$ is weighted by the inverse distance, this measure is more sensitive to small distances s than to large distances. As example, a simulated gap of 10 m compared to a gap of 5 m in the empirical data results in a large error of 100%, whereas the same deviation of 5 m leads to an error of 5% only for a spacing of, for instance, 100 m which is typical for large velocities. In contrast, the absolute error $\mathcal{F}_{\text{abs}}$ is less sensitive to small gaps because the denominator in Eq. (7) is averaged over the whole time series interval. However, the absolute error measure is more sensitive to large differences in the numerator, i.e., for large distances s. As the absolute error systematically overestimates errors for large gaps (at high velocities), while the relative error systematically overestimates deviations of the observed headway in the low velocity range, we have also defined a "mixed" error $\mathcal{F}_{\text{mix}}$. Note that the error measures are normalized in order to make them independent of the duration ΔT of the considered time series allowing for a direct comparison of different datasets.

2.5 Nonlinear Optimization Algorithm

Finding an optimal parameter set for a car-following model with a nonlinear acceleration function such as (1) and (3) corresponds to a *nonlinear optimization problem* which has to be solved numerically. For finding an approximative solution to the nonlinear optimization problem, we will apply a genetic algorithm as a search heuristic. The implemented genetic algorithm proceeds as follows: An "individual" represents a parameter set of a car-following model and a "population" consists of N such sets. In each generation, the *fitness* of each individual in the population is determined via one of the objective functions (6), (7) or (8). Pairs of two individuals are stochastically selected from the current population based on their fitness score and recombined to generate a new individual. Except for the best individual (which is kept without any modification to the next generation), the "genes" of all individuals, i.e., their model parameters, are varied randomly corresponding to a mutation that is controlled by a given probability. The resulting new generation is then used in the next iteration. The termination criterion is implemented as a two-step process: Initially, a fixed number of generations is evaluated. Then, the evolution terminates after convergence which is specified by a constant best-of-generation score for at least a given number of generations.

For an efficient search, we have applied constraints for the minimum and maximum values to restrict the parameter space to positive and reasonable parameter values. Furthermore, we have to take into account that some regions of the OVM parameter space lead to collisions. In order to make these "solutions" unattractive to the optimization algorithm, we have added a large crash penalty value to the objective function.

3 Calibration and Validation Results

3.1 Optimal Model Parameters

By applying the optimization method described in the previous section, we have found the best fit of the IDM and the OVM to the empirical data. The calibration results for the three datasets and the considered three objective functions (6), (7) and (8) are summarized in Table 1. Furthermore, Fig. 1 compares the dynamics of the simulated (using model parameters calibrated with respect to the mixed error measure (8)) with the empirically measured trajectories. The obtained errors are in the range between 11% and 34% which is consistent with typical error ranges obtained in previous studies [1, 2].

Obviously, the calibrated model parameters vary from one dataset to another because of different driving situations and drivers. Furthermore, a model that fits best a certain driver not necessarily does so for a different driver: In dataset 3, the IDM performs considerably better than the OVM, while hardly any difference is found for set 2. Moreover, the calibrated model parameters

Table 1. Calibration results for the Intelligent Driver Model (IDM) and the Optimal Velocity Model (OVM) for three different datasets and three different error measures $\mathcal{F}[s]$ regarding deviations in the gap s to the externally given leading car. Notice that the three datasets represent three different drivers and corresponding driving styles.

IDM	Dataset 1			Dataset 2			Dataset 3		
Measure	$\mathcal{F}_{\rm rel}[s]$	$\mathcal{F}_{\rm mix}[s]$	$\mathcal{F}_{\rm abs}[s]$	$\mathcal{F}_{\rm rel}[s]$	$\mathcal{F}_{\rm mix}[s]$	$\mathcal{F}_{\rm abs}[s]$	$\mathcal{F}_{\rm rel}[s]$	$\mathcal{F}_{\rm mix}[s]$	$\mathcal{F}_{\rm abs}[s]$
Error [%]	24.0	20.7	20.7	28.7	26.2	25.6	18.0	13.0	11.2
v_0 [m/s]	70.0	69.9	70.0	69.8	69.9	69.9	16.1	16.1	16.4
T [s]	1.07	1.12	1.03	1.51	1.43	1.26	1.30	1.30	1.39
s_0 [m]	2.41	2.33	2.56	2.63	2.82	3.40	1.61	1.52	1.04
a [m/s^2]	1.00	1.23	1.40	0.956	0.977	1.06	1.58	1.56	1.52
b [m/s^2]	3.21	3.20	3.73	0.910	0.994	1.11	0.756	0.633	0.614

OVM	Dataset 1			Dataset 2			Dataset 3		
Measure	$\mathcal{F}_{\rm rel}[s]$	$\mathcal{F}_{\rm mix}[s]$	$\mathcal{F}_{\rm abs}[s]$	$\mathcal{F}_{\rm rel}[s]$	$\mathcal{F}_{\rm mix}[s]$	$\mathcal{F}_{\rm abs}[s]$	$\mathcal{F}_{\rm rel}[s]$	$\mathcal{F}_{\rm mix}[s]$	$\mathcal{F}_{\rm abs}[s]$
Error [%]	25.0	23.6	23.2	29.3	26.7	25.7	33.6	29.4	33.0
v_0[m/s]	13.7	13.5	18.51	13.2	11.5	12.5	14.0	14.8	29.9
τ[s]	1.25	1.10	0.584	1.47	1.68	1.40	1.50	2.78	0.393
$l_{\rm int}$[m]	3.28	3.70	5.81	4.37	3.84	4.93	5.52	3.81	31.3
β[m]	2.96	2.65	1.99	2.61	2.74	2.26	2.69	4.68	0.133

also depend considerably on the underlying objective function. The IDM parameter values show a significantly smaller variation for a considered dataset than the OVM. This finding is relevant for benchmarking traffic models: It is not sufficient to consider only the fit errors, but the quality of the traffic model is also determined by the consistency and robustness of the calibrated parameters. In the following section, we will therefore systematically study the models' parameter spaces. In Sec. 4, we will discuss the influencing factors for the deviations between empirical and simulated car-following behavior.

3.2 Sensitivity Analysis

Starting from the optimized model parameters summarized in Table 1, it is straightforward to vary a single model parameter while keeping the other parameters constant. The resulting one-dimensional scan of the parameter space gives a good insight in the model's parameter properties and sensitivities. Furthermore, the application of different objective functions (cf. Sec. 2.4) can be seen as a benchmark of the "robustness" of the model calibration.

Figure 2 shows the resulting error measures of the dataset 3. Remarkably, all error curves for the IDM are smooth and show only one minimum. Consequently, the optimal parameters are easy to determine by the optimization algorithm. As the datasets mainly describe car-following situations in obstructed traffic and standstills, the IDM parameters T, s_0 and a are particularly significant and show distinct minima for the three proposed error

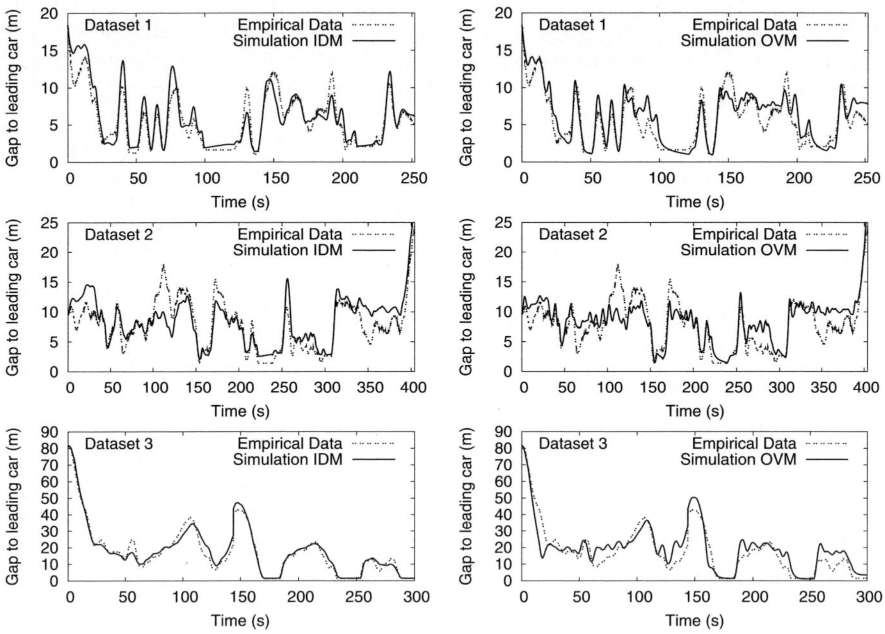

Fig. 1. Comparison of simulated and empirical trajectories. The model parameters are calibrated according to Table 1 for the "mixed" error measure (8).

measures while the values of v_0 were hard to determine exactly from the datasets 1 and 2 where the desired velocity is never approximated. The comfortable deceleration b is also not very distinct (not shown here). The solutions belonging to different objective functions are altogether in the same parameter range. This robustness of the IDM parameter space is an important finding of this study.

The results for the OVM imply a less positive model assessment: The calibration results strongly vary with the chosen objective function indicating a strong sensitivity of the model parameters. Furthermore, too high values of the desired velocity lead to vehicle collisions in the simulation as indicated by abrupt raises in the error curves.

3.3 Microscopic Flow-Density Relations

The car-following dynamics shown in the trajectories of Fig. 1 can alternatively be represented by the microscopic flow-density relations. A translation from the microscopic gap s into the density ρ is given by the *micro-macro relation* $\rho(s) = 1/(s + l)$ where l is the vehicle length (which we fix to 5 m). The flow is then given by $Q(s, v) = v\rho = v/(s + l)$. In Fig. 3, the flow-density points, $(Q(t), \rho(t))$, are plotted for each recorded time step t of the empirical data and the simulated trajectories. Furthermore, we have also plotted

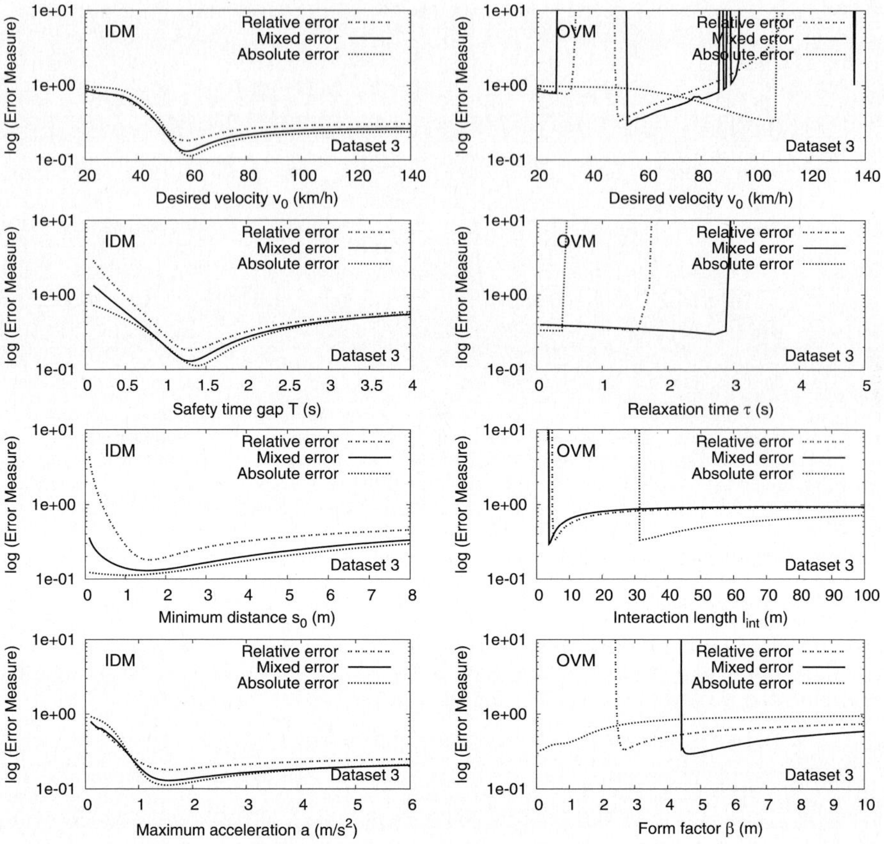

Fig. 2. Systematic variation of one model parameter while keeping the other parameters at the optimal values listed in Table 1. The diagrams show the considered error measures (6), (7) and (8) for the IDM (left column) and OVM (right column) using dataset 3. The errors are plotted in logarithmic scale.

the "fundamental diagrams" describing states of homogeneous and stationary traffic. As equilibrium traffic is defined by vanishing velocity differences and accelerations, the modeled drivers drive at a constant velocity v_e which depends on the gap to the leading vehicle. For the OVM, this equilibrium velocity is directly given by the optimal velocity function (4). For the IDM under the conditions $\dot{v} = 0$ and $\Delta v = 0$, only the inverse, i.e., the equilibrium gap s_e as a function of the velocity, can be solved analytically. However, the fundamental diagrams of the IDM can numerically be obtained by parametric plots varying v.

These representations depicted in Fig. 3 provide a good overview of the recorded traffic situations. While the sets 1 and 2 mainly contain car-following behavior at distances smaller than 20 m (corresponding to densities larger than

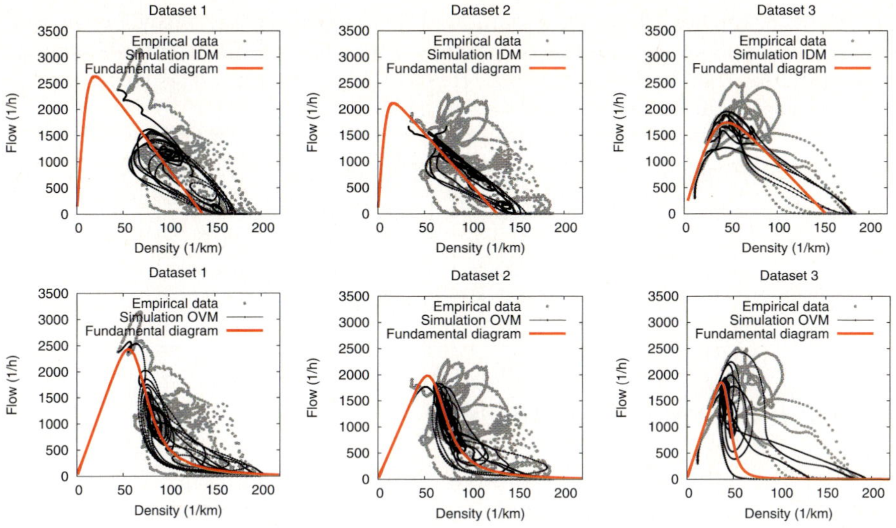

Fig. 3. Microscopic flow-density relations derived from the given and simulated gaps $s(t)$ and velocities $v(t)$. Additionally, the equilibrium flow-density relation (fundamental diagram) is plotted as well. This representation offers an alternative view to Fig. 1 on the empirical and simulated data.

50 /km), the dataset 3 also features a non-restricted driving situation with a short period of "free" accelerating corresponding to the branch with densities lower than 30/km of the flow-density plot.

3.4 Validation

Let us finally validate the obtained calibrated parameters by applying these settings to the other datasets, i.e., using the parameters calibrated on the basis of another dataset. We use the three optimal parameter sets listed in Table 1 and restrict ourselves to the "mixed" error measure (8). The obtained errors can be found in Table 2. Moreover, we have evaluated the arithmetic average of the calibrated parameters which are listed in the caption of Table 2. This *cross-comparison* allows to check for the reliability of the obtained parameters and reflects the variance of the calibrated parameter values. For the IDM, the obtained errors for the cross-compared simulation runs are of the same order as for the calibrated parameter sets. Therefore, the car-following behavior of the IDM turned out to be robust with respect to reasonable changes of parameter settings. This is, however, not the case for the OVM. Small changes in the parameter values lead to large errors and some settings even led to collisions. This low stability of the OVM results from the fact that velocity differences are not taken into account.

Table 2. Cross-comparison of the calibrated parameters for the "mixed" error measure (8) by applying the calibrated parameter sets for the IDM and OVM to the other datasets. The underlined errors refer to the parameter values corresponding to the best calibration results.

Model	Dataset	Calib. Set 1	Calib. Set 2	Calib. Set 3
IDM	Set 1	<u>20.7%</u>	28.8%	28.7%
	Set 2	35.2%	<u>26.2%</u>	40.1%
	Set 3	41.1%	27.0%	<u>13.0%</u>
OVM	Set 1	<u>23.6%</u>	collision	collision
	Set 2	30.5%	<u>26.7%</u>	48.9%
	Set 3	collision	collision	<u>29.4%</u>

4 Discussion and Conclusions

Let us finally discuss three qualitative influences which contribute to these deviations between observations and reproduction. A significant part of the deviations between measured and simulated trajectories can be attributed to the *inter-driver variability* as it has been shown by cross-comparing sets 1 and 3 with either model: The errors of the calibrated sets (representing both the model error and the intra-driver variations, see below) typically are only half or less of the errors of the cross-compared sets to which the inter-driver variations contribute as well. For set 2, however, the intra-driver variation and model errors seem to play the dominating role. Note that microscopic traffic models can easily cope with this kind of heterogeneity because different parameter values can be attributed to each individual "driver-vehicle unit". However, in order to obtain these distributions of calibrated model parameters, more trajectories have to be analyzed.

A second contribution to the overall calibration error results from a non-constant driving style of human drivers which is also referred to as *intra-driver variability*: Human drivers do not drive constantly over time, i.e., they change their characteristic behavior while driving. For a first estimation, we have compared the distances at three standstills in the dataset 3 with the minimum distance s_0 as direct model parameter of the IDM. The gaps are $s_{\text{stop},1} = 1.39\,\text{m}$, $s_{\text{stop},2} = 1.42\,\text{m}$ and $s_{\text{stop},3} = 1.64\,\text{m}$, respectively. An analytical minimization of s_0 results in $s_0^{\text{opt}} \approx 1.458\,\text{m}$. This optimal solution defines a theoretical lower bound (based on about 15% of the data of the considered time series) for a relative error of 7.9%. Therefore, the intra-driver variability accounts for a large part of the deviations between simulations and empirical observations. This influence could be captured by considering *time-dependent* model parameters reflecting driver adaptation processes as for example proposed in [7, 8].

Finally, *driver anticipation* contributes to the overall error as well but is not incorporated in simple car-following models. This is one possible cause for a *"model error"*, i.e., the residual difference between a perfectly time-independent driving style and a model calibrated to it. More complex microscopic traffic models aim at taking those aspects into account including an explicit reaction time [9]. Multi-leader anticipation, however, requires trajectory data because the data recording using radar sensors of single "floating" cars is limited to the immediate predecessor.

References

1. E. Brockfeld, R. D. Kühne, and P. Wagner, "Calibration and validation of microscopic traffic flow models," Transportation Research Record: Journal of the Transportation Research Board **1876,** 62–70 (2004).
2. V. Punzo and F. Simonelli, "Analysis and comparision of microscopic flow models with real traffic microscopic data," Transportation Research Record: Journal of the Transportation Research Board **1934,** 53–63 (2005).
3. S. J. Ossen and S. P. Hoogendoorn, "Car-following behavior analysis from microscopic trajectory data," Transportation Research Record: Journal of the Transportation Research Board **1934,** 13–21 (2005).
4. Deutsches Zentrum für Luft- und Raumfahrt (DLR), "Clearing house for transport data and transport models,", `http://www.dlr.de/cs/` – Access date: May 5, 2007.
5. M. Treiber, A. Hennecke, and D. Helbing, "Congested traffic states in empirical observations and microscopic simulations," Physical Review E **62,** 1805–1824 (2000).
6. M. Bando, K. Hasebe, A. Nakayama, A. Shibata, and Y. Sugiyama, "Dynamical model of traffic congestion and numerical simulation," Physical Review E **51,** 1035–1042 (1995).
7. M. Treiber and D. Helbing, "Memory effects in microscopic traffic models and wide scattering in flow-density data," Physical Review E **68,** 046119 (2003).
8. M. Treiber, A. Kesting, and D. Helbing, "Understanding widely scattered traffic flows, the capacity drop, and platoons as effects of variance-driven time gaps," Physical Review E **74,** 016123 (2006).
9. M. Treiber, A. Kesting, and D. Helbing, "Delays, inaccuracies and anticipation in microscopic traffic models," Physica A **360,** 71–88 (2006).

A Queueing Theory Approach for a Multi-Speed Exclusion Process

Cyril Furtlehner[1] and Jean-Marc Lasgouttes[2]

[1] INRIA Futurs, Project Team TAO – Bat. 490 – Université Paris-Sud – 91405 Orsay Cedex, France cyril.furtlehner@inria.fr

[2] INRIA Paris-Rocquencourt, Project Team IMARA – Domaine de Voluceau – BP. 105 – 78153 Le Chesnay Cedex, France jean-marc.lasgouttes@inria.fr

Summary. We consider a one-dimensional stochastic reaction-diffusion generalizing the totally asymmetric simple exclusion process, and aiming at describing single lane roads with vehicles that can change speed. To each particle is associated a jump rate, and the particular dynamics that we choose (based on 3-sites patterns) ensures that clusters of occupied sites are of uniform jump rate. When this model is set on a circle or an infinite line, classical arguments allow to map it to a linear network of queues (a zero-range process in theoretical physics parlance) with exponential service times, but with a twist: the service rate remains constant during a busy period, but can change at renewal events. We use the tools of queueing theory to compute the fundamental diagram of the traffic, and show the effects of a condensation mechanism.

1 A Multi-Speed Exclusion Process

The totally asymmetric exclusion process (TASEP) is a popular statistical physics model of one-dimensional interacting particles particularly adapted to traffic modeling. This is due to its simple definition, and to the non-trivial exact solutions which have been unveiled in the stationary regime [1]. One important shortcoming of this model is that it does not allow particles to move at different speeds. Cellular automata like the Nagel-Schreckenberg model [2] address this issue, leading to very realistic though still simple simulators. However, these models are difficult to handle mathematically beyond the mean field approximation [3] and an approximate mapping with the asymmetric chipping model suggests that the jamming phenomenon takes place as a broad crossover rather than a phase transition [4]. In this paper, we are interested in analyzing the nature of fluctuations in the fundamental diagram (FD), that is the mean flow of vehicles plotted against the traffic density. To address this question, we propose to extend the TASEP in a different way, more convenient for the analysis albeit less realistic from the point of view of traffic.

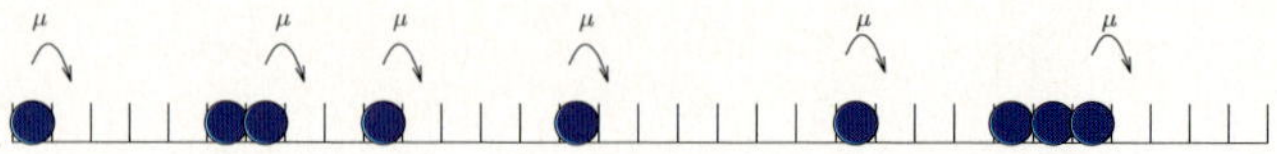

Fig. 1. The basic 1-dimensional TASEP model.

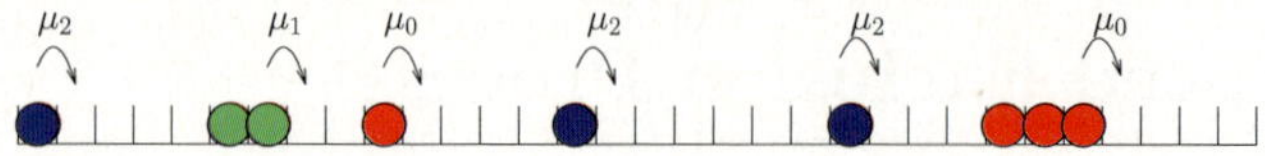

Fig. 2. The multi-speed 1-dimensional TASEP model.

The elementary 1-dimensional TASEP (Fig. 1) model is defined on a discrete lattice (e.g. a finite ring with boundary periodic conditions, a segment with edges or an infinite line), where each site may be occupied with at most one particle. Each particle moves independently to the next site (say, to the right), at the times of a Poisson process with intensity μ. Therefore, the model is a continuous time Markov process, which state is the binary encoded sequence $\sigma_t \in \{O, V\}^N$ of size N (the size of the system), where the letter V (resp. O) denotes a vehicle (resp. an empty space) at site i. Each transition involves two consecutive letters when a particle moves from site i to site $i+1$:

$$VO \xrightarrow{\mu} OV.$$

In order to encode various speed levels, we propose to extend the basic TASEP by allowing the particle to jump at different possible rates which themselves vary stochastically in time (Fig. 2). Assuming for now a finite number of $n-1$ speed levels, the Markov chain that we consider is a sequence encoded into a n-alphabet $\{O, A, B, \ldots\}$

$$\sigma_t = \{V_i, \ldots, i = 1 \ldots N\}, \qquad V_i \in \{O, A, B, \ldots\}$$

where O is again an empty site, and A, $B, \ldots$ represent occupied sites with jump rates $\mu_a, \mu_b, \ldots$. In our model, the transitions remain local: the particle may jump to the next site only if it is empty, we allow the final state to be conditioned by the site after the next. More precisely, we assume that any transition involves three consecutive letters, and distinguish between two cases:

$$\begin{cases} \ldots XOY \ldots \xrightarrow{\mu_x} \ldots OYY \ldots, & Y \neq O \quad \text{rule 1,} \\ \ldots XOO \ldots \xrightarrow{\mu_x} \ldots OZO \ldots, & Z \neq O \quad \text{rule 2.} \end{cases}$$

In the second case, the type (or equivalently the jump rate) of the particle is chosen randomly according to a distribution F. As a limiting case, we will consider a general continuous distribution F on $\mathbb{R}^+$. In other words, a particle at site i with rate μ_x jumps to site $i+1$ and acquires a new rate μ_z which is a random function of V_{i+2}.

The basic assumption is that if a car gets in close contact to another one, it will adopt its rate. Conversely, if it arrives at a site not in contact with any other car, the new rate will be freely determined according to some random distribution. This models the acceleration or deceleration process in an admittedly crude manner. This setting is different from usual exclusion processes with multi-type particles, each having its own jump rate. It is more in line with the Nagel-Schreckenberg model, with the difference that only local jumps are allowed and speed is replaced by jump rate.

2 L-Stage Tandem Queue Reformulation

In the context of exclusion processes, jams are represented as cluster of particles. Clustering phenomena can be analyzed in some cases by mapping the process to a tandem queueing network (i.e. a zero range processes in statistical physics terms). For the simple TASEP on a ring two dual mappings are possible:

- the queues are associated to empty sites and the clients are the particles in contact behind this site,
- the queues are associated with particles and the clients are the empty sites in front of this particle.

By using one of these mappings, the TASEP is equivalent to a closed cyclic queueing network, with fixed service rates equal to the jumping rate μ of the particles. Steady states of such queueing network have been analyzed thoroughly (see for ex. Kelly [5]) in terms of a simple product form structure which we expose now.

Consider an open L-stage tandem queue, with arrival rate λ and a common service rate μ: L queues with service rate μ are arranged in successive order (the departures from a given queue coincide with the arrivals to the next one) and the arrival process of the first queue is Poisson with intensity λ. Each queue is stable when $\lambda < \mu$, transient when $\lambda > \mu$. It is then well known that the distribution of the number of clients $X_1, \ldots, X_L$ in the queues is

$$P(\{X_i = x_i, i = 1 \ldots L\}) = \prod_{i=1}^{L} P_\lambda(X_i = x_i), \tag{1}$$

where

$$P_\lambda(X_i = x) = (1-\rho)\rho^x, \qquad \text{with} \qquad \rho \stackrel{\text{def}}{=} \frac{\lambda}{\mu}.$$

If the network is closed (the last queue is connected to the first one in the ring geometry), then expression (1) remains valid, with the constraint that the total number of clients $\sum_{i=1}^{L} X_i$ is fixed. In this case, λ can be chosen arbitrarily, as long as each queue in isolation remains ergodic.

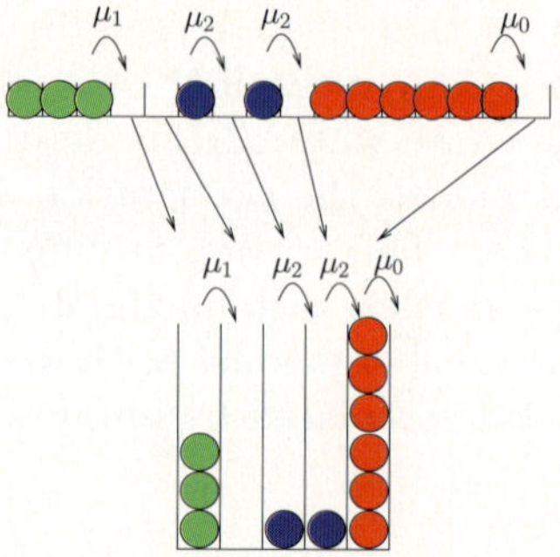

Fig. 3. Mapping of the variable-speed TASEP to a tandem queue.

It can be shown easily [6] that, for a plain TASEP on a ring, the size of the jams is asymptotically a geometric random variable with parameter ρ (when $N = kL$, $k = 1, 2 \ldots$). In the open geometry, the arrival rate is an external parameter which can be set between 0 and μ. When it becomes comparable to the service rate, i.e. when $\rho \simeq 1$, large queues may form and a random walk first time return calculation yields a realistic scaling behavior for the lifetime distribution of jams [7]

$$P(t) \simeq t^{-3/2}.$$

In our multi-speed exclusion process, particles are guaranteed by construction to form clusters with homogeneous speed, and the mapping of empty sites to queues is suitable (Fig. 3). The new feature is that the service rate R_i of a given queue can change with time: it is drawn randomly from a distribution with cumulative distribution F when the first customer arrives. It is assumed that there exists a minimal service rate $\mu_0 > 0$ such that $F(\mu_0) = 0$. $\rho_0 = \lambda/\mu_0 \leq 1$ is therefore the maximal possible load. The state is determined by the pair (X, R) and the possible transitions are as follows:

$$(X = x, R = \mu) \xrightarrow{\lambda \mathbb{1}_{\{x>0\}}} (X = x + 1, R = \mu),$$
$$(X = 0, R = \mu) \xrightarrow{\lambda F(d\mu')} (X = 1, R = \mu'),$$
$$(X = x, R = \mu) \xrightarrow{\mu \mathbb{1}_{\{x>0\}}} (X = x - 1, R = \mu).$$

Since these transitions form a tree (see Fig. 4(a)), each queue in isolation is a reversible Markov process and its stationary distribution reads:

$$P_\lambda(X = x, R \in d\mu) = P_\lambda(X = 0) F(d\mu) \left(\frac{\lambda}{\mu}\right)^x,$$

with

$$P_\lambda(X = 0) = \left(\int_{\mu_0}^{\infty} F(d\mu) \frac{\mu}{\mu - \lambda}\right)^{-1}.$$

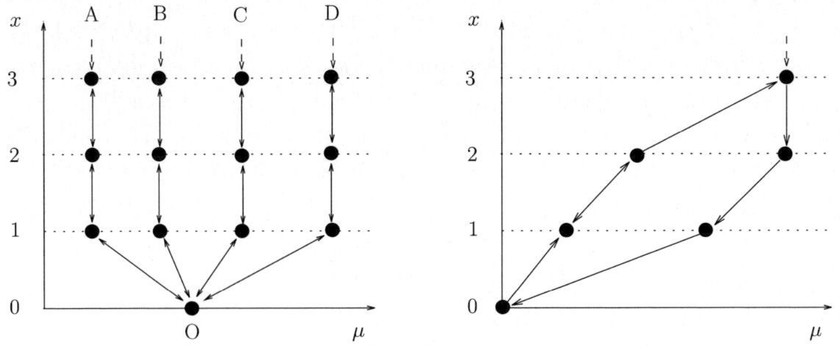

(a) (b)

Fig. 4. State graph for isolated queues in the case of (a) the multi-speed process definition, and (b) a non-reversible queue with hysteresis. States are represented by black dots and transitions by arrows.

The distribution of the number of customers in the queue is therefore no longer geometric:

$$P_\lambda(X=x) = \int_{\mu_0}^{\infty} P_\lambda(X=x, R \in d\mu) = P_\lambda(X=0) \int_{\mu_0}^{\infty} F(d\mu) \left(\frac{\lambda}{\mu}\right)^x . \quad (2)$$

Nevertheless, the product form expression (1) for the invariant measure remains valid, because of the reversibility of the individual queues taken in isolation (see again [5]). The stationary distribution of the L-stage tandem queue takes the form

$$P_\lambda(S) = \prod_{i=1}^{L} P_\lambda(X_i, R_i),$$

for any sequence $S = \{(X_i, R_i), i = 1, \ldots, L\}$. While each queue has a different service rate at a given time, all the queues have globally the same distribution. Our model is therefore encoded in the single queue stationary distribution $P_\lambda(X_i, R_i)$.

3 The Fundamental Diagram

As announced in Section 1, we turn now to the fundamental diagram (FD), that is the plot of the mean flow of vehicles against the traffic density. By nature, the fluctuations in the FD are associated to the jam formation. Schematically, three main distinct regimes or traffic phases have been identified by empirical studies [8]: one for to free-flow, and two congested states, the "synchronized flow" and the "wide moving jam".

In the case of the basic TASEP, it is well known, and rigorously proved in some cases, that an hydrodynamic limit can be obtained by rescaling both the spatial variable $x = i/N$ and the jumping rate according to $\mu(N) = NV_0$, where N is a rescaling which we let to ∞ and V_0 is a constant. The corresponding coarse grained density ρ satisfies the inviscid Burger equation

$$\frac{\partial \rho}{\partial t} = V_0 \frac{\partial}{\partial x}\Big[\rho(1-\rho)\Big].$$

The FD at this scale is deterministic, since

$$J(\rho) = V_0\rho(1-\rho),$$

and symmetric w.r.t. $\rho = 1/2$ because of the particle-hole symmetry. V_0 represents the free velocity of cars, when the density is very low.

In practice, points plotted in experimental FD studies are obtained by averaging data from static loop detectors over a few minutes (see e.g. [8]). This is difficult to do with our queue-based model, for which a space average is much easier to obtain. The equivalence between time and space averaging is not an obvious assumption, but since jams are moving, space and time correlations are combined in some way [7] and we consider this assumption to be quite safe. In what follows, we will therefore compute the FD by considering either the joint probability measure $P_\lambda(d,\phi)$ for an open system, or the conditional probability measure $P_\lambda(\phi|d)$ for a closed system, where

$$\begin{cases} d = \dfrac{N}{N+L}, \\[2ex] \phi = \dfrac{\Phi}{N+L}, \end{cases} \qquad \text{with} \qquad \begin{cases} L & \text{number of queues} \\[1ex] N = \displaystyle\sum_{i=1}^{L} X_i & \text{number of vehicles} \\[1ex] \Phi = \displaystyle\sum_{i=1}^{L} R_i \mathbb{1}_{\{X_i>0\}} & \text{integrated flow} \end{cases}$$

are spatial averaged quantities and represent respectively the density and the traffic flow. We perform the analysis in the ring geometry: this avoids edge effects, fixes the numbers N of vehicles and L of queues, and finally makes sense as an experimental setting. In the statistical physics parlance, the fact that N is fixed means that we are working with the canonical ensemble. As a result this constraint yields the following form of the joint probability measure:

$$P(S) = \frac{1}{Z_L[N]} \prod_{i=1}^{L} P_\lambda(x_i,\mu_i),$$

with the canonical partition function

$$Z_L[N] \stackrel{\text{def}}{=} \sum_{\{x_i\}} \prod_{i=1}^{L} P_\lambda(x_i)\delta\big(N - \textstyle\sum_{i=1}^{L} x_i\big).$$

These expressions are actually independent of λ in this specific ring geometry. The density-flow conditional probability distribution takes the form

$$P(\phi|d) = \frac{Z_L[N,\Phi]}{Z_L[N]}, \tag{3}$$

with

$$N = \frac{d}{1-d}L, \qquad \Phi = \frac{\phi}{1-d}L,$$

and

$$Z_L[N,\Phi] \stackrel{\text{def}}{=} \sum_{\{x_i\}} \int \cdots \int \prod_{i=1}^{L} P_\lambda(x_i, d\mu_i)\delta\big(N - \textstyle\sum_{i=1}^L x_i\big)\delta\big(\Phi - \textstyle\sum_{i=1}^L \mu_i \mathbb{1}_{\{x_i>0\}}\big).$$

Note (by simple inspection, see e.g. [5]) that $P(\phi|d)$ is independent of λ.

4 Phase Transition and Condensation Mechanism

The connection between spontaneous formation of jams and the Bose-Einstein condensation has been analyzed in some specific models, with e.g. quenched disorder [9], where particles are distinguishable with different but fixed hopping rates attached to them. In the present situation, all particles are identical, but hopping rates may fluctuate, which is related to annealed disorder in statistical physics. The condensation mechanism for zero range processes within the canonical ensemble has been clarified in some recent work [10]. Let us translate in our settings the main features of the condensation mechanism. Assume that the number of clients X of an isolated queue has a long-tailed distribution

$$P(X=x) \underset{x\gg 1}{\propto} \frac{1}{x^\alpha}, \qquad \alpha > 1.$$

The empirical mean queue size reads

$$\bar{X} = \frac{1}{L}\sum_{i=1}^{L} X_i, \qquad \text{and} \qquad \mathbb{E}\bar{X} = \mathbb{E}_\lambda(X) \stackrel{\text{def}}{=} \frac{\int_{\mu_0}^{\infty} F(d\mu)\frac{\mu^2}{(\mu-\lambda)^2}}{\int_{\mu_0}^{\infty} F(d\mu)\frac{\mu}{\mu-\lambda}},$$

where $\mathbb{E}_\lambda(X)$ is the expected number of clients in an isolated queue, when the arrival rate is λ. Within the canonical ensemble, $\bar{X}$ is fixed, while for the grand canonical ensemble, only the expectation $\mathbb{E}(\bar{X})$ is fixed. In both cases, for $\alpha > 2$ there exists $\bar{X}_c$ such that, when $\bar{X} > \bar{X}_c$ ($\mathbb{E}(\bar{X}) > \bar{X}_c$ in the grand canonical ensemble), one of the queues condenses, i.e. carries a macroscopic number of particles. When $\alpha > 2$, there is a condensate with probability weight $O(L^{1-\alpha})$.

This condensation corresponds to a second order phase transition, and occurs at a critical density d_c which is the same in the canonical and grand-canonical formalism. To determine d_c, first consider

$$\bar{d}(\lambda) \stackrel{\text{def}}{=} \frac{\mathbb{E}_\lambda(X)}{1+\mathbb{E}_\lambda(X)}.$$

$\mathbb{E}_\lambda(X)$ increases monotonically with λ, which cannot exceed μ_0 (see Section 2). Therefore, if $\mathbb{E}_{\mu_0}(X) = x_c < \infty$, then there exists a critical density

$$d_c = \frac{x_c}{1+x_c},$$

such that for $d \geq d_c$ one of the queues condenses. The interpretation is that $N_c = Lx_c$ is the maximal number of clients that can be in the queues in the fluid regime, and the less costly way to absorb the excess $N - N_c$ is to put it in one single queue. Let us give an example, by specifying the joint law through

$$P(\mu_0 \leq R \leq \mu_0 y) = F(\mu_0 y) = \Big(\frac{y-1}{r-1}\Big)^\alpha, \qquad 1 \leq y \leq r, \tag{4}$$

where $r > 1$ is ratio between the highest and lowest speed. In that case, using (2) we have the following asymptotic as $\xi \to \infty$

$$P_\lambda(X = x) \propto \Big(\frac{\lambda}{\mu_0}\Big)^x \int_1^r (y-1)^{\alpha-1} y^{-x} dy \sim \frac{1}{x^\alpha}\Big(\frac{\lambda}{\mu_0}\Big)^x,$$

and $\mathbb{E}_{\mu_0}(X) < \infty$ when $\alpha > 2$, which yields the possibility of condensation above the critical density

$$d_c(\alpha, r) = \frac{(r-1)^{\alpha-2}}{\alpha - 2 + (r-1)^{\alpha-2}}. \tag{5}$$

5 Numerical Results

The analysis of (3) in the ring geometry can in principle be performed by means of saddle point techniques [10, 11], which we postpone to another work. Instead we present a numerical approach: the FD presented in Fig. 5(a)-(c) is obtained by solving the recursive relation

$$Z_L[N, \Phi] = P_\lambda(X=0) Z_{L-1}[N, \Phi] + \sum_{x=1}^{N} \int P_\lambda(x, d\mu) Z_{L-1}[N-x, \Phi - \mu \mathbb{1}_{\{x>0\}}],$$

up to some value $L_{MAX} = 100$ for the number of queues, with a fixed value of $\lambda < \mu_0$. Although in principle one arbitrary value of λ should suffice, in practice, the results for different values of λ have to be superposed in the diagram to get significant results. Since this recursion is only tractable with a finite number of possible velocities, the distribution F used here is concentrated to two values μ_0 and μ_1. The presence of a discontinuity in the fundamental diagram for small values of $\eta \stackrel{\text{def}}{=} P(R = \mu_0)/P(R = \mu_1)$ is a finite size effect, which disappears when the system size is increased while η is kept

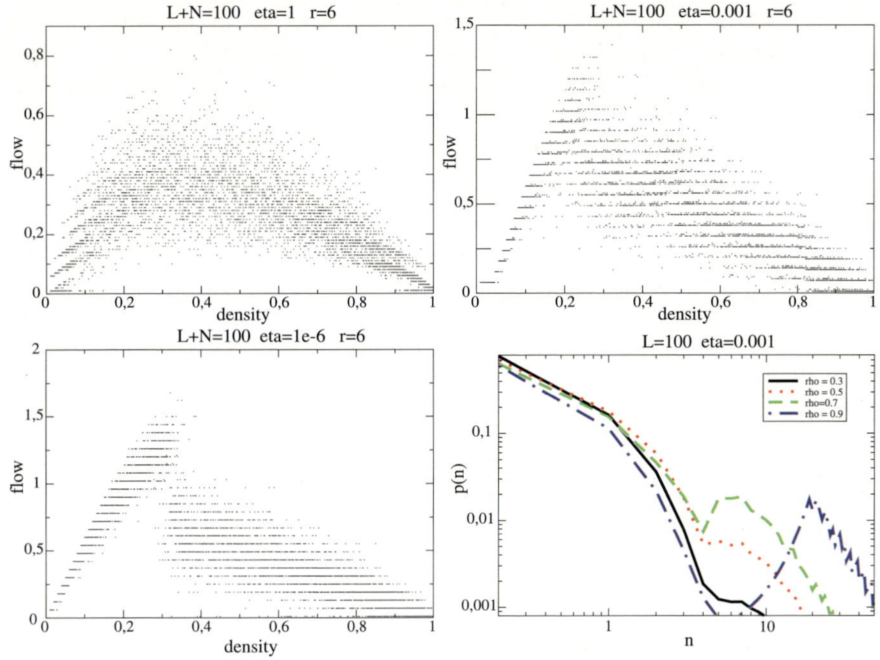

Fig. 5. Probabilistic fundamental diagram on the ring geometry ($L + N = 100$) for a two-level speed distribution with $\eta = 1$ (a), 0.001 (b) and 10^{-6} (c); corresponding single queue distribution as a function of the density for $\eta = 0.001$ (d).

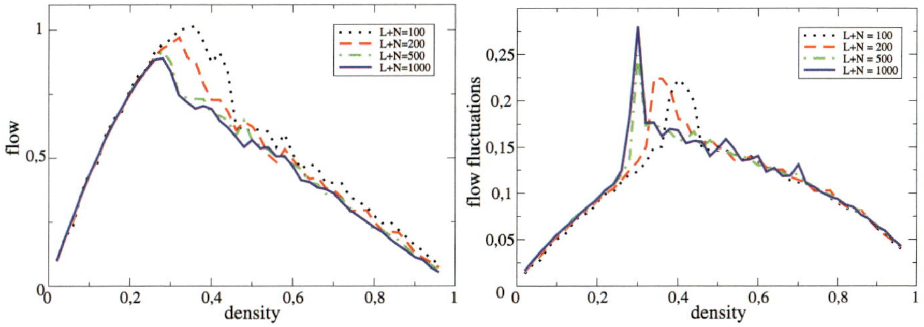

Fig. 6. Mean flow as a function of density for a continuous speed distribution ($\alpha = 3$) on the ring geometry with varying sizes $L + N$ (left) and corresponding standard deviation rescaled by $\sqrt{L}$ (right).

fixed. Nevertheless, the direct simulation of the closed L-stage tandem queues, with continuous distribution (4), indicates as expected a second order phase transition when $\alpha > 2$ (Fig. 6). This transition is related to the formation of a condensate, which is marked by the apparition of a bump in the single queue distribution at the critical density (see Fig. 5(d)). This condensation mechanism is responsible for the slope discontinuity. Fluctuations scale like $1/\sqrt{L}$, as expected from the Central Limit Theorem. Note however that the critical density is different than the one given by (5) for the open system.

6 Perspectives

In this work, we analyze the fluctuations in the fundamental diagram of traffic by considering models from statistical physics and using probabilistic tools. We propose a generalization of the TASEP by considering a multi-speed exclusion process which is conveniently mapped onto an L-stage tandem queue. When the individual queues are reversible, general results from queueing network theory let us obtain the exact form of the steady state distribution. This measure is conveniently shaped to compute the FD. Depending on the speed distribution, it may present two phases, the free-flow and the congested ones, separated by a second order phase transition. This transition is associated to a condensation mechanism, when slow clusters are sufficiently rare.

In practice, it is conjectured [8] that there are three phases in the FD, separated by first order phase transition. A large number of possible extensions of our model are possible, by playing with the definition of the state graph of a single queue (Fig. 4(a)). This graph accounts either for the dynamics of single vehicle clusters, when queues are associated to empty sites, or to the behavior of single drivers when queues are associated to occupied sites. In order to obtain first order phase transitions, we will consider in future work models where the single queues are not reversible in isolation, for example because of an hysteresis phenomenon (Fig. 4(b)).

References

1. B. Derrida, M. R. Evans, V. Hakim, and V. Pasquier. *J. Phys. A: Math. Gen.*, 26:1493–1517, 1993.
2. K. Nagel and M. Schreckenberg. *J. Phys. I*, 2, pages 2221–2229, 1992.
3. D. Chowdhury, L. Santen, and A. Schadschneider. *Phys. Rep.*, 329:199, 2000.
4. E. Levine, G. Ziv, L. Gray, and D. Mukamel. *J. Stat. Phys.*, 117:819–830, 2004.
5. F. P. Kelly. *Reversibility and stochastic networks.* John Wiley & Sons Ltd., 1979.
6. O. Pulkkinen and J. Merikoski. *Phys. Review E.*, 64(5):56114, 2001.
7. K. Nagel and M. Paczuski. *Phys. Rev. E*, 51(4):2909–2918, 1995.
8. B. S. Kerner. *Phys. Rev. Lett.*, 81(17):3797–3800, 1998.
9. M. R. Evans. *Europhys. Lett.*, 36(1):13–18, 1996.
10. M. R. Evans, S. N. Majumdar, and R. K. P. Zia. *J. Stat. Phys.*, 123(2):357–390, 2006.
11. G. Fayolle and J.-M. Lasgouttes. *Markov Proc. Rel. Fields*, 2:317–348, 1996.

A Stochastic Macroscopic Traffic Model Devoid of Diffusion

Megan M. Khoshyaran[1] and Jean-Patrick Lebacque[2]

[1] ETC Economics Traffic Clinic, 35 avenue des Champs Elysées, F 75008 Paris, France etclinic@wanadoo.fr

[2] INRETS-GRETIA, 2 avenue du Général Malleret Joinville, F 94114 Arcueil, France lebacque@inrets.fr

1 Introduction

Macroscopic traffic flow models are based on the *continuum hypothesis*, which postulates the existence of macroscopic functions $\rho(x,t)$ (the density), $v(x,t)$ (the speed), $q(x,t)$ (the flow) describing the state of traffic at any location x and time t.

The macroscopic dynamics of traffic flow are constrained by the local supply and demand of traffic [1] as well as by infrastructure capacity and current regulations. This explains why most macroscopic models are formulated in terms of deterministic systems of partial differential equations which account correctly for most macroscopic phenomena. But the macroscopic dynamics of traffic are affected by the intrinsic stochastic character of driver interaction, as various phenomena attest (spontaneous jam formation, traffic breakdown, etc). Driver behavior is partly stochastic, a feature which is largely accounted for in most microscopic models.

Kinetic models generally contain stochastic components (lane change, desired distributions, etc) which are integrated and become deterministic elements of the model (refer to [2] and the references therein). Similarly some first order models such as SSMT [3] contain integrated stochastic building blocks (for instance conflicts between movements in intersections).

Other efforts have aimed at modeling directly some stochastic processes of traffic flow [4, 5].

Very few researches have introduced directly stochastic processes into macroscopic equations of traffic flow (see [6] for an early model). The main difficulties are computational intractability and diffusive effects. The model described in this paper introduces a model which is non-diffusive, tractable numerically, and readily interpretable from a physical point of view.

2 Basic Properties of the Generic 2nd Order Model

2.1 Model Formulation

The generic second order model (GSOM) has been introduced in [7]. It is essentially a LWR 1st order traffic flow model [8] with variable (driver-dependent) fundamental diagram. The model can be expressed as follows:

$$\begin{array}{ll} \partial_t \rho + \partial x(\rho v) = 0 & \text{Conservation of vehicles} \\ \partial_t \rho I + \partial x(\rho v I) = 0 & \text{Conservation of the driver attribute } I \\ v = \Im(\rho, I) & \text{Driver dependent fundamental diagram (FD)} \end{array} \quad (1)$$

I is a driver attribute that characterizes the behaviour of each driver. A driver with attribute I behaves according to a speed-density relationship $v = \Im(\rho, I)$ and a flow-density relationship

$$q = \rho v = \Re(\rho, I) \stackrel{def}{=} \rho \Im(\rho, I)$$

$\Re$ is assumed to be concave with respect to ρ, for all values of I. The ARZ model [9] and the 1-phase Colombo model [7] are special cases of the GSOM (1).

2.2 Model Properties

The system (1) is a system of two conservation laws. It admits two characteristic wave speeds,

$$\begin{aligned} \lambda_1 &= \frac{\partial \Re}{\partial \rho} \\ \lambda_2 &= v \end{aligned} \quad (2)$$

The 1-waves are kinematical waves which affect density (deceleration - shock-waves and acceleration - rarefaction waves), with invariant I. The 2-waves are contact discontinuities which propagate the discontinuity of I at the speed of traffic (v is the invariant).

Since by combining the first two equations of (1) we deduce:

$$\dot{I} \stackrel{def}{=} \partial_t I + v \partial_x I = 0 \quad (3)$$

it follows that I is conserved along vehicle trajectories, a result in keeping with the fact that 2-waves propagate discontinuities of I at the speed of traffic.

The model (1) is best understood by noticing the following. If the initial condition is piecewise constant with respect to I, then the discontinuities of I propagate at the velocity of the traffic flow and model (1) reduces to a simple first-order LWR model wherever I is constant, i.e. between discontinuity points.

2.3 Examples of GSOM

A first example of GSOM model is given by the ARZ model [9], in which the function $\Im$ is defined by:

$$\Im(\rho, I) = V_e(\rho) + I \tag{4}$$

with $V_e(\rho)$ the fundamental speed-density diagram. This model requires some adjustment in order for the density to stay within physically sound bounds i.e. to stay bounded by the jam density. The reader is referred to [10] for a discussion of these problems. The corresponding functions $\Im$ and $\Re$ are illustrated by the following (with triangular flow density relationship):

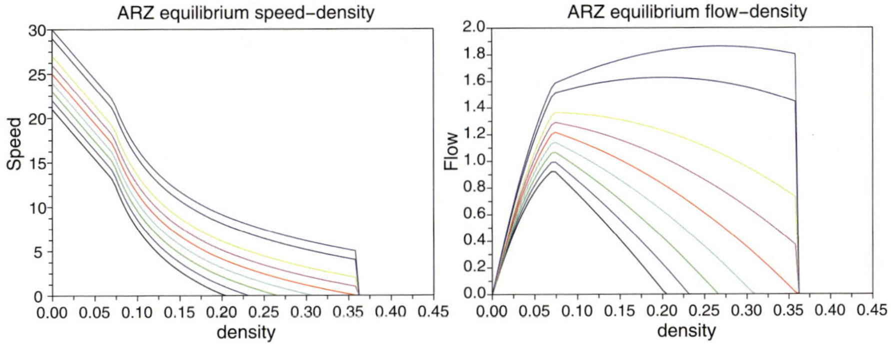

Fig. 1. ARZ model: equilibrium speed and flow relationships.

A second example of GSOM model is given by the 1-phase Colombo model [7], which is defined by the following $\Im$ function

$$\Im(\rho, I) = \begin{cases} v_f(\rho) & \text{if } \rho \le \rho_{crit}(I) \\ \left(I + \frac{q_*}{\rho}\right)\nu_0(\rho) & \text{if } \rho \ge \rho_{crit}(I) \end{cases} \tag{5}$$

with

$$\left| \begin{array}{ll} v_f(\rho) & = V_{max} - \beta\rho \\ \beta & = \dfrac{V_{max} - V_{crit}}{Q_{max}/V_{crit}} \\ \nu_0(\rho) & = 1 - \rho/\rho_{max} \end{array} \right.$$

and

$$\rho_{crit}(I) = \frac{1}{2\left(\beta - \dfrac{I}{\rho_{max}}\right)} \left[V_{max} - I + \frac{q_*}{\rho_{max}} - \sqrt{\left(V_{max} - I + \frac{q_*}{\rho_{max}}\right)^2 - 4q_*\left(\beta - \frac{I}{\rho_{max}}\right)}\right]$$

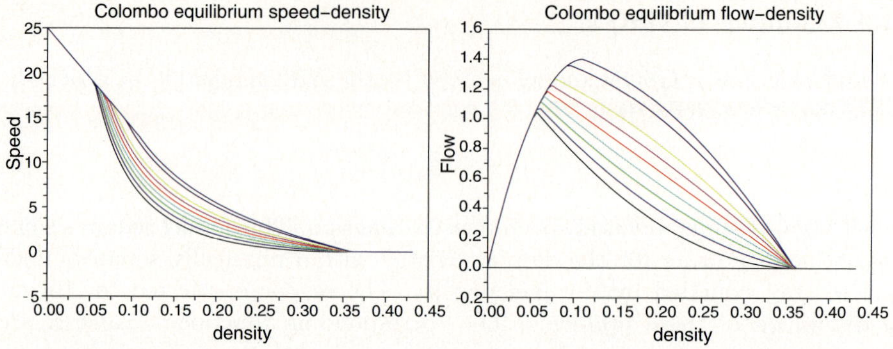

Fig. 2. 1-Phase Colombo model: equilibrium speed and flow relationships.

This model is illustrated by the figure below. The 1-phase Colombo model is devoid of any problems such as analyzed in [10], i.e. that jam density is dependent on I.

2.4 Lagrangian Expression of the GSOM

This section is a generalization of [11]. The first fundamental variable of the model in Lagrangian form is the cumulative density

$$N \stackrel{def}{=} \int_x^{+\infty} \rho(\xi, t) d\xi \tag{6}$$

which is completed by a second variable which is time-like:

$$T \stackrel{def}{=} t$$

N can be interpreted intuitively as a vehicle index. From these definitions it follows:

$$\left| \begin{array}{l} \partial_x N = -\rho \\ \partial_t N = q \end{array} \right.$$

and some easy algebra yields the following expressions for coordinate change:

$$\left| \begin{array}{l} \partial_x = -\rho \partial_N \\ \partial_t = \partial_T + q \partial_N \end{array} \right. \Leftrightarrow \left| \begin{array}{l} \partial_N = -r \partial_x \\ \partial_T = \partial_t + v \partial_x \end{array} \right. \tag{7}$$

with

$$r \stackrel{def}{=} 1/\rho$$

the spacing (note that $v = rq$).

Let us rewrite (1). The conservation equation can be restated as:

$$\partial_t \rho + \partial_x (\rho v) = 0 \Leftrightarrow \partial_T r + \partial_N v = 0$$

with

$$v = \Im(\rho, I) \overset{def}{=} \mathcal{V}(r, I) \tag{8}$$

The driver attribute equation can be restated as:

$$\partial_t \rho I + \partial x(\rho v I) = 0 \Leftrightarrow \partial_T I = 0$$

This result is logical: I is conserved along trajectories. Let us check the result:

$$\begin{aligned}
&\partial_t(\rho I) + \partial_x(\rho v I) = 0 \\
&\partial_T(\rho I) + q\partial_N(\rho I) - \rho\partial_N(\rho v I) = 0 \\
&r\partial_T(\rho I) + v\partial_N(\rho I) - \partial_N(\rho v I) = 0 \\
&r\partial_T(\rho I) - \rho I \partial_N v = 0
\end{aligned}$$

Now using $\partial_T r + \partial_N v = 0$ it follows

$$\begin{aligned}
&r\partial_T(\rho I) + \rho I \partial_T r = 0 \\
&\partial_T(r\rho I) = 0 \\
&\partial_T I = 0
\end{aligned}$$

It can be shown that the transformation (7) is admissible, in the sense that it respects shockwaves (shockwaves in (x, t) and (N, T) coordinates are equivalent).

Thus the GSOM system (1) is equivalent to the following system in Lagrangian coordinates:

$$\begin{array}{ll}
\partial_T r + \partial_N v = 0 & \text{Conservation of vehicles} \\
\partial_T I = 0 & \text{Conservation of the driver attribute } I \\
v = \mathcal{V}(r, I) & \text{Driver dependent fundamental diagram}
\end{array} \tag{9}$$

3 The Stochastic Generic 2nd Order Model

3.1 Model Formulation

The basic idea of the model is to consider the driver attribute I as a variable which is stochastic, as the result of random interactions of the driver with other drivers. Thus we assume that the dynamics of the attribute I are described by:

$$\dot{I} = \Phi\left(I, \frac{dB_t}{dt}\right) \tag{10}$$

with B_t a Brownian process (and $\frac{dB_t}{dt} = W_t$ the corresponding white noise process). It should be noted that more complicated interactions than (10) could be considered, with the effect of interactions being dependent on the density for instance.

The resulting stochastic model in natural (x, t) coordinates is:

$$\begin{array}{lll} \partial_t \rho + \partial x(\rho v) = 0 & \text{Conservation of vehicles} & \\ \partial_t \rho I + \partial x(\rho v I) = \rho \Phi(I, \frac{dB_t}{dt}) & \text{Dynamics of the driver attribute } I & (11) \\ v = \Im(\rho, I) & \text{Driver dependent FD} & \end{array}$$

The model is better specified in the coordinates (N, T) (refer to (6)). Let $(\Omega, \mathbb{P})$ be the underlying probability space; I is a random variable depending on the vehicle index N (it is a driver attribute) and on the random event $\omega \in \Omega$:

$$I = I(N, T; \omega)$$

The white noise perturbation is also vehicle-specific, thus

$$W_T = W_T(N, \omega)$$

Finally the stochastic GSOM model can be expressed in the Lagrangian (N, T) coordinates as:

$$\begin{array}{lll} I = I(N, T; \omega) & & \\ W_T = W_T(N, \omega) & & \\ \partial_T r + \partial_N v = 0 & \text{Conservation of vehicles} & (12) \\ \partial_T I = \Phi(I, W_T) & \text{Dynamics of the driver attribute } I & \\ v = \mathcal{V}(r, I) & \text{Driver dependent FD} & \end{array}$$

The idea for solving (12) is the following:

- fix $\omega \in \Omega$
- integrate $\partial_T I = \Phi(I, W_T)$ for each N, yielding a solution $I(N, T; \omega)$,
- solve $\partial_T r + \partial_N \mathcal{V}(r, I(N, T; \omega)) = 0$, yieldig a solution $r(N, T; \omega)$.

If $I(N, T; \omega)$ is regular enough (continuous for almost all $\omega \in \Omega$), $\partial_T r + \partial_N \mathcal{V}(r, I(N, T; \omega)) = 0$ can be solved for almost all $\omega \in \Omega$ and can be approximated by a Godunov scheme.

3.2 Godunov (Particle) Discretization of (12)

Let us first consider the Godunov scheme for solving (12). The elements of the Godunov scheme are described in [1]; we use here the basic elements, notably the supply/demand approach. The scheme developed below is just an adaptation for the (N, T) coordinate system of the [1] template.

The "cells" of the scheme are packets of vehicles counting ΔN vehicles. In the sequel we simply consider 1 vehicle packets that is $\Delta N = 1$. Note that in the case of the LWR model [8] (same attribute I for all drivers) the resulting Godunov scheme is similar to the particle discretization of the INTEGRATION model [12]. Time is discretized into time-steps ΔT, the size of which is defined by the CFL condition (Courant-Friedrichs-Lewy). The reader is referred to [7] and to (22).

The flux function in (12) is $\mathcal{V}(r, I)$, which is increasing with respect to r (see figure 3 below).

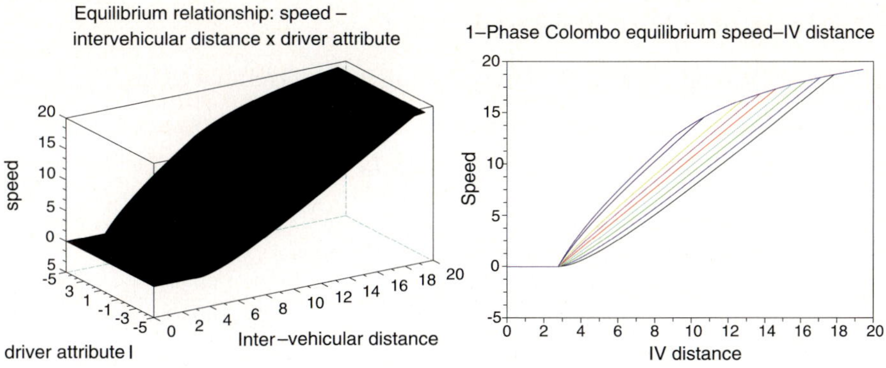

Fig. 3. Flux function in (12): speed as a function of inter-vehicular distance.

Thus the demand associated to the flux function $\mathcal{V}(r, I)$ is this function itself, whereas the supply is the maximum speed. In the sequel we shall only consider homogeneous sections, therefore by the min formula we can ignore the supply. It follows that for a one-vehicle cell (n), with inter-vehicular distance $r_n(t) \stackrel{def}{=} x_{n-1}(t) - x_n(t)$ at time step $(t) = [t\Delta t, (t+1)\Delta t]$, the discretized conservation equation can be stated as:

$$\begin{aligned} &r_n(t+1) = r_n(t) + \Delta t\,(-v_n(t) + v_{n-1}(t)) \\ &v_n(t) = \mathcal{V}(r_n(t), I_n(t)) \end{aligned} \tag{13}$$

It follows trivially that the vehicle positions $x_n(t)$ are updated according to

$$\begin{aligned} &x_n(t+1) = x_n(t) + \Delta t v_n(t) \\ &v_n(t) = \mathcal{V}(r_n(t), I_n(t)) \end{aligned} \tag{14}$$

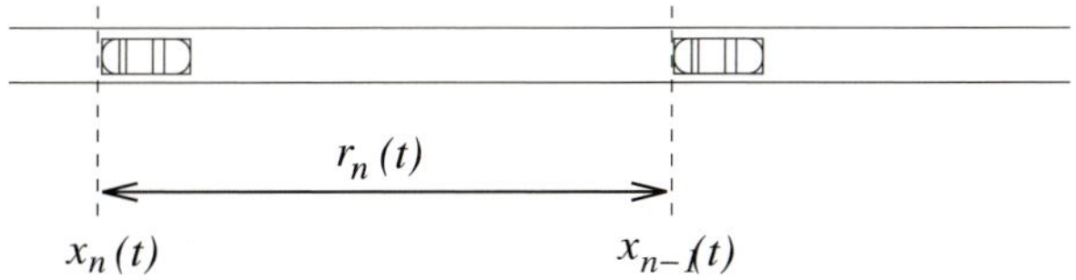

Fig. 4. Particle discretization.

In order to evaluate $I_n(t+1)$, it is necessary to integrate $\partial_T I = \Phi(I, W_T)$. Thus if we denote by $\mathcal{L}(I^0, t_0, t_1; \omega)$ the solution at t_1 of

$$\begin{aligned} &\dot{I}(s) = \Phi(I(s), W_s(n, \omega)) \\ &I(t_0) = I^0 \end{aligned}$$

it follows:

$$I_n(t+1) = \mathcal{L}(I_n(t), t, t + \Delta t; \omega) \tag{15}$$

The complete Godunov discretization is summarized by (14), (15) and (22).

3.3 Microscopic Formulation of (12)

By letting Δt tend towards 0, the model can be reduced to a follow-the leader (FTL) model:

$$\left| \begin{array}{l} \dot{x_n}(t) = v_n(t) = \mathcal{V}(r_n(t), I_n(t)) \\ \dot{I}_n(t) \ = \Phi(I_n(t), W_t(n, \omega)) \end{array} \right. \quad \forall \omega \in \Omega \tag{16}$$

which can also be expressed using the semi-group $\mathcal{L}$:

$$\left| \begin{array}{l} \dot{x_n}(t) = v_n(t) = \mathcal{V}(r_n(t), I_n(t)) \\ \dot{I}_n(t) \ = \mathcal{L}(I_n(0), 0, t; \omega) \end{array} \right. \quad \forall \omega \in \Omega \tag{17}$$

The numerical viscosity in the Godunov discretization scheme (14), (15) is minimal (i.e. the shock-waves are fitted most exactly) when the CFL condition (22) is an equality. Thus the FTL model (16) smoothens shockwaves.

4 Examples of Stochastic Models

4.1 Ornstein-Uhlenbeck Process for *I*

The simplest model for (10) is the linear model:

$$\Phi(I, W) = -\alpha I + \sigma W \tag{18}$$

with α and σ two parameters and $W_t = \frac{dB_t}{dt}$ a white noise process: the Brownian motion process has independent normal Gaussian increments $B_t - B_s$ of variance $t - s$. Let us recall also that the trajectories $t \to B_t(\omega)$ are continuous for nearly all $\omega \in \Omega$. The idea of expression (18) is simple: the dynamics of I_n results from two competing processes: a relaxation process and a white noise perturbation process.

The expression of the semi-group $\mathcal{L}$ is easily deduced

$$\mathcal{L}(I(t), 0, t; \omega) = \sigma \int_0^t e^{-\alpha(t-s)} dB_s(\omega) + I(0)e^{-\alpha t} \tag{19}$$

(19) defines the solution of (18) as an Ornstein-Uhlenbeck process; the mean of $I(t, \omega)$ is equal to $I(0)e^{-\alpha t}$ and its variance to

$$\sigma^2 \frac{1 - e^{-2\alpha t}}{2\alpha}$$

The integral in (19) is the standard Wiener integral. These processes are illustrated in figure 5. Asymptotically, that is at the limit $t \to +\infty$, $I(t, \omega)$ is Gaussian with variance $\sigma^2/2\alpha$. Thus, the model predicts that the occurrence of arbitrarily large values of the driver attribute I is possible, albeit with vanishingly small probability. Such driver behavior is not in accordance with observations.

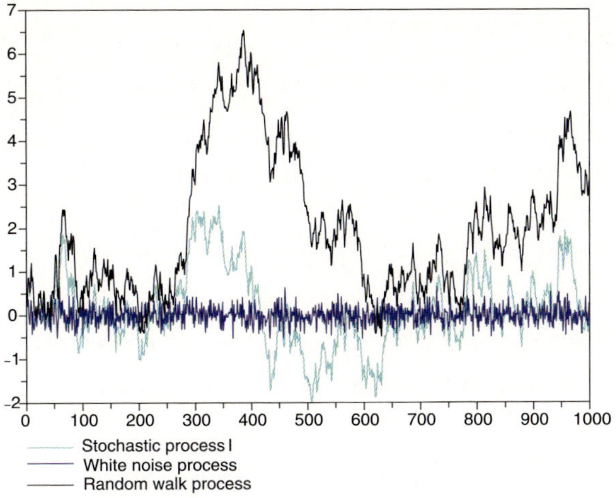

Fig. 5. Processes: W_t, B_t, I_t.

4.2 Transformed Ornstein-Uhlenbeck Process for *I*

The idea is to apply a non-linear transformation on an Ornstein-Uhlenbeck process in order to constrain I within physically acceptable bounds. We will consider a transformation

$$\mathcal{F} : [-I_*, I_*] \to \mathbb{R}$$

and a process I defined by

$$\dot{\mathcal{F}(I)}(t) = -\alpha \mathcal{F}(I(t)) + \sigma \frac{dB_t}{dt}$$

For instance, $\mathcal{F}(I) = \mathrm{Argth}(I/I_*)$ or $\mathcal{F}(I) = \mathrm{tg}(\pi I/I_*)$ constitute two possible choices. We can deduce from (19) that

$$I(t,\omega) = \mathcal{F}^{-1}\left(\sigma \int_0^t e^{-\alpha(t-s)} dB_s(\omega) + \mathcal{F}(I(0))e^{-\alpha t}\right) \tag{20}$$

The process I is illustrated by the following figure 6. The model above is defined by the following function Φ:

$$\Phi(I,W) = -\alpha \frac{\mathcal{F}(I)}{\mathcal{F}'(I)} + \sigma \frac{W}{\mathcal{F}'(I)}$$

4.3 Discretization

We apply the Godunov discretization scheme (14), (15), but replace (15) by an implicit Euler iteration. This iteration is derived from:

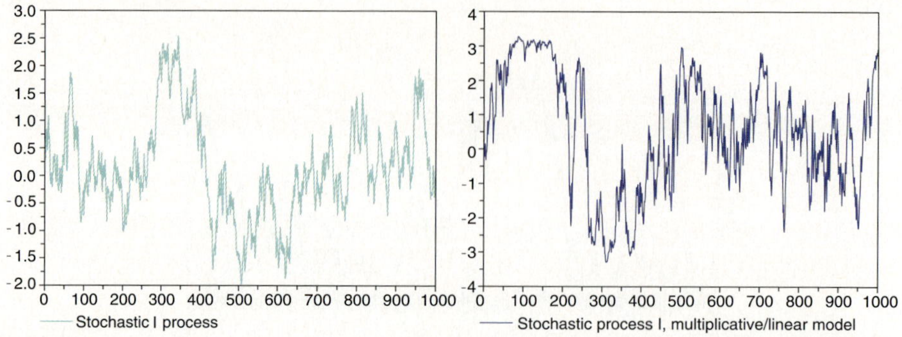

Fig. 6. Examples of O-U process (left) versus transformed O-U process.

$$\mathcal{F}(I_n^{t+1}) = \mathcal{F}(I_n^t) - \alpha\mathcal{F}(I_n^{t+1})\Delta t + \sigma\Delta B_t$$

In this iteration ΔB_t is Gaussian, centered, of variance t and the successive increments ΔB_t are independent.

Thus the Godunov (particle) discretization can be expressed as:

$$\left| \begin{array}{l} x_n^{t+1} = x_n^t + \Delta t\, v_n^t \\ v_n^t = \mathcal{V}(r_n^t, I_n^t)) \\ I_n^{t+1} = \mathcal{F}^{-1}\left(\left[\mathcal{F}(I_n^t) + \sigma\Delta B_t\right]/[1+\alpha\Delta t]\right) \end{array} \right. \tag{21}$$

The CFL condition for (21) can be obtained by expressing that if the distance between particle n and $n-1$ is greater than the minimum headway $r_{min}(I_n^t)$, it stays so after one iteration:

$$r_{min}(I_n^t) + \Delta t\, \mathcal{V}(r_n^t, I_n^t)) \leq x_{n-1}^t - x_n^t$$

The following constraints results, which is applied in the simulation runs:

$$\Delta t \leq \text{Max}_{\left[\begin{array}{c} r \geq r_{min}(I) \\ -I_* \leq I \leq I_* \end{array}\right]} \left(\frac{1 - r_{min}(I)/r}{\mathcal{R}(1/r, I)}\right) \tag{22}$$

4.4 Some Simulation Results

As expected, on a free section the random fluctuations of the trajectories induce clustering and generate spontaneous random perturbations which once created behave approximately in conformity with LWR theory.

Supply reduction are modelled by limiting the outflow, that is to say the speed of the lead vehicle. Demand is modelled by introducing new vehicles at a rate equal to demand, but a distance from the last vehicle which is compatible with the section supply. This is the reason for the accumulation of vehicles at the entry point of the section in figure 7.

Features such as shock waves and rarefaction waves are also recaptured, but include the expected amount of stochastic variability.

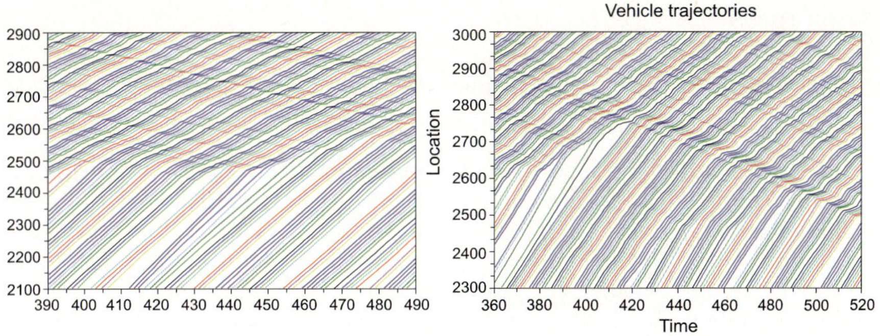

Fig. 7. Examples of particle trajectories, with downstream supply reduction and upstream variable demand (increasing).

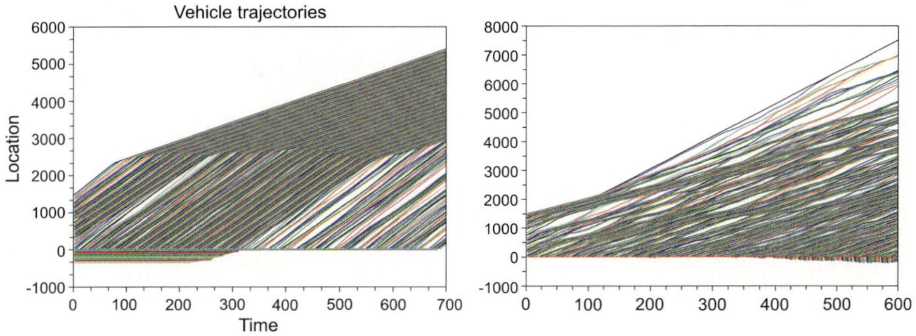

Fig. 8. Examples of classical traffic dynamics. Left: shock wave + variable demand (decreasing), right: constant demand + supply increase.

5 Conclusion

The stochastic feature introduced into the GSOM model (11) does not induce any diffusion effects, as results trivially from (21): the vehicle speed for instance cannot be negative. The Godunov particle discretization (21) constitutes a convergent discretization for almost all $\omega \in \Omega$, and provides an intuitive description of the model solutions. The model reproduces the variability of trajectories, as described in [13]. Ongoing research addresses the following

- deduce the resulting stochastic properties of flow and density, in order to apply the model to stochastic traffic control,
- extend the model to intersection modeling [14], in order to check whether the model (11) predicts breakdown of traffic and similar phenomena.

References

1. Lebacque, J.P. (1996). *The Godunov scheme and what it means for first order traffic flow models.* In: Transportation and Traffic Theory, proceeding of the 13th ISTTT (J.B. Lesort, ed.), 647-677, Elsevier.
2. Ngoduy, D. (2006). *Macroscopic Discontinuity Modeling for Multiclass Multilane Traffic Flow Operations.* PhD thesis, Delft University of Technology, Netherlands.
3. Lebacque, J.P. (1984). *Semimacroscopic simulation of urban traffic.* Int. 84 Minneapolis Summer Conference. AMSE.
4. Kühne, Mahnke R. (2005). *Controlling traffic breakdowns.* In: Transportation and traffic theory, Proceedings of the 16th ISTTT (H.S. Mahmassani, ed.), 229-244.
5. Dundon, N., Sopasakis, A., (2007). *Stochastic modeling and simulation of multilane traffic.* In: Transportation and traffic theory 2007, Proceedings of the 17th ISTTT, pp 661-689.
6. Weits, E. (1992). *Stationary freeway traffic flow modelled by a linear stochastic partial differential equation.* Transportation Research B, 26, 2, pp 115-126.
7. Lebacque, J.P., Mammar, S., Haj-Salem, H. (2007). *Generic second order traffic flow modeling.* In: Transportation and traffic theory 2007, Proceedings of the 17th ISTTT, pp 755-776.
8. Lighthill, M.H. and Whitham, G.B. (1955). *On kinematic waves II: A theory of traffic flow on long crowded roads.* Proceedings of the Royal Society (London) A 229, 317-345.
   Richards, P.I. (1956). *Shock-waves on the highway.* Operations Research, 4, 42-51.
9. Aw, A. and Rascle, M. (2000). *Resurrection of second order models of traffic flow.* SIAM Journal of Applied Mathematics, 60(3), 916-938.
   Zhang, H.M. (2002). *A non-equilibrium traffic model devoid of gas-like behavior.* Transportation Research, 36, 275-290.
10. Lebacque, J.P., Mammar, S. and Haj-Salem, H. (2007). *The Aw Rascle and Zhangs model: Vacuum problems, existence and regularity of the solutions of the Riemann problem.* Transportation Research Part B 41, 710-721.
11. Aw, A., Klar, A., Materne, T. and Rascle, M. (2002). *Derivation of continuum traffic flow models from microscopic follow-the-leader models.* SIAM Journal of applied Mathematics, 63, 259-278.
12. Van Aerde, M. (1994) *INTEGRATION: A model for simulating integrated traffic networks.* Transportation Systems Research Group. Queens University Kingston, Ontarion.
13. Treiterer, J. and Myers, J.A. *The Hysteresis Phenomenon in Traffic Flow.* In: Procs. 6th International Symposium on Transportation and Traffic Theory. ed. by D.J. Buckley (A.H. & A.W. Reed, London 1974) pp. 1338.
14. Lebacque, J.P. and Khoshyaran, M.M. (2005). *First order macroscopic traffic flow models: intersection modeling, network modeling.* In: Transportation and traffic theory, Proceedings of the 16th ISTTT (H.S. Mahmassani, ed.), 365-386.
15. Colombo, R. (2002). *Hyperbolic phase transitions in traffic flow.* SIAM Journal of Applied Mathematics, 63(2), 708-721.

A Multiclass Car-Following Rule Based on the LWR Model

Ludovic Leclercq[1] and Jorge A. Laval[2]

[1] Laboratoire Ingénierie Circulation Transports - ENTPE / INRETS - Université de Lyon, Vaulx-en-Velin, France leclercq@entpe.fr
[2] Georgia Institute of Technology, Atlanta, USA jorge.laval@ce.gatech.edu

1 Introduction

The Lighthill-Whitham and Richards (LWR) model [1, 2] is well known for its simplicity, parsimony and its robustness to replicate basic traffic features. However, it considers traffic as a homogeneous flow. This can be a serious limitation when the traffic stream is composed of radically different vehicle classes, such as cars and heavy trucks near an uphill grade.

Extensive research has been conducted to introduce heterogeneity in the LWR model; e.g. [3–9]. All these extensions are based on the same principle: disaggregating the heterogeneous traffic flow into homogeneous and continuum vehicle classes that obey a conservation law with a specific fundamental diagram (FD). These models are solved numerically in Eulerian coordinates with methods such as the Godunov scheme [10]. However, these schemes are known to be very diffusive for hyperbolic systems of conservation laws [4].

Recent developments in traffic flow theory have led to efficient numerical schemes for solving the LWR model. They are derived in Lagrangian framework rather than in traditional Eulerian one; see for example [11–14]. Additionally, variational theory [15–17] and its extension in Lagrangian coordinates [14] make it possible to prove that these schemes are exact for the LWR model when the FD is triangular. This is an important leap forward since current methods introduce numerical errors that can be devastating in practice. The aim of this paper is to extend the framework in [14] in order to incorporate multiple vehicle types, each one with a different car-following rule. In this way, the free-flow speed, the jam density and the wave-speed can be defined for each individual vehicle class. Note that the one-class car-following rule has already been coupled with a lane-changing model [18] and thus the proposed extension is fully compatible with the latter.

The sketch of the paper is as follows: section 1 will recall the Lagrangian formulation of the LWR model and its numerical resolution using (i) the Godunov scheme and (ii) the variational theory. Section 2 will introduce the proposed extension, the resulting numerical schemes using (i) and (ii) above, and the conditions for the exactness of the numerical schemes. Finally, section 3 will present some numerical results focusing on the representation of a bimodal flow of trucks and cars.

2 The Homogeneous LWR Model in Lagrangian Coordinates

Traditionally, the LWR model has been formulated in (x,t) coordinates. Introducing the cumulative count function $N(x,t)$, which represents the number of vehicles that have crossed position x by time t, allows to define a new coordinate system (N,t). These Lagrangian coordinates are fixed to a given fluid particle and move with it in space-time. The purpose is no longer to determine the local density $k(x,t)$ but the position $X(n,t)$ of vehicle n. In the remainder of the paper, capital N (respectively X) will stand for the $N(x,t)$ (respectively $X(n,t)$) function while n (respectively x) will define a specific value taken by this function. This section briefly presents the Lagrangian formulation of the LWR model. Details can be found in [14] or in [19].

2.1 Continuum Formulation

Lagrangian Conservation Law

The LWR model in (n,t) coordinates can be expressed as a conservation law:

$$\partial_t s + \partial_n v = 0 \tag{1}$$

This model is fully described by the spacing s which corresponds to $1/k$. The speed v can be derived from the FD $v{=}V(s)$. Therefore, the LWR model can be described by the following hyperbolic equation in s:

$$\partial_t s + \partial_n V(s) = 0 \tag{2}$$

Lagrangian Variational Principle

The reader should refer to [15] and [16] for a complete description of variational theory in Eulerian coordinates. Indeed, the transformation to Lagrangian coordinates preserves the nature of the problem: the LWR model can be expressed as the Hamilton-Jacobi equation 3 derived from the FD.

$$\partial_t X = V(-\partial_n X) \tag{3}$$

All the results proven in [15] and [16] can thus be applied to the Lagrangian variational formulation of the LWR model. Notably, the value of X at a point P in the (n,t) plane, X_P, can be expressed as a least-cost path problem:

$$X_P = \min(B_\wp + C(\wp) : \forall \wp \in \mathbf{V} \cap \mathbf{P}_P), \text{ where}$$
$\mathbf{V}$: set of all valid paths
$\mathbf{P}_P$: set of all paths from boundary condition to P (4)
$B_\wp$: X value at the beginning of the path
$C(\wp)$: cost of path $\wp$

Analogously to [15], "waves" in (n,t) coordinates are characteristics where s is constant. They have slopes $u = \partial_s V(s)$ representing a passing rate. We define two types of passing rates: (a) u is a "possible passing rate" if there exists s such that $u = \partial_s V(s)$; (b) $\hat{u}$ is an "allowable passing rate" if $\min \partial_s V(s) \leq \hat{u} \leq \max \partial_s V(s)$. "Valid paths" are continuous and piecewise differentiable paths $n(t)$ in the (n,t) plane whose slopes $n'(t)$ are allowable passing rates. "Wave paths" are valid paths whose slopes are possible passing rates and are thus composed of a succession of waves.

The cost rate r on a wave path is given by $\mathrm{d}_t X$. The scalar r represents the speed of the Eulerian characteristic associated to the passing rate u.

$$r = \mathrm{d}_t X = \partial_t X + \partial_n X \partial_t n = v - su \tag{5}$$

As 3 holds and V is concave, one can express r only as a function $R(u)$ using the Legendre transformation as in [15]:

$$r = R(u) = \sup_s \{V(s) - su\} \tag{6}$$

The cost on a Lagrangian valid path P from B to P is thus:

$$C(\wp) = \int_{t_B}^{t_P} R(n'(t))dt \tag{7}$$

In the next section, we will show how the Lagrangian variational principle makes it possible to construct a numerical scheme which is exact under few restrictive assumptions.

2.2 Numerical Resolution

Godunov Scheme

In the Godunov scheme, the N-function is discretized in cells $i = 1, 2, \ldots$ of size Δn and the spacing s is approximated by a constant value, s_i^t, which is updated at every time step Δt; see Figure 1. Since the flux function V in 2 is non-decreasing in s, the characteristic speed is always non-negative (traffic anisotropy). The Godunov method reduces then to the upwind method:

$$s_i^{t+\Delta t} = s_i^t + \frac{\Delta t}{\Delta n}\left(V(s_i^t) - V(s_{i-1}^t)\right) \tag{8}$$

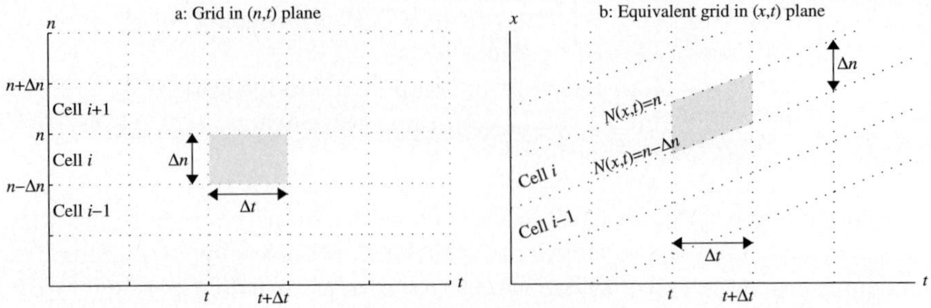

Fig. 1. Lagrangian grid.

The Courant-Friedrich-Lewy's (CFL) condition 9 should hold to guaranty the stability and the convergence of 8.

$$\Delta n \geq \max_s |\partial_s V(s)| \, \Delta t \tag{9}$$

The Lagrangian Godunov scheme can also be expressed in terms of $X(n,t)$ by noting that the flux $V(s_i^t)$ at a boundary n of a cell i is:

$$\frac{X(n,t+\Delta t) - X(n,t)}{\Delta t} = V(s_i^t) = V\left(\frac{X(n-\Delta n,t) - X(n,t)}{\Delta n}\right) \tag{10}$$

Let us suppose now that the FD is triangular when expressed in terms of flow with respect to density:

$$V(s) = \min(v_m, w(\kappa s - 1)) \tag{11}$$

where v_m is the free-flow speed, w, the wave speed and κ the jam density; see Figure 2b. After simplification 10 becomes:

$$X(n,t+\Delta t) = \min(X(n,t)+v_m\Delta t, (1-\alpha)X(n,t)+\alpha X(n-\Delta n,t)-w\Delta t) \tag{12}$$

where $\alpha = w\kappa\Delta t/\Delta n$. If the CFL condition 9 is satisfied as an equality, $\alpha = 1$ and the scheme produces exact results [14]. In that case, one should fix either the time step Δt or the quantity of vehicle inside a Lagrangian cell Δn. For practical reason, it is easier to set Δn and to deduce $\Delta t = \Delta n/w\kappa$. Equation 12 reduces then to:

$$X\left(n, t + \frac{\Delta n}{w\kappa}\right) = \min\left(X(n,t) + v_m\frac{\Delta n}{w\kappa}, X(n-\Delta n,t) - \frac{\Delta n}{\kappa}\right) \tag{13}$$

Note that when n is an integer and $\Delta n = 1$ then $X(n,t)$ corresponds to the position x_n^t of vehicle n at time t and $X(n\text{-}1,t)$ to the position x_{n-1}^t of its leader at the same time.

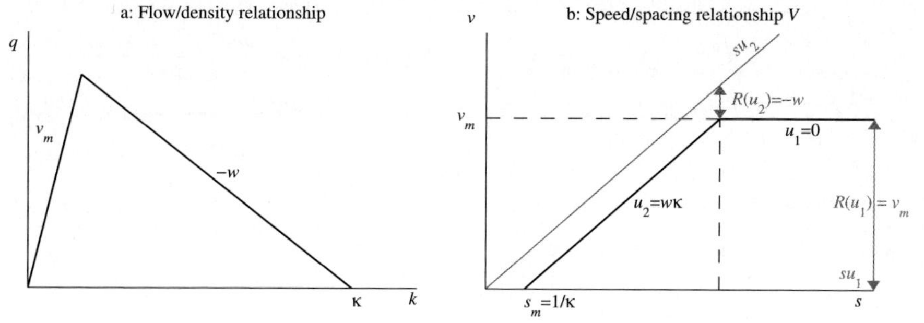

Fig. 2. Triangular fundamental diagram.

Variational Principle

Daganzo [16] proposed efficient methods to solve the LWR model using the concept of "sufficient networks". A "network" is defined as a directed graph where arcs are valid paths. A network is "sufficient" when the least-cost path through the network between every pair of nodes is optimum according to the continuous formulation of the model. In a sufficient network, the solution is exact at every node provided that the initial data is linear between two consecutive initial nodes. Otherwise, numerical errors are introduced as all optimum paths are not necessarily included in the network.

In Lagrangian coordinates, a sufficient network may easily be constructed when the FD is triangular. In this case, waves have only two possible velocities: $u_1 = 0$ (free-flow wave) and $u_2 = w\kappa$ (congestion wave). The resulting cost rates 6 are: $R(u_1) = v_m$ and $R(u_2) = -w$; see Figure 2b. Any geometric network formed by two families of parallel equidistant lines with slopes u_1 and u_2 and separated by Δn_1 and Δn_2 is sufficient; see Figure 3a. Therefore, with appropriate initial data, the solution at nodes is exact.

Since $u_1 = 0$ nodes are always lined-up along rows where n values are constant. Furthermore, if one sets $\Delta n_1 = \Delta n_2 = \Delta n$ nodes also line-up along "time-columns"; see Figure 3b. This defines a rectangular lattice in the (n,t) plane with $\Delta t = \Delta n/(w\kappa)$, which is very practical for computational implementation. Furthermore, with only two incoming arcs per node, the computation of 4 at each node is straightforward; i.e.:

$$\begin{aligned} &X(n, t + \Delta t) \\ &= \min\left(X(n,t) + \Delta_{(n,t)\rightarrow(n,t+\Delta t)}, X(n - \Delta n, t) + C_{(n-\Delta n,t)\rightarrow(n,t+\Delta t)}\right) \\ &= \min\left(X(n,t) + v_m \Delta t, X(n - \Delta n, t) - w\Delta t\right) \end{aligned} \tag{14}$$

where $C_{(n,t)\rightarrow(n',t')}$ is the cost of the arc between (n,t) and (n',t').

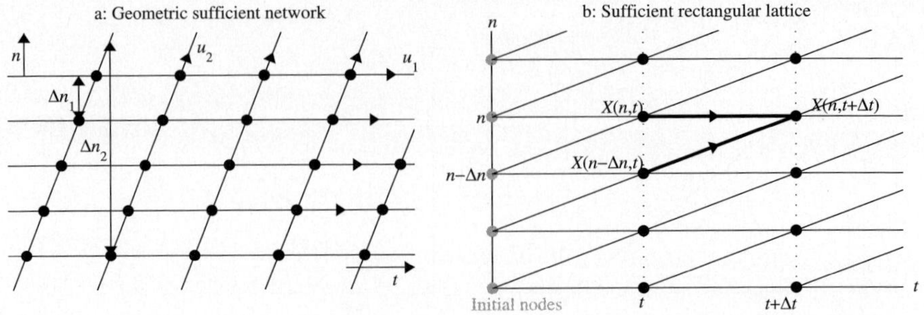

Fig. 3. Geometric networks associated to the Lagrangian variational principle.

Notice that 14 and 13 are identical. Therefore, the Godunov and the variational schemes are equivalent when the FD is triangular and the CFL condition 9 is satisfied as an equality; i.e., when $\Delta t = \Delta n/(w\kappa)$.

3 Introducing Different Vehicle Characteristics

3.1 Principle

The only way to distinguish different vehicle characteristics in Eulerian coordinates is to separate the global flow into several continuous homogeneous classes. The Lagrangian coordinate system makes this easier. Indeed, it allows for tracking vehicles and therefore applying a specific FD to them.

In the sequel, we suppose that m vehicle classes flow in proportions r_j, $j = 0, 1, \ldots m-1$. Each class j is characterised by a triangular FD V_j with the following parameters: $v_{m,j}$, the free-flow speed, κ_j, the jam density and w_j the wave-speed. The free-flow speed $v_{m,j}$ depends both on vehicles' mechanical performances and legislation. The jam-density κ_j can be estimated as the inverse of the mean distance between two stopped vehicles of the same class j. This distance, l_j, is roughly equal to the mean vehicle length plus some buffer distance between vehicles (about one meter). The wave-speed w_j can be estimated as the ratio between l_j and the mean time between two successive departures inside a queue formed upstream from a traffic signal. Indeed, the wave-speed corresponds to the velocity of the starting wave observed when a traffic signal turns green. Note that the index $j = 0$ will always represent the class in which the mean vehicle length is minimum. Typically, it corresponds to passenger cars. For this latter class, the subscript j will be omitted.

For simplicity, we assume that the behavior of a given vehicle class is defined by the FD of its own class. We will see in the next section how the car-following rule is modified to account for different FDs.

3.2 Numerical Resolution

The Lagrangian cell size Δn will now be set to 1, so that cells contain only one vehicle characterized by its specific FD. We can therefore drop the index i and refer to cells by the vehicle number, n. Numerical schemes will be derived either from the Godunov method or the variational principle.

Godunov Scheme

Lebacque in [20] has demonstrated that the Godunov scheme can be applied with different FDs in each Eulerian cell. This also holds in Lagrangian coordinates. Thus, if the vehicle in cell n belongs to class j, 12 becomes:

$$X(n,t+\Delta t) = \min\big(X(n,t)+v_{m,j}\Delta t, (1-\alpha_j)X(n,t)+\alpha_j X(n-1,t) - w_j\Delta t\big) \tag{15}$$

where $\alpha_j = w_j\kappa_j\Delta t$. The CFL condition imposes that:

$$\forall k \in [0, m-1] \quad \Delta t \leq \frac{1}{w_j\kappa_j} \tag{16}$$

This condition cannot be satisfied as an equality for all classes. Thus, the numerical scheme is no longer exact. Numerical viscosity appears except for the class where the equality holds.

3.3 Variational Principle

Considering a specific diagram for each vehicle does not modify the slope of free-flow waves in the Lagrangian network. Indeed, this slope $u_{1,j}$ is always equal to 0. Only the cost depends on vehicle class: $R(u_{1,j}) = v_{m,j}$. This can be easily accounted for as it does not change the structure of the network.

The slope of congestion wave, however, which depends on vehicle class: $u_{2,j} = w_j\kappa_j$. Cost on congestion waves is $R(u_{1,j}) = -w_j$. Thus it is no longer possible to build a sufficient network in the general case. To reduce numerical errors, one should set $\Delta t = 1/(w\kappa)$. Passenger car positions will then be exactly calculated. Congestion waves for all other classes will reach position $X_c(t)$ at time t; see Figure 4a. The location $X_c(t)$ is bounded by $X(n,t)$ and $X(n-1,t)$, and should be estimated as this point does not belong to the network. This will introduce numerical errors. If we suppose that the spacing is uniform in each Lagrangian cell, $X_c(t)$ can be deduced by:

$$X_c(t) = (1-\alpha_j)X(n,t) + \alpha_j X(n-1,t) \tag{17}$$

with $\alpha_j = w_j\kappa_j\Delta t$. The variational scheme is then:

$$X(n,t+\Delta t) = \min\left(X(n,t) + v_{m,j}\Delta t, X_c(t) - w_j\Delta t\right) \tag{18}$$

Note that 18 is fully equivalent to the Godunov scheme 15. This explains why this latter scheme is no longer exact for heterogeneous flow.

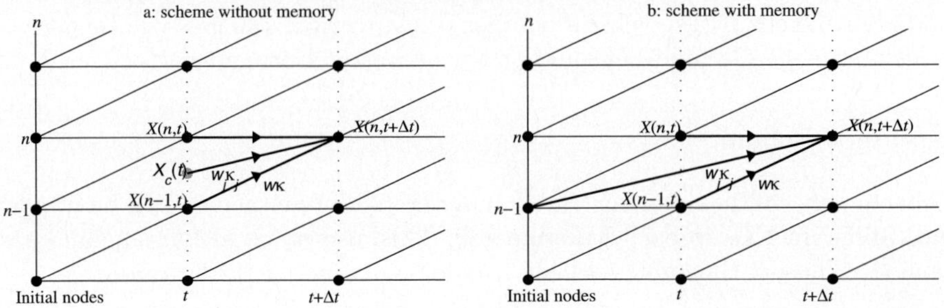

Fig. 4. Networks to account for different vehicle characteristics.

Under some specific assumptions, it is still possible to have an exact numerical scheme. These assumptions are:

- the wave speed is the same for each j: $w_j = w$;
- the ratio l_j (mean vehicle length) over l (mean passenger car length) is an integer for each j.

Under these assumptions, the ratio $w\kappa/(w_j\kappa_j)$ is always an integer. Thus, each congestion wave can join a network node among the previous time steps; see Figure 4b. One has just to store the X values at the T previous time steps with $T = \max(w\kappa/(w_j\kappa_j))$. The numerical scheme becomes:

$$\begin{aligned} & X(n, t+\Delta t) \\ & = \min\left(X(n,t) + v_{m,j}\Delta t, X\left(n-1, t-\left(\tfrac{\kappa}{\kappa_j}-1\right)\Delta t\right) - w_j\tfrac{\kappa}{\kappa_j}\Delta t\right) \end{aligned} \tag{19}$$

Note that the above assumptions are not too restrictive. Indeed, many authors assume that vehicle's behaviours are almost the same in congestion [4, 6, 8]. Therefore a constant wave-speed for all classes is a reasonable assumption.

4 Numerical Example

We are now going to focus on the bimodal case: light trucks and passengers cars. The FD parameters for both classes are:

- passenger cars: $v_m = 20$ m/s; $w = 5$ m/s; $\kappa = 0.2$ veh/m ($l = 5$ m);
- light trucks: $v_m, 1 = 12$ m/s; $w_1 = 5$ m/s; $\kappa_1 = 0.1$ veh/m ($l_1 = 10$ m).

Figure 5 presents the simulation results for a one-lane arterial road with a traffic signal located 400 m after the entrance. The cycle time is 90 s with 60 s of green time. The incoming flow is equal to 1080 veh/h. The ratio of light trucks r_1 is equal to 10% during 150 s and then switches to 40% until time $t = 340$ s which corresponds to the end of the simulation.

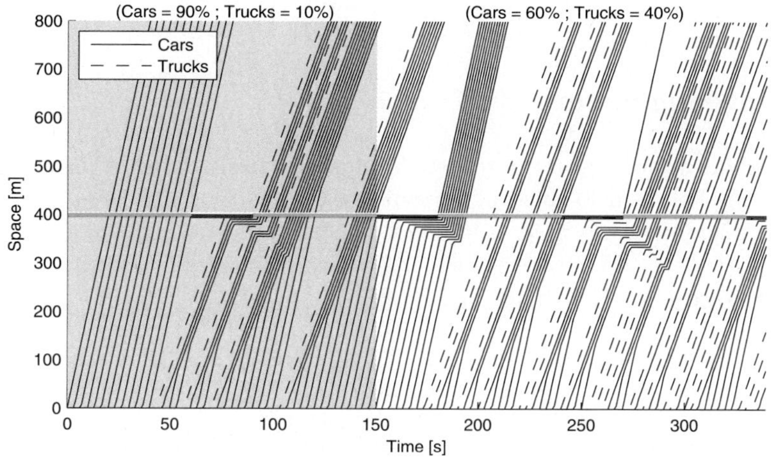

Fig. 5. Numerical example - exact variational scheme.

The simulation results show that the proposed numerical scheme reproduces the different vehicle behaviour as expected. Notice how shockwaves propagate upstream as sharp discontinuities in vehicle speeds, regardless of the vehicle class it crosses. This appealing property is a consequence of the numerical method being exact for both vehicle classes.

5 Conclusion

This paper proposes an original method to account for a traffic heterogeneity in the LWR model. Contrary to previous approaches, this extension is introduced in Lagrangian coordinates rather than in classical Eulerian ones, which enable distinguishing vehicle characteristics via class-specific FDs. Furthermore, numerical resolution can be addressed with simple and effective schemes. The main advantage is that these schemes are exact under few restrictive assumptions.

For simplicity, we made the assumption that the behavior of a given vehicle class is defined by the FD of its own class. In particular, this amounts to assuming that the class of the vehicle immediately downstream of a given vehicle will not affect its behavior. While this is certainly true in freely-flowing traffic, in congested conditions this may or may not be the case. Empirical evidence should be used to discern this matter. Fortunately, it is straightforward to extend the proposed model to capture such correlations: all one needs to do is define the FD for each pair of vehicle types.

The car-following rules proposed in this paper can be straightforwardly coupled with the lane-changing model in [18], yielding a complete microscopic multilane and multiclass model. Further research is being undertaken to validate this complete model with empirical data.

Acknowledgements

The authors are very grateful to Estelle Chevalier for her interesting comments and discussions and for helping to improve the final form of the paper.

References

1. Lighthill M.J, Whitham J.B (1955) On kinematic waves II: A theory of traffic flow in long crowded roads. Proceedings of the Royal Society, A229:317–345.
2. Richards P.I (1956) Operations Research, 4:42–51.
3. Benzoni-Gavage S, Colombo R.M (2003) European Journal on Applied Mathematics 14:587–612.
4. Chanut S (2005) First-order macroscopic model for mixed traffic including moving bottleneck effects. In: Mahmassani H.S (eds) 16th ISTTT. Elsevier, London.
5. Chanut S, Buisson C (2003) TRR 1852:209–219.
6. Logghe S (2003) Dynamic modeling of heterogenous vehicular traffic. PhD Thesis, Kathlieke Universiteit Leuven, Leuven.
7. Wong G.C.K, Wong S.C (2002) Transportation Research part A 36(9):827–841.
8. Zhang H.M, Jin W.L (2002) TRR, 1802:197–204.
9. Zhu Z, Chang G.L, Wu T (2003) TRR, 1852:201–208.
10. Godunov S.K (1959) Mat. Sb. 47:271–290.
11. Newell G.F (2002) Transportation Research part B 36(3):195–205.
12. Daganzo C.F (2006) Transportation Research part B 40(5):396–403.
13. Leclercq L (2007) Transportation Research part B 41(7):701–709.
14. Leclercq L, Laval J.A, Chevallier E (2007) The Lagrangian coordinates and what it means for first order traffic flow models. In: Allsop R.E, Bell M.G.H, Heydecker B.G (eds) 17th ISTTT. Elsevier, London.
15. Daganzo C.F (2005) Transportation Research part B 39(2):187–196.
16. Daganzo C.F (2005b) Transportation Research part B 39(10):934–950.
17. Daganzo C.F, Menendez M (2005) A variational formulation of kinematic waves: bottlenecks properties and examples. In: Mahmassani H.S (eds) 16th ISTTT. Elsevier, London.
18. Laval J.A, Leclercq L (2007) A microscopic theory of lane-changing. Submitted for publication. Available as Research Report LICIT 06-05, INRETS, Bron.
19. Daganzo C.F (2006b) Networks and Heterogeneous Media 1(4):601–619.
20. Lebacque J.P (1996) The Godunov scheme and what it means for first order traffic flow models. In: Lesort J.B (eds) 13th ISTTT. Elsevier, London.

A Cross Entropy Based Multi-Agent Approach to Traffic Assignment Problems

Tai-Yu Ma and Jean-Patrick Lebacque

INRETS/GRETIA - 2, Avenue du General-Malleret Joinville, F-94114 Arcueil, France tai-yu.ma@inrets.fr, lebacque@inrets.fr

Summary. In this paper, we propose a Cross Entropy (CE) [1] based multiagent approach for solving static/dynamic traffic assignment problems (TAP). This algorithm utilizes a family of probability distributions in order to guide travelers (agents) to network equilibrium. The route choice probability distribution depends on the average network performance experienced by agents on previous days. Based on the minimization of cross entropy concept, optimal probability distributions are derived iteratively such that high quality routes are more attractive to agents. The advantage of the CE method is that it is based on a mathematical framework and sampling theory, in order to derive the optimal probability distributions guiding agents to the dynamic system equilibrium. Interestingly, we demonstrate that the proposed approach based on CE method coincides with dynamic system approaches. Numerical studies illustrate both nonlinear and bimodal static traffic assignment problems. A comparative study of the proposed method and the dynamic system approach is provided to justify the efficiency of proposed method.

1 Introduction

Solving both static/dynamic traffic assignment problems based on realistic traffic flow models is an important issue in transportation research. Following the Wardrop equilibrium principle, drivers are assumed to swap to cheaper cost routes until the costs of all routes actually used are equal and are less than the costs of unused routes. The traffic assignment problems can be usually formulated as variational inequality, non-linear complementarity or fixed point problems [2]. For one user class static traffic assignment problems, route travel cost is evaluated based on a performance function under monotone, symmetric or separable assumptions. Based on these assumptions, classical convex optimization techniques can be applied for solving these problems without difficulty. However, for dynamic or multimodal traffic assignment problems, the classical solution methods can be no more applicable, given that the analytical form of link cost function is generally difficult to derive and also that the monotonicity assumption is no longer respected in the case of multimodal network [3].

To address this issue, there exist many variant approaches in the literature. We are particularly interested in dynamical system approaches [4–6] for comparative study with the proposed CE based method. The dynamical system approaches are based on a day-to-day route flow adjustment process, for which network flows between the same OD pair are adjusted following some deterministic or stochastic procedure until (Wardrop) user equilibrium is achieved. However, the adjustment process is problematic, especially when there exist more than one class flows between each OD pair. More explicitly, there is no systematical way to adjust all class flows with consideration of the asymmetric effect caused by other class users on the network. Another issue is that the dynamical system approach cannot be applied when there are multiple equilibrium in multimodal network. Indeed, there is no Lyapunov function in this case. On the other hand, the dynamical system approach could be applied to each mode in sequence.

In this work, we derive iteratively the optimal route choice probability distributions towards network equilibrium based on the CE Method, which is a stochastic optimization technique for combinatorial and continuous optimization problems. Based on the minimization of cross entropy concept, optimal probability distributions are derived iteratively such that high quality routes are more attractive to agents. We demonstrate that the proposed CE method can be seen as a stochastic version of the dynamical system approach, which is based on importance sampling and rare event theory. Two numerical examples are illustrated for one and multiple user class static traffic assignment problems. A comparaison study of the proposed method with dynamical system approach is also shown in order to evaluate the efficiency of proposed method.

2 The Cross Entropy Based Multi-Agent Method

The CE method is an adaptive algorithm for estimating probabilities of rare events in complex systems. Based on this method, we can modelize user equilibria as rare events depending on complex stochastic behavior of all travellers. Consider a static traffic assignment or network equilibrium problem, for which a network consists in an origin-destination (OD) pair connected by a set of routes. Agents are located at origins and choose a route stochastically to a destination based on a route choice probability distribution. Note that the probability distribution is related to each OD pair and utilized by agents of the same OD pair. Let C_i denote travel cost of route i. For simplicity of the exposition we assume here that C_i is a function of its demand, i.e. $C_i = C_i(d_i), \forall i \in I$; and I denotes the set of all routes connecting the same OD pair. Let p_i denote the choice probability of route i and $H_i(\gamma)$

denote the performance function of route i, defined by Boltzmann distribution [7],

$$H_i(\gamma) = e^{-C_i(d_i)/\gamma}, \quad \forall i \in I, \tag{1}$$

where γ is the control parameter or temperature. We note that γ controls the swapping force pushing flow to shift from more expensive routes to cheaper ones. Note that as γ decreases, the swapping force increase. The overall expected performance based on the choice probability distribution $\mathbf{p}$ is the expectation of all route performance, defined as:

$$h(\mathbf{p}, \gamma) = \sum_{i \in I} p_i H_i(\gamma), \tag{2}$$

subject to:

$$\sum_{i \in I} p_i = 1, \quad \forall p_i \geq 0 \tag{3}$$

As mentioned above, agents swap from costly routes to cheaper ones in order to minimize their own cost. Following Rubinstein [1], the optimal choice probability distribution towards to cheaper routes at next iteration is the solution of the following optimization problem which minimize the Kullback-Liebler relative entropy between two consecutive probability distributions,

$$\mathbf{p}^{w+1} = \max_{\mathbf{p}} \mathbb{E}_{\mathbf{p}^w}[H(\gamma) \ln \mathbf{p}], \tag{4}$$

subject to equation (3), where w is the index of iteration.

The optimization problem of equation (4) is equivalent to

$$\mathbf{p}^{w+1} = \arg\min_{\mathbf{p}} \sum_{i \in I} p_i^w e^{\frac{C_i(d_i)}{\gamma}} \ln p_i, \tag{5}$$

subject to equation (3).

We can derive the solution of the above equation as:

$$p_i^{w+1} = p_i^w \times \frac{e^{-C_i^w/\gamma^w}}{\sum_{j \in I} p_j^w e^{-C_j^w/\gamma^w}}, \quad \forall i \in I \tag{6}$$

We note that the derived probability distribution depends on the control parameter γ, which must be estimated adaptively in order to converge to the user equilibrium.

As mentioned above, keep γ small will cause more agents swapping to routes which are cheap. However, swapping flow too early will cause traffic congestion on the chosen routes and make them unattractive since the cost-flow relationship is nonlinear. Thus keeping γ too small may renforce routes that turn out to be inefficient. An adaptive parameter estimation technique to obtain optimal γ is to minimize γ, provided that the sum over the differences between two consecutive probability distributions is bounded. This bound is decreased iteratively towards 0 like the cooling technique utilized in simulated annealing method [8]. Thus, we consider the following adaptive estimation of the control parameter γ:

$$\min \gamma^w \text{s.t.} \sum_{i \in I} |p_i^{w+1} - p_i^w| \leq \alpha^w, \tag{7}$$

where p_i^w is defined by equation (5), $\alpha^w = \frac{C}{w}$ a numerical divergent series, the generic term of which converges to 0, and C a constant for setting initial value of γ.

If the first order approximation is applied to exponential function, i.e.

$$e^{-C_i^w/\gamma^w} \simeq 1 - \frac{C_i^w}{\gamma^w}, \quad \text{if } \gamma^w \to \infty \tag{8}$$

We can obtain the following dynamical system which is the same as the system introduced in [6]:

$$\dot{\mathbf{p}} = -\mathbf{p}(C(\mathbf{p}) - \overline{C}(\mathbf{p})), \tag{9}$$

where $\overline{C}(\mathbf{p}) = \mathbb{E}_{\mathbf{p}} C(\mathbf{p})$ denotes the expectation of travel cost with respect to the probability distribution $\mathbf{p}$.

We note that at Wardrop equilibrium, the derived probability distribution is stable. i.e.

$$p_i^{w+1} = p_i^w, \quad \text{if } C_i^w = \min_{j \in I} C_j^w, \tag{10}$$

which coincides with the result obtained by the dynamical system approach. Note that the classical CE method has been tailored for cost functions independent of demand, which is why it diverges from the adaptation presented here.

Finally, let us investigate its impact on the derived probability changes of γ tending to ∞ and 0, respectively.

In the case of $\gamma^w \to \infty$, we obtain the following results

$$\lim_{\gamma^w \to \infty} (p_i^{w+1} - p_i^w) = 0, \quad \forall w \tag{11}$$

On the contrary, when $\gamma^w \to 0$, we obtain

$$\lim_{\gamma^w \to 0} (p_i^{w+1} - p_i^w) = -p_i^w \left(1 - \frac{1}{\sum_{j \in I} p_j^w e^{\frac{C_i^w - C_j^w}{\gamma^w}}} \right) = \begin{cases} -p_i^w & \text{if } C_i^w > \min_{j \in I} C_j^w \\ -\infty & \text{otherwise} \end{cases} \tag{12}$$

3 The Algorithm Based on CE Method

The algorithm for solving general TAPs consists of an iterative procedure which updates adaptively the probability distributions and the control parameters until the desired result is obtained. The algorithm is described as follows:

Main algorithm

1. *Initialization*
   a) Initialize the parameter α^w according to the magnitude of route choice set and set iteration index $w = 0$.
   b) Initialize route choice probability according to a uniform probability $p_i^w = 1/|I_k|, \forall i \in I_k, \forall k \in K$, where I_k is the set of routes connecting the same O/D pair k, K is the set of all OD pairs.
2. *Compute travel cost*
   Compute travel costs of all routes with respect to agents' choice. In the static case, travel cost can be evaluated by cost-flow functions. However, in the dynamic case, agents' experienced travel costs can be computed with respect to analytical or simulation-based traffic flow models. Different traffic flow models can be utilized, such as first-order model [9, 10], the generic second-order traffic flow model [11, 12] or the cell transition model [13]. Moreover, different alternatives with related cost specifications can be taken into account, e.g. mode choice, departure time choice or destination choice etc.
3. *Update the choice probability distribution*
   For adaptive estimation of γ, we utilize the interpolation method by bounding $0 \leq \gamma \leq M$, where M is a constant. The optimal γ can be deduced from equation 7. The derived optimal probability distribution can be thus evaluated by equation 6.
4. *Stopping criteria*
   When $w = w_{max}$ or the derived choice probability distributions are stable, stop; otherwise go to step 2.

We note that for multimodal traffic assignment problems, there is one user equilibrium per class of users, given the other classes of users' choice. Accordingly, the probability distributions utilized by agents of different classes should be distinguished.

4 Numerical Studies

In this section, we study two static traffic assignment problems with respect to one and multiple classes. The first example is shown in Fig. 1 consisting of an OD pair with three links [6]. The nonlinear cost-flow functions are the following:

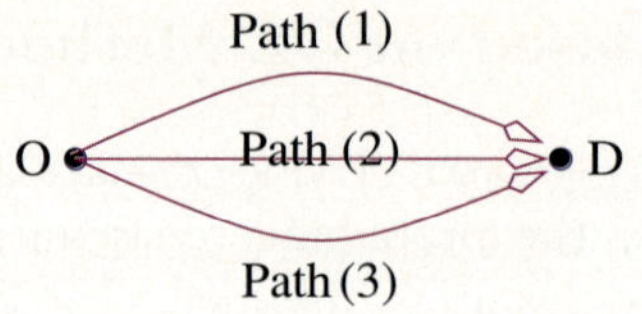

Fig. 1. A static mono-modal network.

$$
\begin{aligned}
t_1 &= 10(1.0 + 0.15(\frac{x_1}{2})^4) \\
t_2 &= 20(1.0 + 0.15(\frac{x_2}{4})^4) \\
t_3 &= 25(1.0 + 0.15(\frac{x_3}{3})^4)
\end{aligned}
\tag{13}
$$

$$x_1 + x_2 + x_3 = 10$$

where x_i is the route flow on link/path i and the travel demand is 10. We initialize the probability distribution for route (link) choice as uniform. The evolution of the probability distribution is shown in Fig. 2. We see that after 15 iterations, the probability distribution for link choice becomes stable, and the link cost over all links is almost equal, as shown in Fig. 3. The estimates of γ oscillate after the solution nears the network equilibrium, as shown in Fig. 4.

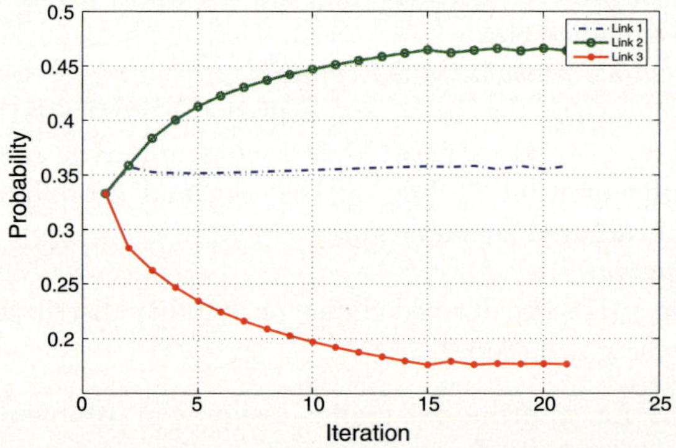

Fig. 2. Evolution of link choice probability.

The second example concerns a bimodal traffic assignment problem and is described in Fig. 5. Two classes of users, e.g. car and bus, are present in a two link/path network. The linear flow-cost function is given by [3]:

$$
\begin{aligned}
t_{ai} &= 1.5x_{ai} + 5x_{bi} + 30, \forall i \in \{1,2\} \\
t_{bi} &= 1.3x_{ai} + 2.6x_{bi} + 28, \forall i \in \{1,2\}
\end{aligned}
\tag{14}
$$

$$d_a = 16, d_b = 4$$

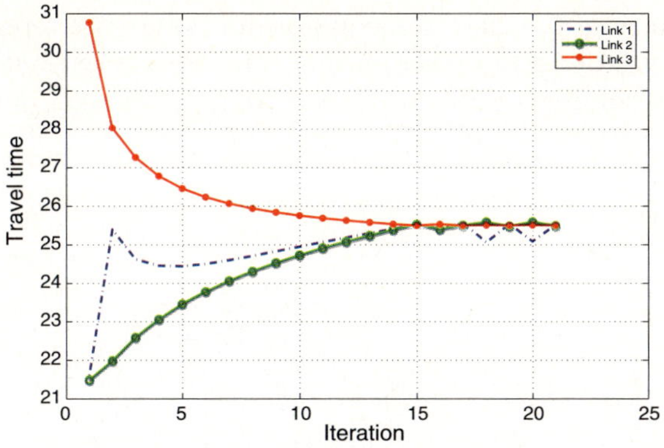

Fig. 3. Evolution of link travel time.

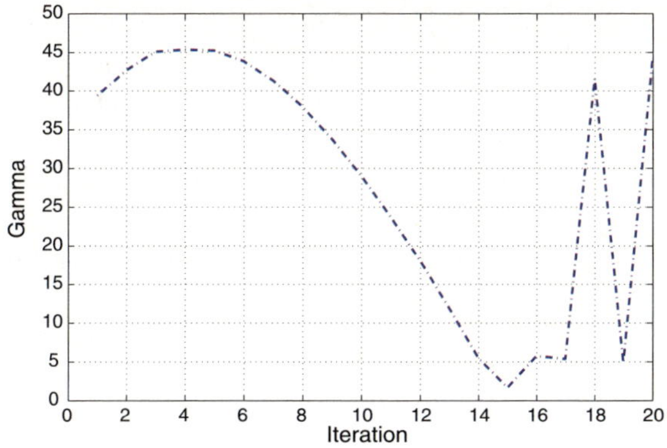

Fig. 4. Evolution of control parameter γ.

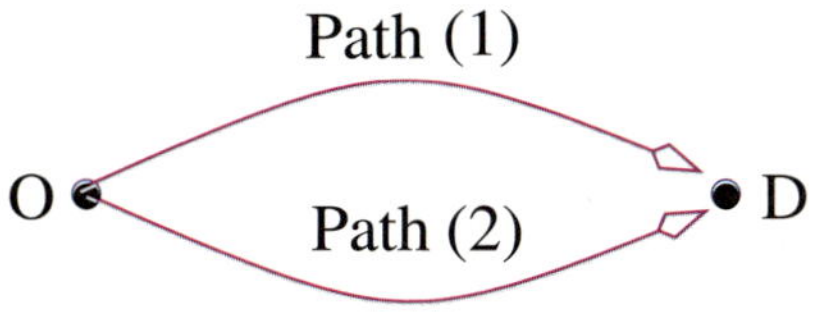

Fig. 5. A static bimodal network.

where t_{ki} is the travel time of class k on link i, x_{ki} is the number of class k on link i. d_k is the demand of class k. To treat this example, we introduce two probability distributions with respect to the two user classes. Accordingly, two different control parameters γ need to be estimated. We initialize the probability distribution as a uniform distribution for each class. Similarly to the previous example, we obtain convergence towards a user equilibrium. The Nagurney field has as components the differences between path costs per mode on link (1) and (2) as functions of the modal flows on link (1). The Nagurney field is shown in Fig. 6, where the equilibrium points are represented by red stars. The flow vector far from equilibrium converges to the stable equilibria at the up-left corner or down-right corner according to different initial conditions. The unstable equilibrium can be found at the center with flows being 8 and 2 for class a and b.

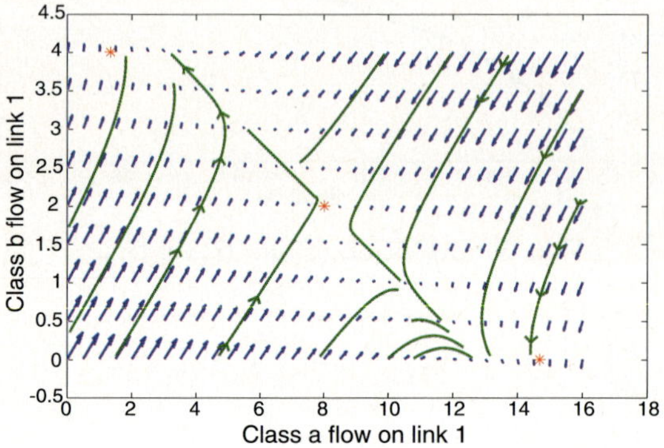

Fig. 6. The velocity vectors associated to the Nagurney field on link 1.

5 Comparison with Dynamical System Approach

In this section, we are interested in comparing our method with other approaches. Note that the projection-type method [5] attempts iteratively to a find a fixed point but its convergence speed is slow (linear or less). For this reason, we choose the dynamical system approach to compare our proposed CE method. The network is composed of 8 nodes and 12 links with two origins and two destinations, as shown in Fig. 7. The OD travel demand equals to 1000 for each OD. The nonlinear cost functions are defined as follows:

$$\begin{aligned} t_1 &= 20(1.0+0.15(\frac{x_1}{4})^4) & t_7 &= 10(1.0+0.15(\frac{x_7}{2})^4) \\ t_2 &= 15(1.0+0.15(\frac{x_2}{4})^4) & t_8 &= 20(1.0+0.15(\frac{x_8}{4})^4) \\ t_3 &= 25(1.0+0.15(\frac{x_3}{3})^4) & t_9 &= 25(1.0+0.15(\frac{x_9}{3})^4) \\ t_4 &= 10(1.0+0.15(\frac{x_4}{2})^4) & t_{10} &= 90(1.0+0.15(\frac{x_{10}}{2})^4) \\ t_5 &= 20(1.0+0.15(\frac{x_5}{4})^4) & t_{11} &= 20(1.0+0.15(\frac{x_{11}}{4})^4) \\ t_6 &= 5(1.0+0.15(\frac{x_6}{3})^4) & t_{12} &= 35(1.0+0.15(\frac{x_{12}}{3})^4) \end{aligned} \quad (15)$$

$$d_k = 1000, \forall k \in K,$$

where x_i denotes number of travelers, k an OD pair. The network equilibrium solutions obtained by these two approaches (CE and dynamic system) are very close for all OD pairs. The evolution of path costs, connecting the same OD pair (O1-D1), obtained by these two methods is shown in Fig. 8. The CE method converges to equilibrium faster than the dynamical system approach.

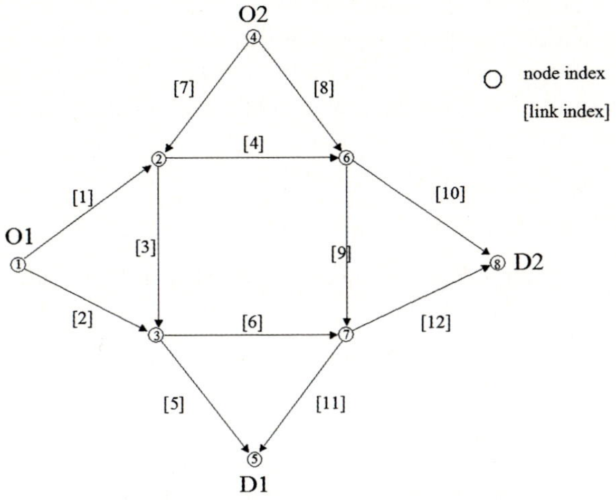

Fig. 7. The multiple OD network.

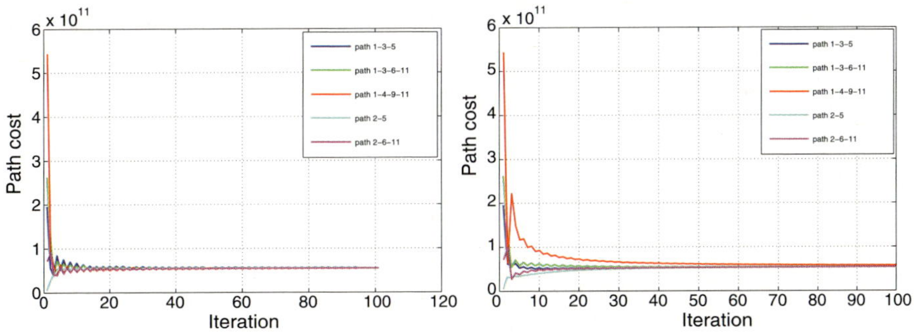

Fig. 8. Evolution of path costs connecting origin O1 and destination D1 based on CE method (left) and dynamical system approach [4] (right).

6 Conclusion

In this study, a fast convergent CE based multi-agent approach for solving the general static/dynamic traffic assignment problem has been proposed. Based on the CE method, the adaptive route flow shifts are derived, resulting from travel cost realized on previous day. We have also shown that in a multimodal network, given that the monotonicity of the travel cost function does not hold, multiple user equilibrium solutions can be found quickly with different initial conditions. The results with comparison to the dynamical system approach show that the proposed CE method approach converges more quickly. It is also better adapted to complex cost functions, as in the dynamic assignment problems based on traffic flow models.

We have provided the main algorithm for solving general static/dynamic traffic assignment problems. Further applications in large scale multimodal network based on realistic traffic flow models are actually under study.

References

1. Rubinstein RY (1999) The Cross-Entropy Method for Combinatorial and Continuous Optimization. Methodology and Computing in Applied Probability, 1(2):127–190.
2. Patriksson M (1994) The Traffic Assignment Problem–Models and Methods. VSP, Utrecht, The Netherlands.
3. Wynter L (2001) A convergent algorithm for the multimodal traffic equilibrium problem. Research report. No. 4125, INRIA, France.
4. Smith MJ (1984) The Stability of Dynamic Model of Traffic Assignment: An Application of Method of Lyapunov. Transportation Science, 18(3):245–252.
5. Nagurney A, Zhang D (1996) Projected Dynamical Systems and Variational Inequalities with Applications. Kluwer, Boston.
6. Jin WL (2007) A dynamical system model of the traffic assignment problem. Transportation Research B 41:32–48.
7. Helvik BE, Wittner O (2001) Using the Cross-Entropy Method to Guide/Govern Mobile Agent's Path Finding in Networks. Proc. of 3rd International Workshop on Mobile Agents for Telecommunication Applications.
8. Aarts EHL, Korst JMH (1989) Simulated Annealing and Boltzmann Machines: A Stochastic Approach to Combinatorial Optimization and Neural Computing. John Wiley & Sons.
9. Buisson C, Lebacque JP, Lesort JB (1995) Macroscopic modelling of traffic flow and assignment in mixed networks. In: Pahl PJ, Werner H (eds) Proc. of the Berlin ICCCBE Conf. 1367-1374.
10. Lebacque JP, Khoshyaran MM (1999) Modelling vehicular traffic flow on networks using macroscopic models. In Proc. of the FVCA II, 551-558.
11. Lebacque JP, Mammar S, Haj-Salem, H (2007) Generic second order traffic flow modelling. In Proc. of the 17th ISTTT (International Symposium on Transporation and Traffic Theory), London.
12. Lebacque JP, Mammar S, Haj-Salem H (2006) Second-order Traffic Flow Modeling : Comments on Rascle and Zhang's Models. Future appears in Transportation Research B.
13. Daganzo CF (1994) The cell transmission model: a simple dynamic representation of highway traffic. Transportation Research B 28:269–287.

Handling of Contacts in Crowd Motion Simulations

Bertrand Maury and Juliette Venel

Laboratoire de Mathématiques, Université Paris-Sud, 91405 Orsay Cedex, France
Bertrand.Maury@math.u-psud.fr, Juliette.Venel@math.u-psud.fr

1 Introduction

This paper is concerned with the modelling of crowd motion in highly packed configurations. We are especially interested in describing evacuation situations: people want to escape from a room, a building, a railway station or a plane, that may contain obstacles (walls, seats, tables, ...).

Many strategies have been followed to model crowd motions: macroscopic models [1–3], Cellular Automata [4–6], queuing models [7, 8] and microscopic models [9, 10]. The possibility that people may actually get into contact is usually handled indirectly (repulsive forces or adapted cut-off for microscopic models, exclusion principle for Cellular Automata or queuing models). We propose here to integrate a strong non-overlapping constraint in a ODE framework, in the spirit of granular flow models.

Our approach rests on two principles. On the one hand, we define a spontaneous velocity, which corresponds to the velocity each individual would like to have in the absence of other people. On the other hand, individuals (which are identified to rigid discs) must obey a non-overlapping constraint. Those two principles lead us to define the actual velocity field as the projection of the spontaneous velocity over the set of admissible velocities (regarding the non-overlapping constraints). To perform this projection, we put the problem in a saddle-point form, which leads us to introduce a collection of Lagrange multipliers. Those Lagrange multipliers can be interpreted as interaction pressure between people which are in contact.

2 Modelling

We consider N persons identified with rigid disks of common radius r, in a room represented by a domain $\Omega \subset \mathbb{R}^2$. The centre of the i-th disk is denoted by $\mathbf{q}_i$. We define the set of configurations:

$$Q = \{\mathbf{q} = (\mathbf{q}_1, \mathbf{q}_2, \ldots, \mathbf{q}_N) \in \mathbb{R}^{2N}\}.$$

Moreover, we introduce a spontaneous velocity field

$$\mathbf{U} = (\mathbf{U}_1, \ldots, \mathbf{U}_N),$$

where $\mathbf{U}_i$ represents the velocity which person i would like to have if he were alone. As a first step, we consider here the simplest model one may think of: all individuals have the same behaviour, and they do not elaborate complex strategies to escape. We therefore introduce a global spontaneous velocity field $\mathbf{x} \mapsto \mathbf{U}_0(\mathbf{x})$, and we write

$$\mathbf{U}(\mathbf{q}) = (\mathbf{U}_0(\mathbf{q}_1), \ldots, \mathbf{U}_0(\mathbf{q}_N)).$$

In this hard-sphere approach, overlapping is strictly forbidden, which leads to the following set of feasible configurations:

$$Q_0 = \{\mathbf{q} \in Q, D_{ij}(\mathbf{q}) = |\mathbf{q}_j - \mathbf{q}_i| - 2r \geq 0 \quad \forall i \neq j\}.$$

As overlapping is forbidden, two persons already in contact can only increase their distance: the set of feasible velocities is

$$\mathcal{C}_{\mathbf{q}} = \{\mathbf{v}\,,\ \mathbf{G}_{ij}(\mathbf{q}) \cdot \mathbf{v} \geq 0 \quad \text{as soon as} \quad D_{ij}(\mathbf{q}) = 0\}\,,$$

with (see Fig. 1)

$$\mathbf{G}_{ij} = \nabla D_{ij} = (0, \ldots, 0, -\mathbf{e}_{ij}, 0, \ldots, 0, \mathbf{e}_{ij}, 0, \ldots, 0) \in \mathbb{R}^{2N}.$$

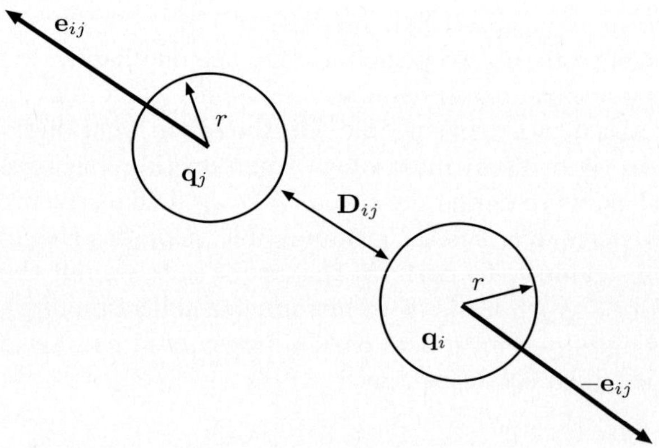

Fig. 1. Notations.

We may now state the basic form of the model: the actual velocity field is the feasible field which is the closest to $\mathbf{U}$ for the euclidean distance, which writes

$$\frac{d\mathbf{q}}{dt} = P_{\mathcal{C}_\mathbf{q}}\mathbf{U}(\mathbf{q}),$$

where $P_{\mathcal{C}_\mathbf{q}}$ denotes the euclidean projection onto the closed convex cone $\mathcal{C}_\mathbf{q}$.

Remark 1. We focus in this note on the handling of contacts, limiting our scope to the simplest behavioral model, but more sophisticated strategies can be integrated in the approach. In this spirit, the spontaneous velocity of a person can be made dependent upon its own characteristics (like personality or physical force), and the position of its neighbours (according to some self-optimization strategy, or social rules). It amounts to write the global velocity field in the general form

$$\mathbf{U}(\mathbf{q}) = (\mathbf{U}_1(\mathbf{q}), \ldots, \mathbf{U}_N(\mathbf{q})).$$

Note that the dependence is likely to be non-smooth (it involves discrete decision processes), so that the mathematical results presented in the next section could be affected. From the numerical point of view though, those modifications can be implemented straightforwardly, and do not change the computational times.

3 Mathematical Framework

Despite its formal simplicity, this model does not fit directly into a standard framework. Let us reformulate the problem by introducing $\mathcal{N}_\mathbf{q}$, the outward normal cone to the set of feasible configurations Q_0

$$\mathcal{N}_\mathbf{q} = \mathcal{C}_\mathbf{q}^o = \{\mathbf{w}\,,\ (\mathbf{w}, \mathbf{v}) \leq 0 \quad \forall \mathbf{v} \in \mathcal{C}_\mathbf{q}\}\,.$$

Thanks to Farkas' Lemma, this cone can be expressed

$$\mathcal{N}_\mathbf{q} = \left\{-\sum \lambda_{ij}\mathbf{G}_{ij}\,,\ \lambda_{ij} \geq 0\,,\ D_{ij}(\mathbf{q}) = 0 \Longrightarrow \lambda_{ij} = 0\right\}.$$

Now using the classical orthogonal decomposition of a Hilbert space as the sum of mutually polar cones (see [11]), we obtain

$$\frac{d\mathbf{q}}{dt} = P_{\mathcal{C}_\mathbf{q}}\mathbf{U}(\mathbf{q}) = \mathbf{U}(\mathbf{q}) - P_{\mathcal{N}_\mathbf{q}}\mathbf{U}(\mathbf{q}).$$

As a consequence,

$$\frac{d\mathbf{q}}{dt} + \mathcal{N}_\mathbf{q} \ni \mathbf{U}(\mathbf{q}). \tag{1}$$

The problem takes the form of a differential inclusion, which has motivated a huge amount of papers in the last decades. Let us first consider a special situation where standard theory can be applied. Consider N individuals in a corridor (see Fig. 2). In that case, as people are not likely to leap accross each

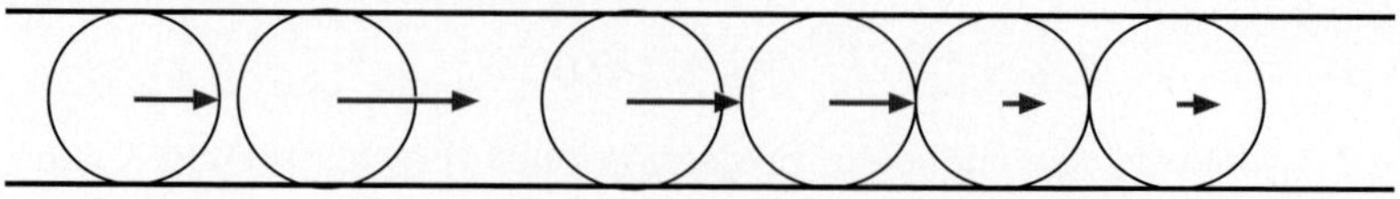

Fig. 2. One dimensional situation.

other, it is natural to restrict the set of feasible configurations to one of its connected components:

$$Q_0 = \{\mathbf{q} = (q_1, \ldots, q_N)\,,\; q_{i+1} - q_i \geq 2r\}.$$

In this very situation, as Q_0 is closed and convex, the multivalued operator $\mathbf{q} \longmapsto \mathcal{N}_{\mathbf{q}}$ identifies to the subdifferential of the indicatrix function of Q_0:

$$\partial I_{Q_0}(\mathbf{q}) = \{\mathbf{v}, I_{Q_0}(\mathbf{q}) + (\mathbf{v}, \mathbf{h}) \leq I_{Q_0}(\mathbf{q}+\mathbf{h}) \quad \forall \mathbf{h}\}\,,\; I_{Q_0}(\mathbf{q}) = \left| \begin{array}{ll} 0 & \text{if } \mathbf{q} \in Q_0 \\ +\infty & \text{if } \mathbf{q} \notin Q_0 \end{array} \right.$$

therefore $\mathbf{q} \longmapsto \mathcal{N}_{\mathbf{q}}$ is maximal monotone. As soon as the spontaneous velocity is regular (i.e. Lipschitz), standard theory (see e.g. Brezis [12]) ensures well-posedness.

In the general case, Q_0 is not convex, so that $\mathbf{q} \longmapsto \mathcal{N}_{\mathbf{q}}$ is not maximal monotone. Despite the formal simplicity of the model, its analysis calls for sophisticated abstract tools developed recently by Edmond and Thibault [13, 14]. The well-posedness rests on the fact that the set of feasible configurations is prox-regular (see [15]), which means that the projection onto Q_0 is well-defined in its neighbourhood. As a consequence, it can be established that the problem is well-posed.

4 Numerical Simulation

4.1 Time-Stepping Scheme

The numerical scheme we propose is based on a first order expansion of the constraints expressed in terms of velocities. We introduce a uniform sequence of time steps

$$t^0 = 0 < t^1 < \cdots < t^p = T\,,\; t^{n+1} - t^n = T/p,$$

and we denote by $\mathbf{q}^n$ the approximation of $\mathbf{q}(t^n)$. The next configuration is obtained as

$$\mathbf{q}^{n+1} = \mathbf{q}^n + h\mathbf{u}^{n+1},$$

where $\mathbf{u}^{n+1}$ minimizes

$$\frac{1}{2}\,|\mathbf{v} - \mathbf{U}(\mathbf{q}^n)|^2 \text{ over } \mathcal{C}^h_{\mathbf{q}^n}\,, \quad \text{with}$$

$$\mathcal{C}_{\mathbf{q}}^h = \{\mathbf{v}\,,\ D_{ij}(\mathbf{q}) + h\mathbf{G}_{ij}(\mathbf{q}) \cdot \mathbf{v} \geq 0\}.$$

In other words,

$$\mathbf{u}^{n+1} = P_{\mathcal{C}_{\mathbf{q}^n}^h}(\mathbf{U}(\mathbf{q}^n)).$$

This scheme can be shown to converge to the exact solution:

Theorem 1. *Let us denote by $\mathbf{q}_h$ the continuous, piecewise linear function associated to the numerical scheme. Then $\mathbf{q}_h$ goes to $\mathbf{q}$ uniformly in $[0,T]$, where $t \mapsto \mathbf{q}(t)$ is the exact solution.*

Proof. The proof is technical and lengthy, and we shall detail it in a forthcoming paper.

The scheme can be interpreted in the following way. Let us introduce the set

$$Q_0^h(\mathbf{q}) = \{\tilde{\mathbf{q}}\,,\ D_{ij}(\mathbf{q}) + \mathbf{G}_{ij}(\mathbf{q}) \cdot (\tilde{\mathbf{q}} - \mathbf{q}) \geq 0\},$$

which can be seen as an inner convex approximation of Q_0. Note that Q_0^h does not depend on h. Yet we keep this notation to emphasize the fact that it is an approximation of Q_0. It is straightforward to check that

$$\frac{\mathbf{q}^{n+1} - \mathbf{q}^n}{h} + \partial I_{Q_0^h(\mathbf{q}^n)}\left(\mathbf{q}^{n+1}\right) \ni \mathbf{U}(\mathbf{q}^n),$$

so that the scheme can be seen as a semi-implicit discretization of (1), where $\partial I_{Q_0^h(\mathbf{q}^n)}$ approximates $\mathcal{N}_{\mathbf{q}^n}$.

4.2 Projection Step

The costly part of the algorithm lies in the projection of the spontaneous velocity onto the approximated set of feasible velocities. We propose here to solve this projection by a Uzawa algorithm (note that any algorithm could be used to perform this task). This algorithm (see e.g. [16]) is based on a reformulation of the projection problem onto a saddle-point form:

$$\begin{cases} \mathbf{u} + B^\star \boldsymbol{\lambda} & = \mathbf{U} \\ B\mathbf{u} & \leq \mathbf{D} \\ (B\mathbf{u} - \mathbf{D}, \boldsymbol{\lambda}) & = 0. \end{cases}$$

with

$$\mathbf{U} = \mathbf{U}(\mathbf{q}^n)\,,\ B\mathbf{v} = (-h\mathbf{G}_{ij}(\mathbf{q}^n) \cdot \mathbf{v})_{i<j}\,,\ \mathbf{D} = (D_{ij}(\mathbf{q}^n))_{i<j},$$

$$B^\star \boldsymbol{\lambda} = -h \sum_{i<j} \lambda_{ij} \mathbf{G}_{ij}(\mathbf{q}^n).$$

The Uzawa algorithm produces a sequence $\boldsymbol{\lambda}^k$ according to

$$\boldsymbol{\lambda}^{k+1} = \Pi_+ \left(\boldsymbol{\lambda}^k + \rho\left(B(\mathbf{U} - B^\star\boldsymbol{\lambda}^k) - \mathbf{D}\right)\right),$$

where Π_+ is the euclidean projection onto the cone of vectors with nonnegative components (a simple cut-off in practice), and $\rho > 0$ is a fixed parameter. The algorithm can be shown to converge as soon as $0 < \rho < 2/\|B\|^2$ (see [16]).

Remark 2. The auxiliary quantities λ_{ij} appear as numerical ways to handle the corresponding constraints. They quantify the way $\mathbf{U}$, the spontaneous velocity field, does not fit the constraints, and as such they can be interpreted in terms of pressures undergone by individuals. Although it would be presumptuous at this stage to assimilate λ_{ij} to an actual measure of the discomfort experienced by persons i and j, it is obvious that high values for those Lagrange multipliers can be expected on zones where people are likely to be crushed.

5 Numerical Experiments

5.1 Counter flowing crowds

The first set of experiments is based on the following situation: we consider two populations of individuals in a periodic bidimensional domain. They are represented in black (B) and white (W) in the figures. B-individuals want to go to the right, and W-individuals tend to go in the opposite direction (with the same desired speed 1). The simplest model we proposed at the beginning of this paper, without any avoiding strategies, makes it possible to recover the so-called fingering pattern (see [10, 17]). The number of inviduals in each population is 750. Figure 3 represents snapshots of the two populations, at times 0 (random distribution), 25, 75, and 100. Note on the second figure from the top, during the transitory period, the apparition of white and black clusters, due to absence of any social force or auto-optimization strategy in this test.

5.2 Exit from a Building

The second experiment we performed correspond to the situation of 3000 persons which are randomly distributed in a room at initial time. The spontaneous velocity field has constant modulus 1, and is directed along streamlines of the geodesic distance to a safe spot, far away from the room. We represented (see Fig. 4), at times 3, 7, 65, and 110 the current configuration and the corresponding network of interaction pressures. For any couple of discs in contact, we represent the segment between centers, having its color (from white to black) depend upon the (positive) Lagrange multiplier which handles the corresponding constraint. We recover the apparition of arches upstream the exit (see [17, 18]).

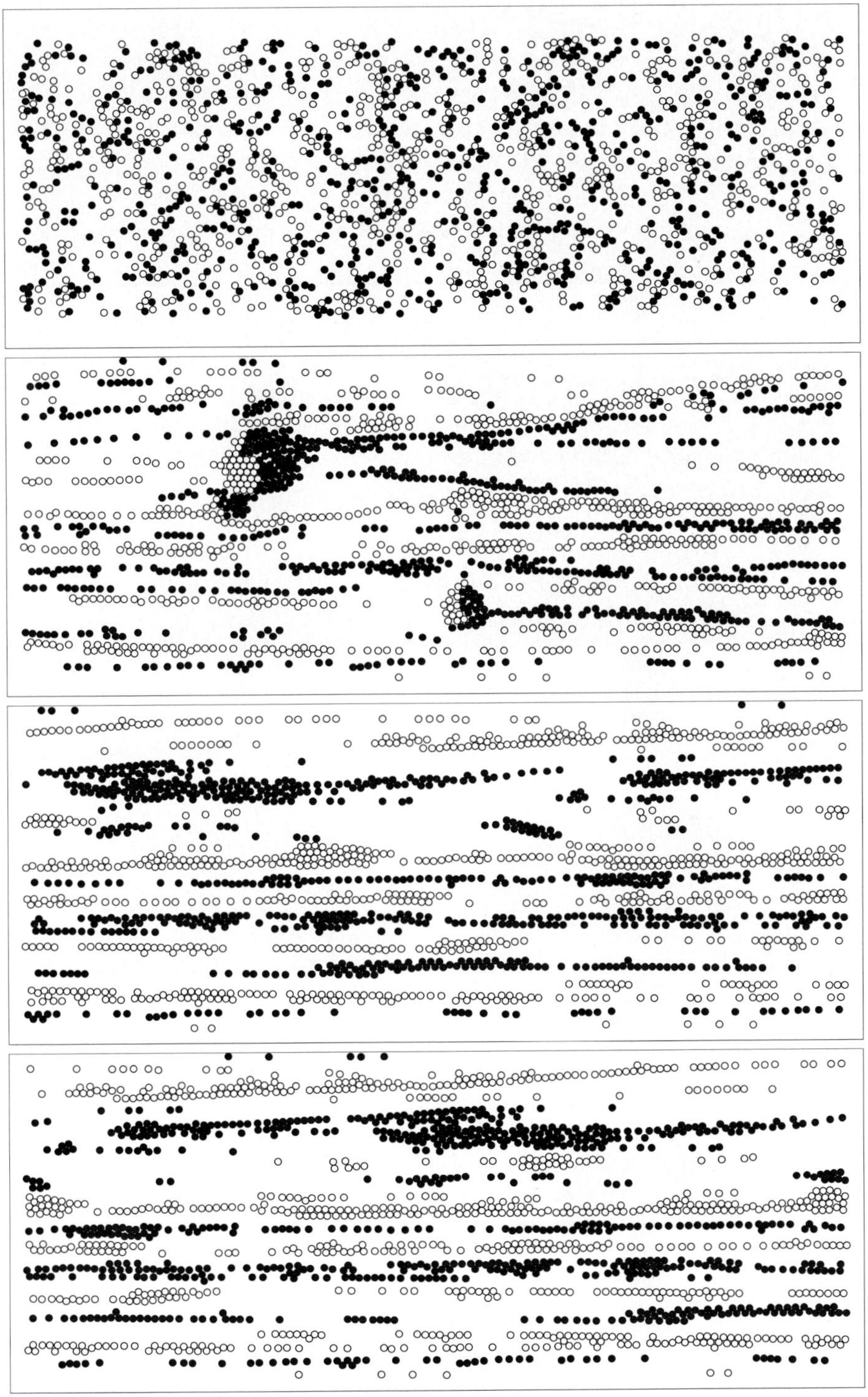

Fig. 3. Counter flowing crowds.

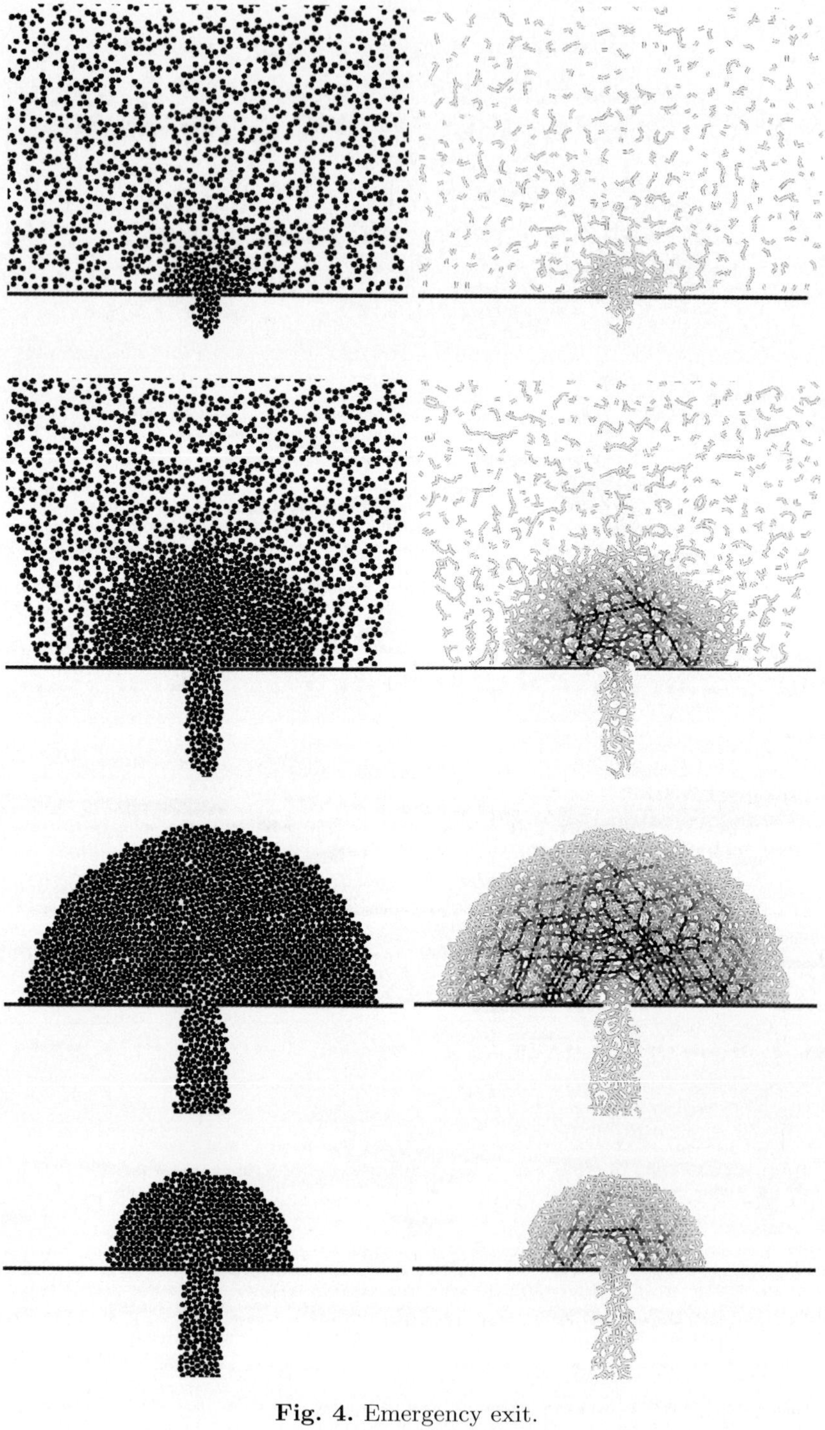

Fig. 4. Emergency exit.

References

1. L.F. Henderson, *On the fluid mechanics of human crowd motion*, Transp. Res. 8, 509-515, 1974.
2. D. Helbing, *A fluid-dynamic model for the movement of pedestrians*, Complex Systems 6, 391-415, 1992.
3. S.P. Hoogendoorn, P.H.L. Bovy, *Dynamic user-optimal assignment in continuous time and space*, Transp. Res. B 38, 571-592, 2004.
4. V. Blue, J.L. Adler, *Cellular automata microsimulation for modeling bi-directional pedestrian walkways*, Transp. Res. B 35, 293-312, 2001.
5. A. Kirchner, A. Schadschneider, *Simulation of evacuation processes using a bionics-inspired cellular automaton model for pedestrians dynamics*, Physica A 312, 260-276, 2002.
6. C. Burstedde, K. Klauck, A. Schadschneider, J. Zittartz, *Simulation of pedestrian dynamics using a two-dimensional cellular automaton*, Physica A 295, 507-525, 2001.
7. G.G. Lovas, *Modelling and simulation of pedestrian traffic flow*, Transp. Res. B 28, 429-443, 1994.
8. S.J. Yuhaski, J.M. MacGregor Smith, *Modelling circulation systems in buildings using state dependent queueing models*, Queueing Systems 4, 319-338, 1989.
9. P.G. Gipps, B. Marksjö, *A micro-simulation model for pedestrian flows*, Math. Comp. Simul. 27, 95-105, 1985.
10. D. Helbing, P. Molnár, *Social force model for pedestrians dynamics*, Phys. Rev. E 51, 4282-4286, 1995.
11. J.J. Moreau, *Décomposition orthogonale d'un espace hilbertien selon deux cônes mutuellement polaires*, C. R. Acad. Sci. 255, Série I, 238-240, 1962.
12. H. Brezis, *Opérateurs Maximaux Monotones et Semi-groupes de contractions dans les espaces de Hilbert*, 1973.
13. J.F. Edmond, L. Thibault, *Relaxation of an optimal control problem involving a perturbed sweeping process*, Math. Program. 104, no. 2-3, Ser. B, 347-373, 2005.
14. J.F. Edmond, L. Thibault, *BV solutions of nonconvex sweeping process differential inclusion with perturbation*, J. Differential Equations 226, no.1, 135-179, 2006.
15. G. Colombo, M.D.P. Monteiro Marques, *Sweeping by a continuous prox-regular set*, J. Differential Equations 187, no. 1, 46-62, 2003.
16. P.G. Ciarlet, *Introduction à l'analyse numérique matricielle et l'optimisation*, Masson, Paris, 1990.
17. D. Helbing, T. Vicsek, *Optimal self-organization*, New J. Phys. 1, 13.1-13.17, 1999.
18. D. Helbing, I. Farkas, T. Vicsek, *Simulating dynamical features of escape panic*, Nature 407, 487, 2000.
19. D. Helbing, M. Isobe, T. Nagatani, K. Takimoto, *Lattice gas simulation of experimentally studied evacuation dynamics*, Phys. Rev. E 67, 067101, 2003.
20. B. Maury, *A time-stepping scheme for inelastic collisions*, Numerische Mathematik 102, no. 4, 649-679, 2006.
21. T. Meyer-König, H. Klüpfel, and M. Schreckenberg, *Assessment and analysis of evacuation processes on passenger ships by microscopic simulation*, Pedestrian and Evacuation Dynamics (PED), Springer, Berlin, 2002.

22. J.J. MOREAU, *Evolution Problem associated with a moving convex set in a Hilbert space*, J. Differential Equations 26, no. 3, 347-374, 1977.
23. L. THIBAULT, *Sweeping Process with regular and nonregular sets*, J. Differential Equations 193, no. 1, 1-26, 2003.

Optimal Velocity Model Via Considering Multi-Velocity Difference Ahead

Mo Yeliu[1], Yu Xue[1,2], Shi Wei[1], and Liu Zhang[1]

[1] Institute of physical science and engineering, Guangxi University, Nanning, 530004, China yuxuegxu@gxu.edu.cn
[2] Shanghai Institute of Applied Mathematics and Mechanics, Shanghai University, Shanghai, 200072, China

1 Introduction

All the time, traffic problems have attracted extensive attention of scientists because of the appearance of complex phenomena. i.e. there exist the observed non-equilibrium phase transitions and various nonlinear dynamical phenomena in the process of traffic congestion. There have been many attempts to construct traffic models, such as cellular automaton models, car-following models, gas kinetic models and hydrodynamic models, to explain the mechanism of jams from the viewpoint of physics [1]. The classic optimal velocity (OV) model is a kind of car-following model proposed by Bando *et al.* [2] in 1995. The dynamical equation of the optimal velocity model is obtained as follows

$$\ddot{x}_j(t) = a(V(\Delta x_j(t)) - \dot{x}_j(t)) \tag{1}$$

where $x_j(t)$ is position of the jth car at time t, $\Delta x_j(t)$ is the headway of the jth car at time t and a is the driver's sensitivity. The optimal velocity model assumed that a driver can obtain the vehicle's acceleration via adjusting a vehicle's optimal velocity $V(\Delta x_j(t))$. The OV model is a very simple car-following model and has succeeded in reproducing the dynamical formation of congestion. However, OV model can not avoid crash at values of the sensitivity smaller than about one, $a \leq 1s^{-1}$ for it shows unrealistic values of the acceleration for various optimal functions [3]. Recently, Lenz *et al.* [4] and Wilson *et al.* [3] analyzed the characteristics of the optimal velocity model and proposed the multiple look-ahead models with many-neighbor and non-locality to implement the OV model's drawbacks, in which Wilson *et al.* [3] presented three multiple look-ahead models with many-neighbor and non-locality. In what follows, for simplicity, we call the three models as the model A, B and C, respectively.
Model A:

$$\ddot{x}_j(t) = a(\sum_{l=1}^{n} \gamma_l V(\Delta x_{j+l-1}(t)) - \dot{x}_j(t)) \tag{2}$$

Model B:

$$\ddot{x}_j(t) = a(\sum_{l=1}^{n} \gamma_l V(\frac{1}{l}\sum_{p=1}^{l} \Delta x_{j+p-1}(t)) - \dot{x}_j(t)) \tag{3}$$

Model C:

$$\ddot{x}_j(t) = a(V(\sum_{l=1}^{n} \gamma_l \Delta x_{j+l-1}(t)) - \dot{x}_j(t)) \tag{4}$$

where γ_l is a weight factor to denote the driver to notice the behaviours of many vehicle ahead in the different manner. In addition, Ge *et al.* [5] has also proposed a multiple look-ahead model similar to equation (4) in the form of difference equation via considering the effect of multi-headway ahead on optimal velocity. Their studies indicates that the properties with many-neighbor and non-locality is beneficial to enhance the stability of traffic flow. However, we have carefully performed the simulation at values of the sensitivity smaller than about one, $a < 1s^{-1}$ and found that the multiple look-ahead model can not avoid a negative velocity or to move backward, which it shows unrealistic and unallowed in real traffic. We will discuss these drawbacks and remove the defects via introducing multiple relative velocity in section 3. Although Herman *et al.* [6] has ever observed that variation of multiple car relative velocity ahead affected the current drivers in real traffic, the OV model (1) and all of the multiple look-ahead models do not contain a driver response to the velocity difference $\Delta v_j(t)$ (relative velocity) with respect to the preceding car. As early as 1961, Gazis *et al.* [7] have proposed the following stimulus-response equation via considering the velocity difference in a single lane.

$$\ddot{x}_j(t+T) = \kappa \frac{v_j(t+T)^{\eta}}{\Delta x_j^{\rho}} \Delta v_j(t) \tag{5}$$

Here, κ is the driver's sensitivity to relative velocity, η and ρ is an exponent of velocity and headway, respectively. But, this equation can not avoid to obtain the larger acceleration, which exceeds one in real traffic [8]. Helbing *et al.* have considered that the relative velocity $\Delta v_j(t) = v_{j+1}(t) - v_j(t)$ enables to enhance the stability of traffic flow in Ref. [8]. Xue *et al.* [9] also extended the OV model to take the effect of the relative velocity into account and its stability is superior to the original OV model.

In this paper, we will combine the multiple look-ahead models with Gazis et al.'s stimulus-response equation by introducing multi-velocity-differences ahead.We would like to address whether or not the extended model displays marked improvement of stability of traffic flow. The linear stability analysis indicates that the effect of multiple velocity-ahead differences enhance the stability of traffic in great measure, and is beneficial to reduce the excess energy consumption of the whole flow in respect to energy consumption.

2 Model and Stability Analysis

Based on multiple look-ahead model, the stimulus-response equation proposed by Gazis et al. and symmetrical consideration, the extended model to consider multiple velocity difference ahead is proposed as follows

$$\ddot{x}_j(t) = a(\sum_{l=1}^{n} \gamma_l V(\Delta x_{j+l-1}(t)) - \dot{x}_j(t)) + a\sum_{l=1}^{n} \frac{v^{\eta}_{j+l-1}(t)}{\Delta x^{\rho}_{j+l-1}} \Delta v_j(t), \quad (6)$$

$$\ddot{x}_j(t) = a(\sum_{l=1}^{n} \gamma_l V(\frac{1}{l}\sum_{p=1}^{l} \Delta x_{j+p-1}(t)) - \dot{x}_j(t)) + a\sum_{l=1}^{n} \frac{v^{\eta}_{j+l-1}(t)}{\Delta x^{\rho}_{j+l-1}} \Delta v_j(t), \quad (7)$$

$$\ddot{x}_j(t) = a(V(\sum_{l=1}^{n} \gamma_l \Delta x_{j+l-1}(t)) - \dot{x}_j(t)) + a\sum_{l=1}^{n} \frac{v^{\eta}_{j+l-1}(t)}{\Delta x^{\rho}_{j+l-1}} \Delta v_j(t). \quad (8)$$

For simplicity, κ is approximately chosen as a Δv_{j+l-1} and Δx_{j+l-1} is respectively velocity difference (relative velocity) and headway between jth car and $(j+l-1)$th car ahead at time t. γ_l $(l = 1, 2...n)$ is a weight factor chosen according to the different variations of many-vehicle states ahead and has the following properties [10]. When $n > 1$,

$$l \neq n, \gamma_l = \frac{m-1}{m_l}; l = n, \gamma_l = \frac{1}{m^{n-1}} \quad (9)$$

m is a parameter in the weight factors and $m \geq 3$. Phenomenological parameters η and ρ to be fixed by comparison with empirical data are $\eta = 0.953$ and $\rho = 3.05$ or $\eta = 0.8$ and $\rho = 2.8$ [11, 12]. The optimal velocity $V(\Delta x_j)$ is

$$V(\Delta x_j(t)) = (V_{max}/2)(tanh(\Delta x_j(t) - h_c) + tanh(h_c)), \quad (10)$$

where V_{max} is the maximal velocity, h_c is the safety distance.

To apply the linear stability analysis to analyze the stability of the model described by Eq. (6), (7) and (8), the stable conditions are respectively as

$$\frac{V'(h)}{a} = \sum_{l=1}^{n} \gamma_l \frac{2l-1}{2} + n\frac{(V(h))^{\eta}}{h^{\rho}}, \quad (11)$$

$$\frac{V'(h)}{a} = \sum_{l=1}^{n} \frac{\gamma_l}{2l} \sum_{p=1}^{l} (2p-1) + n\frac{(V(h))^{\eta}}{h^{\rho}}, \quad (12)$$

$$\frac{V'(h)}{a} = \sum_{l=1}^{n} \gamma_l \frac{2l-1}{2} + n\frac{(V(h))^{\eta}}{h^{\rho}}, \quad (13)$$

where $V'(h) = dV(\Delta x_j)/d\Delta x_j|_{\Delta x_j = h}$. Obviously, the stability criterion Eq. (11), (12) and (13) is determined by the value of the weighted function γ_l. We take the extended model A,C for example to compare its stability with those of Wilson's model A,C and choose the parameter of weight functions $m = 3$. Fig. 1 (a) and (b) show the neutral stability lines of model A,C and their extended model as $n = 1, 2, 3$, and 5, respectively. In fig. 1, the apex of each curve denotes the critical point (h_c, a_c), the traffic flow above each curve is stable and the traffic jam does not appear. Below each curve, the traffic flow loses its

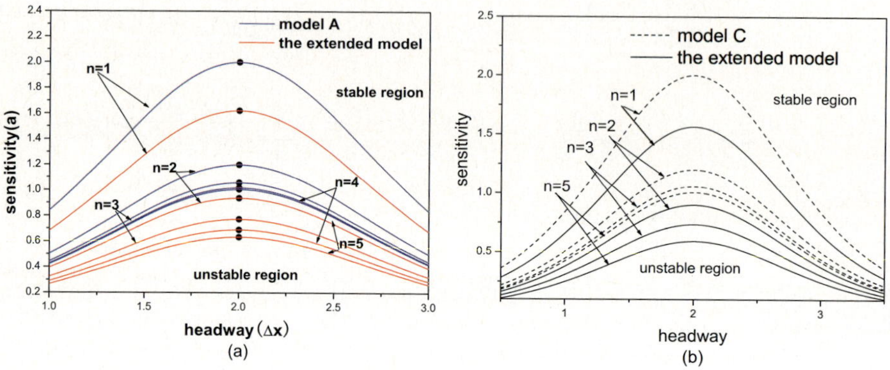

Fig. 1. Neutral stability curves in the headway-sensitivity space, where $v_{max} = 2.0$, $h_c = 2.0$. The neutral stability curves in Wilson's model A,C (the dash line) and their extended model (the solid line) as $m = 3$, $n = 1, 2, 3$, and 5, respectively.

stability and the density wave emerges. From fig. 1, it can be found that with the consideration of more cars ahead, the critical points and the neutral stability curves are lowered, which means the traffic flow more stable. Moreover, just only considering two cars ahead and their velocity differences in extended model is more stable than those taking an arbitrary number of cars ahead in Wilson's model A,C into account. That is to say, the stability of the extended model has been strengthened and traffic jamming is suppressed efficiently.

3 Numerical Simulation and Analysis

In order to demonstrate how the stability of traffic flow is enhanced in the extended model, we carry out the computer simulation for the extended model Eq. (6)–(8) and Wilson's model Eq. (2)–(4) using the forth order Runge-Kutta scheme under initial condition of homogeneous distribution and periodic boundary condition. Initially, 100 vehicles are homogeneously distributed on a straight line with the length 200, the mean headway $h = 2$ and the initial mean velocity is $tanh(2.0)$. The safety distance chosen is $h_c = 2.0$, the time-step is 0.01 and the exponent of velocity and headway $\eta = 0.953$ and $\rho = 3.05$. In the following, we study the stability and properties of two models through the neutral stability curves, consumption of energy and distribution of energy consumption of each vehicle, respectively.

First, the disturbance applies to the leading car to make its larger headway than the others. Fig. 2 exhibits the snapshots of velocity configuration of 100 vehicles of two models in the unstable condition $a = 0.35$ at time step $t = 1500$. From fig. 2, we can find that the negative speed appears in the Wilson's model

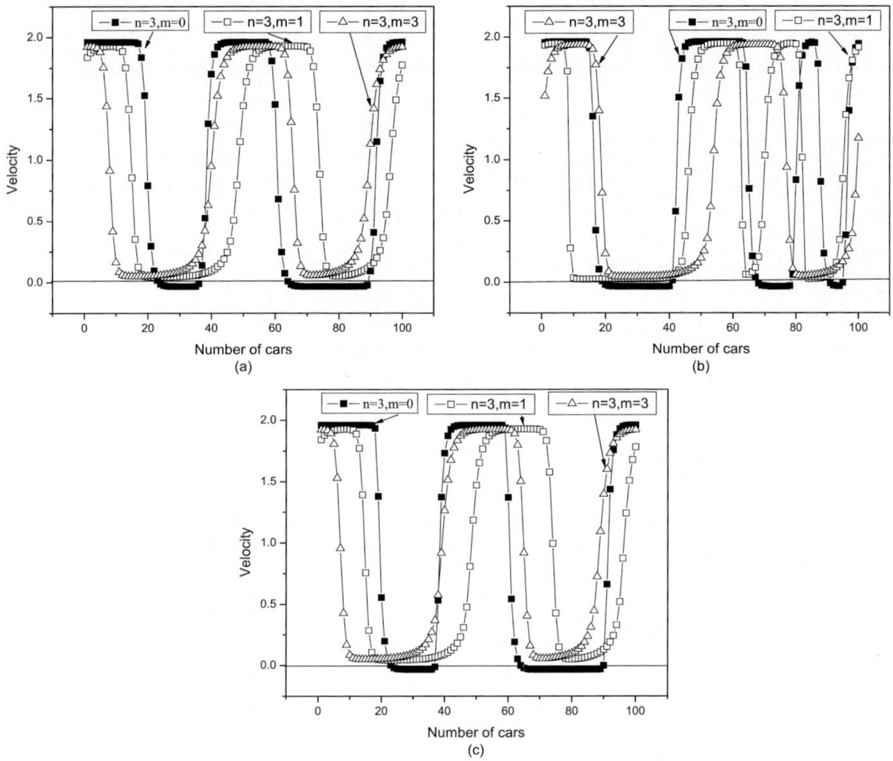

Fig. 2. A snapshot of velocities of the 100 vehicles of the Wilson's model A,B,C and their extended model in the unstable condition $a = 0.35$ at $t = 500$.

A,B,C while the extended model are still positive speed. It means the Wilson's model can not avoid crash as the sensitivity is $a \leq 0.35\ s^{-1}$. It is clear that the multiple velocity differences has an important effect on the stability of the traffic flow and is consistent with the results of stability analysis..

On the other hand, we study the stable properties of the models from energy consumption. Under the stable condition, Nakayama *et al.* [13] presented the following quantity to estimate such additional energy consumption for all vehicles,

$$E = \sum_{cars} \sum_{waves} \frac{1}{2}(V_{max}^2 - V_{min}^2). \tag{14}$$

where waves denotes the summation of all periods of oscillation until the disturbance disappears. Shi *et al.* [10] analyzes additional energy consumption of some typical car-following models and shows that minimizing the energy consumption depends on the stability of model. Fig. 3 exhibits the additional energy consumption for Wilson's model A,C and their extended model as $n = 1, 2, 3, 4, 5$ and the parameter of weighting function $m = 3$. One can find in fig. 3 that the consumption of the whole additional energy decreases with the increasing value of sensitivity (a) and will diverge when sensitivity (a)

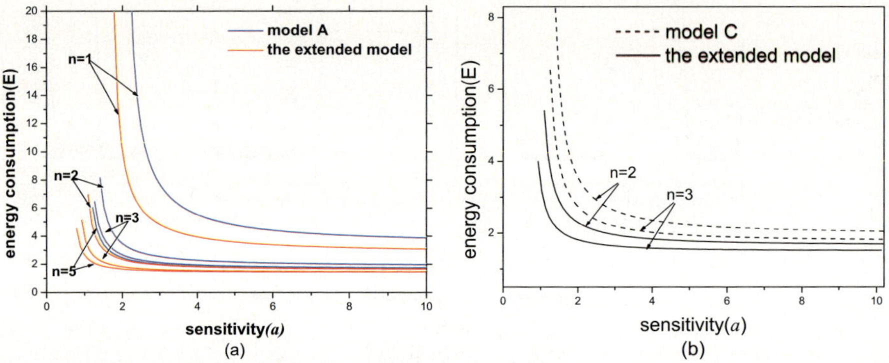

Fig. 3. The energy consumption for Wilson's model A, C (the dashed line) and their extended (the solid line) model as $m = 3$, $n = 2, 3$, respectively.

approaches or exceeds the critical point, it means the traffic flow is in the case of unstable state and the perturbation on the traffic flow will induce stop-and-go traffic. And we notice that the energy consumption curves will be reduced with increasing value of the parameter n in each model. In addition, the energy consumption in the extended model with $n = 2$ is lower than that in Wilson's model A,C with $n = 3$. Thus, it can be inferred that the flows in the extended model with $n = 2$ have more stability than that in Wilson's model A,C in any value of n and a, which confirms the results obtained in stability analysis above. Under the unstable condition, the disturbance to induce stop-and-go traffic in the form of oscillation will not disappears. Eq. (14) is not suitable to describe the whole energy change of traffic flow, but each vehicle's energy change is more important to understand formation of traffic jamming. The change in kinetic energy ΔE_j for the vehicle j between the two successive time step is defined as

$$\Delta E_j = \frac{1}{2}[v_j^2(t) - v_j^2(t-1)] \tag{15}$$

where $v_j(t)$ and $v_j(t-1)$ are the velocity of vehicle j in the two successive time-steps, respectively. Fig. 4 shows the profile of distribution on energy consumption in the extended model A, where the green solid denotes the case

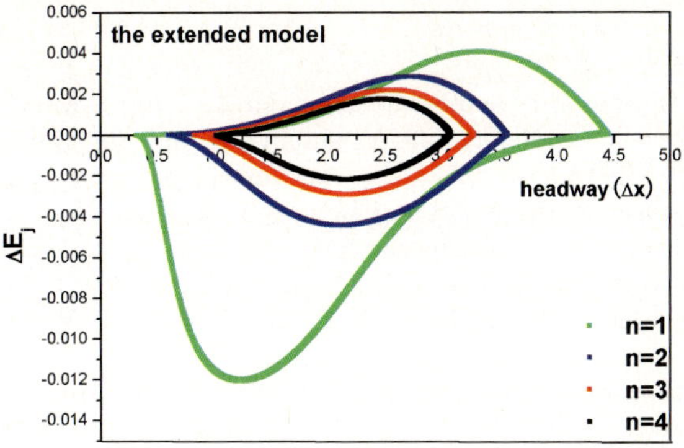

Fig. 4. Profile of energy consumption of the extended model as $m = 3$ and $n = 1$, 2, 3, 4, respectively, and $a = 0.5$.

as $n = 1$, the blue solid denotes the case as $n = 2$, the red solid represents the case as $n = 3$ and the black solid represents case as $n = 4$. The profile of energy consumption in each case is loop, and each loop is divided into two regions with $\Delta E_j > 0$ and $\Delta E_j < 0$. $\Delta E_j > 0$ describes the vehicles that will leave congestion region in the state of acceleration process. $\Delta E_j < 0$ indicates that vehicles enter into the congestion region in way of deceleration. It is obviously noticeable that the area of acceleration process and deceleration process are not symmetrical. The latter is larger than the former, which indicates that the energy consumption of entering into congestion is larger than those of leaving congestion. In addition, the whole areas of each loop are not equivalent. The area of loop diminished with increasing n. It can be inferred that the loop of free flow (or in stable region) will shrink into a dot$(h, 0)$. That is to say, the traffic is free flow. As a result, for smaller loop of the energy consumption profiler, traffic flow is more stable.

4 Summary

We have presented an extended model with many-neighbors and non-locality via considering the effect of multiple velocity differences ahead on the optimal velocity model. The stability of extended model has been strengthened. Based on the intelligent transport system, drivers can conveniently receive information of headway and multiple velocity changes ahead. It is possible to improve traffic, suppress the emergence of a traffic jam, and reduce the energy consumption.

This work was supported by the National Natural Science Foundation of China, the National Basic Research Program of China, the Natural Science Foundation of Guangxi and the Special Foundation for the New Century Talents Program of Guangxi Zhuang Autonomous Region.

References

1. Chowdhury D, Santen L, and Schadschneider A (2000) Phys. Rep. 329:199-329.
2. Bando M, Hasebe K, Nakayama A, Shibata A and Sugiyama Y (1995) Phys. Rev. E 51:1035-1042.
3. Wilson R E, Berg P, Hooper S and Lun S (2004) Eur. Phys. J. B 39:397-408.
4. Lenz H, Wagner C K and Sollacher R (1999) Eur. Phys. J. B 7:331-335.
5. Ge H X, Dai S Q, Dong L Y and Xue Y (2004) Phys. Rev. E 70:066134-066140.
6. Herman R, Montroll E W, Potts R B and Rothery R W (1959) Oper. Res. 7:86-106.
7. Gazis D C, Herman R and Rothery R W (1961) Oper. Res. 9:545-567.
8. Helbing D (2001) Rev. Mod. Phys. 73:1067-1141.
9. Xue Y, Dong L Y, Yuan Y W and Dai S Q (2002) Acta. Phys. Sin. 51(3):492-499 (in Chinese).
10. Shi Wei, Xue Yu (2007) Physica A 381:399-406.
11. May A D and Keller H E M (1967) Highw. Res. Rec. 199:19.
12. Hoefs D H, Untersuchung des Fahrverhaltens (1972) in Fahrzeugkolonnen (Bundesministerium für Verkehr, Abt. Straßenbau, Bonn-Bad Godesberg).
13. Nakayama A, Sugiyama Y and Hasebe K (2001) Phys. Rev. E 65:016112-016118.

Empirical Data for Pedestrian Flow Through Bottlenecks

Armin Seyfried[1], Bernhard Steffen[1], Andreas Winkens[2], Tobias Rupprecht[2], Maik Boltes[1], and Wolfram Klingsch[2]

[1] Jülich Supercomputing Centre, Research Centre Jülich, 52425 Jülich, Germany
a.seyfried@fz-juelich.de

[2] Institute for Building Material Technology and Fire Safety Science, University of Wuppertal, Germany

Summary. The number of models for pedestrian dynamics has grown in the past years, but the experimental data to discriminate between these models is still to a large extent uncertain and contradictory. To enhance the data base and to resolve some discrepancies discussed in the literature over one hundred years we studied the pedestrian flow through bottlenecks by an experiment performed under laboratory conditions. The time development of quantities like individual velocities, densities, individual time gaps in bottlenecks of different width and the jam density in front of the bottleneck is presented. The comparison of the results with experimental data of other authors supports a continuous increase of the capacity with the bottleneck width. The most interesting results of this data collection is that maximal flow values measured at bottlenecks can exceed the maxima of empirical fundamental diagrams significantly. Thus either our knowledge about empirical fundamental diagrams is incomplete or the common assumptions regarding the connection between the fundamental diagram and the flow through bottlenecks need a thorough revision.

1 Introduction

Studies of the dependence between the capacity and the width of a bottleneck for pedestrian flow can be traced back to the beginning of the last century [1, 2]. But up to now it is discussed controversially whether it increases stepwise or continuously with width. In the following the basic assumptions of these two positions are introduced.

Commonly the flow equation in combination with empirical measurements is used to calculate maximal flow values through bottlenecks [3–6]. With the width of the pedestrian facility, b, the flow equation can be written

$$J = J_s\, b \quad \text{with} \quad J_s = \rho\, v. \tag{1}$$

The specific flow, J_s, gives the flow per unit-width and is the product of the average density, ρ, and the average speed, v, of a pedestrian stream. The empirical relation between flow and density, $J = J(\rho)$, is called the fundamental diagram and with a given fundamental diagram the capacity of a facility is defined as the maximum of this function. In general it is assumed that for a given facility (e.g. corridors, stairs, doors) the fundamental diagrams for different b merge into one diagram for the specific flow J_s. Consequently the capacity, C, is assumed to be a linear function of the width, b.

Contrary to this, Hoogendoorn and Daamen [7, 8] claim that the capacity is growing in a step-wise manner. This statement is based on their observation that inside a bottleneck the formation of lanes occurs, resulting from the zipper effect during entering the bottleneck. The data in [7, 8] indicate that the distance between these lanes is independent of the bottleneck-width. This would imply that the capacity increases only when an additional lane can develop, i.e. that this would occur in a stepwise manner with increasing width [8]. Consequently, either the specific flow would decrease between the values where the steps occur or the flow equation in combination with the concept of a specific flow would not hold. One goal of this work is to examine this claim.

To resolve the discrepancies an experiment is arranged where the density and the velocity and thus the flow inside the bottleneck is measured while a jam occurs in front of the bottleneck. Our experiment is performed under laboratory conditions with a homogeneous group of test persons and equal initial conditions for the density and position of the test persons in front of the bottleneck. Exclusively the width of the bottleneck and the number of the pedestrians are varied. For a detailed discussion the results are compared with experimental data of other studies. In this article we concentrate on unidirectional pedestrian movement through bottlenecks under normal conditions. The term movement under normal means that panic or in particular non adaptive behavior which can occur in critical situation or under circumstances including rewards [9] are excluded. This contribution summarizes parts of an articles and two diploma thesis. The reader may consult [10–12] for more detailed discussions and additional results.

2 Experimental Setup

The experiment was arranged in the auditorium 'Rotunde' at the Jülich Supercomputing Centre (JSC) of the Research Centre Jülich. The configuration is shown in Figure 1. The group of test persons was composed of students and ZAM staff. The boundary of the corridor in front of the bottleneck and the bottleneck was arranged from desks. The height of the bottleneck assured a constant width from the hips to the shoulders of the test persons. The length of the bottleneck amounted to $l_{bck} = 2.8\ m$. The holding areas ensured an equal initial density of the pedestrian bulk in front of the bottleneck for each run. The distance from the center of the first holding area to the entrance of the bottleneck was three meter.

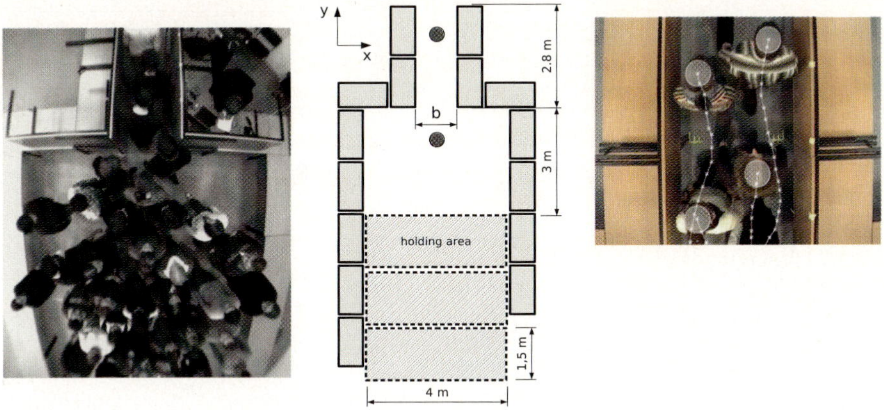

Fig. 1. Experimental setup. In the drawing the position of the video cameras are marked with circles. The holding areas are hatched. The left photo shows a picture of the camera in front of the entrance in the bottleneck. The right is a snapshot from the camera above the bottleneck. The trajectories are determined by marking the center of the head of each person manually.

A stepwise increase of the flow due to lane formation is expected more pronounced for small numbers of lanes. Thus the width of the bottleneck was increased from the minimal value of $b = 0.8\ m$ in steps of $0.1\ m$ to a maximal value of $b = 1.2\ m$. For every width runs are performed with $N = 20$, 40 and 60 pedestrians in front of the bottleneck. At the beginning of each run N test persons were placed in the holding areas with a density of $\rho_{ini} = 3.3\ m^{-2}$. They were advised to move through the bottleneck without haste but purposeful. It was emphasized not to push and to walk with normal velocity. The test persons started to move after an acoustic signal. The whole cycle of each run was filmed by two cameras, one situated above the center of the bottleneck and the other above the entrance of the bottleneck.

3 Data Analysis

3.1 Jam Density in Front of the Bottleneck

For the analysis of the jam density in front of the bottleneck only the runs with $N = 60$ are used. After selecting and extracting the pictures made by the camera located a half meter in front of the entrance, people are detected manually with the help of the software tool Censys 3D^{TM} [13]. To get an overview of the time dependence and the local values of the density this procedure was repeated for every second of the run. The measurement area of $1\ m^2$ was chosen directly in front of the bottleneck.

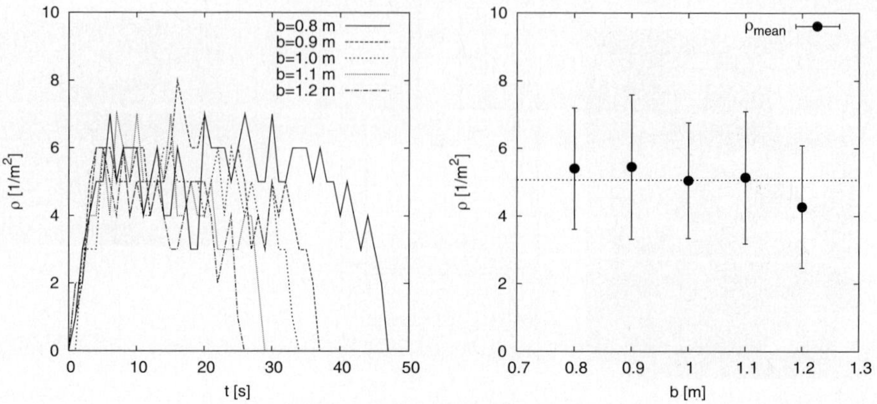

Fig. 2. Time dependence of density in an observation area of 1 m^2 located directly in front of the entrance (left). Different lines refer to different bottleneck widths. The resulting mean values of the stationary state as a function of the bottleneck width (right).

In Figure 2 one observes for every width b the following qualitative development of the density in time. In the first five seconds the pedestrian stream reaches the entrance to the bottleneck and the density increases rapidly. It follows a stationary phase with large fluctuations around a constant value. The length of the stationary phase decreases with increasing width. In the last ten seconds of every run the density decreases to zero. The large fluctuation in the second phase ranging from $\rho = 3$ to $8\ m^{-2}$ can be explained by the small observation area of 1 m^2. However these fluctuations oscillate about a width independent mean value of $\rho = 5\ m^{-2}$. For the calculation of the mean value only the data of the second phase are used and as shown in the left figure they are consistent with the assumption that the width of a bottleneck has no influence on the density in front of the bottleneck. Indeed the fluctuations are very large and do not allow a conclusive judgement. Moreover it can not be excluded that this independence is restricted to small N and $b \geq 0.8\ m$.

3.2 Trajectories and Probability Distributions in the Bottleneck

The investigation of the flow insight the bottleneck is done by means of the trajectories (x_{ij}, y_{ij}, t_j). The index i marks the pedestrian, while j marks the sequence of the points in time. For the determination of the trajectories a manual procedure based on the standard video recordings of a camera above the bottleneck is used, see Fig. 1. For details of this procedure and how flow values, densities and velocities are extracted from the trajectories to study their time dependence we refer to [10].

In Figure 3 we have collected for the runs with $N = 60$ the trajectories, the probability distribution to find a pedestrian at the position x averaged over y and the probability distribution of the individual time gaps between

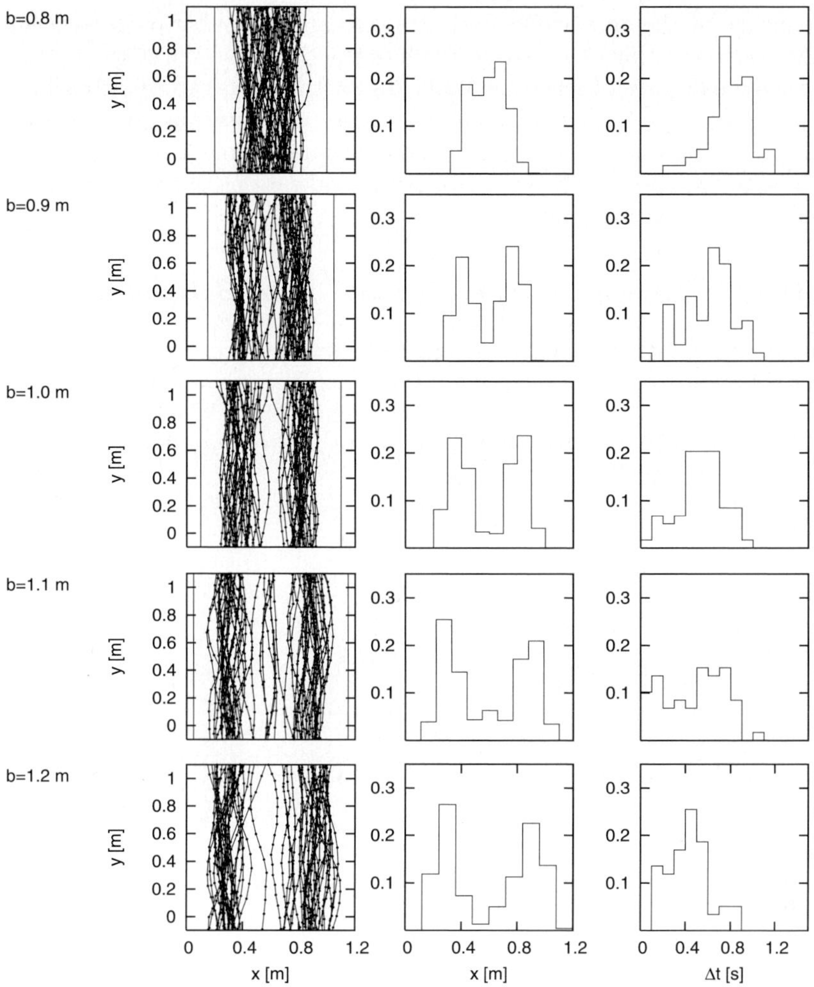

Fig. 3. For the runs with $N = 60$ and from top to bottom with increasing b: The trajectories (left), the probability to find a pedestrian at position x (middle) and probability distribution of the time gaps Δt_i at $y = 0.4\ m$ (right). For $b \geq 0.9\ m$ the formation of lanes is observable. However the distance between the lanes increases continuously with b leading to a continuous decrease of time gaps between two following pedestrians. Thus no indications of a stepwise change of the flow can be found.

the crossing of two adjacents pedestrians, Δt_i, at the center of the bottleneck at $y = 0.4\ m$. The double peak structure in the probability distribution for $b \geq 0.9\ m$ of the positions indicates the formation of lanes. The separation of the lanes is continuously growing with the width of the bottleneck. As a consequence of the zipper effect one expects also a double peak distribution for Δt. However this is not as articulated as in the separation of lanes in space. One can only observe a broadening of the time gap distribution with increasing b and a drift to smaller values. It is important to note that all changes as a function of the width are continuous except for the transition from one to two lanes and thus there are no indications of a stepwise increase or decrease in any observable.

3.3 Time Dependence of ρ, v_i, and Δt_i in the Bottleneck

For the first pedestrian in a run passing the bottleneck the velocity and density will be different from the velocity and the density of the following pedestrians. One expects that the density will increase while the velocity will decrease in time. A systematic drift to a stationary state, where only fluctuation around a constant value will occur, is expected.

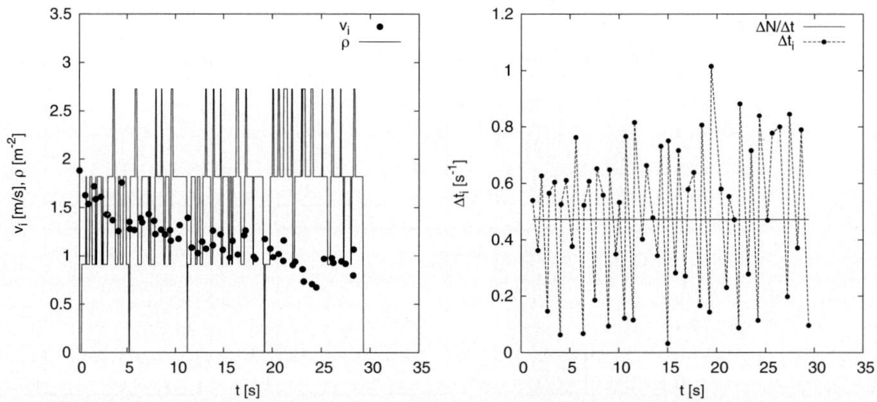

Fig. 4. Run with $N = 60$ and $b = 1.1\ m$. Development of individual velocity and density (left). While the velocity decreases the density increases. Development of the individual time gaps (right).

Figure 4 shows the time-development of the individual velocities and the density for the run with $N = 60$ and $b = 1.1\ m$. Plots for other runs can be found in [11]. The concept of a momentary density in this small observation area is problematic because of the small (1-4) number of persons involved and leads to large fluctuations in the density, see also Sect. 3.1. But one can clearly identify the decrease of the velocity and the increase of the density. For the individual time gaps a time dependence or a trend to a stationary state is hard to identify because the velocity decrease and the density increase compensate largely. A possible time dependence is hidden by large and regular jumps from small to high time gaps caused by the zipper effect.

To find stationary values for the velocity and density by means of regression analysis the tool *MINUIT* [14] for function minimization is used with the following model function borrowed from relaxation processes $f(t) = f_{stat} + A \exp -\frac{t}{\tau}$ for $f(t) = v_i(t)$ and $f(t) = \rho(t)$. The relaxation time τ characterizes the time in which a stationary state will be reached. The amplitude A gives the difference between the stationary state and the initial velocity or density. The velocity or density at the stationary state is labeled f_{stat}. For the fit we use the data of all three runs for one width with different N. Note, that the model function for the regression only describes the overall decrease in time and does not account for the density-fluctuations due to the small observation area or the fluctuations of the velocity in a stable state. Consequently we do not quote an error margin in Table 1.

Table 1. Results for the fit to $v_i(t)$ and $\rho(t)$

b $[m]$	v_{stat} $[m/s]$	A_v $[m/s]$	τ_v $[s]$	ρ_{stat} $[m^{-2}]$	A_ρ $[m^{-2}]$	τ_ρ $[s]$
0.8	1.18	0.354	3.55	1.42	−1.82	0.24
0.9	1.22	0.604	3.00	1.50	−1.20	0.95
1.0	1.17	0.485	3.83	1.59	−1.87	0.31
1.1	0.94	0.745	7.33	1.73	−1.30	2.10
1.2	0.99	0.836	5.63	1.70	−1.28	1.45

The results of the regression analysis are collected in Table 1. For $b \geq 1.0$ m even with $N = 60$ the stationary state is not reached, see e.g. Fig. 4. The results for A and τ indicate that the relaxation into the stationary state is almost independent of the width. However, for a final judgment more data or a larger number of test persons would be necessary. Nevertheless, the results are accurate enough to check at which position of the fundamental diagram the stationary state will be located. Again, the increase of the stationary values for the density ρ_{stat} can be explained by means of the zipper effect in combination with boundary effects.

4 Combined Analysis with Data from Other Experiments

4.1 Comparison with the Data of Other Experiments

In Figure 5 we have collected experimental data for flows through bottlenecks (left) and show how far our measurements fit into common fundamental diagrams (right). All measurements for bottleneck flows were performed under laboratory conditions. The amount of test persons ranged from $N = 30$ to 180 persons. The influence of panic or pushing can be excluded as the collection is

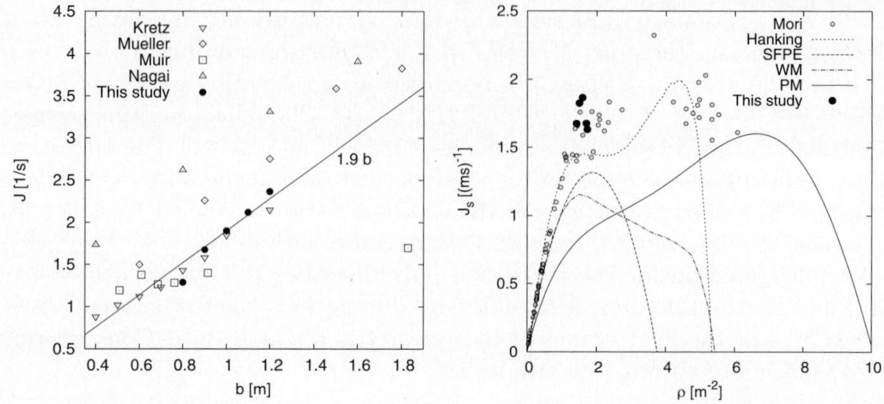

Fig. 5. Influence of the width of a bottleneck on the flow (left). Experimental data from other authors at different types of bottlenecks and initial conditions in comparison with the results of the above described experiment. Experimental data of the flow and the associated density in the bottleneck (right) in comparison with experimental data for the fundamental diagram of unidirectional pedestrian streams (Mori [15], Hanking [16]) and the specifications for the fundamental diagram according to the SFPE Handbook [6] (SFPE) and the guidelines of Weidmann [5] (WM) and Predtechenskii and Milinskii [3] (PM).

limited to measurements where the test persons were asked to move normally. However, the experimental arrangements under which this data were taken differ in many details which provide possible explanations for the discrepancies. Significant differences concern first the geometry of the bottleneck, i.e. its length and position with respect to the incoming flow, and second the initial conditions, i.e. initial density values and the initial distance between the test persons and the bottleneck. The flow measurements of [17] show a leveling off at $b > 0.6\ m$. But the range of the flat profile from $b = 0.6\ m$ to $b = 1.8\ m$ indicates that obviously the passage width is not the limiting factor for the flow in this setup. The data of [18] and [19] are shifted to higher flows in comparison with the data of [17, 20] and our data. The height of the flows in the experiments of Müller and Nagai can be explained by their use of much higher initial densities which amount to $\rho_{ini} \approx 5\ m^{-2}$. That the initial density has this impact is confirmed by the study of Nagai et al., see Fig. 6 in [18]. There it is shown that for $b = 1.2\ m$ the flow grows from $J = 1.04\ s^{-1}$ to $3.31\ s^{-1}$ when the initial density is increased from $\rho_{ini} = 0.4\ m^{-2}$ to $5\ m^{-2}$. The agreement between our data and the results obtained by Kretz indicates the minor importance of the bottleneck-length. This collection suggests that details of the bottleneck geometry and position play a minor role only, while the initial density in front of the bottleneck has a major impact.

4.2 Linear Dependence of Flow and Bottleneck-Width

As mentioned in the introduction one goal of this work is to examine if the flow or the capacity is a linear function of the width, b, of a bottleneck or if it grows in a stepwise manner, as suggested by [8]. Such a stepwise growth would question the validity of the specific flow concept used in most guidelines, see Sect. 1. However, the previous section has argued for the coherence of our data set and previous measurements. All of these results are compatible with a linear and continuous increase of the flow with the width of the bottleneck. Only around $b = 0.7\ m$ the data of [20] show a small edge. The edge is located exactly at the width where the zipper effect can begin to act, i.e. provides no evidence for a stepwise behavior in general. Moreover does the alleged stepwise increase of the flow follows from the assumption that inside a bottleneck the formation of lanes with constant distance occurs. In [8] this assumption is based on flow measurements at two different bottlenecks at $b = 1\ m$ and $b = 2\ m$. It is doubtful whether this results can be extrapolated to intermediate values of the width. In fact our data show no evidence for the appearance of lanes with constant distance (see Sect. 3.2, in particular Fig. 3).

4.3 Connection Between Bottleneck Flow and Fundamental Diagrams

The above results can be used to address a crucial question in pedestrian dynamics, namely the criteria for the occurrence of a jam and thus the connection between bottleneck flow and the fundamental diagram. Commonly it is assumed that jamming happens when the incoming flow exceeds the capacity of the bottleneck. Here the capacity of the bottleneck is defined as the maximum of the fundamental diagram for the specific flow, $J_s(\rho)$, times its width. Moreover most authors assume that in case of a jam the flow through the bottleneck persist on the capacity. However the comparison in Fig. 5 of the collected flow values (left) and fundamental diagrams (right) suggest a more complicated picture and cast doubt on assumptions outlined before. Our results from Section 3 can be used to examine which density and flow inside the bottleneck is present for a situation where a jam occurs in front of the bottleneck. In Sect. 3.1 it was shown that directly in front of the bottleneck the density fluctuates around $5\ m^{-2}$. Inside the bottleneck we found a density of $\rho \approx 1.8\ m^{-2}$ (see Tab. 1). Fig. 5 (right) indicates that the value for the stationary density is exactly located at the position where the fundamental diagram according to the SFPE-Handbook and the guideline of Weidmann show the maximum of the flow while the absolute value of the flow exceeds the predicted values. This seems to support the common jam-occurrence criteria. However, two observations cast doubt on this conclusion. Already when discussing the data of Müller and Nagai we have mentioned that higher initial densities result in higher flow values, i.e. that the maximal flow can not be near $\rho = 1.8\ m^{-2}$. In addition do the fundamental diagrams of Mori [15],

Hanking [16] and PM [3] display a completely different shape. According to Mori and Hanking and in agreement with the specification of PM the flow will increase from $\rho \approx 1.8\ m^{-2}$ or stay constant with increasing density. Moreover does the level of the flow measured in our experiment conforms much better with their specifications. The most important conclusion which can be drawn from the data collection for fundamental diagrams and bottleneck flow is that the high flow values reached by increasing the initial density in front of the bottleneck can not be explained by the maxima of common fundamental diagrams. Moreover does the complicated picture of density values in front and inside the bottleneck suggest a revision of the common assumptions for bottleneck flow.

5 Summary

We have studied experimentally the flow of unidirectional pedestrian streams through bottlenecks under normal conditions. The jam-density in front of the bottleneck shows large fluctuations around a mean value of $\rho = 5\ m^{-2}$ independent of the width. The analysis of the trajectories inside the bottleneck shows that the density tunes around $\rho = 1.8\ m^{-2}$. For a small variation of the width quantities like the time gap distribution or the lane distance change continuously if the zipper effect is acting. The comparison of our data with flow measurements through bottlenecks of different types and lengths suggests that the exact geometry of the bottleneck is of only minor influence on the flow. Regarding the increase of the flow with the width all collected data are compatible with a continuous and linear increase, except for the edge at $b \approx 0.7\ m$ due to zipper effect is beginning to act. The linear dependency between the flow and the width holds for different kinds of bottlenecks and initial conditions. Hence it seems that the basic flow equation in combination with the use of the specific flow concept is justified for facilities with $b > 0.7\ m$. However, the rise of the flow through the bottleneck due to an increase of the initial density in front of the bottleneck from $\rho = 1.8\ m^{-2}$ to $5\ m^{-2}$ and the resulting high flow values through the bottleneck can not be explained by the maxima of common fundamental diagrams. Thus either the available measurements of density flow relation for pedestrian traffic are incomplete or the connection between bottleneck flow and fundamental diagram need a rigorous revision.

References

1. D. Dieckmann. *Die Feuersicherheit in Theatern.* Jung (München), 1911. in German.
2. Herbert Fischer. *Über die Leistungsfähigkeit von Türen, Gängen und Treppen bei ruhigem, dichtem Verkehr.* Dissertation, Technische Hochschule Dresden, 1933. in German.

3. V. M. Predtechenskii and A. I. Milinskii. *Planing for foot traffic flow in buildings.* Amerind Publishing, New Dehli, 1978. Translation of: Proekttirovanie Zhdanii s Uchetom Organizatsii Dvizheniya Lyuddskikh Potokov, Stroiizdat Publishers, Moscow, 1969.
4. J. J. Fruin. *Pedestrian Planning and Design.* Elevator World, New York, 1971.
5. U. Weidmann. Transporttechnik der Fußgänger. Schriftenreihe des IVT 90, ETH Zürich, 1993.
6. H. E. Nelson and F. W. Mowrer. Emergency movement. In P. J. DiNenno, editor, *SFPE Handbook of Fire Protection Engineering*, chapter 14, page 367. National Fire Protection Association, Quincy MA, third edition, 2002.
7. S. P. Hoogendoorn, W. Daamen, and P. H. L. Bovy. Microscopic pedestrian traffic data collection and analysis by walking experiments: Behaviour at bottlenecks. In E. R. Galea, editor, *Pedestrian and Evacuation Dynamics '03*, pages 89–100. CMS Press, London, 2003.
8. S. P. Hoogendoorn and W. Daamen. Pedestrian behavior at bottlenecks. *Transportation Science*, 39 2:0147–0159, 2005.
9. A. Mintz. Non-adaptive group behaviour. *The Journal of abnormal and social psychology*, 46:150–159, 1951.
10. A. Seyfried, T. Rupprecht, O. Passon, B. Steffen, W. Klingsch, and M. Boltes. New insights into pedestrian flow through bottlenecks. *arXiv:physics/0702004*, 2007.
11. T. Rupprecht. Untersuchung zur Erfassung der Basisdaten von Personenströmen. diploma thesis, Bergische Universität Wuppertal, 2006. www.fz-juelich.de/jsc/JSCpeople/seyfried/teaching.
12. A. Winkens. Analyse der lokalen Dichte in Fußgängerströmen vor Engstellen. diploma thesis, Bergische Universität Wuppertal, 2007. www.fz-juelich.de/jsc/JSCpeople/seyfried/teaching.
13. Censys3D™. Point Grey Research Inc., `www.ptgrey.com`.
14. F. James. MINUIT - Function Minimization and Error Analysis, 1994. CERN Program Library entry D506.
15. M. Mori and H. Tsukaguchi. A new method for evaluation of level of service in pedestrian facilities. *Transp. Res. Part A*, 21A(3):223–234, 1987.
16. B. D. Hankin and R. A. Wright. Passenger flow in subways. *Operational Research Quarterly*, 9:81–88, 1958.
17. H. C. Muir, D. M. Bottomley, and C. Marrison. Effects of motivation and cabin configuration on emergency aircraft evacuation behavior and rates of egress. *The International Journal of Aviation Psychology*, 6(1):57–77, 1996.
18. R. Nagai, M. Fukamachi, and T. Nagatani. Evacuation of crawlers and walkers from corridor through an exit. *Physica A*, 367:449–460, 2006.
19. K. Müller. *Die Gestaltung und Bemessung von Fluchtwegen für die Evakuierung von Personen aus Gebäuden.* dissertation, Technische Hochschule Magdeburg, 1981.
20. T. Kretz, A. Grünebohm, and M. Schreckenberg. Experimental study of pedestrian flow through a bottleneck. *J. Stat. Mech.*, page P10014, 2006.

Modeling Synchronized Flow at Highway Bottlenecks

Florian Siebel[1], Wolfram Mauser[2], Salissou Moutari[3], and Michel Rascle[3]

[1] Sebastian-Andre-Weg 8, D-82362 Weilheim, Germany
`florian.siebel@gmx.de`
[2] Department of Geography, University of Munich, Luisenstraße 37, D-80333 Munich, Germany `w.mauser@iggf.geo.uni-muenchen.de`
[3] Laboratoire J. A. Dieudonné, UMR CNRS N° 6621, Université de Nice-Sophia Antipolis, Parc Valrose, 06108 Nice Cedex 2, France
`{salissou, rascle}@math.unice.fr`

Summary. Many experimental studies have shown the appearance of synchronized flow at highway bottlenecks. We study highway bottlenecks within the macroscopic BVT model. The BVT model describes traffic flow as a hyperbolic system of balance laws. It generalizes the traffic model of Aw, Rascle and Greenberg by introducing in the momentum equation a new source term, which can become negative due to the finite reaction and relaxation times of drivers. The model is capable of reproducing multivalued fundamental diagrams, the metastability of free traffic flow at the onset of instabilities and wide moving jams. Based on previous work we describe the coupling conditions for the Riemann problem of the system and apply them to highway bottlenecks. We focus our study on the situation where the bottlenecks are either caused by the reduction of the number of lanes or by on-ramps or off-ramps. Our numerical simulations reproduce the appearance of synchronized flow at these highway bottlenecks. The analysis of the lane reduction setup shows that the outflow from the synchronized flow region in front of the bottleneck is constant and below the maximum free flow. This observation can be understood from the study of the static solutions within the model. As a consequence of the coupling conditions static solutions have to cross the jam line, one of the additional equilibrium solutions within the BVT model. This crossing determines the flow value of the static solution.

1 Balanced Vehicular Traffic: The BVT Model

The BVT model (balanced vehicular traffic model, see [1, 2]) generalizes the traffic model of Aw, Rascle and Greenberg (often called Aw, Rascle and Zhang model, too, see [3–5]) by prescribing a more general source term to the pseudo-momentum equation. The evolution of traffic density ρ and dynamical velocity v is described by the following hyperbolic system of balance laws

$$\frac{\partial \rho}{\partial t} + \frac{\partial(\rho v)}{\partial x} = 0, \tag{1a}$$

$$\frac{\partial(\rho(v - u(\rho)))}{\partial t} + \frac{\partial(\rho v(v - u(\rho)))}{\partial x} = \beta(\rho, v)\rho(u(\rho) - v), \tag{1b}$$

where t denotes the time coordinate and x the space coordinate. The function $u(\rho)$ denotes the equilibrium velocity, which is monotonously decreasing, $\beta(\rho, v)$ is the effective relaxation coefficient. Due to finite *reaction times* and finite *relaxation times*, an effective relaxation coefficient results with negative values in the neighborhood of the equilibrium flow curve $u(\rho)$ for medium to high traffic densities ρ [1]. As a consequence of the negative relaxation coefficients, there are two additional equilibrium velocity curves, the so-called *high-flow branch* and the *jam line*. The high-flow branch is metastable for medium traffic densities and unstable for high traffic densities, whereas the jam line is unstable for medium traffic densities and metastable for high traffic densities, see [2, 6] for a detailed discussion. We stress, that the BVT model, like the model of Aw, Rascle and Greenberg, fulfills the anisotropic condition, the characteristic speeds of the system $\lambda_1 = v - \rho u'(\rho)$ and $\lambda_2 = v$ are bounded from above by the vehicle speed v. Hence the model does not show the unphysical behavior pointed out by Daganzo [7]. Moreover, the subcharacteristic condition [8] is essential for the stability properties above and can explain the form of the reversed λ observed in the fundamental diagram of traffic flow [9].

Here we are interested in synchronized flow at highway bottlenecks [10]. We applied the coupling conditions for the Riemann problem of the system and studied highway bottlenecks either caused be the reduction of the number of lanes or by on- and off ramps [11]. Our results show that the BVT model can explain the observation of a capacity drop at highway bottlenecks, i.e. the outflow from the bottleneck regions is substantially lower than the maximum capacity of free flow (see e.g. [12, 13]).

2 Coupling Conditions at Intersections

In order to model bottlenecks in the BVT model one has to describe the coupling conditions at intersections or junctions. These coupling conditions give rise to the boundary values for vehicle and pseudo-momentum fluxes of each road section, which are necessary to solve the corresponding initial value problem in each section.

Although the source term in the pseudo-momentum equation, i.e. the right-hand side of Eqn. (1b), plays an essential role for the traffic dynamics on road sections, it can be neglected for the analysis of the Riemann problem at intersections, since it is never a delta-function. For the homogeneous system without this source term, i.e. the Aw-Rascle model [3], there is a large amount of theoretical work on the coupling conditions at intersections [14–18]. For comments on the difference between these approaches, see [11].

For a given junction n, let us denote by δ_n^- and δ_n^+, respectively the set of all incoming roads to n (indexed i in the sequel) and the set of all the outgoing roads from n (indexed k in the following). We require the Eqs. (1a)–(1b) to hold on each road of $\delta_n^- \cup \delta_n^+$. The percentage of cars on the road i intending to go to the road k are denoted by α_{ik}, such that $\forall i \in \delta_n^-, \sum_k \alpha_{ik} = 1$. These coefficients are assumed to be known.

Let $U_i^- = (\rho_i^-, \rho_i^- v_i^-)$, $\forall i \in \delta_n^-$ and $U_k^+ = (\rho_k^+, \rho_k^+ v_k^+)$, $\forall k \in \delta_n^+$, respectively the boundary values at the junction on the incoming and outgoing roads. We denote by $\hat{\beta}_{ik}$, such that $\forall k \in \delta_n^+, \sum_i \hat{\beta}_{ik} = 1$, the proportion of the flow on the road k coming from the road i. We set

$$\hat{\beta}_{ik} = \frac{\alpha_{ik} d_i(\rho_i^-)}{\sum_{i \in \delta_n^-} \alpha_{ik} d_i(\rho_i^-)} \quad \forall i \in \delta_n^-, \ \forall k \in \delta_n^+, \tag{2}$$

with the demand functions d_i defined as

$$d_i(\rho) = \begin{cases} \eta_{di}(\rho), & \text{if } \rho \leq \tilde{\rho}_i, \\ \eta_{di}(\tilde{\rho}_i), & \text{if } \rho > \tilde{\rho}_i, \end{cases} \tag{3}$$

where the functions η_{di} on the incoming roads i are given by

$$\eta_{di}(\rho) = \rho u_i(\rho) + \rho w_i(U_i^-), \tag{4}$$

and the maximum of the corresponding curves corresponds to $\tilde{\rho}_i$. The *distance from equilibrium* for each road section for arbitrary $i \in \delta_n^-$ (and $k \in \delta_n^+$, see below) can be determined from the state $U = (\rho, \rho v)$ by

$$w_i(U) = v - u_i(\rho). \tag{5}$$

On the outgoing roads near the junction we define the *homogenized distance from equilibrium*

$$w_k^* = \sum_{i \in \delta_n^-} \hat{\beta}_{ik} w_i(U_i^-), \quad \forall k \in \delta_n^+. \tag{6}$$

The supply functions s_k for arbitrary $k \in \delta_n^+$ read

$$s_k(\rho) = \begin{cases} \eta_{sk}(\tilde{\rho}_k), & \text{if } \rho < \tilde{\rho}_k, \\ \eta_{sk}(\rho), & \text{if } \rho \geq \tilde{\rho}_k. \end{cases} \tag{7}$$

Here, the functions η_{sk} are defined as

$$\eta_{sk}(\rho) = \rho u_k(\rho) + \rho w_k^*, \tag{8}$$

which reaches its maximum value at $\tilde{\rho}_k$.

For all $k \in \delta_n^+$, the intermediate state of density $\rho_k^\dagger$ on the outgoing road is given by the intersection point of the curves $v = v_k^+$ and $w_k(U) = w_k^*$.

To obtain the downstream fluxes on incoming road section i

$$\hat{f}_i^- = q_i \left(\frac{1}{w_i(U_i^-)} \right) \tag{9}$$

and the upstream fluxes on outgoing road section k

$$\hat{f}_k^+ = q_k \left(\frac{1}{w_k^*} \right), \tag{10}$$

we have to solve the following optimization problem

$$\max \sum_{k \in \delta_n^+} q_k \text{ subject to} \tag{11a}$$

$$0 \leq q_i \leq d_i(\rho_i^-), \forall i \in \delta_n^-; \tag{11b}$$

$$0 \leq q_k \leq s_k(\rho_k^\dagger), \forall k \in \delta_n^+; \tag{11c}$$

$$q_i = \sum_{k \in \delta_n^+} \hat{\beta}_{ik} q_k, \forall i \in \delta_n^-; \tag{11d}$$

$$q_k \leq \sum_{i \in \delta_n^-} \alpha_{ik} d_i(\rho_i^-), \forall k \in \delta_n^+. \tag{11e}$$

Details on the solution of this optimization problem for selected bottlenecks can be found in [11].

3 Synchronized Flow at a Lane Drop Bottleneck

We first analyze the traffic dynamics for the setup depicted in Fig. 1. The highway under study consists of two 7 km long road sections, with a lane drop from three lanes in section 1 to two lanes in section 2 at 7 km. Note that in the mathematical description the transition from three to two lanes is immediate, the length of the merging segments is neglected. We further prescribe periodic boundary conditions. Hence, the evolution is fully determined by the initial data on the two road sections. Throughout this work we use the equilibrium velocity function of Newell

$$u(\rho) = u_m \Big(1 - \exp\Big(- \frac{\lambda}{u_m} \Big(\frac{1}{\rho} - \frac{1}{\rho_m} \Big) \Big) \Big) \tag{12}$$

with parameter values $u_m = 160$ km/h, $\lambda = 3600$ [1/h/lane], $\rho_m = 160$ [1/km/lane] and an effective relaxation coefficient

$$\beta(\rho, v) = \begin{cases} \frac{a_c}{u-v}, & \text{if } \tilde{\beta}(\rho, v)(u(\rho) - v) - a_c \geq 0, \\ \frac{d_c}{u-v}, & \text{if } \tilde{\beta}(\rho, v)(u(\rho) - v) - d_c \leq 0, \\ \tilde{\beta}(\rho, v), \text{ else}, \end{cases} \tag{13}$$

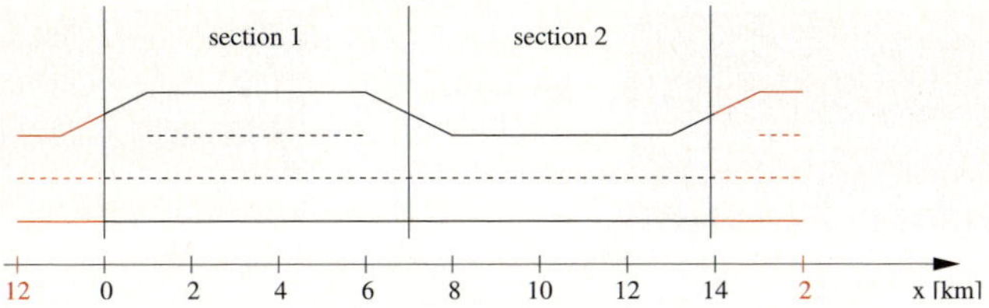

Fig. 1. Sketch of the highway under study. The highway consists of two road sections of 7 km length. The road section 1 consists of three lanes whereas the road section 2 consists of two lanes. We use periodic boundary conditions, i.e. the road section 1 is also located downstream of the road section 2.

$$\tilde{\beta}(\rho, v) = \frac{1}{\hat{T} u_m}\Big(|u(\rho) - v + a_1 \Delta v| + a_2 \Delta v\Big) \tag{14}$$

and

$$\Delta v(\rho) = \tanh\Big(a_3 \frac{\rho}{\rho_m}\Big)\Big(u(\rho) + c\rho_m\Big(\frac{1}{\rho} - \frac{1}{\rho_m}\Big)\Big), \tag{15}$$

with parameters $a_c = 2$ m/s^2, $d_c = -5$ m/s^2, $\hat{T} = 0.1$ s, $a_1 = -0.2$, $a_2 = -0.8$, $a_3 = 7$ and $c = -14$ km/h. Thus the maximum density of the road section 1 (2) is 480 vehicles/km (320 vehicles/km). The initial data to start the numerical simulations consists of equilibrium data on the two road sections. We prescribe a constant vehicle density ρ_0 in both road sections, setting the initial velocity to $v = u(\rho_0)$. We choose the constant ρ_0 to be independent of the number of the lanes, the corresponding scaled densities in each road section follow from dividing ρ_0 by the number of lanes of that road section.

Figure 2 shows the simulation results for the density (left column) and velocity (right column) for two simulations with initial density $\rho_0 = 50$ [1/km] (first row) and $\rho_0 = 100$ [1/km] (second row) covering an evolution time of two hours. Note that, although the initial data are in equilibrium in each road section, the coupling conditions at the interface between the two road sections do not guarantee the equilibrium during the evolution.

For a density $\rho_0 = 50$ [1/km], a small region of higher density and lower velocity forms between about 5.5 km and 7 km. This region corresponds to data located in the fundamental diagram on and scattered around the jam line. Clearly, this congested region is fixed at the bottleneck and therefore cannot correspond to a wide moving jam. Instead it corresponds to synchronized flow. For a density $\rho_0 = 100$ [1/km], the dynamics becomes more complicated, but finally a synchronized flow region of extended width (ranging from about 2 km to 7 km) forms. Only in a small region of the road section 1 (between 0 km and 2 km) traffic is in free flow.

For both initial data, we observe the formation of a static solution during the simulation. These static solutions can be analyzed analytically as follows. Steady-state solutions of the system (1a) and (1b) are solutions for which, in an appropriate coordinate system $(\tilde{t}, z) = (t, x - \hat{w}t)$ with a constant speed

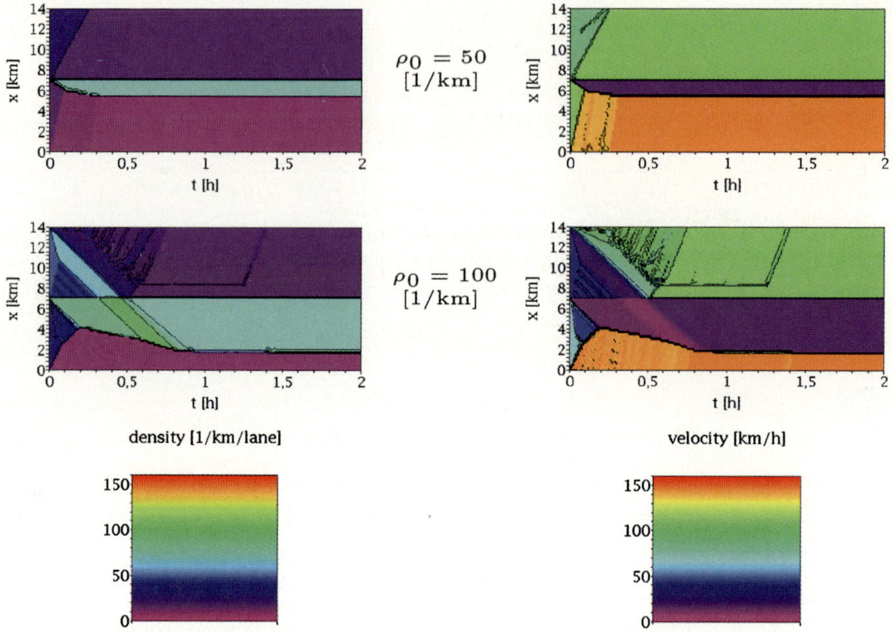

Fig. 2. Traffic dynamics at the bottleneck caused by the reduction of the number of lanes on a highway. The column on the left shows the evolution of the vehicle density in units [1/km/lane], the column on the right the corresponding evolution of the velocity in units [km/h]. The first row corresponds to the simulation run for an initial density $\rho_0 = 50$ [1/km], the second row to an initial density $\rho_0 = 100$ [1/km].

$\hat{w}$, all time derivatives with respect to $\tilde{t}$ vanish. It follows from (1a) and (1b) that steady-state solutions fulfill the following set of equations (see [2, 6])

$$q = \rho(v - \hat{w}), \tag{16}$$

$$\frac{d\rho}{dz} = \frac{\rho^2}{q} \frac{\beta(\rho, v)(\rho\hat{w} + q - \rho u(\rho))}{q + u'(\rho)\rho^2}, \tag{17}$$

with a constant q. For static solutions, moreover, we have $\hat{w} = 0$ and $z = x$.

Figure 3 shows the numerical results for the static solutions in more detail. We note that apart from the location of the shock discontinuity at about 5.5 km and about 1.8 km respectively, the solutions for initial data $\rho_0 = 50$ [1/km] and $\rho_0 = 100$ [1/km] are identical. The right panel of Fig. 3 shows the solution for $\rho_0 = 50$ [1/km] in the flow-density-diagram. In addition we indicated for each highway section the equilibrium flow curve $\rho u(\rho)$ and the additional equilibrium curves, which correspond to the zeros of the effective relaxation coefficient $\beta(\rho, v)$, i.e. the high-flow branch and the jam line. As a result of [2, 6], these curves can be subdivided into stable, metastable and unstable branches, which are shown as solid, dashed and dotted curves in this panel.

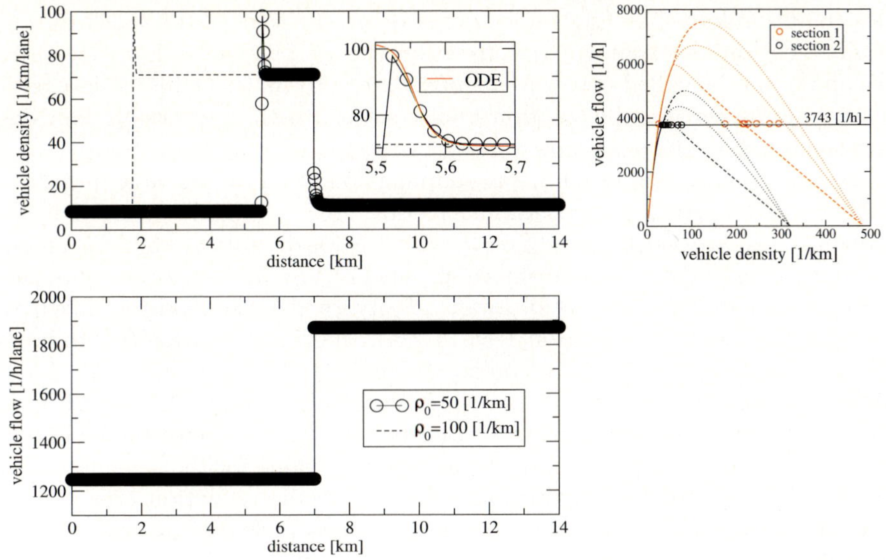

Fig. 3. Static solution with synchronized flow in front of the bottleneck. The two panels on the left show the vehicle density and flow profiles for the static solutions, which finally form for an initial density $\rho_0 = 50$ [1/km] (solid line with circles) and $\rho_0 = 100$ [1/km] (dashed line). Up to the location of the shock discontinuity, the static solution is independent of the initial data. Note that for the vehicle flow, the two curves for initial densities $\rho_0 = 50$ [1/km] and $\rho_0 = 100$ [1/km] fully overlap. The peaks located at about 5.5 km for $\rho_0 = 50$ [1/km] and at about 1.8 km for $\rho_0 = 100$ [1/km] in the left upper panel correspond to sections of nontrivial static solutions. This can be seen in the inset of the left upper panel, where we showed for comparison the solution of the ordinary differential equations (16)–(17). The situation in the flow-density-diagram is shown in the right panel, where the circles correspond to the static data for $\rho_0 = 50$ [1/km]. Up to discretization errors the constant flow value 3743 [1/h] agrees with the theoretical value determined by the crossing of the jam line of section 2.

Let us first focus on the jump discontinuities (shocks) at 5.5 km and 1.8 km, respectively. In systems of conservation laws, shocks have to fulfill the Rankine-Hugoniot jump conditions, which reduce for the BVT model to the conservation of flow

$$(\rho v)_R = (\rho v)_L, \tag{18}$$

and the conservation of the *distance from equilibrium*

$$(v - u(\rho))_R = (v - u(\rho))_L. \tag{19}$$

Here the subscripts L and R denote the states left and right of the shock. The flow conservation (18) is fulfilled everywhere for the static solutions. Moreover, left (upstream) of the shock, traffic is in equilibrium $v = u(\rho)$. Hence, this condition has to be fulfilled for the state on the right of the shock (downstream) as well. In analogy to detonation theory, this state downstream of the shock can be called a *von Neumann state* [19]. As this state is an unstable equilibrium state, the solution finally settles to the metastable jam line corresponding to the same flow value.

Second, we are interested in the coupling conditions at the lane drop bottleneck at 7 km. Analogous conditions to Eqs. (18)–(19) lead to a state downstream of the bottleneck in section 2, which is located above the jam line in the fundamental diagram (see the right panel of Fig. 3). In order to link this state to an equilibrium state of lower density (which then can be linked to the equilibrium state in section 1 at 14 km), the static solution has to cross the jam line (see [20, 21] for a similar discussion). This is only possible provided that the denominator of the right hand side of Eqn. (17) vanishes on the jam line (the numerator trivially vanishes, as $\beta(\rho, v) = 0$ for the jam line). From this condition, we obtain a theoretical flow value for the static solution of 1889.9 [1/h/lane], which we reproduce, up to discretization errors, with our numerical simulations. Note that this value is independent of the initial data, and moreover is far below the maximum capacity of free flow. Due to the existence of the jam line, which follows from negative effective relaxation coefficients, we thus can explain a capacity drop within the BVT model.

We further note that for higher densities in the initial data, wide moving jams form for the above lane drop setup.

4 On- and Off-Ramps

In this setup we analyze a two-lane highway with an on-ramp and an off-ramp (see Fig. 4). For our numerical simulations we chose a length of 7 km for two-lane road sections 1 and 3 each, and a length of 10 km for the one-lane road section 2. For the parameterization of the equilibrium velocity curve and the

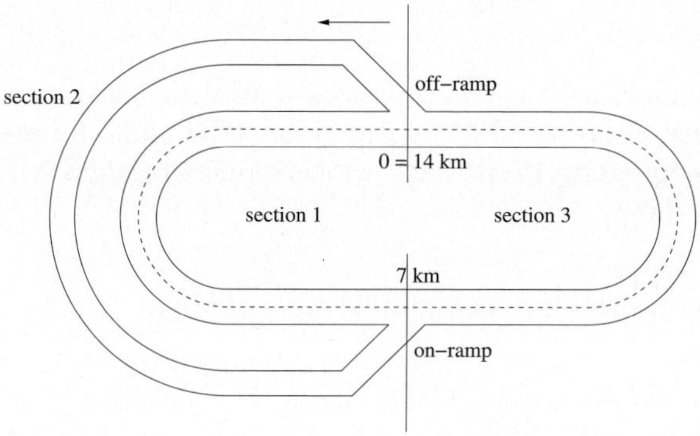

Fig. 4. Sketch of the simulation setup of a two-lane highway with an on-ramp between the road sections 1 and 3 at 7 km (bottom) and an off-ramp between the road sections 3 and 1 at 0 km = 14 km (top).

effective relaxation coefficient, we use again the values given in Eqs. (12)–(15). We start the simulations with a constant density in equilibrium on all road sections of $\rho_0 = 50$ [1/km/lane] and vary the percentage of cars $\alpha = \alpha_{31}$ aiming to enter the road section 1 from the road section 3. Some numerical results are presented in Fig. 5. For simplicity, we do not present the simulation results on road section 2. Nevertheless, the evolution of section 2 influences the evolution of the other sections due to the boundary conditions at the ramps. For a value $\alpha = 0.5$, a narrow stretch of synchronized flow develops in front

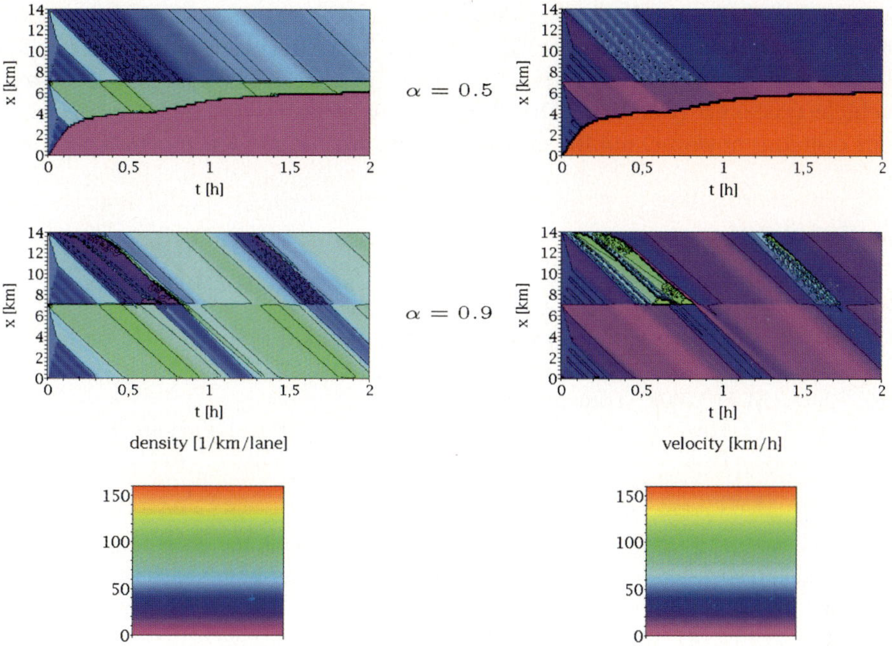

Fig. 5. Traffic dynamics at highway bottlenecks caused by on-ramps and off-ramps. The plot shows the traffic dynamics on the two-lane highway, the road section 1 corresponds to the region between 0 and 7 km, the road section 3 corresponds to the region between 7 and 14 km. We vary the percentage of cars α intending to enter road section 1 from the road section 3 (different rows). The column on the left shows the evolution of the vehicle density in units [1/km/lane], the column on the right the corresponding evolution of the velocity in units [km/h].

of the on-ramp in the road section 1. For a parameter value $\alpha = 0.9$ a region of narrow moving jams forms in front of the off-ramp in the road section 3.

5 Conclusion

We studied synchronized flow at highway bottlenecks within the macroscopic BVT model. The bottlenecks were either caused by the reduction of the number of lanes or by an on-ramp and an off-ramp. We observed a constant outflow from the synchronized flow region in front of the lane drop bottleneck. This flow is independent of the width of the synchronized flow region and has a value far below the maximum free flow in the downstream section, thus reproducing a capacity drop. It follows from the coupling conditions at intersections and in particular from the crossing of the corresponding static solution with the jam line in the downstream section.

We stress that the BVT model can reproduce multi-valued fundamental diagrams, the metastability of free flow at the onset of traffic instabilities, synchronized flow, wide moving jams and the capacity drop at lane-drop bottlenecks. All these aspects of traffic flows can be reproduced without modeling lane changes, stochasticity of traffic flow or different driver characteristics and vehicle types. Of course, all this plays a role in observed traffic patterns. In our opinion, however, nonlinear instabilities are the most important factor.

References

1. Siebel, F, Mauser, W (2006) SIAM J Appl Math 66:1150–1162.
2. Siebel, F, Mauser, W (2006) Phys Rev E 73:066108.
3. Aw, A, Rascle, M (2000) SIAM J Appl Math 60:916–938.
4. Greenberg, J (2001) SIAM J Appl Math 62:729–745.
5. Zhang, H M (2002) Transp Res B 36:275–290.
6. Siebel, F, Mauser, W (2007) Stability of Steady State Solutions in Balanced Vehicular Traffic. In: Schadschneider A et al (eds) Traffic and Granular Flow '05. Springer 559–564.
7. Daganzo, C (1995) Transp Res B 29:277–286.
8. Whitham, G B (1974) Linear and Nonlinear Waves. John Wiley, New York.
9. Koshi, M, Iwasaki, M, Ohkura, I (1983) Some findings and an overview on vehicular flow characteristics. In: Hurdle V, Hauer E, Stuart G (eds) 8th International Symposium on Transportation and Traffic Theory, 403–424.
10. Kerner, B (2004) The Physics of Traffic. Springer, Berlin.
11. Siebel, F, Mauser, W, Moutari, S, Rascle, M (2006) submitted to Physica A.
12. Bertini, R, Leal, M (2005) J Transp Engin 131:397–407.
13. Laval, J, Cassidy, M, Daganzo, C (2007) Impacts of lane changes at merge bottlenecks: a theory and strategies to maximize capacity. In: Schadschneider A et al (eds) Traffic and Granular Flow '05. Springer 577–586.
14. Garavello, M, Piccoli, B (2006) Comm Part Diff Equat 31:243–275.
15. Herty, M, Rascle, M (2006) SIAM J Math Anal 38:595–616.
16. Herty, M, Moutari, S, Rascle, M (2006) Networks and Heterogenous Media 1:275–294.
17. Haut, B, Bastin, G (2005) A second order model for road traffic networks. In: 8th International IEEE Conference on Intelligent Transportation Systems. 178–184.
18. Haut, B, Bastin, G (2007) Networks and Heterogeneous Media 2:227–253.
19. LeVeque, R (2001) Some Traffic Flow Models Illustrating Interesting Hyperbolic Behavior. http://www.amath.washington.edu/~rjl/pubs/traffic.
20. Zhang, P, Wong, S C (2006) Phys Rev E 74:026109.
21. Zhang, P, Wong, S C, Dai, S Q (2006) Chinese Physics Letter 23:516–519.

Modeling Lane-Changing Decisions with MOBIL

Martin Treiber and Arne Kesting

Technische Universität Dresden, Institute for Transport & Economics,
Andreas-Schubert-Straße 23, D-01062 Dresden, Germany

Summary. We present the general model MOBIL ("Minimizing Overall Braking Induced by Lane Changes") to derive lane-changing rules for a wide class of car-following models. Both the utility of a given lane and the risk associated with lane changes is determined in terms of longitudinal accelerations calculated with microscopic traffic models. This allows for the formulation of compact and general safety and incentive criteria both for symmetric and asymmetric passing rules. Moreover, anticipative elements and the crucial influence of velocity differences of the longitudinal traffic models are automatically transferred to the lane-changing rules. While the safety criterion prevents critical lane changes and collisions, the incentive criterion takes into account not only the own advantage but also the (dis-)advantages of other drivers associated with a lane change via a "politeness factor". The parameter allows to vary the motivation for lane-changing from purely egoistic to a more cooperative driving behavior. This novel feature allows first to prevent change lanes for a marginal advantage if this obstructs other drivers, and, second, to let a "pushy" driver induce a lane change of a slower driver ahead in order to be no longer obstructed. In a more general context, we show that applying the MOBIL concept without politeness to simple car-following models and cellular automata results in lane changing models already known in the literature.

1 Introduction

In the past, single-lane car-following models have been successfully applied to describe traffic dynamics [1]. Particularly, collective phenomena such as traffic instabilities and the spatiotemporal dynamics of congested traffic can be well understood within the scope of single-lane traffic models. But real traffic consists of different types of vehicles, e.g., cars and trucks. Therefore, a realistic description of heterogeneous traffic streams is only possible within a multi-lane modeling framework allowing faster vehicles to improve their driving conditions by passing slower vehicles. Hence, freeway lane changing has recently received increased attention [2–4].

The modeling of lane changes is typically considered as a multi-step process. On a *strategic* level, the driver knows about his or her route in a network which influences the lane choice, e.g., with regard to lane blockages, on-ramps, off-ramps, or other mandatory merges [5]. In the *tactical* stage, an intended lane change is prepared and initiated by advance accelerations or decelerations of the driver, and possibly by cooperation of drivers in the target lane [6]. Finally, in the *operational* stage, one determines if an immediate lane change is both safe and desired [7].

In this contribution, we model only the operational decision process. When considering a lane change, we assume that a driver makes a trade-off between the expected own advantage and the disadvantage imposed on other drivers. In particular, our model includes the follower on the target lane in the decision process. For a driver considering a lane change, the subjective *utility* of a change increases with the gap to the new leader on the target lane. However, if the velocity of this leader is lower, it may be favorable to stay on the present lane despite of the smaller gap. A criterion for the utility including *both* situations is the difference of the accelerations after and before the lane change, at least, if the acceleration of the longitudinal model is sensitive to velocity differences. Consequently, the utility of a given lane increases with the acceleration possible on this lane: The higher the acceleration, the nearer it is to the "ideal" acceleration on an empty road and the more attractive it is to the driver. Therefore, the basic idea of our lane-changing model is to formulate the anticipated advantages and disadvantages of a prospective lane change in terms of single-lane accelerations.

Compared to explicit lane-changing models, the formulation in terms of accelerations of a longitudinal model has several advantages. First, the assessment of the traffic situation is transfered to the acceleration function of the car-following model, which allows for a compact and general model formulation with only a small number of additional parameters. In contrast to the classical gap-acceptance approach, critical gaps are not taken into account explicitly. Second, it is ensured that both longitudinal and lane-changing models are consistent with each other. For example, if the longitudinal model is collision-free, the combined models will be accident-free as well. Third, any complexity of the longitudinal model such as anticipation is transfered automatically to a similarly complex lane-changing model. Finally, the braking deceleration imposed on the new follower on the target lane to avoid accidents is an obvious measure for the (lack of) safety. Thus, safety and motivational criteria can be formulated in a unified way.

The contribution is structured as follows: In Sec. 2, the safety and the incentive criteria of the lane-changing model MOBIL will be formulated for symmetric lane-changing rules. In Secs. 2.3 and 2.4, the general rules will be applied to simple car-following models leading to lane-changing models already known in the literature. Asymmetric lane-changing rules will be presented in Sec. 3. We will conclude with a discussion in Sec. 4.

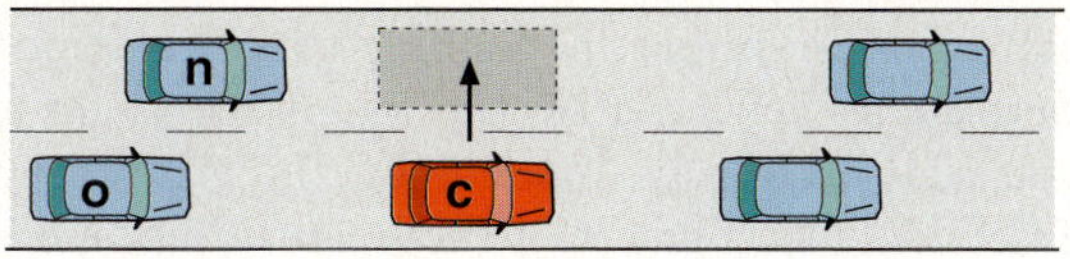

Fig. 1. Sketch of the nearest neighbors of a central vehicle c considering a lane change to the left. The new and old successors are denoted by n and o, respectively.

2 Lane-Changing for Symmetric Passing Rules

In the following, we will formulate the lane-changing model MOBIL for the class of car-following models which are defined by an acceleration function of the general form

$$a_\alpha := \frac{\mathrm{d}v_\alpha}{\mathrm{d}t} = a(s_\alpha, v_\alpha, \Delta v_\alpha). \tag{1}$$

That is, the motion of a single driver-vehicle unit α depends on its velocity v_α, the gap s_α to the front vehicle $(\alpha - 1)$ and the relative velocity $\Delta v_\alpha = v_\alpha - v_{\alpha-1}$. Generalizations to models taking into account more than one predecessor or an explicit reaction time are straightforward [8].

A specific lane change, e.g., from the right lane to the left lane as shown in Fig. 1, generally depends on the leader and the follower on the present and the target lane, respectively. In order to formulate the lane-changing criteria, we use the following notation: For a vehicle c considering a lane change, the followers on the target and present lane are represented by n and o, respectively. The acceleration a_c denotes the acceleration of vehicle c on the actual lane, while $\tilde{a}_c$ refers to the prospective situation on the target lane, i.e., to the expected acceleration of vehicle c on the target lane for the same position and velocity. Likewise, $\tilde{a}_o$ and $\tilde{a}_n$ denote the acceleration of the old and new followers after the lane change of vehicle c. Note that the leader on the target lane is the nearest vehicle on this lane for which the position is $x > x_c$. Likewise for the followers for which $x < x_c$. This also applies for the case where the vehicles on neighboring lanes are nearly side by side and a possible change would lead to negative gaps. In this case, the longitudinal model must return a very high braking deceleration such that lane changes are excluded by the criteria to be discussed below.

2.1 Safety Criterion

The safety criterion checks the possibility of executing a lane change by considering the effect on the follower n in the target lane, cf. Fig. 1. Formulated in terms of longitudinal accelerations, the safety criterion guarantees that, after the lane change, the deceleration $\tilde{a}_n$ of this vehicle does not exceed a given safe value b_{safe}, i.e.,

$$\tilde{a}_n \geq -b_{\text{safe}}. \tag{2}$$

Although formulated as a simple inequality, this condition implicitly contains all the dependencies reflected by the longitudinal car-following model, as the acceleration $\tilde{a}_n(t)$ typically depends on the gap, the velocity and the approaching rate, cf. Eq. (1). That is, if the longitudinal model has a built-in sensitivity with respect to *velocity differences*, this dependency is inherited to lane-changing decisions. In this way, larger gaps between the following vehicle in the target lane and the own position are required to satisfy the safety constraint if the following vehicle is faster than the changing vehicle. In contrast, smaller gaps are acceptable if the following vehicle is slower. Compared to conventional gap-acceptance models, this approach depends on gaps only indirectly, via the dependence on the longitudinal acceleration.

By formulating the criterion in terms of safe braking decelerations of the longitudinal model, collisions due to lane changes are *automatically* excluded. For realistic longitudinal models, $b_{\rm safe}$ should be well below the maximum possible deceleration $b_{\rm max}$ which is about $9\,{\rm m/s^2}$ on dry road surfaces. Increasing the value for $b_{\rm safe}$ generally leads to stronger perturbations due to individual lane changes. This is relevant in traffic simulations due to the fact that performing a lane change implies a discontinuous change in the input parameters in the acceleration function of the new follower.

2.2 Incentive Criterion

An actual lane change is only executed if, besides the safety criterion (2), the incentive criterion is simultaneously fulfilled. The *incentive criterion* typically determines whether a lane change improves the individual local traffic situation of a driver. In the presented model, we propose an incentive criterion that includes a consideration of the immediately affected neighbors as well. A *politeness factor p* determines to which degree these vehicles influence the lane-changing decision of a driver. For symmetric overtaking rules, we neglect differences between the lanes and propose the following incentive criterion for a lane-changing decision of the driver of vehicle c:

$$\underbrace{\tilde{a}_c - a_c}_{\text{driver}} + p\,\big(\underbrace{\tilde{a}_n - a_n}_{\text{new follower}} + \underbrace{\tilde{a}_o - a_o}_{\text{old follower}}\big) > \Delta a_{\rm th}. \tag{3}$$

The first two terms denote the advantage (*utility*) of a possible lane change for the driver him- or herself, where $\tilde{a}_c$ refers to the new acceleration for vehicle c after a prospective lane change, and a_c to the acceleration in the present situation. The considered lane change is attractive if the driver can accelerate more. The third term with the prefactor p is an innovation of the presented model. It denotes the total advantage (acceleration gain – or loss, if negative) of the two immediately affected neighbors, weighted with the politeness factor p. It can of course be argued to take into account only the new follower, at least to give him more weight than the old follower, who will anyway find him- or herself in an advantageous situation after the lane change of the leading vehicle. However, it is straightforward to adapt Eq. (3) accordingly. Finally, the switching threshold Δa_{th} on the right-hand side of Eq. (3) models a certain inertia and prevents lane changes if the overall advantage is only marginal compared to a "keep lane" directive.

In summary, the incentive criterion is fulfilled if the own advantage (acceleration gain) is greater than the weighted sum of the disadvantages (acceleration losses) of the new and old successors augmented by the threshold Δa_{th}. Note that the threshold Δa_{th} influences the lane-changing behavior *globally*, while the politeness parameter affects the lane-changing behavior locally, i.e., with respect to the involved neighbors. As is the case for the safety constraint (2), our incentive criterion is more general than a simple gap-based rule. If the longitudinal model is sensitive to velocity differences, there may be an incentive for a lane change even if the gap on the new lane is smaller – provided that the leader on the new lane is faster. The generalization to traffic on more than two lanes per direction is straightforward. If, for a vehicle on a center lane, the safety and incentive criteria are satisfied for both neighboring lanes, the change is performed to the lane where the incentive is larger.

Since the disadvantages of other drivers and the own advantage are balanced via the politeness factor p, the lane-changing model contains typical strategic features of classical game theory. The value of p can be interpreted as the degree of altruism. It can vary from $p = 0$ (for selfish lane-hoppers) to $p > 1$ for altruistic drivers, who do not change if that would deteriorate the traffic situation of the followers. They would even perform disadvantageous lane changes if this would improve the situation of the followers sufficiently. By setting $p < 0$, even malicious drivers could be modeled who accept own disadvantages in order to thwart others. Together with the parameter b_{safe} of the safety criterion (2), a classification of different driver types is depicted in Fig. 2. By means of simulation, we found that realistic lane-changing behavior results for politeness parameters in the range $0.2 < p < 0.5$ [9]. In the special case $p = 1$ and $\Delta a_{\mathrm{th}} = 0$, the incentive criterion simplifies to

$$\tilde{a}_c + \tilde{a}_n + \tilde{a}_o > a_c + a_n + a_o. \tag{4}$$

Thus, lane changes are only performed, when they increase the sum of accelerations of all involved vehicles which corresponds to the concept of *"Minimizing Overall Braking Induced by Lane Changes"* (MOBIL) in the strict sense. When identifying the safe braking threshold b_{safe} to the desired braking deceleration of the underlying car-following model, the strict MOBIL strategy corresponding to $p = 1$ has no free parameters and might therefore be considered as a "minimal model" for lane-changing decisions. In the general case, MOBIL contains three parameters, b_{safe}, p, and Δa_{thr}.

2.3 Application to the Optimal Velocity Model

For reasons of illustration, we will now apply the lane-changing rules (2) and (3) to the Optimal Velocity Model [10] as a simple representative of a

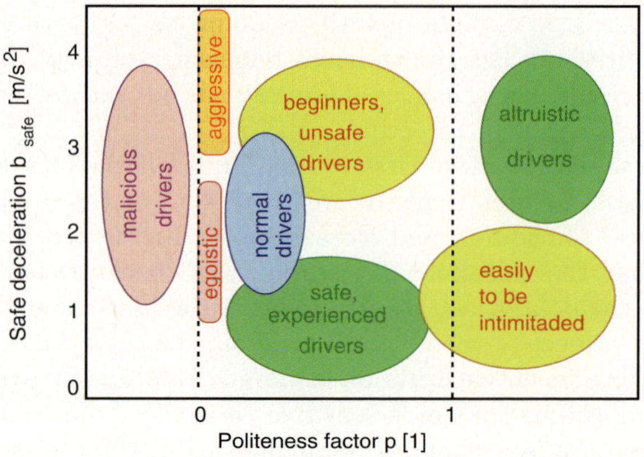

Fig. 2. Classification of different driver types with respect to the safe deceleration parameter and the politeness factor. While the safety criterion prevents critical lane changes and collisions, the incentive criterion also takes into account the (dis-) advantages of other drivers associated with a lane change. Most other lane-changing models implicitly adopt an egoistic behavior ($p = 0$), and often do not allow any interaction with the new follower ($b_{\text{safe}} = 0$). For $p = 1$, lane changes always lead to an increase of the average accelerations of the vehicles involved (MOBIL principle).

car-following model. The acceleration equation of the Optimal Velocity Model for a vehicle α can be written in the form

$$a_\alpha(t) = \frac{\mathrm{d}v_\alpha}{\mathrm{d}t} = \frac{V_{\text{opt}}(s_\alpha(t)) - v_\alpha(t)}{\tau}, \tag{5}$$

where $V_{\text{opt}}(s)$ represents the "optimal velocity function", i.e., the equilibrium velocity for a given spatial vehicle gap s. Defining the inverse $s_{\text{opt}}(v)$ of this function, i.e., the equilibrium distance for a given velocity v, the safety criterion (2) implies for the new follower n on the target lane a minimum safe distance given by

$$\tilde{s}_n > s_{\text{opt}}\left(v_n - \tau b_{\text{safe}}\right). \tag{6}$$

The incentive criterion (3) without politeness factor ($p = 0$) implies

$$\left(V'_{\text{opt}}(s_c) > 0\right) \quad \text{AND} \quad \left(\tilde{s}_c > s_c + \frac{\Delta a_{\text{thr}}\tau}{V'_{\text{opt}}(s_c)}\right), \tag{7}$$

where a first-order Taylor expansion of the optimal velocity function has been assumed. This approximation is justified by the small values of $\Delta a_{\text{thr}}\tau$ which are $0.1\,\text{m/s}$ for the chosen parameters (see below).

The resulting lane-changing rules define a simple gap-acceptance model: The safety criterion is fulfilled if the gap $\tilde{s}_n$ to the back vehicle on the target lane is larger than the equilibrium gap for the actual velocity v_n reduced by τb_{safe}. The incentive criterion is satisfied if there is an interaction at all ($V'_{\text{opt}}(s) > 0$), and if the gap to the front vehicle $\tilde{s}_c$ on the other lane is larger by the amount of $\Delta a_{\text{thr}}\tau/V'_{\text{opt}}(s_c)$. The decision model has two parameters: The safe deceleration with a typical value of $b_{\text{safe}} = 3\,\text{m/s}^2$, and a lane-changing threshold of the order of $\Delta a_{\text{thr}} = 0.1\,\text{m/s}$. Assuming a typical value $\tau = 0.5\,\text{s}$ for the OVM velocity adaptation time, and typical values for the gradient $V'_{\text{opt}}(s)$ of the optimal-velocity function of the order of 1/s, we have $\tau b_{\text{safe}} = 1.5\,\text{m/s}$, and $\Delta a_{\text{thr}}\tau/V'_{\text{opt}}$ of the order of 0.1 m. If both terms are neglected, the OVM safety criterion simply states that the new lag gap must be at least equal to the "optimal" gap, while an incentive to change lanes is given if the lead gap on the new lane is larger than that on the present lane.

2.4 Application to the Nagel-Schreckenberg Model

Now, we will apply the lane-changing criteria (2) and (3) to the deterministic part of the Nagel-Schreckenberg model [11] as generic representative of cellular automata in traffic modeling. Its update rule is defined by

$$v_\alpha(t+1) = \min(v_\alpha + 1, v_0, s_\alpha). \tag{8}$$

Here, the time t is given in seconds, v_α is the velocity of vehicle α in units of 7.5 m/s, v_0 the maximum velocity (in the same units), and s_α the gap measured by the number of empty cells of 7.5 m length. This rule may be interpreted as a discretized version of the car-following equation

$$\frac{dv_\alpha}{dt} = \min(1, v_0 - v_\alpha, s_\alpha - v_\alpha). \tag{9}$$

Applying the rules (2) and (3) (with $p = 0$ and $\Delta a_{thr} < 1$) leads to the safety criterion

$$\tilde{s}_n > v_n - b_{\text{safe}}, \tag{10}$$

and the incentive criterion

$$s_c < \min(v_0, \tilde{s}_c). \tag{11}$$

Remarkably, for $b_{\text{safe}} = 0$, these rules are identical to one of the set of rules proposed by Wagner et al. [12]. In summary, the MOBIL scheme produces purely gap-oriented lane-changing rules when applied to the OVM and the Nagel-Schreckenberg model, i.e., the required gap sizes depend on the own velocity but not on velocity differences. These (not very realistic) results reflect the fact that the underlying longitudinal models do not depend on the velocity difference themselves. In contrast, when applying the MOBIL principle to longitudinal models that are sensitive to velocity differences, the resulting lane-changing models depend on velocity differences as well [9].

3 Lane-Changing for Asymmetric Passing Rules

In most European countries, the driving rules for lane usage are restricted by legislation. We now formulate an asymmetric lane-changing criterion for two-lane freeways and assume, without loss of generality, that the right lane is the default lane, i.e., we implement a "keep-right" directive. Specifically, we presuppose the following "European" traffic rules: (i) *Passing rule*: Passing on the right-hand lane is forbidden, unless traffic flow is bound or congested, in which case the symmetric rule (3) applies. We treat any vehicle driving at a velocity below some suitably specified velocity v_{crit}, e.g., $v_{\text{crit}} = 60\,\text{km/h}$, as driving in bound or congested traffic. (ii) *Lane usage rule*: The right lane is the default lane. The left lane should only be used for the purpose of overtaking. The passing rule is implemented by replacing the longitudinal dynamics on the right-hand lane by the condition

$$a_c^{\text{Eur}} = \begin{cases} \min(a_c, \tilde{a}_c) & \text{if } v_c > \tilde{v}_{\text{lead}} > v_{\text{crit}}, \\ a_{\text{c}} & \text{otherwise}, \end{cases} \tag{12}$$

where $\tilde{a}_c$ corresponds to the acceleration of the considered vehicle if it were on the left lane (at the same longitudinal coordinate), and $\tilde{v}_{\text{lead}}$ denotes the velocity of the front vehicle on the left-hand lane. The passing rule influences the acceleration on the right-hand lane only (i) if there is no congested traffic ($\tilde{v}_{\text{lead}} > v_{\text{crit}}$), (ii) if the front vehicle on the left-hand lane is slower ($v_c > \tilde{v}_{\text{lead}}$) and (iii) if the acceleration $\tilde{a}_c$ for following this vehicle would be lower than the single-lane acceleration a_c in the actual situation. Note that the condition $v_c > \tilde{v}_{\text{lead}}$ prevents that vehicles on the right-hand lane brake whenever they are passed.

The "keep-right" directive of the lane-usage rule is implemented by a constant bias Δa_{bias} in addition to the threshold Δa_{th}. Furthermore, we neglect the disadvantage (or advantage) of the successor in the right lane in Eq. (3) because the left lane has priority. This does not mean that this vehicle will be ignored, because the safety criterion is applied in any case, see Fig. 3. Explicitly speaking, the resulting asymmetric incentive criterion for lane changes from left to right reads

$$\tilde{a}_c^{\text{Eur}} - a_c + p\,(\tilde{a}_o - a_o) > \Delta a_{\text{th}} - \Delta a_{\text{bias}}, \tag{13}$$

while the incentive criterion for a lane change from right to left is given by

$$\tilde{a}_c - a_c^{\text{Eur}} + p\,(\tilde{a}_n - a_n) > \Delta a_{\text{th}} + \Delta a_{\text{bias}}. \tag{14}$$

Again, the quantities with a tilde refer to the new situation after a prospective lane change. While the parameter Δa_{bias} is small, it clearly has to be larger than the threshold Δa_{th}. Otherwise, the switching threshold would prevent changes to the right-hand lane even on an empty road.

Neglecting the follower on the right-hand lane for the incentive criterion allows one to model the following situation: Via the politeness factor p, a driver on the right lane considering a lane change to the left takes into account the *disadvantage* of the approaching vehicle in the target lane. This can prevent the considered lane change, even if the lane change is not critical which is assured by the safety criterion (2). This feature of the MOBIL

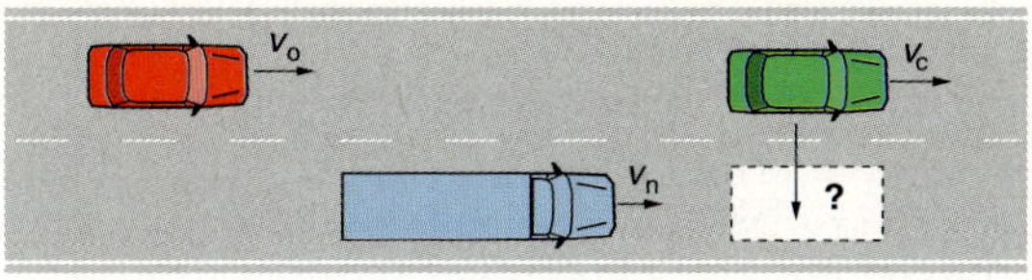

Fig. 3. The asymmetric incentive criterion additionally includes only the following vehicle in the (left) passing lane. The sketch illustrates the "dynamic pressure" which is imposed by a fast follower o to the vehicle c. The succeeding driver may induce a lane change of vehicle c to the right lane if the disadvantage (of being hindered) exceeds the own disadvantage in the right lane. This "passive cooperation" of the subject c is frequently observed in countries with asymmetric lane-changing rules, e.g., after having passed a slow truck.

lane-changing model realistically reflects a perceptive and anticipative driving behavior, as commonly observed for asymmetric passing rules. Furthermore, by taking into account only the follower on the faster (left) lane via the politeness factor p, one models a selective *dynamic pressure* to change lanes that faster (possibly tailgating) drivers on the fast (left) lane exert on their slower predecessors, see Fig. 3. This is a frequently observed behavior on European freeways, particularly on Germany freeways with their wide distribution of desired velocities.

4 Discussion and Conclusions

We have presented the general concept MOBIL (*"Minimizing Overall Braking Induced by Lane Changes"*) defining lane-changing models for a broad class of car-following models. The basic idea of MOBIL is to measure both the attractiveness of a given lane, i.e., its utility, and the risk associated with lane changes in terms of accelerations. This means, both the incentive criterion and the safety constraint can be expressed in terms of the acceleration function of the underlying car-following model, which allows for an efficient and compact formulation of the lane-changing model with only a small number of additional parameters. As a consequence, the properties of the car-following model, e.g., any dependence on relative velocities or the exclusion of collisions are transfered to the lane-changing behavior. By virtue of the acceleration-based decisions, the lane changes are more anticipative as that of gap-based models. For example, if a leading vehicle on a possible target lane is faster than the own leader, MOBIL in combination with models that are sensitive to velocity differences such as the Gipps model [13] or the Intelligent Driver Model [14], can suggest a lane change even if the lead gap on the target lane is smaller than that on the actual lane. In a way, MOBIL *anticipates* that the gap will be larger in the future. In contrast, we have shown that MOBIL produces purely gap-oriented lane-changing rules for the Optimal Velocity Model and the Nagel-Schreckenberg cellular automaton.

Furthermore, our model takes into account other drivers via a *politeness factor* p. The politeness factor characterizes the degree of "passive" cooperativeness among drivers, i.e., the subject vehicle makes a decision by considering its effects on other drivers. More specifically, even advantageous lane changes will not be performed if the personal advantage is smaller than the disadvantage to the traffic environment, multiplied by p. Furthermore, a "pushy" driver is able to initiate a lane change of his or her leader, which is a commonly observed driving behavior in countries with asymmetric lane-changing rules and dedicated passing lanes.

Finally, extensions of the proposed acceleration-based concept to other discrete decision processes of drivers are possible as well. For example, when approaching a traffic light that switches from green to amber, one has to decide whether to stop in front of the signal or to pass it with unchanged speed. In the framework of MOBIL, the "stop" decision will be based on the safe braking deceleration b_{safe}. Similar considerations apply when deciding whether it is safe enough to cross an unsignalized intersection, entering a priority road, or to start an overtaking maneuver on the opposite lane of a two-way rural road [15].

References

1. D. Helbing, "Traffic and related self-driven many-particle systems," Reviews of Modern Physics **73,** 1067–1141 (2001).
2. J. Laval, M. Cassidy, and C. Daganzo, "Impacts of Lane Changes at Merge Bottlenecks: A Theory and Strategies to Maximize Capacity," in *Traffic and Granular Flow '05*, A. Schadschneider, T. Pöschel, R. Kühne, M. Schreckenberg, and D. E. Wolf, eds., (Springer, Berlin, 2007), pp. 577–586.
3. B. Coifman, S. Krishnamurthy, and X. Wang, "Lane-changing maneuvers consuming freeway capacity," in *Traffic and Granular Flow '03*, S. Hoogendoorn, S. Luding, P. Bovy, M. Schreckenberg, and D. Wolf, eds., (Springer, Berlin, 2005), pp. 3–14.
4. K. Nagel, D. Wolf, P. Wagner, and P. Simon, "Two-lane traffic rules for cellular automata: A systematic approach," Physical Review E **58,** 1425–1437 (1998).
5. T. Toledo, C. F. Choudhury, and M. E. Ben-Akiva, "Lane-Changing Model with Explicit Target Lane Choice," Transportation Research Record: Journal of the Transportation Research Board **1934,** 157–165 (2005).
6. P. Hidas, "Modelling vehicle interactions in microscopic traffic simulation of merging and weaving," Transportation Research Part C: Emerging Technologies **13,** 37–62 (2005).
7. P. G. Gipps, "A model for the structure of lane-changing decisions," Transportation Research Part B: Methodological **20,** 403–414 (1986).
8. M. Treiber, A. Kesting, and D. Helbing, "Delays, inaccuracies and anticipation in microscopic traffic models," Physica A **360,** 71–88 (2006).
9. A. Kesting, M. Treiber, and D. Helbing, "MOBIL – A general lane-changing model for car-following models," Transportation Research Record: Journal of the Transportation Research Board (in print) (2007).

10. M. Bando, K. Hasebe, A. Nakayama, A. Shibata, and Y. Sugiyama, "Dynamical model of traffic congestion and numerical simulation," Physical Review E **51,** 1035–1042 (1995).
11. K. Nagel and M. Schreckenberg, "A cellular automaton model for freeway traffic," J. Phys. I France **2,** 2221–2229 (1992).
12. C. Wagner, K. Nagel, and D. Wolf, "Lane-changing rules in two lane traffic simulation using cellular automata," Physica A **234,** 687 (1997).
13. P. G. Gipps, "A behavioural car-following model for computer simulation," Transportation Research Part B: Methodological **15,** 105–111 (1981).
14. M. Treiber, A. Hennecke, and D. Helbing, "Congested traffic states in empirical observations and microscopic simulations," Physical Review E **62,** 1805–1824 (2000).
15. M. Treiber and A. Kesting, "Acceleration-Based Model for Discrete-Choice Situations in Vehicular Traffic," In *Submitted to the Annual Meeting 2008*, (Transportation Research Board, Washington, D.C., 2008).

Kinetic Derivation for a Traffic Flow Model

Alma R. Méndez and Rosa M. Velasco

Department of Physics, Universidad Autónoma Metropolitana, 09340, México
armr@xanum.uam.mx, rmvb@xanum.uam.mx

Summary. A macroscopic traffic model for aggressive drivers is developed from a kinetic equation by means of a maximization procedure and Grad's method. The macroscopic equations thus obtained contain a kind a of viscosity effect as well as density gradients. An stability analysis is developed and stable regions are observed. The simulation results are shown for one initial condition and they are in general agreement with traffic flow characteristics.

1 Introduction

Macroscopic model equations to study traffic flow characteristics have been studied for a long time. In the evolution of such models several lines of thought have been developed, going from those derived from some phenomenological considerations up to their derivation from microscopic assumptions [1, 2]. Derivations based on Kinetic Theory have also been presented, in this kind of approach it is assumed that the single distribution function for a vehicle in a highway satisfies a kinetic equation. In this work, the starting point is the Paveri-Fontana equation [3], this kinetic equation considers a phase space in which the individual desired velocity of drivers is considered explicitly, consequently the distribution function depends on such variable. To our knowledge, the complete Paveri-Fontana equation has not been solved in an analytical way, though numerical solutions exist for some special cases [4]. Then, Paveri-Fontana equation has been used to construct macroscopic equations to generate models based on a kinetic support [5–7]. As noticed above, the Paveri-Fontana distribution function contains the individual desired velocity of drivers as an independent variable, it is very difficult to give a prescription about that because it depends on the drivers disposition. As an alternative procedure we have found that the integrated equation can be solved in an exact way for the homogeneous steady state, if we introduce a model for the averaged desired velocity [8, 9]. Thinking in this way, firstly in this work we solved the Paveri-Fontana equation in the homogeneous steady state for a specific model of the average desired velocity of drivers. By means of a maximization procedure [10], a local distribution function is also obtained. Such a distribution function is taken as a basis to develop the Grad's method [11] which allows us to expand any distribution function in terms of a complete set of orthonormal polynomials. The distribution function obtained through this method contains the density and the average velocity of vehicles as macroscopic variables to describe the time evolution in the system. In order to consider the spatial inhomogeneities, it was assumed a relaxation behavior in the perturbed distribution function [8], as a result we obtained the contribution of the gradients in the density and velocity. In the following step we used the distribution function to calculate the traffic pressure which gave us the closure relation to be introduced in the macroscopic equations. The closed macroscopic model is simulated with periodic boundary conditions and one initial situation. The simulation results show a general agreement [1] with the characteristics of traffic flow and the corresponding behavior is shown.

2 The Kinetic Equation

To start with the kinetic model we will develop here, let us recall that the complete Paveri-Fontana equation considers a distribution function $g(v, w, x, t)$ which depends on the instantaneous velocity v, the individual desired velocity of drivers w, position x and time t. The desired velocity of drivers, in principle, carries the information about drivers' disposition. This characteristic makes the problem somewhat uncontrollable, in such a way that a model for the desired velocity is necessary to go further. On the other hand the complete Paveri-Fontana equation can be used to construct an evolution equation for $f(v, x, t)$ which does not contain explicitly the desired velocity. Such an equation is constructed by means of the integration over the individual desired velocity,

$$\frac{\partial f}{\partial t} + v\frac{\partial f}{\partial x} + \frac{\partial}{\partial v}\left(f\frac{V_0 - v}{\tau}\right) = f\int_0^{\infty}(1-p)(v' - v)f(x, v', t)dv', \quad (1)$$

where $f(v, x, t)dvdx$ is the distribution function which gives us the probability to find a vehicle with instantaneous velocity v in $(v, v+dv)$, position in $(x, x+dx)$ at time t. The left hand side of (1) represents the drift in the phase space (v, x) and the right hand side corresponds to the interaction between vehicles, p is the probability of overtaking. Then we have a kinetic equation in which it is the average desired velocity $V_0(v, x, t)$ the quantity which plays a role, it is defined by $V_0(v, x, t) = \int_0^{\infty} w\frac{g(x,v,w,t)}{f(v,x,t)}dw$. This desired velocity is an explicit function of the instantaneous velocity, position and time. In this work we will assume a model for aggressive drivers, hence

$$V_0(v, x, t) = \omega v, \quad \omega > 1, \quad (2)$$

where ω is considered as a constant. With this model, equation (1) can be solved exactly for the equilibrium distribution function $f_e(v)$ given by

$$f_e(v) = \frac{\alpha}{\Gamma(\alpha)}\frac{\rho_e}{V_e}\left(\frac{\alpha v}{V_e}\right)^{\alpha-1} exp\left(-\frac{\alpha v}{V_e}\right), \tag{3}$$

where the density ρ_e and the velocity V_e correspond to the equilibrium state and the numerical value of the equilibrium velocity is determined by the fundamental diagram, $\Gamma(\alpha)$ is the gamma function and α is a dimensionless quantity given by $\alpha = \frac{\rho_e(1-p)V_e\tau}{\omega-1}$, which is a constant including information about the equilibrium state, as well as the model parameters (p,ω,τ). With the distribution function we have found (3), it is possible to calculate the velocity variance $\Theta = V_e^2/\alpha$ and it is important to emphasize that this expression for the velocity variance, in this special state, can be compared with experimental data [12].

3 Entropy Maximization

The distribution function $f_e(v)$ describes the equilibrium state for a special traffic situation and to go further we need a distribution function representing the behavior of the system out of this reference state. We obtain that distribution function using the informational entropy as introduced by Shannon [13]

$$s(x,t) = -\int_0^\infty f^{(0)}(v,x,t) Ln\left(\frac{f^{(0)}(v,x,t)}{f_e(v)v}\right) dv, \tag{4}$$

where $f^{(0)}(v,x,t)$ will be the distribution function corresponding to this nonhomogeneous nonsteady state, which will be determined by the maximization of the entropy with the restriction of the density given by

$$\rho(x,t) = \int_0^\infty f^{(0)}(v,x,t)dv. \tag{5}$$

The maximization procedure gives us the next result

$$f^{(0)}(v,x,t) = \frac{\alpha}{\Gamma(\alpha)}\frac{\rho(x,t)}{V_e}\left(\frac{\alpha v}{V_e}\right)^{\alpha-1} exp\left(-\frac{\alpha v}{V_e}\right). \tag{6}$$

In this case the local state is determined by $\rho(x,t)$, $V(x,t) = V_e(\rho(x,t))$, $\Theta(x,t) = \frac{V_e^2(\rho(x,t))}{\alpha}$.

Inhomogeneities

To obtain a distribution function with the density $\rho(x,t)$ and the average velocity $V(x,t)$ as independent variables we will develop Grad's method. We consider that a general distribution function can be expanded in terms of a complete set of orthonormal polynomials generated by the local distribution function $f^{(0)}$ in such a way that

$$f(v,x,t) = f^{(0)} \sum_{n=0}^{\infty} B_n^{\alpha}(x,t) P_n^{\alpha}(y), \tag{7}$$

this expansion in the distribution function is infinite and we need a hypothesis to cut it in a finite number of terms. In order to have the average velocity as an independent variable we consider $B_2^{\alpha}(x,t) = B_3^{\alpha}(x,t) = \cdots = 0$, now the distribution function is

$$f^{(G)}(v,x,t) = f^{(0)} \left[1 + \alpha \left(\frac{V(x,t)}{V_e} - 1 \right) \left(\frac{v}{V_e} - 1 \right) \right]. \tag{8}$$

The distribution function given in (8) depends only on the local values of the relevant variables $\big(\rho(x,t), V(x,t)\big)$ and does not have the contribution of the corresponding gradients in order to measure the corresponding spatial inhomogeneities. To go further it is necessary to consider a modification of the distribution function which takes into account the contribution of the gradients of the local variables. Then we assume that

$$f(v,x,t) = f^{(G)}(v,x,t) + f^{(1)}(v,x,t) + \cdots , \tag{9}$$

where the correction to the distribution function $f^{(1)}(v,x,t)$ must satisfy the compatibility conditions

$$\int_0^{\infty} f^{(1)}(v,x,t)\, dv = 0, \quad \int_0^{\infty} v f^{(1)}(v,x,t)\, dv = 0. \tag{10}$$

These conditions imply that the density and the average velocity will be always determined by the distribution function $f^{(G)}(v,x,t)$. To complete the proposal we will assume that the interaction term in (1) can be approximated by a kind of collective relaxation, hence

$$\frac{\partial f^{(G)}}{\partial t} + v \frac{\partial f^{(G)}}{\partial x} + \frac{\partial}{\partial x} \left(\frac{W - v}{\tau} f^{(G)} \right) = \rho(1-p)(V-v) f^{(G)} - \frac{1}{\tau_0} f^{(1)}, \tag{11}$$

where τ_0 is the collective relaxation time and $W(x,t) = \omega V(x,t)$ according to our model for the average desired velocity. The substitution of $f^{(G)}$ in (11) allows us to obtain $f^{(1)}$.

4 The Traffic Pressure

The distribution function $f^{(1)}$ we have found with (11) contains the spatial gradients of the density and the velocity, and this fact takes into account the spatial inhomogeneities in the system, as measured by the corresponding gradients. The complete distribution function (9) will reproduce the values of the macroscopic variables density and velocity, however the velocity variance and consequently the complete traffic pressure are affected by the approximation we have made, a straightforward calculation shows that

$$\rho(x,t)\Theta(x,t) = \mathcal{P}(x,t) = \mathcal{P}_{local} - \rho V_e^2 \tau^* \left\{ \frac{1}{\alpha}\frac{\partial V}{\partial x} + \rho \frac{V_e}{\alpha}\frac{\partial \rho}{\partial x}\left(\frac{V}{V_e} - 1\right)\right\}, \quad (12)$$

where

$$P_{local} = P^{(G)}(x,t) - \rho V_e^2 \tau^* \left(\frac{V}{V_e} - 1\right)\left(\frac{\rho V_e}{\alpha}(1-p) - \frac{(\omega - 1)}{\tau}\right) \quad (13)$$

and $\tau^* = 2\tau_0 \frac{\alpha+1}{\alpha}$ is an effective collective relaxation time. It is clear that this new expression for the traffic pressure is modified by the presence of the gradients in the macroscopic variables. Certainly some other cumulants of the distribution function can be calculated, however the traffic pressure is enough to construct the macroscopic model, as we will see in the next section.

The expression for the traffic pressure (12), deserves some comments: Firstly, we observe a term proportional to the velocity gradient in such a way that a kind of viscosity coefficient can be identified as $\eta = \frac{\rho V_e^2 \tau^*}{\alpha}$. Secondly, there is also a term proportional to density gradient. Finally, the terms in $P_{local}(x,t)$ do not contain any gradients, in such a way that it can be identified with a kind of hydrostatic pressure and it depends on the model parameters.

5 The Macroscopic Equations

The evolution equations for the macroscopic variables can be obtained from (1) by multiplying by 1 and v and integrating over the speed in the interval $(0,\infty)$. Then, the two first equations of motion can be written as

$$\frac{\partial \rho}{\partial t} + \frac{\partial(\rho V)}{\partial x} = 0, \quad (14)$$

$$\rho\left(\frac{\partial V}{\partial t} + V\frac{\partial V}{\partial x}\right) = -\frac{\partial \mathcal{P}}{\partial x} + \rho\frac{W - V}{\tau} - \rho(1-p)\mathcal{P}. \quad (15)$$

It is clear that equations (14–15) need a closure hypothesis in order to have a closed set of equations for the density and velocity. In this model the closure hypothesis is given by the traffic pressure calculated, (12).

On the other hand, the evolution of the velocity (15) contains the usual drift terms and it becomes of second order in the derivatives of the density and the velocity. As we said above the presence of those terms give account of inhomogeneities effects as well as viscosity in the system. These characteristics allow us to classify this model as a Navier-Stokes-like one.

6 Stability

As another point in the analysis of this model, we will consider a small perturbation of the homogeneous steady state in order to study the stability regions. Hence, we assume that the density and velocity can be written as

$$\begin{aligned}\rho(x,t) &= \rho_e + \hat{\rho}\; exp(ikx + \gamma t),\\ V(x,t) &= V_e + \hat{V}\; exp(ikx + \gamma t),\end{aligned} \tag{16}$$

where $\hat{\rho}$, $\hat{V}$ do not depend on (x,t) and $\hat{\rho} \ll \rho_e$, $\hat{V} \ll V_e$. The quantity k corresponds to the wave vector and γ can be a complex function of the wave vector.

The stability condition is given by $Re\ \gamma < 0$, otherwise the homogeneous steady state is unstable with respect to small perturbations. To determine the stability regions we linearize the macroscopic equations (14–15) around ρ_e, V_e taking into account that the probability to overpass is given by $p = \exp(-10\ \frac{\rho}{\rho_{max}})$ [14]. The direct substitution of (16) in the macroscopic equations (14–15) allows us to construct a dispersion relation from which $Re\ \gamma(k, \rho_e)$ can be obtained. The results are shown in Fig. 1(a) for $\frac{\tau_0}{\tau} = 3$ and in Fig. 1(b) when $\frac{\tau_0}{\tau} = 9$ respectively. The regions where $Re\ \gamma(k, \rho_e) > 0$ correspond to the unstable case. We notice that the instability regions shrink when the quotient between the collective relaxation and the individual relaxation times grows, meaning that the value of the collective relaxation time is a stabilizing factor. In Fig. 1 the equilibrium density as well as the wave vector are taken in units of ρ_{max}.

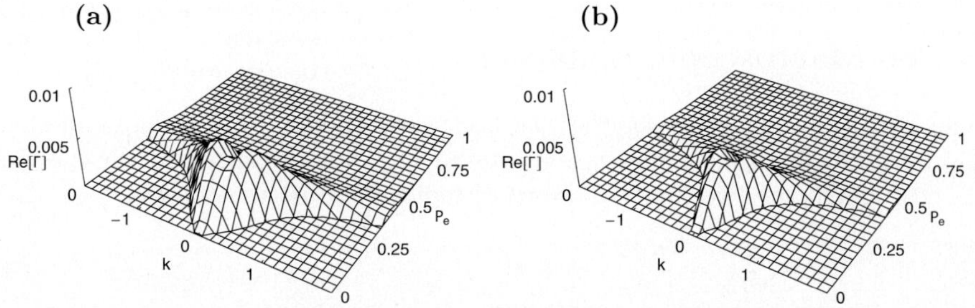

Fig. 1. Stability regions for (a) $\frac{\tau_0}{\tau} = 3$ and (b) $\frac{\tau_0}{\tau} = 9$.

7 Simulation Results

The complete set of equations must be solved by numerical means. The first step along this line is given by writing the equations in a conservative form,

$$\frac{\partial \mathbf{u}}{\partial t} + \frac{\partial \mathbf{F}(\mathbf{u})}{\partial x} = \mathbf{S}(\mathbf{u}) \tag{17}$$

where

$$\mathbf{u} = \begin{pmatrix} \rho \\ \rho V \end{pmatrix},$$

$$\mathbf{F}(\mathbf{u}) = \begin{pmatrix} \rho V \\ \rho V^2 + P \end{pmatrix},$$

$$\mathbf{S}(\mathbf{u}) = \begin{pmatrix} 0 \\ \rho \frac{W-V}{\tau} - \rho(1-p)P \end{pmatrix}. \tag{18}$$

Secondly, we make $x_i = i\ \Delta x$, $t^n = n\ \Delta t$ in such a way that $\mathbf{u}_j^n = \mathbf{u}(x_j, t^n)$. The MacCormack method [15, 16], is now written as follows

$$\begin{aligned} \tilde{\mathbf{u}}_j^n &= \mathbf{u}_j^n - \frac{\Delta t}{\Delta x}(\mathbf{F}_j^n - \mathbf{F}_{j-1}^n) + \Delta t\ \mathbf{S}_j^n, \\ \mathbf{u}_j^{n+1} &= \frac{1}{2}\left[\tilde{\mathbf{u}}_j^n + \mathbf{u}_j^n - \frac{\Delta t}{\Delta x}(\tilde{\mathbf{F}}_{j+1}^n - \tilde{\mathbf{F}}_j^n) + \Delta t\ \tilde{\mathbf{S}}_j^n\right]. \end{aligned} \tag{19}$$

The reference values for the density and velocity are taken as $\rho_{max} = 140\ veh/km$, $V_{max} = 120\ km/h$, the individual relaxation time $\tau = 30\ s$ and the collective relaxation time $\tau_0 = 90\ s$. The probability of immediate overtaking is taken according with the values given by Helbing [14] $p = exp\left(-10\ \frac{\rho}{\rho_{max}}\right)$. We will take periodic boundary conditions, so $\rho(0,t) = \rho(L,t)$ and $V(0,t) = V(L,t)$, where L is the length of the highway. As initial condition we will take the next,

$$\begin{aligned} \rho(x,0) &= \rho_e, \\ V(x,0) &= V_e(\rho_e)\left\{1 + \delta V\ sin\left(\frac{2\pi x}{L}\right)\right\}, \end{aligned} \tag{20}$$

where we have taken $L = 12\ km$, $\rho_e = 28\ veh/km$, $V_e(\rho_e) = 84\ km/h$ according to the fundamental diagram and $\delta V = 0.84\ km/h$.

The simulation results for the density and the velocity are shown in Figs. 2(a) and 2(b) where we see that a small variation in the drivers velocity produce a growing in the density which propagates along the highway. The corresponding reduction in the velocity follows the behavior of the density. We can observe that the variance calculated with (12) has a maximum in an advanced position with respect to the position in the maximum of the density as shown in Fig. 2(c). This characteristic is the usual one in real traffic.

8 Concluding Remarks

The macroscopic model we have presented is based on a kinetic equation and the Grad's method of solution as well as the approximation made have provided us with a closure in the macroscopic equations. The variance which is

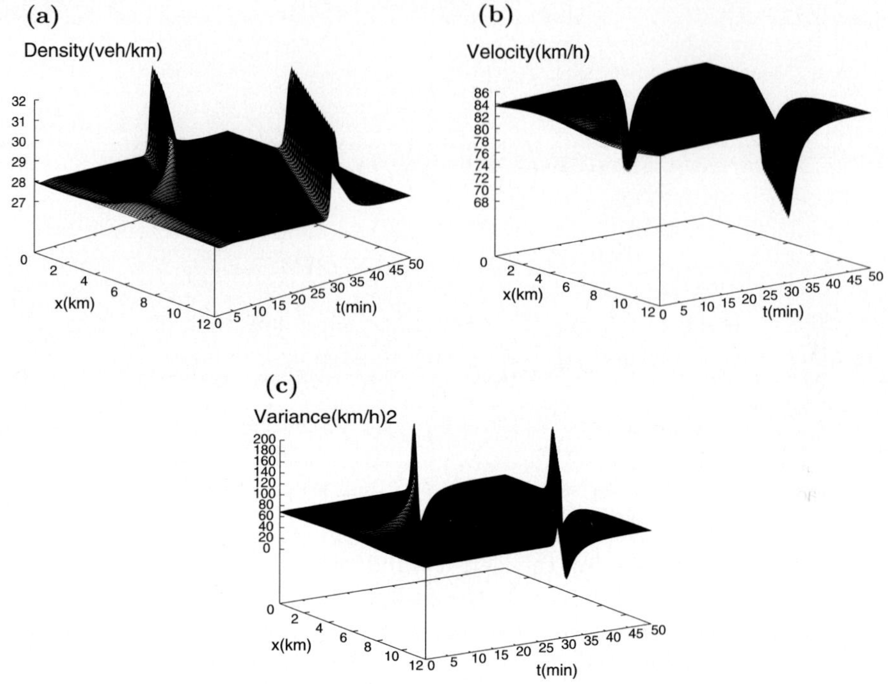

Fig. 2. Spatial and temporal behavior of the (a) density, (b) average velocity and (c) velocity variance for the initial condition given in (20).

a measure of the traffic pressure can be calculated and its behavior is illustrated in the corresponding figure. As general comments we can say that this model has the main characteristics of usual traffic models, in the sense that the density and velocity behavior are coupled in their maximums and minimums. The maximum in the variance occurs in an advanced position with respect to the maximum in the density. The initial condition considered in the simulation may correspond to a velocity disorder in the drivers. Also we can observe a propagating perturbation in which the front wave is almost undisturbed. It is worth noticing that in this model there are only two parameters: the aggression of drivers measured through the parameter ω and the collective relaxation time $\frac{\tau_0}{\tau}$. A comment on the value of ω must be made, because in our calculations its value has been taken from several experimental data. In fact, we calculated it taking the value of α from the variance prefactor, the equilibrium density and velocity from the fundamental diagram and the probability p from Helbing's data [14]. Such a calculation shows that ω is greater than one in a very small proportion ($<5\%$), a characteristic which can be interpreted as an average behavior of drivers.

References

1. Helbing D (2001) Rev. Mod. Phys. 73:1067–1141.
2. Chowdhury D, Santen L, Schadschneider A (2000) Phys. Rep. 329:199–329.
3. Paveri-Fontana S L (1975) Transp. Res. 9:225–235.
4. Wagner C (1998) J. Stat. Phys. 90:1251–1275.
5. Helbing D (1996) Phys. Rev. E 53:2366–2381.
6. Wagner C, Hoffmann C, Sollacher R, Wagenhuber J, Schürmann B (1996) Phys. Rev. E 54:5073–5085.
7. Helbing D, Treiber M (1998) Granular Matter 1:21–31.
8. Velasco R M, Marques Jr (2005) Phys. Rev. E 72:046102.
9. Velasco R M, Saavedra P Physica D 228:153–158.
10. Velasco R M, Méndez A R (2005) The informational entropy in traffic flow. In: Statistical Physics and Beyond: 2nd Mexican Meeting on Mathematical and Experimental Physics, Uribe F J, García Colín L S, Díaz-Herrera E (eds) AIP Conference Proceedings 757, New York.
11. Grad H (1949) Commun. Pure & Appl Math 2:325–330.
12. Shvetsov V, Helbing D (1999) Phys. Rev. E 59:6328.
13. Shannon C E (1984) Bell System Tech J. 27:379–623.
14. Helbing D (1998) arXiv:cond-mat/9805136v1.
15. Helbing D, Treiber M (1999) Comp. Sci. & Eng. Sept-Oct:89–99.
16. Press W H, Teukolsky S A, Vetterling W T, Flannery B P (1992) Numerical Recipes: The art of Scientific Computing. Cambridge University Press.

Qualitative Change of Car-Following Behavior Observed in Real Traffic

Yasushi Yokoya, Yoichi Asano, and Nobuyuki Uchida

Japan Automobile Research Institute, 2530 Karima, Tsukuba, Ibaraki 305-0822, Japan yyokoya@jari.or.jp, http://www.jari.or.jp

Summary. Microscopic dynamical properties of traffic flow are studied from the aspects of vehicle behavior and driver operation. We studied the nature of fluctuations around the critical region in real traffic by analyzing a time series of variations of velocity obtained from single-vehicle data measurement. We found that the probability density function calculated from the time series of velocity variations is transformed, while a Gaussian distribution transitions into a stable symmetrical Lévy distribution. The power-law tail in the Lévy distribution indicated that the time series of velocity variation exhibits critical fluctuations. The power-law tail in the probability density function suggests that dynamical processes of vehicular traffic are related to a time-discrete stochastic process driven by random amplification with additive external noise. In contrast, the empirical data of deceleration in a car-following situation obtained from the driving simulator experiment indicated a large dispersion of perceptual quantities of a driver when operating the brake pedal. The result suggests that the algorithm for operating the brake pedal is closely related to the random amplification in the discrete stochastic process.

1 Introduction

The phase transition of vehicular traffic is a universal feature of traffic flow [1–3]. One of the empirical facts characterizing the dynamical process in the phase transition of vehicular traffic is given by the metastable branch derived from the flux-density relation (i.e., the fundamental diagram), where spontaneous formation of high-density flow and its decay are observed in measurement of variations of flux with density. In this relation, flux increases in proportion to density under the critical point. However, a discontinuous reduction of flux occurs beyond the critical point, and jams eventually emerge through a mixed state of freely flowing traffic and jams (e.g., stop-and-go traffic).

In typical measurements of traffic using roadside detectors, the assumption of constant velocity and the following distance of individual vehicles on a stretch of road, which is divided by sensors, restricts the resolving power of data collection [2–5]. The restriction hinder the direct comparison of empirical data on real traffic and the results of the numerical simulations or the analytical calculations especially in the critical-density region with strong fluctuations of velocity [6]. Therefore, it is important to interpret the flux-density relation given by empirical data of the time series of the velocity and the headway of individual vehicles.

In this paper, microscopic dynamical properties of traffic flow are studied from the aspects of both vehicle behavior and human driver operation. The real-time behavior of a vehicle in real traffic is measured by an onboard apparatus. The analysis based on single-vehicle data potentially enables us to analyze the relation between the microscopic processes in the various dynamical phases and the whole surrounding situation (e.g., environmental factors, road geometry, and conditions of vehicle control). Furthermore, deceleration is measured by a driving simulator, which provides us the origin of the dynamical properties of traffic flow.

This paper is organized as follows. Section 2 presents dynamical features of microscopic traffic flow observed in real traffic. Section 3 describes driver behavior in car-following situations obtained from a driving simulator experiment. Finally, Section 4 presents the conclusions.

2 Critical Flow in Real Traffic

2.1 Data Collection

Single-vehicle data is obtained directly from an onboard measurement apparatus. Figure 1 schematically illustrates the measurement apparatus, which was made up of sensors for the vehicle states and a video camera for a forward-view image. Using sensors and laser radar, we measured acceleration, velocity, and distance headway. We also recorded a forward-view image simultaneously with a video camera installed in the front of the vehicle, in order to understand the relation between the measuring vehicle and the preceding vehicles. The motion of a vehicle in real traffic can be affected by not only the leading vehicle but also by a few vehicles ahead of the leading vehicle.

In this system, the resolving power of the data collection is determined by its sampling rate. The sampling rate in our measurement system was 10 [Hz]. Since this measurement system enabled us to calculate local density and flow directly in real time, we could analyze the states of traffic even in the critical region where strong fluctuations developed. The traffic was measured on several urban roads in Ibaraki Prefecture. These roads are primary distributors with one or two lanes, in which the average traffic flow is 30,000 vehicles per day. Traffic jams occur on these roads in the morning and during the evening rush hour.

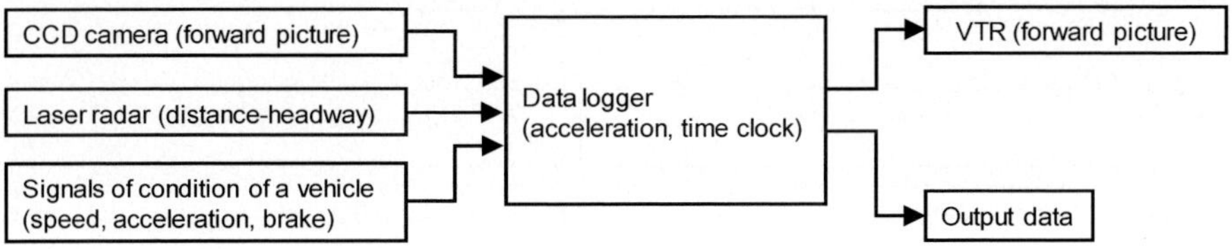

Fig. 1. Schematic diagram of measurement apparatus.

2.2 Time Series of Single-Vehicle Data

Figures 2 (a) and (d) depict successive variations of velocity as

$$\Delta v(t) \equiv \bar{v}_T(t) - \bar{v}_T(t - \Delta t). \tag{1}$$

Here, Δt is the sampling rate, and we adopted $\Delta t = 0.1$ [sec]. Our measurement apparatus contains noise (about 3 [Hz]) primarily due to the performance of the sensor detecting a pulse signal from a turning axle and variation of tire radius due to fine structures of the road surface. Therefore, we eliminated the noise by calculating the moving average of time series of velocity with time interval $T = 0.3$ [sec] as follows.

$$\bar{v}_T(t) = \frac{1}{T} \sum_{\tau=t-T+1}^{t} v(\tau). \tag{2}$$

Figures 2 present the observed time series of Δv, $\bar{v}_T$, and the following distance h for typical stable and unstable runs. Figure 2 (a) reveals small fluctuations. We can confirm that the relation between velocity and the following distance obeys the general car-following model, e.g., the GM model. However, we find large fluctuations in Fig. 2 (d), where vehicle density increases and the vehicle runs unstably. This period includes the transition process from free flow to congested flow. A particular relation between velocity and the following distance no longer appears to exist in this period.

2.3 Probability Density Function

Since we can confirm that the observed time series of Δv is stationary, it is possible to calculate the probability density function (PDF) of Δv. We introduce the PDF of Δv as

$$p\left(\Delta v(t)\right) = \frac{n(\Delta v(t), \Delta v(t) + \delta)}{N\delta}. \tag{3}$$

The quantities n, δ, and N denote the number of data Δv included in an interval of $[\Delta v(t), \Delta v(t) + \delta]$, size of time intervals, and the total number of data Δv.

Figure 3 presents a semilogarithmic plot of $p(\Delta v)$ obtained for three different time series of stable runs, one of which (open circles) has already been

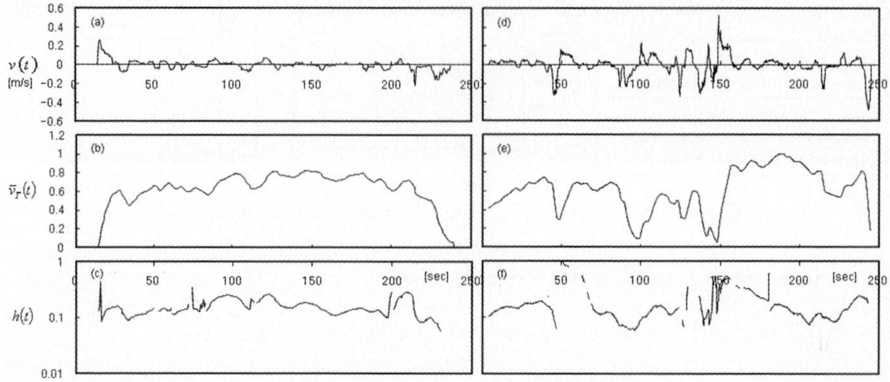

Fig. 2. Time series of single-vehicle data of stable (unstable) runs in left column (right column). (a) ((d)) Successive variation of velocity. (b) ((e)) Velocity. (c) ((f)) Following distance. Velocity (b) and following distance (c) are normalized by the maximum values of (e) and (f).

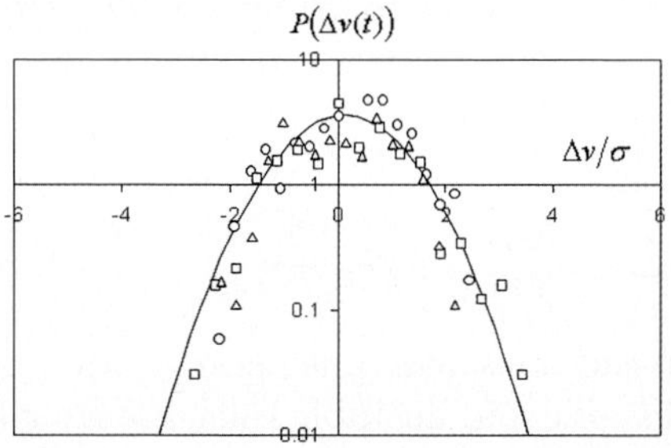

Fig. 3. Semilogarithmic plot of the probability density function Δv obtained for three different time series of stable runs. The solid line represents a Gaussian distribution.

presented in Fig. 2 (a). We find that the distributions are roughly symmetrical with respect to Δv, which is scaled by the standard deviation σ in each distribution. In Fig. 3, the solid line denotes a Gaussian distribution with standard deviation σ calculated from one of the three distributions (open circles). We find that $p(\Delta v)$ of the stable runs are well fitted with a Gaussian distribution.

Figure 4 presents distributions obtained for three different time series of the unstable runs. A time series that gives one of the distributions (open circles) has already been depicted in Fig. 2 (d). These distributions with sharp peaks and long tails are more leptokurtic than Gaussian.

The long tails of the distribution of the unstable runs can by identified by the cumulative distribution defined as

$$p\left(\geq |\Delta v(t)|\right) = \int_{-\infty}^{-|\Delta v|} p(x)dx + \int_{|\Delta v|}^{\infty} p(x)dx. \tag{4}$$

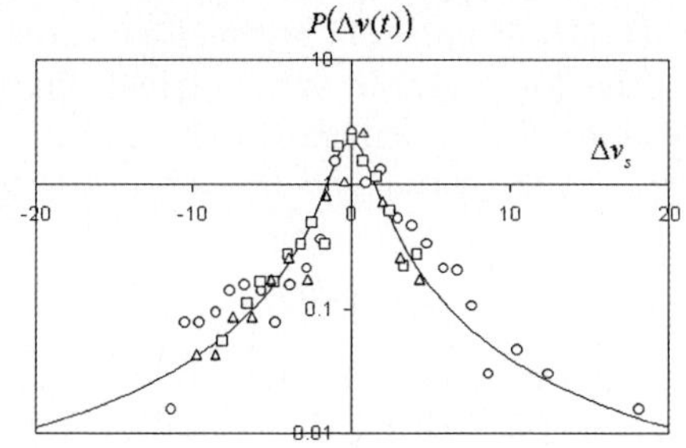

Fig. 4. Semilogarithmic plot of the probability density function of Δv obtained for three different time series of unstable runs. The solid line represents the Lévy stable symmetrical distribution.

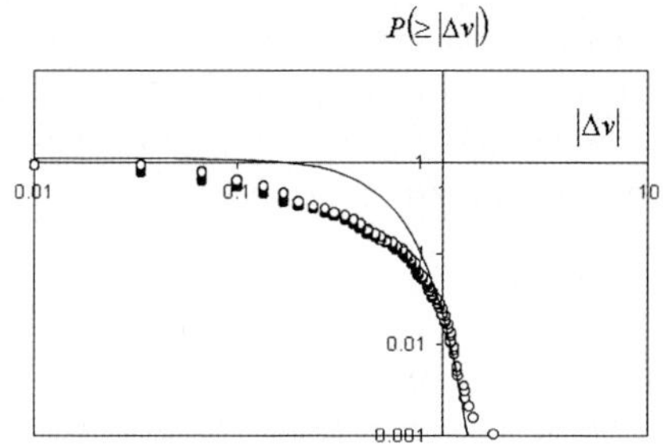

Fig. 5. Cumulative distribution for one of the three distributions given in Fig. 5. The solid line represents a Gaussian distribution.

Figure 5 plots the cumulative distribution for one of the three distributions given in Fig. 4 (open circles) with a Gaussian distribution (solid line). Figure 5 illustrates a power-law tail with an exponent α in the cumulative distribution:

$$p\,(\geq |\Delta v|) \propto |\Delta v|^{-\alpha}\,. \tag{5}$$

We find that the slope of the cumulative distribution given in Fig. 5 is fitted by a power-law distribution with an exponent $\alpha \approx 1.02$.

Taking into account the leptokurtic distribution with a sharp peak and long tails depicted in Fig. 4, the distribution is described by a Lévy stable symmetrical distribution [7, 8].

$$p\,(\Delta v) = \frac{1}{\pi} \int_0^\infty dk \cos(k\Delta v) \exp(-\gamma \Delta t k^\alpha). \tag{6}$$

Here, index α refers to the slope of the cumulative distribution given in Eq. (4). Index γ is a scale factor at $\Delta t = 1$. Scale factor γ is given by the Gamma function as

$$\gamma \Delta t = \exp\left[\alpha \ln\left(\frac{\Gamma(1/\alpha)}{\pi \alpha p(0)}\right)\right]. \tag{7}$$

Figure 4 compares the PDF $p(\Delta v_s)$ obtained for three different time series of unstable runs with the Lévy stable symmetrical distribution of $\alpha = 1.02$ and $\gamma \Delta t = 0.13$ (solid line). The horizontal axis of Δv is scaled as

$$\Delta v_s \equiv \frac{\Delta v}{(\Delta t)^{1/\alpha}}. \tag{8}$$

The index α of the PDF obtained for each time series is unique. Figure 4 demonstrates good agreement with the Lévy stable distribution in the whole region.

2.4 Origin of Power-Law Tail

It is well established that empirical observations around the critical region exhibit a remarkable repetitive property [3, 5, 9], where the velocity of the n-th vehicle evolves in the same way as that of its preceding $(n-1)$-th vehicle except for a certain time delay, which is given as

$$x_{n-1}(t) = x_n(t+T) + v_B T. \tag{9}$$

Here, v_B represents a constant velocity, at which the global pattern moves backward. In this condition, we can reduce the general car-following equation (Eq. (10)) to a one-body equation (Eq. (11)).

$$\ddot{x}_{n-1}(t+T) = \kappa \frac{\dot{x}_n(t) - \dot{x}_{n-1}(t)}{x_n(t) - x_{n-1}(t)} + f(t), \tag{10}$$

$$\ddot{x}(t+T) = K(t)\ddot{x}(t) + f(t), \tag{11}$$

where $K(t) = \kappa/(\dot{x}(t) + v_B)$, and κ is the sensitivity to the stimulus and $f(t)$ represents random additive noise (e.g., the so-called acceleration noise that is often given by white noise [1]).

A discrete stochastic process involving random amplification a_i with additive external noise b_i, which is given by Eq. (12), exhibits a power-law distribution [10, 11].

$$X_{i+1} = a_i X_i + b_i. \tag{12}$$

In Eq. (12), a stationary process emerges for the condition $\langle \ln a_i \rangle < 0$, where $\langle \cdots \rangle$ denotes an average over realizations. In addition, a power-law tail is obtained when a_i changes intermittently or randomly; for example, a_i sometimes takes values exceeding 1, and b_i needs to be so finite that it is susceptible to intermittent amplifications [11].

The contrast between Eq. (11) and Eq. (12) suggests that the power-law tail of the PDF of $\Delta v(t)$ is closely related to the nature of the driver's response to the stimulus.

3 Deceleration Process

A driver analyzes various information of road and traffic conditions as well as the relation between the preceding vehicle and his/her own vehicle (e.g., headway distance and relative speed) in order to maintain and stabilize vehicle following. The information characterizes the interaction between vehicles, which provides the fundamental aspects of traffic phenomena from the perspective of statistical physics. Therefore, it is important to determine what kind of information drivers base their acceleration or deceleration decisions on. For a simple example, drivers change their acceleration, when perceptual quantities, e.g., headway distance to a vehicle ahead, exceed some criteria.

3.1 Driving Simulator Experiment

We studied driving behavior when the preceding vehicle decelerates. The deceleration is recorded by the wide-view driving simulator. Basic specifications for the driving simulator are listed in Table. 1. The experiment subjects were 20 healthy adults, 16 males and 4 females (six persons aged 20 to 29 years old, 10 persons aged 30 to 39 years old, two persons aged 40 to 49 years old, and two persons aged 50 to 59 years old). We conducted the experiment on a single-lane straight road. Subjects were instructed to follow the preceding vehicle running ahead at 60 [km/h] and to allow a safe headway distance. The vehicle preceding the subject vehicle was programmed to decelerate 48 times in a 36 [sec] period at six different deceleration rates ($-0.02G, -0.05G, -0.075G, -0.1G, -0.125G$, and $-0.15G$ ($G = 9.8$ [m/s^2])) with equal probability.

Table 1. Driving simulator specifications

Vehicle model	3-DOF, 4 wheels, Load-shiftable model (2000cc class passenger car)
Horizontal screen size	237° (Include curved surface screen 150°)
Vertical screen size	34°

3.2 Deceleration Operations

Deceleration by a driver consists of releasing the accelerator pedal and pressing the brake pedal. Figure 6 plots a typical time evolution of deceleration recorded in the experiment. We find that small deceleration is generated by engine braking due to releasing the accelerator pedal and that large deceleration is generated by consecutively pressing the brake pedal.

Deceleration is triggered by certain changes in the visual angle of the preceding vehicle [12], which includes the time headway. Figure 7 plots the time

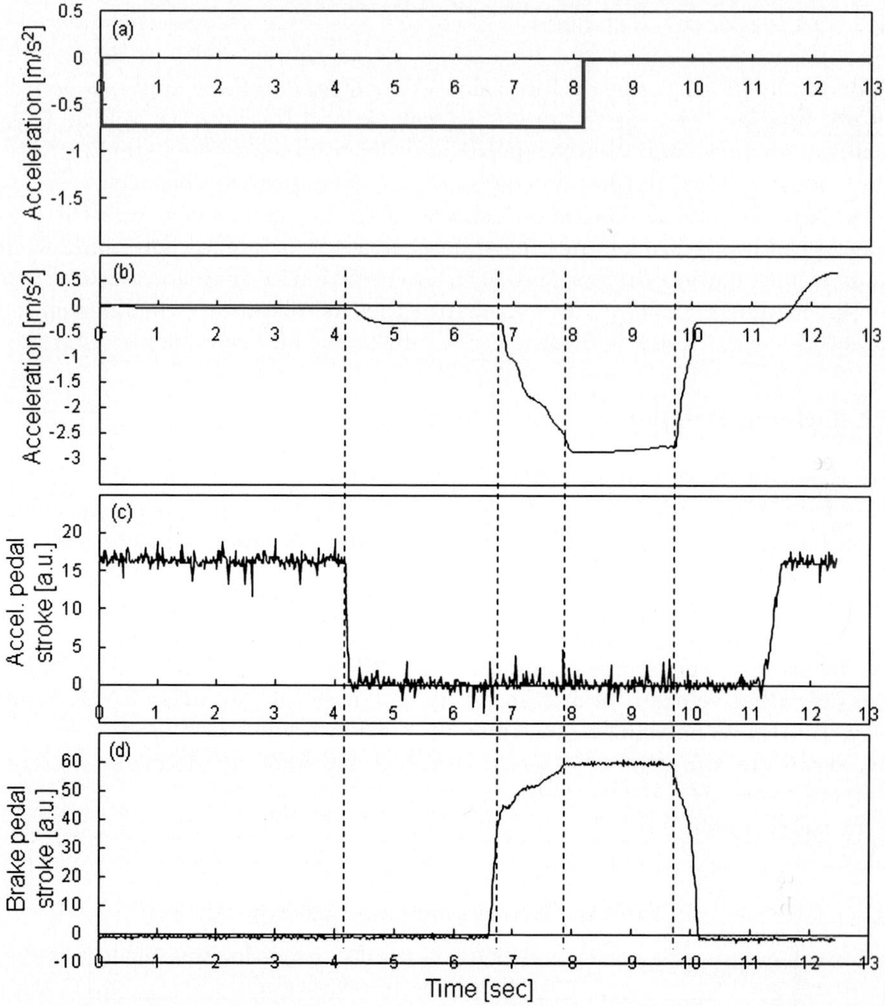

Fig. 6. Deceleration of (a) the preceding vehicle, (b) the following (subject) vehicle. (c) ((d)) indicates the voltage proportional to pedal stroke of accelerator (brake).

headway $\bar{t}_h$ at two operation points averaged over all subjects ($n = 20$) for six different a_p values (Fig. 7(a)) and the standard deviation of $\bar{t}_h$ (Fig. 7(b)). We can find a_p dependence of $\bar{t}_h$ for pressing the brake pedal but can find hardly any dependence for releasing the accelerator. Furthermore, the dispersion of $\bar{t}_h$ for pressing the brake pedal exceeds that for releasing the accelerator pedal. We can also confirm the same tendency in the dispersion of other quantities such as the headway distance and the relative speed. These results indicate that pressing the brake pedal is not decided simply by the relation between the preceding vehicle and the following (own) vehicle at the moment, while

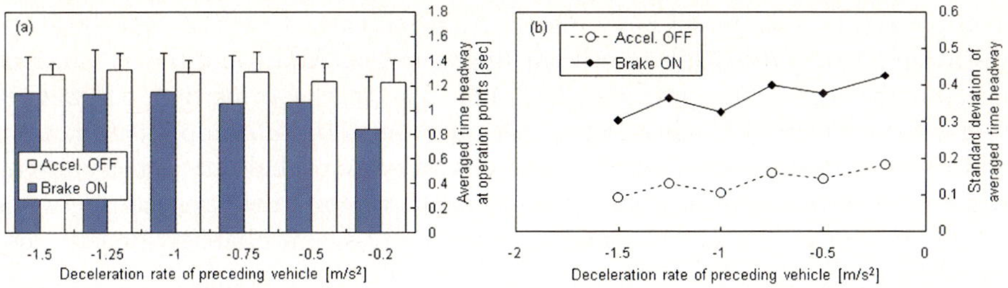

Fig. 7. (a) Time headway averaged over all subjects and (b) its standard deviation at two operation points (accelerator and brake) for six different deceleration rates of the preceding vehicle.

releasing the accelerator pedal is almost completely decided by the relation between vehicles, such as time headway.

The empirical results obtained from the driving simulator experiment suggest that the algorithms of the two deceleration operations (i.e., releasing the accelerator pedal and pressing the brake pedal) qualitatively differ from each other. Therefore, it is necessary to deal with the deceleration process by superposing physiological and psychological factors [13–15] in order to determine the algorithms behind the deceleration operations of drivers.

Our empirical results indicate that the algorithm for releasing the accelerator pedal can be constructed mainly from physiological factors, such as visual perception and simple reaction time. In contrast, psychological factors, such as anticipation based on the traffic situation in the past and driving attitude, seem to play important roles in determining when to press the brake pedal [16]. As regards the formulation of psychological factors, it is not clear at this stage whether the large dispersion of physical variables observed in the relation between vehicles can be accounted for simply by the so-called random noise terms incorporated in conventional models of traffic flow.

4 Conclusion

We measured time series of single-vehicle data in urban traffic with an onboard apparatus and found the nature of fluctuations of variations of velocity around the nonequilibrium phase transition of local vehicular traffic. At low vehicle density, small fluctuations due to the uncertainty of the driver's reaction or various external conditions on the roads add statistical noise to the time series of the states of local vehicular traffic. An increased density leads to amplification of the fluctuations as a result of interaction between vehicles, and instability of traffic flow is induced at the critical density.

We found that the probability-density function calculated from the time series of variation of velocity is transformed around the critical density, where a Gaussian distribution changes into a Lévy stable symmetrical distribution.

The power-law tail in the Lévy distribution indicates that the time series of variation of velocity exhibits critical fluctuations. We attempted to analyze microscopic dynamical processes of traffic flow in the transition, in connection with a discrete stochastic process involving random amplification with additive external noise, which exhibits a power-law probability density distribution of the variation of velocity. As a result, we found the relation between the driver's response to the stimulus and a power-law probability density distribution.

The empirical data of deceleration in a car-following situation obtained from the driving simulator experiment indicated a large dispersion of perceptual quantities about the relations with a preceding vehicle during operation of the brake pedal. These results suggest that the property of brake pedal operation that contributes mainly to the deceleration of a vehicle is closely related to the random amplification in the discrete stochastic process (Eq. (12)). In addition, the experiment conditions of car-following set up in the driving simulator include same density region (Fig. 2 (d)). Therefore, the empirical data of deceleration is expected to give us some clues for understanding the transition mechanisms of traffic flow.

Acknowledgements

We would like to thank T. Fujikawa, T. Fujii and H. Uno for helpful discussions and comments. We are also grateful to T. Kobayashi for setting up the onboard measurement system and K. Fujita for technical support of the driving simulator experiment.

References

1. Chowdhury D, Santen L, Schadschneider (2000) A Phys. Rep. 329: 199.
2. Helbing D (2001) Rev. Mod. Phys. 73: 1067.
3. Koshi M, Iwasaki M, Ohkura I (1983) Proc. 8th International Symposium Transportation Traffic Theory, p. 403, Hurdle V. F, Stewart G. N (eds) Elsevier, Amsterdam, The Netherlands, and Toronto, Ontario, Canada.
4. Gerlough D. L, Huber M. J (1975) Traffic flow theory, Special report 165, Transportation Research Board, National Research Council, Washington, DC.
5. Leutzbach W (1988) Introduction to the Theory of Traffic Flow. Springer, Berlin.
6. Neubert L, Santen L, Schadschneider A, Schreckenberg M (1999) Phys. Rev. E 60: 6480.
7. Bouchaud J.-P, A. Georges (1990), Phys. Rep. 195, 127.
8. Shlesinger M. F, Zaslavsky G. M, Klafter J (1993) Nature 363: 31.
9. Treiterer J (1975) Ohio State Technical Report No. PB 246 094; Treiterer J, Myers J. A (1974) The hysteresis phenomenon in traffic flow – From Proc. 6th ISTT. p.13. In: Buckley D. J (eds) Elsevier, Sydney, New York.

10. Kesten H (1973) Acta Math. 131: 207.
11. Sornette D, Cont R (1997) J. Phys. I 7: 431; Sornette D (1998) Phys. Rev. E 57: 4811.
12. Koppa R. J (1992) Human factors – From Traffic Flow Theory, Chap. 3. In: Gartner N, Messner C. J, Rathi A. J (eds) Transportation Research Board Special Report, Vol.165. Washington, D.C.
13. Brackstone M, McDonald M (1999) Transportation Research Record 1689: 9-17.
14. Zhang Y, Owen L. E, Clark J. E (1998) Transportation Research Record 1644: 103-115.
15. Treiber M, Hennecke A, Helbing D (2000) Phys. Rev. E 62: 1805.
16. Knospe W, Santen L, Schadschneider A, Schreckenberg M (2001) Phys. Rev. E 65: 015101.

Phase Transition Induced by Homogeneous Initial Distribution of Cars

Xue Yu[1,2], Yanfang Wei[2,3], and Shi Wei[1]

[1] Institute of physical science and engineering, Guangxi University, Nanning, 530004, China yuxuegxu@gxu.edu.cn

[2] Shanghai Institute of Applied Mathematics and Mechanics, Shanghai University, Shanghai, 200072, China

[3] Department of physics and information science, Guangxi Yulin Normal University, Yulin, 537000, China

1 Introduction

A very simple microscopic model for single-lane traffic is the cellular automaton model presented by Nagel and Schreckenberg in 1992, called the NaSch model [1]. Recent experiments have shown that the flow-density in the fundamental diagram is more complicated. The different scenario for the jam formation was proposed by Kerner and Konhauser at DaimlerChrysler in 1994 [2]. Moreover, hysteresis effects encountered in empirical observations are related to the existence of metastable states in certain density regions, which is not observed in the NaSch model. For a better understanding of such complex traffic phenomena, a variety of modifications to the NaSch model have been proposed by introducing the slow-to-start rules, among which are the VDR model [3], the TT model [4], the BJH model [5] and so on. They are able to reproduce metastable states and exhibit the clear separation of the congestion and free-flow regions in a spatiotemporal plot. However, the nature of phase transition to induce jamming formation has always attracted considerable arguments. Roters *et al.* [6] considered the stochastic NaSch model displays criticality via investigating the dynamical structure factor of the nondeterministic NaSch model. Conversely, Chowdhury *et al.* [7] showed the existence of a crossover instead of a critical point. Recently, Levine *et al.* [8] also discussed the same question by studying existence of a jamming phase transition in the asymmetric chipping model and obtained that the system exhibits a smooth crossover between free flow and jammed states, as the car density is increased. Cheybani *et al.* [9] have study the transition from freely to jammed and super-jammed traffic in the stochastic NaSch model using the spatiotemporal correlation function. They have turned out that both the transition from freely moving to jammed traffic and from jammed to super-jammed traffic in the stochastic NaSch model is not sharp but rather like a crossover. In theory of car-following model, however, traffic jamming will occur when traffic flow loses its stability starting from uniform initial distribution of cars. Thus, we can infer that complicated phase transition to induce jamming formation is not only related to the rule of model but also to the initial distribution and the maximum velocity of cars. In this paper, we systematically study the phase transition induced by initial condition to the typical cellular automaton traffic model, such as the NaSch, VDR (BJH) and TT model. The several definition of order parameter is used to analyze the characteristics of phase transition of the several models above mentioned under the homogeneous initial condition.

2 Model and Method of Simulation

In one-dimensional cellular automata traffic model, road is divided into discrete lattice of length L and states of each cell is determined by $v_{max} + 1$ velocity of cars $v \in \{0, 1, 2, ..., v_{max}\}$. The state of the system at the time $t+1$ could be obtained from the state at the time t by applying the following rules to all cars at the same time (i.e., the parallel update rules):

(1) Acceleration:
$v_n \rightarrow \min(v_n + 1, v_{max})$
(2) Deterministic deceleration to avoid accidents:
$v_n \rightarrow \min(v_n, d_n)$
(3) Randomization:
$v_n \rightarrow \max(v_n - 1, 0)$ with probability p
(4) Update of positions:
$x_n(t+1) \rightarrow x_n + v_n$

The NaSch model is a minimum model in the sense that all the four steps are necessary to reproduce the basic features of real traffic. In the NaSch model, step 3 combined three different behavioral patterns into one computational rule: fluctuations at maximum speed, retarded acceleration and overreactions at braking. In the VDR model, the delay probabilities were first considered as velocity-dependent in contrast to the constant randomization in the NaSch model. A car will be slow to start when its velocity equals zero, otherwise accelerate to advance, i.e. the randomization parameter is

$$p(v) = \begin{cases} p_0 & v = 0 \\ p & v > 0 \end{cases}$$

Probability p_0 depicts the behavior of standing car slow to start and probability p determines uncertain behaviors of other motion cars. The slow-to-start rule can lead not only to metastable state and hysteresis, but also to phase separated states at high densities. In the TT model, a standing car with exactly one empty cell in front accelerates with the probability $q_t = 1 - p_t$ and all other cars accelerate deterministically. In our simulation, a single lane is

represented a one-dimensional lattice of L = 10^4 cells with periodic boundary condition, which corresponds to the length of the actual road 7.5 km. Each cell is either empty, or occupied by just one vehicle with discrete velocity v. Velocity ranged between 0 and $v_{max} = 5$, where v_{max} is the speed limit to correspond to 135 km/h in real traffic. The time step $t = 1s$ corresponds to the order of the reaction time for a driver. Let v_n and x_n denote the current velocity and position of the nth vehicle respectively. We denote $d_n(t)$ by $d_n(t) = x_{n+1} - x_n - 1$, which is the number of empty cells in front of the nth vehicle. The computational formulas are given as follows:
Average density:

$$\rho = N/L \tag{1}$$

Mean velocity:

$$V = \sum_{i=1}^{n} v_i(t)/N = \sum_{t=t_0}^{T+t_0-1} v_i(t)/T \tag{2}$$

Flow:

$$J = \rho V \tag{3}$$

The numerical simulation was performed according to the above rules. For each simulation, each run is first conducted 1×10^4 time steps in order to remove the transient effects and then the data are recorded in successive 4×10^4 time steps. The fundamental diagram is obtained by averaging over 50 runs of simulations.

3 Fundamental Diagram, Solution and Order Parameters

3.1 Fundamental Diagram

The most important empirically measured quantities in traffic flow are usually shown in the so-called fundamental diagram, which reflects the transit capacity for a one-lane traffic flow model. The fundamental diagrams of the deterministic and stochastic NaSch model are independent on the initial condition, wherever the initial condition is homogeneous and stochastic initial distribution. However, for models with slow-to-start rule, such as VDR model, BJH model and TT model, the fundamental diagrams of these models under the different initial condition are different and dependent on the initial condition. For example, if $p = 0$, the fundamental diagrams of VDR model is shown in Fig. 1(a). There are three piecewise curves in Fig. 1(a). From velocity versus density of Fig. 1(b), the three piecewise curves respectively correspond to free flow, low velocity and traffic jamming. The homogeneous initial distribution of identical vehicles with the equivalent gap ranges from $\rho = 0$ to $\rho = 0.5$. If the density exceeds 0.5, the homogeneous initial distribution is impossible to realize. It will lead to drop of flow rate in the fundamental diagram. It is found that the curve with smaller than density $\rho = 0.5$ coincides with those of deterministic NaSch model, and drops to the branch of traffic jamming while the

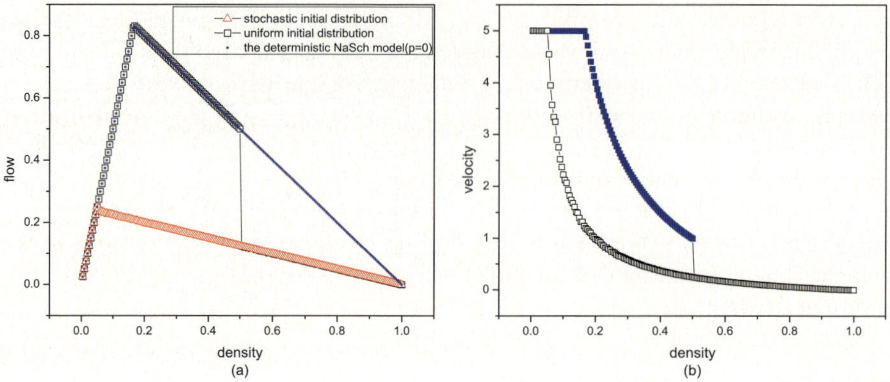

Fig. 1. Fundamental diagram and velocity versus density of the VDR model ($v_{max} = 5$, $p_0 = 0.75$, $p = 0$).

density exceeds 0.5. It is obvious that at least two cars in this case have gap less than other's gap and probability p_0 acts on blocking traffic and leads to drop of flow rate, in which the curve is coincident with the branch of traffic jamming of VDR model with same slow-to-start probability p_0 from stochastic initial distribution. Using the same method of simulation, we obtain the fundamen-

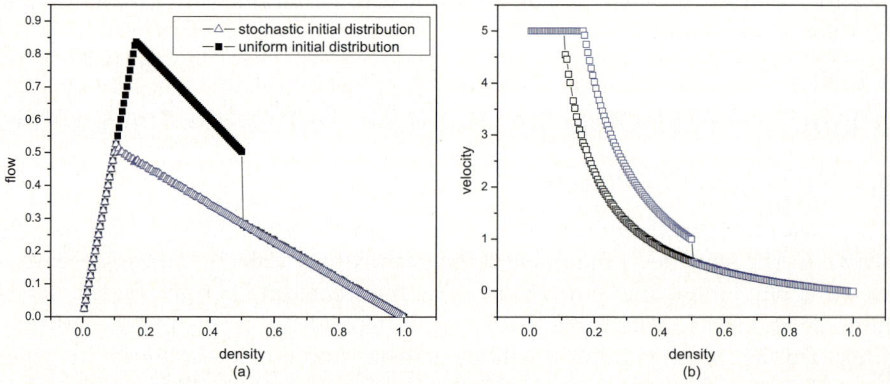

Fig. 2. Fundamental diagram of the TT model ($v_{max} = 5$, $p_t = 0.75$, $p = 0$).

tal diagram of TT model in Fig. 2 from the different initial distribution while probability p equals zero and have similar results in the VDR model. From above discussion, there are two common characteristics that the first transition point locates at $\rho_{c1} = 1/(v_{max}+1)$ and the second point at $\rho_{c2} = 0.5$. The first transition point means that the criticality occurs, because the traffic flow really evolves according to the deterministic rule ($p = 0$) and the deterministic NaSch model has a criticality [10]. The second point at density $\rho = 0.5$ has a discontinuous skipping in the fundamental diagram. It means that traffic occur the first order phase transition from low velocity to jamming.

3.2 Solution

From the above analysis, when the density is less than 0.5, the free flow and low velocity flow are defined in terms of the deterministic NaSch model. If the density exceeds 0.5, the branch of traffic jamming is determined via making use of solution of VDR model [3]. The solution of VDR model with $p = 0$ from the homogeneous initial distribution is as follows.

$$J = \begin{cases} \rho v_{max} & \rho < \rho_{c1} = \frac{1}{v_{max}+1} \\ 1-\rho & \rho_{c1} = \frac{1}{v_{max}+1} \leq \rho < \rho_{c2} = \frac{1}{2} \\ (1-p_0)(1-\rho) & \rho \geq \rho_{c2} = \frac{1}{2} \end{cases}$$

The solution of VDR model from the stochastic initial distribution is [3]

$$J = \begin{cases} \rho(v_{max} - p) & \rho < \rho_c = \frac{1-p}{v_{max}+1-2p} \\ (1-p_0)(1-\rho) & \rho \geq \rho_c = \frac{1-p}{v_{max}+1-2p} \end{cases}$$

It is obvious that both solutions are different. Similarly, the approximate solutions of TT model from the homogeneous initial condition and the stochastic initial distribution are obtained, respectively,

$$J \approx \begin{cases} \rho v_{max} & \rho < \rho_{c1} = \frac{1}{v_{max}+1} \\ 1-\rho & \rho_{c1} = \frac{1}{v_{max}+1} \leq \rho < \rho_{c2} = \frac{1}{2} \\ (1-p_0)(\frac{1}{2} + \frac{1-p_t}{2}) & \rho \geq \rho_{c2} = \frac{1}{2} \end{cases}$$

and

$$J \approx \begin{cases} \rho(v_{max} - p) & \rho < \rho_c = \frac{1-p}{v_{max}+1-2p} \\ (1-p_0)(1-\rho)(\frac{1}{2} + \frac{1-p_t}{2}) & \rho \geq \rho_c = \frac{1-p}{v_{max}+1-2p} \end{cases}$$

3.3 Order Parameters

In statistic physics, the order parameter is an important quantity to describe phase transition. Many authors have introduced many definitions of order parameter [10–12]. Eisenblatter's definition [10]:

$$m = \frac{1}{L} \sum_{i=1}^{L} n_i(t) n_{i+1}(t), \tag{4}$$

where for an empty cell $n_i = 0$ and for a cell occupied by a car $n_i = 1$.

Boccara's definition [12]:

$$m = v_{max} - \langle v \rangle \tag{5}$$

Vilar *et al.* [11] and Eisenblätter *et al.* [10] have ever investigated the order parameter of the deterministic NaSch model and have found that the order parameter vanishes below the transition density $\rho_c = \frac{1}{v_{max}+1}$. The reason is that every car has at least v_{max} empty sites in front and propagates with v_{max}. Otherwise, the order parameter is nonzero. In this section, we just discuss order parameter Eq. (4) and (5) in order to more understand the features of traffic transition. The order parameter of VDR model under the homogeneous and stochastic initial conditions is shown in Fig. 3(a) and (b). Similarly, the order parameter of TT model from the homogeneous and stochastic initial conditions is shown in Fig. 4(a) and (b), respectively. From these plots of order parameter, it is found that there is a criticality at $\rho_c = \frac{1}{v_{max}+1}$ and

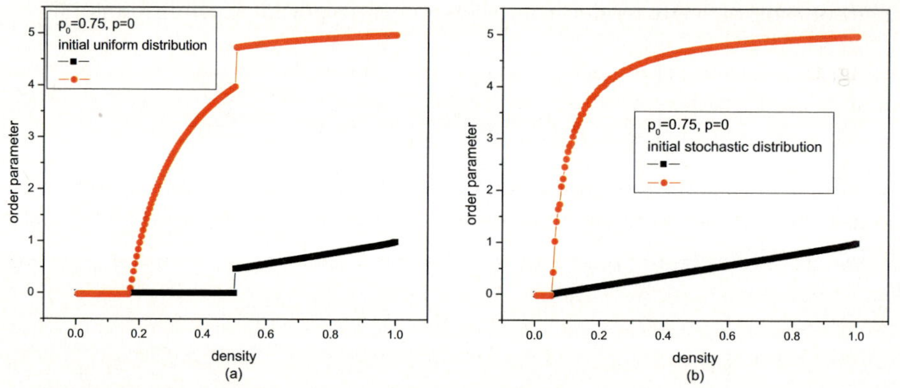

Fig. 3. Order parameter of VDR model in the different initial conditions ($v_{max} = 5$, $p_0 = 0.75$, $p = 0$), where the dash of red is obtained from Eq. (5) and the dash of gray from Eq. (4).

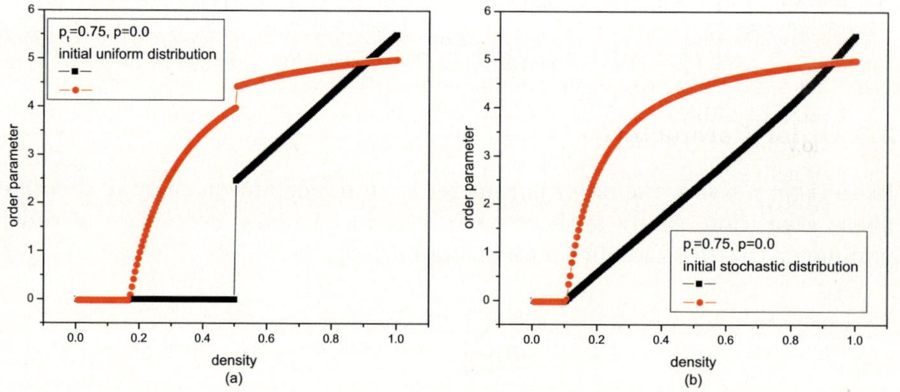

Fig. 4. Order parameter of TT model from the different initial conditions ($v_{max} = 5$, $p_t = 0.75$, $p = 0$), where the dash of red is obtained from Eq. (5) and the dash of gray from Eq. (4).

a discontinuous jump, which reflects the occurrence of the first order phase transition. These results are consistent with the above discussions. Comparing the several plots, we consider that order parameter of Eq. (4) is not better to reflect the phase transition under the homogeneous initial conditions while Eq. (5) can depict traffic transition of both initial conditions.

4 Summary

There exist three different traffic phases: free flow, low velocity flow and traffic jamming in fundamental diagrams of VDR and TT models without the stochastic delay probability under the homogeneous initial condition. The behaviour from free flow to low velocity flow ($0 < \rho \leq \rho_c = \frac{1}{v_{max}+1}$) is similar to those of the deterministic NaSch model. From low velocity flow to traffic jamming, the first order phase transition occurs, where the probability p_0 or p_t of slow-to-start plays an important role in traffic jamming. The computation of the order parameter confirms the characteristics of phase transition. The analytical solutions corresponding to these models are obtained, respectively.

This work was supported by the National Natural Science Foundation of China, the National Basic Research Program of China, the Natural Science Foundation of Guangxi and the Special Foundation for the New Century Talents Program of Guangxi Zhuang Autonomous Region.

References

1. Nagel K and Schrekenberg M (1992), J. Physique I 2:2221-2229.
2. Kerner B S (2004), The physics of traffic. Springer, Berlin Heidelberg New York.
3. Barlovic R, Santen L, Schadschneider A *et al.* (1998), Eur. Phys. J. B 5:793-800.
4. Benjamin S C, Johnson N F, Hui P M (1996), J. Phys. A: Math. Gen. 29:3119-3127.
5. Takayasu M and Takayasu T (1993), Fractals 1:860-864.
6. Roters L, Lübeck S, Usadel K D (1999), Phys. Rev. E 59:2672-2676.
7. Chowdhury D, Kertész J, Nagel K *et al.* (2000), Phys. Rev. E 61:3270-3271.
8. Levine E, Ziv G, Gray L *et al.* (2004), J. Stat. Phys. 117:819-830.
9. Cheybani S, Kertész J and Schreckenberg M (1998), J. Phys. A: Math. Gen. 31:9787-9799.
10. Eisenblätter B, Santen L, Schadschneider A *et al.* (1998), Phys. Rev. E 57:1309-1314.
11. Vilar L C Q and de Souza A M C (1994), Physica A 211:84-92.
12. Boccara N and Fukś H (2000), J. Phys. A: Math. Gen. 33:3407-3415.

Similarity Between Temporal and Spatial Structures in Pattern Formation of Dissipative Non-Equilibrium System

Rintarou Akiyama, Hiroshi Watanabe, and Yūki Sugiyama

Department of Complex Systems Science, Nagoya University, Nagoya 464-8601, Japan sugiyama@phys.cs.is.nagoya-u.ac.jp

1 Introduction

Similarity between the temporal structure of emerging patterns and the spatial structure of stable patterns is observed in nature, such as biological development, growth of trees, forming livers, etc. Can these phenomena be originated in a general property in dissipative non-equilibrium systems? Traffic flow is also an example of the cases. The phenomenon is observed in real traffic, the experiment of circuit and the flow of highway traffic upstream of a bottleneck. The phenomenon is simulated using a mathematical model such as OV model.

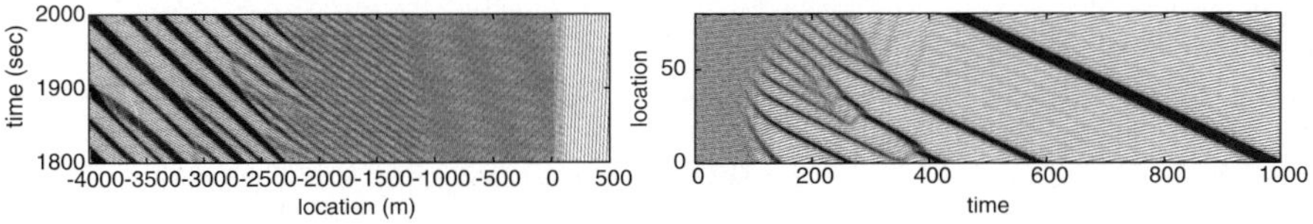

Fig. 1. (Left:) The stable spatial patterns of traffic flow on the upstream of a bottleneck after relaxation time. The bottleneck is placed at $0m$-point, and the length is $100m$. Vehicles move from left to right. (Right:) Space-time plot of jam formation on periodic boundary condition (a part). The vertical axis is location on the circuit. The horizontal axis is time evolution.

Figures 1 are the result of simulation of OV model, which shows the stable structure of traffic flow upstream of a bottleneck [1] (left), and the process of forming a jam [2] (right). The three distinct patterns of flow are observed, such as uniform flow, "oscillatory wave flow", and jam flow. Comparing the results of two simulations, the similarity is clearly observed. For the purpose of analytical studying of the similarity, we investigate "the delay-OV (dOV) model" instead of original OV model. The model is expressed as the following formula,

$$\frac{d}{dt}x_n(t+\tau) = V(x_{n+1}(t) - x_n(t)), \tag{1}$$

where x_n is a position of nth particle (vehicle), and τ is a real positive constant called "delay" as a control parameter. The dOV model shows the very similar dynamical behavior as the original OV model in case $V(x) = \tanh(x)$, which is the reasonable choice for traffic flow [2]. Eq. (1) has been popular in many physical phenomena of relaxation towards an optimal equilibrium state, such as relaxation effect in gases, chemical reactions and synchronization problem. The formula has been first introduced for traffic flow model with another choice of OV function [3, 4]. We rewrite Eq. (1) for the headway distance of two successive particles, $h_n(t) = x_{n+1}(t) - x_n(t)$, as

$$\frac{d}{dt}h_n(t+\tau) = V(h_{n+1}(t)) - V(h_n(t)). \tag{2}$$

2 Linear Wave Solution and its Stability

We first derive the linear wave solution, assuming that the amplitude $h_n(t)$ is infinitesimal and $V(h_n) = V(0) + V'(0)h_n$. Eq. (2) becomes

$$\frac{d}{dt}h_n(t+\tau) = h_{n+1}(t) - h_n(t). \tag{3}$$

Here we set $V'(0) = 1$ without loss of generality. We obtain the linear solution as

$$h_n(t) = \exp \pm i\alpha\left(n + \frac{t}{2\tau}\right) \tag{4}$$

where α is given as a solution of

$$\frac{\sin \alpha/2}{\alpha/2} = \frac{1}{2\tau}, \tag{5}$$

which has a solution if τ is larger than 1/2, meaning that $\tau = 1/2$ is critical delay. Eq. (4) represents a traveling wave solution with the velocity $1/(2\tau)$ in the space of index n, which is treated as a continuous variable, moving backward against the numbering direction. It appears as the traveling wave in the real space moving opposite direction in the flow of particles [4, 6].

The linear stability of the above solution is analyzed by the behavior of the small deviation b_n beyond the solution, as

$$\frac{d}{dt}b_n(t+\tau) = b_{n+1}(t) - b_n(t). \tag{6}$$

We suppose $b_n = \exp(iqn + zt)$, and the dispersion relation $z(q)$ by long wave-length expansion, $q \sim 0$ is

$$z(q) = iq + q^2(2\tau - 1) + o(q^3), \tag{7}$$

which indicates that the solution exists for $\tau > 1/2$, but it is always unstable.

3 Elliptic Solutions

In general, traveling wave solutions are expressed in the form as $h_n = f(n+vt)$, which should satisfy Eq. (2) as

$$vf'(n + v(t + \tau)) = V(f(n + 1 + vt)) - V(f(n + vt)). \tag{8}$$

We set $v = 1/(2\tau)$ and introduce $u = n + vt + 1/2$, which is the condition of integrability [6]. Actually, the traveling waves propagate with the velocity $1/(2\tau)$, just the same as the linear wave. The solutions f are exactly obtained as the solutions for the following difference-differential equation [6]

$$v\frac{dG(u)/du}{1 - G(u)^2} = G\left(u + \frac{1}{2}\right) - G\left(u - \frac{1}{2}\right), \tag{9}$$

where $G = V(f)$. In our case $V(x) = \tanh(x)$, which is the choice for corresponding to OV model. A solution of Eq. (9) is

$$G(u) = \beta\,\mathrm{sn}(\alpha u, k), \tag{10}$$

where sn is Jacobi's elliptic function with modulus k [6], and β is

$$\beta = \pm k\frac{\alpha}{4\tau}. \tag{11}$$

The parameter α is determined by

$$\frac{\mathrm{sn}(\alpha/2, k)}{\alpha/2} = \frac{1}{2\tau}, \tag{12}$$

where Eq. (12) has a real solution only if $\tau > 1/2$. In the case of $k = 0$, Eq. (12) reduces to the result of linear theory, Eq. (5). The modulus k is a free parameter of the solution, which indicates the existence of many solutions traveling with the same velocity $v = 1/(2\tau)$. The variation of the solutions depending on k gives the different kind of traveling waves in terms of amplitude and dispersion relation, from linear wave ($k = 0$) to cluster solutions ($k \sim 1$).

4 Linear Stability of Elliptic Solutions

We investigate the linear stability of elliptic solutions for small amplitudes, $f \sim$ small in Eq. (8). The equation of motion for small deviation b_n beyond a traveling wave solution up to the second order of f, as

$$\frac{d}{dt}b_n(t + \tau) = \left\{1 - f^2(n + 1 + vt)\right\} b_{n+1}(t) - \left\{1 - f^2(n + vt)\right\} b_n(t). \tag{13}$$

For $f \sim$ small leads $G = V(f) = \tanh f \sim f$, then

$$\begin{aligned}\frac{d}{dt}b_n(t + \tau) = &\left\{1 - \beta^2\mathrm{sn}^2(\alpha(n + 1 + vt), k)\right\} b_{n+1}(t)\\ &- \left\{1 - \beta^2\mathrm{sn}^2(\alpha(n + vt), k)\right\} b_n(t).\end{aligned} \tag{14}$$

We assume $k \ll 1$, which is consistent with small f. Here, $sn(\alpha u, k)$ is expanded in terms of modulus k. Up to the second order of k, Eq. (14) is

$$\frac{d}{dt}b_n(t+\tau) = \left\{1 - \left(\frac{k\alpha}{4\tau}\right)^2 \sin^2(\alpha(n+1+vt))\right\} b_{n+1}(t)$$
$$- \left\{1 - \left(\frac{k\alpha}{4\tau}\right)^2 \sin^2(\alpha(n+vt))\right\} b_n(t). \tag{15}$$

The evolution of the time-dependent coefficients in the equation for b_n is slowly comparing with the time development of b_n. Then we can ignore the time-dependence of the coefficients, and use the smallness of the solution α in Eq. (12) for small k. As the result, we can estimate that $\sin^2(\alpha(n+vt)) \sim \sin^2(\alpha n) \sim (\alpha n)^2$. Then we derive the linearized equation for the small deviation beyond the elliptic solution for the relatively small $k < 1$ as

$$\frac{d}{dt}b_n(t+\tau) = \left\{1 - \left(\frac{\alpha^2}{4\tau}\right)^2 k^2(n+1)^2\right\} b_{n+1}(t) - \left\{1 - \left(\frac{\alpha^2}{4\tau}\right)^2 k^2 n^2\right\} b_n(t). \tag{16}$$

In the following subsections we investigate the stabilities of elliptic solutions in the two different boundary conditions corresponding to the boundary of the open system with bottleneck and the periodic condition with no bottleneck.

4.1 Stability of Traveling Wave Solutions on the Fixed Boundary

First we investigate the stability of elliptic solutions on the fixed boundary condition that any small deviation is forbidden, which condition is regarded as a bottleneck [7]. On this condition we assume that $b_n = 0, (n \geq N+1)$ meaning that the gate of a bottleneck is set between the particle number N and $N+1$ at a time. Then the equation of the deviation for N is written as

$$\frac{d}{dt}b_N(t+\tau) = -\left\{1 - \left(\frac{\alpha^2}{4\tau}\right)^2 k^2 N^2\right\} b_N(t). \tag{17}$$

It is easily solved for b_N, and the solution for the oscillation mode $b_{N-1}(\omega)$ is expressed using the solution $b_N(\omega)$ by solving Eq. (16) for $n = N-1$.

$$b_{N-1}(\omega) = \frac{1 - k^2(\alpha^2/4\tau)^2 N^2}{k^2(\alpha^2/4\tau)^2(2N-1)} b_N(\omega). \tag{18}$$

In general the solution for $b_n(\omega)$ can be determined by the recursive relation for the sequence $b_N, b_{N-1}, \ldots, b_{n+1}, b_n, \ldots$ upstream of the bottleneck. For small k the recursive relation is simply expressed as

$$b_{n-1}(\omega) \sim \frac{1}{k^2} b_n(\omega). \tag{19}$$

This relation shows that the amplitude of the deviation grows rapidly enhanced both as far as upstream of a bottleneck and as small as the modulus k. The result indicates the elliptic solution of modulus k is more and more unstable as far from a bottleneck with small k. Inversely say, as far from a bottleneck the cluster (jam) solution ($k \sim 1$) is stable comparing that the small modulus ($k < 1$) solution is relatively stable not so far from a bottleneck. This characteristic property of instability depending on modulus-parameter connected with the spatial position for the solutions, is consistent with the spatial structure of flow upstream of a bottleneck.

4.2 Stability of Traveling Wave Solutions on the Periodic Boundary

Next, we investigate the stability of elliptic solutions on the periodic boundary without bottleneck. We set $b_{N+n} = b_n$ on the periodic boundary condition, where N is a total number of particles on a periodic lane. We suppose that $k^2N^2 \sim$ finite for large N together with small k. On this condition, the n-dependence of the coefficient of the equation for b_n Eq. (16), can be neglected and reduced to the given N as,

$$\frac{d}{dt}b_n(t+\tau) = \left\{1 - \left(\frac{\alpha^2}{4\tau}\right)^2 k^2N^2\right\} b_{n+1}(t) - \left\{1 - \left(\frac{\alpha^2}{4\tau}\right)^2 k^2N^2\right\} b_n(t). \quad (20)$$

Using the above formula we can estimate the k-dependence of the stability for elliptic solutions. In the same way as for the analysis for the linear wave, the dispersion relation for the Eq. (20) is

$$z(q) = iq + q^2 \left[2\tau \left\{1 - \left(\frac{\alpha^2}{4\tau}\right)^2 k^2N^2\right\} - 1\right] + o(q^3). \quad (21)$$

For given large-N approximation, we can estimate the stability depending on k. In the case $k = 0$ the above dispersion relation is reduced to that of the deviation beyond the linear wave, Eq. (7). If given τ is larger than 1/2, the elliptic solutions exist. But the solution with modulus k is unstable under the condition

$$\tau \geq \frac{1}{2\left(1 - (\alpha^2/4\tau)^2 k^2N^2\right)}, \quad (22)$$

where the inequality is meaningful as long as the denominator is positive. Then, the solution for $k < k_c$ is unstable, while that for $k > k_c$ is stable, where k_c is given by the equality on Eq. (22).

The exact form of dispersion can be derived. For small q

$$\omega \propto q^2 + \frac{(\alpha^2/4\tau)^4 k^4N^4}{1 - (\alpha^2/4\tau)^2 k^2N^2}, \quad (23)$$

which shows that the deviation wave of disturbance propagates in long range as "massless particle" for $k \sim 0$, meaning that the solution of small k is easily decayed. For larger k the deviation propagates in short range as "massive particle", which means the solution is hard to decay.

As a result, the smaller modulus k solution the earlier it decays and the larger k solution is more stable and remains after the optimization time. This characteristic property of modulus-dependence of decay is consistent with the temporal structure in the phenomenon of flow on periodic boundary with no bottleneck.

5 Similarity Between Temporal and Spatial Structures

Summarizing up the results of previous sections, we can provide the explanation for the main subject, the similarity between temporal and spatial structures. We observe the family of traveling wave solutions for particle-flow in OV type models, such as a linear wave flow, an oscillatory wave flow and a cluster (jam) flow. In dOV model these solutions are characterized by modulus parameter ($0 \leq k \leq 1$) in elliptic solutions. The flow patterns of temporal and spatial structures are commonly considered as the stability of each solution depending on modulus parameter. As for the spatial stability upstream of a bottleneck, the smaller of modulus k, the more unstable a solution is, when it exists far from the bottleneck. This property results in the stable spatial pattern of flow. As for the temporal stability on periodic boundary condition, the smaller of modulus k, the earlier a solution decays, while a cluster flow as the solution $k \sim 1$ is stable. This property results in the emerging temporal pattern of flow. Namely, the temporal and spatial patterns are originated in the common concept, that is the modulus-dependency of stability, which leads to the appearance of the similarity of patterns.

References

1. Y. Sugiyama and A. Nakayama, *Modeling, simulation and observations for freeway traffic and pedestrians*, 'Computational Physics of Transport and Interface Dynamics', (eds) H. Emmerich, B. Nestler and M. Schreckenberg, (Springer), (2003) 406-421.
2. M. Bando, K. Hasebe, A. Nakayama, A. Shibata and Y. Sugiyama, Phys. Rev. E **51**, 1035 (1995); Japan J. of Ind. and Appl. Math. **11**, 203 (1994).
3. G.F. Newell, Oper. Res. **9**, 2209 (1961).
4. G.B. Whitham, Proc. R. Soc. Lond. A **428**, 49 (1990).
5. Y. Igarashi, I. Itoh, K. Nakanishi, K. Ogura, K. Yokokawa, Phys. Rev. Lett. **88**, 718 (1999).
6. K. Hasebe, A. Nakayama, Y. Sugiyama, Phys. Lett. A **259**, 135 (1999).
7. N. Mitarai, H. Nakanishi, Phys. Rev. Lett. **85**, 1766 (2000).

Pattern Formation in Traffic Microscopic Model

Tomas Apeltauer, Petr Holcner, and Jiri Macur

Brno University of Technology, Faculty of Civil Engineering, Brno, CZ-602 00, Czech Republic apeltauer.t@fce.vutbr.cz, holcner.p@fce.vutbr.cz, macur.j@fce.vutbr.cz

Summary. The behavior of vehicle following model was investigated in the paper from the point of view of non-linear dynamics. When the mean density of the traffic flow exceeds some critical limit the formation of stable dissipative structures can be observed. This pattern-producing system shows further properties typical of complex system as well.

1 Traffic Flow Model

Simple single lane car following model (CFM) [1] was used for simulation experiments. From the point of view of dynamic systems theory several common properties can be expected for relatively wide class of the models.

In the CFM vehicle acceleration may be expressed as following function

$$a = a(v, \Delta v, \Delta x) \tag{1}$$

where v is the vehicle velocity in the traffic flow, Δv the relative velocity with regard to a vehicle driving before and Δx is the distance from a vehicle driving ahead.

In the system of vehicles assuming the finite driver reaction interval T_R accelerations of vehicles emerge retarded motion equations system

$$a_i(t) = a(v_i(t - T_R), \Delta v_i(t - T_R), \Delta x_i(t - T_R)) \tag{2}$$

Due to its simplicity we picked the Intelligent Driver Model (IDM) [2] with finite reaction time. In this model the acceleration is proposed in the form

$$a_{i\,\mathrm{acc}} = a_{i\,0} \left[1 - \left(\frac{v_i}{v_{i\,0}} \right)^{\delta} \right] \tag{3}$$

where a_{i0} is the vehicle acceleration for a low velocity and v_{i0} is the maximal vehicle velocity. The coefficient δ is calibrated to the value $\delta \approx 4$ according to measurements. Then the vehicle acceleration drops monotonously with a growing velocity and after reaching the maximal velocity it becomes zero.

The proposed deceleration contribution is valid in any distance Δs_i (distance of a frontal bumper and a back bumper of the previous driving vehicle, i.e. the gap between the vehicles) and is of the following form

$$a_{i\,\mathrm{dec}} = -a_{i\,0}\left(\frac{\Delta s_{i\,\mathrm{opt}}}{\Delta s_i}\right) \tag{4}$$

where $\Delta s_{i\,\mathrm{opt}}$ is the optimal distance from the previous vehicle with the given velocity

$$\Delta s_{i\,\mathrm{opt}} = \Delta s_{i\,0} + \Delta v_i T_i + \frac{v_i \Delta v_i}{2\sqrt{a_{i\,0} b_{i\,0}}} \tag{5}$$

Parameter s_{i0} is the distance at zero velocity, T_i is the optimal time distance from the previous vehicle, Δv_i is the relative velocity with regard to the previous vehicle. The third term in the relation (5) expresses an "intelligent" trend to maintain a distance which enables the vehicle to accelerate in a maximal way in a quickly changing situation and at the same time not to brake more intensively than the comfort deceleration expressed by the parameter b_{i0}.

2 Model Verification

To verify and calibrate the parameters of IDM model the GPS measurement of vehicles position was used. Standard static measurement is not useful because of its poor accuracy even in case of DGPS (differential GPS) method with corrections. The RTK (Real Time Kinematic) measurement method [3] of the carrier phase of GPS signal measurement is more suitable. Wave length of these signals is about 20 cm, thus measuring with 1% phase accuracy makes possible the millimeter distance accuracy. However this determination is disrupted of difficult removable ambiguity represented by unknown integer multiple of carrier wavelength (phase ambiguity). Therefore the reference receiver with known position in the monitored area was used for fast elimination of ambiguities. Conjoined compensation of refraction and other errors is an advantage of this method.

Position measurement of two vehicles with centimeter accuracy and sampling interval 0.1 s allowed us to determine all required properties of vehicles (position, velocity and acceleration).

Good correspondence between measuring and simulating the movement of the following vehicle using model described above was reached (see Fig. 1). Maneuvers of the tested driver are evidently not as fluent as in case of IDM simulation however general behavior is described by the model very realistically. It is rather surprising if we take complex strategy of the following driver into consideration: acceleration using full power of the engine, gentle deceleration using the engine, braking and approach maneuvers with gap and velocity corrections.

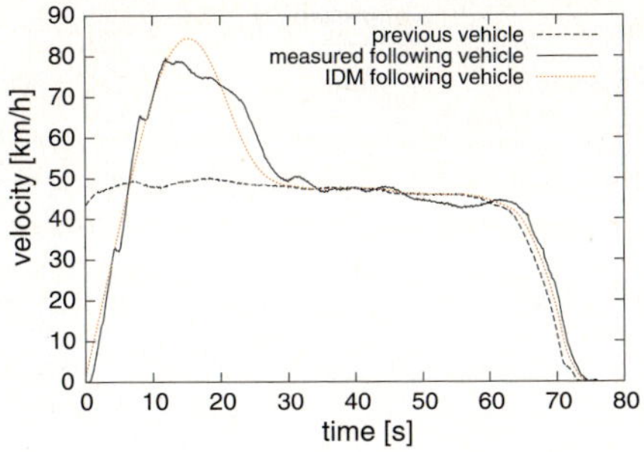

Fig. 1. Measurement of the first reference vehicle together with measured and simulated following vehicle.

Reasonable values of the IDM parameters assumed for the simulation: $a_0 = 2.0\ \mathrm{ms}^{-2}$, $b_0 = 4.0\ \mathrm{ms}^{-2}$, $v_0 = 30\ \mathrm{ms}^{-1}$, $T = 1.5$ s, $T_R = 0.5$ s.

Non-linear regression methods were not sufficiently stable for parameters calibrating, furthermore their exact values are not relevant for our qualitative analysis.

3 Equilibrium Properties of the Traffic Flow

For a traffic flow simulation it is advantageous to use the cyclic border conditions which are commonly used for a systems of interacting particles — in this case it means to put vehicles on a circular road with sufficiently large radius. In case of such closed deterministic system the deformed behavior due to the correlation between the first and the last vehicle can be anticipated.

The 6.28 km circle lane with 500 identical vehicles (with the same parameters mentioned above) can be seen on Fig. 2. Initially homogenously placed vehicles started synchronously and in a small time period spontaneous perturbations appeared. The system reaches steady state after some transient time where identical congestions move upstream the flow. This "excessively regular" behavior is not as unrealistic as could be assumed from the simulation experiment arrangement:

1. Integer number of stationary congestions in the circle due to the cyclic border conditions is irrelevant in case of sufficient large radius of the lane (see Fig. 3)
2. The same amplitude of all congestions is not the result of identical vehicles or system simplicity. The congestion velocity (analogy of the group velocity in wave physics) depends on the congestion shape in particular on the maximal density fluctuation. Thus the congestions of different shape move with various velocities — this process eliminates the system steady state.

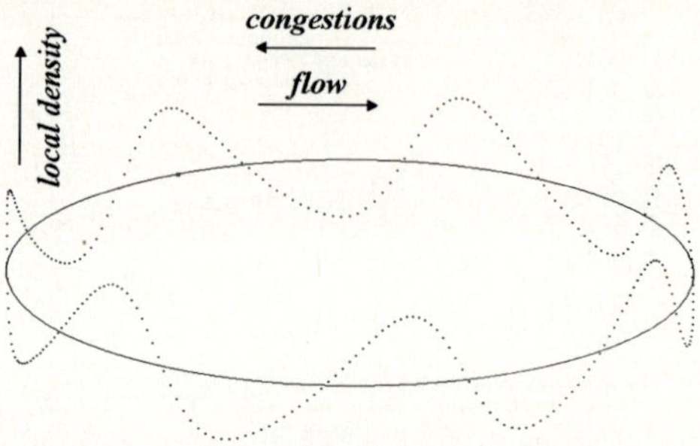

Fig. 2. Spontaneous periodic congestions in steady state of the initial homogenous flow. Solid line marks the value of the constant global density.

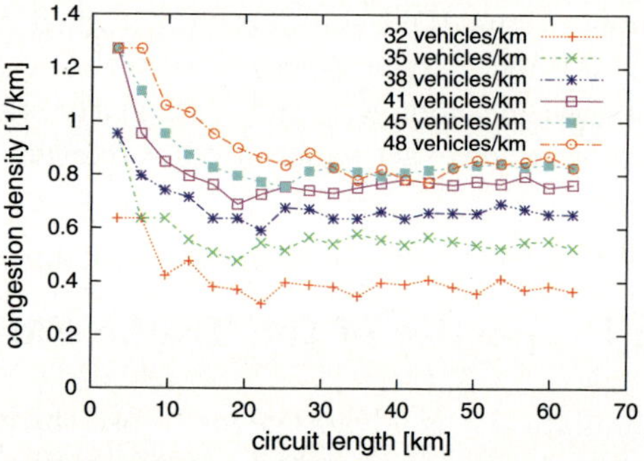

Fig. 3. Convergence of congestions number for long circle lanes. Behavior for different average densities of the flow can be seen.

Density of the congestions asymptotically converges to the value dependent only on the mean density of the flow (cyclic enclosure of the system is irrelevant for the radius greater than 4 km — see Fig. 3).

Figure 4 illustrates other common properties of the single lane steady state flow:

1. The density of the congestions (number per km) increases with the mean flow density
2. The congestions amplitude (maximal local density) monotonously decreases with the mean flow density (and also with the density of congestions)
3. The product of the congestions density and amplitude depends on the mean flow density linearly.

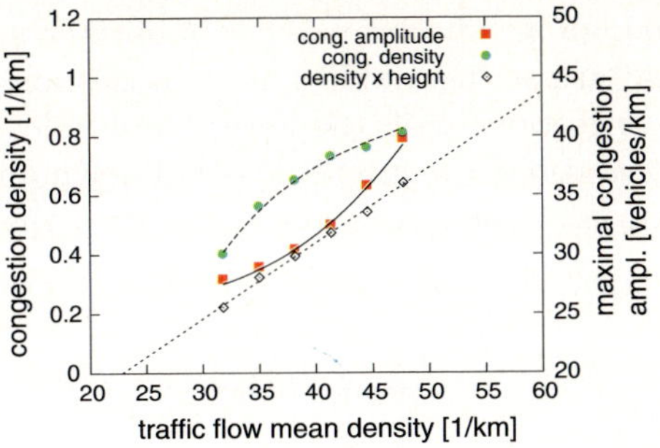

Fig. 4. Properties of congestions in system steady state depending on average (global) flow density.

4 Non-Linear Phenomena of Congestion Behavior

Together with the equilibrium number of congestions we can investigate also their "group" velocity. The diagram on Fig. 5 demonstrates non-smooth dependence of congestion velocity on the mean flow density.

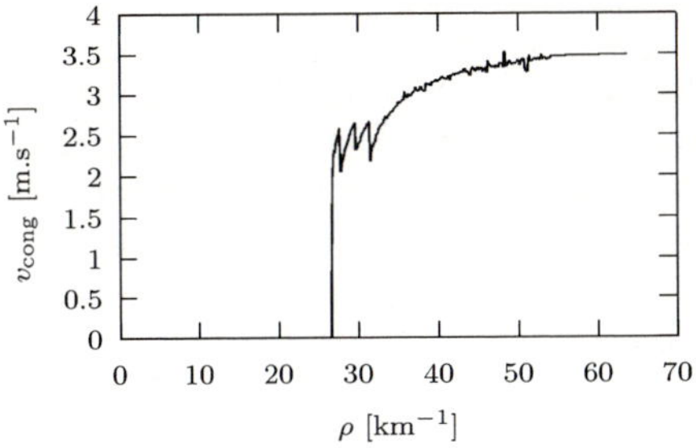

Fig. 5. Velocity of steady state congestions depending on the mean density of vehicles.

Variations of congestion speed at higher densities originate from non-linear effects: the system with certain number of congestions is stable and with a change of density jumps into the different stable states. In case of generation of a new congestion their velocity drops abruptly. It is evident that it would be possible to introduce a phenomenological view on the investigated system, in which growing "tension" is released by the emergency of a new congestion. Figure 5 can be assumed also as an analogy of the bifurcation diagram of low dimensional dynamic system.

Investigated system can exist in various stable states depending only on various initial conditions. For 1000 vehicles on the lane with 4 km radius we can "place" the system into a number of different steady states through the initial periodic density fluctuations. However the stable number of the congestions is limited: an insufficient number of initial congestions is unstable because of their great attracting amplitude — congestions will decay into the greater number of congestions with the lower amplitude. In case of too large number of initial congestions the amplitudes fall down and any perturbation translates the system into the new stable state with the smaller number of congestions.

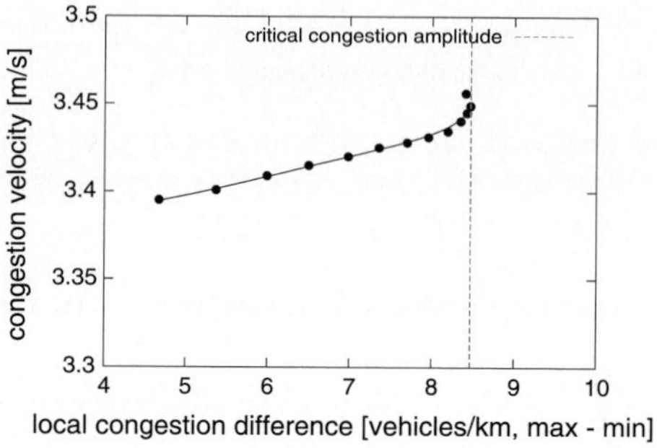

Fig. 6. Congestion velocity depending on the congestion amplitude (difference between the maximal and minimal local density).

Figure 6 illustrates overall linear relation between the amplitude of the congestion and its group velocity. This relation is not perturbed by various attributes of the system: all investigated densities of congestions are stable in the same system. Some critical amplitude of the congestion can be seen: at higher amplitudes the group velocity increases abruptly and the system loses its stability. Demonstrated dependence can be assumed as the "dispersion relation" known in wave theory [4] and is in charge of identical shape of congestions in steady state.

References

1. Haberman R (1998) Mathematical Models: Mechanical Vibrations, Population Dynamics and Traffic Flow. Society for Industrial and Applied Mathematics, Philadelphia.
2. Treiber M, Hennecke A, Helbing D (2000) Phys Rev E, 62:1805–1824.
3. Rizos C (2003) Network RTK Research and Implementation - A Geodetic Perspective. Journal of Global Postioning System, Vol. 1, No. 2.
4. Born M, Huang K (1998) Dynamical Theory of Crystal Lattice, Oxford University Press Inc.

Stability of Flow on a Ring with Three Links *

Alexander P. Buslaev[1], Alexander G. Tatashev[1], and Marina V. Yashina[2]

[1] Moscow State Automobile and Road Technical University, 64, Leningradsky pr., Moscow, Russia apal2006@yandex.ru
[2] Moscow Technical University of Communications and Informatics, 8a, Aviamotornaya str., Moscow, Russia yash-marina@yandex.ru

Summary. The qualitative properties of solutions of nonlinear differential equation system that describe traffic flow on a ring are developed. The ring consists of three links. The stationary points of the system have been found. The flow behavior in the neighborhood of this point has been considered. The stability of the stationary points is studied. The behavior of the solution near the boundary is considered.

1 Three-Chain

The qualitative properties of solutions of nonlinear differential equations system that describe traffic flow on a ring with three links are developed. We will call a ring with three links *a three-chain*, Fig. 1.

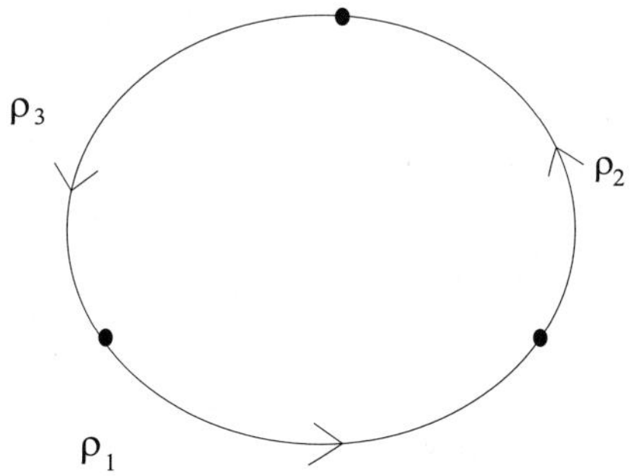

Fig. 1. Symmetrical three-chain.

Let $\rho(t) = (\rho_1(t), \rho_2(t), \rho_3(t))$ be flow densities on each link, $0 \le \rho_i \le \rho_{max}$, $i = 1, 2, 3$.

* Supported by Grant No. RP07-04 of National Taiwan University of Science and Technology

The flow dynamics is described by the differential equations system

$$\begin{cases} \frac{d\rho_1}{dt} = f(\rho_3) - f(\rho_1), \\ \frac{d\rho_2}{dt} = f(\rho_1) - f(\rho_2), \\ \frac{d\rho_3}{dt} = f(\rho_2) - f(\rho_3), \end{cases} \tag{1}$$

where $f(\rho) = a\rho^2 + b\rho + c = \rho(\rho_{\max} - \rho)$ is flow intensity.

Then

$$\begin{cases} \frac{d\rho_1}{dt} = (\rho_3 - \rho_1)(-(\rho_3 + \rho_1) + \rho_{\max}), \\ \frac{d\rho_2}{dt} = (\rho_1 - \rho_2)(-(\rho_1 + \rho_2) + \rho_{\max}), \\ \frac{d\rho_3}{dt} = (\rho_2 - \rho_3)(-(\rho_2 + \rho_3) + \rho_{\max}). \end{cases} \tag{2}$$

2 The Number of Stationary Points

Let C be the flow mass, $C \equiv \rho_1 + \rho_2 + \rho_3$.

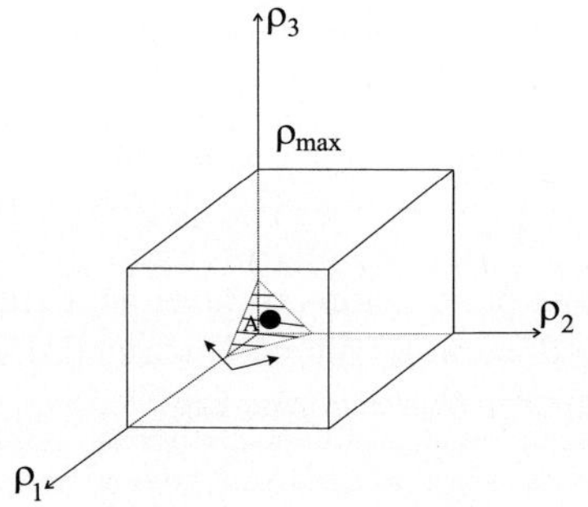

Fig. 2. Domain of definition (DD), $C < \rho_{max}$.

2a) Figure 2 shows the domain boundary on phase space in case of $C < \rho_{\max}$.

Lemma 1. *For $C < \rho_{\max}$ there is only one stationary point $A = (\rho_1 = \rho_2 = \rho_3 = C/3)$.*

Figure 3 shows the projection of the velocity field of the differential equations system (2) to the hyperplane $\rho_1 + \rho_2 + \rho_3 = C$.

Lemma 2. *For $C = \rho_{\max}$ there are four stationary points A^*, B_1^*, B_2^*, B_3^*. $A^* = (\rho_1 = \rho_2 = \rho_3 = \rho_{\max}/3)$, $B_1^* = (\rho_1 = \rho_{\max},\ \rho_2 = \rho_3 = 0)$, $B_2^* = (\rho_2 = \rho_{\max},\ \rho_1 = \rho_3 = 0)$, $B_3^* = (\rho_3 = \rho_1 = 0,\ \rho_2 = \rho_{\max})$.*

2b) Let us consider the case $\rho_{\max} < C < \frac{3}{2}\rho_{\max}$.

Let be true $\rho_1 = \rho_2 < \rho_3$. We have

$$\rho_1 + \rho_3 = \rho_2 + \rho_3 + \rho_{\max}, \tag{3}$$

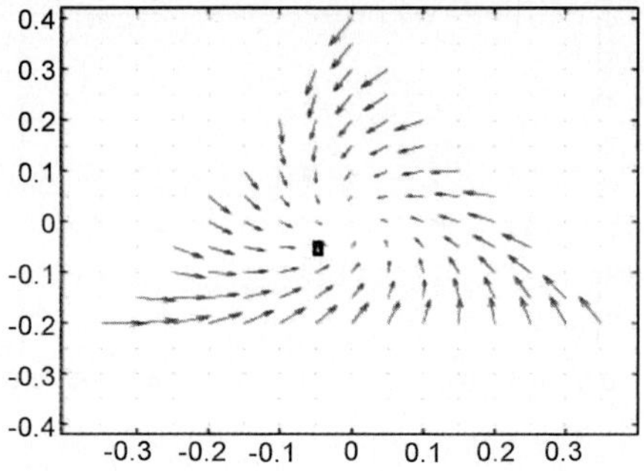

Fig. 3. The velocity field for $C < \rho_{\max}$, $\rho_{\max} = 1$.

$$\rho_3 = 2\rho_{\max} - C \geq 0, \tag{4}$$

$$\rho_1 = \rho_2 = C - \rho_{\max} \geq 0. \tag{5}$$

From (3)–(5) we conclude $2C < 3\rho_{\max}$. So because of the symmetry it is true the following statement.

Lemma 3. *For*

$$\rho_{\max} < C < \frac{3}{2}\rho_{\max}$$

there is the stationary points $B_3 = (\rho_1 = \rho_2 = C - \rho_{\max},\ \rho_3 = 2\rho_{\max} - C)$, $A = (\rho_1 = \rho_2 = \rho_3 = \frac{C}{3})$, $B_1 = (\rho_2 = \rho_3 = C - \rho_{\max},\ \rho_1 = 2\rho_{\max} - C)$, $B_2 = (\rho_3 = \rho_1 = C - \rho_{\max},\ \rho_2 = 2\rho_{\max} - C)$.

Figure 4 shows the points A, B_1, B_2, B_3.

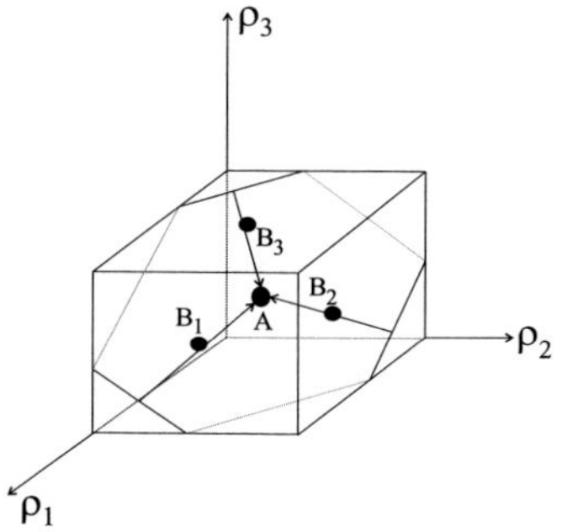

Fig. 4. DD, $\rho_{\max} < C < \frac{3}{2}\rho_{\max}$.

Figure 5 shows the projection of velocity field of the system of the differential equations (2) on hyperplane $\rho_1 + \rho_2 + \rho_3 = C$.

2c) Suppose that $\frac{3}{2}\rho_{\max} < C < 2\rho_{\max}$.

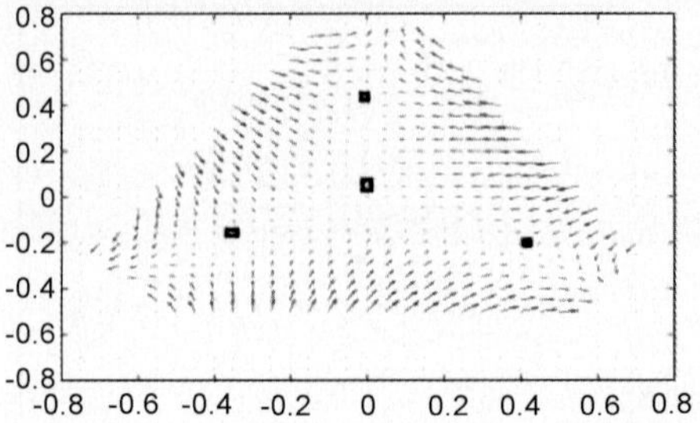

Fig. 5. The velocity field for $\rho_{\max} < C < \frac{3}{2}\rho_{\max}$.

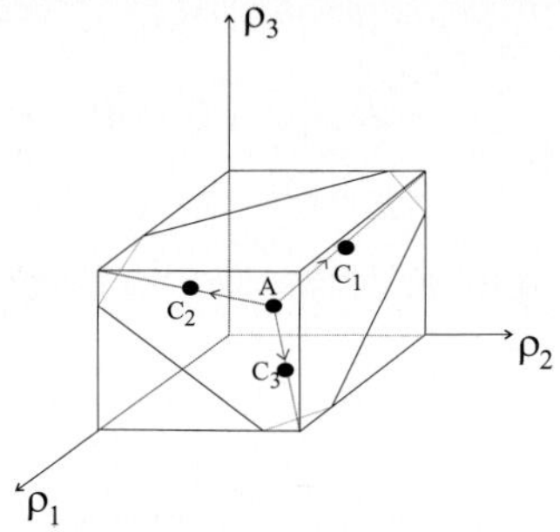

Fig. 6. DD, $\frac{3}{2}\rho_{\max} < C < 2\rho_{\max}$.

The following stationary points exist: $A = (\rho_1 = \rho_2 = \rho_3 = \frac{C}{3})$, $C_1 = (\rho_1 = 2\rho_{\max} - C,\ \rho_2 = \rho_3 = C - \rho_{\max})$, $C_2 = (\rho_2 = 2\rho_{\max} - C,\ \rho_1 = \rho_3 = C - \rho_{\max})$, $C_3 = (\rho_3 = 2\rho_{\max} - C,\ \rho_1 = \rho_2 = C - \rho_{\max})$.

Figure 7 shows the projection of the velocity field of the system of the differential equation (2) to the hyperplane $\rho_1 + \rho_2 + \rho_3 = C$ for $\frac{3}{2}\rho_{\max} < C < 2\rho_{\max}$.

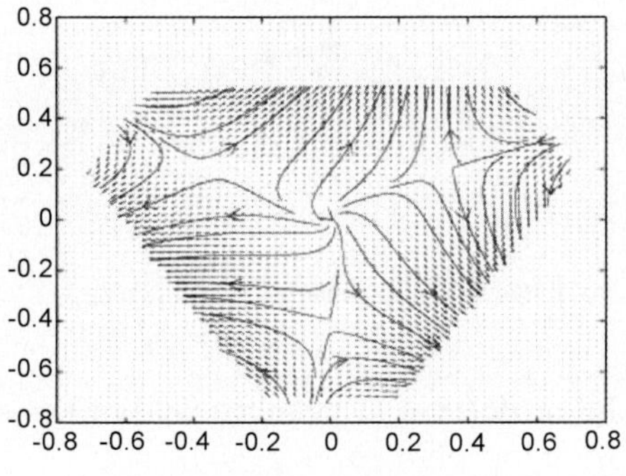

Fig. 7. The velocity field for $\frac{3}{2}\rho_{\max} < c < 2\rho_{\max}$, $(C = 1.75\rho_{\max})$.

So *in the case when* $C \to \frac{3}{2} - 0$ *the points* B_i, $i = 1, 2, 3$, *converge to point* A *and then points* C_i *diverge. For* $C = \frac{3}{2}\rho_{\max}$ *the stationary points* A, B *and* C *coincide.*

Theorem 1. *If* $\rho_{\max} < C < \frac{3}{2}\rho_{\max}$ *or if* $\frac{3}{2}\rho_{\max} < C < 2\rho_{\max}$ *then there are four stationary points*

$$A = (\rho_1 = \rho_2 = \rho_3 = \frac{C}{3}),$$

$$B_1 = (\rho_3 = \rho_2 = C - \rho_{\max},\ \rho_1 = 2\rho_{\max} - C),$$

$$B_2 = (\rho_1 = \rho_2 = C - \rho_{\max},\ \rho_3 = 2\rho_{\max} - C),$$

$$B_3 = (\rho_1 = \rho_2 = C - \rho_{\max},\ \rho_3 = 2\rho_{\max} - C).$$

If $C = \frac{3}{2}\rho_{\max}$ *then there is the only one stationary point*

$$\rho_1 = \rho_2 = \rho_3 = \frac{\rho_{\max}}{2}.$$

Let us suppose $C = 2\rho_{\max}$. In this case besides the stationary points

$$A = (\rho_1 = \rho_2 = \rho_3 = \frac{2\rho_{\max}}{3})$$

there exist also the following stationary points:
$C_3 = (\rho_1 = \rho_2 = \rho_{\max},\ \rho_3 = 0)$, $C_1 = (\rho_2 = \rho_3 = \rho_{\max},\ \rho_1 = 0)$, $C_2 = (\rho_3 = \rho_1 = \rho_{\max},\ \rho_2 = 0)$.

2d) In the case $2\rho_{\max} < C < 3\rho_{\max}$ there is the only one stationary point $\rho_1 = \rho_2 = \rho_3 = \frac{C}{3}$, (Fig. 8).

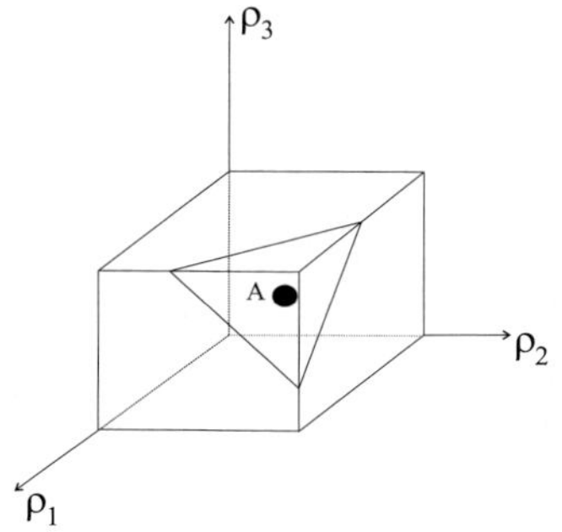

Fig. 8. DD, $2\rho_{\max} < C < 3\rho_{\max}$.

In case of $C = 3\rho_{\max}$ *there is only one stationary point which is also the only one admitted point of the phase space* $\rho_1 = \rho_2 = \rho_3 = \rho_{\max}$.

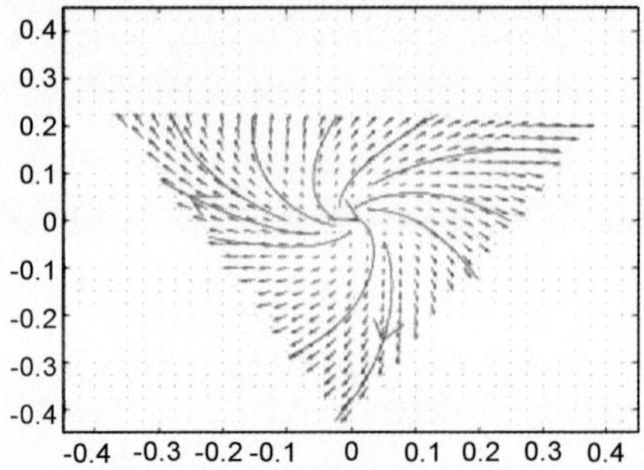

Fig. 9. The velocity field for $2\rho_{\max} < C < 3\rho_{\max}$.

3 Linearization

Let $(\rho_1^*, \rho_2^*, \rho_3^*)$ be some stationary point where $\rho_1^* = \rho_2^* = \rho_3^*$. Let us study the flow behavior in the neighborhood of this point. We have

$$f(\rho_i^* + u_i) = f(\rho_i^*) + f'(\rho_i^*)u_i + o(u_i),\ u_i \sim 0,\ i = 1, 2, 3.$$

When $u_i \to 0$ the system can be described as

$$\begin{cases} \frac{du_1}{dt} = f'(\rho_3^*)u_3 - f'(\rho_1^*)u_1, \\ \frac{du_2}{dt} = f'(\rho_1^*)u_1 - f'(\rho_2^*)u_2, \\ \frac{du_3}{dt} = f'(\rho_2^*)u_2 - f'(\rho_3^*)u_3, \\ u_1 + u_2 + u_3 = 0. \end{cases} \tag{6}$$

Theorem 2. *The stationary point $\rho_1 = \rho_2 = \rho_3 = \frac{C}{3}$ is stable then and only then when the flow mass C is less than half of maximum possible value.*

Proof. Let M be the matrix of the system of the system of the differential equations (6):

$$M = \begin{pmatrix} -f'(\rho_1^*) - f'(\rho_3^*) & -f(\rho_3^*) \\ f'(\rho_1^*) & -f(\rho_2^*) \end{pmatrix}.$$

Let E be the unit matrix. The characteristic equation of the system is

$$0 = |M - \lambda E| = \lambda^2 + \lambda(f'(\rho_1^*) + f'(\rho_2^*) +$$

$$+ f'(\rho_3^*)) + (f'(\rho_1^*)f'(\rho_3^*) + f'(\rho_1^*)f'(\rho_2^*) + f'(\rho_2^*)f'(\rho_3^*)),$$

$$\lambda^2 + 3\lambda f'(\frac{C}{3}) + 3(f'(\frac{C}{3}))^2 = 0.$$

For the strict stability of the local stationary point $\rho_1 = \rho_2 = \rho_3 = \frac{C}{3}$ it is necessary and sufficiently that the real parts of the roots of the characteristic equation are negative. It is in the case when $f'(\frac{C}{3}) > 0$, that is, when $\frac{C}{3} < \frac{\rho_{\max}}{2}$. The roots of the characteristic equation are imagine because of the discriminant of the characteristic equation is negative. *In the case* $\frac{C}{3} < \frac{\rho_{\max}}{2}$ *the point* $A = (\rho_1 = \rho_2 = \rho_3 = C/3)$ *is the stable focus. In the case* $\frac{C}{3} > \frac{\rho_{\max}}{2}$ *this point is the unstable focus.*

Theorem 3. *The stationary point* B_1, B_2, B_3, *for* $\rho_{\max} < C < \frac{3}{2}\rho_{\max}$, *and the stationary points* C_1, C_2, C_3, *for* $\frac{3}{2}\rho_{\max} < C < 2\rho_{\max}$, *are unstable saddles.*

Proof. Suppose $\rho_{\max} < C < \frac{3}{2}\rho_{\max}$. Then the point $\rho_1 = \rho_2 = C - \rho_{\max}$, $\rho_3 = 2\rho_{\max} - C$ is the stationary point. We have $f'(2\rho_{\max} - C) = -f(C - \rho_{\max})$ and then $f'(C - \rho_{\max}) + f'(C - \rho_{\max}) + f'(2\rho_{\max} - C) = f'(C - \rho_{\max}) > 0$. Characteristic equation of the system is

$$\lambda^2 + \lambda f'(C - \rho_{\max}) - (f'(C - \rho_{\max}))^2 = 0.$$

The roots of the characteristic equation are real and their signs are different. *In this case the type of point is the "saddle".*

Suppose $C = \rho_{\max}$. As it has been stated in the section 2, in case of $C = \rho_{\max}$ three additional stationary points exist $\rho_1 = \rho_2 = 0$, $\rho_3 = \rho_{\max}$, $\rho_3 = \rho_1 = 0$, $\rho_2 = \rho_{\max}$, $\rho_2 = \rho_3 = 0$, $\rho_1 = \rho_{\max}$. We have $f'(0) = \rho_{\max}$, $f'(\rho_{\max}) = -\rho_{\max}$ and hence $\lambda^2 - \lambda\rho_{\max} - (\rho_{\max})^2 = 0$. The roots of the characteristic equation are real and their signs are different. So, *for* $C = \rho_{\max}$ *the stationary points* B_1, B_2, B_3 *are unstable.*

Suppose $\frac{3}{2}\rho_{\max} < C < 2\rho_{\max}$. The point $B_3 = (\rho_1 = \rho_2 = C - \rho_{\max}$, $\rho_3 = 2\rho_{\max} - C)$ is a stationary point. We have $f'(2\rho_{\max} - C) = 2 - 3\rho_{\max} = -(3\rho_{\max} - 2C) = -f(C - \rho_{\max})$ and so $f'(C - \rho_{\max}) + f'(C - \rho_{\max}) + f'(2\rho_{\max} - C) = f'(C - \rho_{\max}) > 0$. Characteristic equation of the system (3) is $\lambda^2 + \lambda f'(C - \rho_{\max}) - (f'(C - \rho_{\max}))^2 = 0$. The roots of the characteristic equations are real and their signs are different. The stationary point is the saddle. The stationary points $\rho_3 = \rho_1 = C - \rho_{\max}$, $\rho_2 = 2\rho_{\max} - C$ and $\rho_2 = \rho_3 = C - \rho_{\max}$, $\rho_2 = 2\rho_{\max} - C$ are also saddles.

Suppose $C = 2\rho_{\max}$. As it has been stated in the section 2, in the case $C = \rho_{\max}$ there are three additional stationary points $C_1 = (\rho_2 = \rho_3 = \rho_{\max}$, $\rho_1 = 0)$, $C_2 = (\rho_3 = \rho_1 = \rho_{\max}$, $\rho_2 = 0)$, $C_3 = (\rho_1 = \rho_2 = \rho_{\max}$, $\rho_3 = 0)$.

The roots of the characteristic equation are real and their signs are different. So, *for* $C = 2\rho_{\max}$ *the stationary points* C_1, C_2, C_3 *are unstable. Each of this points is the saddle.*

4 Behaviour of the Solution Near the Boundary

Consider the behaviour of the solution near the boundary of the admitted domain of the phase space that is in the points for which at least one variable has a value near 0 or near $\rho_{\max}$.

a) Consider the case $C < \rho_{\max}$. Suppose that $\rho_3 = 0$, $\rho_1 + \rho_2 = C$. Vector $\mathbf{n} = (-1, -1, 2)$ belongs to the domain of admitted values and is normal for the domain boundary in the considered point. This vector is directed to the inside of the domain. The scalar product of this vector and the vector of the field velocity equals

$$f(\rho_1) - f(\rho_2) + f(\rho_2) - f(\rho_3) + 2f(\rho_1) - 2f(\rho_3) = f(\rho_3) > 0.$$

On account of symmetry it is sufficient to consider one part of the boundary only. Hence the flow in the neighborhood of the boundary is uncritical. Suppose that $\rho_{\max} < C < \frac{3}{2}\rho_{\max}$. Again on account of symmetry, it is sufficient to consider the preceding fragment of the boundary and the neighbor one $\rho_2 = \rho_{\max}$, $\rho_1 + \rho_3 = C - \rho_{\max}$. In the first case the inequality is true still. Therefore we have for the second part $\mathbf{n} = (1, -2, 1)$ and $(\mathbf{v}, \mathbf{n}) = f(\rho_2) - f(\rho_1) - 2f(\rho_3) + f(\rho_1) - f(\rho_3) = -3f(\rho_3) < 0$. Hence the boundary part appearing anew is a zone of attraction.

b) Suppose $\rho_{\max} < C < 2\rho_{\max}$. Consider the case $\rho_1 \sim 0$, $\rho_1 < \rho_3 < \rho_{\max} - \rho_1$. Then the right part of the first equation of the system (3) is positive and so $\frac{d\rho_1}{dt} > 0$. Solution of the system removes from the boundary. Consider the point of the phase space for which $\rho_{\max} - \rho_1 \sim 0$, $\rho_1 < \rho_3 < \rho_{\max} - \rho_1$. Then the right part of the first equation of system (3) is positive and so $\frac{d\rho_1}{dt} > 0$. *The solution of the system approaches to the boundary and the value of the variable ρ_1 attains the critical value $\rho_{\max}$.*

c) Consider the case $2\rho_{\max} \leq C < 3\rho_{\max}$.

Suppose that for some point of the phase space it is true $\rho_{\max} - \rho_1 \sim 0$, $\rho_3 < \rho_{\max} - \rho_1$. We have $C < \rho_{\max}$, $\rho_1 > 0$ and hence $f(\rho_3) > f(\rho_1)$ that is the right part of the first equation of the system (3) positive and therefore $\frac{d\rho_1}{dt} > 0$. *In this case the solution of the system approaches to the boundary and attains the value $\rho_{\max}$.*

References

1. Buslaev A.P., Novikov A.V., Prikhodko V.M., Tatashev A.G., Yashina M.V. (2003) Stochastic and Imitation Approach to Traffic Movement. Mir, Moscow. 286 p.
2. Buslaev A.P., Tatashev A.G., Yashina M.V. (2006) Stability of Flows on Networks. In: Proceedings of International Conference "Traffic and Granular Flows -2005". Springer, Berlin Heidelberg New York: 427–435.
3. Bugaev A.S., Buslaev A.P., Tatashev A.G. (2006) On properties of solutions of one class of system of non-linear differential equations on graphs. J. Imamod RAS 18, No.12: 19–34.
4. Buslaev A.P., Tatashev A.G., Yashina M.V. (2004) On properties of a class of systems of non-linear differential equations on graphs. Vladikavkaz Math. J. 6 No. 4: 17–24.
5. Buslaev A.P., Tatashev A.G., Yashina M.V. (2006) Stability of flows on graphs. In: Korobejnik U.F., Kusraev A.G. (Eds) Compex Analyses. Mathematical Modelling. Vladikavkaz Scientific Center RAS, Vladikavkaz: 263–284.

Improving Pedestrian Micro-Simulations with Event Steps

Mario C. Campanella, Serge P. Hoogendoorn, and Winnie Daamen

Delft University of Technology, Stevinweg 1, 2628 CN Delft, The Netherlands
m.c.campanella@tudelft.nl, s.p.hoogendoorn@tudelft.nl,
w.daamen@tudelft.nl

Summary. Microscopic pedestrian models describe individual pedestrian behavior and the interaction of pedestrians with other pedestrians and obstacles. Continuous time models generally calculate the acceleration of pedestrians due to repulsive or attractive interactions using for instance distances to other pedestrians and obstacles within a two-dimensional influence area. Two problems that usually arise with these types of models when simulating very large crowds are extreme accelerations that occur due to very short distances to other pedestrians and obstacles and large computational times to assess and to calculate the accelerations for individual pedestrians. This paper presents a hybrid pedestrian management algorithm that combines a traditional optimized time-based simulation and an event-driven simulation. This way, the task of assessing the surroundings and the task of dealing with interactions on very short distances are each treated in an optimized way leading to more reliable accelerations in high densities as well as shorter calculation times.

1 Introduction

Although the computational power available in a personal computer grew several orders of magnitude in the last decades, performing micro-simulations involving several thousands of pedestrians still requires a significant amount of time. This may effectively hinder the application of these models in a design process in which an architect wants to assess several layouts of public areas or large buildings. In the next section we will show that most simulation models, among which the NOMAD model [1], apply general approaches to search pedestrians' surroundings to identify other pedestrians and obstacles. This search is one of the most time consuming parts of micro-simulation models computations and it has to be performed each simulation step. In this contribution we propose an optimized spatial search to improve the quality and the efficiency of the NOMAD model.

2 The NOMAD Model

NOMAD is a normative model proposed by Hoogendoorn and Bovy [1]. Being a microscopic simulation model, it has a discrete time step. The movement of a pedestrian p in terms of velocity and acceleration can be described as a system of differential equations. Let $\mathbf{r}_p(t)$, $\mathbf{v}_p(t)$ and $\mathbf{a}_p(t)$ respectively denote the location, the velocity and the acceleration at time instant t, then the following is valid:

$$\frac{d}{dt}\mathbf{r}_p(t) = \mathbf{v}_p(t) \tag{1}$$

$$\frac{d}{dt}\mathbf{v}_p(t) = \mathbf{a}_p(t) \tag{2}$$

The acceleration is the result of different factors affecting the speed of pedestrian p, namely the desire to stay as close as possible to a desired trajectory leading towards the pedestrian's destination, avoiding other pedestrians, avoiding obstacles, the physical forces when pedestrians collide and some random component or noise:

$$\mathbf{a}_p(t) = \frac{\mathbf{v}_p^0(t) - \mathbf{v}_p(t)}{T_p} - A_p \sum\nolimits_q e^{-\frac{\mathbf{d}_{pq}(t)}{R_p^p}} - A_p^o \sum\nolimits_o e^{-\frac{\mathbf{d}_{po}(t)}{R_p^o}} + \mathbf{b}_p(t) + \varepsilon_p(t) \tag{3}$$

where:

$\mathbf{v}_p^0$: desired velocity
$\mathbf{d}_{pq}$: vector pointing from pedestrian p towards pedestrian q
$\mathbf{d}_{po}$: vector pointing from pedestrian p towards obstacle o
T_p : acceleration time of pedestrian p
A_p : interaction strength between pedestrian p and other pedestrians
A_p^o : interaction strength between pedestrian p and obstacles o
R_p : interaction distance between pedestrian p and other pedestrians
R_p^o : interaction distance between pedestrian p and obstacles o
$\mathbf{b}_p(t)$: physical acceleration due to contact forces
$\varepsilon_p(t)$: random term

A pedestrian is only affected by the pedestrians and obstacles present within his influence area. The latter can be represented by different forms (isotropic influence areas, anisotropic influence areas, etc.).

The physical forces $\mathbf{b}_p(t)$ are calculated as follows [2, 3] (Eq. (4) describes normal forces, while Eq. (5) describes tangential forces):

$$\mathbf{b}_p^{(n)}(t) = \kappa_p \delta_{pq}(t) \tag{4}$$

where:

$\mathbf{b}_p^{(n)}$: physical acceleration in direction of centers of p and q
κ : restitution coefficient
$\delta_{pq}(t)$: deformation of pedestrians p and q

$$\mathbf{b}_p^{(t)}(t) = \upsilon_p \left| v_p^{(t)}(t) - v_q^{(t)}(t) \right| \delta_{pq}(t) \tag{5}$$

where:

$\mathbf{b}_p^{(t)}$	: tangential physical acceleration in direction of centers of p and q
υ	: tangential viscosity factor
$v_p^{(t)}(t)$	: tangential speed of pedestrian p
$v_q^{(t)}(t)$	: tangential speed of pedestrian q
$\delta_{pq}(t)$	: deformation of pedestrians p and q

3 Improving Efficiency of Micro-Simulation Models

The simulation model NOMAD is capable of reproducing emergent spatial patterns. However, two problems are potentially encountered due to the formulation of the acceleration term in Eq. (3):

- The physical forces that arise when pedestrians are very close to each other or close to obstacles can be very large in extreme densities due to large deformations of pedestrians.
- The computation times to calculate the distance vectors $\mathbf{d}_{pq}(t)$ and $\mathbf{d}_{po}(t)$ at every simulation step can grow excessively, impairing the total time needed to run a simulation, especially when many pedestrians are present in dense situations.

There are several ways to solve these problems, such as using dedicated data representation structures to reduce look-up times for pedestrians in the direct vicinity of pedestrian p. However, this only partially eliminates the problem, since it is still dependent on the size of the pedestrian population. Another way is to use isotropic (circular) influence areas. Usually, it is possible to implement a simplified query algorithm for this geometry that does not take the heading of the pedestrian into consideration to assess the distances. Again, this is not a complete solution, because such an area may produce unrealistic behavioral results, and it still requires a query at every simulation step.

The second problem of possibly high interaction forces can be tackled by using another formulation such as an exponential spring model for the physical forces:

$$\mathbf{b}_p^{(n)}(t) = c_1 e^{c_2 \delta_{pq}(t)} \tag{6}$$

Physical accelerations progress in a manner that could prevent too large accelerations to occur (in case of small deformations). It would not solve the problem when density is very high and large deformations will occur. The most effective way to prevent high deformations is to reduce the time step dt. A sufficiently small dt prevents pedestrians to deform too much (due to numerical issues) since the acceleration terms are calculated more frequently.

However, a small dt increases the simulation time substantially, which has already been identified as a problem in the previous section.

4 Hybrid Pedestrian Management

In a simulation, in each time step the walking behavior is calculated for each individual pedestrian present on the walking area. In this contribution we call pedestrian management the part of the simulation model responsible to calculate this walking behavior. Hybrid pedestrian management, presented in this contribution, is a combination of a time based simulation step and an event based simulation step, where most models are restricted to a time based simulation step. A time based simulation step considers discrete (constant) time steps, during which every pedestrian moves the distance calculated considering the constant speed in the time step (Fig. 1a).

To overcome the high physical forces in high densities, the hybrid management will distinguish the pedestrians that are very close to each other (categorized as pedestrians "in-collision"). These pedestrians are handled as colliding particles in an event-based simulation [4] (Fig. 1b), while for the other pedestrians the traditional time-based simulation is applied. The latter are further distinguished in two categories: those being sufficiently far away from other pedestrians and obstacles in order not to collide during the next time step ("in-range") and those being even further away from pedestrians and obstacles guaranteeing that their influence area will be not interfered ("isolated"). This way of distinguishing three degrees of isolation allows the hybrid management to optimize the amount of pedestrians that have to calculate the repulsion terms caused by physical contact.

4.1 Event Step

The hybrid way of handling collisions that may occur during a simulation step is to subdivide the simulation step into *smaller collision event steps of variable time lengths.* The smallest distance that any pedestrian "in-collision" can walk before he collides with another pedestrian or an obstacle will determine the next event step. If the time for the next collision event is smaller than the simulation time step, the pedestrians involved are moved until they collide, while other pedestrians in the event step will move according to this event step. The process is then repeated for the next smallest event time step. This process stops when the added length of the realized event steps including the next calculated event step is larger than the simulation time step. At this moment the last collision event is not realized and the residual time (see Eq. (7)) will be used to move the pedestrians to their final positions in this step without collisions:

$$R_{step}(t) = \sum_{i=1}^{n} E^{i}_{step}(t) - S_{step} \tag{7}$$

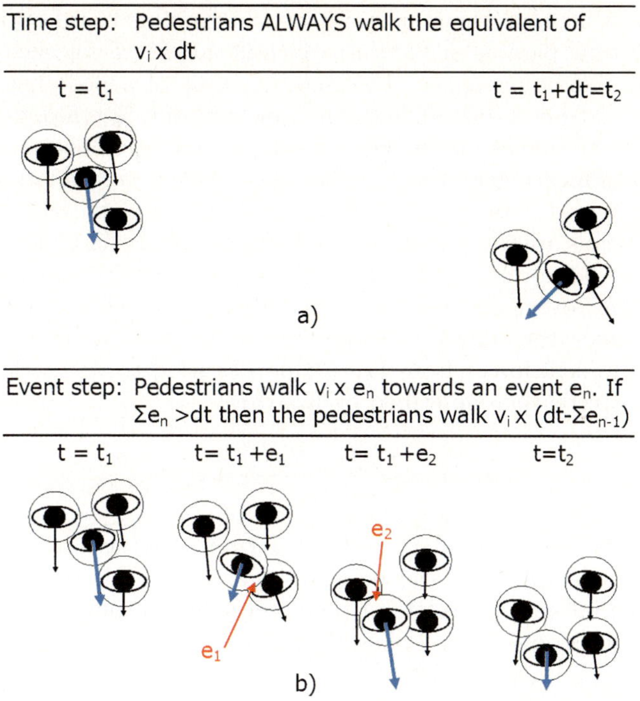

Fig. 1. a) and b) Comparison between a simulation time step and an event step.

where:

$R_{step}(t)$	: residual time step
$E^i_{step}(t)$	: event time step
S_{step}	: simulation time step

To handle the collisions that occur at each event step we assume that there a pedestrian has a maximum deformation in the normal direction (δ_{max}). With this assumption an event becomes a collision when at least one pedestrian has $\delta_p(t) = \delta_{max}$. With the introduction of this condition the hybrid management makes sure that the collision forces will never be larger then a certain calibrated value independent of the value of dt.

4.2 Optimized Time Based Simulation Step

The time based simulation step is an optimized step in which the necessity to assess the influence areas by the pedestrians is minimized. Preferably, all pedestrians are calculated using the optimized time based simulation step and only those pedestrians that are too close to other pedestrians or walls will perform the event step. The main parameter in the time based optimization is the isolation time parameter ($T^{ip}(t)$) which is calculated for each pedestrian during his initialization phase. The isolation time of a pedestrian indicates the amount of time a pedestrian can walk without interfering with other pedestrians or obstacles, since the nearest pedestrian or obstacle is too far away to risk appearing in his influence area. When the isolation time has expired, the hybrid management will recalculate it and the process restarts. These pedestrians are said to be "isolated" from other pedestrians whilst their isolation time did not expire. By definition an "isolated" pedestrian does not need to look for other pedestrians in the current time step. To calculate his walking behavior the repulsion terms due to pedestrians and obstacles as well as the physical forces from Eq. (3) do not need to be recalculated. His acceleration can be then described by Eq. (8):

$$\mathbf{a}_p(t) = \frac{\mathbf{v}_p^0(t) - \mathbf{v}_p(t)}{\mathrm{T}_p} + \varepsilon_p(t) \tag{8}$$

To calculate the lower boundary of $T^{ip}(t)$ for a pedestrian p the hybrid management calculates the distances between him and all pedestrians around him and assumes that each pedestrian is in direct collision route regardless of the real walking directions. Fig. 2 shows the above mentioned worst case situation.

To be sure that pedestrian q will not reach the influence area of pedestrian p, the hybrid management uses the maximum speeds (V_{max}) and maximum body radii (R_{max}) for both pedestrians. It also must assume all possible direc-

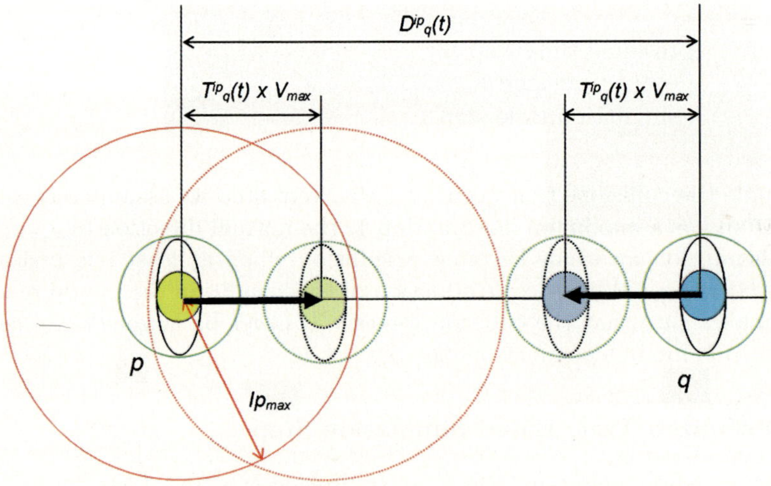

Fig. 2. Pedestrian q entering the influence area of pedestrian p assuming a conflicting trajectory.

tion changes by both pedestrians. Therefore it uses a circle with radius Ip_{max} to calculate $T_q^{ip}(t)$ and not the real influence area shape. From Fig. 2 we can derive the expressions for the distance between two pedestrians in terms of the maximum constants and $T_q^{ip}(t)$ and the expression for $T_q^{ip}(t)$ (Eq. (10)).

$$D_q^{ip}(t) = 2 \cdot \left(T_q^{ip}(t) \cdot V_{\max}\right) + Ip_{\max} + R_{\max} \tag{9}$$

$$T_q^{ip}(t) = \left(\frac{D_q^{ip}(t) - Ip_{\max} - R_{\max}}{2 \cdot V_{\max}}\right) \tag{10}$$

The isolation parameter time $T^{ip}(t)$ for a pedestrian is obtained as the minimum of all $T_q^{ip}(t)$ that are calculated according to Eq. (11):

$$T^{ip}(t) = \min\left(\frac{D_q^{ip}(t) - Ip_{\max} - R_{\max}}{2 \cdot V_{\max}}\right) \forall q \neq p \tag{11}$$

If the value obtained in Eq. (11) is larger than the simulation step dt, then the pedestrian is "isolated" from other pedestrians. For obstacles, a similar approach is implemented.

A pedestrian p is "in-range" if the nearest pedestrian q is sufficiently distant to guarantee that no collision will occur, but too close to guarantee that pedestrian q will not fall within pedestrian p's influence area. In this case the acceleration of pedestrian p takes into account all terms of the acceleration formula in Eq. (3) except the physical term since by definition there is no chance of collision with any pedestrian (Eq. (12)).

$$\mathbf{a}_p(t) = \frac{\mathbf{v}_p^0(t) - \mathbf{v}_p(t)}{T_p} - A_p^p \sum_{q \in P} e^{-\frac{\mathbf{d}_{pq}(t)}{R_p^P}} - A_p^o \sum_{o \in O} e^{-\frac{\mathbf{d}_{po}(t)}{R_p^O}} + \varepsilon_p(t) \tag{12}$$

To calculate the in-range time ($T_{rp}(t)$) that a pedestrian is guaranteed to be free of collisions from other pedestrians we apply the same assumptions used for the isolation time: maximum parameter values and a straight collision path. We can deduce the value of $T_{rp}^q(t)$ that is the in-range time that pedestrian q would take to collide with pedestrian p in a straight trajectory with full speed:

$$T_q^{rp}(t) = \left(\frac{D_q^{rp}(t) - 2 \cdot R_{\max}}{2 \cdot V_{\max}}\right) \tag{13}$$

The maximum time during which pedestrian p will certainly not collide with other pedestrians is shown in Eq. (14). We define:

$$T^{rp}(t) = \min\left(\frac{D_q^{rp}(t) - 2 \cdot R_{\max}}{2 \cdot V_{\max}}\right) \forall q \neq p \tag{14}$$

$$T^{ip}(t) \leq dt \tag{15}$$

If the value obtained in Eq. (14) is larger than the simulation step dt and his $T_{ip}(t)$ is smaller than dt then the pedestrian is "in-range" from other pedestrians. For obstacles, again a similar approach is used.

If a pedestrian is neither "isolated" nor in "in-range", than his nearest pedestrian (or obstacle) is so close that there is a theoretical possibility that they may collide. In this case this pedestrian is "in-collision" and will perform the event step.

5 Conclusions and Future Work

In this contribution we presented a new hybrid pedestrian management approach that combines an optimized time-based simulation step and an event-based simulation step. This hybrid pedestrian manager differentiates pedestrians in micro-simulations according to three isolation degrees: "isolated", "in-range" and "in-collision". With this, the hybrid management is able to optimize the time to compute pedestrians' accelerations and prevent unrealistic repulsive accelerations in very dense situations. This hybrid manager is expected to improve calculation times and accuracy of accelerations during high densities in spatially distributed micro-simulations. The next step in this research is to implement and calibrate the hybrid model including the event step and compare the quality and the performance of the results with the traditional management model.

References

1. Hoogendoorn S P, and P H L Bovy (2003) Simulation of Pedestrian Flows by Optimal Control and Differential Games. Optimal control applications & methods, 24(3), pp. 153-172.
2. Helbing D, and P Molnar (1995) Social force model of pedestrian dynamics. Physical Review, E 51, 4282-4286.
3. Helbing D, I Farkas, and T Vicsek (2000) Simulating dynamical features of escape panic. Nature, 407, 487-490.
4. Luding S (1998) Collisions & Contacts between two particles. In H.J. Herrmann, J.-P. Hovi, and S. Luding (Eds.): Physics of dry granular Media. Kluwer Academic Publishers, Dordrecht.

Road Traffic Models Using Petri Nets and Minplus Algebra

Nadir Farhi, Maurice Goursat, and Jean-Pierre Quadrat

INRIA - Rocquencourt, Domaine de Voluceau 78153 Le Chesnay, Cedex, France
(Nadir.Farhi, Maurice.Goursat, Jean-Pierre.Quadrat)@inria.fr

Summary. We present in this paper a traffic model based on Petri nets and minplus algebra. A Petri net is represented by two matrices one in standard algebra and the other in minplus algebra. Then a system point of view is developed based on a matrix product combining these two algebras. Introducing inputs and outputs on transitions and places we can link two Petri nets by associating outputs with inputs of the two systems. This linking corresponds to a contraction operator. Combining elementary systems with this contraction operator we can build large systems. This point of view is used to define the traffic dynamics of a regular town from three elementary Petri nets.

1 Introduction

Two modeling approaches of road traffic can be distinguished: – the macroscopic point of view where the traffic is seen as a gas with a comportment law given by a diagram called the fundamental traffic diagram, – the microscopic point of view where each individual dynamics is given. The cellular automata approach is a microscopic point of view with very simplified individual dynamics where the fundamental diagram emerges from large simulations.

In this paper, we present a very simplified microscopic traffic model based on Petri nets [1] with dynamics represented by generalized matrices and a generalized product mixing standard and minplus matrix (see [2]) product. Our approach is closed to the cellular automata one given for example in [3]. As in statistical physics, we study the individual vehicle movements to derive macroscopic laws. The effectiveness of this type of model in determining the fundamental traffic diagrams has been shown on a circular road [4] and on traffic networks [5, 6].

To be able to compute the fundamental traffic diagram of towns, we focus our attention on the system point of view, that is, connecting deterministic Petri nets with inputs and outputs. In particular we introduce a unique combinator called contraction which generalizes series, parallel and feedback composition. An application is given to traffic where three elementary Petri nets are given from which we can derive the traffic dynamics of a complete regular town (with traffic lights) using only this contraction operator.

2 Minplus Algebra

The minplus algebra [2] is the structure $\mathbb{R}_{\min} = (\mathbb{R} \cup \{+\infty\}, \oplus, \otimes)$ where the set $\mathbb{R} \cup \{+\infty\}$ is endowed with the operations min (denoted by $\oplus$, called minplus *sum*) and + (denoted by $\otimes$, called minplus *product*). The element $\varepsilon = +\infty$ is the *zero* element, and the element $e = 0$ is the *unity* one.

This minplus structure on scalars induces an idempotent semiring structure on $m \times m$ square matrices with the element-wise minimum and the matrix product defined by $(A \otimes B)_{ik} = \min_j (A_{ij} + B_{jk})$, where the zero and the unit matrices are still denoted by ε and e.

Event Graphs (Petri nets with places having only one input and one output arcs) are linear dynamical systems in this algebra and can be represented by a triple (A, B, C) of minplus matrices. The throughput of the event graph is then given by the minplus eigenvalue of the matrix A (see [2]).

Traffic on a circular road can be described by an event graph whose dynamics is given by a minplus linear system. Determining the fundamental diagram giving the average flow as a function of the car density corresponds to compute the eigen value of a minplus matrix (see [4]).

3 Petri Net Dynamics

A Petri net $\mathcal{N}$ is a graph with two sorts of nodes: the *transitions* $\mathcal{Q}$ and the *places* $\mathcal{P}$ and two sorts of arcs, the *synchronization arcs* (from places to transitions) and the *production arcs* (from transitions to places). In this paper we consider deterministic Petri nets where all the places have only one downstream arc.

A minplus $|\mathcal{P}| \times |\mathcal{Q}|$ matrix D, called *synchronization (decision) matrix* is associated to the synchronization arcs. $D_{pq} = a_p$ if there exists an arc from the place $p \in \mathcal{P}$ to the transition $q \in \mathcal{Q}$, and $D_{pq} = \varepsilon$ elsewhere, where a_p is the *initial marking* and is represented graphically by tokens in the places. We suppose here that the sojourn time in all the places is one time unit.

A standard $|\mathcal{Q}| \times |\mathcal{P}|$ matrix H, called *production (hazard) matrix* is associated with the production arcs. It is defined by $H_{qp} = m_{qp}$ if there exists an arc from q to p, and 0 elsewhere, where m_{qp} is the *multiplicity* of the arc.

Therefore a deterministic Petri net is characterized by the quadruple:

$$(\mathcal{P}, \mathcal{Q}, H, D).$$

It is a dynamic system in which the token evolution is partially defined by the transition firings saying that a transition can fire as soon as all its upstream places contain a positive quantity of tokens (fluid) having stayed at least one time unit. When a transition fires, it consumes a quantity of tokens equal to the minimum of all the available quantities being in the upstream places. Cumulating the firings done up to present time defines the *cumulated firing* of the transition. The firing produces a quantity of tokens in each downstream place equal to the firing of the transition multiplied by the multiplicity of the corresponding production arc. If the multiplicity of a production arc, going from q to p, is negative, the firing of q consumes tokens of p.

In the case of a *deterministic* Petri net, the dynamics is well defined, that is, there is no token consuming conflict between the downstream transitions. Then, denoting $Q = (Q_q^k)_{q\in\mathcal{Q},k\in\mathbb{N}}$ the vector of sequences of cumulated firing numbers of the transitions and $P = (P_p^k)_{p\in\mathcal{P},k\in\mathbb{N}}$ the vector of sequences of cumulated tokens numbers arrived in the places we have:

$$\begin{bmatrix} P^{k+1} \\ Q^{k+1} \end{bmatrix} = \begin{bmatrix} 0 & H \\ D & \varepsilon \end{bmatrix} \boxtimes \begin{bmatrix} P^{k+1} \\ Q^k \end{bmatrix} \overset{def}{=} \begin{bmatrix} HQ^k \\ D \otimes P^{k+1} \end{bmatrix} . \tag{1}$$

We can generalize the autonomous Petri nets to the input-output Petri nets. For that we partition the transition and the place sets in three parts: $(\mathcal{V}, \mathcal{Q}, \mathcal{Z})$ for the transitions and $(\mathcal{U}, \mathcal{P}, \mathcal{Y})$ for the places. These parts represent the inputs, the states, and the outputs respectively. The inputs are the transitions [resp. places] without upstream arcs and the outputs are the ones without downstream arcs. The dynamics (1) can be rewritten:

$$\begin{bmatrix} P^{k+1} \\ Q^{k+1} \\ Y^{k+1} \\ Z^{k+1} \end{bmatrix} = \begin{bmatrix} 0 & A & 0 & B \\ C & \varepsilon & D & \varepsilon \\ 0 & E & 0 & 0 \\ F & \varepsilon & \varepsilon & \varepsilon \end{bmatrix} \boxtimes \begin{bmatrix} P^{k+1} \\ Q^k \\ U^{k+1} \\ V^k \end{bmatrix} \overset{def}{=} \begin{bmatrix} AQ^k + BV^k \\ C \otimes P^{k+1} \oplus D \otimes U^{k+1} \\ EQ^k \\ F \otimes P^{k+1} \end{bmatrix} . \tag{2}$$

This dynamics, denoted S and defined by the matrices (A, B, C, D, E, F), associates to the input signals $(U^k, V^k)_{k\in\mathbb{N}}$ the output signals $(Y^k, Z^k)_{k\in\mathbb{N}}$. We write: $(Y, Z) = S(U, V)$. On these systems many operations such as parallel, series, feedback and contraction can be defined [7]. Here, we discuss only the contraction operator which will be used later and which generalize the three standard ones.

To define the *contraction* we have first to partition the inputs (denoted in the case of a partition in two sets by (i, i')) and the outputs (denoted (o, o')) of a system. Then, a system S with the previous partition of inputs and outputs is denoted $S_{oo'}^{ii'}$.

Given two partitioned systems $S_{oo'}^{ii'}$ and $\bar{S}_{i'o''}^{o'i''}$ we define the contracted system $\hat{S}_{oo''}^{ii''} \overset{def}{=} S_{oo'}^{ii'} \bar{S}_{i'o''}^{o'i''}$ as the solution in Y, Y'', Z, Z'' of the system:

$$\begin{cases} (YY', ZZ') = S(UU', VV'), \\ (U'Y'', V'Z'') = \bar{S}(Y'U'', Z'V''), \end{cases}$$

where the partition of the input places associated to the input partition (i, i') is denoted UU' with the analogous notations for the outputs YY' and for the input and output transitions VV' and ZZ'. The dynamics of $\hat{S}$ is defined by:

$$\hat{A} = \begin{bmatrix} A & 0 & 0 & B' \\ 0 & \bar{A} & \bar{B}' & 0 \\ E' & 0 & 0 & 0 \\ 0 & \bar{E}' & 0 & 0 \end{bmatrix}, \hat{B} = \begin{bmatrix} B & 0 \\ 0 & \bar{B}'' \\ 0 & 0 \\ 0 & 0 \end{bmatrix}, \hat{C} = \begin{bmatrix} C & \varepsilon & \varepsilon & D' \\ \varepsilon & \bar{C} & \bar{D}' & \varepsilon \\ F' & \varepsilon & \varepsilon & \varepsilon \\ \varepsilon & \bar{F}' & \varepsilon & \varepsilon \end{bmatrix}, \hat{D} = \begin{bmatrix} D & \varepsilon \\ \varepsilon & \bar{D}'' \\ \varepsilon & \varepsilon \\ \varepsilon & \varepsilon \end{bmatrix},$$

$$\hat{E} = \begin{bmatrix} E & 0 & 0 & 0 \\ 0 & \bar{E}'' & 0 & 0 \end{bmatrix}, \hat{F} = \begin{bmatrix} F & \varepsilon & \varepsilon & \varepsilon \\ \varepsilon & \bar{F}'' & \varepsilon & \varepsilon \end{bmatrix},$$

where the input matrices of S and $\bar{S}$ have been partitioned in: $\left[BB'\right]$, $\left[\bar{B}'\bar{B}''\right]$, $\left[DD'\right]$, $\left[\bar{D}'\bar{D}''\right]$, and the output matrices have been partitioned in: $\begin{bmatrix} E \\ E' \end{bmatrix}, \begin{bmatrix} \bar{E}' \\ \bar{E}'' \end{bmatrix}, \begin{bmatrix} F \\ F' \end{bmatrix}, \begin{bmatrix} \bar{F}' \\ \bar{F}'' \end{bmatrix}$.

4 Traffic Modelling

The traffic is modeled using the following assumptions:

- Each road is made up of sections which can contain only one vehicle.
- A junction is a section which can contain at most one vehicle.
- If a vehicle is not hampered by another one, it stays one time unit in a section, before moving.
- The vehicles move forward without overtaking.
- At a junction, one goes forward, the next on right or left depending of the junction type.
- At each junction there is a traffic light giving the moving authorizations.

4.1 The Junction Modelling

Before building the Petri net of the complete regular city, let's explain the junction modeling on a system of two circular roads with one junction controlled by a traffic light (Figure 1-Right).

Each road is modeled by a set of sections, and each section is represented by two places a and $\bar{a}$ in the Petri net. If $a = 1$, the section is occupied by one vehicle and we have $\bar{a} = 0$, and if $a = 0$, the section is free, and $\bar{a} = 1$ gives the permission to enter to this section.

The traffic light is described on the right-side of Figure 1 by the transitions q_N, q_E and the places connected to these transitions. At the initial time, $a_c = a_g$ and $\bar{a}_c = \bar{a}_g = 1 - a_c$. If $a_g = 1$, then $\bar{a}_g = 0$, and the light is green for the vehicles coming from the north, and red for the vehicles coming from the east. We assume that the phase times of the traffic light takes one time unit[1]. At the end of the first phase, q^E fires and consumes the token in a_c and products one token in $\bar{a}_c$, one token in $\bar{a}_g$, and minus one token in a_g (it consumes the token). The light commute to the other phase where

[1] In fact the sojourn times in a_c and $\bar{a}_c$.

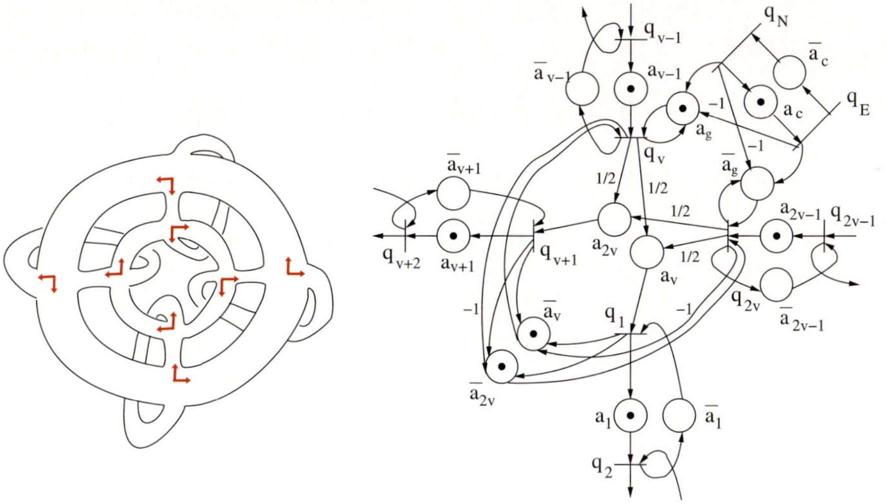

Fig. 1. Left. Regular city on a torus, **Right.** The Petri net model of a junction.

$\bar{a}_g = \bar{a}_c = 1$, which corresponds to a green light for the vehicles coming from the east, and red light for the vehicles coming from the north. This phase takes also one unit of time, then q^N fires and the system come back to the first phase.

Before entering in the crossing we have to know if it is free, this is done by the tokens in the places $\bar{a}_\nu$ and $\bar{a}_{2\nu}$ which must be counted correctly. The negative multiplicities is necessary for that. Thanks to the negative multiplicity, each time when a vehicle enter by the North (for example), the authorization by the east is consumed until the vehicle leaves the crossing. In the right-side of Figure 1, when $a_\nu = 1$ (resp. $a_{2\nu} = 1$), the intersection is occupied by a vehicle which wants to go South (resp. West), and we have $\bar{a}_\nu = \bar{a}_{2\nu} = 0$. When $a_\nu = a_{2\nu} = 0$, the intersection is free, and $\bar{a}_\nu = \bar{a}_{2\nu} = 1$ gives the permission to enter into the junction.

We consider that half of the vehicles entered in the crossing want to go West (by entering in the place $a_{2\nu}$) and the other half want to go South (by entering in the place a_ν). In fact, by rounding in appropriate way the number of vehicles in these two places, the even vehicles entered in the junction go towards the South, and the odd ones go towards the West.

The dynamic of this system is then written as follows:

$$\begin{cases} Q_i^{k+1} = \min\{a_{i-1} + Q_{i-1}^k, \, \bar{a}_i + Q_{i+1}^k\}, \qquad i = 2, \dots, \nu-1, \nu+2, \dots, 2\nu-1, \\ Q_\nu^{k+1} = \min\{\bar{a}_\nu + Q_1^k + Q_{\nu+1}^k - Q_{2\nu}^k, \, a_{\nu-1} + Q_{\nu-1}^k, \, a_g + Q_\nu^k + Q_N^k - Q_E^k\}, \\ Q_{2\nu}^{k+1} = \min\{\bar{a}_{2\nu} + Q_1^k + Q_{\nu+1}^k - Q_\nu^k, \, a_{2\nu-1} + Q_{2\nu-1}^k, \, \bar{a}_g + Q_{2\nu}^k + Q_E^k - Q_N^k\}, \\ Q_1^{k+1} = \min\{a_\nu + \lceil (Q_\nu^k + Q_{2\nu}^k)/2 \rceil, \, \bar{a}_1 + Q_2^k\}, \\ Q_{\nu+1}^{k+1} = \min\{a_{2\nu} + \lfloor (Q_\nu^k + Q_{2\nu}^k)/2 \rfloor, \, \bar{a}^{\nu+1} + Q_{\nu+2}^k\}, \\ Q_N^{k+1} = \bar{a}_c + Q_E^k, \qquad Q_E^{k+1} = a_c + Q_N^k. \end{cases}$$

The two roads being circular, the system above is closed and it can be written in the form (1). The average density of vehicles d in this system is given by:

$$d = \left[\sum_{i=1}^{\nu-1} a_i + \sum_{i=\nu+1}^{2\nu} a_i + (a_\nu + a_{2\nu})/2\right] /(2\nu).$$

The average flow of vehicles f in every section of the network is given by:

$$f = \lim_{k\to\infty} \frac{1}{k} Q_i^k, \ \forall i = 1, 2, \ldots, 2\nu.$$

The fundamental diagram is then obtained by computing numerically the flows for a discretized set of densities.

4.2 Regular City Modeling

To model the traffic of a regular city, we use the following elementary systems:

- A traffic section denoted by $\mathcal{T}_{Z\bar{Z}}^{V\bar{V}}(a, b)$,
- A junction entry denoted by $\mathcal{E}_{ZZ_sZ_wZ_0Z_1Z_2}^{VV_nV_eV_0V_1V_2}(a, b, c, d)$,
- A junction exit denoted by $\mathcal{X}_{Z_sZ_wZ}^{V_nV_eV}(a, b)$,

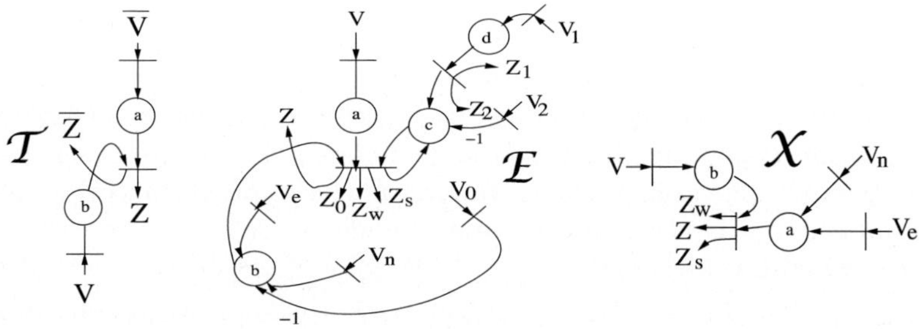

Fig. 2. Elementary systems.

More details of this construction can be found in [7].

References

1. T. Murata: *Petri Nets: Properties, Analysis and Applications* Proceedings of the IEEE, Vol. 77, No. 4, pp. 541-580, 1989.
2. F. Baccelli, G. Cohen, G.J. Olsder, and J.P. Quadrat: *Synchronization and Linearity*, Wiley, 1992 available `http://maxplus.org`.
3. D. Chowdhury, L. Santen, A. Shadschneider: *Statistical physics of vehicular traffic and some related systems.* Physics Report 329, pp. 199-329, 2000.
4. P. Lotito, E. Mancinelli, J.P. Quadrat: *A Min-Plus Derivation of the Fundamental Car-Traffic Law*, IEEE-AC V.50, N.5, pp. 699-705, 2005.
5. N. Farhi, M. Goursat, and J.P. Quadrat: *Derivation of the fundamental traffic diagram...*, 44th IEEE-CDC-ECC December 2005, Seville.
6. N. Farhi, M. Goursat, J.-P. Quadrat: *Fundamental Traffic Diagram of Elementary Road Networks*, in Proceedings ECC, 2006, Kos.
7. N. Farhi: *PhD thesis*, to appear, Paris 1, 2008.

Vehicular Traffic Flow at a Non-Signalised Intersection

M. Ebrahim Fouladvand and Somayyeh Belbasi

Department of Physics, Zanjan University, P.O. Box 45196-313, Zanjan, Iran
fooladQiasbs.ac.ir

1 Introduction

Besides various theoretical efforts aiming to understand the basic principles governing the spatial-temporal structure of traffic flow, considerable attempts have been made towards realistic problems involving optimization of vehicular traffic flow. While the existing results in the context of highway traffic seem to need further manipulations in order to find direct applications, researches on *city traffic* have more feasibility in practical applications [1–4]. We believe that optimisation of traffic flow at a single intersection is a substantial ingredient for the task of global optimisation of city networks. Isolated intersections are fundamental operating units of complex city networks and their thorough analysis would be inevitably advantageous not only for optimisation of city networks but also for local optimization purposes. Recently, physicists have paid notable attention to controlling traffic flow at intersections and other traffic designations such as roundabouts [5–9]. In this respect, our objective in this paper is to study another aspect of conflicting traffic flow at intersections. In principle, the vehicular flow at the intersection of two roads can be controlled via two distinctive schemes. In the first scheme the traffic is controlled without traffic lights. In the second scheme, signalized traffic lights control the flow. In the former scheme, approaching car to the intersection yield to traffic at the perpendicular direction by adjusting its velocity to a safe value to avoid collision. According to driving rules, the priority is given to the nearest car to the intersection. It is evident that this scheme is efficient if the density of cars is low. When the density of cars increases, this method fails to optimally control the traffic and long queues may form which gives rise to long delays. At this stage the implementation of the second scheme is unavoidable. Therefore it is a natural and important question to find out under what circumstances the intersection should be controlled by traffic lights? More concisely, what is the critical density beyond which the non-signalised schemes begins to fail.

2 Description of the Problem

We now present our CA model. Consider two perpendicular one dimensional closed chains each having L sites. The chains represent urban roads accommodating unidirectional vehicular traffic flow. They cross each other at the sites $i_1 = i_2 = \frac{L}{2}$ on the first and the second chain respectively. With no loss of generality we take the direction of traffic flow in the first chain from south to north and in the second chain from east to west (see Fig. 1 for illustration). The discretisation of space is such that each car occupies an integer number of cells denoted by L_{car}. The car position is denoted by the location of its head cell. Time elapses in discrete steps of Δt *sec* and velocities take discrete values $0, 1, 2, \ldots, v_{max}$ in which v_{max} is the maximum velocity of cars.

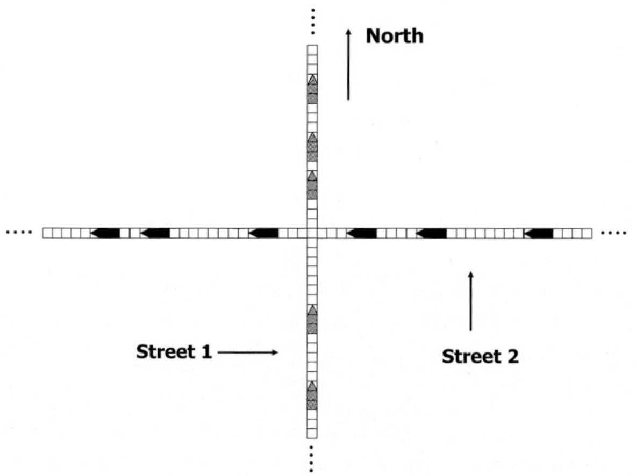

Fig. 1. Intersection of two uni-directional streets.

To be more specific, at each step of time, the system is characterized by the position and velocity configurations of cars. The system evolves under the Nagel-Schreckenberg (NS) dynamics. The yielding dynamics in the vicinity of the intersection is implemented by introducing a safety distance D_s. The approaching cars (nearest cars to the crossing point $i = \frac{L}{2}$) should yield to each other if their distances to the crossing point, denoted by d_1 and d_2 for the first and second street respectively, are both less than the safety distance D_s. In this case, the movement priority is given to the car which is closer to the crossing point. This car adjust its velocity as usual with its leading car. On the contrary, the further car, which is the one that should yield, brakes irrespective of its direct gap. The simplest way to take into this cautionary braking is to adjust the gap with the crossing point itself. This implies that the yielding car sees the crossing point as a hindrance. In this way, the model is collision-free. Figure two illustrates the situation.

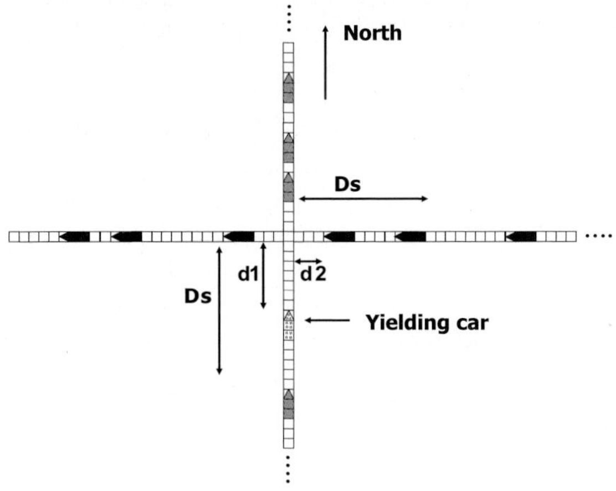

Fig. 2. Two approaching cars to the intersection yield to each other. $d_1 > d_2$.

Let us now specify the physical values of our time and space units. The spatial grid Δx equals to $\frac{4.5}{L_{car}}$ m. We take the time step $\Delta t = 1$ s. Furthermore, we adopt a speed-limit of 75 km/h. The value of v_{max} is so chosen to give the speed-limit value 75 km/h or equivalently 21 m/s. In addition, each discrete increments of velocity signifies a value of $\Delta v = \frac{4.5}{L_{car}} m/s$ which is also equivalent to the acceleration. For $L_{car} = 5$ this gives the comfort acceleration $0.9\ \frac{m}{s^2}$. Moreover, we take the value of random breaking parameter at $p = 0.1$. In the next section, the simulation results of the above-described dynamics is presented.

3 Monte Carlo Simulation

The streets sizes are equally taken as $L_1 = L_2 = 1350$ m and the system is update for 10^6 time steps. After transients, two streets maintain steady-state currents, defined as the number of vehicles passing from a fixed location per a definite time interval, denoted by J_1 and J_2. They are functions of the global densities $\rho_1 = \frac{N_1 \times L_{car}}{L_1}$ and $\rho_2 = \frac{N_2 \times L_{car}}{L_2}$ where N_1 and N_2 are the number of vehicles in the first and the second street respectively. We kept the global density at a fixed value ρ_2 in the second street and varied ρ_1. Figure 3 exhibits the fundamental diagram of the first street i.e.; J_1 versus ρ_1.

It is observed that for small densities ρ_2 up to 0.05, J_1 rises to its maximum value, then it undergoes a short rapid decrease after which a lengthy plateau region, where the current is independent of ρ_1, is formed. Intersection of two chains makes the intersection point appear as a site-wise dynamical defective site. It is a well-known fact that a local defect can affect the low dimensional non-equilibrium systems on a global scale [10–13]. This has been confirmed not only for simple exclusion process but also for cellular automata models

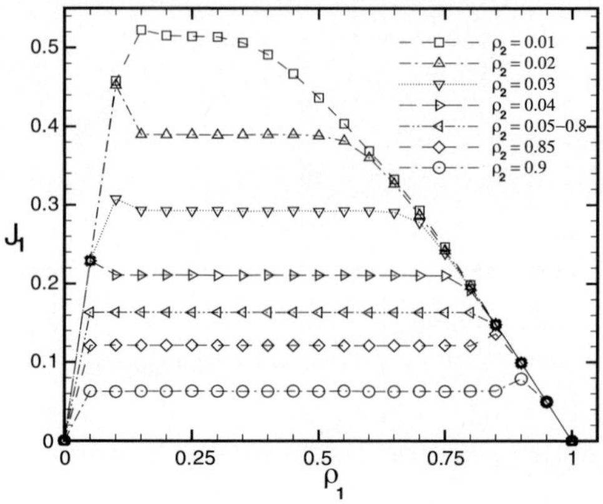

Fig. 3. J_1 vs ρ_1 for various values of ρ_2. $D_s = 25\ m$ and $L_{car} = 5$ cells. The road length is 1350 metre.

describing vehicular traffic flow. Analogous to static defects, in our case of dynamical impurity, we observe that the effect of the site-wise dynamic defect is to form a plateau region $\rho \in [\rho_-, \rho_+]$ in which $\Delta = \rho_+ - \rho_-$ is the extension of the plateau region in the fundamental diagram. The larger the density in the perpendicular chain is, the more strong is the dynamic defect. For higher ρ_2, the plateau region is wider and the current value is more reduced. After the plateau, J_1 exhibits linear decrease versus ρ_1 in the same manner as in the fundamental diagram of a single road. In this region which corresponds to $\rho_1 > \rho_+$ the intersecting road imposes no particular effect on the first road. Increasing ρ_2 beyond 0.05 gives rise to substantial changes in the fundamental diagram. Contrary to the case $\rho_2 < 0.05$, the abrupt drop of current after reaching its maximum disappears for $\rho_2 > 0.05$ and J_1 reaches its plateau value without showing any decrease. The length and height of the plateau does not show significant dependence for $\rho_2 \in [0.05, 0.8]$. This marks the efficiency of the non signalised controlling mechanism in which the current of each street is highly robust over the density variation in the perpendicular street. When ρ_2 exceeds 0.8, the plateau undergoes changes. Its length increases whereas its height decreases. We now consider the flow characteristics in the second street. Although the global density is constant in street 2 its current J_2 is affected by density variations in the first street. In Figure 4 we depict the behaviour of J_2 versus ρ_1.

For each value of ρ_2, the current J_2 as a function of ρ_1 exhibits three regimes. In the first regime in which ρ_1 is small, J_2 is a decreasing function of ρ_1. Afterwards, J_2 reaches a plateau region (second regime) which is approximately extended over the region $\rho_1 \in [0.1, 0.8]$. Eventually in the third regime, J_2 exhibits decreasing behaviour towards zero. Analogous to J_1, the existence of wide plateau regions indicates that street 2 can maintain a constant flow capacity for a wide range of density variations in the first street.

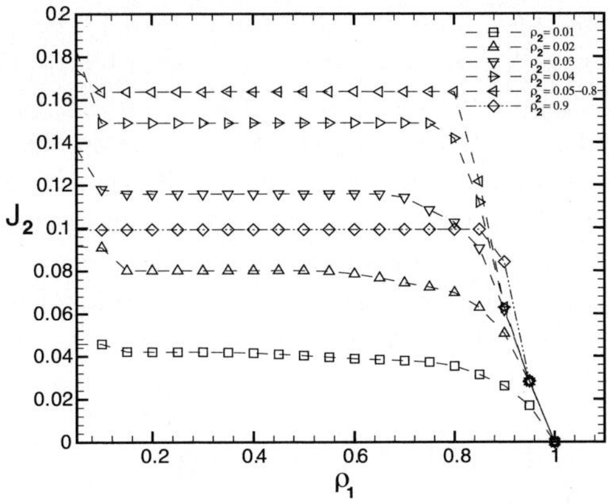

Fig. 4. J_2 vs ρ_1 for various values of ρ_2. $Ds = 25\ m$ and $L_{car} = 5$ cells. The road length is 1350 metre.

The other feature is that in fixed ρ_1, J_2 is an increasing function for small values of ρ_2. This is natural since the current in street 2 has not reached its maximal value. This increment persists up to $\rho_2 = 0.05$. Beyond that, for each ρ_1, J_2 saturates. In the plateau region, the saturation value is slightly above 0.16. The current saturation continues up to $\rho_2 = 0.8$ above which J_2 again starts to decrease. We note that the behaviours depicted in $J_1 - \rho_1$ and $J_2 - \rho_1$ diagrams are consistent to each other. Due to the existence of $1 \rightleftharpoons 2$ symmetry, the $J_2 - \rho_2$ diagram is identical to $J_1 - \rho_1$ and $J_1 - \rho_2$ is identical to $J_2 - \rho_1$. In order to find a deeper insight, it would be illustrative to look at the behaviour of total current $J_{tot} = J_1 + J_2$ as a function of density in one of the streets. Fig. 5 sketches this behaviour.

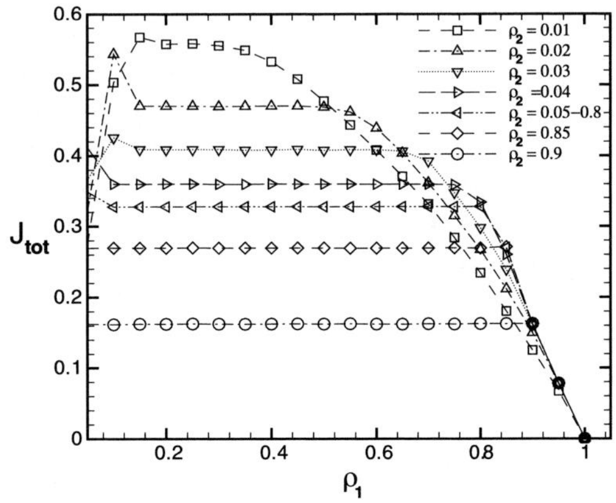

Fig. 5. Total current J_{tot} vs ρ_1 for various values of ρ_2. $D_s = 25\ m$ and $L_{car} = 5$ cells.

For $\rho_2 < 0.05$, the maximum of J_{tot} lies at $\rho_1 = 0.1$. However, for $\rho_2 > 0.05$, the maximum shifts backward to $\rho_1 = 0$. According to the above graphs, after a short increasing behaviour, J_{tot} enters into a lengthy plateau region. Evidently for optimisation of traffic one should maximize the total current J_{tot}. The existence of a wide plateau region in J_{tot} suggests that yielding mechanism can be regarded as an efficient method in the plateau range of density in the first street. Finally, we have also examined the effect of varying the safety distance D_s. Our simulations do not show any significant dependence on D_s. This is due to unrealistic deceleration in the NS model.

4 Summary and Concluding Remarks

We have explored the flow characteristics of a non signalised intersection by developing a Nagel-Schreckenberg cellular automata model. Our findings show yielding of cars upon reaching the intersection gives rise to formation of plateau regions in the fundamental diagrams. This is reminiscent of the conventional role of a single impurity in the one dimensional out of equilibrium systems. We remark that our approach is open to serious challenges. The crucial point is to model the yielding braking as realistic as possible. Empirical data are certainly required for this purpose. We expect the system characteristics undergo substantial changes if realistic yielding declaration is taken into account.

References

1. O. Biham, A. Middleton and D. Levine, *Phys. Rev. A*, **46**, R6124 (1992).
2. T. Nagatani, *J. Phys. Soc. Japan*, **63**, 1228 (1994).
3. S. Tadaki, *Phys. Rev. E*, **50**, 4564 (1994).
4. D. Chowdhury and A. Schadschneider, *Phys. Rev. E*, **59**, R 1311 (1999).
5. M.E. Fouladvand and M. Nematollahi, *Eur. Phys. J. B*, **22**, 395 (2001).
6. S. Lämmer, H. Kori, K. Peters and D. Helbing, *Physica A*, **363**, 39 (2006).
7. B. Ray and S.N. Bhattacharyya, *Phys. Rev. E*, **73**, 036101 (2006).
8. C. Rui-Xiong, Bai Ke-Zhao and L. Mu-Ren, *Chinese Physics*, 15, July 2006.
9. S.-B. Cools, C. Gershenson and B.D. Hooghe, arXive: nlin.AO/0610040
10. S. Janowsky and J. Lebowitz, *Phys. Rev. A*, **45**, 618 (1992).
11. G. Tripathy and M. Barma, *Phys. Rev. Lett.*, **78**, 3039 (1997).
12. A.B. Kolomeisky, *J. Phys. A: Math, Gen.*, **31**, 1153 (1998).
13. G. Lakatos, T. Chou and A.B. Kolomeisky, *Phys. Rev. E*, **71**, 011103 (2005).

Traffic Anticipation Effect in the Lattice Hydrodynamic Model

Hong-xia Ge

Faculty of science, Ningbo University, Ningbo 315211, China
gehongxia@nbu.edu.cn

Summary. A lattice hydrodynamic model is proposed by incorporating the anticipation effect. The linear and nonlinear methods are used to analyze the model, and the results show that considering "backward looking" effect leads to the stabilization of the system. Numerical simulation confirms that our model is more reasonable for describing the real traffic flow.

1 Introduction

Since the seminal work of Lighthill, Whitham and Richards (LWR model) [1, 2] on kinematic waves in vehicular traffic flow, Payne [3] introduced a high-order continuum traffic flow model including a dynamic equation in 1971, which was derived from the car-following theory:

$$\frac{\partial \rho}{\partial t} + \frac{\partial(\rho v)}{\partial x} = 0, \tag{1}$$

$$\frac{\partial v}{\partial t} + v\frac{\partial v}{\partial x} = \frac{V_e(\rho) - v}{\tau} - \frac{\mu}{\rho\tau}\frac{\partial \rho}{\partial x}, \tag{2}$$

where ρ the traffic density, v the space mean speed, τ the relaxation time, $V_e(\rho)$ the equilibrium speed and $\mu = -0.5\partial V_e/\partial \rho$ the anticipation coefficient. The first term on the right-hand side represents a relaxation to equilibrium state. The second term is called the anticipation term, which reflects a driver's reaction to the preceding car. Recently, Nagatani [4, 5] first proposed the simplified version of the LWR model for the convenient to analyze continuum model. The governing equations are described as follows:

$$\partial_t \rho + \rho_0 \partial_x(\rho v) = 0, \tag{3}$$

$$\partial_t(\rho v) = a\rho_0 V(\rho(x+\delta)) - a\rho v, \tag{4}$$

where $a = 1/\tau$ the sensitivity of a driver, ρ_0 the average density, $\rho(x+\delta)$ the local density at position $x + \delta$ at time t, and δ the average headway, which means $\delta = 1/\rho_0$. Local density $\rho(x+\delta)$ is related with the inverse of headway $h(x,t)$: $\rho(x+\delta) = 1/h(x,t)$. The idea is that a driver adjusts the car velocity according to the observed headway $h(x,t)$(or density ahead $\rho(x+\delta)$). Equation (4) corresponds to Eq. (2), but the anticipation term is omitted for simplicity.

According to real traffic, the anticipation term is essential. In this paper, it will be included into the simplified continuum model, which leads to the "backward looking" effect in the lattice hydrodynamic model. The linear stability theory is given and the neutral stability condition is obtained. The analytical result shows that this model could stabilize the traffic flow. The nonlinear analysis is given to derive the mKdV equation, which could describe the propagation of density waves near the critical point. Numerical simulation is conducted to validate the above fact.

2 The Anticipation Simplified Continuum Model

First we add the corresponding anticipation term of Eq. (2) into the right hand of Eq. (4). The motion equation reads

$$\partial_t(\rho v) = a\rho_0 V(\rho(x+\delta)) - a\rho v + ra\frac{\partial V(\rho(x+\delta))}{\partial x}. \tag{5}$$

Here r is a constant, which equals 0.5 in the continuum model. As $r = 0$, it recovers the original model Eq. (4). Then we modify the continuity Eq. (3) with dimensionless space x. Let $\tilde{x} = x/\delta$, and $\tilde{x}$ is indicated as x hereafter. Including Eq. (3), the motion equation of the anticipation simplified continuum model is given below:

$$\partial_t(\rho v) = a\rho_0 V(\rho(x+1)) - a\rho v + ra\rho_0[V(\rho(x+1)) - V(\rho(x))]. \tag{6}$$

The lattice version of the above system is rewritten as the following form.

$$\partial_t\rho_j + \rho_0(\rho_j v_j - \rho_{j-1}v_{j-1}) = 0, \tag{7}$$

$$\partial_t\rho_j v_j = a\rho_0 V(\rho_{j+1}) - a\rho_j v_j + ra\rho_0[V(\rho_{j+1}) - V(\rho_j)], \tag{8}$$

where j denotes site j on the one-dimensional lattice, and $\rho_j(t)$, $v_j(t)$ represent the local density and the local average velocity on site j at time t respectively. So we call it anticipation lattice hydrodynamic (short for, ALH) model. As $r = 0$, the model is called lattice hydrodynamic (short for, LH) model.

By eliminating velocity in Eqs. (7) and (8), one obtains the density equation of the ALH model as follows:

$$\begin{aligned}\partial_t^2\rho_j + a\partial_t\rho_j + a\rho_0^2[V(\rho_{j+1}) - V(\rho_j)]\\ + ra\rho_0^2[V(\rho_{j+1}) - 2V(\rho_j) + V(\rho_{j-1})] = 0.\end{aligned} \tag{9}$$

It is amazing that the anticipation effect in the continuum model changes into "backward looking" effect in the lattice hydrodynamic model. The optimal velocity is selected as that proposed by Nagatani [6], that is,

$$V(\rho_j) = \tanh\left(\frac{2}{\rho_0} - \frac{\rho_j(t)}{\rho_0^2} - \frac{1}{\rho_c}\right) + \tanh\left(\frac{1}{\rho_c}\right), \tag{10}$$

where ρ_c is the inverse of the safety distance h_c in the optimal velocity model. This function has the turning point at $\rho_j = \rho_0 = \rho_c$. So we could derive the modified Korteweg-de Vries equation (the mKdV equation, for short) near the inflection point, that is, the critical point.

3 Linear and Nonlinear Analysis

We know that the uniform traffic flow with constant density ρ_0 and constant optimal velocity $V(\rho_0)$ is the steady state solution for Eqs. (7) and (8), given as: $\rho_j(t) = \rho_0 v_j(t) = V(\rho_0)$. Then suppose $y_j(t)$ to be a small deviation from the steady state density of the jth vehicle: $\rho_j(t) = \rho_0 + y_j$. Substituting them into Eq. (9), linearizing it, and expanding y_j in the Fourier-modes: $y_j(t) = exp(ikj + zt)$, where $z = z_1(ik) + z_2(ik)^2 + \cdots$. Finally, we have the neutral stability condition:

$$a = -\frac{2\rho_0^2 V'(\rho_0)}{1+2r}, \tag{11}$$

where $V'(\rho_0) = dV(\rho_j)/d\rho_j|_{\rho_j=\rho_0}$. According to the linear stability theory, the condition for instability is

$$a < -\frac{2\rho_0^2 V'(\rho_0)}{1+2r}. \tag{12}$$

The neutral stability lines according to different r in the parameter space (ρ, a) are shown in Fig. 1 for the ALH model. It is obvious from Eq. (12) that the region is unstable under the neutral stability lines. Figure 1 shows that the critical points (the apexes of the neutral stability lines) and the neutral stability lines decrease with increasing the value of r, that is to say, once the anticipation effect is considered, the stable region is enlarged, and the traffic flow is more stable. As $r = 0$, it reduces to the LH model.

Then we use the reductive perturbation method to derive the mKdV equation from the ALH model for describing the traffic jam around the critical point (ρ_c, a_c). We introduce slow scales for space variable j and time variable t, and define the slow variables X and T [7, 8]

$$X = \varepsilon(j + bt) \quad T = \varepsilon^3 t, \quad 0 < \varepsilon \ll 1, \tag{13}$$

where b is a constant to be determined. Let

$$\rho_j(t) = \rho_c + \varepsilon R(X,T). \tag{14}$$

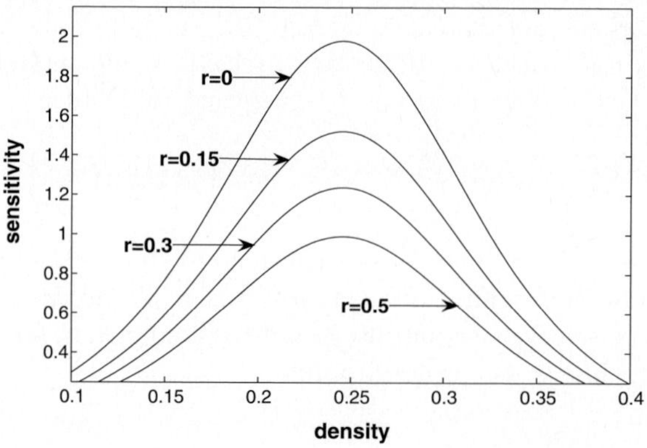

Fig. 1. The phase diagram of the ALH model according to different parameter r.

Substituting Eqs. (13)–(14) into Eq. (9) and making the Taylor expansions to the fifth order of ε lead to the expression:

$$\begin{aligned}&\varepsilon^2 f_1\partial_X R + \varepsilon^3 f_2\partial_X^2 R + \varepsilon^4(\partial_T R + f_3\partial_X^3 R + f_4\partial_X R^3)\\&+\varepsilon^5(f_5\partial_X\partial_T R + f_6\partial_X^4 R + f_7\partial_X^2 R^2) = 0,\end{aligned}\tag{15}$$

Table 1. The Coefficients f_i of the Three Models

f_1	f_2	f_3	f_4	f_5	f_6	f_7
$b+\rho_c^2V'$	$\frac{b^2/a+(1+2r)\rho_c^2V'}{2}$	$\frac{\rho_c^2V'}{6}$	$\frac{\rho_c^2V'''}{6}$	$\frac{2b}{a}$	$\frac{(1+2r)\rho_c^2V'}{24}$	$\frac{(1+2r)\rho_c^2V'''}{12}$

where $V' = dV(\rho_j)/d\rho_j|_{\rho_j=\rho_c}$ and $V''' = d^3V(\rho_j)/d\rho_j^3|_{\rho_j=\rho_c}$. V', V''' correspond to $V'(\rho_c)$, $V'''(\rho_c)$ in the above equation and hereafter respectively. Near the critical point $a = (1-\varepsilon^2)a_c$, taking $b = -\rho_c^2V'$ and eliminating the second- and third-order terms of ε from Eq. (15) result in the simplified equation:

$$\varepsilon^4(\partial_T R + g_1\partial_X^3 R + g_2 R\partial_X R) + \varepsilon^5(g_3\partial_X^2 R + g_4\partial_X^4 R + g_5\partial_X^2 R^3) = 0,\tag{16}$$

where

Table 2. The Coefficients g_i of the Three Models

g_1	g_2	g_3	g_4	g_5
$\frac{\rho_c^2V'}{6}$	$\frac{\rho_c^2V'''}{6}$	$\frac{-(1+2r)\rho_c^2V'}{2}$	$\frac{-(1+2r)\rho_c^2V'}{8}$	$\frac{-(1+2r)\rho_c^2V'''}{4}$

We make the following transformations to obtain the standard mKdV equation with higher order correction

$$T' = -g_1T,\quad R = \sqrt{\frac{-g_1}{g_2}}R'.$$

So we get the regularized mKdV equation with $O(\varepsilon)$ correction term

$$\partial_{T'} R' - \partial_X^3 R' + \partial_X R'^3 + \varepsilon M[R'] = 0, \tag{17}$$

where

$$M[R'] = \frac{-g_3}{g_1}\partial_X^2 R' - \frac{g_4}{g_1}\partial_X^4 R' + \frac{g_5}{g_2}\partial_X^2 R'^3.$$

Analogously, we obtain the mKdV soliton solution

$$\rho_j(t) = \rho_c + \varepsilon\sqrt{\frac{-5V'}{V'''}} tanh\sqrt{\frac{5}{2}}\left[j + \left(-\rho_c^2 V' + \frac{5\rho_c^2 V'}{6}\varepsilon^3\right)t\right]. \tag{18}$$

4 Numerical Simulation and Conclusion

For the convenience of simulation, Eq. (9) could be rewritten as the difference equation by transforming time derivatives to asymmetric forward differences:

$$\begin{aligned}\rho_j(t+2\tau) - \rho_j(t+\tau) + \tau\rho_0^2[V(\rho_{j+1}(t)) - V(\rho_j(t))]\\ + \tau r \rho_0^2[V(\rho_{j+1}(t)) - 2V(\rho_j(t)) + V(\rho_{j-1}(t))] = 0.\end{aligned} \tag{19}$$

Computer simulation is performed for the difference ALH model described by Eq. (19). The boundary condition selected is a periodic one. The initial conditions are chosen as follows: $\rho_j(0) = \rho_0 = 0.3$, $\rho_j(1) = \rho_j(0) = 0.3$ for $j \neq 50, 51$, $\rho_j(1) = 0.3 - 0.05$ for $j = 50$, and $\rho_j(1) = 0.3 + 0.05$ for $j = 51$, where the total number of sites is $N = 100$, the safety density is $\rho_c = 0.25$ and the sensitivity $\tau = 0.5$.

Figure 2 shows the space-time evolution of the density for various r at a fixed time $t = 10300$ in the ALH model. The pattern (a) corresponds to the original LH model. The patterns (b), (c) and (d) exhibit the density profile as $r = 0.15, 0.25, 0.5$, respectively. In patterns (a) and (b), due to the "backward looking" effect, the maximum value of density decreases, that is to say, the degree of vehicles aggregation is reduced. In patterns (c) and (d), large cluster disappears compared to the pattern (a)–(b), and the wide amplitude congestion changes into stop-and-go traffic flow, which demonstrates that the "backward looking" effect is important to describe the nature of density wave near the critical point in the ALH model. It is obvious that, the stop-and-go density wave appears more frequently than wide amplitude congestion in real traffic, so our model is reasonable. Moreover, the bigger r, the stronger of the density oscillation frequency, and the phenomena of the kink-anti-kink soliton is not clear. The parameter r selected rightly will make the model describe the traffic flow effectively.

We have proposed the ALH model of traffic flow for the purpose of constructing a cooperative driving system. Taking into account the anticipation effect in the continuum model, we give a new motion equation, describing

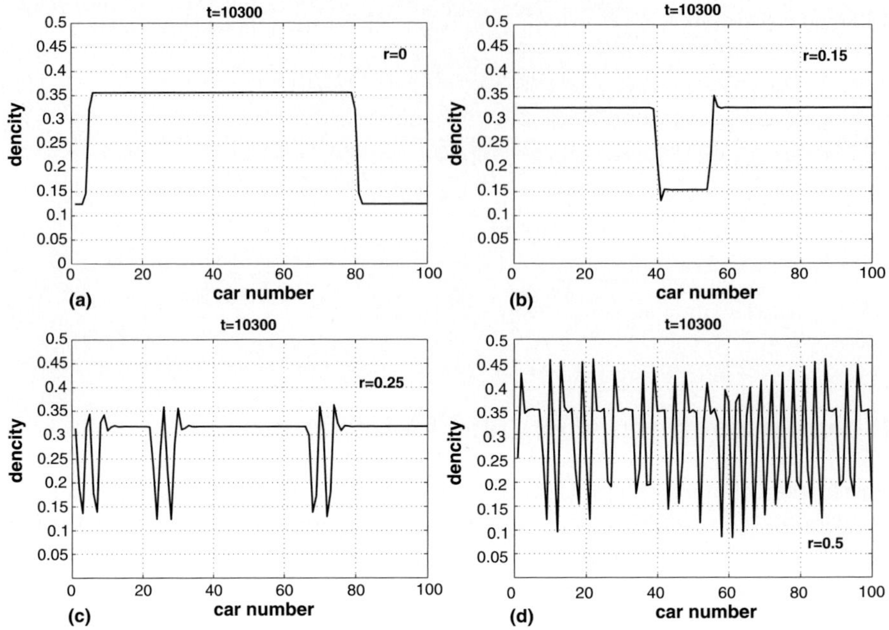

Fig. 2. Space-time evolution of the density at $t = 10300$ for the ALH model.

the "backward looking" effect in the lattice hydrodynamic model. The traffic nature has been analytically analyzed by using the linear analysis. It has been shown that consideration of "backward looking" effect could further stabilize traffic flow. The simulation outcomes confirm that, on one hand, the stability of the ALH model is improved compared to the LH model, on the other hand, the model could describe the real traffic flow more accurately.

Acknowledgements

This work was supported by the National Basic Research Program of China (Grant No. 2006CB705500), the National Natural Science Foundation of China (Grant Nos. 10532060, 10602025) Scientific Research Fund of Zhejiang Provincial Education Department (Grant Nos. 20061634), and K.C.Wong Magna Fund in Ningbo University.

References

1. Lighthill MJ, Whitham GB (1955) In: Proceedings of Royal Society, Vol 229 (1178) of A. pp. 317.
2. Richards PI (1956) Oper Res 4:42.

3. Payne HJ (1971) Simulation Councils Proc. Series: Mathematical Models of Public Systems, Vol 1, No 1, ed. George A. Bekey.
4. Nagatani T (1998) Physica A 261:599.
5. Nagatani T (1999) Physica A 264:581.
6. Nagatani T (1999) Phys Rev E 59:4857.
7. Dai SQ (1982) Advan in Mech 12(1):2 (in Chinese).
8. Nagatani T (1998) Phys Rev E 58:4271.

Is the Nearest Neighbor Distribution Enough to Describe the Statistical Behavior of a Domain System?

Diego Luis González and Gabriel Téllez

Departamento de Física, Universidad de Los Andes, A. A. 4976 Bogotá, Colombia
die-gon1@uniandes.edu.co, gtellez@uniandes.edu.co

Summary. In many cases complex systems are mapped onto more simple systems with the only criteria of the nearest neighbor distribution similitude. Some of these simplified systems are coalescing and interacting random walk, another common simplification is the independent interval approximation. However, we found that the nearest neighbor distribution does not contain enough information about the statistics of the system: several different systems could share the same nearest neighbor distribution.

1 Introduction

This article summarize the most important result presented in Ref. [1], where the statistical behavior of many non equilibrium systems is studied. In particular, we study of the statistical behavior of two non-equilibrium systems. The first one is a quasi one-dimensional gas introduced in Ref. [2]. There, the authors studied the biased diffusion of two species in a fully periodic $2 \times L$ rectangular lattice half filled with two equal number of two types of particles (labeled by their charge $+$ or $-$). An infinite external field drives the two species in opposite directions along the x axis (long axis). The only interaction between particles is an excluded volume constraint, i.e., each lattice site can be occupied at most by only one particle. The authors map this system onto two simplified models: the coalescing random walk (CRW) where the particles diffuse and interact with the reaction $A + A \to A$, the other one is the interacting random walk (IRW) where the particles additionally experiment an interaction which decrease with the distance between particles. The second system is a one-dimensional spin system introduced in Ref. [3], where the authors consider a chain of L Ising spins with nearest neighbor ferromagnetic interaction J. The chain is subject to spin-exchange dynamics with a driving force E that favors motion of up spins to the right over motion to the left.

The gas and spin systems evolve in a similar way. For each system there are two kind of domains: in the spin case, there are domains of up spins and domains of down spins and in the gas case, there are domains filled with particles and empty domains. Both systems evolve with formation of domains in such way that the average length of domains grow in time. For intermediate times where the size of the domains is much smaller than the total size L of the system, the domain size distribution exhibit a dynamic scaling. For later times, in both systems, it is possible to find a non equilibrium steady state (NESS). For the gas system, in this regime there is a only one macroscopic domain with an approximate size of $L/2$. This macroscopic domain it is not in equilibrium because there are particles (travellers) which leak out from one end of this domain and travel along the lattice until reach the other end of this domain. In the spin system, for later times there are two macroscopic domains. One of them with up spins and the other one with down spins, because of the external field, they move in opposite directions along the lattice.

We are interested in the statistical properties of these domains regardless of their type (filled or empty, up or down) in the scaling regime. We use the same definitions for the spacing distributions as Ref. [1], then, $P^{(n)}(S,t)$ is the probability density function that the distance between the external borders of $n+1$ consecutive domains is S at time t, $\langle S\rangle$ is the average of S and the relative spacing between borders is $s = S/\langle S\rangle$. Finally, the scaling probability density function is given by

$$p^{(n)}(s) = \langle S\rangle\, P^{(n)}(s\,\langle S\rangle\,,t)\,, \tag{1}$$

in particular $p^{(0)}(s)$ is the scaling nearest neighbor edges distribution, i.e. the domain size distribution. In the scaling regime, $p^{(n)}(s)$ does not depend on the time t. Several numerical simulations of both systems, see [1–4], show that the scaling domain size distribution function for both systems in the scaling regime is well fitted by

$$p^{(0)}(s) = \frac{\pi}{2} s\, e^{-\pi s^2/4}\,. \tag{2}$$

This distribution also it appears in the random matrices theory, it is known as the Wigner surmise, and it describes the spacing distribution between eigenvalues in the Gaussian and circular orthogonal ensembles (GOE and COE respectively) [5]. The GOE and COE ensemble are equivalents to a system of particles on a circle interacting through a log-potential. For more details about these systems see [1–5].

2 The Nearest Neighbor Spacing Distribution

In order to verify the results obtained by [2–4], we performed simulations for the gas, the spin, the CRW, the IRW and the circular orthogonal ensemble (COE) of random matrices with the same parameters used in Ref. [1]. Figure 1 shows ours results for the scaling domain size distribution $p^{(0)}(s)$ of these systems, in all cases the nearest neighbor distribution is well fitted by the Wigner surmise. This fact poses a question: Are these systems equivalents for $n > 0$, as occurs for $n = 0$?

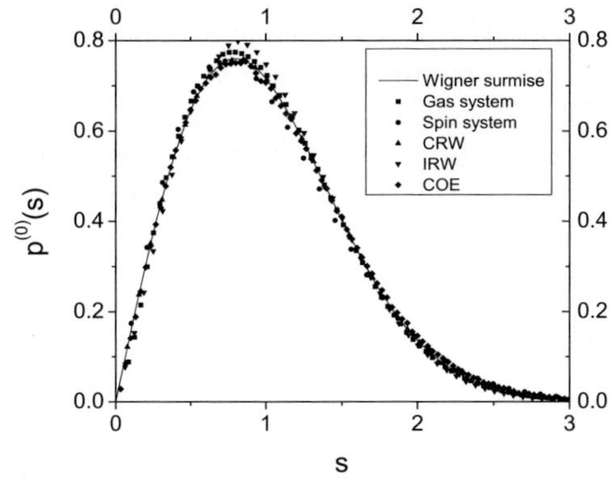

Fig. 1. Nearest neighbor spacing distributions for the gas, spin, IRW, CRW and COE and Wigner surmise.

3 The Spacing Distributions for $n > 0$

From our simulations, we computed the other spacing distributions $p^{(n)}(s)$ and the correlation function $g(r)$ of all these systems in order to compare them. The results are presented in the following subsections.

3.1 The Quasi One-Dimensional Gas and its Approximate Models: the IRW and the CRW

The CRW and IRW were use in Ref. [2] as approximate models for the quasi one-dimensional gas. We now compare the other spacing distribution functions and the correlation function. The results are shown in figure 2. As we already know, the CRW and IRW reproduce correctly the nearest neighbor distribution, but they fail describing the other spacing distribution functions for $n > 0$ and as consequence, they also fail describing the two-point correlation

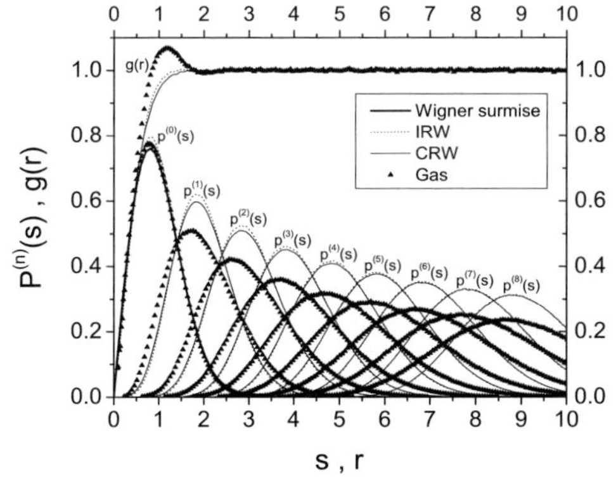

Fig. 2. Spacing distributions and the pair correlation function in the gas system, IRW and CRW for a lattice with $N = 500$ and $N = 1000$ sites respectively.

function $g(r)$. The most important difference it happens in the correlation function $g(r)$. In the gas system there is an oscillation around $r = 1$ (and probably more for higher values of r) but in the CRW and IRW systems this oscillation does not exist. The spacing distributions $p^{(n)}(s)$ for $n \geq 1$ of the IRW and CRW are somehow similar, but they are different from the ones of the gas system. We conclude that the CRW and IRW systems are not statistically equivalent to the quasi one-dimensional gas.

3.2 The Circular Orthogonal Random Matrix Ensemble (COE) and the CRW and IRW

In figure 3(a), we compare the correlation functions and spacing distribution functions of both the CRW and IRW systems with COE. The correlation function $g(r)$ of the CRW and IRW differs from the one of the COE. Although the three correlation functions have no oscillations, the correlation function of the COE is smaller than the one of the IRW and the CRW. The first two distribution functions $p^{(0)}(s)$ and $p^{(1)}(s)$ of these systems are very similar. However from $n \geq 2$, the COE spacing distributions $p^{(n)}(s)$ start to differ from the corresponding ones of the CRW and the IRW. The most important conclusion is that the CRW and the IRW systems are not statistically equivalent to the COE. Additionally, the CRW and the IRW show a very similar statistical behavior, therefore, the interaction proposed in Ref. [2] for the IRW do not change considerably the statistical behavior of this system from the CRW in the scaling regime.

3.3 The Gas and Spin Systems and the Circular Orthogonal Random Matrix Ensemble (COE)

Now, in figure 3(b), we compare the spacing distributions $p^{(n)}(s)$ and the correlation function $g(r)$ of both the gas and spin systems with the ones for

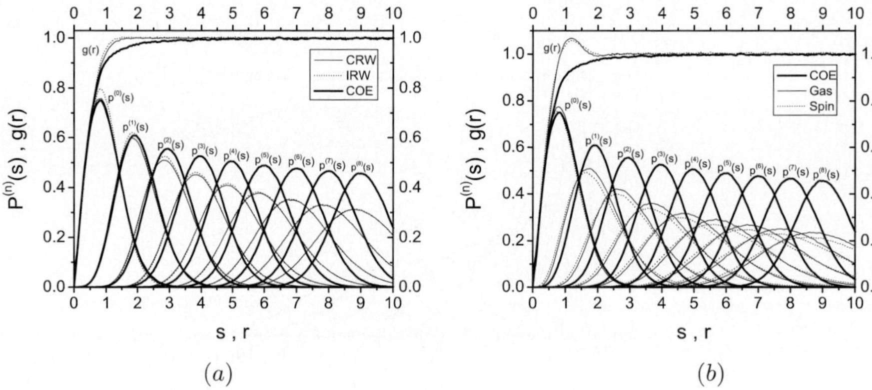

Fig. 3. (a) Comparison between the CRW, IRW and COE for a lattice with $N = 1000$ sites. (b) Comparison between the gas, spin system ($\mu = 0.5$) and COE for a lattice with $N = 1000$ sites. In the COE we took 20000 matrices of 200×200.

the random matrix COE. The statistical behavior of the gas and spin system is very similar. In particular, the correlation functions of both systems are almost identical and they have the oscillation near $r = 1$ mentioned above. The spacing distributions are somehow similar for the gas and the spin system, with some small differences between them. But both systems, gas and spin, have a very different statistical behavior from the COE. It is possible to conclude that the gas and spin systems have a similar statistical behavior but this behavior is different from the COE. They only coincide in the nearest neighbor distribution.

3.4 The Independent Interval Approximation (IIA)

The independent interval approximation is commonly used in this kind of systems in order to find analytical results, in this approximation, $p^{(n)}(s)$ is given by the convolution product of $n+1$ nearest neighbor distribution factors. Because of that it is possible to calculate the distributions $p^{(n)}(s)$, for any $n \geq 1$ and the pair correlation function $g(r)$, see Refs. [1, 6, 7]. In figure 4, we compare the gas and spin system correlation functions and spacing distributions with the theoretical predictions from the independent interval approximation. We notice that this approximation reproduce more closely the spacing distributions and the correlation function. The IIA correlation function has an oscillatory behavior near at $r = 1$ as it occurs in the gas and spin system. The fact that the independent interval approximation reproduce much better the correlation function and spacing distributions of the gas and the spin system than the other approaches considered (CRW, IRW and COE), suggest that in the gas and spin systems the domains are not strongly correlated.

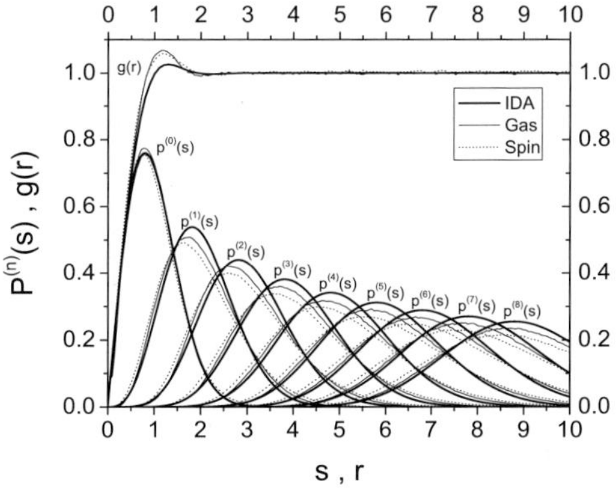

Fig. 4. Comparison between the gas, spin system and IIA for a lattice with $N = 1000$ sites.

4 Conclusion

We studied the statistical behavior of several equilibrium and non equilibrium systems, the last ones in the scaling regime. In all studied systems the nearest neighbor distribution function $p^{(0)}(s)$ is well fitted by the Wigner surmise, Eq. (2). However, by comparing the other spacing distribution functions $p^{(n)}(s)$ for $n > 0$ and the correlation function $g(r)$, we find that those systems are not equivalents and have different statistical behaviors. This result show us that the nearest neighbor distribution contains limited information about the statistic of the system and the finer details and differences between them are contained in the correlation function and the other spacing distributions. Then we must be cautious when we map complex systems onto more simple systems with the only criteria of the nearest neighbor distribution similitude.

Acknowledgements

This work was partially supported by an ECOS Nord/COLCIENCIAS action of French and Colombian cooperation and by the Faculty of Sciences of Los Andes University.

References

1. González D, Téllez G (2007) Phys. Rev. E **76**:011126.
2. Mettetal J, Schmittmann B, Zia R (2002) Europhysics Lett. **58**:653–659.
3. Cornell S, Bray A (1996) Phys. Rev. E **54**:1153–1160.
4. Spirin V, Krapivsky P, Redner S (1999) Phys. Rev. E **60**:2670–2676.
5. Mehta M (1991) Random matrices 2^{ed}. Academic press.
6. Alemany P, ben-Avraham D (1995) Phys. Lett. A **206**:18–25.
7. Derrida B, Hakim V and Zeitak R (1996) Phys. Rev. Lett. **77**:2871–2874.

Statistical Properties of Disordered Driven Lattice Gases with Open Boundaries

Philip Greulich[1] and Andreas Schadschneider[1,2]

[1] Institut für Theoretische Physik, Universität zu Köln, 50937 Köln, Germany
pg@thp.uni-koeln.de
[2] Interdisziplinäres Zentrum für komplexe Systeme, 53117 Bonn, Germany
as@thp.uni-koeln.de

Summary. We investigate driven lattice gases with open boundary conditions in presence of randomly distributed defect sites with reduced hopping rate [1]. These systems can be used as models for intracellular transport systems impurified by immobile blocking molecules. In contrast to equilibrium, even macroscopic quantities in disordered non-equilibrium systems depend sensitively on the defect sample. We show that the leading behaviour in the disordered system is determined by the longest stretch of consecutive defect sites. Using results from extreme value statistics [2] this single-bottleneck approximation gives accurate results for the expectation values of the maximum current at small defect densities. Corrections from bottleneck interactions can be taken into account systematically by a perturbative expansion.

1 Introduction

Driven diffusive systems play an important role in non-equilibrium statistical physics. They are used as models for transport processes like vehicular traffic [3], and biological transport by motor proteins [4–7]. The paradigmatic model is the totally asymmetric simple exclusion process (TASEP) which was first introduced to describe protein polymerization in ribosomes [8]. The TASEP exhibits some generic properties like *boundary induced phase transitions* [9] that also occur in more complex driven systems. It was solved exactly [10, 11] and results can be used for qualitative and quantitative approaches to other systems.

In contrast to the homogeneous TASEP, the TASEP with site-dependent hopping rates is not fully understood. There have been some numerical and analytical approaches [1, 12–14], while exact solutions for even single sites with lower hopping rates are still pending.

Here we consider systems with disorder, i.e. two possible hopping rates randomly assigned to the sites. Sites with a lower hopping rate q are called *defect sites*, while sites with the higher hopping rate p are *non-defect sites.*

In open systems the main effect of defects is a decrease of the flow capacity compared to the homogeneous system. In contrast to this, below critical system parameters, the system seems rather unaffected by the disorder on a macroscopic scale [15]. Nonetheless, there are small deviations from the homogeneous system even in the low current regimes if there are defects near the boundaries [16]. If the system current reaches the maximum current, phase separation occurs which marks the transition to a *maximum current* or *phase separated* phase, which is larger than the corresponding phase of the pure TASEP. Inside this phase, system properties are independent of the boundary conditions.

The maximum current, which is a macroscopic quantity, depends sensitively on microscopic details of the defect distribution. For applications to real systems macroscopic parameters and quantities are most relevant. Therefore we are mainly interested in determining statistical properties, e.g. probability distributions and expectation values, of relevant quantities taking an ensemble of systems rather than looking at single samples.

2 Model

We consider a TASEP consisting of L sites which can either be empty or occupied by one particle. With each site i we associate a hopping rate p_i which corresponds to the rate at which a particle at this site will move to its right neighbour $i+1$ if this is empty. At the boundary sites $i = 1$ and $i = L$ particles can be inserted and removed, respectively. If site 1 is empty a particle will be inserted there with rate α. On the other hand, if site L is occupied this particle will be removed with rate β. Here we will use a random-sequential update corresponding to continuous-time dynamics.

The hopping rates p_i can take two different values $p_i = p$ and $p_i = q$ with $q < p$. Sites with hopping rate q are called *defect* or *slow* sites, while sites with hopping rate p are *non-defect* or *fast* sites. In the following we will take $p = 1$ which can always be achieved by rescaling time. We consider systems with randomly distributed defects by assigning hopping rates to sites according to

$$p_i = \begin{cases} q & \text{with probability } \phi \\ p & \text{with probability } 1-\phi \end{cases} . \tag{1}$$

We have simulated the TASEP with defects to illustrate some of the special properties of the system. The results of the Monte-Carlo (MC) simulations can be seen in fig. 1. A sketch of the phase diagram of a specific, randomly chosen disorder realization can be seen in fig. 2.

3 Expectation Values of the Maximum Current

The crucial issue for treating disordered systems is the observation that in the disordered TASEP with periodic boundary conditions the longest stretch

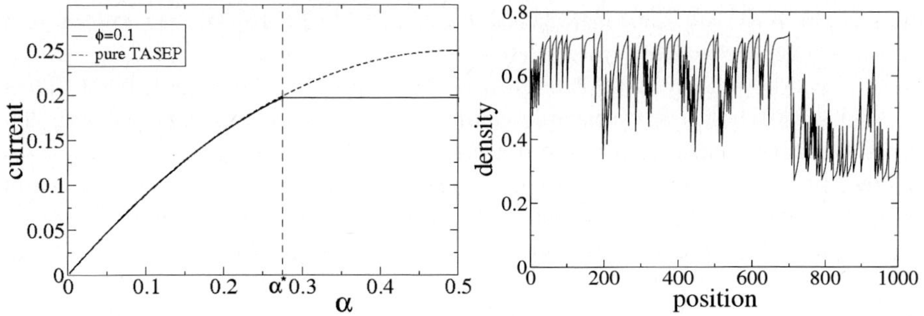

Fig. 1. left: Current as function of the entry rate α for fixed exit rate $\beta = 0.9$ in the pure and disordered TASEP. Below the critical value α^*, the currents are almost identical. For $\alpha > \alpha^*$ the current in the disordered TASEP saturates at the maximum current J^*. **right**: Density profile of the disordered TASEP above the critical entry rate α^*, here $\alpha = 0.8$. One observes phase separation exhibiting a macroscopic high density region on the left and a low density region on the right.

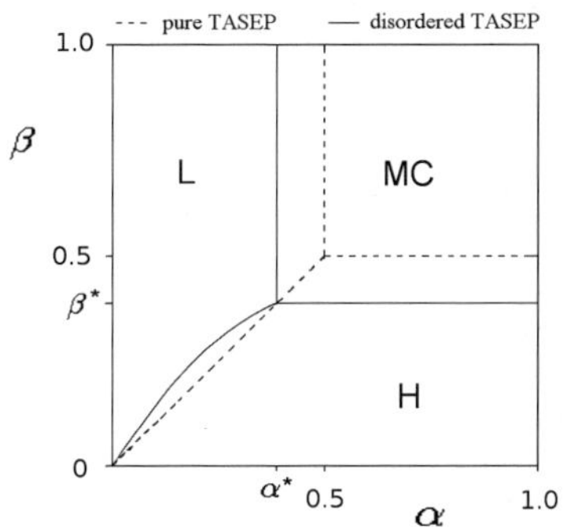

Fig. 2. Sketch of the phase diagram of the TASEP for one specific realization of randomly distributed defects, compared with that of the pure TASEP. The maximum current phase is larger than in the pure TASEP (here denoted as PS because of phase separation) due to decreased maximum current. The transition line between high and low density phase is distorted due to defects near the boundaries [16].

of defects is the limiting factor for the current [1, 14]. This remains true for open boundaries [17], as one can see in table 1. Here we call a stretch of l consecutive defects a *bottleneck of size (length) l*. This leads to the *Single Bottleneck Approximation (SBA)*:

> *In a system with many defects, the maximum current is approximately the same as in a system with a single bottleneck which has the same size as the longest one.*

The SBA reduces the problem to the much simpler one of a single bottleneck in a TASEP. For such systems, efficient methods have been developed recently, namely the *finite segment mean field theory (FSMFT)* [18] and the *interaction subsystem approximation (ISA)* [19].

The validity of the SBA is confirmed by results displayed in table 1. Thus we can state that a macroscopic quantity (the maximum current) can depend sensitively on microscopic details (the longest bottleneck), in contrast to equilibrium systems. Since we are interested in macroscopic properties, and we can control only macroscopic parameters like the defect density ϕ, we focus on disorder averages and obtain probability distributions and expectation values.

Table 1. SBA results and Monte-Carlo-simulations for the maximum current in the disordered TASEP for fixed slow hopping rate $q = 0.6$. Column 3: length of the longest bottleneck. Columns 4 and 5: distance and length of the bottleneck next to the longest one. Column 6: Results by MC-simulations. Column 7: SBA-results by simulating a system with a *single* bottleneck whose size corresponds to the value in column 3.

L	ϕ	l^*	next bn.	length	J^*_{MC}	J^*_{SBA}
1000	0.05	2	2	1	0.2174	0.2294
1000	0.1	3	12	1	0.2080	0.2131
1000	0.2	3	2	2	0.1963	0.2131
3000	0.1	3	4	1	0.2048	0.2131
3000	0.2	5	5	1	0.1866	0.1925

Since the longest bottleneck is the dominant part of the system we need the probability distribution $\mathcal{P}(l^*)$ of the length l^* of the longest bottleneck as a function of the defect density ϕ. Extreme value statistics provide the tools to solve this problem [2]. The probability that a single bottleneck has size l is given by $P(l) = \phi^l(1-\phi)$. This leads to a Gumbel-like extreme value distribution for the probability of the longest bottleneck l^* [17]

$$\mathcal{P}(l^*) = -L(1-\phi)\phi^{l^*+\frac{1}{2}} \ln\phi \exp(-\phi^{l^*+\frac{1}{2}}(1-\phi)L) . \tag{2}$$

This distribution is plotted in fig. 3. It can be used to calculate the sample averaged expectation value of the maximum current in SBA:

$$\langle J^*_{SBA}(q,\phi)\rangle = \frac{\sum_{l^*=1}^{\infty} J^*(q,l)\,\mathcal{P}(l^*)}{\sum_{l^*=1}^{\infty} \mathcal{P}(l^*)} \tag{3}$$

Here $J^*(q,l)$ is the maximum current in a system with a single bottleneck of size l. This quantity can be obtained by computer simulations of the corresponding system or by analytical approximations [18, 19]. The results are displayed in table 2. In practice, $\langle J^*(q,\phi)\rangle$ is calculated by truncating the sum at a value where $\mathcal{P}(l^*)$ becomes negligibly small.

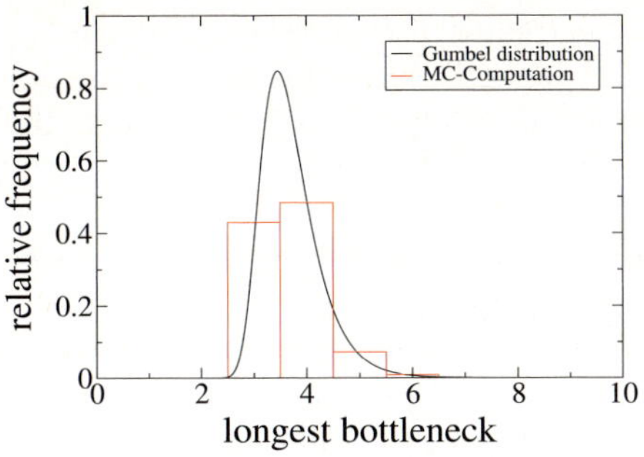

Fig. 3. Frequency of the longest bottleneck for 500 samples, system size $L = 10000$ defect density $\phi = 0.1$. The average size of the longest bottleneck is 3.65, the statistical standard deviation is 0.03. Using (3) one obtains the expectation value $\langle l^* \rangle = 3.70$.

4 Perturbative Expansion

In order to improve the results and to control the error of the SBA we systematically take into account the effect of defects in the vicinity of the longest bottleneck on the maximum current. For this purpose a kind of perturbative series expansion is derived.

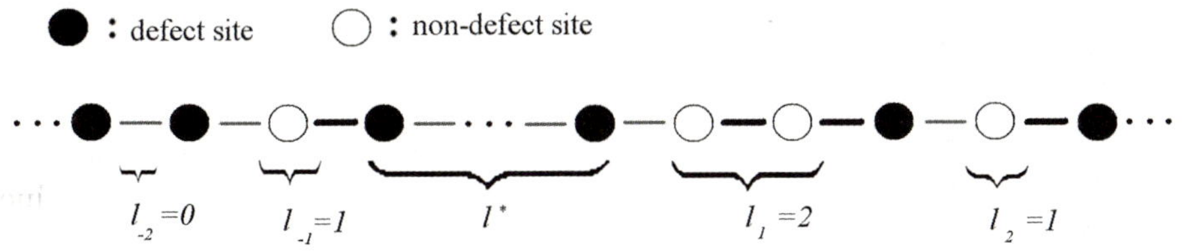

Fig. 4. Illustration of the vicinity of the longest bottleneck (length l^*). The distances between defects to the left are denoted by $l_{-1}, l_{-2}, \ldots$, and $l_1, l_2, \ldots$ to the right.

Consider the vicinity of the longest bottleneck as displayed in fig. 4. We assume the defects on the left and right of the bottleneck being screened, i.e. the perturbation by a defect on the right does not depend on the distribution of defects on the right of the bottleneck. For a given longest bottleneck of size l^*, the current in a system with N defects can then be decomposed as

$$J^*_{l^*}(l_1, l_2, \ldots) = J^*_{l^*} - \sum_{k=1}^{N} \Delta^k J^+_{l^*}(l_1, \ldots, l_k) - \sum_{k=1}^{N} \Delta^k J^-_{l^*}(l_{-1}, \ldots, l_{-k}) \quad (4)$$

where $\Delta^k J^{\pm}_{l^*}(l_1, \ldots, l_k)$ is the contribution of the k-th defect ("defect-defect interaction") on the left/right of the longest bottleneck to the decrease of the maximum current. This quantity still depends on the defects that are nearer to the bottleneck than the k-th defect, as the arguments suggest.

Taking the expectation value over all defect distributions (length of bottleneck fixed) yields a function of the defect density ϕ that can be expanded formally into a Taylor series with a remainder term [17]. Investigation of this remainder term reveals that the expansion converges if the functions $\Delta^k J^+_{l^*}(l_1, \ldots, l_k)$ decay faster than any power law. If these functions decay not slower than l_i^{-3}, at least a first order asymptotic expansion is possible, while the expansion can be taken to higher orders if the decay is faster. This confirms the validity of the SBA if the defect-defect interaction is weak enough. Note that although the estimate is rather crude, MC simulations indicate a wider range of validity of the expansion.

For illustration we explicitly expand the series up to second order

$$\langle \Delta J_{l^*} \rangle = \phi \left(\sum_{l_1=1}^{\infty} \Delta^1 J_X(l_1) \right) + \phi^2 \left(- \sum_{l_1=1}^{\infty} l_1 \, \Delta^1 J_{l^*}(l_1) + \sum_{l_1,l_2=1}^{\infty} \Delta^2 J_{l^*}(l_1, l_2) \right) + \mathcal{O}(\phi^3) \, . \tag{5}$$

The full expectation value is obtained by averaging over all l^* analogous to (3). In table 2, the results of the Taylor expansion are compared with numerical results from MC simulations. The MC results of the SBA serve as reference points.

Table 2. Results by SBA and pertubative expansion for the expectation value of the maximum current compared with average values of MC simulations ($q = 0.6$). Column 4: MC-results. Column 5: SBA-results. Columns 6 and 7: first and second order perturbative expansion around the SBA solution (5).

L	ϕ	samples	$\langle J^* \rangle_{MC}$	$\langle J^*_{SBA} \rangle_{MC}$	1.order	2.order
500	0.1	200	0.2099	0.2244	0.2126	0.2077
1000	0.2	100	0.1918	0.2024	0.1922	0.1899
3000	0.1	100	0.2018	0.2110	0.2041	0.2025
3000	0.2	50	0.1866	0.1960	0.1884	0.1880

References

1. Barma M (2006) Physica A 372:22–33.
2. Sornette D (2000) Critical Phenomena in Natural Sciences. Springer, Berlin Heidelberg New York.
3. Chowdhury D, Santen L, Schadschneider A (2000) Phys. Rep. 329:199.
4. Parmeggiani A, Franosch T, Frey E (2003) Phys. Rev. Lett. 90:086601.
5. Klumpp S, Nieuwenhuizen T M, Lipowsky R (2005) Physica E: 29:380.
6. Nishinari K, Okada Y, Schadschneider A, Chowdhury D (2005) Phys. Rev. Lett. 95:118101.

7. Greulich P, Garei A, Nishinari K, Schadschneider A, Chowdhury D (2007) Phys. Rev. E 75:041905.
8. MacDonald J T, Gibbs J H, Pipkin A C (1968) Biopolymers 6:1.
9. Krug J (1991) Phys. Rev. Lett. 67:1882.
10. Derrida B, Evans M R, Hakim V, Pasquier V (1993) J. Phys. A 26:1493.
11. Schütz G M, Domany E (1993) J. Stat. Phys. 72:277.
12. Janowsky S A, Lebowitz J L (1992) Phys. Rev. A 45:618.
13. Janowsky S A, Lebowitz J L (1993) J. Stat. Phys. 77:35.
14. Krug J (2000) Braz. Jrl. Phys. 30:97.
15. Tripathy G, Barma M (1997) Phys. Rev. Lett. 78:3039.
16. Enaud C, Derrida B (2004) Europhys. Lett. 66:83.
17. Greulich P, Schadschneider A (2007) in preparation.
18. Chou T, Lakatos G (2004) Phys. Rev. Lett. 93: 198101.
19. Greulich P, Schadschneider A (2007) submitted to Physica A.

Control of Traffic Congestion in the Modified Coupled Map Car-Following Model Based on Intelligent Transportation System

Han Xianglin[1,2], Ge Hongxia[3], Jiang Changyuan[2], Li Xingli[1], and Dai Shiqiang[1]

[1] Shanghai Institute of Applied Mathematics and Mechanics, Shanghai University, Shanghai 200072, China xlhan653@163.com; lxl326@163.com; sqdai@shu.edu.cn
[2] School of Science, Huzhou University, Huzhou 313000, China jcy@hutc.zj.cn
[3] Institute of Science, Ningbo University, Ningbo 315211, China ghxmocy@tom.com

Summary. With the application of intelligent transportation system (ITS), a modified coupled map car-following model is proposed to describe the dynamic characteristics of one-way traffic flow and the control of traffic congestion. It is concluded that the proposed control strategy is more effective in suppressing the formation of traffic congestion.

1 Introduction

Traffic congestion (traffic jam) leads to not only traffic safety problems but also the waste of considerable traffic transportation time and the pollution of the environment. A variety of approaches have been applied to describe the serious problem [1–11]. It is noticeable that when suppressing the congestion, coupled map (CM) car-following model is important for it, as a discrete version of car-following mode, not only explains many traffic phenomena but also suits to describe complicated traffic control based on its simple arithmetic in numerical simulation. Our investigations are based on a coupled map car-following traffic model [12], which consists of a leading vehicle and following vehicles, under open boundary condition. The dynamics of the ith vehicle can be given as:

$$v_i(n+1) = \alpha_i \left[v_i^{op}(y_i(n)) - v_i(n) \right] T + v_i(n) \tag{1}$$

$$y_i(n+1) = v_0 T - v_i(n) T + y_i(n) \tag{2}$$

where $x_i(n) > 0$ is the position of the ith vehicle at time $t = nT$, $T > 0$ is the sampling time, $v_i(n)$ is the velocity of the ith vehicle at time $t = nT$, N is the total number of vehicles, α_{i1} and α_{i2} are constants which denote respectively the sensitivity to preceding and following vehicles of ith vehicle

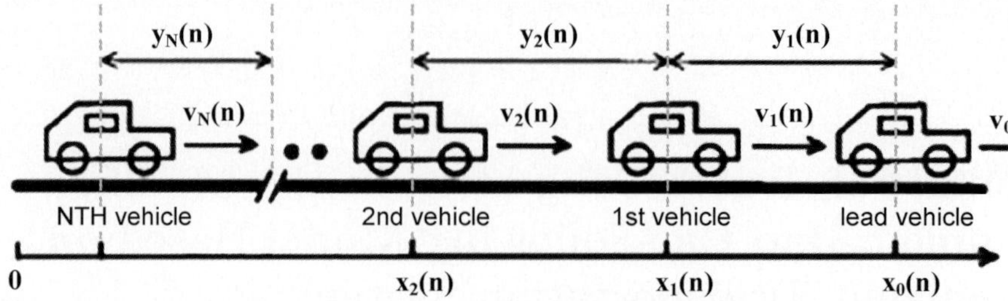

Fig. 1. Illustrations of car-following model.

driver. $v_i^{op}(y_i(n))$ is the optimal velocity (OV) function, $y_i(n) = x_{i-1}(n) - x_i(n)$. p and q are non-negative constants which stand for the relative effect on considered vehicle coming from the preceding and following ones, respectively, as well as $p + q = 1$. Namely, q is the probability of $y_{i+1}(n)$ less than safety distance h_c. The OV function has been given by [12]

$$v_i^{op}[y_i(n)] = \frac{v_i^{\max}}{2}\left[1 + \bar{H}_{sat}(2\frac{y_i(n) - \eta_i}{\xi_i})\right]$$

where the saturation function $\bar{H}_{sat}$ is described as

$$\bar{H}_{sat}(\rho) = \begin{cases} +1 & \text{if} \quad \rho > 1 \\ \rho & \text{if} \quad -1 \leq \rho \leq 1 \\ -1 & \text{if} \quad \rho < -1 \end{cases}$$

$v_i^{\max} > 0$ is the maximum speed, $\eta_i > 0$ is the neutral headway distance, and $\xi_i > 0$ is the parameter. The steady state of system (1)–(2) is

$$[v_i^* \ y_i^*]^T = [v_0 \quad \frac{v_0}{r_i} - \frac{\xi_i}{2} + \eta_i]^T \tag{3}$$

where $r_i = \frac{v_i^{\max}}{\xi_i}$.

By the control theory, Ref. [12] derived the no traffic congestion conditions of (1)–(2). Then, they proposed a control scheme, which is called KKH model, for suppression of traffic jam in the model. A decentralized delayed-feedback control signal term is added to system (1), this term is $\bar{u}_i(n)$, which is given as:

$$w_i(n+1) = k_i^a w_i(n) + k_i^b\left[v_i(n) - v_i(n-1)\right]$$
$$u_i(n+1) = k_i^c w_i(n) + k_i^d\left[v_i(n) - v_i(n-1)\right]$$

where k_i^a,k_i^b,k_i^c and $k_i^d \in R$ are the feedback gains.

Based on KKH model, Xiaomei Zhao and Ziyou Gao [13] modified $\bar{u}_i(n)$ in the KKH model as

$$\bar{u}_i(n) = k\left[v_{i-1}(n) - v_i(n)\right], i = 1, 2, \ldots, N$$

the feedback gain has only one, we call this model as ZG model.

In the next section, we will propose simpler control schemes to suppress traffic jam.

2 Delayed-Feedback Control Under ITS

We suppose all the vehicles are equipped with ITS (i.e. Intelligent Transportation System), and the ith vehicle can obtain the information's of velocities of all vehicles before it. Thence, we designate a feedback control signal term $u_i(n)$ as follows:

$$u_i(n) = \sum_{l=1}^{s} k_l[v_{i-1}(n) - v_i(n)], s \geq 2, i = 1, 2, \ldots, N$$

where $k_l > 0(l = 1, 2, \ldots, s)$ are the feedback gains. Let $\bar{y}_i(n) = \sum_{l=1}^{s} k_l y_{i-l+1}(n)$, add the control signal term to (1), i. e.

$$v_i(n+1) = \alpha_i[v_i^{op}(\bar{y}_i(n)) - v_i(n)]T + v_i(n) + u_i(n) \tag{4}$$

In order to suppress the traffic jam in the system (1) and (4), we design the feed back gains k_l which satisfies: $p_i(z)$ is stable; and $\max_{|z|=1} |G_i(z)| \leq 1, i = 1, 2, \ldots, N$.
where

$$G_i(z) = [(k_1 - k_2)(z-1) + \alpha_i r_i k_1 T^2, (k_2 - k_3)(z-1), \ldots, (k_{s-1} - k_s)(z-1), k_s(z-1), \alpha_i r_i T k_2(z-1), \alpha_i r_i T k_3(z-1), \ldots, \alpha_i r_i T k_{s-1}(z-1), \alpha_i r_i T k_s(z-1)]/p_i(z)$$

and $p_i(z) = z^2 + a_i z + b_i$, $a_i = \alpha_i T + k_1 - 2$, $b_i = 1 - k_1 - \alpha_i T + \alpha_i r_i k_1 T^2$.
We adopt all assumes of Ref. [12], and suppose that [9]

$$k_l = \frac{2}{3^l} R \ (l = 1, 2, \ldots, s), \quad k_s = \frac{1}{3^{s-1}} R, \quad R > 0;\ 0 < T \leq 0.5.$$

The following theorem provides one of the sufficient conditions of system (1) and (4) for no traffic jams on how to design the feedback gains k_l.

Theorem. *Supposing the system (1) and (4) satisfies all of the above assumptions, and* $0 < \alpha_i r_i T^2 < 1$*, then there is no traffic jam, if* $A_1 = -\frac{2}{9} + \frac{7}{9^s} + \alpha_i^2 r_i^2 T^4 (\frac{1}{18} + \frac{1}{2 \times 3^{2s-2}}) - \frac{8}{27}\alpha_i r_i T^2 < 0$, $\Delta_2 > 0$ *and* R *satisfies*

$$\{R > R_{11} or R < R_{12}\} \cap \left\{ \max\left\{ \frac{3(1-\alpha_i T)}{2(1-\alpha_i r_i T^2)},\ 0 \right\} < R < \frac{3(4-2\alpha_i T)}{2(2-\alpha_i r_i T^2)} \right\} \tag{5}$$

where $R_{21} = \frac{-B_2 - \sqrt{\Delta_2}}{2A_1}$, $R_{22} = \frac{-B_2 + \sqrt{\Delta_2}}{2A_1}$, $B_2 = \frac{2}{3}(4 - \alpha_i r_i T^2 + \alpha_i^2 r_i T^3 - 2\alpha_i T)$, $C_2 = -4 + 4\alpha_i T - \alpha_i^2 T^2$.

3 Numerical Simulation Results

Our simulations are based on the CM car-following with ITS under open boundary. The parameters of all vehicles are the same and set as Ref. [12], the method of simulation is also same as Ref. [12]. We will compare our model with KKH model and ZG model: How does the external disturbance effect on the traffic jam by three method?

According to Ref. [9], $s = 3$ is the optimal value. Substituting all of the parameters to (5), we can obtain: $1.23558 < R < 1.5944$.

We chose $R = 1.59$, then we have
The second vehicle: $k_1 = 1.06, k_2 = 0.53$.
The ith $(s \geq i > 2)$ vehicle: $k_1 = 1.06, k_2 = 0.286, k_3 = 0.144$.

Fig. 2(a), 3(a), 4(a) and 5(a) show the spatio-temporal pattern of the traffic flow about uncontrolled system, KKH model, our model and ZG model after $nT = 90s$, respectively. Fig. 2(b), 3(b), 4(b) and 5(b) show the temporal velocity of the first, 25th, and 50th vehicles corresponding to Fig. 2(a), 3(a), 4(a) and 5(a), respectively. It should be noted that there is smaller oscillating amplitudes in the system controlled by our method, compared with KKH model and ZG model. The simulation results indicate that all of the above three control models can be used to suppress the traffic jam, and our model is better.

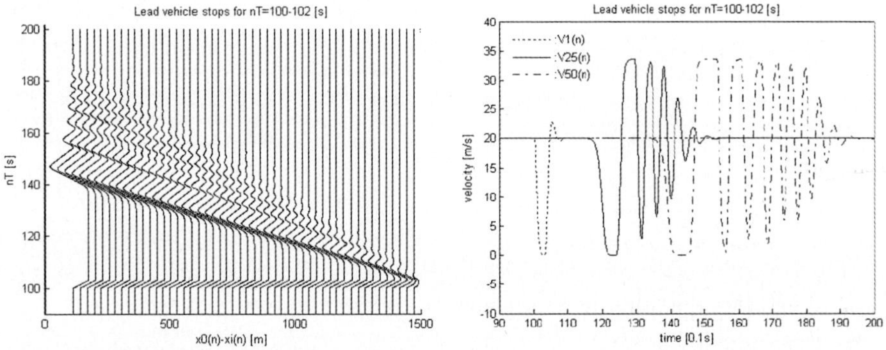

Fig. 2. (a).Spatio-temporal pattern of the traffic flow about uncontrolled system; (b).Temporal velocity of the first, 25th, and 50th vehicles corresponding to Fig. 1(a).

4 Conclusions

In this paper, a modified coupled map car-following model for the suppression of the traffic jam is proposed, based on the pioneering work of Ref. [1] and the application of intelligent transportation system (ITS) to describe the dynamic

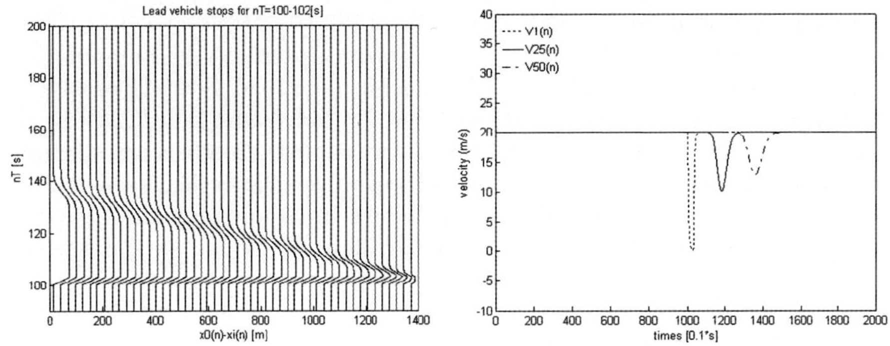

Fig. 3. (a).Spatio-temporal pattern of the traffic flow about KKH model; (b). Temporal velocity of the first, 25th, and 50th vehicles corresponding to Fig. 2(a).

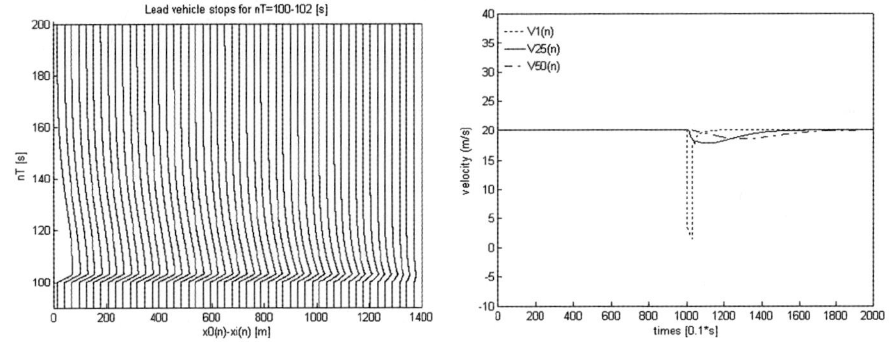

Fig. 4. (a).Spatio-temporal pattern of the traffic flow about (4)–(5) model; (b). Temporal velocity of the first, 25th, and 50th vehicles corresponding to Fig. 3(a).

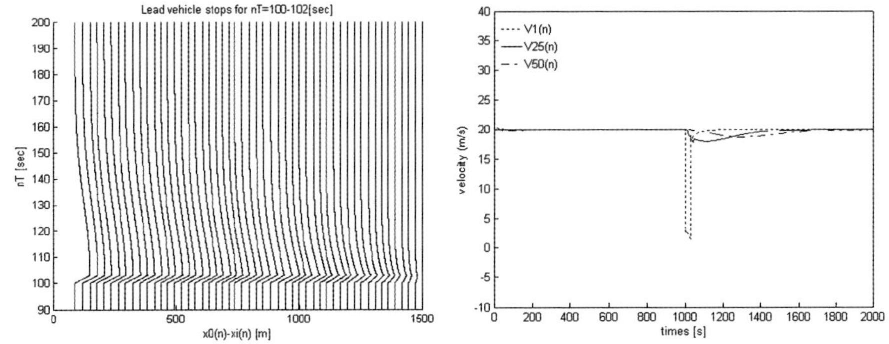

Fig. 5. (a).Spatio-temporal pattern of the traffic flow about ZG model; (b).Temporal velocity of the first, 25th, and 50th vehicles corresponding to Fig. 4(a).

characteristics of one-way traffic flow. The stability criteria are given as the speed of the lead vehicle changes. By theoretical analysis and simulations, it can concluded that (i) the information on more leading vehicles of each vehicle could lead to a stabilization effect for the traffic flow, that is, the stability conditions could be considerably weakened, (ii) the corresponding numerical simulations confirm the correctness of the theoretical analysis, (iii) the proposed control strategy is more effective in suppressing the formation of traffic congestion by applying and controlling the induction information provided by the intelligent transportation system.

Acknowledgements

This paper was partly supported projects of the National Basic Research Program of China (Grant No. 2006CB705500), the National Natural Science Foundation of China (Grants Nos. 10532060, 10471039,10602025) and the Natural Science Foundation of Zhejiang Province (Grants No. Y606268)

References

1. Bando M et al (1995) Phys. Rev. E 51:1035–1042.
2. Nagatani T and Nakanishi K (1998) Phys. Rev. E 57:6415–6421.
3. Yukawa S and Kikuchi M (1995) J. Phys. Soc. Jpn. 64:35–38.
4. Tadaki S et al (1998) J. Phys. Soc. Jpn. 67:2270–2276.
5. Nagatani T (1999) Phys. Rev. E 60:6395–6401.
6. Lenz H, Wagner C K and Sollacher R (1999) Eur. Phy. J. B 7:331–335.
7. Hasebe K, Nakayama A and Sugiyama Y (2004) Phys. Rev. E 69:017103.
8. Han X L, Jiang C Y, Ge H X and Dai S Q (2007) Phys. Acta 56:4383–4392.
9. Ge H X and Dai S D (2004) Phys. Rev. E, 70:066134.
10. Nakayama A, Sugiyama Y and Hasebe K (2001) Phys. Rev. E 65:016112.
11. Hasebe K, Nakayama A and Sugiyama Y (2003) Phys. Rev. E 68:026102.
12. Konishi K, Kokame H and Hirata K (1999) Phys. Rev. E 60:4000–4007.
13. Xiaomei Zhao and Ziyao Gao (2006) Physica A 366:513–522.

Generic Driving Behavior Modeling by Differential Game Theory

Serge P. Hoogendoorn and Piet Bovy

Delft University of Technology, Faculty of Civil Engineering and Geosciences, Stevinweg 1, 2628 CN Delft, The Netherlands s.p.hoogendoorn@tudelft.nl, p.h.l.bovy@tudelft.nl

Summary. In the last few decades, a number of driving models aiming at modeling the longitudinal and lateral driving tasks have been put forward based on the analogy with self-driven many particle systems. Examples of such models are the social-forces type models for car-following behavior [1], the IDM (Intelligent Driver Model) and its modifications [2], and the MOBIL model [3] describing lane-changing behavior. Although these models can describe many phenomena in motorway traffic flow operations, a clear behavioral foundation has however been lacking so far.

This contribution puts forward a new generic theory of driving behavior, based on the principle of least effort. In this theory, drivers are assumed to minimize the *predicted subjective perceived effort* of their control actions, including for instance acceleration towards the free speed, car-following and lane changing. In this game-theoretic approach, drivers may or may not anticipate the reactions of the other drivers on their control decisions. Also non-cooperative and cooperative driving rules can be incorporated using the flexibility of the modeling approach.

In this contribution, we present the main behavioral assumptions, the model derivation, and the resulting car-following and lane changing models for the *non-cooperative* case. The workings of the model will be illustrated by means of a simple example.

1 Introduction

Microscopic driver models aim to predict traffic flow operations using mathematical models describing individual driving behavior. In general, these models distinguish between:

- the *longitudinal driver tasks*, such as choosing the free speed and accelerating towards it, and car-following, and
- the *lateral driver tasks*, such as lane-changing and overtaking.

For an elaborate overview of driving tasks, we refer to [4].

This contribution generalizes many of these modeling approaches by adopting the concept of effort minimization: drivers are assumed to minimize the predicted subjective effort of their control actions, taking into account the anticipated actions of the other drivers. This can be achieved in a cooperative (working together to achieve some common goal) or a non-cooperative way (possibly taking account expected counteractions of the 'opponents'). The remainder of this contribution considers the main behavioral principles that will be used for model derivation.

2 Driving and Effort Minimization

Several authors (e.g. [5] and [6]) have proposed using a utility (or cost) optimization model to conceptually model execution of the driving task. Clearly, this process, its parameters, and the resulting control decisions will differ between drivers, due to differences in the control objectives, preferences, vehicle characteristics, etc. Differences in driving style are reflected by different preferences for the desired driving speed, desired comfortable deceleration levels, minimum gap distances, etc.; some driver give priority to safe driving; others prefer driving at a high speed, accepting smaller headways and increased risk.

In general, we assume that the objective of drivers will be a subset of the following:

1. Maximize safety and minimize risks
2. Minimize lane-changing maneuvers
3. Maximize travel efficiency (restricting deviations from the desired driving speed)
4. Maximize smoothness and comfort
5. Minimize stress, inconvenience, fuel consumption, accelerations, decelerations, etc.

The importance of each of these objectives will vary among the individuals, given the possibilities of their vehicle in terms of maximum speed, braking capabilities, etc.

In the remainder of this contribution, we will derive an operational model for driving behavior based on these conceptual notions. Effectively, this implies that we formalize the control objectives into a cost function (predicted disutility) a driver aims to minimize. Before doing so, let us briefly reflect on the main behavioral assumptions made in this contribution.

2.1 Behavioral Assumptions

Underlying the mathematical model that will be derived in the ensuing is a behavioral theory, describing how we assume driver p to act and react to other drivers q – referred to as the opponents – on the road. These assumptions pertain to the way drivers scan their environment, and act upon these observations, given their control objectives. The following list provides an overview of the main behavioral assumptions:

1. Drivers minimize the generalized predicted costs of their control actions. These generalized costs reflect the different (and subjective) control objectives (e.g. risk, travel time, stress, smoothness, etc.).
2. Drivers reconsider their operational control decisions at non-equidistant discrete time instants t_k based on a) observations of the prevailing traffic conditions and b) predictions of the impacts of their control actions, possibly including the expected reactions on the other drivers (see 4).
3. Car-drivers are to a large extent anisotropic, implying that they will generally only consider stimuli in front of them. Drivers will however not only consider the driver directly ahead or only drivers on the same lane, but may also take into account other drivers further downstream and on other lanes.
4. Drivers anticipate on the behavior of other drivers by predicting their driving behavior according to either non-cooperative or cooperative strategies.
5. Drivers have limited prediction possibilities. This is reflected by discounting the costs both over time and space.
6. Drivers will minimize the predicted cost stemming from: a) not driving at the desired state, described by e.g. the free speed, the desired distance, and the desired lane, b) driving too close to other drivers, c) acceleration and braking and d) changing lanes (lane switching costs).
7. Observing, decision making, and maneuvering take time and are prone to errors.
8. Control objectives may change over time (adaptation effects).

All these behavioral assumptions are supported by empirical evidence for different sources. It is beyond the scope of this contributions to provide a complete overview of all literature providing evidence for these assumptions. In the remainder, we will focus on the principle of least effort and how this can be formalized to yield a microscopic driver model.

2.2 Driving as an Optimal Control Cycle

We assume that decisions to adapt speed or change lanes are made at discrete time instants, say t_k, with $k = 1, 2, \ldots, K$. Let us assume that at each of these instants t_k, driver i has information on the positions and speeds of all cars influencing his or her behavior. This information stems from (erroneous) observations made by the driver which are used in conjunction with the drivers experience, mental model, etc., to form an estimate of the current state of the system.

The estimate of the state of the system available at t_k is used as the starting conditions for the (conditional) predictions of the evolution of the system. We assume that the driver has some mental model allowing him or her to make such a prediction, given the candidate control decision (longitudinal acceleration or changing lanes). He or she will then choose the control decision that minimizes the predicted effort (or cost, or disutility) and apply this control. The decision will be reconsidered when new information about the state of the system becomes available (i.e. at t_{k+1}).

3 Control Model Formulation

The presented model distinguishes two components, namely 1) the physical model and 2) the control model. In this contribution, we focus on the latter as this describes the most important component for the purpose of our study.

3.1 Conceptual Driving Task Model

A driver is assumed to use an internal mental model for determining appropriate control decisions. Experience and knowledge have skilled the driver in performing the driving task, and hence the assumption of subjective optimal behavior appears justified. The optimization however includes the limitations of the driver in terms of observation, information processing, internal state estimation, as well as processing times and reaction times. Moreover, the process includes the driving strategy.

3.2 Mental Model for State Predictions

For the longitudinal dimension, we will use the following kinematic equations describing the dynamics of the location $x_i(t)$ of driver i at instant t:

$$\frac{d}{dt}x_i = v_i \text{ and } \frac{d}{dt}v_i = a_i = u_i \tag{1}$$

Driver i will use the same model to predict the behavior of the opponents $j \neq i$ (the opponents). In particular the assumption driver i makes with respect to the accelerations a_j of the opponents are relevant, since this describes the anticipated reaction of the opponents j to the actions of i.

For the lateral dimension, the variable $y_i(t) \in \{1, 2, \ldots, m\}$ denotes the lane number. In the model, a lane change is modeled by an instantaneous change in this variable. However, generalizations to continuous lane positions are straightforward. Note that lane changes are assumed to happen instantaneously, and will be considered separately in the ensuing of the contribution.

3.3 Cost Specification

The principle of least effort stipulates that drivers minimize some cost function. We hypothesize that this cost can be described by the following cost functional:

$$J(t_k, \mathbf{z}(t_k)) = \int_{t_k}^{\infty} e^{-\eta s} L(\mathbf{z}(s), \mathbf{u}(s)) ds \tag{2}$$

$$\text{s.t.} \ \frac{\mathrm{d}}{\mathrm{dt}} \mathbf{z} = f(\mathbf{z}, \mathbf{u}) \tag{3}$$

In Eq. (2), $\mathbf{z}$ denotes the state of the system from the perspective of driver i: it contains the locations x_j, the speeds v_j and the lane y_j of all drivers j; $\mathbf{u}$ denotes the control, including the acceleration and braking (denoted by $a_i(t)$), and the lane changing behavior. The parameter $\eta > 0$ denotes the so-called discount factor (see [7]). Note that without loss of generality, we may assume that $t_k = 0$.

Note that the cost function $J(t_k, \mathbf{z}(t_k))$ describes the expected cost given the current state of the system $\mathbf{z}(t_k)$ (as perceived by the driver), the control actions of the driver and the evolution of the system, starting from time t_k onward. At time t_{k+1}, the decision will be re-assessed using the most current observations or estimates of the system state (at time t_{k+1}).

In Eq. (2), L denotes the running costs. These costs include cost of acceleration, driving too close to the preceding vehicles, and the cost of not driving at the desired speed v_0. For now, we propose [7]:

$$L(\mathbf{z}, \mathbf{u}) = \frac{1}{2} a_i^2 + l(\mathbf{z}) = \frac{1}{2} a_i^2 + c_1 \sum_{j \in O_i} e^{-|x_j - x_i|/R_0} + \frac{c_2}{2} (v_0 - v)^2 \tag{4}$$

where O_i denotes the set of opponents of driver i. Note that other, more involved, specifications of the running costs are also possible (e.g. using speed differences, etc.). Lane changes will be considered separately in the remainder of the paper, based on the trade-off between expected costs of driving in either of the lanes and lane changing costs.

4 Non-Cooperative Driving

In this section, we will present the optimal control model describing driving acceleration, given that driver i remains on his / her current lane. We will only consider driver strategies for the *non-cooperative feedback case*. More specifically, we hypothesize that drivers predict the behavior of the other drivers assuming that they maintain driving at the same speed. Furthermore, we will calculate the perceived cost of driving on the current lane, which in the next sections will be used to determine whether a lane change will take place or not. The whole approach is based on two concepts for optimal control theory: Pontryagin's minimum principle and the Hamilton Jacobi Bellman equation for the discounted cost problem. Let us first recall both of them.

4.1 Dynamic Programming

Let us consider a control problem of the form (2). We can show that the so-called value function $V(t_k, \mathbf{z})$ describing the cost functional J under optimal control (i.e. the minimum cost) satisfies [8]:

$$V(t_k, \mathbf{z}) = e^{-\eta t_k} W(\mathbf{z}) \tag{5}$$

where $W(\mathbf{z})$ satisfies the so-called *Hamilton Jacobi Bellman equation* for the *discounted cost infinite horizon problem*:

$$\eta W(\mathbf{z}) + H\left(\mathbf{z}, \mathbf{u}^*, \partial W/\partial \mathbf{z}\right) = 0 \tag{6}$$

with the optimal control $\mathbf{u}^*$ defined by

$$\mathbf{u}^* = \arg\min_{\mathbf{u}\in U} \left\{H\left(\mathbf{z}, \mathbf{u}, \partial W/\partial \mathbf{z}\right)\right\} \tag{7}$$

and where the Hamilton function H is defined by:

$$H(z, u, \lambda) = L(\mathbf{z}, \mathbf{u}) + \lambda' \mathbf{f}(\mathbf{z}, \mathbf{u}) \tag{8}$$

We can easily show that when using the cost specification Eq. (4), the optimal control (i.e. the acceleration of driver i) satisfies:

$$\frac{\partial H}{\partial u} = 0 \Rightarrow \mathbf{u}^* = a_i^* = -\partial W/\partial v_i \tag{9}$$

Let us note that:

$$\boldsymbol{\lambda} = \frac{\partial W}{\partial \mathbf{z}} \tag{10}$$

are generally referred to as the *co-states* or the *marginal costs* of the state $\mathbf{z}$.

Finally, let us note that the optimal cost W satisfies:

$$W(\mathbf{z}) = -\eta^{-1} H\left(\mathbf{z}, \mathbf{u}^*, \partial W/\partial \mathbf{z}\right) \tag{11}$$

Eq. (11) shows how we can compute the incurred cost when applying the optimal control $\mathbf{u}^*$. This expression can be used to compare the perceived costs on the different lanes and hence to decide whether a lane change is beneficial or not.

4.2 Derivation of Optimal Acceleration Control Law

Suppose that driver i assumes that all other drivers j will not accelerate or decelerate. I.e., we have $a_j = 0$ for all $j \neq i$. In this case, when driver i is predicting the behavior of the other drivers, he assumes no direct reaction of the other drivers until new information becomes available at the next decision time instant t_{k+1}.

To determine the optimal control, Pontryagin's minimum principles uses the so-called Hamiltonian, defined by:

$$H(\mathbf{z},\mathbf{u},\boldsymbol{\lambda}) = L(\mathbf{z},\mathbf{u}) + \boldsymbol{\lambda}'\mathbf{f}(\mathbf{z},\mathbf{u}) = L(\mathbf{z},\mathbf{u}) + \lambda_i^v a_i + \sum_j \lambda_j^x v_j \tag{12}$$

where $\boldsymbol{\lambda} = (\boldsymbol{\lambda}^x, \boldsymbol{\lambda}^y)$ denotes the vector of marginal costs with respect to the location and the speed.

The so-called *stationary conditions* allow computation of the optimal control:

$$\frac{\partial H}{\partial a_i} = 0 \quad \Rightarrow \quad a_i^* = -\lambda_i^v \tag{13}$$

Where λ_i^v denotes the marginal cost of the speed v_i of vehicle i. The equation states that when the marginal cost of the speed is negative, the driver will accelerate (thus increasing the speed and reducing the cost). Vice versa, when the marginal cost of the speed is positive, the driver will decelerate.

Pontryagin's principle also allows us to determine an expression for the marginal cost. It can be shown easily that:

$$\eta\lambda = \frac{\partial H}{\partial \mathbf{z}} \tag{14}$$

Using this expression, it turns out that:

$$\lambda_i^x = \frac{1}{\eta}\frac{\partial H}{\partial x_i} = \frac{1}{\eta}\frac{\partial L}{\partial x_i} \tag{15}$$

and that:

$$\lambda_i^v = \frac{1}{\eta}\frac{\partial H}{\partial v_i} = \frac{1}{\eta}\left(\frac{\partial L}{\partial v_i} + \lambda_i^x\right) = \frac{1}{\eta^2}\left(\eta\frac{\partial L}{\partial v_i} + \frac{\partial L}{\partial x_i}\right) \tag{16}$$

Combining these results, we get the following expression for the optimal acceleration:

$$a_i^* = -\frac{1}{\eta^2}\left(\eta\frac{\partial L}{\partial v_i} + \frac{\partial L}{\partial x_i}\right) \tag{17}$$

If we use the specification Eq. (4), we get:

$$a_i^* = \frac{v_0 - v}{\tau} - A_i \sum_{j\in O_i} e^{-|x_j - x_i|/R_0} \tag{18}$$

where:

$$\frac{1}{\tau} = \frac{c_2}{\eta} \text{ and } A_i = \frac{c_1}{\eta^2 R_0} \tag{19}$$

Eq. (19) and the fact that $\lambda_i^v = -a_i^*$ show that the acceleration is determined by two terms. One term describes the acceleration towards the free speed v_0, while the second term describes deceleration when a driver is driving too close to the vehicles ahead.

Expected Driving Costs

Using Eq. (11), we can now compute the expected cost of driving in the current lane:

$$W(\mathbf{z}) = -\eta^{-1} H(\mathbf{z}, -\lambda_i^v, \lambda) \tag{20}$$

$$= \frac{1}{\eta}\left(\frac{1}{2}(a_i^*)^2 - l(\mathbf{z}) - \frac{1}{\eta}\sum_j \frac{\partial L}{\partial x_j} v_j\right) \tag{21}$$

Now, note that under the assumptions of anisotropy, the running cost L is only determined by drivers j *ahead* of i (as reflected by the set of so-called *opponents* O_i). This implies that $x_j > x_i$ for all $j \in O_i$, and that the running cost decrease as the position x_j is increased:

$$\frac{\partial L}{\partial x_i} > 0 \text{ and } \frac{\partial L}{\partial x_j} < 0 \text{ with } j \in O_i \tag{22}$$

Now, let us rewrite:

$$L = \frac{1}{2}a_i^2 + l(\mathbf{z}) = \frac{1}{2}a_i^2 + \mu(v_i) + \sum_{j \in O_i} \sigma(x_j - x_i) \tag{23}$$

where $\sigma = \sigma(r)$ denotes the running cost component due to the spacing r between two vehicles j and i. Then, we find:

$$\sum_j \frac{\partial L}{\partial x_j} v_j = \sum_j v_j \frac{\partial}{\partial x_j}\left(\sum_{k \in O_i} \sigma(x_k - x_i)\right) = \sum_{j \in O_i} (v_j - v_i)\sigma'(x_j - x_i) \tag{24}$$

This expression shows nicely that when $v_j > v_i$, the contribution of this term to the total cost Eq. (20) is (using the specification of the costs proposed in Eq. (4)):

$$W(\mathbf{z}) = \frac{1}{\eta}\left(\frac{1}{2}(a_i^*)^2 - \frac{c_2}{2}(v_0 - v)^2 - \left(c_1 + \frac{v_j - v_i}{\eta R_0}\right)\sum_{j \in O_i} e^{-\frac{|x_j - x_i|}{R_0}}\right) \tag{25}$$

Note that if we would compare the costs of two lanes, the middle term expressing the costs due to not driving at the free speed (at the current time instant) is the same for all lanes.

4.3 Lane Changing Modeling

Using the expected cost of driving on a certain lane Eq. (25), decisions to change from the current lane to another lane can be determined easily. We assume that the lane change is a *discrete decision*, yielding a change from lane y_i to lane $y_i' = y_i \pm 1$ (i.e. to the left or right lane). Furthermore, we assume that a driver changes lanes if the expected costs on the target lane is substantially smaller than on the current lane.

For a lane change from the current lane to the left lane, we assume the following criterion for a lane change

$$W(x_i, y_i + 1) < W(x_i, y_i) - \delta_i^+ \tag{26}$$

In Eq. (26), δ_i^+ denotes the costs of switching from lane y_i to the lane y_{i+1} on the left. In case of 'keep your lane' driving rules, an equivalent relation holds for lane-changes to the right lane y_{i-1}. For 'drive on the right, overtake to the left' traffic rules, this will apply only for sufficiently small speeds. For larger speeds, the switching costs δ_i^- may be zero or even negative, to ensure that drivers have the natural tendency to drive on the rightmost lane.

Note that besides the costs of switching, the driver decides to change lanes based on:

- The expected acceleration a_i^* on the target lane
- The expected changes in the proximity costs.

The latter is expressed by the rightmost term in Eq. (25). Note that this term in a decreasing function of the relative speeds $v_j - v_i$, implying that when the vehicles on the target lane have a higher speed than the considered vehicle, the expected costs are reduced (and vice versa). Note that this term is not present in the MOBIL model [3].

Let us illustrate the workings of the model by means of a simple example shown in Fig. 1 and Fig. 2. Consider a three lane motorway. On each lane of the motorway, a platoon of 5 vehicles is present. The platoon leader of lane 1 (right lane), lane 2 and lane 3 respectively have desired speeds of 16 m/s, 24 m/s and 32 m/s. All other vehicles have a desired speed of 32 m/s. Initially, driver 9 is driving on lane 3 without being disturbed by the fact that further downstream, the platoon leader is driving at 16 m/s. However, once he / she has caught up with this slow driving vehicle, driver 9 starts braking (round 140 s) and the cost of driving on lane 3 is increasing.

5 Summary and Outlook

In this contribution, we have put forward a microscopic theory of driving behavior based on the concept of subjective effort minimization. More importantly, we have shown how the theory can serve as a basis for advanced driving model derivation. The approach is generic, and can be used to include different kinds of driving, including cooperative and non-cooperative behavior. In the example shown in the contribution, the theory was used to derive a joint car-following and lane-changing model. The model turns out to behave realistically.

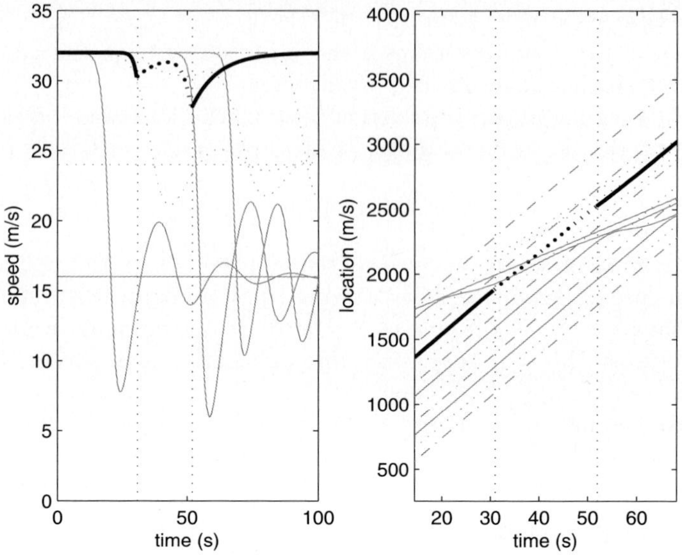

Fig. 1. Vehicle speeds (left) and positions (right) for platoon driving according to optimal control law.

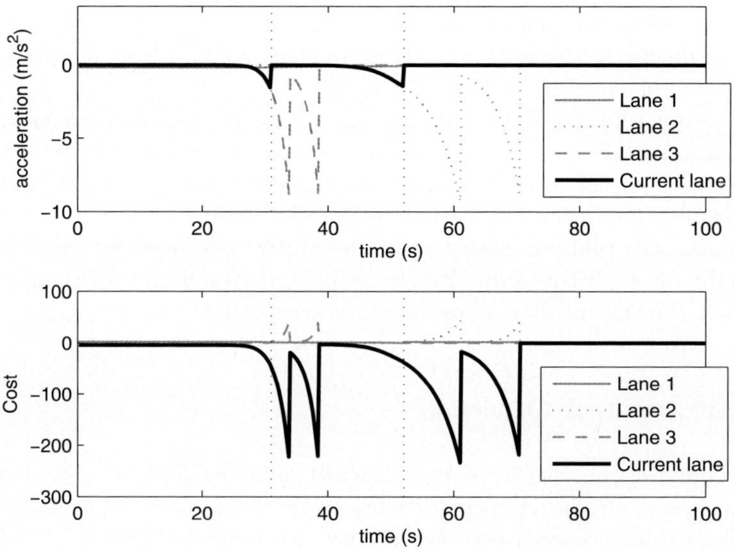

Fig. 2. Expected acceleration and costs (as a function of time) of vehicle 9, initially starting on lane 1 (right lane).

Acknowledgements

The research presented in this paper is part of the research programme "Tracing Congestion Dynamics - with Innovative Traffic Data to a better Theory", sponsored by the Dutch Foundation of Scientific Research MaGW-NWO.

References

1. Helbing D (1997) *Verkehrsdynamik.* Springer-Verlag.
2. Treiber M, A Kesting, and D Helbing (2006) *Delays, inaccuracies, and anticipation in microscopic traffic models.* Physica A **360**, 71–88.
3. Kesting A, M Treiber, and D Helbing (2006) *MOBIL: General Lane-Changing Model for Car-Following Models.* Transportation Research Board Annual Meeting 2006.
4. Michon J A (1993) *Generic Intelligent Driver Support.* Taylor & Francis.
5. Stanton N (1995) *Back to the future: Analysing Driver Performance with Potential IVHS.* In: Proceedings of the Second Annual World Congress on Intelligent Transportation Systems, Japan.
6. Minderhoud M (1999) *Supported Driving: Impacts on Motorway Traffic Flow.* PhD thesis, Delft University of Technology.
7. Hoogendoorn S P and P H L Bovy (2002) *Normative pedestrian behaviour theory and modelling.* In: Proceedings of the 15th International Symposium on Transportation and Traffic Theory. Adelaide, Australia.
8. Fleming W H and H M Soner (1993) *Controlled Markov Processes and Viscosity Solutions.* Applications of Mathematics 25. Springer-Verlag.
9. Bando M, K Hasebe, A Nakayama, A Shibata, and Y Sugiyama (1995) *Dynamical model of traffic congestion and numerical simulation.* Physical Review E **51**, 1035–1042.
10. Kerner B S (2002) *The Physics of Traffic: Empirical Freeway Pattern Features.* Springer, Berlin.

A Model for City Traffic

Ding-wei Huang and Wei-neng Huang*

Department of Physics, Chung Yuan Christian University, Chung-li, Taiwan
*wnhuang@phys.cycu.edu.tw

Summary. We propose a simple cellular automaton to simulate city traffic. Vehicles on road are treated as particles hopping on a lattice. In the morning rush hours, downtown parking lots are taken as the sinks for vehicles moving from suburban; in the afternoon rush hours, the parking lots become the sources for vehicles moving out of town. The traffic is regulated by the operation of a series of traffic lights on each street block. Both the traffic flow (from global perspective) and the travel time (from personal perspective) are studied. It is found that even in maintaining the same traffic flow, the tuning of traffic lights can significantly affect the travel time, and correspondingly the duration of rush hours.

1 Introduction

In the time scale of most interest, traffic problem is intrinsically an inequilibrium problem. As a mean of transportation, vehicles move from one place to another, and rarely did a roadway keep same traffic flow all day. The situation is no more apparent than in city traffic, where large number of vehicles move in and out of the city in relatively short rush hours according to the rhythm of city life.

Cellular automaton has been widely used in the study of highway traffic problems [1, 2]. The method proves very successful in simulating and explaining various highway traffic patterns. There have been attempts to extend the method to city traffic, with mixed results [3–9]. Unlike the highway traffic, which is mostly determined by the vehicle density, city traffic pattern obviously is also influenced by other factors, such as operation of traffic lights and the street layout. The highway traffic is characterized by the traffic flow, i.e. the ability to move vehicles from one place to another in certain time. But for city traffic, city itself is the destination, and the ability to absorb and dissipate traffic is crucial, and that is determined not only by traffic flow but by travel time as well.

To keep the essence of the city traffic characteristics with as few parameters as possible, we propose a cellular automaton model for the city traffic. For the rush hours in the morning, heavy traffic moves from suburban area to downtown center. In the afternoon, the process is reversed. We study the influence of the traffic light control to regulate the traffic. We suspect that the optimization to achieve the highest traffic flow on road might not lead to the shortest travel time for individual, and this in turn will affect the time it takes to absorb or dissipate all the vehicles getting in or out of the city.

2 Model

We assume a symmetric city street layout, with equal block lengths and similar traffic movements in all directions. The afternoon rush hours traffic is considered the morning traffic in reverse direction. The system consists of a single-lane roadway with open boundaries, as shown in Fig. 1.

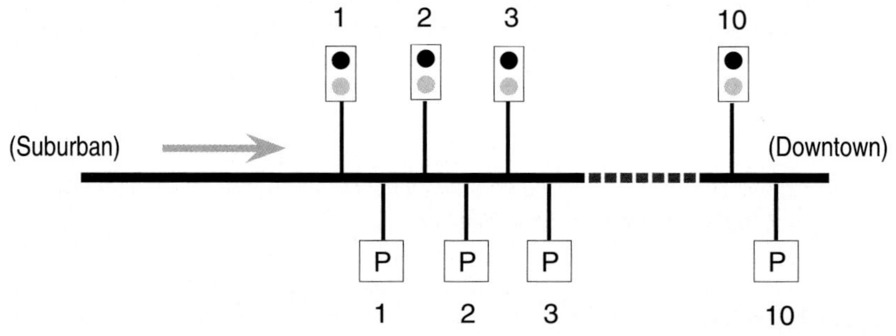

Fig. 1. Configuration of the open system.

Vehicles move in one direction only, i.e. from left (suburban) to right (downtown). Along the roadway, ten traffic lights are set up with equal distance $L = 100$ to simulate the layout of the downtown area. In between the traffic lights, ten parking lots act as the destinations of the in-bound traffic. The vehicles on road are mainly injected from the left boundary (suburban) and prescribed by a stochastic rate α_0. For each newly injected vehicle, one of the ten parking lots is assigned randomly as its destination. The vehicle will remain on the roadway until it reaches the designated parking lot, then it is removed from the system. The traffic lights operate in alternating green and red phases. To simulate the traffic turning in from the perpendicular direction, new vehicles are also injected right at the intersection when the traffic light is in the red phase. For example, at the intersection of the first traffic light, new vehicles are injected with a rate α_1 in the red phase, and their destinations are randomly assigned from parking lot 1 to 10. Similarly, at the intersection of the second traffic light, the injection rate is α_2, and the destinations are randomly assigned from parking lot 2 to 10. The vehicle injected from the i-th intersection is forbidden to have a destination at the j-th paring lot with

$j < i$, because it would be counted as part of the symmetric right-left traffic. To simplify the parameterization, we assume $\alpha_1/10 = \alpha_2/9 = \alpha_3/8 = \cdots$. In summary, we have two parameters to prescribe the injection of vehicles: α_0 for the main road and α_1 for the perpendicular direction. We also assume that the ten traffic lights are operated at the same green phase and the same red phase, and these two phases are set equal. Then the operation of traffic lights can be characterized by two parameters: the period T and the delay δ, which prescribed the time delay between two consecutive traffic lights.

The movement of vehicles is modeled by the well-known cellular automaton, the Nagel-Schreckenberg traffic model [10, 11]. The dynamics is controlled by two parameters: speed limit v_{max} and noise p. Comparing to the setting of highway traffic at $v_{max} = 5$ and $p = 0.5$, we adopt a setting at $v_{max} = 3$ and $p = 0.1$ for the city traffic. Both the speed limit and the noise are reduced.

3 Traffic Flow

First, we measure the traffic flow in the downtown area, i.e. the traffic flow after the first intersection. In this simple model, a decreasing traffic flow along the roadway can be expected. We observe that, when the perpendicular traffic is absent, the average traffic flow is essentially not affected by the operation of traffic lights. The typical results are shown in Fig. 2(a).

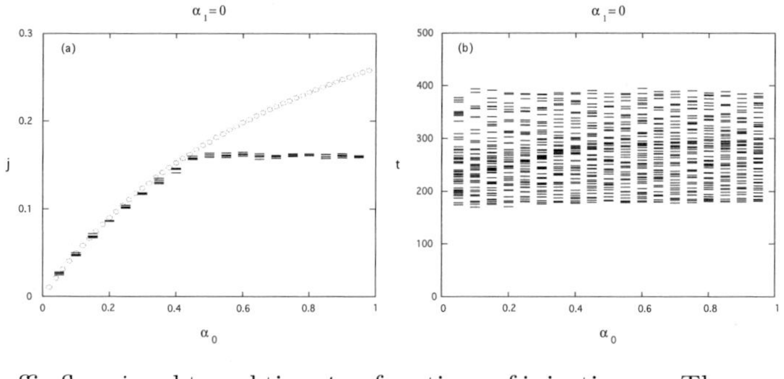

Fig. 2. Traffic flow j and travel time t as functions of injection α_0. The perpendicular traffic is cut off ($\alpha_1 = 0$). The period of traffic lights is fixed at $T = 100$. Different bars at each α_0 show the variation caused by different delay δ. (a) traffic flow, the gray circles show the results when the traffic lights are absent; (b) travel time.

A traffic light only introduces local disturbance to the system. Not all the vehicles were forced to stop when the lights turn red; only those approaching the intersection will be blocked. When the vehicles are scarce on the road, it is plausible that most vehicles will remain moving in the red phase of a traffic light, thus the global traffic flow remains constant when the two parameters T and δ are varied. When the traffic demand increases, which was prescribed by an increasing α_0, the flow increases accordingly and then saturates to its capacity. In this open system, the downtown traffic always remains in the free flow regime in the absence of perpendicular traffic, because the possible vehicle queuing is caused by the first traffic light.

As the perpendicular traffic is allowed to turn in, the global traffic flow does respond to the operation of traffic lights. The typical results are shown in Fig. 3(a) and 3(b).

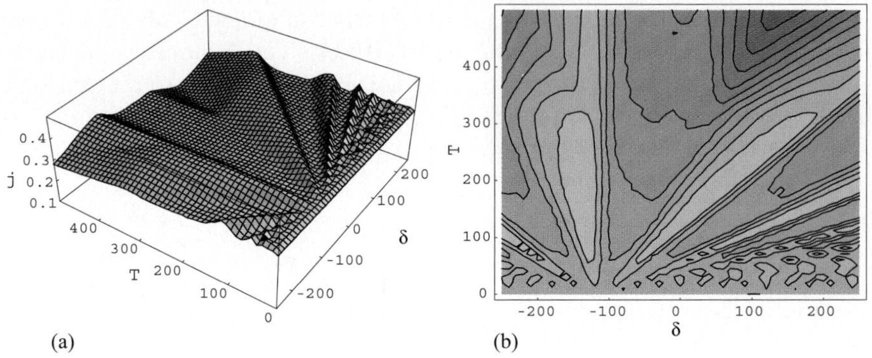

Fig. 3. Traffic flow j as function of (T, δ). Parameters are fixed at $(\alpha_0, \alpha_1) = (0.3, 0.3)$. (a) 3D plot; (b) contour plot.

Such a traffic pattern can be associated with the emergence of congestion. Basically, the radial structure can be understood as the repetition of a negative delay at $\delta \sim -110$, which characterizes the backward propagation of traffic jams [6]. To achieve a high flow, the parameters T and δ should be locked in these radial wave.

4 Travel Time

Next, we study the influence of traffic lights to the travel time, which can be taken roughly as a counter indicator of the traffic flow. Basically, a higher traffic flow would imply a shorter travel time. However, these two measurements can have different implications. While traffic flow reflects the global perspective of how many vehicles are moving at an instant, which might be the most concerned issue in traffic management, travel time, which measures how soon one vehicle can go through its journey, can be much more relevant from a personal perspective.

We record the travel time for those vehicles from suburban to downtown. Specifically, a clock is turned on when the vehicle crossed the first intersection, and the reading is registered as that vehicle went to its designated parking lot. The travel time for those vehicles turning from the perpendicular direction is not recorded. The typical results for the averaged travel time are shown in Figs. 4(a) and 4(b), respectively for free flow and congestion.

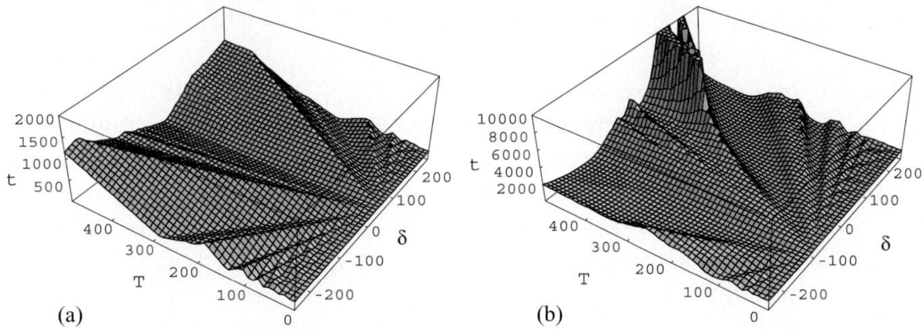

Fig. 4. Travel time t as function of (T, δ). (a) free flow regime at $(\alpha_0, \alpha_1) = (0.3, 0.1)$; (b) congestion regime at $(\alpha_0, \alpha_1) = (0.3, 0.3)$.

Contrary to the traffic flow, the travel time shows a clear dependence on the traffic lights in the free flow regime. Even when $\alpha_1 = 0$, the operation of traffic lights still has a significant influence on the travel time. The typical results are shown in Fig. 2(b). By adjusting the delay δ, the travel time can vary by as much as a factor of 2. In contrast, the variation of traffic flow is quite limited, as can be seen in Fig. 2(a).

In the congestion regime, both traffic flow and travel time are greatly affected by the operation of traffic lights. The patterns for travel time and traffic flow can be easily mapped into each other, and is observed to be inversely related.

From a driver's perspective, the increased time it takes to reach destination is an indication of rush hours. By doubling α_0 for a short period before returning to its normal value, and observing the consequent rise and fall of travel time, the duration of rush hours can be estimated. It is found that longer travel time does prolong rush hours, a clear indication that more time is needed to absorb all the traffic. A result is shown in Fig. 5.

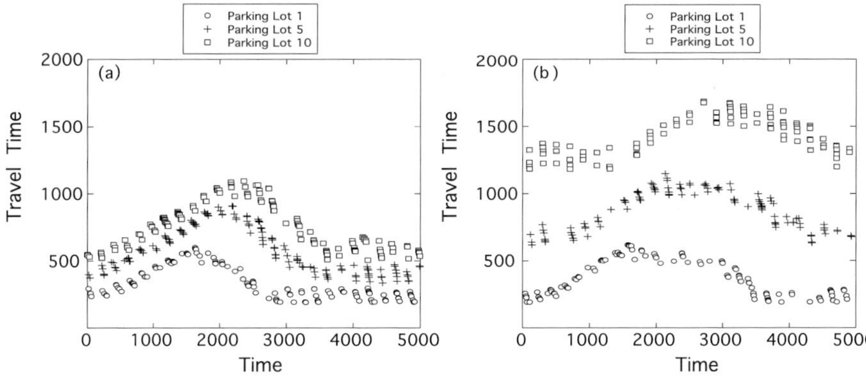

Fig. 5. Travel time t for individual vehicles to reach first, fifth and tenth parking lots. The setting used is $(\alpha_0, \alpha_1) = (0.3, 0.1)$ and $T = 200$. Travel time and duration of rush hours is influenced by the tuning of phase δ. (a) $\delta = 35$; (b) $\delta = 120$.

5 Discussions

In this simple city traffic model with as few parameters as possible, we found that the first traffic light acts as a regulator, only allowing limited number of vehicles into the city, thus creating a congestion-free downtown, in the case vehicles from perpendicular direction are cut off. When the perpendicular traffic is allowed to turn in, congestion may arise, and the traffic pattern is greatly affected by the operation of traffic lights. In the free flow regime, while traffic flow remains more or less constant, the travel time may change by as much as a factor of 2 by tuning of traffic lights. This seemingly uncorrelated traffic flow and travel time indicates that the extra time is mostly caused by the traffic light delay, i.e. the forced stop at most intersections. With same traffic flow but longer travel time in the city, this also means that it takes longer to absorb all the vehicles coming in from suburban area, resulting in an extended rush hours. In the congestion region, the traffic flow and travel time are almost inversely related; the maintenance of a high traffic flow would yield a shorter travel time, and can be achieved by proper tuning of traffic lights.

We conclude that in city traffic, the operation of traffic lights is most important in achieving a high traffic flow and low travel time. And while they may be related, the traffic flow is not the sole indicator of city traffic. The ability to absorb or dissipate all the vehicles quickly is crucial, and is directly related to travel time.

References

1. D. Chowdhury, L. Santen, and A. Schadschneider, Phys. Rep. **319**, 199 (2000).
2. D. Helbing, Rev. Mod. Phys. **73**, 1067 (2001).
3. E. Brockfeld, R. Barlovic, A. Schadschneider, and M. Schreckenberg, Phys. Rev. **E64**, 056132 (2001).
4. M. Sasaki and T. Nagatani, Physica **A325**, 531 (2003).
5. D. W. Huang and W. N. Huang, Int. J. Mod. Phys. **C14**, 539 (2003).
6. D. W. Huang and W. N. Huang, Phys. Rev. **E67**, 056124 (2003).
7. H. L. Tan, C. Y. Zhang, L. J. Kong, and M. Liu, Int. J. Mod. Phys. **B18**, 2658 (2004).
8. T. Nagatani, Physica **A350**, 563 (2005).
9. T. Nagatani, Physica **A361**, 619 (2006).
10. K. Nagel and M. Schreckenberg, J. Phys. I (France) **2**, 2221 (1992).
11. M. Schreckenberg, A. Schadschneider, K. Nagel and N. Ito, Phys. Rev. **E51**, 2939 (1995).

Traffic Behaviors of Mixed Bicycle System in the Multi-Value Cellular Automata Model

Bin Jia[1], Xin-Gang Li[1,2], Rui Jiang[3], and Zi-You Gao[1,2]

[1] School of Traffic and Transportation, Beijing Jiaotong University, Beijing 100044, P.R.China bjia@bjtu.edu.cn
[2] State Key Laboratory of Railway Traffic Control and Safety, Beijing Jiaotong University, Beijing 100044, P.R.China zygao@center.njtu.edu.cn
[3] School of Engineering Science, University of Science and Technology of China, Hefei 230026, P.R.China rjiang@ustc.edu.cn

Summary. In this paper, the traffic behaviors of mixed bicycle flow are investigated by using the multi-value cellular automata (CA) model. Two types of bicycles with different maximum speed are considered in the system. The system of mixed bicycles is investigated under both periodic and open boundary conditions. As to periodic boundary condition, it is shown that under the deterministic case there appear multiple states both in congested flow and free flow regions. Analytical analysis is carried out and is in good agreement with the simulation results. In the stochastic case, the multiple states effect disappears only when both slow and fast bicycles are randomized. As to open boundary condition, the flux in saturated state will not change with different proportion of slow bicycle if only the slow bicycle has randomization effect, but in other cases, they decrease as the proportion of slow bicycle increasing. Spacetime plots are presented to show the evolution of mixed bicycle flow.

1 Introduction

In the past few decades, traffic problems have attracted much attention [1–3]. Traffic flow is a kind of many-body system of strongly interacting vehicles, and it can exhibit very complex behavior. In order to understand the mechanisms in traffic flow, many theoretical models have been proposed [4–9]. Among them we will focus on Cellular automata (CA) models. CA has become an efficient tool for simulating traffic flow, for it is conceptually simple and can be easily implemented on computers for numerical investigations.

The rule-184 [10] CA is a prototype of all CA models for traffic flow. In 1992, Nagel and Schreckenberg proposed the well-known Nagel-Schreckenberg (NS) model [7]. As an extension of rule-184 CA, higher speeds are allowed in the NS model. The NS model is the minimal model to reproduce the basic properties of real traffic flow. It can simulate some basic phenomena, such as spontaneous jams, stop-and-go wave. However, it cannot explain such empirically observed phenomena as metastable states, capacity drop and synchronized flow. Therefore, several improved versions of the NS model were proposed, such as the anticipation model [8], the slow-to-start model [9], etc.

Recently, Nishinari and Takahashi proposed a family of multi-value CA models [11–13]. Different from previous cases, in these models each site is assumed to hold L vehicles at most. The basic version of the family is obtained from an ultradiscretization of the Burgers equation, so it is called the Burger cellular automata (BCA). Its evolution equation is

$$U_j(t+1) = U_j(t) + \min(U_{j-1}(t), L - U_j(t)) - \min(U_j(t), L - U_{j+1}(t)) \tag{1}$$

where $U_j(t)$ represents the number of vehicles at site j and time t. If it is assumed that the road is an L-lane freeway, then the model can describe the multi-value traffic without explicitly considering the lane-changing rule.

The maximum speed of vehicles in BCA is 1. Nishinari and Takahashi had extended BCA for the case of maximum speed 2 and presented extended BCA (EBCA) models [13]. Matsukidaira and Nishinari have investigated the Euler-Lagrange correspondence of cellular automata models for traffic flow [14, 15] and proposed the generalized BCA (GBCA) model with high speed and long perspective [16]. In our previous work [17], the EBCA models were used to model bicycle flow, and stochastic randomization was introduced into the models.

In real traffic, bicycles do not have the same maximum velocity due to differences in the personalities of riders. For example, young riders always tend to ride at high speed, while old riders usually ride at low speed.

In this work, mixed bicycle flow is investigated using the EBCA model. There are two kinds of bicycles: slow bicycles with maximum speed 1 and fast bicycles with maximum speed 2. Both the periodic and open boundary conditions are considered. As to periodic boundary condition, it is found under the deterministic case that the multiple states effect occurs both in the free flow and congested flow regions. In the stochastic case, the multiple states effect disappears only when both slow bicycles and fast bicycles are randomized. As to the open boundary condition, the flux in saturated state keep the same value with different proportion of slow bicycle if only the slow bicycle has randomization effect, but in other cases, the flux in saturated state decreasing as the proportion of slow bicycle increasing.

This paper is organized as follows: the model for mixed bicycle flow is proposed in section 2. Next the simulation results are presented and analyzed in section 3. The conclusion is given in section 4.

2 The Multi-Value CA Model for Mixed Bicycle Flow

In the EBCA model[4], bicycle movement from t to $t+1$ consists of the following two successive procedures:

(a) bicycles move to their next site if the site is not fully occupied;
(b) only bicycles which moved in procedure (a) can move a further one site if their next site is not fully occupied after procedure (a).

Therefore, the evolution equation of the EBCA is given by:

$$U_j(t+1) = U_j(t) + b_{j-1}(t) - b_j(t) + c_{j-2}(t) - c_{j-1}(t). \qquad (2)$$

Here $b_j(t) = \min(U_j(t), L - U_{j+1}(t))$ represents the number of moving bicycles at site j and time t in procedure (a); $c_j(t) = \min(b_j(t), L - U_{j+2}(t) - b_{j+1}(t) + b_{j+2}(t))$ represents the number of bicycles that can move in procedure (b).

To investigate mixed bicycle flow, two kinds of bicycles, slow bicycles with maximum speed 1 and fast bicycles with maximum speed 2, are considered in the system. The number of slow bicycles and fast bicycles at site j and time t are $U_j^s(t)$ and $U_j^f(t)$ respectively. The updating procedures are changed to:

(1) all bicycles move to their next site if the site is not fully occupied, and the fast bicycles have priority over slow bicycles;
(2) only the fast bicycles moved in procedure (1) can move a further one site if their next site is not fully occupied after procedure (1).

The number of fast and slow bicycles that move one site at site j and time t in procedure (1) are $b_j^f(t)$ and $b_j^s(t)$ respectively. $c_j(t)$ represents the number of fast bicycles that move two sites at site j and time t. The randomization effect of slow bicycles is introduced as: $b_j^s(t+1)$ decreases by 1 with probability p_s if $b_j^s(t+1) > 0$. The randomization effect of fast bicycles is: $c_j(t+1)$ decreases by 1 with probability p_f if $c_j(t+1) > 0$. The updating rules are as follows:

Step 1: calculation of $b_j^f(t+1)$, $b_j^s(t+1)$ and $b_j(t+1)$ $(j = 1, 2, \ldots, K)$:
$b_j^f(t+1) = \min(U_j^f(t), L - U_{j+1}(t))$;
$b_j^s(t+1) = \min(U_j^s(t), L - U_{j+1}(t) - b_j^f(t+1))$,
if $rand() < p_s$, $b_j^s(t+1) = \max(b_j^s(t+1) - 1, 0)$;
$b_j(t+1) = b_j^f(t+1) + b_j^s(t+1)$.
$b_j^f(t+1)$ is calculated first because the fast bicycles have priority over slow bicycles.

Step 2: calculation of $c_j(t+1)$:
$c_j(t+1) = \min(b_j^f(t+1), L - U_{j+2}(t) - b_{j+1}(t+1) + b_{j+2}(t+1))$;
if $rand() < p_f$, then $c_j(t+1) = \max(c_j(t+1) - 1, 0)$.

[4] Here we consider only the EBCA1 model, the extension of the EBCA2 model to mixed bicycle flow will be carried out in a future study.

Step 3: updating $U_j(t+1)$, $U_j^s(t+1)$ and $U_j^f(t+1)$:
$U_j^f(t+1) = U_j^f(t) - b_j^f(t+1) + b_{j-1}^f(t+1) - c_{j-1}(t+1) + c_{j-2}(t+1)$;
$U_j^s(t+1) = U_j^s(t) - b_j^s(t+1) + b_{j-1}^s(t+1)$;
$U_j(t+1) = U_j^f(t+1) + U_j^s(t+1)$.

Here $rand()$ is an uniformly distributed random number between 0 and 1.

In the simulations, $L = 4$ and $K = 100$ are selected, where K is the length of the road. Let ρ represents the density and R is the proportion of slow bicycles.

Under periodic boundary condition, slow bicycles and fast bicycles are randomly distributed on the road in the initial state. The number of fast (slow) bicycles in the system is $N_f = (1-R)\rho LK$ ($N_s = R\rho LK$). We note that the new model with $R = 0.0$ is equivalent to the stochastic EBCA1 model, and that with $R = 1.0$ corresponds to the stochastic BCA model. The average density ρ and flow Q over all sites are defined by

$$\rho = \frac{1}{KL}\sum_{j=1}^{K} U_j(t) \qquad Q(t) = \frac{1}{KL}\sum_{j=1}^{K}(b_{j-1}(t) + c_{j-2}(t)) \tag{3}$$

Under open boundary condition, we check the first cell at each time step. If the first cell is not full filled with bicycles, a bicycle is inserted with the probability (inflow rate) α, and the bicycle is a slow one with probability R. This procedure should be done $(L-n)$ times at each time step. Here n represents the number of bicycles at the first cell. The bicycles with positions larger than K are deleted. The flow Q is defined by the average number of bicycles driving out of the system per time step per lane.

3 Simulation Results

3.1 Periodic Boundary Condition

Deterministic Case

In this subsection, we focus on the deterministic case, i.e., $p_s = 0.0$ and $p_f = 0.0$. The fundamental diagrams are shown in Fig. 1. In the case of $R = 0.0$, the multiple states effect is observed only in the congested flow region (Fig. 1(a)). In other words, there is only one branch with positive slope but there are several branches with negative slope. The detailed analysis of the multiple states has been given in Ref. [13].

When $0 < R < 1$, the multiple states effect occurs both in the congested flow and free flow regions, i.e., there are several branches with positive slope as well as several branches with negative slope. With increasing R, the number of branches with positive slope decreases. Simultaneously, the multiple states effect in the congested flow region is also suppressed. When $R = 1$, the multiple states effect disappears.

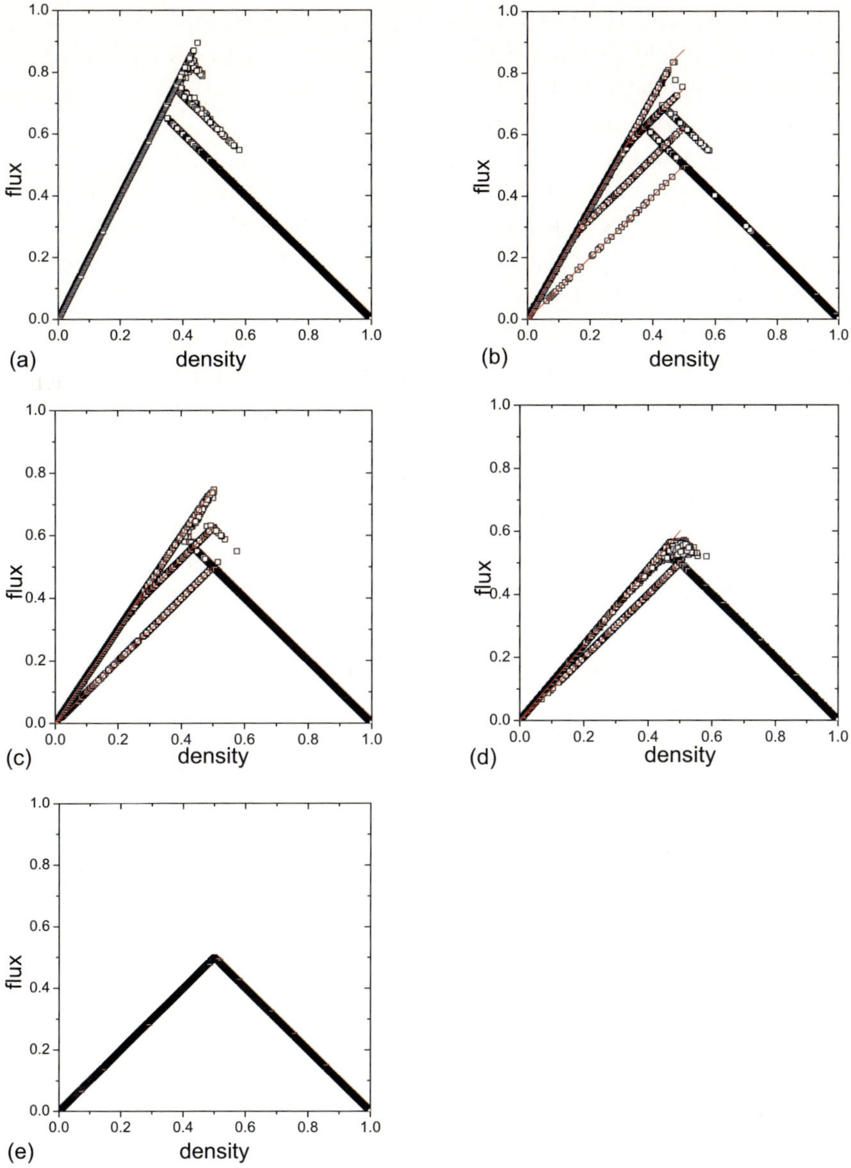

Fig. 1. The fundamental diagrams in the case of (a) $R = 0.0$ (b) $R = 0.2$ (c) $R = 0.5$ (d) $R = 0.8$ (e) $R = 1.0$. The red lines correspond to the analytical values.

Next we focus on the branches with positive slope. Our simulations show that in the free flow region, every slow bicycle can move in each time step. Therefore, the average velocity of the system is given by

$$\bar{v} = \frac{N_s \times 1 + N_f \times \bar{v}_f}{N_s + N_f}.$$

Here N_s and N_f are the number of slow and fast bicycles respectively; $\bar{v}_f$ is the average velocity of the fast bicycles. Since $L = 4$ is used in the simulations, next four subcases are classified:

1. There exist four slow bicycles that occupy one site. In this case, these four slow bicycles form a moving bottleneck by moving side by side. Fast bicycles cannot overtake them and have to follow behind them. As a result, $\bar{v}_f = 1$ and $\bar{v} = 1$. The flow rate is, therefore, $Q = \rho$. This gives the lowest free flow branch in Figs. 1(b)–(e).
2. There exist at most three slow bicycles at each site. In this case, there will be one fast bicycle overtaking the moving bottleneck formed by the three slow bicycles at each time step. Due to the parallel update rule, this allows at most $K/2$ fast bicycles to move with velocity 2. As a result,

$$\bar{v}_f = \frac{(N_f - K/2) \times 1 + K/2 \times 2}{N_f} \text{ for } N_f > K/2$$

and

$$\bar{v}_f = 2 \text{ for } N_f \leq K/2.$$

Therefore, the flow rate is

$$Q = \rho \left[\frac{N_s \times 1 + N_f \times 2}{N_s + N_f} \right] \text{ for } N_f \leq K/2 \tag{4}$$

and

$$Q = \rho \left[\frac{N_s \times 1 + N_f \times \frac{(N_f - K/2)\times 1 + K/2 \times 2}{N_f}}{N_s + N_f} \right] \text{ for } N_f > K/2. \tag{5}$$

Substituting $N_s = \rho KLR$ and $N_f = \rho KL(1 - R)$ into Eqs. (4) and (5), one has

$$Q = \rho(2 - R) \text{ for } \rho \leq \frac{1}{2L(1-R)} \tag{6}$$

and

$$Q = \rho + \frac{1}{2L} \text{ for } \rho > \frac{1}{2L(1-R)}. \tag{7}$$

Eq. (7) gives the second lowest free flow branch in Figs. 1(b) and (c), and Eq. (6) gives the high free flow branch in Fig. 1(d).

3. There exist at most two slow bicycles at each site. In this case, there will be two fast bicycle overtaking the two slow bicycles each time step. This allows at most $K/2\times 2 = K$ fast bicycles to move with velocity 2. Similarly to case 2, we have

$$Q = \rho(2 - R) \text{ for } \rho \leq \frac{1}{L(1-R)} \tag{8}$$

and

$$Q = \rho + \frac{1}{L} \text{ for } \rho > \frac{1}{L(1-R)}. \tag{9}$$

Eq. (9) gives the second highest free flow branch in Fig. 1(b), and Eq. (8) gives the high free flow branch in Fig. 1(c).

4. There exists at most one slow bicycle in each site. Similarly to cases 2 and 3, we have

$$Q = \rho(2 - R) \text{ for } \rho \leq \frac{3}{2L(1-R)} \tag{10}$$

and

$$Q = \rho + \frac{3}{2L} \text{ for } \rho > \frac{3}{2L(1-R)}. \tag{11}$$

Eq. (10) gives the highest free flow branch in Fig. 1(b). The states determined by Eq. (11) are unstable and will transit into congested states.

From Fig. 1, we can see that the simulation results are consistent with the analytical results, except that some high flux states are not reached in the simulation. In real bicycle flow, a group of cyclists may ride together because that they are class mates, friends or colleagues. They will occupy a large part of the cycle lane, thus the bicycles that follow are blocked. We argue that the multiple states effect in the free flow region surely occurs in real bicycle flow.

Stochastic Case

In this subsection, we investigate the effect of randomization in the mixed bicycle flow. To this end, two typical values of R are chosen: $R = 0.2$ and $R = 0.5$.

Firstly we study the case that only slow bicycle has randomization effect, i.e., $p_f = 0$ and $p_s > 0$. Fig. 2 shows that the multiple states effect does not occur in the free flow region but still exists in the congested region. This is because with the consideration of the randomization effect of slow bicycles, a stationary moving bottleneck will not exist. A large moving bottleneck may split into several small ones and several small moving bottlenecks may merge into one large one. This breaks the mechanism of multiple branches shown in the previous subsection.

The fundamental diagram only slightly depends on the value of p_s when the ratio of slow bicycles is small (Fig. 2(a)). However, when the ratio of slow

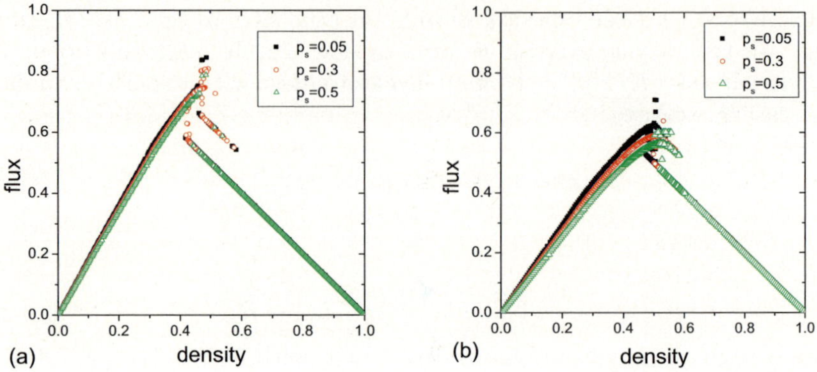

Fig. 2. The fundamental diagrams with different values of p_s when $p_f = 0.0$. (a) $R = 0.2$ and (b) $R = 0.5$.

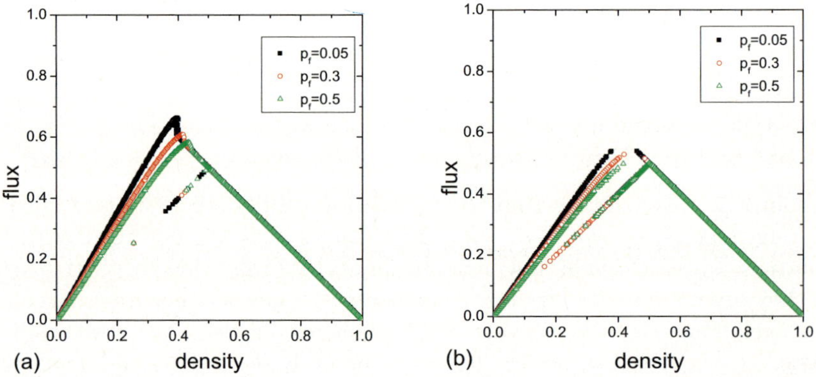

Fig. 3. The fundamental diagrams with different values of p_f when $p_s = 0.0$. (a) $R = 0.2$ and (b) $R = 0.5$.

bicycles increases, the maximum flow rate notably decreases with the increase of p_s (Fig. 2(b)).

Next we consider the case $p_f > 0$ and $p_s = 0$. In Fig. 3, one can see that the multiple states effect does not occur in the congested flow region. However, there always exist two branches in the free flow region. The branch $Q = \rho$ (the low branch) exists because once four slow bicycles appear in one site, this moving bottleneck will not dissolve since no randomization is exerted on slow bicycles. The high branch exists in the situations that four slow bicycles never appear in one site. When the ratio of slow vehicles is small, the chance that four slow bicycles appear in one site is small. As a result, the data points on the low branch are sparse (Fig. 3(a)).

The value of p_f only affects the high branch. As p_f increases, the flux in the free flow region decreases. We also note that there is a gap between the high branch and the congested branch in the case of $R = 0.5$. This is because in the intermediate range of density, the appearance of four slow bicycles in one site will always occur sooner or later. When four slow bicycles accumulate on one site, a moving bottleneck forms and it will not dissolve, so it hinders the movement of fast bicycles. Consequently, a flow rate drop appears.

In contrast, the probability that four slow bicycles appear in one site is small in the intermediate range of density in the case of $R = 0.2$. Therefore, there is not a gap between the high branch and the congested branch.

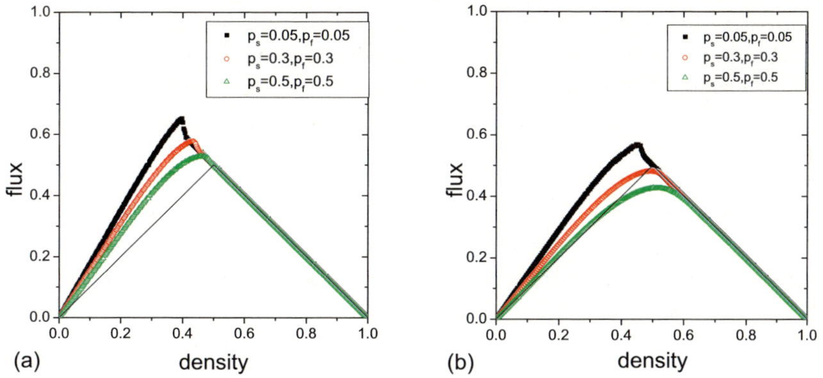

Fig. 4. The fundamental diagrams with different values of p_s and p_f. (a) $R = 0.2$ and (b) $R = 0.5$. The solid line corresponds to the case of $R = 1.0$ with $p_s = 0.0$.

Figs. 4(a) and (b) show the results with randomization parameter $p_s > 0$ and $p_f > 0$. One can see that the multiple states effect disappears even if p_s and p_f are slightly larger than zero. As p_s and p_f increase, the maximum flow rate decreases and the density corresponding to the maximum flow rate increases.

We note that the properties of mixed bicycle flow are different from that of mixed vehicle flow. In mixed vehicle flow, the flux is constrained by the plug formed by slow vehicles [18, 19]. The fundamental diagram will be essentially the same as in the case of $R = 1.0$ even if only a small number of slow vehicles are introduced into the system. However, in mixed bicycle flow, introducing a small proportion of slow bicycles will not affect the flow rate so much. This can be seen from Fig. 4, in which the flow rate in the case of $R = 0.2$ is significantly larger than in the case of $R = 1.0$. We believe this difference is mainly because the bicycle lanes are not so clearly separated from each other as vehicle lanes.

3.2 Open Boundary Condition

Under open boundary condition, the bicycle enters into the road from the left edge with inflow rate α, and drives off the road from the right edge with probability 1. The flux as a function of α with different R are shown in Fig. 5. Fig. 5(a) corresponds to the deterministic case, in which we choose $p_s = 0.0$ and $p_f = 0.0$. We know that the fast bicycle has velocity expectation effect in procedure (2), because the fast bicycles moved in procedure (1) can move to a further one site if their next site is not fully occupied after procedure (1). When $R = 0.0$, all the fast bicycles can move with velocity 2, and there is a linear relationship between Q and α. When $R > 0$, the liner relationship remains at small values of α, because the number of slow bicycles is so small that bottlenecks can not be formed and the fast bicycles can move freely. As α increasing, the curve bends down, because some bottlenecks formed and the flux is saturated. We note that the maximum flux decreasing as R growing.

Figs. 5(b), (c) and (d) show the results with stochastic cases. In Fig. 5(b), we choose $p_s = 0.3$ and $p_f = 0.0$. When $R = 0.0$, it is equivalent to deterministic case. When $R > 0$, the maximum flux become smaller in comparison with deterministic case, because the slow bicycles have randomization effect. In Fig. 5(c), the parameters are $p_s = 0.0$ and $p_f = 0.3$. When $R = 0.0$, the bicycle also becomes saturated with large α for randomization effect. When $R > 0$, the bottlenecks with 4 slow bicycles always happen in saturated flow, but they could not dissolve because the slow bicycles have no randomization effect. Thus the flux are the same with different R in saturated flow. Fig. 5(d) corresponds the case $p_s = 0.3$ and $p_f = 0.3$. We note that as R increasing, the flux in saturated flow decreasing.

4 Conclusion

In this paper, mixed bicycle flow is investigated using the multi-value CA model. Both the periodic and open boundary condition are studied.

As to periodic boundary condition, the fundamental diagrams and the spacetime plots are analyzed in detail.

In the deterministic case, the multiple states effect exists both in the free flow and congested regions. The site which contains the largest number of slow bicycles forms a moving bottleneck to the system and the flux is constrained by this bottleneck. The simulation results are consistent with the analytical ones.

In the stochastic case, the multiple states effect disappears in the free flow region when $p_s > 0.0$, and it disappears in the congested region when $p_f > 0.0$. The multiple states effect disappears when both slow bicycles and fast bicycles are randomized.

As to the open boundary condition, the flux in saturated state keep the same value with different R if only the slow bicycle has randomization effect,

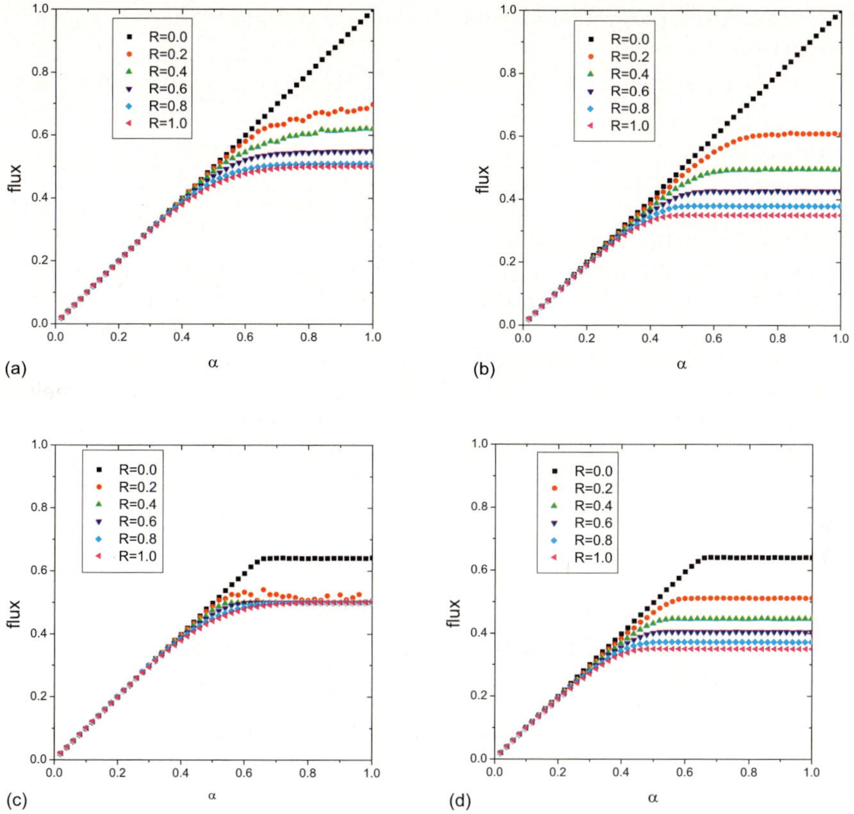

Fig. 5. The flux as a function of α with different R. (a) $p_s = 0.0$, $p_f = 0.0$; (b) $p_s = 0.3$, $p_f = 0.0$; (c) $p_s = 0.0$, $p_f = 0.3$; (d) $p_s = 0.3$, $p_f = 0.3$.

but in other cases, the flux in saturated state decreasing as the proportion of slow bicycle increasing.

In our future work, we will extend the EBCA2 model to study mixed bicycle flow and compare the results of the two models. The calibration and verification of the models using real bicycle flow data will also be performed.

Acknowledgements

This work is partially supported by the National Basic Research Program of China under No. 2006CB705500, the National Natural Science Foundation of China under Key Project No. 70631001, 10532060 and Project Nos. 70501004, 70471088 and 10404025, Special Program for the Preliminary Research of the Momentous Fundamental Research under No. 2005CCA03900, and the

Science and Technology Foundation of Beijing Jiaotong University under No. 2004SM026.

References

1. Chowdhury D, Santen L, Schadschneider A (2000) Phys. Rep. 329:199–329.
2. Helbing D (2001) Rev. Mod. Phys. 73:1067–1141.
3. Nagatani T (2002) Rep. Prog. Phys. 65:1331–1386.
4. Bando M, Nakayama A, Shibata A, Sugiyama Y (1995) Phys. Rev. E 51:1035–1042.
5. Lee H Y, Lee H W, Kim D (1998) Phys. Rev. Lett. 81:1130–1133.
6. Kerner B S, Klenov S L, Wolf D E (2002) J. Phys. A: Math. Gen. 35:9971–10013.
7. Nagel K, Schreckenberg M (1992) J. Physique I 2:2221–2229.
8. Li X B, Wu Q S, Jiang R (2001) Phys. Rev. E 64:066128.
9. Benjamin S C, Johnson N F, Hui P M (1996) J. Phys. A: Math. Gen. 29:3119–3127
   Barlovic R, Santen L, Schadschneider A (1998) Eur. Phys. J. B 5:793–800.
10. Wolfram S (1986) Theory and Applications of Cellular Automata. World Scientific, Singapore.
11. Nishinari K, Takahashi D (1998) J. Phys. A: Math. Gen. 31:5439–5450.
12. Nishinari K, Takahashi D (1999) J. Phys. A: Math. Gen. 32:93–104.
13. Nishinari K, Takahashi D (2000) J. Phys. A: Math. Gen. 33:7709–7720.
14. Nishinari K (2001) J. Phys. A: Math. Gen. 34:10727–10736.
15. Matsukidaira J, Nishinari K (2003) Phys. Rev. Lett. 90:088701.
16. Matsukidaira J, Nishinari K (2004) Int. J. Mod. Phys. C 15:507–515.
17. Jiang R, Jia B, Wu Q S (2004) J. Phys. A: Math. Gen. 37:2063–2072.
18. Chowdhury D, Wolf D E, Schreckenberg M (1997) Physica A 235:417–439.
19. Knospe W, Santen L, Schadschneider A, Schreckenberg M (1999) Physica A 265:614–633.

Empirical Differences Between Time Mean Speed and Space Mean Speed

Victor Knoop, Serge P. Hoogendoorn, and Henk van Zuylen

Delft University of Technology, Faculty of Civil Engineering and Geosciences, Stevinweg 1, 2628 CN Delft, The Netherlands v.l.knoop@tudelft.nl, s.p.hoogendoorn@tudelft.nl, h.j.vanzuylen@tudelft.nl

Summary. Insight into traffic flow characteristics is often gained using local measurements. To determine macroscopic flow characteristics, time aggregation of microscopic information is required.

Usually, a data collection system stores values averaged over time. However, it is well known that a time mean average overestimates the influence of faster vehicles, and consequently overestimates the mean speed. As a direct result, densities, computed from flow and speed, are underestimated.

This paper compares the time mean speed and space mean speed, using data of individual car passages on a motorway road stretch. We show that the differences between time mean and space mean averages are substantial, up to a factor four. In particular in the lower speed regions the error is big. We indicate the considerable consequences for the jam density and shock wave speed. Finally, a fundamental diagram based on correctly averaged microscopic data can be fitted much better.

1 Introduction

The most common data available are aggregated (dual) loop detector counts. The loop detects the passage of a vehicle and its speed. The data are usually presented in aggregated values for a time period, for instance, 1 or 15 minutes.

It has been known that there is a difference between the space mean speeds and the time mean speeds [1, 2]. In studies, usually mean speeds are used, as opposite from space mean speeds. Helbing discusses the use and meaning of both mean speeds [3]. It is well known that traffic density is better computed from a space mean speed (e.g., [4]).

Rakha and Zhang [5] discuss the difference and possible conversions between the two mean speeds. Under the assumption of stationary road way conditions, the space mean speed equals the harmonic average of the time mean speed [6], which always is higher than the arrhythmic mean. This raises the question how large the differences are between these two averages in reality. Van Lint [7] discusses the topic and finds:

$$v_T = \frac{\sigma_M^2}{v_M} + v_M \tag{1}$$

In this equation, v_T is the time mean speed, v_M the space mean speed and σ_M^2 the variance of the space mean speed. Note that it does not give a rule to compute space mean speeds out of time-mean measurements: σ_M^2 is unknown if one has local measurements.

The next section explains how space mean speeds can be approximated using speeds of individual vehicles. We apply this technique on a dataset, thus getting space mean speeds and time mean speeds. Then, we analyze the differences of applying both averaging techniques. We found that the averaging method has a big impact on the mean speeds, up to a factor 4. Using this in calculations to find the density, the difference becomes more important; we also show the consequences for the speed of the propagation of a traffic jam, and the shock wave speed.

2 From Dual Loop Counts to Density

There are many possibilities to measure road traffic. One of the most common methods uses induction loops that are situated in the road surface. Generally, the output of the detectors state the flow and speed aggregated per time period. The flow q and the vehicle density ρ are calculated from the number of passages n and the time interval t_{agg} using the equations:

$$q = \frac{n}{t_{agg}} \tag{2}$$

$$\rho = \frac{q}{v} \tag{3}$$

Eq. 3 will yield the correct, space averaged, density if the space mean speed is used. However, averaging the speed of vehicles $i = 1..n$ at one location will give the local mean speed v_L:

$$v_L = \langle v \rangle_L = \frac{1}{n} \sum_{i=1}^{n} v_i \tag{4}$$

Assuming stationary conditions, we can compute the space mean speed from local measurements, using a harmonic mean of the measurements [4]:

$$v_M = \langle v \rangle_M = \frac{1}{\frac{1}{n} \sum\limits_{i=1}^{n} \frac{1}{v_i}} = \frac{1}{\langle \frac{1}{v} \rangle_L} \tag{5}$$

In this paragraph, we will indicate why this formula will give the space mean speed under stationary assumptions. A detector lies at location x_{det}. Now, let us reconstruct which vehicles will pass in the time of one aggregation period.

For this, the vehicle must be closer to the detector than the distance it travels in the aggregation time t_{agg}:

$$x_{det} - x_i \leq t_{agg} \cdot v_i \tag{6}$$

In this formula, x is the vehicle position on the road. For faster vehicles, this distance is larger. Therefore, Eq. 4 overestimates the influence of the faster vehicles. This holds for every property of traffic, including speed. The computed average speed v_L, being biased to the property of the faster cars, is therefore higher than the space mean speed v_M. To compensate, one could attach a weight of $1/v_i$ to each measurement. This is stated below. After rewriting, it states the same as Eq. 5.

$$v_L = \langle v \rangle_L = \frac{\sum_{i=1}^{n} \frac{1}{v_i} v_i}{\sum_{i=1}^{n} \frac{1}{v_i}} = \frac{\sum_{i=1}^{n} 1}{\sum_{i=1}^{n} \frac{1}{v_i}} = \frac{n}{\sum_{i=1}^{n} \frac{1}{v_i}} = \frac{1}{\frac{1}{n} \sum_{i=1}^{n} \frac{1}{v_i}} \tag{7}$$

3 Data Collection and Analysis

This paper shows the differences in magnitude between the local mean speed and the space mean speed. It also gives the error for the density, which is derived from the speed. Finally, it discusses the consequences of taking the wrong mean for jam density and shock wave speed.

For this research, a dataset of individual vehicle passages was used. The loop detectors are placed at the ring road of Amsterdam, a three lane motorway with a 100 km/h speed limit. Time and speed of each individual vehicle were recorded in the period 16 June 2005 to 11 July 2005. From these data we compute both the local mean speed, v_L, using Eq. 4 and the space mean speed v_M using Eq. 5. The speeds are aggregated over the lanes.

We used different aggregation times (10 and 20 seconds and 1, 2, 5 and 15 minutes) to see the influence of the aggregation time. The necessary assumption that the speed profile does not change over time is more likely to hold over a shorter time. In order not to be influenced by changing traffic states, the aggregation periods both before and after a change were removed. We used 2 possible states (congested, uncongested), which were distinguished by the mean speed (under and over 70 km/h). When within 5 aggregation intervals the traffic state would change again, all intermediate aggregation periods were discarded.

3.1 Computed Densities for the Roadway

Fig. 1 reads the quotient of the time mean speed compared and the space mean speed for different speeds. It thus shows for different speeds how large the difference is between the two averages. We see for lower speeds, the two averages are more distinct. Using Eq. 1, this can only be explained by a low variation of speeds in the higher speed regions. For the same reason, the speed variation in congestion must be higher. Higher aggregation times make the differences grow, so the speed variation grows.

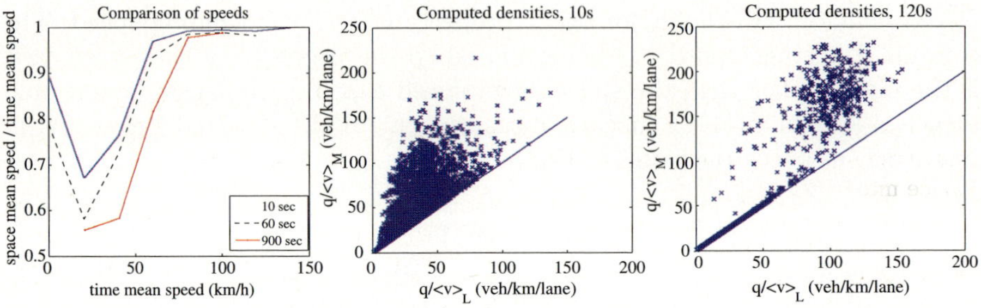

Fig. 1. (left) average speed difference (middle) densities, 10 s aggregation time (right) densities, measurements.

We use flow q and sequentially speeds, v_L and v_M, in Eq. 3 to compute two different densities: $\rho_L = q/v_L$ and $\rho_M = q/v_M$. Fig. 1 shows both densities (ρ_M and ρ_L). The deviations from the line $x = y$ show the differences between the averaging methods.

The flows (Eq. 2) and densities given here are summed over the three lanes. When the aggregation time equals 10 seconds, the density q/v_S can be much (up to fourfold) higher than the estimation q/v_M. This is much higher than the results stated by Rakha and Zhang [5]. They already stated that the results differ per location. Since their measurements were performed in the USA, lower speed differences between trucks and cars are expected, and a lower variation causes a lower difference between time mean speed and space mean speed. Besides, in Europe overtaking is allowed only at the left, so vehicles in the left lane are faster than the right lane. Aggregation over the lanes causes than larger variations. Finally, the number of measurements (4 weeks in our case) results in more points, and therefore more extreme points.

In Fig. 2 we plot the flow versus the density (as illustration the one-minute data), both calculated using the time mean speed and calculated using the space mean speed. The figures show the cloud of measured points. The red line is fit to that cloud (the red line connecting them), according to the shape Wu proposed [8]. Each of the measured points is assigned to one of the branches [9]. It sometimes is unclear to which branch a point belongs, especially for the lower time intervals. This confusion causes a bad fit for the time mean speed. The green line, the shock wave speed, will be discussed in section 3.2.

For each time period, we determine the flow that would be related to the computed density, according to the fit parameters. The errors between these

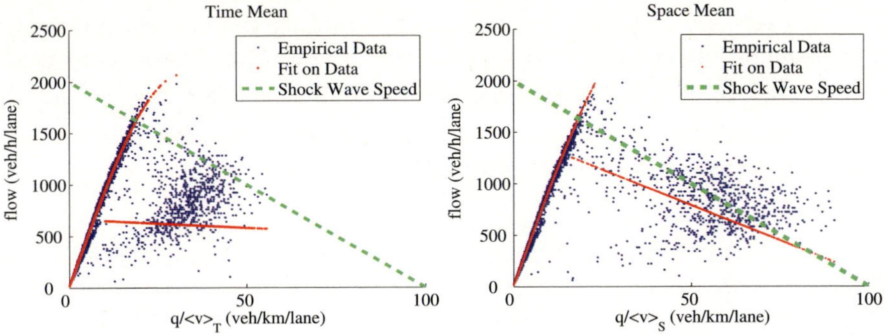

Fig. 2. (top) The fundamental diagrams constructed using time mean speed and space mean speed.

flows and the measured flows are squared and averaged. The errors, stated in Tab. 1, show that using the space mean speed improves the fit.

Table 1. Results of fits of the fundamental diagrams for different aggregation times

Aggregation time (s)	Error on fit time mean $((km/h)^2)$	Error on fit space mean $((km/h)^2)$	Jam Density time mean $(veh/km/lane)$	Jam Density space mean $(veh/km/lane)$
10	112	85	118	166
20	90	66	145	525
60	70	48	158	107
120	65	41	113	249
300	56	39	99	113
900	40	37	87	131

3.2 Consequences for Macroscopic Traffic Models

Two different average speeds lead to two different densities, but the flows are equal. Consequently, the speed of the propagation of traffic jams, the shock wave speed (ω), is calculated differently, since Stokes law states:

$$\omega = \frac{\Delta q}{\Delta \rho} \tag{8}$$

The shock wave speed can also be derived directly from the measurements. The speed of the shockwave is approximately 20 km/h. According to Eq. 8, this speed should equal the slope of the congested branch of the fundamental diagram. The green line in the left plots of Fig. 2 illustrates this speed; note that the line can be shifted and still indicate the same shockwave speed. It fits much better with the fundamental diagram using the space mean speed.

Furthermore, the buffer capacity of a road stretch is estimated completely different, see Table 1. Due to fit a wide cloud of points, outliers can occur (e.g., 525 veh/km, 249 veh/km). When disregarding the outliers, we see an underestimate of the jam density in almost all cases if a time-mean average is made. Using the space mean speed also gives more consistent results.

3.3 Conclusions and Future Research

In this contribution, we discussed two ways to average the speeds of passing vehicles: a local time mean average and a harmonic average, approximating the space-mean averages. Theory says that the space mean speed is to be used in the fundamental relation. Using empirical data, we showed from that the two different mean speeds might differ up to a factor 4.

We show that the space mean speed gives a better fit for a fundamental diagram. The estimates for both jam density and shock wave speed differ considerably between the two mean speeds. A better knowledge of these values can improve traffic models.

This paper assumes stationary traffic conditions. Using reliable trajectory data, one could compare the two mean speeds presented in this paper to a directly observed space mean speed. Aggregation times and aggregation lengths then are an interesting aspect.

References

1. W. Leutzbach (1969) Trans Res 3: 421-428.
2. A. D. May (1990) Traffic Flow Fundamentals. Prentice Hall, New Jersey, Englewood Cliffs.
3. D. Helbing (1997) Verkehrsdynamik. Springer, Heidelberg.
4. S. P. Hoogendoorn, G. Hegeman and T. Dijker (2006) Traffic Flow Theory and Simulation. TU Delft.
5. H. Rakha and W. Zhang (2005) Estimating Traffic Stream Space-Mean Speed and Reliability From Dual and Single Loop Detectors. TRB.
6. C. F. Daganzo (1997) Fundamentals of Transportation and Traffic Operations. Pergamon.
7. H. van Lint (2004) Reliable Travel Time Prediction for Freeways. PhD Thesis, TU Delft.
8. N. Wu (2002) Trans Res A 36: 867-884.
9. H. Heikoop, S. P. Hoogendoorn and G. J. Martens (2007) Onderzoek Verkeersafwikkeling en capaciteitswaarden discontinuteiten Autosnelwegen. AVV, in Dutch.

Anisotropy and Stabilization of Traffic Flow in Extended Cooperative Driving Lattice Hydrodynamic Models Based on Backward-Looking Effect

Li Xingli[1], Li Zhipeng[2], Han Xianglin[1,3], and Dai Shiqiang[1]

[1] Shanghai Institute of Applied Mathematics and Mechanics, Shanghai University, Shanghai 200072, China lxl326@163.com; sqdai@shu.edu.cn
[2] College of Electronics and Information Engineering, Tongji University, Shanghai 201804, China newspring88@126.com
[3] Department of Mathematics, Huzhou Teacher College, Huzhou 750021, China xlhan653@163.com

Summary. Two extended cooperative driving lattice hydrodynamic models are proposed by considering the backward looking effect in traffic flow. The stability conditions for the two models are investigated with the linear stability theory and it is found that new consideration leads to the improvement of the stability of traffic flow. The modified Korteweg-de Vries equations (the mKdV equation, for short) near the critical point are derived by using the nonlinear perturbation method to show that the traffic jam could be described by the kink-antikink soliton solutions for the mKdV equations. Moreover, the anisotropy of traffic flow is further discussed through examining the negative propagation velocity as the effect of following vehicle is involved.

1 Introduction

Over the past decades, traffic problems, especially traffic congestions, have attracted considerable attention. Various traffic models have been suggested, such as the car-following models, cellular automaton models and hydrodynamic models, etc. [1]. Here we focus on the lattice hydrodynamic model, which was proposed by Nagatani [2, 3] in 1998. Its basic idea is that the traffic current can be optimized by the product of the optimal velocity and average density. Using this model, Nagatani obtained a series of important results. After that, many improvements have been made, and accordingly, a lot of relative more reasonable models have been put forward [4–9].

The above models are related to the forward looking effect, but only a few models studied the backward looking effect which has been studied only by employing the car-following models [10–12]. However, all the work only validated the fact that considering the effect of following vehicles could strength the stability of traffic flow based on linear stability analysis and did not perform the nonlinear analysis near the critical points, yet. Especially the anisotropy of traffic flow [13–15] caused by from the effect of following car has not been involved. In this paper, we try to present two cooperative driving lattice hydrodynamic models with the consideration of arbitrary number of preceding vehicles and one following vehicle on a single-lane highway. Furthermore, through the linear and nonlinear stability analysis, we will discuss the stabilization and anisotropy of traffic flow.

2 Brief Description of the Models

Two extended cooperative driving lattice hydrodynamic models are proposed by taking into account an arbitrary number of vehicles ahead and one following vehicle on a single-lane highway. The basic equations are as follows:

$$\partial_t \rho_j + \rho_0(\rho_j v_j - \rho_{j-1} v_{j-1}) = 0 \tag{1}$$

$$\rho_j(t+\tau) - \rho_j(t) + \tau\rho_0(\rho_j v_j - \rho_{j-1} v_{j-1}) = 0 \tag{2}$$

$$\rho_j(t+\tau)v_j(t+\tau) = \rho_0 V_B(\rho_{j-1})(1-p) + \rho_0 V_F(\rho_{j+1}, \rho_{j+1}, \ldots, \rho_{j+n})p \tag{3}$$

We can obtain two models, called model A (Eqs. (1) and (3)) and model B (Eqs. (2) and (3)), respectively. The density equations for them are

$$\begin{aligned} \partial_t \rho_j(t+\tau) + \rho_0^2 \Bigg(& [V_B(\rho_{j-1}(t)) - V_B(\rho_{j-2}(t))](1-p) \\ & + \left[V_F\left(\sum_{l=1}^{n} \beta_l \rho_{j+l}\right) - V_F\left(\sum_{l=1}^{n} \beta_l \rho_{j+l-1}\right)\right]\Bigg) p = 0 \end{aligned} \tag{4}$$

$$\begin{aligned} \rho_j(t+2\tau) - \rho_j(t+\tau) + \tau\rho_0^2 \Bigg(& [V_B(\rho_{j-1}(t)) - V_B(\rho_{j-2}(t))](1-p) \\ & + \left[V_F\left(\sum_{l=1}^{n} \beta_l \rho_{j+l}\right) - V_F\left(\sum_{l=1}^{n} \beta_l \rho_{j+l-1}\right)\right]\Bigg) p = 0 \end{aligned} \tag{5}$$

where p stands for the acting fraction of the two optimal velocity functions and the other notations follow the convention. The two optimal velocity functions are chosen as

$$V_B(\rho_{j-1}(t)) = -\tanh\left(\frac{2}{\rho_0} - \frac{\rho_{j-1}(t)}{\rho_0^2} - \frac{1}{\rho_c}\right) + \tanh\left(\frac{1}{\rho_c}\right) \tag{6}$$

$$V_F(\rho_{j+1}(t), \rho_{j+1}(t), \ldots, \rho_{j+n}(t)) = \tanh\left(\frac{2}{\rho_0} - \frac{\sum_{l=1}^{n} \beta_l \rho_{j+l}(t)}{\rho_0^2} - \frac{1}{\rho_c}\right) + \tanh\left(\frac{1}{\rho_c}\right) \quad (7)$$

where β_l is the weighting function which is same to those in ref. [9].

3 Linear Stability Analysis

We apply the linear stability analysis method to the models A and B described by Eqs. (4) and (5). The neutral stability conditions are given, respectively as

$$\tau = -\frac{pV_F' \sum_{l=1}^{n} \beta_l (l - 1/2) + 3/2(1-p)V_B'}{\rho_0 (V')^2} \quad (8)$$

$$\tau = -\frac{pV_F' \sum_{l=1}^{n} \beta_l (2l - 1) - 3(1-p)V_B'}{3\rho_0 (V')^2} \quad (9)$$

where $V_B'(\rho_0) = \frac{dV_B(\rho_{j-1})}{d\rho} \mid_{\rho_{j-1}=\rho_0}$, $V_F'(\rho_0) = \frac{dV_F(\rho_j)}{d\rho} \mid_{\rho_j=\rho_0}$.

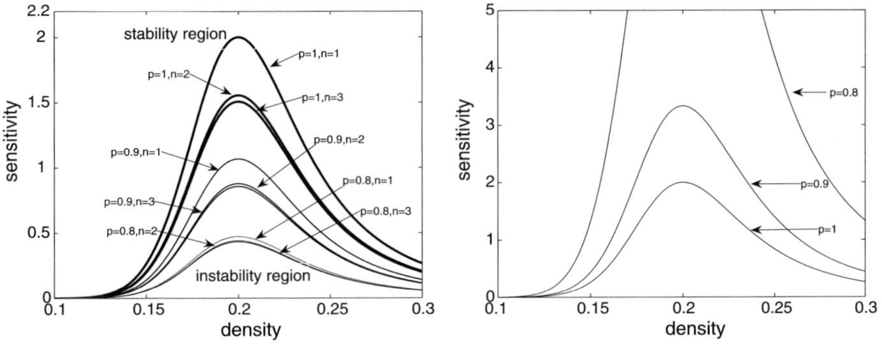

Fig. 1. The neutral stability lines for model A at two cases: the optimal velocity functions (6) and (7) are chosen (left) and the optimal velocity functions are not distinguished (right).

The neutral stability lines of model A in the parameter (ρ, a) space are shown in Fig. 1. There exist critical points (ρ_c, a_c) for neutral stability lines, such that uniform states with any density are always linearly stable for $a > a_c$. From Fig. 1, we can see clearly that considering the backward looking effect will help to strengthen the stability to a certain extent. Besides, we find that the stability decrease greatly when V_B is selected as $V_B(\rho_{j-1}(t)) = \tanh(\frac{2}{\rho_0} - \frac{\rho_{j-1}(t)}{\rho_0^2} - \frac{1}{\rho_c}) + \tanh(\frac{1}{\rho_c})$, which validates that an unreasonable conclusion that considering the following vehicle can weaken the stability of traffic flow in ref. [8] was resulted from have not distinguished V_B from V_F.

4 Nonlinear Analysis

Using the reductive perturbation method to Eqs. (4) and (5) to analyze the system behavior near the critical point (ρ_c, a_c), we find that the nature of kink-antikink solitons can be described by the following mKdV equations:

$$\partial_{T'} R' = \partial_X^3 R' - \partial_X R'^3 - \varepsilon M_1[R'] \tag{10}$$

$$\partial_{T'} R' = \partial_X^3 R' - \partial_X R'^3 - \varepsilon M_2[R'] \tag{11}$$

After some deduction, we obtain the propagation velocity c for the two models.

$$\begin{aligned} c_1 = \frac{5}{3}(2p-1)^3\tau_c \Bigg\{ \frac{2}{3}(2p-1)\Bigg[-\frac{5(2p-1)^4\tau_c^3}{6} - \frac{7(2p-1)(1-p)\tau_c}{3} \\ + \frac{(2p-1)p\tau_c}{3}\sum_{l=1}^{n}\beta_l(3l^2-3l+1) - \frac{15(1-p)}{24} - \frac{p}{24}\sum_{l=1}^{n}\beta_l(4l^3-6l^2+4l-1)\Bigg] \\ +3\left[\frac{1-p}{2} - \frac{2(2p-1)^2\tau_c}{3} + \frac{p}{6}\sum_{l=1}^{n}\beta_l(2l-1)\right] \\ \left[\frac{(2p-1)^3\tau_c^2}{2} + \frac{7(1-p)}{6} - \frac{p}{6}\sum_{l=1}^{n}\beta_l(3l^2-3l+1)\right]\Bigg\}^{-1} \end{aligned} \tag{12}$$

$$\begin{aligned} c_2 = \frac{5}{2}(2p-1)^3\tau_c \Bigg\{ \frac{2}{3}(2p-1)\Bigg[-\frac{23(2p-1)^4\tau_c^3}{8} - \frac{(2p-1)(1-p)\tau_c}{2} \\ + \frac{(2p-1)p\tau_c}{2}\sum_{l=1}^{n}\beta_l(3l^2-3l+1) - \frac{15(1-p)}{24} - \frac{p}{24}\sum_{l=1}^{n}\beta_l(4l^3-6l^2+4l-1)\Bigg] \\ +3\left[\frac{1-p}{2} - (2p-1)^2\tau_c + \frac{p}{6}\sum_{l=1}^{n}\beta_l(2l-1)\right] \\ \left[\frac{7(2p-1)^3\tau_c^2}{6} + \frac{7(1-p)}{6} - \frac{p}{6}\sum_{l=1}^{n}\beta_l(3l^2-3l+1)\right]\Bigg\}^{-1} \end{aligned} \tag{13}$$

5 Results Analysis and Discussion

If we adopt the explicit forms (5) and (6) of the optimal velocity ($-\rho_c^2 V' = 2p - 1, \rho_c^6 V''' = 2(2p-1)$), we can obtain the critical density (ρ_c, a_c) and propagation velocities c_1, c_2 of the kink-antikink solution for models A and B (see table 1). In the following, we discuss the results from four aspects.

Table 1. a_c and c in models A and B for $p = 0.8, 0.9, 1$, respectively

	n	1	2	3	4
A	a_c	0.514286	0.442105	0.433415	0.432202
	c_1	-0.793283	-0.804063	-0.813698	-0.81684
B	a_c	0.771429	0.663158	0.650123	0.648302
	c_2	-0.96888	-0.988985	-1.00223	-1.00679

	n	1	2	3	4
A	a_c	1.06667	0.874381	0.856831	0.853831
	c_1	-2.74286	-2.95832	-3.13782	-3.17641
B	a_c	1.6	1.31765	1.28525	1.28075
	c_2	-3.63636	-3.94645	-4.25826	-4.32755

	n	1	2	3	4
A	a_c	2	1.55556	1.50769	1.50109
	c_1	24	10.6586	9.0225	8.84629
B	a_c	3	2.33333	2.26154	2.25164
	c_2	27	11.6053	9.7383	9.53980

(1) The effect of looking backward on traffic. In both models, a_c decreases with the increase of n, and the stability regions are enlarged when the looking backward effect is considered. It is also found that the information from two preceding vehicles and one following vehicle plays the most important role in the stability. We may consider this state as the optimal one.

(2) Comparison of propagation velocity with that in ref. [10]. From Table 1, it can be seen that the propagation velocities decrease as the increase of n in both models for the case $p = 1$, which is different from the conclusion in Ref. [10], where the propagation velocities are constant for model A or increase with the increasing of n for model B. It is well known that in real traffic, when considering the information of more vehicles is employed, the stability of traffic system will increase, and as a result, the propagation velocities of density wave will decrease. The reason why the conclusion in ref. [10] is contrary to the real traffic is that there existed an error in deriving the mKdV equation.

(3) The negative propagation velocity when the looking backward effect is involved. The occurrence of the negative propagation velocity when the effect of following vehicle is involved constitutes the anisotropy of the models. According to the definition of V_B as the following vehicle is very close to the considered vehicle, the considered one has to accelerate to avoid collision, so in the unstable region, as the following vehicle has a slight density (or headway) fluctuation, if the considered vehicle cannot accelerate on time, collision easily appears. Meanwhile, because both the preceding and following vehicles are considered, the absolute value of propagation velocity is smaller than that in the case of only considering forward vehicles.

(4) The variation of propagation velocity under different inclusions of backward looking effect. It is found that the absolute value of propagation velocity at $p = 0.8$ is smaller that at $p = 0.9$ under other same conditions, which means that more inclusion of backward effect leads to slower propagation of fluctuation, in other words, once the driver pays more attention to the following vehicle, the accelerating probability of considered vehicle increases so as to decrease collision probability. In addition, in the same model, under the constant probability, the absolute value of propagation velocity increases with the increase of n, but the propagation velocity hardly changes as n rises to a certain value. This can be explained by the fact that the following vehicle could influence the considered vehicle, which, corresponding to the nearest-neighbor vehicle, can be as the following vehicle, so when the considered vehicle cannot accelerate on time to avoid collision, the influence will extend to the nearest-neighbor vehicle,and the rest may be deduced by analogy, up to the next-nearest-neighbor vehicle, etc., but the influence will weaken with increasing n. This result can also account for the collision accidents, especially the considered car is pasted on by the following car and several cars are pasted on along a highway.

6 Conclusions

We have proposed two lattice hydrodynamic models involving the effect of looking forward on traffic flow through defining the different optimal velocity functions for forward and backward looking cases. The traffic nature has been analytically investigated by using the linear stability theory and through the nonlinear wave analysis. Furthermore, the negative propagation velocity caused by the effect of backward looking on traffic is analyzed in detail.

Acknowledgements

This work was supported by the National Basic Research Program of China (Grant No. 2006CB705500), the National Natural Science Foundation of China (Grant No. 10532060), the Special Research Fund for the Doctoral Program in Higher Education of China (Grant No. SRFDP 20040280014) and the Shanghai Leading Academic Discipline Project (Grant No. Y0103)

References

1. Chowdhury D., Santen L. and Schadschneider A. (2000) Phys. Rept. 329: 199-329.
2. Nagatani T. (1998) Physica A 261: 599-607.

3. Nagatani T. (1999) Physica A 264: 581-592.
4. Nagatani T. (1999) Physica A 265: 297-310.
5. Nagatani T. (1999) Phys. Rev. E 59: 4857-4864.
6. Nagatani T. (1999) Physica A 271: 200-221.
7. Nagatani T. (1999) Physica A 272: 592-611.
8. Xue Y. (2004) Acta Physica Sinica 53: 25-30.
9. Ge H.X., Dai S.Q., Xue Y. and Dong L.Y. (2005) Phys. Rev. E 71: 066119.
10. Nakayama A., Sugiyama Y. and Hasebe K. (2001) Phys. Rev. E 65: 016112.
11. Hasebe K., Nakayama A. and Sugiyama Y. (2003) Phys. Rev. E 68: 026102.
12. Ge H.X., Zhu H.B. and Dai S.Q. (2006) Eur. Phys. J. B 54: 503-507.
13. Danganzo C.F. (1995) Transp. Res. B 29: 277-286.
14. Zhang H.M. (2002) Transp. Res. B 36: 275-290.
15. Xue Yu and Dai Shiqiang (2003) Phys. Rev. E 68: 066123.

Complex Fundamental Diagram of Traffic Flow in the Deep Lefortovo Tunnel (Moscow)

Ihor Lubashevsky[1], Cyril Garnisov[2], and Boris Lifshits[2]

[1] A.M. Prokhorov General Physics Institute of Russian Academy of Sciences, Vavilov str., 38, Moscow, 119311 Russia ialub@fpl.gpi.ru
[2] Moscow Technical University of Radioengineering, Electronics, and Automation, Vernadsky pros., 78, Moscow, 117454 Russia

Summary. The fundamental diagram for tunnel traffic is constructed based on the empirical data collected during the last two years in the deep long branch of the Lefortovo tunnel located on the 3rd circular highway of Moscow. This tunnel of length 3 km is equipped with a dense system of stationary radiodetetors distributed uniformly along it chequerwise at spacing of 60 m. The data were averaged over 30 s. Each detector measures three characteristics of the vehicle ensemble; the flow rate, the car velocity, and the occupancy for three lanes individually. The conducted analysis reveals an original complex structure of the fundamental diagram.

1 Traffic Flow in Long Tunnels

The properties of traffic flow in long highway tunnels has been under individual consideration since the middle of the last century (see, e.g., Refs [1, 2]). Interest to this problem is caused by several reasons. The first and, may be, main one is safety. Jams in long tunnels are rather dangerous and detecting the critical states of vehicle flow leading to the jam formation is of the prime importance for the tunnel operation. Second, the tunnel traffic in its own right is an attractive object for studying the basic properties of vehicle ensembles on highways. On one hand, it is due to the individual car motion being more controllable inside tunnels with respect to velocity limits and lane changing. On the other hand, long tunnels typically are equipped with a dense system of detectors, which provides a unique opportunity to receive a detailed information about the spacial-temporal structures of traffic flow.

The present work continues the investigation of tunnel traffic properties reported previously [3]. The analysis is based on empirical data collected during the last time in the Lefortovo tunnel located on the 3rd circular highway of Moscow (Fig. 1). It comprises two branches and the upper one is a deep linear three lane tunnel of length about 3 km. Exactly in this branch the analyzed data were collected. The tunnel is equipped with a dense system of stationary radiodetectors (Remote Traffic Microwave Sensor, X model) distributed

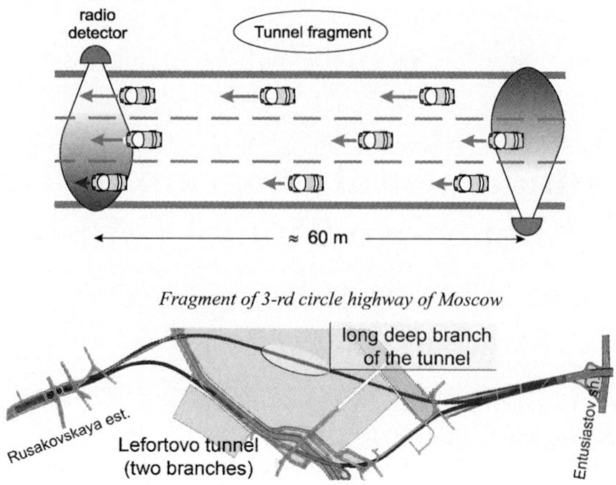

Fig. 1. Structure of the Lefortovo tunnel and the system of car motion detectors.

uniformly along it chequerwise at spacing of 60 m. Because of the technical features of the detectors traffic flow on the left and right lanes is measured at spacing of 120 m whereas on the middle lane the spacial resolution is 60 m. The data were averaged over 30 s.

Each detector measures three characteristics of vehicle ensemble; the flow rate q, the car velocity v, and the occupancy k for three lanes individually. The occupancy is analog to the vehicle density and is defined as the total relative time during witch vehicles were visible in the view region of a given detector within the averaging interval. It is measured in percent. The detectors themselves and their records were analyzed initially to justify the reliability of the collected data.

2 Fundamental Diagram

The fundamental diagram under consideration was constructed as follows. The phase space $\{k, v, q\}$ was divided into cells of size about $1\ \% \times 1$ km/h $\times$ 0.01 car/s. Each 30 seconds a detector contributes unity to one of the cells. Taking into account a certain rather long time interval of traffic flow observation, all the detectors, and then dividing the result by the total number of records we obtain the three-dimensional distribution $P(k, v, q)$ of fixed traffic flow states over this phase space. In order to elucidate the obtained result we present the projection of $P(k, v, q)$ on three phase planes $\{kq\}$, $\{kv\}$, and $\{vq\}$. Besides, in projecting onto the given phase planes some layers can be singled out, for example, the expression

$$P_{DV}(k, q) \propto \int_{v \in DV} dv\, P(k, v, q)$$

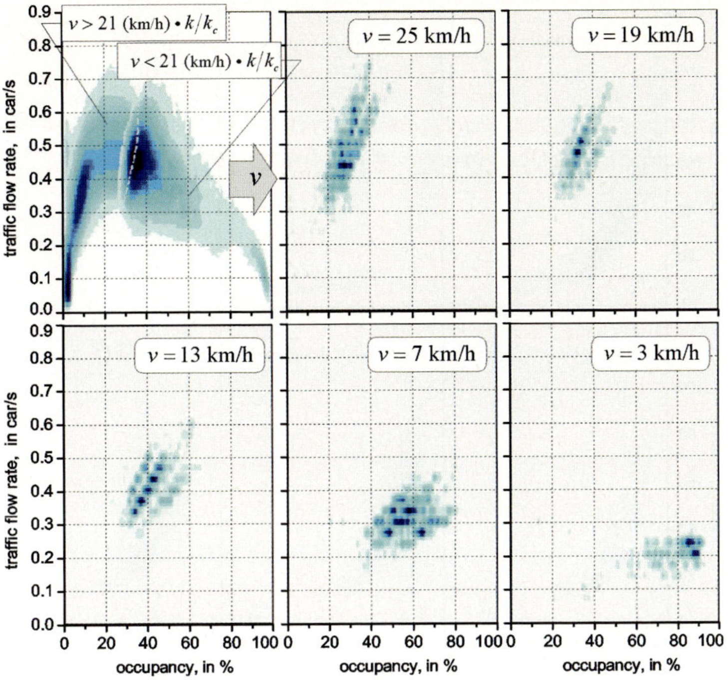

Fig. 2. Projection of the fundamental diagram onto the plane $\{k, q\}$ as well as its slices parallel to this plane.

specifies the projection of the layer $DV = (v_{\min}, v_{\max})$ onto the plane $\{kq\}$ within a constant cofactor normalizing it to unity. Such distributions will be also referred to as slices of the fundamental diagram.

Figure 2 presents the projection of the whole fundamental diagram onto the plane $\{k, q\}$ (the upper left frame) as well as its slices parallel to this plane. In the frame of the whole projection two branches are singled out by the relation $v \lesseqgtr 21$ km/h $\times\, k/k_{c2}$, where the critical occupancy $k_{c2} = 31\%$ according the results to be demonstrated further. The two branches with a small degree of overlap are separated actually by the transition from light to heavy synchronized traffic (see below). The given slices of fixed velocity demonstrate the fact that, at least, three different states of heavy congested traffic were observed. It reflects in the existence of three branches visible well for $v = 19$, 13, 7 km/h. Their additional analysis demonstrated us that these branches are characterized individually by different mean lengths of vehicles. In particular, the higher is a branch in Fig. 2, the shorter, on the average, vehicles forming it. The distribution of the traffic flow states becomes rather uniform for very low velocities matching the jam formation. On the whole fundamental diagram the jammed traffic is described by the region looking like a certain "beak".

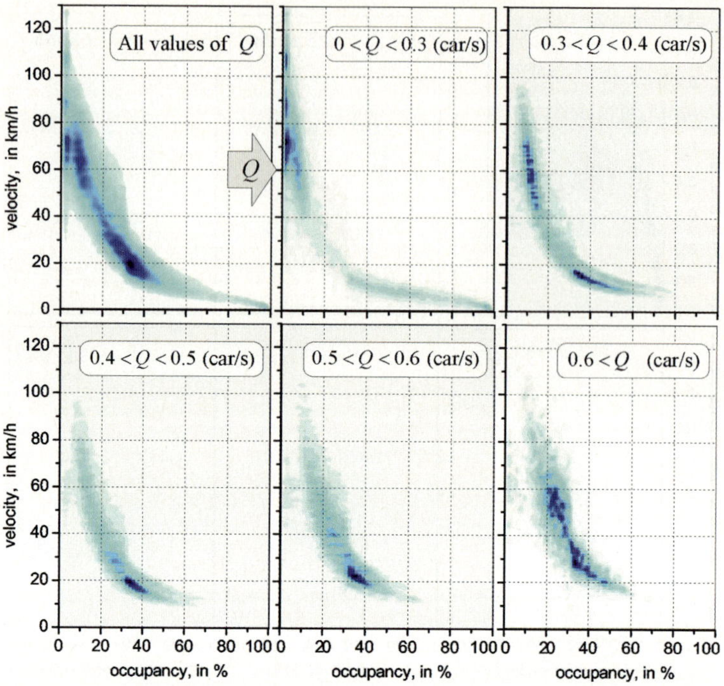

Fig. 3. Projection of the fundamental diagram onto the plane $\{k, v\}$ as well as its slices parallel to this plane that are made up by projecting the noted layers.

Figure 3 depicts a similar projection of the fundamental diagram onto the plane $\{k, v\}$. For low values of the traffic flow rate two states of traffic flow are clearly visible, the free flow and jam. The slice of $0.3 < Q < 0.4$ (car/s) shows actually the light and heavy phases of synchronized traffic flow, with the latter phase splitting into several branches. The final slice corresponding to large values of the traffic flow rate exhibits the phase transition between the two light and heavy states of traffic flow at the critical value of occupancy $k_{c2} = 31\%$, where the velocity drop about 15 km/h is clearly visible. It should be pointed out that the traffic flow states are distributed with the comparable intensity on both the sides of the phase transition at $k = k_{c2}$, which enables us to assume that this phase transition proceeds in the both directions. The whole projection of the fundamental diagram on the plane $\{k, v\}$ also shows this phase transition as well as the existence of two accumulation points of traffic flow states in the region of light synchronized traffic and in the vicinity of phase transition between the two states of synchronized traffic. The latter feature poses a question about the possibility of phenomena like "stop-and-go waves" but based on transitions between different states of the synchronized traffic.

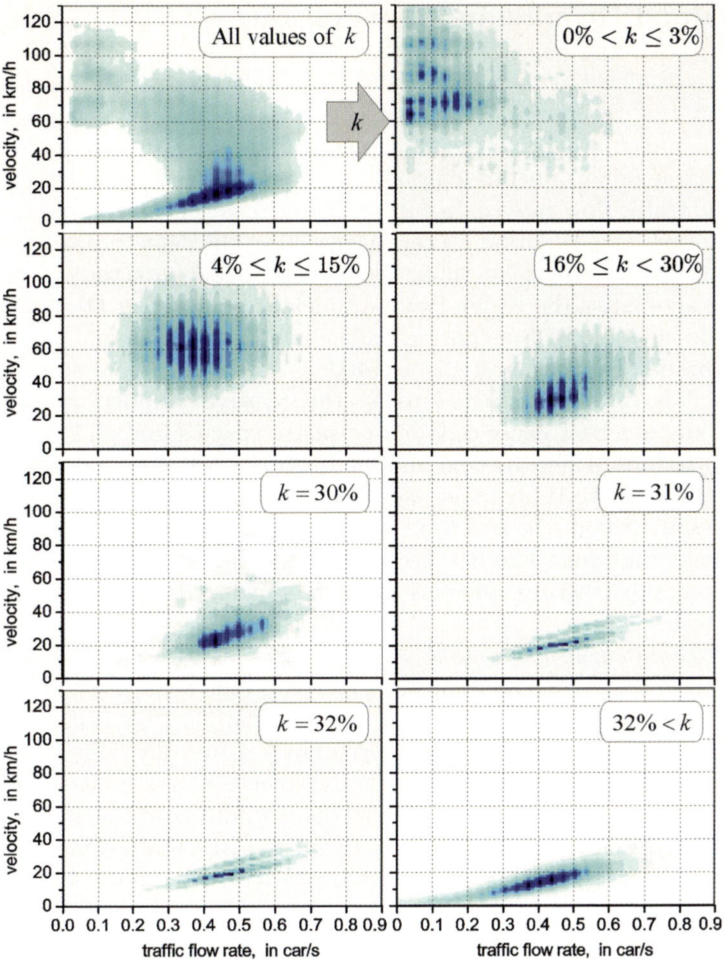

Fig. 4. Projection of the fundamental diagram onto the $\{q, k\}$-plane as well as its several slices parallel to this plane.

Figure 4 exhibits the projection of the fundamental diagram onto the plane $\{q, v\}$ and evolution of its slices for fixed values of the occupancy. In this figure the four different phase states of the analyzed tunnel traffic are visible. The free flow where the overtaking manoeuvres are most feasible corresponds to the three branches that can be related to trucks, passenger cars, and high-speed cars. As the traffic flow rate grows with the occupancy k the three branches terminate and are followed by a structureless two-dimensional domain via a certain phase transition. Then this phase state in turn is followed by a structural domain which itself converts again into the structureless beaked region corresponding to jam.

3 Obtained Results and Conclusion

The paper is devoted to constructing the fundamental diagram for tunnel traffic based on the empirical data collected in the linear higher branch of the Lefortovo tunnel located on the 3rd circular highway of Moscow in 2004–2005. It is the three lane tunnel of length 3 km equipped with radiodetectors measuring the traffic flow rate (q, in car/s), the vehicle velocity (v, in km/h), and the road occupancy (k, in %) averaged over 30 s. The detectors are distributed chequerwise at spacing 60 m along the tunnel. Because of the detector technical characteristics the traffic flow parameters are fixed at 60 m spacing on the middle lane and 120 m spacing on the left and right ones.

The fundamental diagram is treated as the traffic flow state distribution and has been constructed using the relative number of records per $1\,\% \times 1$ km/h $\times$ 0.01 car/s cells in the phase space $\{k, v, q\}$. Analyzing the three projections of this 3D field and its different slices we have demonstrated the fundamental diagram to be complex in structure. Four possible traffic flow states are found, the free flow, light synchronized traffic, heavy synchronized traffic, and jam. The free flow state as well as the heavy synchronized traffic has a substructure, whereas the light synchronized traffic and jam are structureless.

The free flow comprises three branches related to trucks, passenger cars, and high-speed cars. These branches exist while the occupancy is less then a certain critical value, $k < k_{c1} \approx 3\,\%$ and are clearly visible in the projection onto the phase plane $\{q, v\}$. As the occupancy grows the light synchronized traffic appears which is characterized by the structureless region of widely scatted states. When the occupancy exceeds the next critical value $k_{c2} \approx 31\,\%$ the heavy synchronized traffic changes the previous phase state. This transition is accompanied by some jump in the mean velocity. In the projections onto the phase planes $\{k, q\}$, $\{k, v\}$, and $\{q, v\}$ it looks like widely scatted states uniformly distributed inside a certain region. However the corresponding slices of the fundamental diagram demonstrate a substructure of the given phase state. It again comprises, at least, three different branches. The conducted analysis demonstrated that the given branches are characterized, on the averaged, by different lengths of vehicles. The jam phase, as should be expected, can be ascribed with a certain relationship between the traffic flow rate q, the mean velocity v, and the occupancy k, in particular, it is possible to write down a certain function $v = v(k)$.

In addition we should note the following. In spite of the complex structure of the fundamental diagram and the existence of four different phase states the distribution of the detected states is, roughly speaking, bimodal. One its maximum is located at the beginning of the region matching the light synchronized traffic. The other maximum drops on the region corresponding to the transition between the two phases of the synchronized traffic.

Acknowledgements

This work was supported in part by INTAS project 04-78-7185, DFG project MA 1508/8-1, and RFBR grants 06-01-04005, 05-01-00723, and 05-07-90248 as well as RNP 2.1.1.6893.

References

1. H.C. Chin and A.D. May: Examination of the Speed-Flow Relationship at the Caldecott Tunnel. In: *Transportation Research Record* **1320**, Transportation Reserch Board (NRC, Washington, DC, 1991), pp. 1–15.
2. R.W. Rothery: Car following models. In: *Traffic Flow Theory* Transportation Research Board, Special Report **165**, ed. by N. Gartner, C.J. Messer, and A.K. Rathi (1992), Chap. 4.
3. I. Lubashevsky, C. Garnisov, R. Mahnke, B. Lifshits, M. Pechersky: States of Traffic Flow in the Deep Lefortovo Tunnel (Moscow): Empirical Data. In: *Traffic and Granular Flow'05*, A. Schadschneider, T. Pöschel, R. Kühne, M. Schreckenberg, and D.E. Wolf (eds.) (Springer-Verlag, Berlin, 2007) p. 717–723.

Three-Frame Algorithm of Car Path Reconstruction from Airborne Traffic Data

Ihor Lubashevsky[1], Namik Gusein-zade[1], Dmitry Klochkov[1], and Sergey Zuev[2]

[1] A.M. Prokhorov General Physics Institute, Russian Academy of Sciences, Vavilov str. 38, Moscow, 119991, Russia ialub@fpl.gpi.ru

[2] German Aerospace Center, Optimal Information Systems, Rutherfordstrasse 2, D-12489 Berlin, Germany

Summary. The airborne traffic monitoring system forms a novel technology of detecting vehicle motion. An optical digital camera located on an airborne platform produces a series of images which then are processed to recognized the fixed vehicles. In this way the video data are converted into the time sequence of frames containing the vehicle coordinates. In the present work a three-frame algorithm is developed to identify the succeeding vehicle positions. It is based on finding the neighboring points in the frame sequence characterized by minimal acceleration. To verify and optimize the developed algorithm a "Virtual Road" simulator was created. Finally available empirical data are analyzed using the created algorithm.

1 Introduction

The traditional techniques of traffic flow measurements are based on local detectors mounted at fixed places of a road network. Such detectors provide adequate information about traffic flow only for road fragments of a rather simple structure. However in cities there are a large number of complex road intersections, where vehicles moving on different branches interact strongly with one another. To measure appropriately traffic flow features on such "hot areas" the information supplied by the stationary detectors should be complemented at least with the detailed information about the spatial structure of traffic streams on these hot areas. Currently the novel technology of airborne monitoring and detecting the main characteristics of traffic flow is under development. The German Aerospace Center (DLR) with its department of Transportation Studies and department of Optical Information Systems has developed an airborne traffic monitoring system which was successfully used during the Soccer World Championship 2007 in Germany [1–3]. In particular, these and some other DLR airborne datasets are used in the present paper. This system is based on an optical sensor placed on an airborne platform.

The optical sensor produces a series of images of area under. The empirical airborne traffic data obtained by recognizing these images form a set of the frames with individual vehicle coordinates. To use these data and to calculate the main parameters of traffic flow trajectories of individual vehicle motion have to be reconstructed. In this way it becomes possible to measure the traffic flow rate, the mean velocity of vehicles, and their density simultaneously.

2 Vehicle Trajectory Reconstruction

The airborne traffic data under consideration are the time series $\mathcal{F}$ of frames $\{F_t\}$ with the coordinates of recognized vehicles, $\mathcal{F} = \{F_t = \{x_\alpha, y_\alpha\}\}$. The problem is to construct a collection of trajectories $\mathcal{P} = \{x(t), y(t)\}_t$ passing through the points $\{\{x_\alpha, y_\alpha\}\}$ at the corresponding time moments $\{t = n\tau\}$, where n is integer and τ is the time span between successive frames. Some constraints should be imposed on the vehicle trajectories $\mathcal{P}$ to make them smooth. There are several ways to do this, in particular, to bound the "acceleration" (the used approach)

$$\sqrt{\left(\frac{d^2x_\alpha}{dt^2}\right)^2 + \left(\frac{d^2y_\alpha}{dt^2}\right)^2} \leq a_{\max} \tag{1a}$$

or to bound the "velocity"

$$\sqrt{\left(\frac{dx_\alpha}{dt}\right)^2 + \left(\frac{dy_\alpha}{dt}\right)^2} \leq v_{\max}\,. \tag{1b}$$

We note that the acceleration and velocity thresholds, $a_{\max}$ and $v_{\max}$, are not the real characteristics of car motion but internal algorithm parameters whose choice is determined by its efficiency being highest.

To explain the crux of the algorithm implementing these constrains let us assume that the preceding frames $\{\ldots, F_{n-2},\ F_{n-1}\}$ have been analyzed and the vehicle trajectories are reconstructed at the previous time moments $\{\ldots, n-2,\ n-1\}$. Then a car $\{x^n_\alpha, y^n_\alpha\}$ in the frame F_n will be incorporated into the trajectory $\omega = \cup_m\{x^m_\omega, y^m_\omega\}$ if

$$\sqrt{(x^{n-2}_\omega - 2x^{n-1}_\omega + x^n_\alpha)^2 + (y^{n-2}_\omega - 2y^{n-1}_\omega + y^n_\alpha)^2} \leq a_{\max}\,\tau^2\,. \tag{2}$$

Figure 1 illustrates the given three-frame algorithm based on (1a) as well as the relative two-frame algorithm based on (1b) mentioned for comparison only.

Let us discuss the possible errors of the car identification algorithm caused by errors in the car coordinate measurements. There are two main types of the latter ones; the errors of individual vehicle positions, ξ_1, and the errors in the frame reference to GPS, ξ_2. According to the pilot airborne monitoring

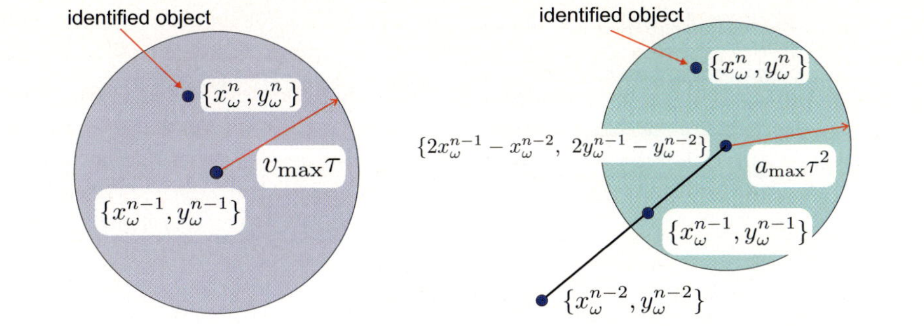

Fig. 1. Illustration of the car identification algorithms. Within the two-frame speed limitation algorithm (left fragment) new objects $\{x^n, y^n\}$ are sought in the circle of radius $r = v_{\max}\tau$ centered at the point $\{x_\omega^{n-1}, y_\omega^{n-1}\}$. In the three-frame algorithm bounding acceleration (right fragment) new objects $\{x^n, y^n\}$ are sought in the circle of radius $r = a_{\max}\tau^2$ centered at the point $\{2x_\omega^{n-1} - x_\omega^{n-2},\ 2y_\omega^{n-1} - y_\omega^{n-2}\}$.

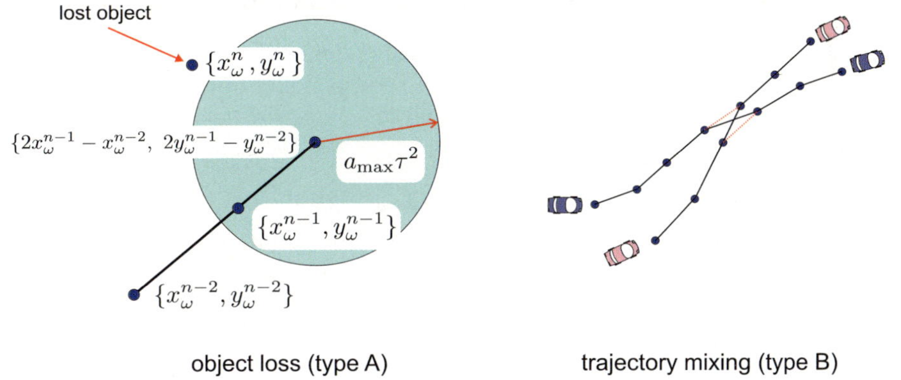

Fig. 2. The main errors of the car identification algorithm.

system (Institute for Traffic Research, DLR, Berlin) [1, 2] the error values can be estimated as $\xi_1 \sim 0.5$ m and $\xi_2 \sim 1$–2 m. The caused errors of the car identification algorithm, i.e. the object loss (type A) and the trajectory mixing (type B) illustrated in Fig. 2 will be quantified in units of

$$A(B) = \frac{\text{The number of lost(mixed) objects}}{\text{Total number of objects}} \cdot 100\%$$

To analyze the feasibilities of the identification algorithm, first, the following kinematic simulator has been used (Fig. 3, upper fragment). The motion of cars is specified by the shown equations where the parameters x_{0i}, y_{0i}, b_{xi} etc. are random and some addition constraints are imposed to prevent the collisions. The noise components $\xi_{x(y)i} = (\xi_1 + \xi_2)_{x(y)i}$ imitate the errors in the measurements of the car coordinates x_i and y_i at the time moments $t = n\tau$. The system parameters and the number of cars are specified so that

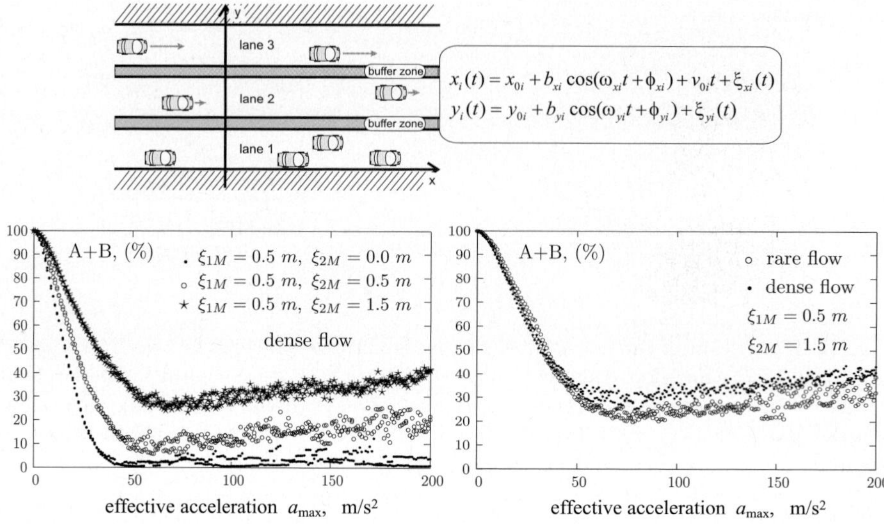

Fig. 3. Airborne data simulator, "Virtual Road", used to verify the vehicle identification algorithm (upper fragment) and the corresponding results of the car identification routine (lower fragment). The subscript M at the noise components ξ_{iM} stands for their amplitudes.

the virtual car ensemble imitate the real characteristics of traffic flow, at least semiquantitatively. The lower fragment in Fig. 3 exhibits the net value of the identification errors, $A + B$, vs. the acceleration threshold $a_{\max}$. As could be expected, the present results demonstrate the fact that the car identification algorithm attains its maximal efficiency when

$$a_{\max} = \frac{4(\xi_{1M} + \xi_{2M})}{\tau^2}, \tag{3}$$

and, in particular, for $\xi_{1M} = 0.5$ m and $\xi_2 = 1.5$ m the optimal acceleration threshold is about $a_{\max} \sim 50\ \text{m/s}^2$.

Second, the developed algorithm has been verified using the available airborne traffic data collected during Soccer Championship 2006 in Stuttgart within the DLR project "Soccer" [3]. These data undergone manual processing, so initially we had a collection of vehicle trajectories regarded as real keeping in mind the human abilities. The vehicle identification algorithm has given the result depicted in Fig. 4. For these data, again, the optimal value of the acceleration threshold meets relation (3). Figure 4 presents also the efficiency of the car identification algorithm for different values of the velocity threshold $v_{\max}$. As seen the imposition of the additional speed limitation constraint reduces the algorithm efficiency. In other words, constraints (1) interfere with each other and, thus, should be used separately. As should be expected the measurement errors and the optimal acceleration threshold are again related by expression (3).

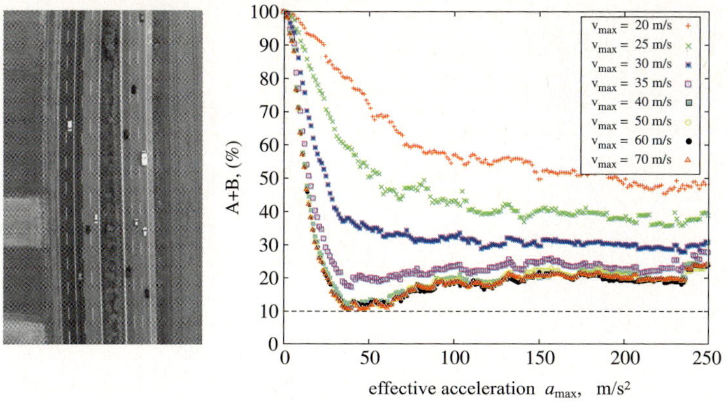

Fig. 4. The airborne data collected during the Soccer Championship 2006 in Stuttgart. Typical example of the analyzed video data (left fragment) and the results of car identification routine vs. the acceleration threshold (right fragment).

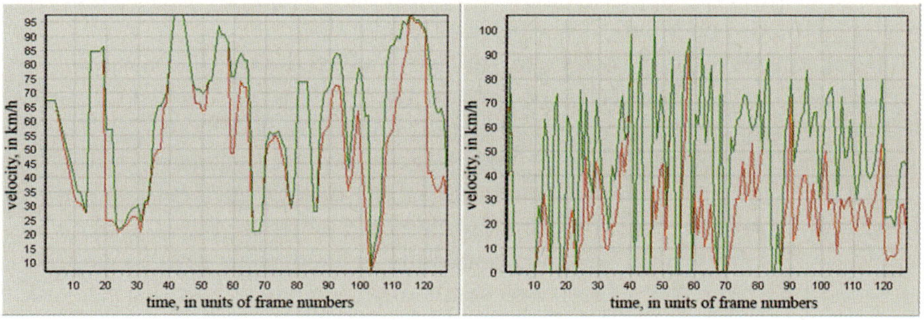

Fig. 5. Time-mean speed (green) and space-mean speed (red) reconstructed using the smoothed vehicle trajectories (left fragment) and the same data obtained using the neighboring points directly. The airborne data of pilot flights in Berlin, 2004, the LUMOS project.

3 Mean Velocity of Traffic Flow and Fundamental Diagram

Because of the considerable errors in the vehicle coordinates the found successive positions of a car cannot be used directly to calculate its velocity. However, the reconstructed trajectories enable one to overcome this problem by fitting a rather smooth trajectory to the found data. In this way we have analyzed the empirical data collected during pilot flights of an airplane in Berlin, 2004 (DLR, LUMOS project). The mean velocity of traffic flow was calculated, first, by smoothing the reconstructed vehicle trajectories via the Savitzky-Golay filter and, then, averaging the velocities over the observed car ensemble. The results are illustrated in Fig. 5. As seen the time pattern of

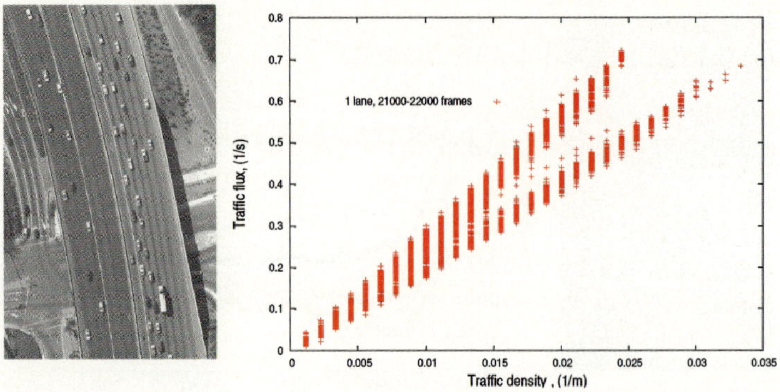

Fig. 6. Fundamental diagram (right fragment) and an example of the analyzed video data (left fragment). The NGSIM "First Prototype" dataset [4].

traffic flow speed obtained by smoothing the reconstructed vehicle trajectories gives much more adequate description of traffic flow.

Using the developed technique we have analyzed also the NGSIM "First Prototype" dataset consisting of vehicle trajectories on a half-mile section of Interstate 80 in Emeryville, California, for one half of hour [4]. The system portrait on the phase plane "traffic flow rate – car density" was drawn using the mean car velocity calculated via the developed technique and car density obtained by just counting the detected objects (Fig. 6). No widely scatted states are visible, instead, branching is fixed. This result again poses the question about the necessity to single out the vehicle flow discreteness in analyzing the states of traffic flow [5]. It should be noted that the developed approach depresses this discreteness effect.

4 Conclusion

We have developed the three-frame-algorithm of reconstructing vehicle trajectories from airborne traffic data, which evaluates the proximity of objects in the neighboring frames in terms of acceleration. The acceleration threshold $a_{\max}$ is determined by the car measurement accuracy. Using the "Virtual Road" simulator and the dataset collected within the project "Soccer" (DLR, Stuttgart, 2006) the value of $a_{\max} \sim 40$–$50\ \mathrm{m/s^2}$ is found and the possibility of decreasing the identification error down to 5–30 % is shown.

Using one of the NGSIM datasets it has been demonstrated that for the observed congested traffic the fundamental diagram splits into two branches when the car density exceeds some value rather than exhibits the widely scatted states. This result again poses a question whether the discreteness of traffic flow could be responsible for the appearance of widely scatted states on the fundamental diagrams. Naturally, the unambiguous answer requires special and detailed investigation.

Acknowledgements

This work was supported in part by INTAS project 04-78-7185, DFG project MA 1508/8-1, and RFBR grants 06-01-04005, 05-01-00723, and 05-07-90248.

References

1. I. Ernst, S. Sujew, K.-U. Thiessenhusen, M. Hetscher, S. Raßmann and M. Ruhé: *LUMOS - Airborne Traffic Monitoring System.* In: Proceedings of the IEEE 6th International Conference On Intelligent Transportation Systems; Shanghai (China), 2003 (CD-ROM).
2. H. Hetzheim, A. Börner: *Vehicle detection from airborne images by separating of texture properties and their fusion.* In: Image and Vision Computing, New Zealand, 2003, Proceedings pp. 48–53 (Palmerston North, 2003).
3. Martin Ruhé, Reinhart Kühne, Ines Ernst, Sergey Zuev, and Eileen Hipp: *Air borne systems and datafusion for traffic surveillance and forecast for the soccer world cup.* In: Proceedings of 86-th Annual Meeting Transportation Research Board, January, 2007, Washington, D.C.
4. Next Generation Simulation Community, `http://ngsim.camsys.com`.
5. I. Lubashevsky, R. Mahnke, P. Wagner, and S. Kalenkov: Phys. Rev. E **66**, 016117 (2002).

Vehicular Motion and Traffic Breakdown: Evaluation of Energy Balance

Christof Liebe[1], Reinhard Mahnke[1], Jevgenijs Kaupužs[2], and Hans Weber[3]

[1] Rostock University, Institute of Physics, D–18051 Rostock, Germany
christof.liebe@uni-rostock.de; reinhard.mahnke@uni-rostock.de
[2] Institute of Mathematics and Computer Science, University of Latvia, LV–1459 Riga, Latvia kaupuzs@latnet.lv
[3] Luleå University of Technology, Department of Physics, SE–97187 Luleå, Sweden
Hans.Weber@ltu.se

Summary. Microscopic traffic models based on follow–the–leader behaviour are strongly asymmetrically interacting many–particle systems. The well–known Bando's optimal velocity model includes the fact that (firstly) the driver is always looking forward interacting with the lead vehicle and (secondly) the car travels on the road always with friction. Due to these realistic assumptions the moving car needs petrol for the engine to compensate dissipation by rolling friction. We investigate the flux of mechanical energy to evaluate the energy balance out of the given nonlinear dynamical system of vehicular particles. In order to understand the traffic breakdown as transition from free flow to congested traffic we estimate the total energy per car at low and high densities and observe the energy of jam formation.

1 Introduction

The formation and growth of clusters is a widely known phenomenon in physics, e.g. condensation of liquid droplets in a supersaturated vapour [1]. The formation of car clusters (jams) at overcritical densities in traffic flow is an analogous phenomenon in the sense that cars can be considered as many asymmetrically interacting particles [2, 3], and the clustering process can be described by similar equations. In particular, the probability that the system has a given cluster distribution at a certain moment in time can be described by the stochastic Langevin equation in both cases. The spontaneous emergence of car clusters has been studied by different authors (see Proceedings *Traffic and Granular Flow* (TGF) for an overview) based on different models and approaches. In spite of the complexity of real traffic, we believe that some general properties of traffic flow, such as headway and velocity importance, exist which can be described and understood by relatively simple models.

In continuation of our contribution to TGF 03 [4] we concentrate again on the microscopic optimal velocity traffic model, first proposed by Bando et al. [5, 6]. Based on investigations about the stability of traffic flow summarized as critical curves in a density c over Bando parameter b diagram [4] we are now evaluating the energy balance as relationship between kinetic and potential energy and incoming and outgoing energy fluxes in different regions of stability [7].

2 Dynamics of Vehicular Particles

Having in mind (from textbooks) the one–particle Newton's equation

$$m\frac{dv}{dt} = F_{cons}(x) + F_{diss}(v) \quad ; \quad \frac{dx}{dt} = v \tag{1}$$

the energy balance out of the given dynamics (1) reads

$$\frac{d}{dt}\,(E_{kin} + E_{pot}) + \Phi = 0 \tag{2}$$

with $E_{kin} = mv^2/2$ as kinetic energy, $E_{pot} = -\int F_{cons}(x)\,dx$ as potential energy and flux term $\Phi = -v\,F_{diss}(v)$ as transformation rate (dissipation or creation) of mechanical energy.

Extending the one–particle case to the many–vehicle system under consideration (N cars moving on a rotary of given length L) the equations of motion (in dimensionless form) are given as follows ($i = 1, \dots, N$)

$$\frac{d}{d\widetilde{t}}\,\widetilde{v}_i = \widetilde{F}_{cons}(\Delta\widetilde{x}_i) + \widetilde{F}_{diss}(\widetilde{v}_i) \quad ; \quad \Delta\widetilde{x}_i = \widetilde{x}_{i+1} - \widetilde{x}_i \tag{3}$$

$$\frac{d}{d\widetilde{t}}\,\widetilde{x}_i = \frac{1}{b}\widetilde{v}_i \qquad \text{with control parameter} \quad b > 0\,. \tag{4}$$

Defining the potential energy of particle i via gradient by

$$\widetilde{F}_{cons}(\Delta\widetilde{x}_i) = -\frac{1}{b}\frac{d\widetilde{E}_{pot}(\Delta\widetilde{x}_i)}{d\widetilde{x}_i} = +\frac{1}{b}\frac{d\widetilde{E}_{pot}(\Delta\widetilde{x}_i)}{d\Delta\widetilde{x}_i} \tag{5}$$

and the kinetic energy of particle i as $\widetilde{E}_{kin}(\widetilde{v}_i) = \widetilde{v}_i^2/2$ the energy balance becomes

$$\frac{d}{d\widetilde{t}}\,\widetilde{E}_{kin}(\widetilde{v}_i) + \frac{d}{d\widetilde{t}}\,\widetilde{E}_{pot}(\Delta\widetilde{x}_i) = \widetilde{F}_{cons}(\Delta\widetilde{x}_i)\,\widetilde{v}_{i+1} + \widetilde{F}_{diss}(\widetilde{v}_i)\,\widetilde{v}_i\,, \tag{6}$$

where the r.h.s. represents the (negative) overall flux term similar to (2).

Leaving the general Newton–like dynamics (3, 4) to be more specific we choose the following forces known from Bando's optimal velocity model

$$\widetilde{F}_{diss}(\widetilde{v}_i) = 1 - \widetilde{v}_i \geq 0 \tag{7}$$

$$\widetilde{F}_{cons}(\Delta\widetilde{x}_i) = \widetilde{v}_{opt}(\Delta\widetilde{x}_i) - 1 \leq 0 \qquad \text{with} \qquad \widetilde{v}_{opt}(\Delta\widetilde{x}_i) = \frac{(\Delta\widetilde{x}_i)^2}{1 + (\Delta\widetilde{x}_i)^2}\,, \tag{8}$$

which can be also interpreted as accelerating (moving fast) and decelerating (avoid collisions) forces.

Taking into account (7, 8) we are able to calculate the potential energy from (5)

$$\widetilde{E}_{pot}(\Delta\widetilde{x}_i) = b\int^{\Delta\widetilde{x}_i} \widetilde{F}_{cons}(y)dy = b\left(\frac{\pi}{2} - \arctan \Delta\widetilde{x}_i\right) \tag{9}$$

with normalization (fixing integration constant) to $\widetilde{E}_{pot}(\Delta\widetilde{x}_i \to \infty) = 0$.

Figure 1 illustrates the time evolution in state space and the behaviour of kinetic and potential energy as well as energy fluxes in time for one particular car (in an ensemble of $N = 60$ vehicles) at low density for two different initial conditions.

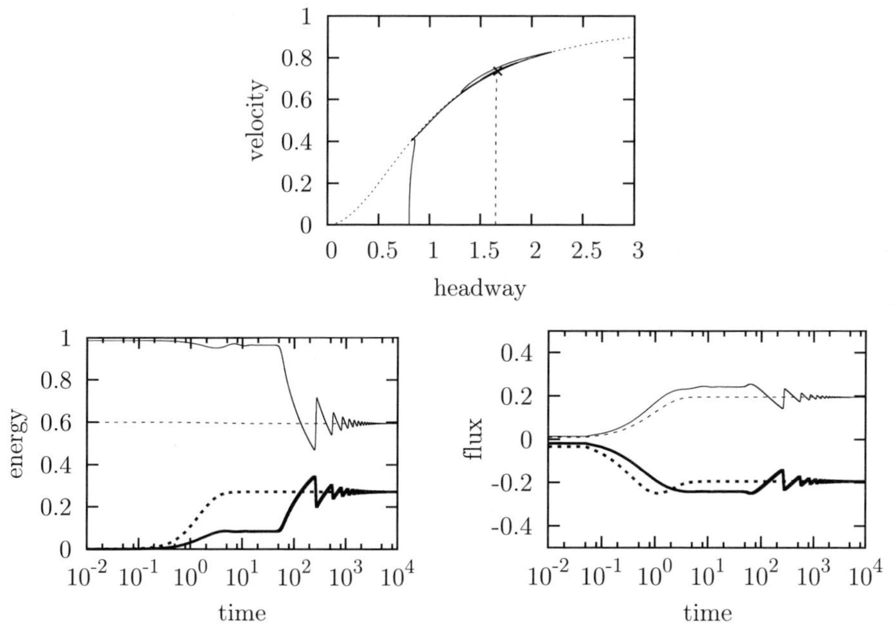

Fig. 1. Numerical integration of equations of motion (3, 4) with forces (7, 8) for $N = 60$ cars on a rotary of dimensionless length $\widetilde{L} = 100$ (at low density $\widetilde{c} = N/\widetilde{L} = 0.6$) and control parameter $b = 1.1$. *Upper picture:* Evolution in state space for one particular car i for two different initial conditions (zero speed and homogeneous spacing by dashed line; zero speed and heterogeneous spacing by full curve). The cross on the optimal velocity function (dotted curve) indicates the stable steady state as free flow solution. *Left picture:* Kinetic energy (starting from zero) and potential energy over time for both given initial situations. *Right picture:* Inflow (positive values) and outflow of energy over time for same initial conditions.

In the following we refer to the whole ensemble consisting of N cars defining overall quantities $\widetilde{E} = \widetilde{E}_{kin} + \widetilde{E}_{pot}$ and $\widetilde{\Phi} = \widetilde{\Phi}_{in} + \widetilde{\Phi}_{out}$ with

$$\widetilde{E}_{kin} = \sum_{i=1}^{N} \widetilde{E}_{kin}(\widetilde{v}_i) = \sum_{i=1}^{N} \frac{1}{2}\widetilde{v}_i^2 \tag{10}$$

$$\widetilde{E}_{pot} = \sum_{i=1}^{N} \widetilde{E}_{pot}(\Delta\widetilde{x}_i) = b\sum_{i=1}^{N}\left(\frac{\pi}{2} - \arctan \Delta\widetilde{x}_i\right) \tag{11}$$

$$\widetilde{\Phi}_{in} = -\sum_{i=1}^{N} \widetilde{F}_{cons}(\Delta\widetilde{x}_i)\,\widetilde{v}_{i+1} \geq 0 \tag{12}$$

$$\widetilde{\Phi}_{out} = -\sum_{i=1}^{N} \widetilde{F}_{diss}(\widetilde{v}_i)\,\widetilde{v}_i \leq 0 \tag{13}$$

and showing energies and fluxes per particle in Fig. 2.

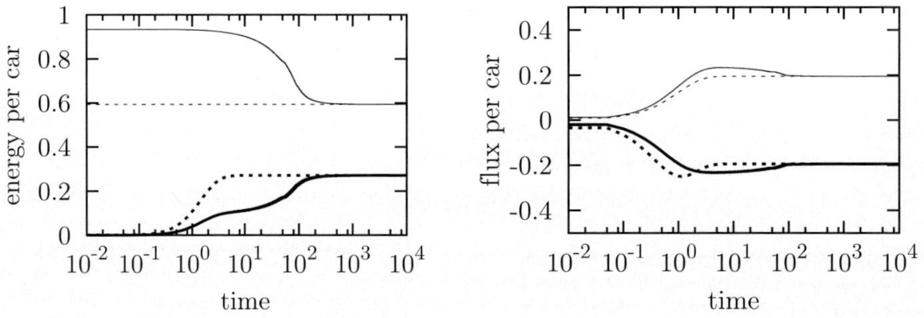

Fig. 2. Temporal development of overall energies (10, 11) (left) and energy flows (12, 13) (right) divided by number of cars N at low density 0.6. Same situation as in previous figure.

3 Traffic Breakdown from Free Flow to Congestion

The steady state solution is given by zero acceleration in (3) characterized by equal velocities and headways for all cars. This homogeneous solution reads

$$\Delta\widetilde{x}_{st} = \frac{\widetilde{L}}{N} = \frac{1}{\widetilde{c}} \quad ; \quad \widetilde{v}_{st} = \widetilde{v}_{opt}(\Delta\widetilde{x}_{st}) = \frac{(1/\widetilde{c})^2}{1+(1/\widetilde{c})^2} \tag{14}$$

and is called vehicular free flow which is stable at low densities $\widetilde{c}$. For the values $\widetilde{c} = 0.6, b = 1.1$ the long–time results (14) are $\widetilde{x}_{st} = 1.667, \widetilde{v}_{st} = 0.735, \widetilde{E}_{kin} = 0.270, \widetilde{E}_{pot} = 0.594, \widetilde{\Phi}_{in} = -\widetilde{\Phi}_{out} = 0.195$ in agreement with the numerical calculations presented in Figs. 1 and 2.

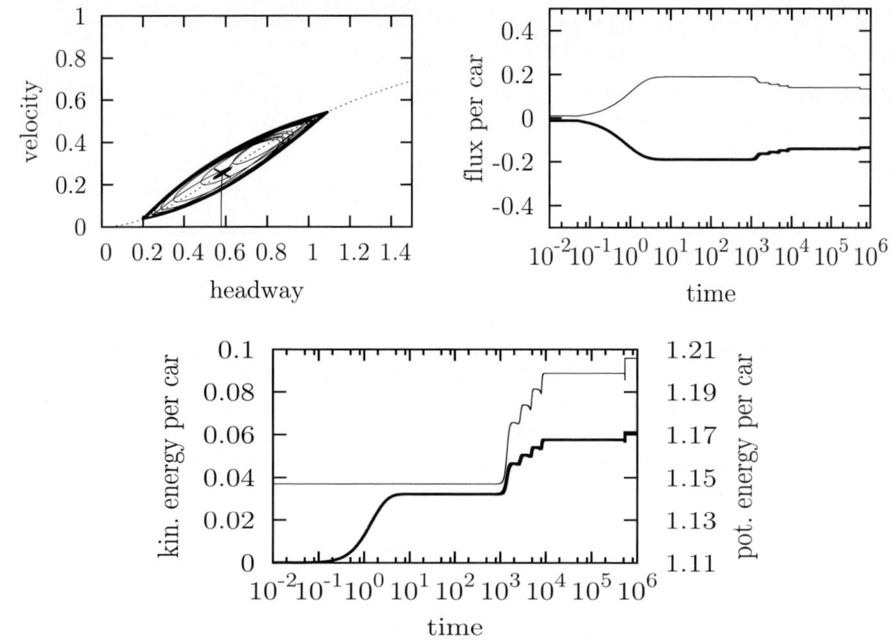

Fig. 3. Numerical integration of equations of motion (3, 4) with forces (7, 8) for $N = 60$ cars on a rotary of dimensionless length $\widetilde{L} = 35$ (at large density $\widetilde{c} = N/\widetilde{L} = 1.714$) and control parameter $b = 1.1$. *Upper picture left:* Evolution in state space for one particular car i for given initial condition (zero speed and homogeneous spacing by full curve). The cross on the optimal velocity function (dotted curve) indicates the unstable steady state. *Upper picture right:* Total energy inflow (positive values) and outflow per car vs. time. *Bottom picture:* Overall kinetic energy (thick curve starting from zero) and overall potential energy per car vs. time.

Figure 3 shows the situation after traffic breakdown as transition from free flow (independent particles with weak interaction) to congested flow where the cars form a dense phase called vehicular cluster. The homogeneous solution (14) becomes unstable and a limit cycle (left upper picture) appears. At large density $\widetilde{c} = 1.714$ (and always $b = 1.1$) the cluster size is $n_{cl} = 30$ vehicles with $\widetilde{x}_{cl} = 0.2, \widetilde{v}_{cl} = 0.04$, $n_{free} = 20$ free particles with $\widetilde{x}_{free} = 1.1, \widetilde{v}_{free} = 0.55$ and $n_{bl} = 10$ cars in the boundary layer (interface between both phases) as long–time limit. Comparing low and high density results presented in Figs. 2 and 3 it is shown that the mean kinetic energy is decreasing from 0.270 to 0.061 whereas the mean potential energy is increasing from 0.594 to 1.206, therefore the total energy per car is changing from 0.864 ($\widetilde{c} = 0.6$) to 1.267 ($\widetilde{c} = 1.714$) summarized in Fig. 4 (left) for the density range $0 \leq \widetilde{c} \leq 5$. The region of congested traffic with a breakdown from homogeneous flow is marked in Fig. 4 (right) by a dashed curve (at $b = 1.1$ from 1.12 to 2.90), whereas the solid curve shows traffic breakdown from heterogeneous flow.

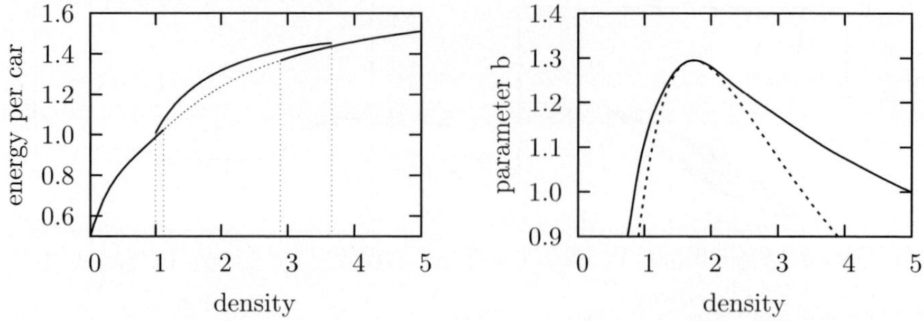

Fig. 4. *Left:* Total energy per car as sum of kinetic and potential energies over vehicular density $\widetilde{c}$ and fixed control parameter $b = 1.1$ Dotted lines mark the region of unstable free flow $1.00 < \widetilde{c} < 3.66$. *Right:* Region of congested traffic below the critical point ($\widetilde{c}_{cr} \simeq 1.714, b_{cr} \simeq 1.295$) bounded by lines of traffic breakdown from heterogeneous flow (solid curve) or homogeneous flow (dashed curve).

4 Appendix: Relationship to Bando's OV Model

Dimensionless variables are original quantities devided by scale units: $\widetilde{x} = x/x_0, \widetilde{v} = v/v_0, \widetilde{t} = t/t_0$ and $\widetilde{E} = E/E_0$ with $E_0 = mv_0^2$ as well as $\widetilde{F} = F/F_0$ with $F_0 = E_0/(t_0 v_0)$ and $\widetilde{\Phi} = \Phi/\Phi_0$ with $\Phi_0 = E_0/t_0$ and control parameter $b = x_0/(t_0 v_0)$.

Dynamics (3, 4)

$$m\,\frac{dv_i}{dt} = F_{cons}(\Delta x_i) + F_{diss}(v_i) \quad ; \quad \Delta x_i = x_{i+1} - x_i \tag{15}$$

$$\frac{dx_i}{dt} = v_i \qquad \text{with control parameter} \quad m > 0 \tag{16}$$

and forces (7, 8)

$$F_{diss}(v_i) = \frac{m}{t_0}\,(v_0 - v_i) \tag{17}$$

$$F_{cons}(\Delta x_i) = \frac{m}{t_0}\,(v_{opt}(\Delta x_i) - v_0) \quad \text{with} \quad v_{opt}(\Delta x_i) = v_0 \frac{(\Delta x_i)^2}{x_0^2 + (\Delta x_i)^2}\,. \tag{18}$$

Energies (9) or (10, 11)

$$E_{kin}(v_i) = \frac{m}{2} v_i^2 \tag{19}$$

$$E_{pot}(\Delta x_i) = \int F_{cons}(\Delta x_i) d(\Delta x_i) = \frac{m x_0 v_0}{t_0}\left(\frac{\pi}{2} - \arctan\left(\frac{\Delta x_i}{x_0}\right)\right) \tag{20}$$

and balance with energy flux (6) or (12, 13)

$$\frac{d}{dt}\,(E_{kin}(v_i) + E_{pot}(\Delta x_i)) + \Phi_i = 0 \tag{21}$$

$$\Phi_i = -\,(F_{cons}(\Delta x_i)\,v_{i+1} + F_{diss}(v_i)\,v_i) \tag{22}$$

with interpretation of parameters: vehicular mass m; maximum speed $v_0 \equiv v_{max}$; interaction distance $x_0 \equiv D$ defined by $v_{opt}(\Delta x = D) = v_{max}/2$ and relaxation time $t_0 \equiv \tau$.

Acknowledgements

We would like to thank the German Science Foundation (DPG) for financial support through grant MA 1508/8 − 1. The authors thank Yuki Sugiyama (Nagoya) for fruitful discussions during the TGF 07 meeting.

References

1. J. Schmelzer, G. Röpke, R. Mahnke: Aggregation Phenomena in Complex Systems, Wiley–VCH, Weinheim, 1999.
2. R. Mahnke, J. Kaupužs, I. Lubashevsky: Physics Reports **408**, 1–130, 2005.
3. D. Helbing: Rev. Mod. Phys. **73**, 1067–1141, 2001.
4. R. Mahnke, J. Kaupužs, J. Tolmacheva: Stochastic Description of Traffic Breakdown: Langevin Approach. In: *Traffic and Granular Flow '03* (Eds.: S. P. Hoogendoorn, S. Luding, P. H. L. Bovy, M. Schreckenberg, D. E. Wolf), Springer, Berlin, 2005, pp. 205–210.
5. M. Bando, K. Hasebe, A. Nakayama, A. Shibata, Y. Sugiyama: Japan J. Indust. and Appl. Math. **11**, 203, 1994; Phys. Rev. E **51**, 1035, 1995.
6. M. Bando, K. Hasebe, K. Nakanishi, A. Nakayama, A. Shibata, Y. Sugiyama: J. Phys. I France **5**, 1389, 1995.
7. R. Mahnke, J. Kaupužs, J. Hinkel, H. Weber: Eur. Phys. J. B **57**, 463, 2007.

Detailed Data of Traffic Jam Experiment

Akihiro Nakayama[1], Minoru Fukui[2], Katsuya Hasebe[3], Macoto Kikuchi[4], Katsuhiro Nishinari[5], Yuki Sugiyama[6], Shin-ichi Tadaki[7], and Satoshi Yukawa[8]

[1] Department of Physics, Meijo University, Nagoya 468-8502, Japan
[2] Nakanihon Automotive College, Sakahogi 505-0077
[3] Faculty of Business Administration, Aichi University, Aichi 470-0296
[4] Cybermedia Center, Osaka University, Toyonaka, Osaka 560-0043
[5] Department of Aeronautics and Astronautics, University of Tokyo, Tokyo 113-8656
[6] Graduate School of Information Science, Nagoya University, Nagoya 464-8601
[7] Computer and Network Center, Saga University, Saga 840-8502
[8] Department of Earth and Space Science, Osaka University, Toyonaka, Osaka 560-0043

Summary. We show detailed data of two traffic jam experiments on a circuit. In the experiments, a traffic jam emerges spontaneously without any bottlenecks. We found the power law nature in time series of average velocity, and also found the homogeneous flow with large velocity is temporarily made before a jam cluster is formed.

1 Traffic Jam Experiment

It is widely believed that the origin of traffic jam on highways is bottlenecks and the jam never appears without bottlenecks. However the recent theories of traffic flow indicate the traffic jam always appears if the car density is sufficiently large [1]. We have carried out a traffic jam experiment to prove that the traffic jam appears without any bottlenecks [2].

The experiment has done on a flat ground and cars run along a circle with the circumference $230m$. The drivers are indicated to follow the preceding car with suitable velocity and headway. We consider the maximum velocity in this circuit will be $30 \sim 40km/h$. From this value, we can estimate the critical density by use of the optimal velocity (OV) model [3]. The car number on a circuit is decided by this critical density and is roughly 20. Then we have carried out several runs by changing car number from 20 to 25 (see Fig. 1a).

The experimental data is extracted from the video record which is taken by 360 degree video camera at the center of the circuit (see Fig. 1b). We obtain positions of each car every 1/3 second (10 frame steps) and other quantities are calculated from this data.

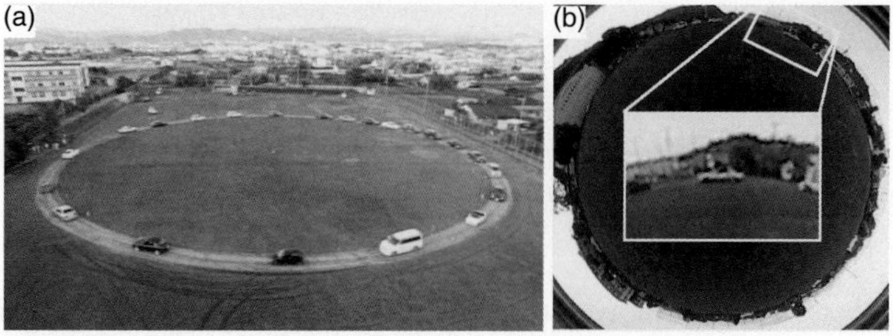

Fig. 1. (a) Snapshot of the experiment. (b) One frame of video records.

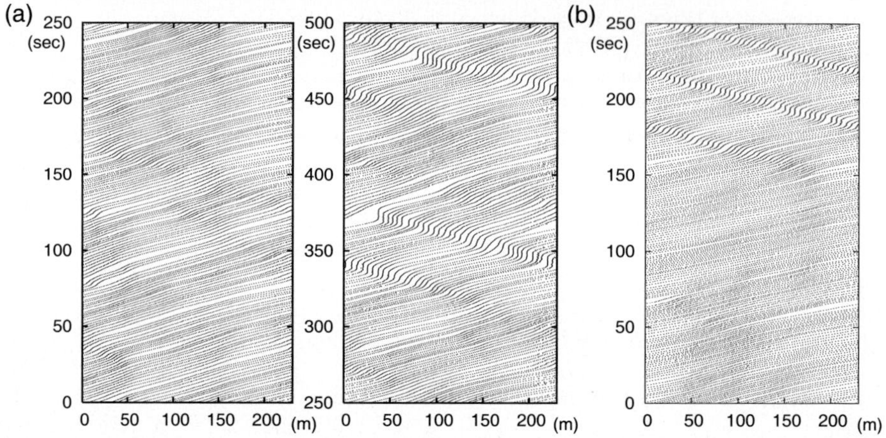

Fig. 2. Spacetime plots of (a) the run of 23 cars and (b) the run of 22 cars. Graphs (a) and (b) show the data during 500 and 250 seconds around the time of jam emergence, respectively. Each dot shows a position of each car.

2 Data of Experiment

In our experiment, no traffic jam emerges in the case that the car number is 20. But a traffic jam emerges if the car number is 22 or 23. Figure 2 shows spacetime plots of such two runs. One is the run of 23 cars (Fig. 2a), and the other is that of 22 cars (Fig. 2b). In both case, a traffic jam emerges spontaneously several minutes after the start. In the former run, a traffic jam is erased by a slow car temporarily, but it reemerges soon. This fact suggests the jam is unavoidable nature of traffic flow, if the density is large. In the latter run, the emergence process of traffic jam is clearly observed, and the details of this run will be shown in section 3. The spacetime plots are quite similar to that of real highway traffic [4]. Also, the velocity of jam cluster moving backward is roughly $20km/h$, which is almost the same as the observed value in real traffic. Then it can be considered that the experiment reveals the feature of real traffic.

Figure 3 shows the average velocity and the variance of velocity. In the first run (Fig. 3a), the fluctuation is relatively large and the transition to congested flow is not clear. On the other hand, the transition is clear in the second run (Fig. 3b). The traffic jam is characterized by the small average velocity and the large variation.

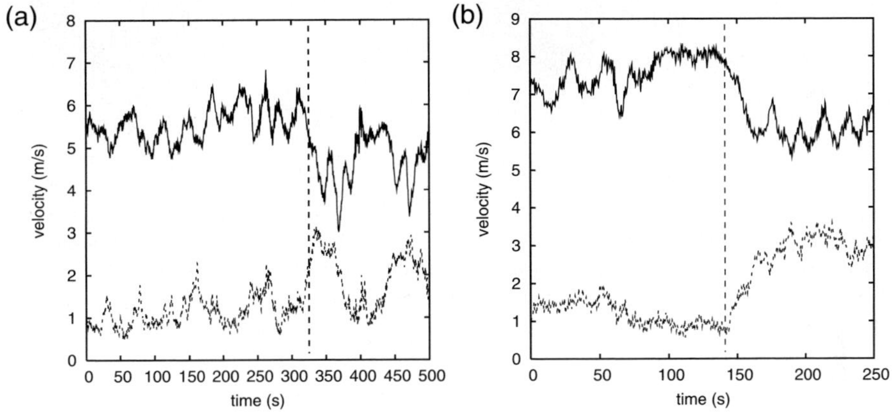

Fig. 3. (a) and (b) show the data corresponding to those of Fig. 2a and Fig. 2b. Solid and dotted lines represent the average velocity and the variance of velocity, respectively. Dashed line indicates the time of jam emergence.

Figure 4 shows the power spectrum of average velocity. We can find the power law nature and some periodicities due to the cyclic boundary condition. It is interesting the experiment has a similar power law nature to the observed one in real highway traffic [5].

We can fix the OV function from the observed data. Figure 5 shows the observed headway-velocity relation and an estimated OV function. We can guess some reasons why the maximum speed of two runs is different each other, for example, the drivers changes their OV function depending on the situation, or the drivers can drive better as they run more times.

3 Jam Formation Process

Here we discuss the details of the jam formation process shown in Fig. 2b. Figure 6 shows the average velocity and headway-velocity relations for three stage of the flow. In the first stage the fluctuation of the velocity is relatively large. A high density area appears many times but disappears soon. In the middle stage the fluctuation becomes small and the average velocity is maintained at a large value. In other words, high density and almost homogeneous flow is realized. The high density area appeared in the homogeneous flow grows up to a true traffic jam cluster. In this experiment the emerges jam cluster is small and can be erased by a single slow car (see Fig. 2a). But the stability

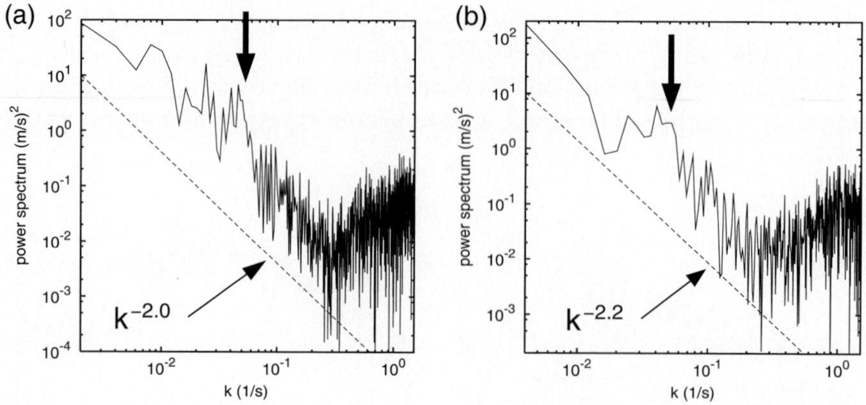

Fig. 4. Power spectrum of average velocity. (a) and (b) show the data corresponding to those of Fig. 2a and Fig. 2b. Solid and dashed lines represent the spectrum and its reference line, respectively. Down arrow indicates the lowest periodicity due to the cyclic boundary condition.

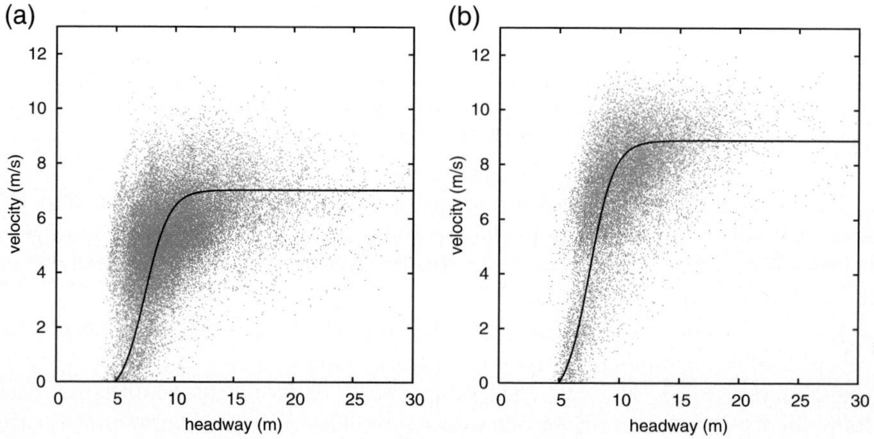

Fig. 5. (a) and (b) show the data of headway-velocity relation corresponding to those of Fig. 2a and Fig. 2b. Gray dots represent the observed data and solid line represents an estimated OV function.

of a jam cluster shown in Fig. 2b indicates that a large cluster as real traffic jam never disappears.

We show the motion of each car in the formation process. Figure 7 shows the motion of three typical cars, (a) fast, (b) standard, and (c) slow cars. Right three graphs show the motion of each car in the three stage corresponding to Fig. 6. In the first and last stages, the difference among the motions of three cars is relatively large. In the middle stage, three cars move at the almost same speed with different headways. Exactly speaking, the flow is in a steady

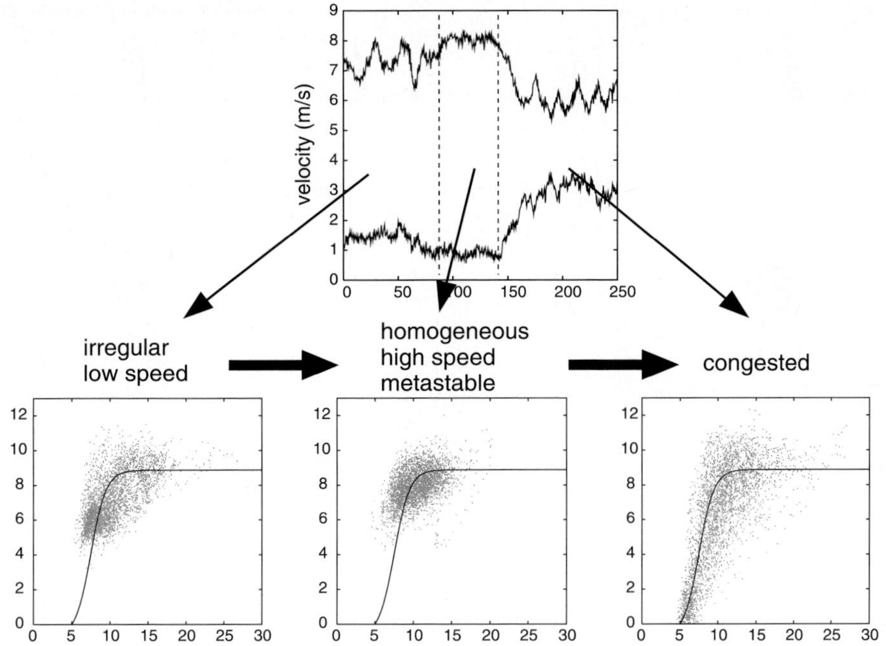

Fig. 6. Clear example of jam formation process. Upper graph is the same as Fig. 3b, and lower three graphs are extracted from Fig. 5b for each period.

state. It may be a necessary condition of the spontaneous emergence of traffic jam in experiment.

4 Summary

We have shown the detailed data of traffic jam experiment on a circuit without any bottlenecks. A traffic jam emerges spontaneously if the car density is large. This fact proves that the density of cars play an essential role of the emergence of traffic jam. It can be considered that the role of bottleneck is to increase the density.

The observed data of real traffic shows a power law in a wide range of time scale. We found the power law nature of average velocity. It suggests that we can experimentally investigate the origin of the power law nature of traffic flow at least at a short time scale.

We also found the homogeneous or steady-state flow with large velocity is temporarily made before a jam cluster emerges. When the irregularity of the density is large, the creation of a traffic jam is prevented by low density part of the flow. A certain length of high density region seems necessary to create a spontaneous traffic jam. It may have a relationship to the property that a small jam cluster is unstable, which is found in numerical simulation.

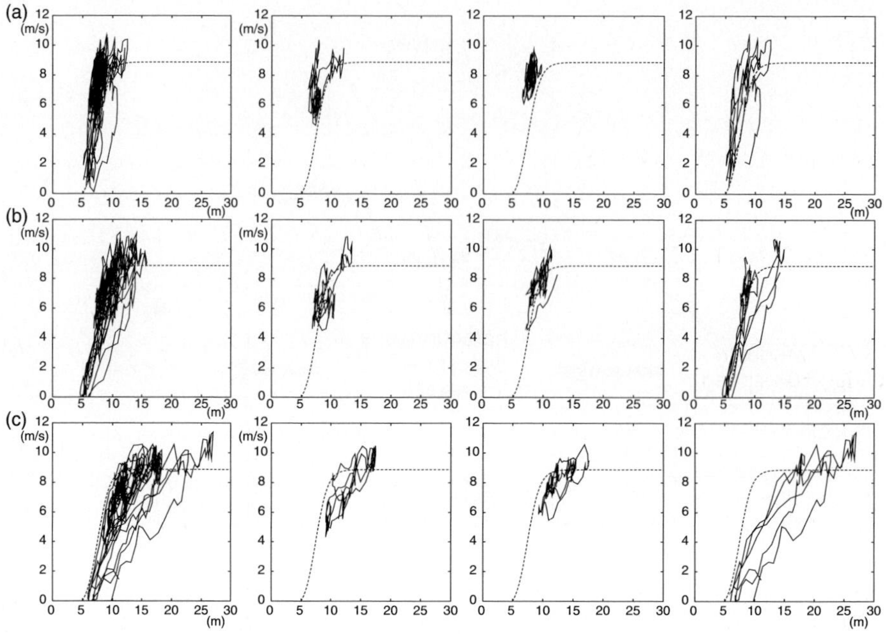

Fig. 7. Motion of each car in headway-velocity plane. The motions of fast, standard, and slow cars are shown in (a), (b) and (c), respectively. Left graph is the plot of all data, and right three graphs are the plots corresponding to three stages of jam formation. Solid and dashed lines represent the motion of car and the reference OV function, respectively.

Acknowledgements

The authors thank Y. Tatetsu, Y. Hagiwara, Y. Terauchi for helping data acquisitions. This work is partly supported by a Grant-in-Aid for Scientific Research (C) (No. 18540409) of the Japanese Ministry of Education, Science, Sports and Culture.

References

1. D. E. Wolf, M. Schreckenberg, and A. Bachem, editors. *Workshop on Traffic and Granular Flow*. World Scientific, Singapore, 1996.
2. Y. Sugiyama, A. Nakayama, M. Fukui, K. Hasebe, M. Kikuchi, K. Nishinari, S. Tadaki,and S. Yukawa. Observation, theory and experiment for freeway traffic as physics of many-body system. In S. P. Hoogendoorn, S. Luding, P. H. L. Bovy, M. Schreckenberg, and D.E. Wolf, editors, *Traffic and Granular Flow '03*, pages 45–58. Springer-Verlag, Berlin, Heidelberg, 2005.
3. M. Bando, K. Hasebe, A. Nakayama, A. Shibata, and Y. Sugiyama. *Phys. Rev.* E, 51:1035–1042, 1995.
4. J. Treiterer and J. A. Myers. *Interin Report EEs*, 278-3:13–38, 1970.
5. S. Tadaki, M. Kikuchi, A. Nakayama, K. Nishinari, A. Shibata, Y. Sugiyama, and S. Yukawa. *J. Phys. Soc. Jpn.*, 75:034002–1–5, 2006.

Prediction Accuracy of Evacuation Times for High-Rise Buildings and Simple Geometries by Using Different Software-Tools

Christian Rogsch[1], Wolfram Klingsch[1], Armin Seyfried[2], and Henning Weigel[1]

[1] University of Wuppertal, Institute for Building Material Technology and Fire Safety Science, Pauluskirchstr. 7, D-42285 Wuppertal, Germany
`christian@rogsch.de; klingsch@uni-wuppertal.de`

[2] Central Institute for Applied Mathematics, Research Centre Jülich, D-52425 Jülich, Germany `a.seyfried@fz-juelich.de`

Summary. Evaluation and optimization of emergency systems can be done by means of several engineering methods, which are entirely different: macroscopic hydraulic models, which can be calculated by hand (so called handcalculation methods), and microscopic computer simulation methods. Both allow forecasting of evacuation-times for various settings. The authors compare results of four commercial software-tools (ASERI, buildingEXODUS, PedGo, Simulex) and some macroscopic handcalculation models with a real evacuation-trial in a high-rise building. Furthermore evacuation times for simple layouts of room and floor are calculated and the results are compared against each other.

1 Analysis of a High-Rise Building

The real evacuation-trial which we compare with results of commercial software-tools was performed [1] in the middle of the 1970 at the Mannesmann-Building, which was built in 1959 in Germany. It consists of three basements, one ground floor, one mezzanine floor and 22 top floors. During the evacuation trial one staircase was closed, thus all people had to use the same staircase. In total, 427 people stayed in the top floors and were evacuated by using the evaluated staircase. A floorplan with measurements is shown in Figure 1, a detailed overview about walking velocity and distribution of people inside the building can be found in [2]. As pre-movement-time we choose 50 sec ± 20 sec, this time based upon the original study [1].

1.1 Results of Commercial Software-Tools

The results show, that commercial software are able to predict a total evacuation time for a high-rise building (see Table 1). If one takes a closer look at

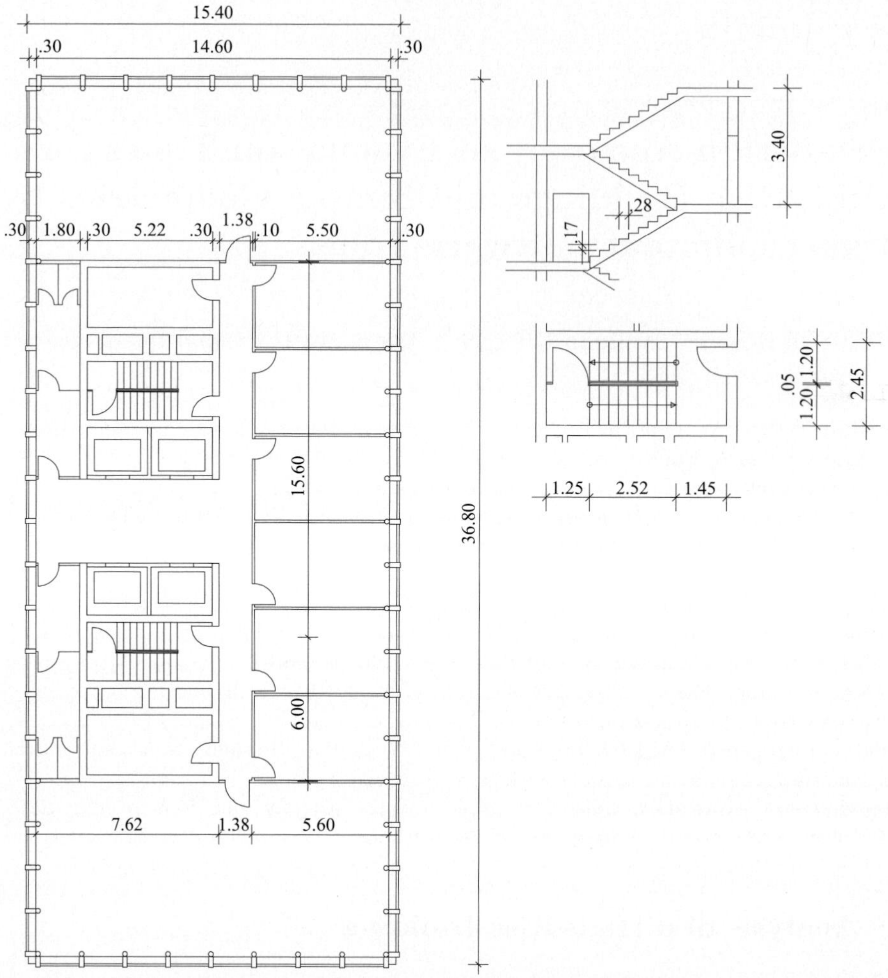

Fig. 1. Floorplan of the Mannesmann-Building [1].

the calculated evacuation times for selected floors, it appears that they differ considerably from the real times (see Table 1). Because of this results we calculate this high-rise buidling with different macroscopic handcalculation methods. These methods should also be able to predict total evacuation times for a high-rise building.

1.2 Results of Macroscopic Handcalculation Methods

The results shown in this section (see Table 2) based upon calculations of ten different macroscopic handcalculation methods. As we can see, some of these handcalculation methods are able to predict "correct" evacuation times for a total building.

Table 1. Comparison of calculated evacuation times with a real evacuation-trial. At selected floors the table shows the first and last person moving into the staircase

	Evacuation-trail	ASERI	buildingEXODUS	PedGo	Simulex
Total Building	8.78 min	Ca. 9 min	ca. 8.5 min	ca. 8 min	ca. 8 min
2nd floor	50–149s	40–82s	38–74s	44–94s	44–82s
4th floor	45–75s	35–86s	49–73s	50–82s	41–86s
5th floor	61–101s	36–87s	35–83s	42–89s	42–90s
6th floor	31–102s	42–82s	35–78s	41–95s	42–85s
7th floor	67–132s	43–96s	37–77s	39–96s	43–95s
10th floor	51–102s	33–117s	41–83s	39–92s	43–90s
15th floor	48–155s	38–83s	38–81s	45–88s	42–80s

Table 2. Comparison of calculated evacuation times by macroscopic handcalculation methods with a real evacuation-trial

Macroscopic handcalculation method	427 people (19 in each floor) with 0.8 min response-time
Predtechenskii and Milinskii, standard method [3]	9,10 min
Predtechenskii and Milinskii, simplified method [4]	9,15 min
SFPE/NFPA-Handbook [5]	14.78 min
Method of W. Müller [6–9]	14.58 min
Method of K. Togawa [10]	5.65 min
Method of M. Galbreath [11]	6.52 min
Effective-Width Model by J. Pauls [12]	7.84 min
Method of Melinek and Booth [13]	7,67 min
Method of E. Kendik [14]	7.98 min
Method of Seeger and John [1]	8.83 min
Real evacuation-trial [1]	8,78 min (without 0.8 min response time: 7.98 min)

2 Analysis of Simple Geometries

We compare the calculation of three different simple geometries (scenarios) against each other. Results are shown in Figure 2 and Figure 3. The results show, that the commercial software-tools are not able tor predict the same evacuation time for such simple geometries. The results differ round about a factor of two at a simple room (Figure 3) and by calculating a linear movement (Figure 2) they might have problems to reproduce a relationship between density and walking velocity. Different boundary conditions (like modelling an extra-room) have also a large influence of the calculated results. At one simulation (without extra-room) people disappear after reaching the time measurement, with extra-room they walk through the extra-room before they disappear at the right door.

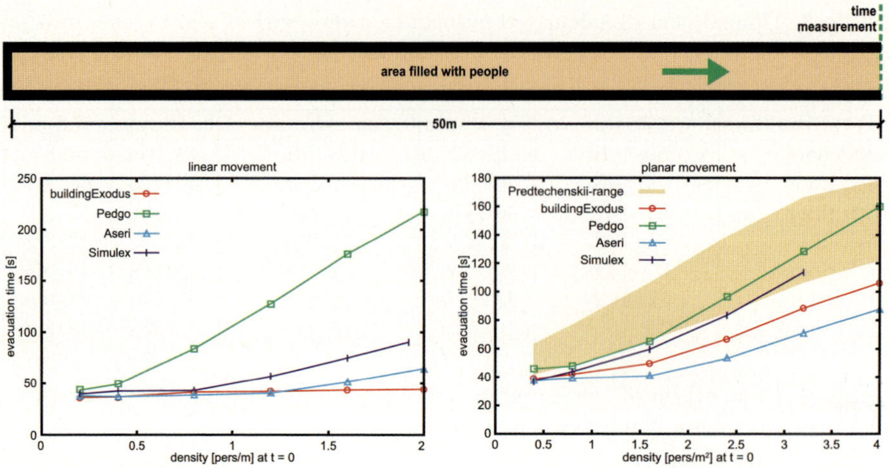

Fig. 2. Linear (left picture; narrow floor, no overtaking is possible) and planar (right picture; floor 2m wide, overtaking is possible) movement of people.

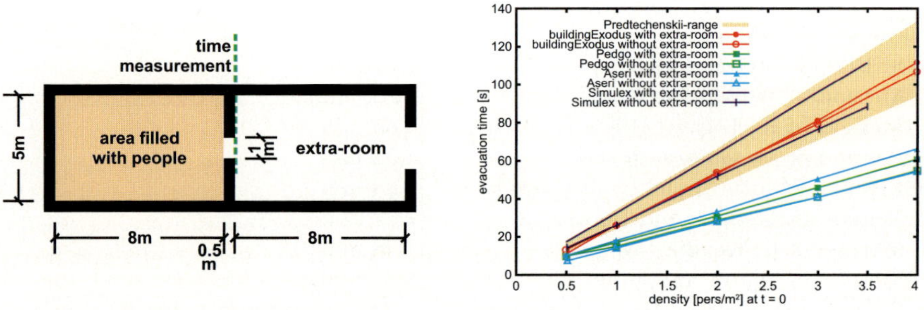

Fig. 3. Evacuation of a simple room with different boundary conditions.

3 Discussion of the Results

The analysis shows that chosen software tools are able to predict total evacuation times for high-rise buildings. However by taking a closer look at evacuation times for selected floors, calculated results differ from the results of the trials. This differences are probably based on the very low density, which is inside each floor (ca. 19P/floor as an arithmetic mean), thus there is no large congestion at the entrance of the staircase. At very low densities like in this case, the response time and the human behaviour has a large influence on the evacuation times for selected floors, thus the people walk "directly" into the staircase without any or only a small congestion, in other words: people are moving with almost free walking velocity through the building. If we have larger densities in these floors, the congestion at the staircase entrance will be still alive when people with high response time reach the staircase.

That means, that human behaviour and response time would have an inferior influence of the evacuation at higher density situations. But this discussion points to another interesting part of pedestrian modelling: the density. If we compare the software-tools about their maximum density there are also large discrepancies: PedGo is able to simulate a maximum density of $6.25\mathrm{P/m}^2$ but buildingEXODUS can only simulate a maximum density of $4\mathrm{P/m}^2$, thus it is quite difficult to say that a congestion will appear or not.

To get a deeper insight we decided to proof the software tools by means of simple geometries, thus compensation effects are excluded and it is possible to recognize weak points of the software and the implemented algorithm. The comparison of the evacuation times of simple geometries reveals differences up to a factor of four among the tools for a linear movement, a problem which is probably based upon the update algorithm used for description of individual movement of pedestrians. The differences at the simple room geometry between buildingEXODUS and PedGo may depend on the different densities which can be simulated by the models. Moreover we found problems in the definition of boundary conditions. Due to the simulation of a simple room geometry with different boundary conditions we achieved results differing up to 20%. Thus it is possible, that only compensation effects provide the correspondence between calculated evacuation times for high-rise buildings and times obtained at the trials. In particular it is obvious that simulated pedestrian flow inside the building is affected by non-negligible uncertainties. It is also obvious that simple macroscopic handcalculation models for the prediction of total evacuation times are as good as more detailed computer simulation tools. But we should not forget that it is very difficult to verify or validate software-tools or models because we do not have very much reliable empirical data, a fact, that we should change as soon as possible in the future.

Acknowledgements

The authors thank I.S.T. GmbH for providing their evacuation software-tool ASERI 3.4c free of charge. We also thank TraffGo HT GmbH for providing their software-tool PedGo 2.1.1 and 2.2.4 free of charge. Furthermore we thank BPK GmbH for using their buildingEXODUS software and IES Ltd. for their student-offer of Simulex. Last but not least the authors want to thank the group of ped-net.org for very useful discussions.

References

1. P. G. Seeger and R. John. Untersuchung der Räumungsabläufe in Gebäuden als Grundlage für die Ausbildung von Rettungswegen, Teil III: Reale Räumungsversuche. Technical Report T395, Forschungsstelle für Brandschutztechik an der Universität Karlsruhe (TH), 1978.

2. C. Rogsch, W. Klingsch, A. Seyfried, and H. Weigel. How reliable are commercial software-tools for evacuation calculation? In *Interflam 2007 - Conference Proceedings*, pages 235–245, 2007.
3. W. M. Predtetschenski and A. I. Milinski. *Personenströme in Gebäuden – Berechnungsmethoden für die Modellierung*. Müller, Köln-Braunsfeld, 1971.
4. W. M. Predtetschenski, W. W. Cholstschewnikow, and H. Völkel. Vereinfachte Berechnung der Umformung von Personenströmen auf Wegabschnitten mit begrenzter Länge. *Unser Brandschutz - wissenschaftlich-technische Beilage*, 6:90–94, 1972.
5. H. E. Nelson and F. W. Mowrer. Emergency movement. In P. J. DiNenno, editor, *SFPE Handbook of Fire Protection Engineering*, chapter 14, page 367. National Fire Protection Association, Quincy MA, third edition, 2002.
6. W. Müller. Untersuchung über zulässige Räumungszeiten und die Bemessung von Rückzugswegen in Gebäuden. 1970.
7. Werner Müller. Die Beurteilung von Treppen als Rückzugsweg in mehrgeschossigen Gebäuden. *Unser Brandschutz - wissenschaftlich-technische Beilage*, 3:65–70, 1966. to be continued in 4/1966.
8. Werner Müller. Die Beurteilung von Treppen als Rückzugsweg in mehrgeschossigen Gebäuden. *Unser Brandschutz - wissenschaftlich-technische Beilage*, 4:93–96, 1966. continuation from 3/1966.
9. Werner L. Müller. Die Überschneidung der Verkehrsströme bei dem Berechnen der Räumungszeit von Gebäuden. *Unser Brandschutz - wissenschaftlich-technische Beilage*, 4:87–92, 1968.
10. K. Togawa. Study on fire escapes basing on the observation of multitude currents. Report of the building research institute, Ministry of Construction, Japan, 1955. in Japanese.
11. M. Galbreath. Time of evacuation by stairs in high buildings. Fire Research Note 8, NRCC, May 1969.
12. J. L. Pauls. Movement of people. In P. J. DiNenno, editor, *SFPE Handbook of Fire Protection Engineering*, chapter 3, page 263. National Fire Protection Association, Quincy MA, second edition, 1995.
13. S. J. Melinek and S. Booth. An analysis of evacuation times and the movement of crowds in buildings. Technical Report CP 96/75, BRE, 1975.
14. Ezel Kendik. *Die Berechnung der Personenströme als Grundlage für die Bemessung von Gehwegen in Gebäuden und um Gebäude*. PhD thesis, TU Wien, May 1984.

Statistical Approach to Traffic Flow

Anton Šurda

Institute of Physics, SAS, Dubravska 9, Bratislava, Slovakia
anton.surda@savba.sk

Summary. Statistical mechanics of a small system of cars on a single-lane road is developed. The system is characterized by a conditional probability of a velocity of a car for a given velocity and distance of the car ahead. Distribution of car velocities for various densities of a group of cars is derived as well as probabilities of density fluctuations of the group for various velocities. Free-flow phase, congested phase, and formation of platoons of cars were found.

1 Introduction

In classical equilibrium statistical mechanics Hamiltonian plays a decisive role in determination of statistical properties of a system of interacting particles. It represents energy as a function of momenta and spatial coordinates of the particles. The system may be treated microcanonically or canonically. In a microcanonical ensemble all the states are equally probable and the number of states of a given energy yields the entropy of the system. In a canonical ensemble the probability of states of a subsystem depends on its energy and their sum is called partition function and determines thermodynamic properties of the system.

In our approach, we do not start from the Hamiltonian of a system of particles, which is not even introduced, but from the knowledge of conditional probabilities of velocities and coordinates of each particle. It is assumed that the conditional probability depends only on the state of the neighbouring particles, and it can be determined from the behaviour of a small system experimentally.

This non-Hamiltonian approach is applied to a system of cars on a single-lane road while the behaviour of the cars is described by conditional probabilities of the velocity of each car depending on the distance and velocity of the car ahead.

Traffic flow of a system of identical cars on a single-lane road has been intensively studied in recent decade using dynamical or kinetic description of car behaviour [1, 2]. The models used were continuous (fluid dynamical models), car-following models, or discrete particle hopping models related to cellular automaton models with stochastic behaviour.

In the approach of Mahnke et al. [3] the group of cars is represented by a grandcanonical ensemble, number of cars in which is not fixed, and its chemical potential is a function of parameters of a master equation.

In the last years we could observe a revival of the microcanonical approach to the problems of statistical mechanics [4]. One of the reasons for that was identification of the region where the entropy of a finite system is convex, instead of the standard concave shape of it, with a point or line in the phase space where the first-order phase transition in the corresponding infinite system takes place. As the number of observed cars in normal traffic is not too large, the techniques developed in statistical physics for small systems are convenient in this case. The term "phase transition" in this paper is used in the sense of the above-cited works.

2 Statistical Mechanics of 1 D Non-Hamiltonian Systems

A single-lane road represents a one-dimensional system so we shall further confine ourselves to 1D systems of particles with nearest neighbour interactions. The conditional probability that the particle has velocity v_i and coordinate r_i while the velocity and coordinate of its neighbouring particles are $v_{i\pm1}$ and $r_{i\pm1}$, respectively, will be denoted as $p(v_i, r_i|v_{i-1}, r_{i-1}, v_{i+1}, r_{i+1}) \equiv p(v_{i-1}, r_{i-1}, v_i, r_i, v_{i+1}, r_{i+1})/p(v_{i-1}, r_{i-1}, v_{i+1}, r_{i+1})$ and taken as an input to the theory.

The probability of velocities and coordinates of all n particles of the system can be easily calculated from slightly different conditional probabilities

$$p(v_1, r_1, \ldots, v_n, r_n) = p(v_1, r_1|v_2, r_2) \ldots p(v_{n-1}, r_{n-1}|v_n, r_n)p(v_n, r_n) \quad (1)$$

where $p(v_i, r_i|v_{i+1}, r_{i+1}) = p(v_i, r_i, v_{i+1}, r_{i+1})/p(v_{i+1}, r_{i+1})$. Probabilities $p(v_i, r_i|v_{i+1}, r_{i+1})$ are generally unknown, but they are closely related to the probabilities $p(v_i, r_i|v_{i-1}, r_{i-1}, v_{i+1}, r_{i+1})$ characterizing our system [5].

In the system of cars on a single-lane road, the driver behaviour is assumed to depend only on the previous car. In such a system the conditional probability of velocity and coordinate of particle 1 depends only on the velocity and coordinate of particle 2 ahead of it, $p \equiv p(v_1, r_1|v_2, r_2)$, and the probabilities of the whole system are given directly by (1). These probabilities will be used for calculations of the sum of probabilities of the system as a function of certain global quantities.

The terms microcanonical and canonical will be used in the case of non-Hamiltonian systems as well, but they will be related rather to the *length* of the system than to its energy, which is not even introduced.

In the canonical ensemble of a Hamiltonian system the usually calculated quantity is the partition function, which is sum of the probabilities of all possible states of the system. In the microcanonical ensemble, the probability to find the system in a state is same for all of them. Logarithm of their sum for some fixed quantities is called entropy. In our non-Hamiltonian approach we calculate sum of the probabilities of states (SPS) for given length of the system and sum of velocities of the particles.

In the microcanonical approach to traffic flow, the length of a group of cars is fixed, and it is influenced by the reservoir only through boundary conditions. The properties of the group are given by the sum of the probabilities of states with the same total velocity.

In the canonical approach, the group of cars represents only a subsystem of a large system whose remaining part is a reservoir. The length of the whole system is constant, and the length of the subsystem is changed only at the expense of the length of the reservoir. The properties of the subsystem are given by sum of the probability of states of the *whole system* having the same total velocity of the cars in the *subsystem.* The canonical approach is able to describe fluctuations of the length of the subsystem around its mean value.

3 Model

The cars are further represented by dimensionless points moving on a discrete one-dimensional lattice and are characterized by 2 quantities: discrete velocity v_i in the interval $\langle 0, v_{\mathrm{max}}\rangle$ and a discrete coordinate (site number) $x_i \in \langle 1, X-1\rangle$. v_{max} is the maximum velocity given by the construction of the car and X is the length of the observed group (subsystem) of cars. The coordinate of each car is measured with respect to the last car of the group. Its coordinate is always 0, i.e., the origin of the coordinate system is moving with it. As the length of the group is X, the coordinate of the last car of the group ahead is also X. Number of cars in the group is N. (The lattice constant is related to the car length). Car velocities and coordinates acquire only integer values.

Kinetics of the system of cars is given by reaction of each driver on the car ahead moving with velocity v_{i+1} at the distance $x_{i,i+1}$. As the driver directly influences only velocity of his car, his reaction is characterized by a conditional probability of a car velocity v_i parametrized by velocity and distance of the car ahead: $p \equiv p(v_i|x_{i,i+1}, v_{i+1})$. The probability could be found experimentally by long observation of two cars at all possible situations. Since such data are not available yet, the probability is calculated from a simple model behaviour of a driver.

The probability distribution is assumed to be peaked around an optimal velocity v_{opt}, which is further chosen as 90% of maximal safe velocity v_{m}. The maximal safe velocity is determined from the requirement that two neighbouring cars, which start to decelerate at the same time with the same deceleration rate a, would stop without crash. Moreover, v_{m} must not be greater than the maximum possible velocity of the car v_{max}, i.e., for every car

$$v_{\rm opt}(v_2, x_{1,2}) = 0.9 v_{\rm m}(v_2, x_{1,2}),$$
$$v_{\rm m}(v_2, x_{1,2}) = -a\tau + \begin{cases} +\sqrt{((a\tau)^2 + 2ax_{1,2} + v_2^2} & \text{if } v_{\rm m} \le v_{\rm max} \\ v_{\rm max} & \text{if } v_{\rm m} > v_{\rm max} \end{cases} \quad (2)$$

where $x_{1,2}$ and v_2 are the distance (headway) and velocity of the car ahead, respectively, τ is the reaction time of the driver of the car 1. As we use only integer values of velocities, the nearest integer value to $v_{\rm opt}$ from 2 is taken for the actual optimal velocity in our further calculations. For very high densities, when discreteness of the lattice is not negligible, another condition $v_1 < v_2 + x_{1,2}$ must be imposed.

The way of driving of the observed drivers is characterized by distribution of probabilities of car velocities around the optimal velocity. Here we use an extremely simple distribution, in which the probability of optimal velocity is p_0, the probabilities of the velocities $v_{\rm opt} \pm 1$ are p_1, while the probability of the car to have any other permitted velocity is p_2. The values of the probabilities for velocities higher then the maximal safe velocity are equal to 0. The sum of all probabilities for each car is equal to 1. The parameters p_0, p_1 and p_2 are the same for every car, and the distribution depends on the headway only by means of the value of optimal velocity.

4 Microcanonical and Canonical Description

The sum of probabilities of states (SPS) with total velocity V of a group of N cars and length X will be denoted as $W(V, X)$. It can be calculated, using (1), recurrently

$$\begin{aligned}
&W_1(V_1, X_1; v_2) = p(V_1|X_1, v_2) \\
&\vdots \\
&W_i(V_i, X_i; v_{i+1}) = \sum_{v_i, x_{i,i+1}} W_{i-1}(V_i - v_i, X_i - x_{i,i+1}; v_i) p(v_i|x_{i,i+1}, v_{i+1}) \\
&\text{for } i = 2, N-1 \\
&\vdots \\
&W_N(V, X; v_{N+1}) = \\
&\quad = \sum_{v_N, x_{N,N+1}} W_{N-1}(V - v_N, X - x_{N,N+1}; v_N) p(v_N|x_{N,N+1}, v_{N+1}) \\
&W(V, X) = \sum_{v_{N+1}} W_N(V, X; v_{N+1}) p(v_{N+1})
\end{aligned} \quad (3)$$

where $0 \le v_j \le v_{\rm max}$, $0 \le V_j \le j v_{\rm max}$, $j \le x_{j,j+1}, X_j \le X - j$. Probability $p(v_{N+1})$ in the last line of (3) is the velocity probability of the last car of a large group ahead (reservoir of length $L_{\rm r}$) of the studied group with the same

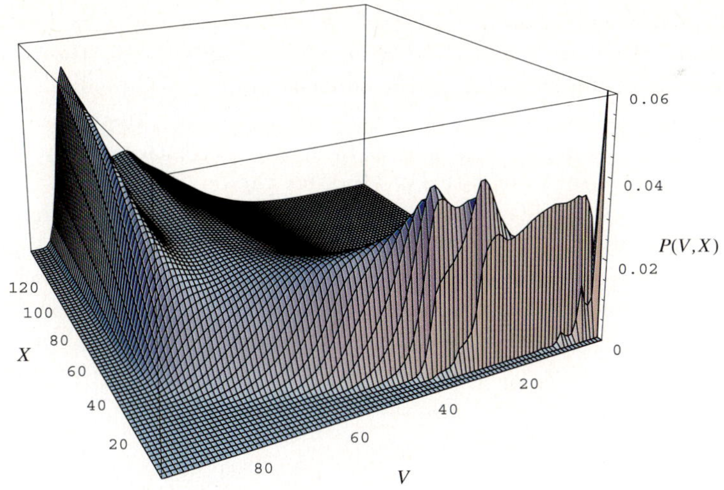

Fig. 1. Probability of the total velocity of a microcanonical ensemble of 5 cars as a function of its total velocity V and length X for $a = 4.0$, $p_2 = 0.025$, $\tau = 0$. The density of the system varies from 0.033 to 1.

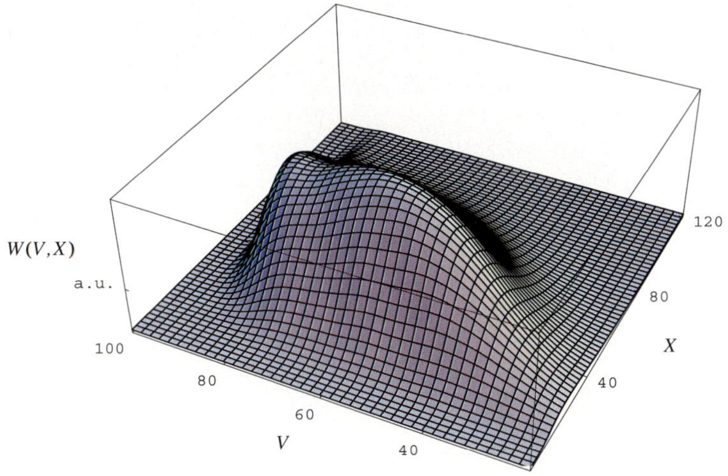

Fig. 2. SPS of a canonical ensemble of 5 cars as a function of its total velocity V and length X for $a = 4.0$, $p_2 = 0.025$, $\tau = 0$, and mean length $\langle X \rangle = 50$, i.e., mean density $\rho = 0.1$. $W(V, X)$ is proportional to the probability of fluctuations of length the group around its mean value.

car density, and it can be calculated using again (1) [5]. In microcanonical approach $W(V, X)$ as a function of V is normalized for each X and the resulting probability $P(V, X)$ of a group of cars to have a total velocity V is shown in the figures below.

In the canonical approach SPS of the whole system with fixed length $L_s = X + L_r$ is calculated, which, after normalization, gives the probability of fluctuations of the length X of the subsystem around its mean value given by the total length L_s. Now, in the last line of Eq. (3), instead of $p(v_{N+1})$, appears SPS of the reservoir of length $L_r = L_s - X$, with velocity of the last car v_{N+1}, summed over velocities of all other cars.

In Figs. 1–3 probability distribution of total velocity and probabilities of length fluctuations of a group of 5 cars are depicted. In Fig. 1 free-flow and congested phases, with first-order phase transitions between them are shown. In a group of cars with inefficient brakes, $a = 0.5$, platoons of cars with the same velocities are formed for small X (high densities) as shown in Fig. 3.

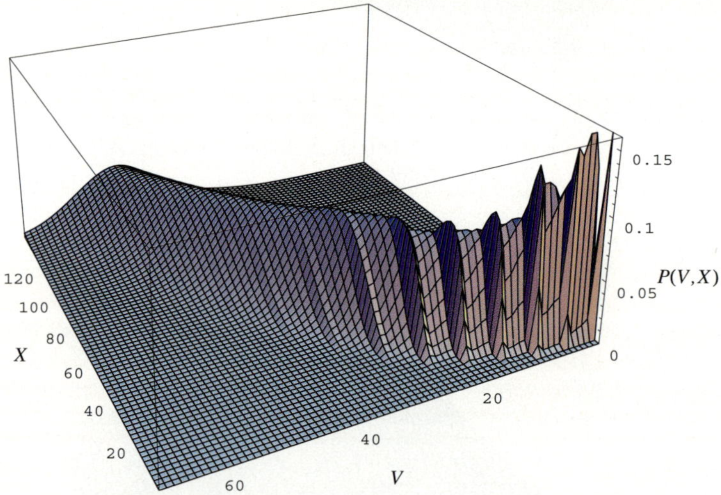

Fig. 3. Probability of the total velocity of a microcanonical ensemble of 5 cars as a function of its total velocity V and length X for $a = 0.5$, $p_2 = 0.005$, $\tau = 0$. The density of the system varies from 0.033 to 1. Maxima at velocities which are integer multiples of 5 show formation of platoons of cars.

References

1. Wolf D E, Schreckenberg M, Bachem A (Eds.) (1996) Traffic and Granular Flow, World Scientific.
2. Nagel K, Schreckenberg M (1992) J. Phys. I (France) 2:2221.
3. Gross D H E (2001) Microcanonical thermodynamics: Phase transitions in "small" systems, Lecture Notes in Physics, 66, World Scientific, Singapore.
4. Mahnke R, Hinkel J, Kaupužs J, Weber H (2006) Thermodynamics of traffic flow, cond-mat/0606509.
5. Šurda A (2007) Statistical mechanics of non-Hamiltonian systems: Traffic flow, (to be published).

A Public Conveyance Model and Analysis on Clustering of Vehicles

Akiyasu Tomoeda[1], Katsuhiro Nishinari[1], Debashish Chowdhury[2], and Andreas Schadschneider[3]

[1] Department of Aeronautics and Astronautics - University of Tokyo, Bunkyo 113-8656 - Japan [tt67055, tknishi]@mail.ecc.u-tokyo.ac.jp
[2] Department of Physics - Indian Institute of Technology, Kanpur 208016 - India debch@iitk.ac.in
[3] Institut für Theoretische Physik - Universität zu Köln, 50937 Köln - Germany as@thp.uni-koeln.de

Summary. A new Public Conveyance Model (PCM) applicable to buses and trains is proposed in this study by using stochastic cellular automaton. We apply our PCM to the bus system and measure the efficiency of the system. By using mean field analysis, we estimate the average velocity and the waiting passengers in the low density limit. We have obtained the theoretical results which are in good agreement with numerical simulations. It is also observed clustering of vehicles which is caused by the time delay effect of passengers when they get on a vehicle. We have found that the big cluster of vehicles is divided into small clusters, by incorporating information of the number of vehicles between successive stops.

1 Introduction

The totally asymmetric simple exclusion process (ASEP) [1–3] is the simplest model of non-equilibrium systems of interacting self-driven particles. Various extensions of this model have been reported in the last few years for capturing the essential features of the collective spatio-temporal organizations in wide varieties of systems, including those in vehicular traffic [4–8]. Traffic of buses and bicycles have also been modeled following similar approaches [9, 10].

In this study, by extending the model in [9], we suggest a new public conveyance model (PCM) [11]. Although we refer to each of the public vehicles in this model as a "bus", the model is equally applicable to train traffic on a given route. In this PCM we can set up arbitrary number of bus stops on the given route. The *hail-and-ride* system in Ref. [9], that is, the passengers could board the bus whenever and wherever they stopped a bus by raising their hand, turns out to be a special case of the general PCM. Moreover, in the PCM the duration of the halt of a bus at any arbitrary bus stop depends on the number of waiting passengers. As we shall demonstrate, the delay in the departure of the buses from crowded bus stops leads to the tendency of the buses to cluster on the route. Furthermore, we also introduce a traffic control system that exploits the information on the number of buses in the "segments" in between successive bus stops; this traffic control system helps in reducing the undesirable tendency of clustering by dispersing the buses more or less uniformly along the route.

2 A Stochastic Cellular Automaton Model for Public Conveyance

In this section, we explain the PCM in detail. For the sake of simplicity, we impose periodic boundary conditions. Let us imagine that the road is partitioned into L identical cells such that each cell can accommodate at most one bus at a time. Moreover, a total of S $(0 \leq S \leq L)$ *equispaced* cells are identified in the beginning as bus stops. Note that, the special case $S = L$ corresponds to the *hail-and-ride* system. At any given time step, a passenger arrives with probability f to the system. Here, we assume that a given passenger is equally likely to arrive at any one of the bus stops with a probability $1/S$. Thus, the average number of passengers that arrive at each bus stop per unit time is given by f/S. In contrast to this model, in ref. [12, 13] the passengers were assumed to arrive with probability f at all the bus stops in every time step.

The symbol H is used to denote the hopping probability of a bus entering into a next cell. In contrast to most of all the earlier bus route models [9], we assume that the maximum number of passengers that can get into one bus at a bus stop is N_{max}. Suppose, N_i denotes the number of passengers waiting at the bus stop i $(i = 1, \ldots, S)$ at the instant of time when a bus arrives there. We assume the form

$$H = \frac{Q}{\min(N_i, N_{\text{max}}) + 1} \tag{1}$$

where $\min(N_i, N_{\text{max}})$ is the number of passengers who can get into a bus which arrives at the bus stop i at the instant of time when the number of passengers waiting there is N_i. The form (1) is motivated by the common expectation that the time needed for the passengers boarding a bus is proportional to their number. If H equals to $Q(= \textit{const.})$, this systems is reduced to ASEP. Fig. 1 depicts the hopping probabilities schematically.

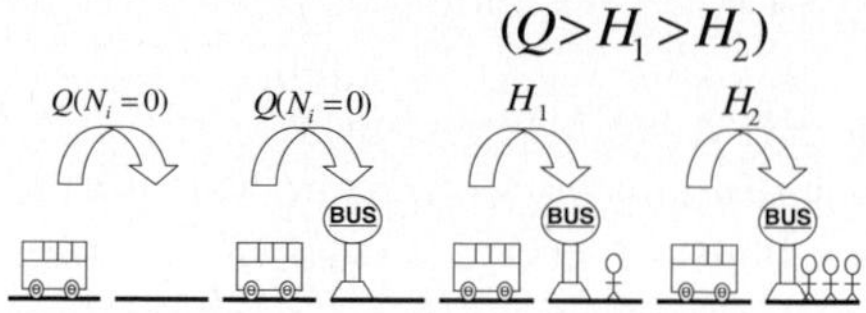

Fig. 1. Schematic illustration of the PCM. The hopping probability to the bus stop depends on the number of waiting passengers.

When a bus arrives at the i-th bus stop cell, the corresponding number N_i of waiting passengers is updated to $\max(N_i - N_{\max}, 0)$ to account for the passengers boarding the bus. Every bus stop has information I_i which is the number of buses in the segment of the route between the stop i and the next stop $i+1$ at that instant of time. This information is updated at each time steps. When one bus leaves the i-th bus stop, I_i is increased to $I_i + 1$. On the other hand, when a bus leaves $(i+1)$-th bus stop, I_i is reduced to $I_i - 1$. The desirable value of I_i is $I_0 = m/S$, where m is the total number of buses, for all i so that buses are not clustered in any segment of the route. We implement a traffic control rule based on the information I_i: a bus remains stranded at a stop i as long as I_i exceeds I_0. This rule restricts the size of such clusters to a maximum of I_0 buses in a segment of the route in between two successive bus stops.

We use the average speed $\langle V \rangle$ of the buses and the number of the waiting passengers $\langle N \rangle$ at a bus stop as two quantitative measures of the efficiency of the public conveyance system under consideration; a higher $\langle V \rangle$ and smaller $\langle N \rangle$ correspond to an efficient transportation system.

3 Computer Simulations of PCM

In the simulations we set $L = 500, Q = 0.9$ and $N_{\max} = 60$. Typical space-time plots are given in Fig. 2. If no information-based traffic control system exists, the buses have a tendency to cluster; this phenomenon is very simular to that observed in the ant-trail model [12, 13]. These quantitative features of the coarsening of the clusters have strong similarities with coarsening phenomena in many other physical systems.

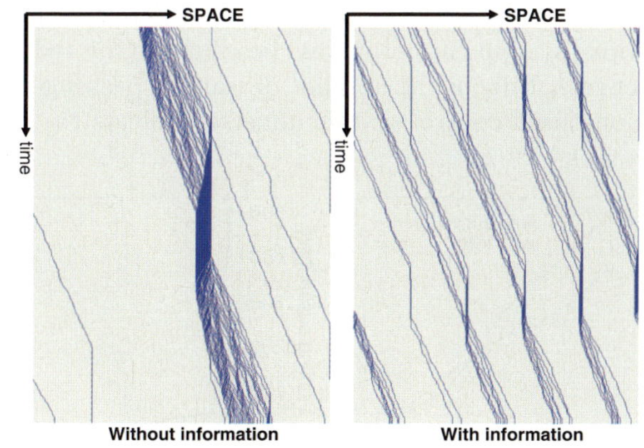

Fig. 2. Space-time plots for the parameter values $f = 0.6, S = 5, m = 30$. The left figure correspond to the case where no traffic control system based on the information is operational. The right figure correspond to the case where the information based bus-traffic control system is operational.

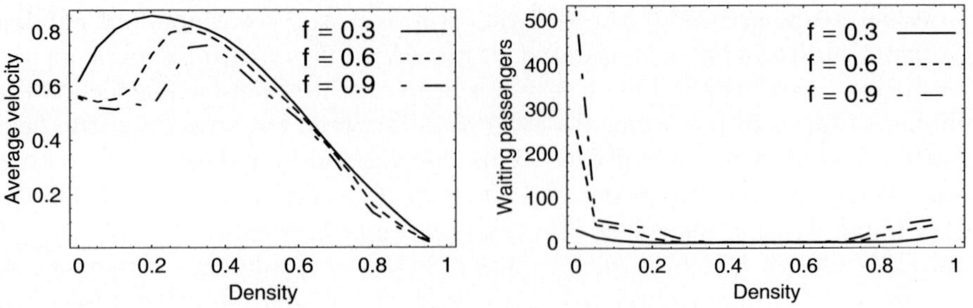

Fig. 3. The plot of $\langle V \rangle$ and $\langle N \rangle$ without information ($S = 5, f = 0.3, 0.6$ and 0.9).

In the Fig. 3 and Fig. 4, we plot $\langle V \rangle$ and $\langle N \rangle$ against the density of buses for several different values of f, where the density of buses is defined by $\rho = m/L$. These figures demonstrate that the average speed $\langle V \rangle$, which is a measure of the efficiency of the bus traffic system, exhibits a *maximum* at low density. As f increases in the system without information, $\langle V \rangle$ is lower by bus clustering. Whereas with information, $\langle V \rangle$ is higher than without information, since the clustering has been resolved. Furthermore, the density corresponding to the peak of $\langle V \rangle$ shifts to lower values when the information-based traffic control system is switched on. The average number of waiting passengers $\langle N \rangle$, whose inverse is another measure of the efficiency of the bus traffic system, is vanishingly small in the region $0.3 < \rho < 0.7$. This is because, in the case with information, bus clustering is occurred and buses do not move smoothly.

The data shown in Fig. 4 establish that implementation of the information-based traffic control system does not necessarily always improve the efficiency of the public conveyance system. In fact, in the region $0.3 < \rho < 0.7$, the average velocity of the buses is higher if the information-based control system is switched off. Comparing $\langle V \rangle$ and $\langle N \rangle$ in Fig. 4, we find that information-based traffic control system can improves the efficiency by reducing the crowd of waiting passengers. But, in the absence of waiting passengers, introduction of the information-based control system adversely affects the efficiency of the

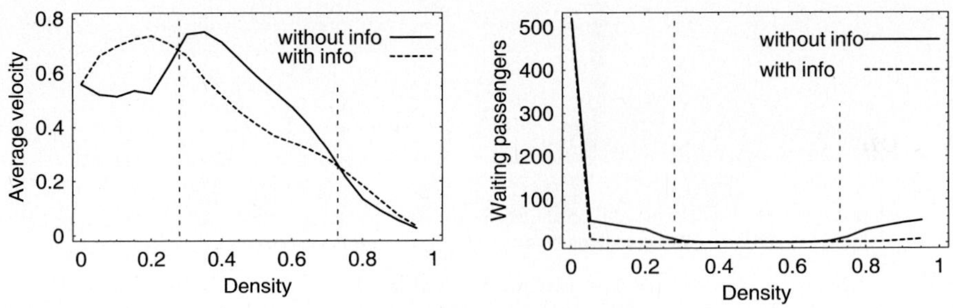

Fig. 4. The comparing plots between with information and without information ($S = 5$ and $f = 0.9$).

public conveyance system by holding up the buses at bus stops when the number of buses in the next segment of the route exceeds I_0.

4 Mean Field Analysis

Let us estimate $\langle V \rangle$ theoretically in the low density limit where m buses run practically unhindered and are distributed uniformly in the system without correlations, that is, this situation are equivalent to $\rho \to 0$. The average number of passengers N waiting at a bus stop, just before the arrival of the next bus, is

$$N = \frac{f}{S}\left(\frac{\frac{L}{S}-1}{Q} + \frac{1}{q}\right)\frac{S}{m}. \tag{2}$$

The first factor f/S on the right hand side of the equation (2) is the probability of arrival of passengers per unit time. The second factor on the right hand side of (2) is an estimate of the average time taken by a bus to traverse one segment of the route, i.e. the part of the route between successive bus stops. The last factor in the same equation is the average number of segments of the route in between two successive buses on the same route.

Suppose, T is the average time taken by a bus to complete one circuit of the route. The number of hops made by a bus with probability q during the time T is S, i.e. the total number of bus stops. Therefore the average period T for a bus and the hopping probability q is well approximated by

$$T = \frac{L-S}{Q} + \frac{S}{q}, \quad q = \frac{Q}{N+1} \tag{3}$$

respectively. Then, we have

$$\langle V \rangle = \frac{L}{T} = Q - \frac{f}{m}. \tag{4}$$

As long as the number of waiting passengers does not exceed $N_{\max}$, i.e. we take sufficiently small f, we have observed reasonably good agreement between the analytical estimates (4) and the corresponding numerical data obtained from computer simulations. However, our mean field analysis breaks down when a bus can not pick up all the passengers waiting at a bus stop.

5 Concluding Discussions

In this study, we have proposed a public conveyance model (PCM) by using stochastic CA. We have obtained quantitative results by using both computer simulations and analytical calculations. In particular, we have introduced two different quantitative measures of the efficiency of the public conveyance system. We have found that the bus system works efficiently in a region of moderate number density of buses; too many or too few buses drastically reduce the efficiency of the bus-transport system. If the density of the buses is lower than optimal, not only large number of passengers are kept waiting at the stops for longer duration, but also the passengers in the buses get a slow ride as buses run slowly because they are slowed down at each stop to pick up the waiting passengers. On the other hand, if the density of the buses is higher than optimal, the mutual hindrance created by the buses in the overcrowded route also lowers the efficiency of the transport system. Moreover, we have found that the average velocity increases, and the number of waiting passengers decreases, when the information-based bus traffic control system is switched on. However, this enhancement of efficiency of the conveyance system takes place only over a particular range of density; the information-based bus traffic control system does not necessarily improve the efficiency of the system in all possible situations.

Acknowledgements

The author (AT) was supported through the 21st Century COE Program, "Mechanical Systems Innovation," by the Ministry of Education, Culture, Sports, Science and Technology.

The work of DC has been supported, in part, by the Council of Scientific and Industrial Research (CSIR), government of India.

References

1. B. Schmittmann and R.K.P. Zia, in: *Phase Transition and Critical Phenomena*, Vol. 17, eds. C. Domb and J.L. Lebowitz (Academic Press 1995).
2. B. Derrida, Phys. Rep. **301**, 65 (1998).
3. G.M. Schütz, in *Phase Transitions and Critical Phenomena*, vol. 19 (Acad. Press, 2001).
4. D. Chowdhury, L. Santen and A. Schadschneider, Phys. Rep. **329**, 199 (2000).
5. D. Helbing, Rev. Mod. Phys. **73**, 1067 (2001).
6. A. Schadschneider, Physica A **313**, 153 (2002).
7. T. Nagatani, Rep. Prog. Phys. **65**, 1331 (2002).
8. D. Chowdhury, K. Nishinari, L. Santen and A. Schadschneider, *Stochastic Transport in Complex Systems*, Elsevier (2008).
9. O.J. O'Loan, M.R. Evans and M.E. Cates, Phys. Rev. E **58**, 1404 (1998).
10. R. Jiang, B. Jia and Q.S. Wu, J. Phys. A: Math. Gen. **37**, 2063 (2004).
11. A. Tomoeda, K. Nishinari, D. Chowdhury and A. Schadschneider, Physica A **384**, 600 (2007).
12. D. Chowdhury, V. Guttal, K. Nishinari and A. Schadschneider, J. Phys. A: Math. Gen. **35**, L573 (2002).
13. A. Kunwar, A. John, K. Nishinari, A. Schadschneider and D. Chowdhury, J. Phys. Soc. Jpn. **73**, 2979 (2004).

Modelling and Simulating Several Time-Delay Mechanisms in Human and Automated Driving

Martin Treiber and Arne Kesting

Technische Universität Dresden, Institute for Transport & Economics,
Andreas-Schubert-Straße 23, D-01062 Dresden, Germany

Summary. In vehicular traffic, reactions to new traffic situations are subject to several mechanisms of time delay. Besides the reaction time of the drivers (or sensors), the finite acceleration capabilities lead to a nonzero "velocity adaptation time" to perform the action itself (e.g. changing the velocity). The commonly used explicit integration schemes for simulating the models introduce the update time as a third delay parameter. By means of numerical simulations with a time-continuous car-following model, we investigate how these times interplay with each other. We show that the three delay times give rise to two types of instabilities: The long-wavelength string instability is mainly driven by the velocity adaptation time while short-wavelength local instabilities arise for sufficiently high reaction and update times. We show that, with respect to stability, there is an 'optimal' adaptation time as a function of the reaction time and draw implications for human *vs.* semi-automated driving.

1 Introduction

An essential feature of human (in contrast to automated) drivers is a considerable reaction time of the order of 1 s. Additional delays in reaching the velocity appropriate for the current traffic situation (and driving style) are due to the finite acceleration and deceleration capabilities. The time step of the numerical update schemes gives rise to a third source of delays. Although of different nature, these sources of delay are often combined. For example, in the car-following models of Newell [1] and Gipps [2], the velocities of each time step are calculated as a function of the situation in the past time step, i.e., the velocity adaptation time is set equal to the numerical update time. Moreover, this time is often denoted as a "reaction time", or as a "safe time gap" (in fact, the time gap to the leading vehicle is a forth characteristic time which is equal to the other times in the models mentioned above). In this work, we investigate the effects of the different delay times separately, thereby clarifying the role of each of the associated mechanisms on stability.

2 Framework and Models

As basis, we consider time-continuous car-following models having an instantaneous *acceleration* function of the following general form:

$$\frac{dv_\alpha}{dt} = a^{\text{mic}}\left(s_\alpha, v_\alpha, \Delta v_\alpha\right), \tag{1}$$

i.e., the acceleration of the considered vehicle α is a function of its velocity v_α, the spatial gap s_α, and the approaching rate Δv_α to the leader.

A reaction time T_r is easily introduced into these models by taking the right-hand side of (1) at the past time $t - T_r$ resulting in a coupled system of delay-differential equations [3]. For the numerical update (fixed timestep Δt), we use following modified Euler scheme:

$$\begin{aligned} v_\alpha(t+\Delta t) &= v_\alpha(t) + a_{\text{mic}}\Delta t, \\ s_\alpha(t+\Delta t) &= s_\alpha(t) + v_\alpha(t)\Delta t + \tfrac{1}{2}a_{\text{mic}}(\Delta t)^2, \end{aligned} \tag{2}$$

where a_{mic} denotes the acceleration (1) of the micromodel at the delayed time $t - T_r$. If T_r is not a multiple of Δt, additional interpolation techniques have been used [3].

Finally, we assume that the drivers are aware of their finite reaction capabilities and adopt temporal anticipation strategies based on constant-velocity and constant-acceleration heuristics, as well as "multi-anticipation" by considering a total of four leaders, see [3] for details. From the point of view of feedback control (Fig. 1), the presented set of models gives rise to three delay mechanisms characterized by the reaction time T_r, the numerical update time Δt, and the adaptation time scale τ_v for reaching a certain target velocity for the actual traffic situation. The latter time (typical values are 5-10 s for city traffic and 10-20 s for freeway traffic) increases with decreasing acceleration and deceleration capabilities.

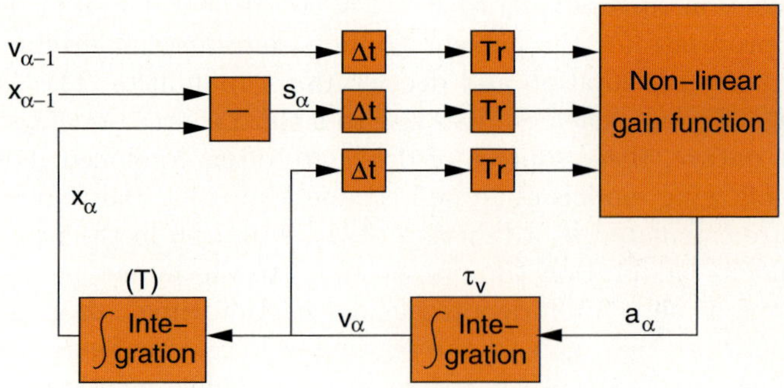

Fig. 1. The acceleration reaction of the general car-following model with delay from the perspective of feedback control.

For the specific implementation of this concept, a specific car-following model of the class given by Eq. (1) has to be selected. In this paper, we use the *intelligent-driver model* (IDM) [4] which, in relation to its simplicity, shows a good agreement with observations [5], and has intuitive model parameters. The acceleration equation of the IDM for vehicle α is given by

$$\frac{dv_\alpha}{dt} = a\left[1 - \left(\frac{v_\alpha}{v_0}\right)^4 - \left(\frac{s^*(v_\alpha, \Delta v_\alpha)}{s_\alpha}\right)^2\right], \tag{3}$$

where the "desired" spatial gap that is appropriate for the actual dynamical situation is given by

$$s^*(v, \Delta v) = s_0 + vT + \frac{v\Delta v}{2\sqrt{ab}}. \tag{4}$$

The IDM parameters and the values used in this paper are the following: Desired velocity $v_0 = 120\,\mathrm{km/h}$, desired time gap in car-following situations $T = 1.5\,\mathrm{s}$, minimum spatial gap (for a complete standstill) $s_0 = 2\,\mathrm{m}$, the comfortable deceleration $b = 2\,\mathrm{m/s^2}$, and the acceleration a in the range between $0.3\,\mathrm{m/s^2}$, and $2.5\,\mathrm{m/s^2}$. To investigate the influencing factors for stability, we will furthermore vary the reaction time T_r, and the numerical update time Δt in the range between 0 s and 2 s.

3 Numerical Investigation of the Stability Regimes

We investigate the stability of traffic flow by simulating two scenarios, (i) a platoon of vehicles following an externally controlled lead vehicle, and (ii) a more realistic open system with a flow-conserving bottleneck.

3.1 Platoon Scenario

In this scenario, a platoon of n vehicles follows an externally controlled vehicle. Initially, the leader drives at $v_{\mathrm{lead}}(0) = 25\,\mathrm{m/s}$, and all following vehicles are in equilibrium, i.e., $v_\alpha(0) = v_{\mathrm{lead}}(0)$, and $s_\alpha(0) = s_e(v_\alpha(0))$ where $s_e(v)$ is defined by $a_{\mathrm{mic}}(s_e(v), v, 0) = 0$. At time $t = 1000\,\mathrm{s}$, the leading vehicle reduces its velocity to $19\,\mathrm{m/s}$ by decelerating for three seconds with $\dot{v}_{\mathrm{lead}}(t) = -2\,\mathrm{m/s^2}$ before continuing with $19\,\mathrm{m/s}$. All platoon vehicles have the same set of IDM parameters and the same values for T_r and Δt.

Figure 2 shows trajectories of selected platoon vehicles for $T_r = 0.9\,\mathrm{s}$, $\Delta t = 0.1\,\mathrm{s}$, and two values of the acceleration parameter. For $a = 0.3\,\mathrm{m/s^2}$ (left column), we observe the classical collective string instability, i.e., each vehicle is locally stable (the perturbation caused by the leading vehicle eventually dies away), but the perturbation grows with increasing vehicle index. Furthermore, the instability is of a long-wavelength type, i.e., the typical wavelengths

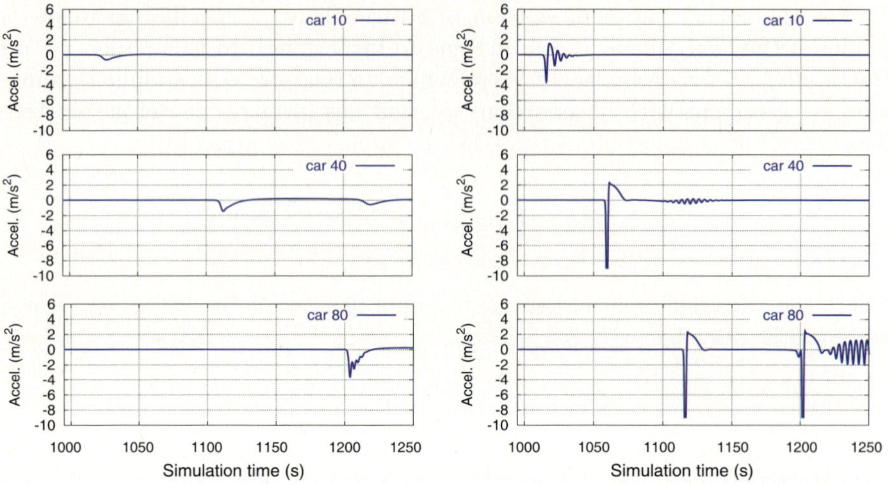

Fig. 2. Trajectories of several vehicles in the platoon for low accelerations (left column, $a = 0.3\,\mathrm{m/s^2}$), and high accelerations (right column, $a = 2.5\,\mathrm{m/s^2}$). Here, "car n" denotes the n-th vehicle behind the leader inducing the perturbation.

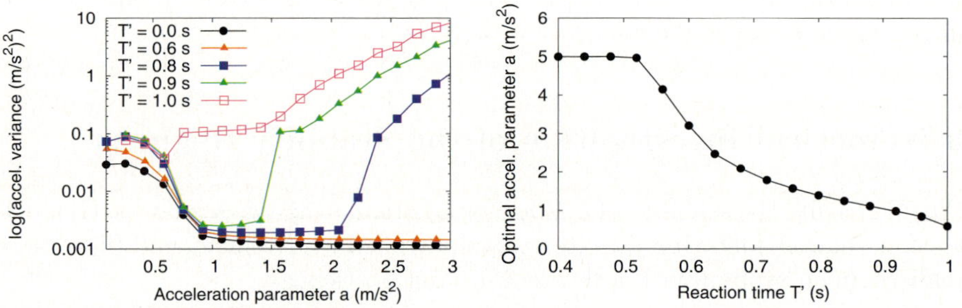

Fig. 3. (Left) Average amplitude of the acceleration changes (logarithmic presentation) as measure of the instability as a function of the reaction time T_r and the IDM acceleration a. (Right) Optimum value for a (upper limit: $5\,\mathrm{m/s^2}$) as a function of the reaction time.

$2\pi/k$ of the growing perturbation satisfy $2\pi/k \gg s_e(v)$. For $a = 1.0\,\mathrm{m/s^2}$, the platoon is string stable (not shown). However, on further increasing the acceleration, the platoon becomes unstable again (right column for $a = 2.5\,\mathrm{m/s^2}$), but the instability starts with short wavelengths that are of the same order of magnitude as the equilibrium gap.

Since this second type of instability does not occur for zero reaction time T_r, it is instructive to investigate its dependence on T_r. Figure 3 (left) displays a continuous measure for the degree of instability (the average acceleration variance) as a function of the acceleration a for several values of T_r. As a remarkable result, the stable range of accelerations is limited from above, and

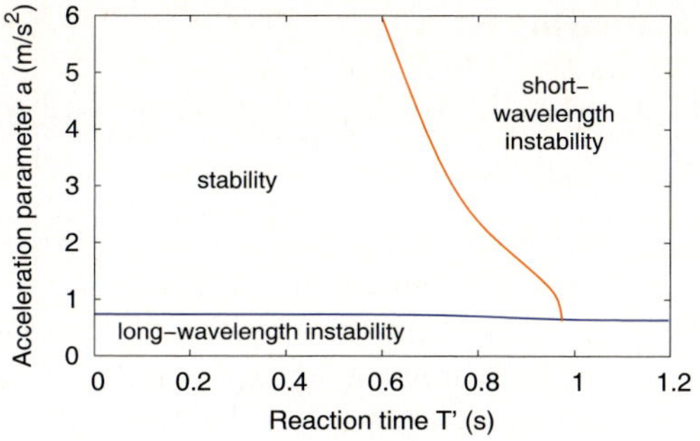

Fig. 4. Phase diagram for the different stability regimes in the space spanned by the IDM acceleration a and the reaction time T_r.

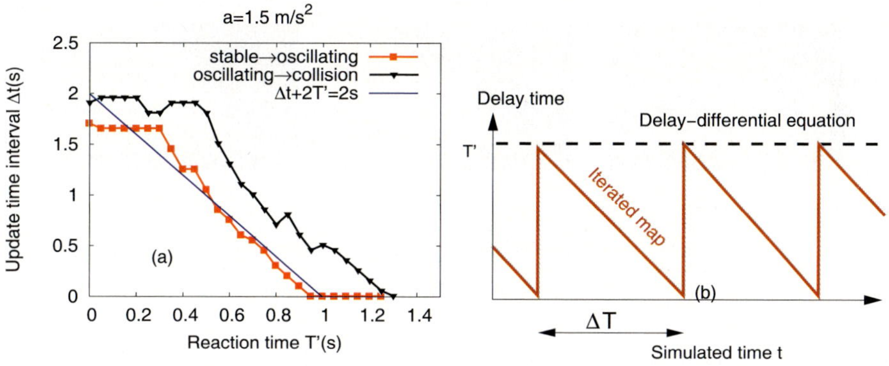

Fig. 5. (Left) stability thresholds in the space spanned by the reaction time T_r and the numerical update time Δt; (right): Visualization of the effective delay time caused by T_r and Δt.

the critical acceleration decreases with increasing reaction time. In contrast, the lower limit is essentially independent from T_r (Fig. 4). Consequently, the "optimal" acceleration with respect to stability decreases with growing values of T_r (Fig. 3 right).

We also compared the destabilizing effects of the reaction time in relation to the effects caused by long update time intervals Δt of the modified Euler integration scheme (2) (Figure 5 left). Remarkably, the destabilizing effect of T_r is comparable to that of $2\Delta t$, i.e., the stability is approximatively constant along the line defined by $T_r + 2\Delta t = C = \text{const}$. This can be explained by observing that, depending on the actual time t, the update time Δt induces effective delays T' between zero and Δt, i.e., $\Delta t/2$, on average (Figure 5 right).

3.2 Bottleneck Scenario

We have simulated an open 15 km long single-lane road section. The traffic demand at the upstream boundary of the initially empty system has been increased from 0 vehicles/h to 1700 vehicles/h in the first half hour and remained at this level, afterwards. A flow-conserving bottleneck was introduced by increasing the time gap parameter T by a factor of 1.5 (i.e. from 1.5 s to 2.25 s) for 10.5 km $\leq x \leq$ 11.5 km thereby locally decreasing the capacity. This can be caused, e.g., by a more defensive driving style adopted on narrow road sections or when encountering increased velocity differences or other perturbations near junctions or intersections [6]. The delay parameters were set to $T_r = 0.6$ s and $\Delta t = 0.5$ s. Figure 6 shows spatiotemporal velocity plots

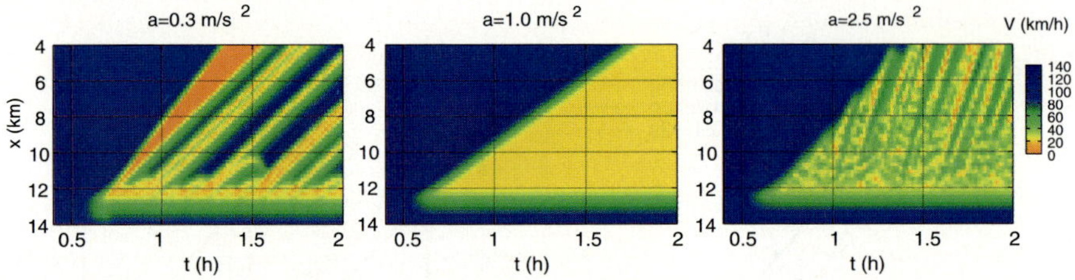

Fig. 6. Spatiotemporal dynamics of the congestion in the "bottleneck" scenario for three values of the IDM acceleration parameter a.

of the simulations. For $a = 0.3\,\mathrm{m/s^2}$, the collective instability displays itself as stop-and-go waves emanating from a permanently congested region at the bottleneck (which is sometimes called the "pinch effect"). The short-wavelength instability at $a = 2.5\,\mathrm{m/s^2}$ also leads to oscillatory congested patterns, but with shorter typical wavelengths. Moreover, the propagation velocity of the perturbations (against the traffic direction) is increased.

4 Discussion and Conclusion

Relevant mechanisms for delays in vehicular traffic dynamics include the reaction time, the velocity adaptation time, and, in the case of automated driving or computer simulations, a finite update time. By separately modelling and simulating each of these delays, we found two qualitatively different instability mechanisms. Besides the common long-wavelength string instability caused mainly by low acceleration capabilities, another short-wavelength instability has been found for finite reaction times and sufficiently *high* accelerations. As a consequence, there is an optimal acceleration that decreases with the reaction time, i.e., people with long reaction times should not accelerate too much, and vice versa. Furthermore, it has been found that the reaction and numerical update times are essentially interchangeable, but the update time step corresponds effectively to only *half* of the reaction time.

References

1. G. Newell, Operations Research **9,** 209 (1961).
2. P. G. Gipps, Transportation Research Part B: Methodological **15,** 105–111 (1981).
3. M. Treiber, A. Kesting, and D. Helbing, Physica A **360,** 71–88 (2006).
4. M. Treiber, A. Hennecke, and D. Helbing, Physical Review E **62,** 1805–1824 (2000).
5. E. Brockfeld, R. D. Kühne, and P. Wagner, Transportation Research Record: Journal of the Transportation Research Board **1876,** 62–70 (2004).
6. M. Treiber, A. Kesting, and D. Helbing, Physical Review E **74,** 016123 (2006).

Multi-Class First Order Traffic Flow Modeling

Hans van Lint[1], Serge P. Hoogendoorn[1], and Marco Schreuder[2]

[1] Traffic and Transport Department, Faculty of Civil Engineering and Geosciences, Delft University of Technology, Stevinweg 1, 2628 CN Delft, The Netherlands j.w.c.vanlint@tudelft.nl, s.p.hoogendoorn@tudelft.nl

[2] AVV Transport Research Centre, Ministry of Transport, Public Works and Water management, 3000 BA, Rotterdam, The Netherlands marco.schreuder@rws.nl

1 Introduction

Modeling and predicting traffic conditions (speeds, travel times, flows, etc) in a heterogeneous traffic network is one of the main research areas in traffic and transport science. Many authors acknowledge that traffic *heterogeneity* constitutes a significant if not dominant factor in accurately modeling freeway traffic flow operations [1, 2]. For example, high truck percentages may induce congestion at much lower volumes than low percentages, and consequently may result in very different network traffic dynamics and hence traffic conditions. This implies that traffic models for real time decision support systems in traffic management centers should provide the means to account for traffic heterogeneity. In this paper we propose a new analytical macroscopic network traffic flow model FASTLANE (FAST multicLass mAcroscopic NEtwork simulation model) which specifically addresses the heterogeneity of network traffic. FASTLANE differs from earlier multi-class models (see e.g. [4] in that it calculates traffic dynamics (of the total heterogeneous traffic flow) in terms of so-called person car equivalents (pce's). The main innovation is that these pce values are state-specific, that is, they are a function of the prevailing class-specific speeds. The next section will first mathematically describe the FASTLANE model. The section thereafter will present some numerical results on a real traffic network to illustrate the main concepts. The paper closes with conclusions and directions for further research.

2 Mathematical Model

Below we briefly describe both the mathematical formulation of FASTLANE as well as a numerical solution to these equations. In this paper a traffic network is described by a directed graph $G = (N, A)$ of nodes $n \in N$ and arcs $a \in A$.

2.1 Traffic Dynamics at a Homogenous Link

Similar to earlier LWR based multiclass traffic flow models (see e.g. [4] for an overview), the core of the FASTLANE traffic flow model is the class-specific conservation of vehicle equation 1. In the ensuing, the subscript u will denote the user-class (e.g. person-cars, trucks). For each class u we have (for each link a)[3]:

$$\frac{\partial k_u}{\partial t} + \frac{\partial q_u}{\partial x} = 0 \tag{1}$$

In equation 1, $k_u = k_u(t,x)$ denotes the class-specific density and $q_u = q_u(t,x) = k_u(t,x) \cdot v_u(t,x)$ denotes the class-specific flow at time instant t and location x. The boundary conditions (inflow at the entry of the link and the outflow at the link exit) are determined respectively by the upstream and the downstream links, expect for the case where a is an origin or a destination link. The class-specific speed is defined by $v_u(t,x) = V_u(K(t,x)), K(t,x) = \kappa(k_1(t,x), \dots, k_U(t,x))$ where V_u denotes the class-specific equilibrium speed as a function of the *effective* density K. $K(t,x)$ is described by a function of the class-specific densities

$$K(t,x) = \sum_{u=1}^{U} pce_u(t,x) k_u(t,x) \tag{2}$$

in which the person-car equivalents pce_u, are state specific and specified by:

$$pce_u(t,x) = \frac{s_u + T_u V_u(t,x)}{s_{u_0} + T_{u_0} V_{u_0}(t,x)} \tag{3}$$

In equation 3, s_u denotes the class-specific gross stopping distance, and T_u denotes the class-specific minimum headway. The equilibrium speed V_u is chosen such that for effective densities K larger than some critical density K_c, the speeds of all classes u are equal to the critical speed v_c. In free-flow conditions, different classes move with different average speeds. More specifically, we have:

$$V_u(K) = \begin{cases} v_u^0 - K\frac{(v_u^0 - v_c)}{K_c} & , K < K_c \\ \frac{v_c K_c}{K}\left(1 - \frac{K - K_c}{K_{jam} - K_c}\right) & , K \geq K_c \end{cases} \tag{4}$$

in which the parameter K_{jam} denotes the effective jam density. Clearly, the parameters K_c and v_c, and the vehicle composition (e.g. percentage of trucks) determine the effective capacity of a link.

2.2 Traffic Dynamics at the Nodes

In the network description, three node types are considered: link-to-link nodes, merges and diverges (Fig. 1a, b and c). A link-to-link node (Fig. 1a) depicts a

[3] We omit the link index a to keep notation simple

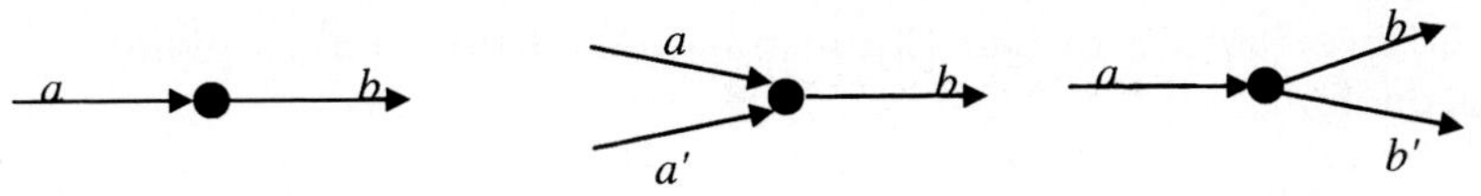

(a) link-to-link node (b) merge (c) diverge

Fig. 1. Normal nodes, merges and diverges.

simple interface between two network links a and b, describing a discontinuity, such as a lane drop, or a change in speed limits. To determine the resulting traffic dynamics, the class-specific demands D_u^a of an incoming link a are compared to the mixed-class supply S^b of the outgoing link b. Put simply, traffic demand is transferred from link a to link b proportional to the traffic composition on link a, and constraint by the maximum possible (total) flow which can enter in link b. This is modeled by means of class-specific shares (proportions) λ_u, which are calculated as follows

$$\lambda_u(t,x) = \frac{pce_u(t,x) \cdot q_u(t,x)}{\sum\limits_u pce_u(t,x) \cdot q_u(t,x)} \tag{5}$$

At the origins, these shares are equal to the traffic composition (e.g. truck percentage) set by the user. As a result, we get the following expression for the class-specific flow between link a and b:

$$q_u^{a \to b} = \frac{1}{pce_u^a} \min\left(D_u^a, \lambda_u^a S^b\right) \tag{6}$$

Merge nodes (Fig. 1b) are more complex. Assuming that we have two incoming links a and a' and one outgoing link b, we need to determine how the supply (so the amount of traffic able to enter outgoing link b) is distributed not only across the classes u through shares λ_u^a, but also across the incoming links a and a' through proportions κ_u^a. In the current version of the model, we make the (rough) assumption that the available supply is distributed according to the effective capacity of the incoming links, that is

$$q_u^{a' \to b} = \frac{1}{pce_u^{a'}} \min\left(D_u^{a'}, \kappa^{a'} \lambda_u^{a'} S^b\right) \tag{7}$$

with

$$\kappa^{a'} = \frac{C^{a'}}{C^a + C^{a'}}, \kappa^a = 1 - \kappa^{a'}$$

Furthermore, if the demand of one of the incoming links is less than the assigned supply, the remaining supply will be assigned to the other link. Finally, the traffic dynamics at a diverge or bifurcation node (Fig. 1c) are described by the turnfractions γ_n^b, which depict the distribution of the total flow over the outgoing links. As a result, for merges we have

$$q_u^{a \to b} = \frac{1}{pce_u^a} \min\left(\gamma_n^b D_u^a, \lambda_u^a S^b\right) \tag{8}$$

2.3 Discretization of the Equations and Numerical Solution Approach

To numerically approximate solutions of the resulting first-order scheme, a new numerical solution approach was developed based on the generalization of the well-known Godunov scheme [3]. As with most numerical solution schemes, each link a is divided into a number of cells i of typically 100–200m. For each cell, the scheme determines a class-specific demand and a mixed-class supply, according to equations 6–8. The mixed-class supply is then distributed over the classes according to the shares in the demands of the upstream cells. The same approach is used to handle the dynamics of the network nodes.

3 Example Application

3.1 Experimental Setup, Network and Parameters

To illustrate multiclass traffic operations produced by FASTLANE under different truck percentages below a testcase is presented. Fig. 2 shows the 15 km A20/A12 freeway corridor, which connects the northern part of the urban beltway around Rotterdam to the A12 freeway corridor from The Hague to Utrecht at the height of the city of Gouda. In this example, two classes of vehicles, trucks and person cars are defined according to the following parameters: For person cars we used: $v_u^0 = 120$ km/h, $s_u = 7.5$m, $T_u = 1.2$s, and for trucks: $v_u^0 = 85$ km/h, $s_u = 15$m, $T_u = 1.8$s. The other mixed-class parameters were chosen as follows: $v_c = 85$ km/h, $K_c = 25$ pce/km and $K_{jam} = 1000/7.5$ pce/km. As a result, the capacity of a single lane equals 2125 (pce/h!). The demand scenario and the set of turnfractions (at the off-ramps) used are derived from real traffic data obtained from inductive loops spaced approximately every 500 meters along this route. We varied the percentage trucks (at the origins) from 0, 5, 10, to 15% respectively.

3.2 Results

Fig. 3 shows the resulting traffic speeds along the 15.8 km route from Rotterdam to Gouda under 4 different truck percentages. In all cases congestion sets in almost simultaneously (a little after 6:15AM) at the onramp "Nieuwerkerk" and the merge between the A20 and A12 motorways further downstream. Clearly, Fig. 3a–d show that both queue length as well as queue duration increase significantly under growing truck percentages. In the (unlikely) case of 0% trucks, congestion resolves around 8 AM (Fig. 3a), while in case of 10% trucks congestion resolves more than an hour later (Fig. 3c). The rationale behind this is intuitive, since in the multi-class case (percentages $> 0\%$), the capacity in veh/h is a (decreasing) function of the percentage of trucks, due to the fact that trucks occupy relatively more space than person cars with

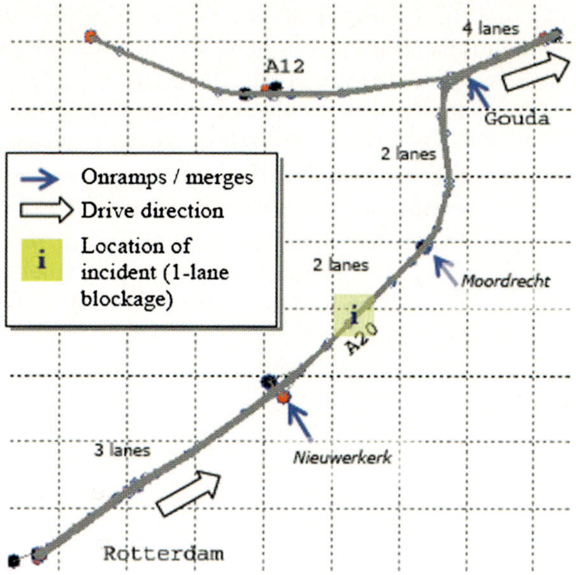

Fig. 2. Network: A20/A12 between Rotterdam and Gouda.

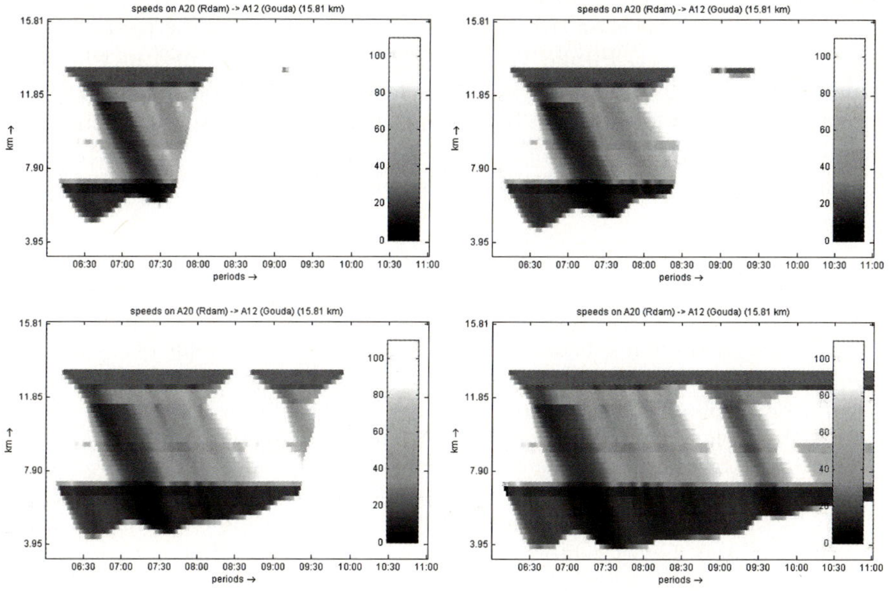

Fig. 3. Resulting speeds under different truck percentages.

decreasing speeds (see 3). A second observation is that the resulting traffic patterns in Fig. 3 exhibit more realistic features than one might expect from first order traffic flow theory. Even in the 0% case (Fig. 3a), a clear backward moving jam is visible which propagates from Gouda (around 6:45AM) to Nieuwerkerk (7AM) and further upstream. These patterns are predominantly the result of using demands, and turnfractions estimated from real data. In this sense, deterministic multi-class first order theory applied on a real traffic network under real (stochastic) demands and turnfractions, is able to reproduce some of the higher order phenomena (e.g. parallel backward moving jams and - to a degree - stop-and-go-traffic) claimed to be reproducible by higher order approaches only.

4 Conclusions

This paper introduced a new multi-class first order traffic flow model FASTLANE. Its main innovation lies in the fact that the dynamics are described in terms person car equivalents, which are state-specific. From results on a realistic test network it appears that the model produces intuitive results. Higher truck percentages yield longer and more severe congestion, which is due to the fact that in congestion, larger vehicles occupy *relatively* more space (gross distance headway). Currently, a first version of the FASTLANE model is developed in the decision support environment BOSS-offline (of the Dutch Ministry of Transport, Public Works and Watermanagement), in which demands, turn fractions and link quantities such as critical speeds and densities are estimated. This system will be rolled out in the five highway traffic management centers in the course of 2008. Further research includes incorporating destination specific flows (OD estimation) and real-time application.

References

1. S. P. Hoogendoorn and P. H. L. Bovy. State-of-the-art of vehicular traffic flow modeling. *Proc. Institution of Mechanical Engineers*, 215(1):283–303, 2001.
2. S. P. Hoogendoorn, P. H. L. Bovy, and J. W. C. Van Lint. Short-term prediction of traffic flow conditions in a multilane multiclass network. In *Proceedings of the 15th International Symposium on Transportation and Traffic Theory*, Sidney, Australia, 2002.
3. J. P. Lebacque. The Godunov scheme and what it means for first order traffic flow models. In J. B. Lesort, editor, *Proceedings of the 13th International Symposium on Transportation and Traffic Theory*, pages 647–677, Lyon, France, 1996.
4. Steven Logghe. *Dynamic Modelling of Heterogeneous Vehicular Traffic.* Faculty of Applied Science, Katholieke Universiteit Leuven, Leuven, 2003.

Dynamics and Thermodynamics of Traffic Flow

Hans Weber[1], Reinhard Mahnke[2], Christof Liebe[2], and Jevgenijs Kaupužs[3]

[1] Luleå University of Technology, Department of Physics, SE–97187 Luleå, Sweden
Hans.Weber@ltu.se
[2] Rostock University, Institute of Physics, D–18051 Rostock, Germany
reinhard.mahnke@uni-rostock.de; christof.liebe@uni-rostock.de
[3] Institute of Mathematics and Computer Science, University of Latvia, LV–1459 Riga, Latvia kaupuzs@latnet.lv

Summary. Application of thermodynamics to traffic flow is discussed. On a microscopic level, traffic flow is described by Bando's optimal velocity model in terms of accelerating and decelerating forces. It allows us to introduce kinetic, potential, as well as a total energy, which is the internal energy of the car system in view of thermodynamics. The total energy is however not conserved, although it has a certain value in any of the two possible stationary states corresponding either to a fixed point or to a limit cycle solution in the space of headways and velocities.

On a mesoscopic level of description, the size n of a car cluster is considered as a stochastic variable in the master equation for the system. Here $n = 0$ corresponds to the fixed point solution with no cluster of the microscopic model, whereas the limit cycle is represented by the coexistence of a car cluster with $n > 0$ and a free flow phase.

The stationary solution obeys a detailed balance condition, which allows us to describe some properties of the model by equilibrium thermodynamics in analogy to the liquid–vapour system. We define the free energy and the chemical potential of the car system. In this sense the behaviour of traffic flow can be described by equilibrium thermodynamics in spite of the fact that it is a driven system.

1 Introduction

It has already been manifested by Mahnke et al. [1] that the thermodynamic approach can be applied to such a many–particle driven system as traffic flow. Investigations of the complex system called vehicular traffic can be done on the microscopic level (dynamical equations like Bando's optimal velocity model), on the mesoscopic level by stochastic equations [2] as well as macroscopically applying equilibrium statistical physics [3]. Recently Krbálek [4] uses the one–dimensional thermodynamical traffic gas to predict clearance (also called spacing or headway) comparing with empirical traffic data. It has been stated that the derived exact formula for thermal–equilibrium spacing distribution of a traffic gas with a short–range repulsive two–body potential

$V(\Delta x) = 1/\Delta x$ is in successful agreement with highway traffic clearance distributions. Depending on vehicular density c the temperature T (or its inverse β), which acts as a fit parameter, changes its value.

2 Bando's Optimal Velocity Model: Dynamics

The dynamical system called Bando's optimal velocity model [5, 6] is a set of $2N$ coupled differential equations acting on a microscopic level, i. e. each car $i = 1, \dots, N$ has its own position x and velocity v equation. The ensemble of cars is subjected to periodic boundary conditions, see Fig. 1.

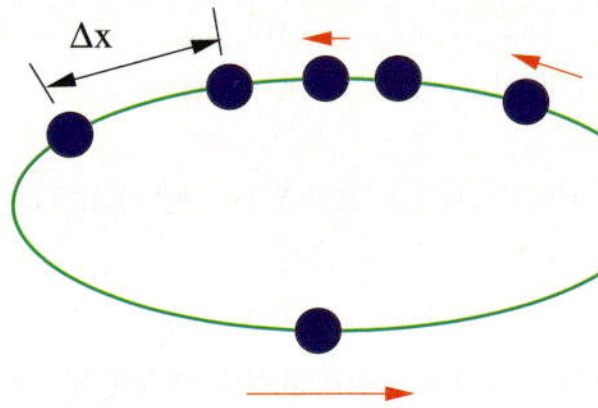

Fig. 1. One–dimensional circular road of length L with periodic boundary conditions. The cars are represented by filled circles, their velocities are marked by arrows and a headway distance is indicated by Δx. Taken from [3].

Denoting headway by $\Delta x_i = x_{i+1} - x_i$ and optimal velocity function by $v_{opt}(\Delta x) = v_{max}(\Delta x)^2/(D^2 + (\Delta x)^2)$, the well–known Bando model [5, 6] reads

$$m\frac{d}{dt}v_i = \frac{m}{\tau}\left\{(v_{max} - v_i) + (v_{opt}(\Delta x_i) - v_{max})\right\} \quad ; \quad \frac{d}{dt}x_i = v_i \ , \qquad (1)$$

where m (mass), v_{max} (maximal velocity), τ (reaction time) and D (interaction distance) are control parameters. The average density of cars on the rotary is $c = N/L$. The corresponding dimensionless version of Bando's model (1) has been investigated in these Proceedings by Liebe et al. [7]. Numerical results showing free flow and car cluster motion (limit cycle regime) at large density $c > c_{cr}$ are presented in Fig. 2.

3 Bando's Optimal Velocity Model: Energetics

The acceleration of cars (1) is split into two contributions

$$m\frac{d}{dt}v_i = F_{acc}(v_i) + F_{dec}(\Delta x_i) \ , \qquad (2)$$

where $F_{acc}(v_i) = \frac{m}{\tau}(v_{max} - v_i) \geq 0$ and $F_{dec}(\Delta x_i) = \frac{m}{\tau}(v_{opt}(\Delta x_i) - v_{max}) \leq 0$.

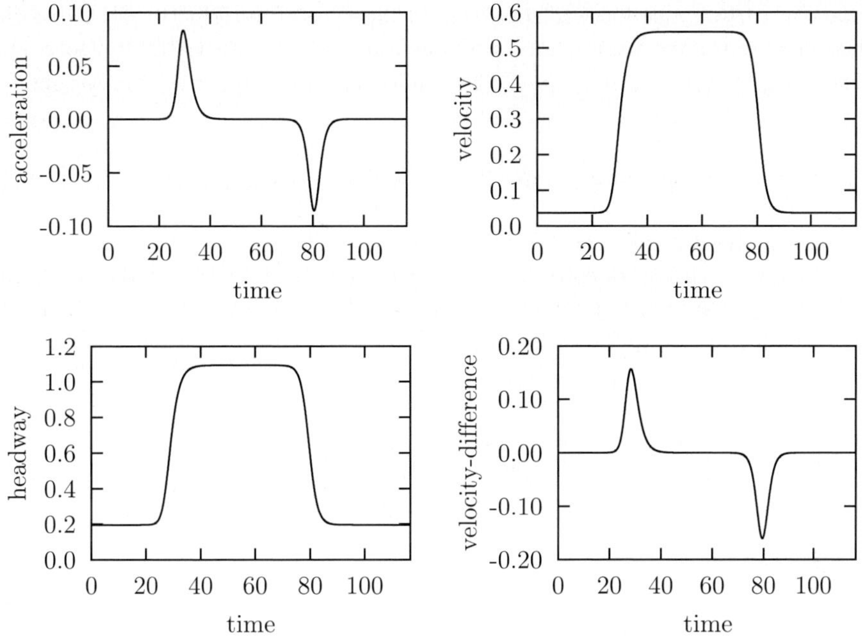

Fig. 2. Numerical integration of (dimensionless) equations of motion (1) for $N = 60$ cars on a rotary of dimensionless length $\widetilde{L} = 35$ (at large density $\widetilde{c} = N/\widetilde{L} = 1.714$) and control parameter $b = D/(\tau v_{max}) = 1.1$. All pictures show one period ($\Delta\tilde{t} = 116.75$) in the limit cycle regime for one single car. *Upper pictures:* Evolution of acceleration and velocity of a particular car i. *Lower pictures:* Evolution of the spatial distance (headway) and the velocity-difference between car i and $i+1$.

Defining kinetic (multiplying (2) by v_i) and potential energy (gradient of headway dependent force),

$$E_{kin}(v_i) = \frac{m}{2} v_i^2 \tag{3}$$

$$E_{pot}(\Delta x_i) = \int F_{dec}(\Delta x_i) d(\Delta x_i) = \frac{m D v_{max}}{\tau} \left(\frac{\pi}{2} - \arctan\left(\frac{\Delta x_i}{D} \right) \right) \tag{4}$$

the energy balance [1, 7, 8] reads

$$\frac{d}{dt} \left(E_{kin}(v_i) + E_{pot}(\Delta x_i) \right) + \Phi_i = 0 \ , \tag{5}$$

$$\Phi_i = - \left(F_{dec}(\Delta x_i)\, v_{i+1} + F_{acc}(v_i)\, v_i \right) \ . \tag{6}$$

It is evident that the total energy $E = T + V$ is not conserved in traffic flow. Due to the energy balance $dE/dt + \Phi = 0$ (compare (5,6) for a single car) there is always power from the engine necessary to compensate dissipation. The performance of a vehicular engine is about 50 kW (= 68 PS [horse power]) or more. The calculated dimensionless energy flux within the Bando model is about $\widetilde{\Phi} = 0.134$, also known from Fig. 3 (upper picture right) in [7]. Using the following exemplary values, for vehicular mass $m = 1$ ton $= 10^3$ kg, maximal speed $v_{max} = 100$ km/h $= \frac{10^2 \cdot 10^3}{3.6 \cdot 10^3}$ m/s and reaction time $\tau = 2$ s, we get for the original energy current (with units)

$$\Phi = \widetilde{\Phi}\,\Phi_0 = \widetilde{\Phi}\,\frac{E_0}{t_0} = \widetilde{\Phi}\,\frac{m v_{max}^2}{\tau} = 51.7 \text{ kW} \tag{7}$$

which is in the order of the power of a passenger car.

Depending on the density $c = N/L$ the vehicular system ends up in one of the two possible states which is either the fixed point $\Delta x_i = \Delta x_{hom}$, $v_i = v_{opt}(\Delta x_{hom})$ or the limit cycle. Fig. 3 shows the evolution of total energy for both cases.

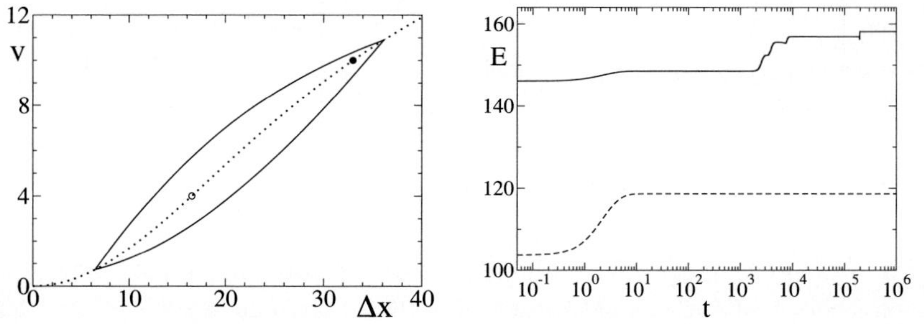

Fig. 3. *Left picture:* Fixed points (circles) and limit cycle (solid line) in the space of headways Δx and velocities v of cars. The solid circle represents the stable fixed point at the car density $c = 0.0303$ m^{-1}. The empty circle is the unstable fixed point at a larger density $c = 0.0606$ m^{-1}, where the long–time trajectory for any car is the limit cycle. The fixed points lie on the optimal velocity curve (dotted line). *Right picture:* Evolution of total energy E (measured in units of $m v_{max}^2/2$) for both cases (solid curve: limit cycle; dashed line: fixed point). The parameters are chosen as $N = 60$, $D = 33$ m, $v_{max} = 20$ m/s, $\tau = 1.5$ s, and $m = 1000$ kg. Taken from [1].

4 Mesoscopic Stochastic Model of Traffic Flow

On higher level of description we will no longer follow individual car trajectories only the number of cars in a jam is of interest. In the simplest model only one cluster on a circular road is considered, and the probability $p(n,t)$ that it contains n cars at time t is given by one–step master equation

$$\begin{aligned}\frac{dp(n,t)}{dt} &= w_+(n-1)\,p(n-1,t) + w_-(n+1)\,p(n+1,t) \\ &- \left[w_+(n) + w_-(n)\right]\,p(n,t) \quad : \quad 0 < n < N\,,\end{aligned} \tag{8}$$

where w_+ and w_- are transition rates. For $n = 0$ (no congestion) and $n = N$ (all cars in one queue) the equations are different. Terms with $p(-1,t)$ and

$p(N+1,t)$ will be absent in this case. Choices for transition rates w_+ and w_- are:

a) $w_-(n) = 1/\tau$ is the rate to leave the queue,
b) $w_+(n) = v_{opt}\left(\Delta x_{free}(n)\right)/\Delta x_{free}(n)$ is the rate to enter the queue.

$\Delta x_{free}(n) = L/(N-n)$ is the mean headway distance in the free flow phase. Some properties of the model, for the above choice of transition rates w_+ and w_- are: (**1**) No large stable cluster forms at low densities of cars, this corresponds to the fixed point solution of the Bando model. (**2**) Above a certain critical density a macroscopic fraction of cars are condensed into the cluster, this corresponds to the limit cycle in the Bando model. (**3**) The stationary solution $dp(n,t)/dt = 0$ obeys the detailed balance condition $p^{st}(n)\,w_+(n) = p^{st}(n+1)\,w_-(n+1)$. (**4**) It allows us to describe some properties of the model by equilibrium thermodynamics in analogy to the liquid–vapour system, in spite of the fact that the traffic flow model is a driven nonequilibrium system.

5 Free Energy of Liquid–Gas System

Taking the relation between transition rates w_+ and w_- and free energy F (if the number of particles is large)

$$\ln\left[\frac{w_+(n)}{w_-(n)}\right] = -\frac{1}{k_BT}\frac{\partial F}{\partial n} \Longrightarrow F = F_0 - k_BT\int_0^n \ln\left[\frac{w_+(n')}{w_-(n')}\right] dn' \tag{9}$$

we get for given ratio of the transition rates w_+ and w_- (see [2]) the following free energy function

$$F = k_BT(N-n)\left[\ln\left(\lambda_0^3(T)\frac{N-n}{V}\right) - 1\right] + f_n(T). \tag{10}$$

This is a well known expression for the free energy of liquid–gas system under isothermal and isochoric conditions. We can calculate the transition rates ratio $w_+(n)/w_-(n)$ and the normalised free energy difference $(F-F_0)/(\widetilde{V}k_BT) = f - f_0$ for three dimensionless densities $\tilde{\rho} = D\rho = DN/L$ as function of n/N density of liquid phase [1], see Fig. 4.

6 Free Energy of Traffic Flow

The ratio of transition rates in this case reads (where $\rho = N/L$)

$$\frac{w_+(n)}{w_-(n)} = \tau\frac{v_{opt}(\Delta x_{free})}{\Delta x_{free}} = v_{max}\tau\rho\,\frac{1-n/N}{1+\rho^2D^2(1-n/N)^2}\,, \tag{11}$$

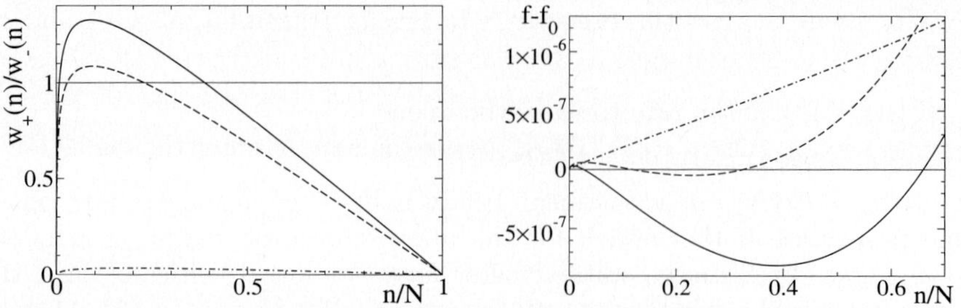

Fig. 4. Calculated $w_+(n)/w_-(n)$ and $(F-F_0)/(\widetilde{V}k_BT)=f-f_0$ for dimensionless densities $\tilde{\rho}=5\cdot10^{-7}$ (dot–dashed line), $\tilde{\rho}=10^{-5}$ (dashed line), and $\tilde{\rho}=1.2\cdot10^{-5}$ (solid line).

where v_{opt} is the optimal velocity function and $\Delta x_{free}=L/(N-n)$ is the headway distance in the free flow phase. After some steps we arrive at the free energy for traffic flow:

$$\frac{F-F_0}{\widetilde{L}T^*}=\tilde{\rho}\left\{\left(1-\frac{n}{N}\right)\ln\left(1-\frac{n}{N}\right)-\frac{n}{N}-\frac{n}{N}\ln\left(\frac{\tilde{\rho}}{\tilde{b}}\right)+\ln\left(1+\tilde{\rho}^2\right)\right.$$
$$\left.-\left(1-\frac{n}{N}\right)\ln\left(1+\tilde{\rho}^2\left[1-\frac{n}{N}\right]^2\right)\right\}+2\arctan\tilde{\rho}-2\arctan\left(\tilde{\rho}\left[1-\frac{n}{N}\right]\right), \quad (12)$$

where $\widetilde{L}=L/D$ is the dimensionless length of the road, $\tilde{b}=D/(v_{max}\tau)$, $\tilde{\rho}=\rho D$ is a dimensionless density, F_0 is the ideal free energy with no clusters: $F_0=-T^*\left[N\ln\left(L/\lambda_0(T^*)\right)-\ln N!\right]\simeq T^*N\left[\ln\left(\rho\lambda_0(T^*)\right)-1\right]$ where $\lambda_0(T^*)=h/(2\pi mT^*)^{1/2}$ and $T^*=k_BT$.

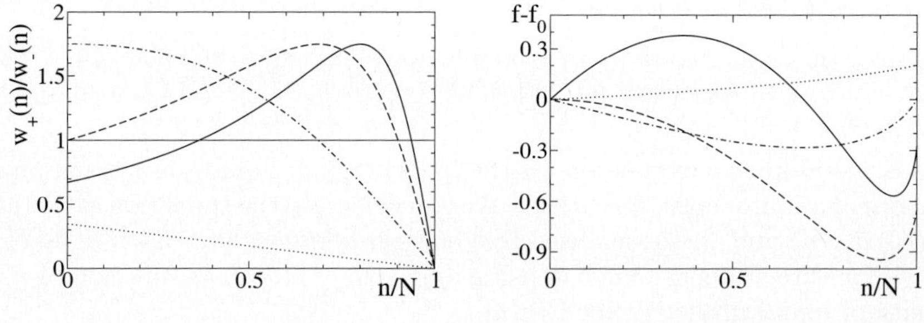

Fig. 5. Transition rates ratio $w_+(n)/w_-(n)$ and normalised free energy difference $(F-F_0)/(\widetilde{V}k_BT)=f-f_0$ depending on the number of congested n/N cars for three dimensionless densities. $\tilde{\rho}=0.1$ (dotted line), $\tilde{\rho}=1$ (dot–dashed line), $\tilde{\rho}=3.186$ (dashed line) and $\tilde{\rho}=5$ (solid line). $D=24$ m, $v_{max}=42$ m/s and $\tau=2$ s.

- We have shown how thermodynamics can be applied to such a many–particle system as traffic flow, based on a microscopic (car–following) as

well as a mesoscopic (stochastic cluster) description, in analogy to equilibrium physical systems like supersaturated vapour forming liquid droplets.
- Distinguishing features between the traffic flow and equilibrium physical systems have been outlined. In particular, we have found that the third Newton's law (action/reaction) is does not hold on the level of "microscopic" equations of motion for individual cars.

Acknowledgements

We would like to thank the German Science Foundation (DPG) for financial support through grant MA 1508/8-2. H. W. acknowledges support from Swedish Research Council Contract No. 621–2001–2545.

References

1. R. Mahnke, J. Kaupužs, J. Hinkel, H. Weber: Eur. Phys. J. B **57**, 463, 2007.
2. R. Mahnke, J. Kaupužs, I. Lubashevsky: Physics Reports **408**, 1–130, 2005.
3. H. Weber, R. Mahnke, J. Kaupužs, A. Strömberg: Models of Highway Traffic and their Connections to Termodynamics, In: Traffic and Granular Flow '05 (Eds.: A. Schadschneider, T. Pöschel, R. Kühne, M. Schreckenberg, D. E. Wolf), p. 545–550, Springer, Berlin, 2007.
4. M. Krbálek: J. Phys. A: Math. Theor. **40**, 5813, 2007.
5. M. Bando, K. Hasebe, A. Nakayama, A. Shibata, Y. Sugiyama: Japan J. Indust. and Appl. Math. **11**, 203, 1994; Phys. Rev. E **51**, 1035, 1995.
6. M. Bando, K. Hasebe, K. Nakanishi, A. Nakayama, A. Shibata, Y. Sugiyama: J. Phys. I France **5**, 1389, 1995.
7. Ch. Liebe, R. Mahnke, J. Kaupužs, H. Weber: Vehicular Motion and Traffic Breakdown: Evaluation of Energy Balance, these Proceedings
8. J. Hinkel: *Applications of Physics of Stochastic Processes to Vehicular Traffic Problems*, PhD–thesis, University of Rostock, 2007.

Phase Transitions and Even/Odd Effects in Asymmetric Exclusion Models

Marko Woelki and Michael Schreckenberg

Universität Duisburg-Essen, Lotharstraße 1, D-47057 Duisburg, Germany
woelki@ptt.uni-due.de

Summary. An asymmetric exclusion process with periodic boundary conditions is investigated. During each time-step a randomly chosen particle moves one site and if possible two sites. This dynamics leads to different gap distributions depending on the parity of the number of holes. Despite the simplicity of the model on a ring there is a phase transition that separates two regimes with different density profiles. For a generalization of the process the steady state is given for two particles on a ring.

1 Introduction

The asymmetric exclusion process (ASEP) has been used to model for example vehicular and pedestrian traffic, granular flow and biological processes. It is generically defined on a discrete one-dimensional lattice comprising exclusive diffusing particles. In its simplest form a particle moves one site to the right if this target site is empty (see e.g. [1] for a review). The steady state for open boundary conditions (injection of particles at the one end and rejection at the other end of the lattice at certain rates) is known to be of the matrix-product form [2] and one observes so-called boundary-induced phase transitions [3]. On the other hand the model with periodic boundary conditions is known to have a uniform stationary measure and of course no phase transitions occur.

However there are some generalizations of the ASEP with periodic boundary conditions with phase transitions: For example, if each of the particles has an individual fixed hopping rate one might observe a phase transition from a fluid into a condensed phase [4]. Another example is the ASEP with a single defect particle with lowered hopping rate that can be overtaken by normal particles at some rate [1, 5]. Finally phase transitions have been studied in asymmetric exclusion models in which the hopping rate depends on the empty space ahead. This models can be related to zero-range processes and the interactions are normally long-ranged when condensation transitions appear [6, 7].

The starting point for this article was to construct a very simple (conserving) process on a ring with one species of particles, no overtaking and short-range interactions that remains exactly solvable beyond a simple factorization and that is on the other hand capable to produce a phase transition. The question is what physical origin this transition might have. In the model to be considered particles have an increased maximum speed (as it is generic for traffic models for example). Under this dynamics all particles tend to have a desired fraction of empty space in front. However if the total amount of empty space is such that not every particle can support this desired fraction finally one particle per time step has a different fraction in front and sometimes his movement rate is decreased. Then this is a sort of defect that might induce phase transitions similarly as in the defect-ASEP mentioned above.

2 Model Definition and Structure of the Steady State

To be precise the model is an asymmetric exclusion process defined on a 1d lattice with L sites, labelled $i = 1, 2, \ldots L$ and periodic boundary conditions (site $L + 1$=site 1) as well as N particles labelled $\mu = 1, 2, \ldots N$. Due to the exclusion rule each site i can be in either of two different states: $\tau_i = 1$, if it is occupied by a particle and $\tau_i = 0$ if it is empty. The system evolves in continuous time (random sequential dynamics) with respect to the following rules:

$$100 \longrightarrow 001, \quad \text{at rate } 1, \tag{1}$$

$$101 \longrightarrow 011, \quad \text{at rate } \beta. \tag{2}$$

It turns out that the stationary distribution of probabilities for the possible configurations depends strongly on the parity of the number $L - N$ of holes (unoccupied sites). In connection with this an important quantity is the number of holes to the right of each particle. We refer to this quantity as the 'gap' to the next particle ahead. The model dynamics is such that odd-numbered gap lengths can not evolve, however they can disappear through transitions in which a configuration $\mathcal{C}(\ldots[\text{any odd number of 0s}]101\ldots)$ turns with rate β to a configuration with two odd gaps less. This processes appear again and again until there rest either no odd gap ($L - N$ even) or exactly one odd gap ($L-N$ odd). So only a certain (absorbing) subspace of the configuration space is visited in the steady state. In the following we are going to consider this two cases separately.

- The model with even number of holes:
  In this case it turns out that all stationary configurations are equally likely what can easily be verified by looking at the master equation for the process what is omitted here. Stationary configurations are those in which the number of holes in front of any particle μ is an even number. A simple argument is that on replacing all gaps by the half then one recovers

the usual ASEP with uniform steady state. Expressing a configuration by the set of occupation numbers leads to the factorization:

$$P(n_1, n_2, \ldots, n_N) = Z_{L,N}^{-1} \prod_{\mu=1}^{N} \delta_{n_\mu \mathrm{mod} 2,0}. \tag{3}$$

where the normalization factor $Z_{L,N}$ was introduced to ensure the probability conservation and δ_{ij} denotes the Kronecker delta. Note that the gap distribution $p(n) \propto \delta_{n \bmod 2,0}$ oscillates between zero (for odd-numbered gap lengths) and a constant value (for even gaps).
Denote the total number of holes by $2M$. Then the normalization is easily obtained combinatorially: For the first particle to place on the lattice there are L possible ways. One can then think of distributing $N-1$ particles and M hole pairs into $N+M-1$ boxes to obtain:

$$Z_{L,N} = \frac{L(L-M-1)!}{N!M!} \delta_{L-N,2M}. \tag{4}$$

Then it is straightforward to calculate correlation functions as

$$\langle \tau_i \rangle = \frac{N}{L}, \quad \langle \tau_i \tau_{i+1} \ldots \tau_{i+m} \rangle = \frac{N}{L} \frac{(N-1)(N-2)\ldots(N-m)}{(L-M-1)\ldots(L-M-m)}. \tag{5}$$

- The model with odd number of holes:
In this case one finds that the stationary configurations are of the matrix-product form. This configurations are those in which exactly one particle has an odd number of holes in front (all other particles have an even number of holes in front). Thus the only non-vanishing probabilities are:

$$P(2n_1, \ldots, 2n_{N-1}, 2n_N + 1) = Z_{L,N}^{-1} \langle W | E^{n_1} D E^{n_2} \ldots D E^{n_N} | V \rangle. \tag{6}$$

Of course similar relations hold if the odd gap is not in front of the Nth particle but in front of any of the others. Here the vectors $\langle W|$, $|V\rangle$ and matrices D, E are the well-known operators from the open-boundary problem and the defect-ASEP on a ring [1, 5] but they have a different meaning: Detect the odd-numbered gap and begin with its rightmost hole. Write this hole together with the following particle as the matrix product $E|V\rangle\langle W|$. Continue to the right on representing all hole pairs by a matrix E (normally the matrix E represents single holes) and represent all particles by a matrix D. Taking finally the trace over the whole word can be rewritten as (6). In fact this is a sort of mapping onto the ASEP with a single defect [1, 5]. The unit 01 corresponds to the defect '2' and hole pairs 00 correspond to usual holes '0' in the defect ASEP. Then the transitions (1) turn into $10 \to 01$ at rate 1, $12 \to 21$ at rate β.
A formal proof of the steady state will be given elsewhere [9], the expressions should here just be made plausible.

Also in the odd case the gap-length distribution oscillates because even gaps are preferred. In fact in the thermodynamic limit the probability for odd gaps goes to zero as in the even case. As for the defect-ASEP [5] the partition function can be derived:

$$Z_{L,N} = \frac{L}{N}\binom{N+M}{N-1}\sum_{m=1}^{\infty} m\binom{N+M-1}{N-m}\left(\frac{1-\beta}{\beta}\right)^{m-1}\delta_{L-N,2M+1} \quad (7)$$

and equivalently a phase transition takes place. The critical density is $\rho_c = 2\beta/(1+2\beta)$. In the following let us in analogy refer to the 01-pair as the defect.

- For $\rho < \rho_c$ the defect behaves as the other particles. In front of the defect the density profile decreases exponentially to its bulk value ρ. The density behind is constant.

- For $\rho > \rho_c$ the defect is similar to a second-class particle [8] that lowers the average speed of the other particles. The density profile decays algebraically to the bulk value. Behind the defect the density is decreased and the profile increases in the same way to its bulk value as in front.

In both cases (even and odd number of holes) the velocity of particles is given by

$$v = 2\frac{1-\rho}{1+\rho} + \mathcal{O}\left(\frac{1}{L+N}\right). \quad (8)$$

Just the special form of the correction differs for even and odd number of holes.

3 Generalization of the Model

An interesting generalization of the model is the following:

$$\begin{aligned} 100 &\to 010, &&\text{with rate } p_1 = 1,\\ &\to 001, &&\text{with rate } p_2,\\ 101 &\to 011, &&\text{with rate } \beta. \end{aligned}$$

This process is in general ergodic what makes it more interesting as the model of the previous section (corresponding to the choice $p_1 = 0$ and $p_2 = 1$). On the other hand here the fully analytical treatment becomes more difficult. The parallel-update version[1] of this general model was considered in [10]. By comparison with the chipping model [11] (for further references see [10]) it was argued that in general no phase transitions should occur for non-deterministic

[1] The random-sequential update in fact is a special limit of the parallel update

rates. For the general process here we considered the solvable case of only two particles on a ring and found the following solution (for $m \geq n$ and $m \geq 1$)

$$P(m,n) = Z^{-1}_{m+n+2,2}\left[c(p_2,\beta,m+n) + (-1)^{n-1}(1-\beta)p_2^n y_{m-n}\right]. \quad (9)$$

The function y_n is given by $y_n := 2^{1-n}\sum_k \binom{n}{2k}(1+4p_2)^k$. The first term $c(p_2,\beta,m+n) = \beta y_{m+n} + (2p_2+1-\beta)y_{m+n-1}$ is constant for given system size. The pre-factor $(-1)^{n-1}$ of the second term indicates that there are in general oscillations except for $\beta = 1$ and/or $p_2 = 0$. This two cases turn out to lead to a simple factorized steady state for arbitrary particle number and have already been studied [12]. However it would be interesting to study the model for other parameter choices and to determine the form of their cluster distributions.

4 Conclusion and Outlook

This article deals with a simple totally asymmetric exclusion process with three-site interaction on a ring. Depending on the parity of the number of holes the steady state takes either a simply factorized form or the matrix-product form. In the latter case the dynamics evolves into an absorbing subspace and a phase transition takes place. This transition separates two regimes with different density profiles. Their forms can be obtained from a relation to the defect-ASEP [1, 5]. An interesting fact is that the gap-length distribution oscillates since even gaps have a higher probability than odd ones. To our knowledge phase transitions and matrix states in such simple asymmetric exclusion processes on a ring (conservative dynamics, one particle species and short-range interactions) are not yet known.

Section 3 generalizes the process to ergodic dynamics by introduction of additional hopping rates. The transition in the non-ergodic process of section 2 is due to the reduction of the configuration space. By a relation to the chipping model it was already argued in [10] that for arbitrary stochastic rates no phase transitions occur. The exact expression for the steady state in the two-particle sector that we obtained shows that the oscillations in the gap distribution remain also in the general process. They seem to be generic for this type of models (with velocity larger than one).

Following related topics are under progress and will soon be published elsewhere [9]: The process under parallel dynamics is perhaps even more interesting than the continuous-time case. For this process we observed qualitatively the same results. The steady state depends on the parity of the number of holes and has a factorized state and matrix-product state, respectively. However it takes a more complicated form. All these models are special cases of a generalized zero-range process (GZRP) - see the review [7] and references therein. They can be mapped by letting the particles become sites in the GZRP that are occupied by the gap to the right of the corresponding particle in the ASEP. While the focus has previously been on deriving a necessary and sufficient condition for GZRPs to have a factorized steady state [7], the process considered here is an example for a GZRP with matrix-product state and factorized steady state depending on the choice of L and N. In fact the factorization condition can be generalized to non-ergodic processes [9]. Finally we have obtained the matrix-product solution for a related two-species zero-range process (ZRP). Particles of species 1 can always move (irrespective of a possible presence of particles of species 2 at the site). Species-2 particles move only (with some other rate) if there is no species-1 particle present at the site. The steady state of the present model in this article corresponds to the steady state of the ZRP with a single particle of species 2. As far as we are aware this are the first matrix-product solutions for GZRPs and two-species zero-range processes.

Acknowledgements

M W thanks the Institut Henri Poincaré in Paris for kind hospitality. We further thank A Schadschneider for helpful discussions as well as B Derrida and M Evans for useful comments.

References

1. Derrida B (1998) Phys Rep 301:65
2. Derrida B, Evans M R, Hakim V, Pasquier V (1993) J Phys A 26:1493–1517.
3. Krug J (1991) Phys Rev Lett 67:1882–1885.
4. Evans M R (1996) Europhys Lett 36:13–18.
5. Mallick K (1996) J Phys A 29:5375–5386.
6. Evans M R (2000) Braz J Phys 30:42.
7. Evans M R, Hanney T (2005) J Phys A 38:R195.
8. Derrida B, Janowsky S A, Lebowitz J L, Speer E R (1993) J Stat Phys 73:5/6.
9. Woelki M, Schreckenberg M (2007) *in preparation.*
10. Levine E, Ziv G, Gray L, Mukamel D (2004) Physica A 340:636.
11. Vigil R D, Ziff R D, Lu B (1988) Phys Rev B 38:942.
12. Klauck K, Schadschneider A (1999) Physica A 271:102.

Simulation and Mean Field Analysis on Pedestrian Dynamics Around an Exit

Daichi Yanagisawa and Katsuhiro Nishinari

Department of Aeronautics and Astronautics, School of Engineering,
The University of Tokyo, Hongo, Bunkyo-ku, Tokyo 113-8656, Japan
[tt66421,tknishi]@mail.ecc.u-tokyo.ac.jp

Summary. Floor field model is a cellular automaton model, which reproduces characteristics of pedestrian dynamics such as an arch phenomenon and lane formation. In the floor filed model, pedestrians move similarly whether they are near an exit or not in egress processes. In reality, pedestrian move fast if they are far from the exit, but move slowly around the exit since they try to avoid conflicts with other pedestrians. The new parameter which represents this phenomenon is introduced to the floor field model. We study egress processes by simulation and cluster approximation and obtained an expression of an average flow of pedestrians going out from the exit when they move to the exit directly. The suitable pedestrian mood and the optimal exit position are also discovered by our analysis.

1 Introduction

Pedestrian dynamics has received growing interest over the last decades from physicists since it shows new collective behaviors such as dynamical phase transitions and spontaneous symmetry breaking [1, 2]. In this paper, we study the floor field (FF) model which is a cellular automaton model, introducing two kinds of FFs, i.e., Static FF (SFF) and Dynamic FF (DFF), to move pedestrians from one cell to another [3–7].

Most of the studies of pedestrian dynamics are based on simulations and there are few analytical results because of the complexity of rules of motion and two-dimensionality. In this paper we present an analytical result on outflow through an exit, which is one of the most important index in evaluating evacuation dynamics. We newly introduce the *bottleneck parameter* β which makes pedestrian behaviors around the exit more realistic. We have succeeded to calculate the average flow $\langle Q \rangle$ as a function of β, μ, and the width of an exit door w by cluster approximation. As far as we know, the analytical expression of the average flow through the exit with arbitrary width is derived for the first time. We also consider how the mood of the pedestrians and a wall beside the exit influence the average flow.

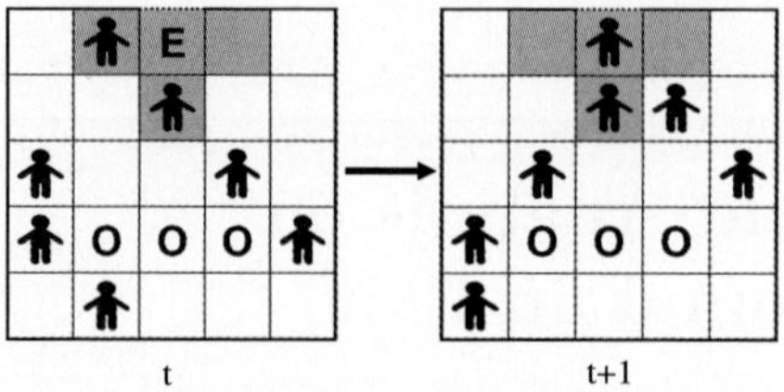

Fig. 1. A schematic view of an evacuation simulation by the FF model. Pedestrians proceed to the exit by one cell at most by one time step.

S=2	1	E 0	1	2
$\sqrt{5}$	$\sqrt{2}$	1	$\sqrt{2}$	$\sqrt{5}$
$2\sqrt{2}$	$\sqrt{5}$	2	$\sqrt{5}$	$2\sqrt{2}$
$2\sqrt{2}+1$	**O**	**O**	**O**	$2\sqrt{2}+1$
$2\sqrt{2}+2$	$2\sqrt{2}+3$	$2\sqrt{2}+4$	$2\sqrt{2}+3$	$2\sqrt{2}+2$

Fig. 2. Static floor field constructed by the exit **E**. The numbers in each cell represent the Euclidean distances from the exit cell.

2 Floor Field Model

We consider a situation that every pedestrian in a room moves to the same exit. The room is divided into cells as given in Fig. 1. Man shaped silhouettes represent pedestrians, an alphabet **E** and alphabets **O** represent the exit cell and obstacle cells, respectively. Each cell contains only a single pedestrian at most. Every time step pedestrians choose which cell to move from 5 cells: a cell which the pedestrian stands now $((i,j) = (0,0))$ and the Neumann neighboring cells $((i,j) = (0,1), (0,-1), (1,0), (-1,0))$. FFs determine the probability of which direction to move. SFF S_{ij} is the shortest distance to the exit cell given by the L^2 norm (Fig. 2). However, when there is an obstacle on the way to the exit, SFF is calculated by making detour of it [6]. Pedestrians move to a cell that has smaller SFF than a cell they occupy and hence go to the exit. We consider only the effect of SFF and ignore DFF for simplicity in this paper. The effect of DFF is written in Ref. [3] in detail. Therefore in this model, the transition probability p_{ij} for a move to a neighbor cell (i,j) is determined by the expression $p_{ij} = N\xi_{ij}\exp(-k_s S_{ij})$. Here the values of the FFs S_{ij} at each cell (i,j) are weighted by a sensitivity parameter k_s with the normalization N. There is a minus sign before k_s since pedestrian move to a cell which SFF decreases. ξ_{ij} returns 0 for an obstacle or a wall cell and returns 1 for other kinds of cells. Note that in our paper a cell occupied by a pedestrian is not regarded as an obstacle cell, thus it affects the normalization N.

Due to the use of parallel dynamics it happens that two or more pedestrians choose the same target cell in the update procedure. Such situations are called conflicts in this paper. To describe the dynamics of a conflict in a quantitative way, *friction parameter* $\mu \in [0,1]$ was introduced in Refs. [4, 5]. This parameter describes clogging and sticking effects between the pedestrians. In a conflict situation, movement of all involved pedestrians remain at their cell with probability μ. One of them is randomly allowed to move to the desired cell with probability $1-\mu$.

3 Introduction of a New Parameter Near the Exit

In real evacuation situations there are few pedestrians near the corner of the room, while there are many pedestrians gathering around the exit. Therefore, pedestrians often conflict with each other around the exit and an arch of pedestrians is likely to be formed in front of the exit due to a friction between them. Since pedestrians know this fact by experience, they walk fast when they are far from the exit, while they walk slowly or give way to each other around the exit. To take into account this situation, we introduced a new parameter $\beta \in [0,1]$ which we call the *bottleneck parameter*. The transition probability of pedestrians who occupy one of the Neumann neighboring cells of the exit cell is described as follows:

$$\begin{cases} p_{ij} = \beta \bar{N} \xi_{ij} \exp(-k_s S_{ij}) & ((i,j) \neq (0,0)) \\ p_{0,0} = (1-\beta) + \beta \bar{N} \exp(-k_s S_{0,0}), & \end{cases} \tag{1}$$

where $\bar{N} = \left[\sum_{i,j} \xi_{ij} \exp(-k_s S_{ij})\right]^{-1}$.

When k_s is large, pedestrians at neighboring cells of the exit move to the exit cell with the probability β and stayed their cells with the probability $1-\beta$. This simplification enables us to analyze the pedestrian behavior theoretically.

4 Cluster Approximation

In this section we focus on the exit cells and the neighbor cells of them and calculate an analytical expression of the average pedestrian flow through the exit by cluster approximation. The flow is defined as the number of evacuated persons per 1 time step thorough an exit. We suppose that a big jam is formed around the exit. This enables us to simplify a situation that only SFF affect pedestrians' motion.

First, we calculate the flow when the width of the exit $w = 1$. The transition probability are defined in Fig. 3. Since we assume a big jam exists at the exit, pedestrians enter into three neighboring cells of the exit with the probability 1. The probability of getting out from the exit cell is set as 1. We define $p_t(0)$ as the probability that a pedestrian is not at the exit cell at time step t and $p_t(1)$ as the probability that a pedestrian is at the exit cell at time step t. The master equations are described as follows:

$$\begin{bmatrix} p_{t+1}(0) \\ p_{t+1}(1) \end{bmatrix} = \begin{bmatrix} 1-r & 1 \\ r & 0 \end{bmatrix} \begin{bmatrix} p_t(0) \\ p_t(1) \end{bmatrix}. \tag{2}$$

Here r represents the probability that a pedestrian enter into the exit cell from the three Neumann neighboring cells, which is described as follows:

$$\begin{aligned} r = {} & \beta_1(1-\beta_2)(1-\beta_3) + \beta_2(1-\beta_3)(1-\beta_1) + \beta_3(1-\beta_1)(1-\beta_2) \\ & + (1-\mu)\{\beta_1\beta_2(1-\beta_3) + \beta_2\beta_3(1-\beta_1) + \beta_3\beta_1(1-\beta_2) + \beta_1\beta_2\beta_3\}. \end{aligned} \tag{3}$$

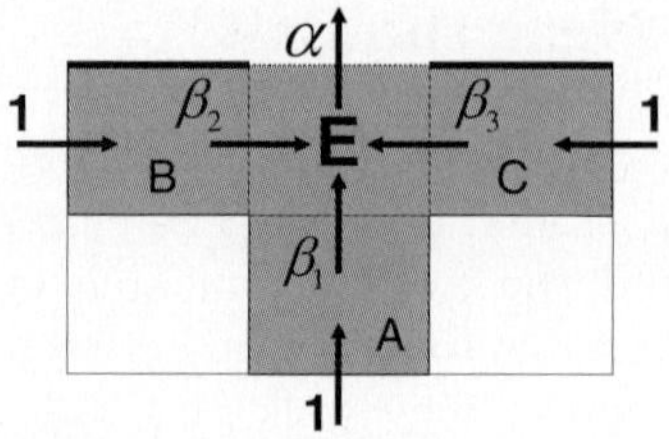

Fig. 3. Cluster approximation at the exit with one cell. 1, β_1, β_2, β_3 represent transition probability.

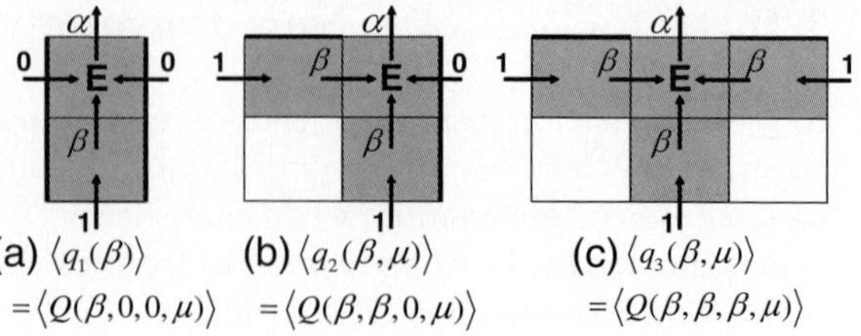

Fig. 4. The three special cases of Fig. 3. We assume that the arrow with transition probability 0 is interpreted as the existence of a wall that blocks pedestrians' motion.

By using (2) and (3) with the normalization condition $p_t(0) + p_t(1) = 1$, we obtain the stationary solution, and the average pedestrian flow through an exit is described as

$$\langle Q(\beta_1, \beta_2, \beta_3, \alpha, \mu)\rangle = \alpha p_\infty(1) = \frac{r}{1+r}. \tag{4}$$

Next we specify (4) by substituting 0 and β for β_1, β_2, β_3 as (a) $\langle q_1(\beta)\rangle$, (b) $\langle q_2(\beta,\mu)\rangle$ and (c) $\langle q_3(\beta,\mu)\rangle$, which describe average flows through an exit with the configuration described in Fig. 4 respectively.

Finally we calculate the average flow of pedestrians through an exit with arbitrary $w \in \mathbf{N}$ width. We represent the average flow through the exit with w cell's width by linear sum of $\langle q_1\rangle$, $\langle q_2\rangle$, and $\langle q_3\rangle$. Here we consider two types of exits: an exit at the center of the wall (Ce-exit) and an exit at the corner of the room (Co-exit). Ce-exit ($w \geq 2$) is divided into $\langle q_1\rangle$-exits, and $\langle q_2\rangle$-exits, and an average flow through an exit with w cell's width $\langle Q_{center,w}\rangle$ is described as:

$$\langle Q_{center,w}\rangle = \begin{cases} \langle q_3\rangle & (w = 1) \\ 2\langle q_2\rangle + (w-2)\langle q_1\rangle & (w \geq 2). \end{cases} \tag{5}$$

In a similar way, $\langle Q_{corner,w}\rangle$ is described as follows:

$$\langle Q_{corner,w}\rangle = \langle q_2\rangle + (w-1)\langle q_1\rangle \quad (w \geq 1). \tag{6}$$

We also define the average flow per 1 cell as $\langle q_{center}\rangle = \langle Q_{center,w}\rangle/w$ and $\langle q_{corner}\rangle = \langle Q_{corner,w}\rangle/w$.

The average flow calculated by analysis and simulation is described in Fig. 5. It shows average pedestrian flows at the exit as a function of β for various μ values. We see that the simulations agree with the analytical results very well. Surprisingly, for $\mu = 0.9$, we clearly find that a maximum flow is attained at a value of β in both simulations and analytical results. For $\mu \to 1$ the number of unsolved conflicts increases as β grows. As a result, pedestrians stick and the average pedestrian flow decreases.

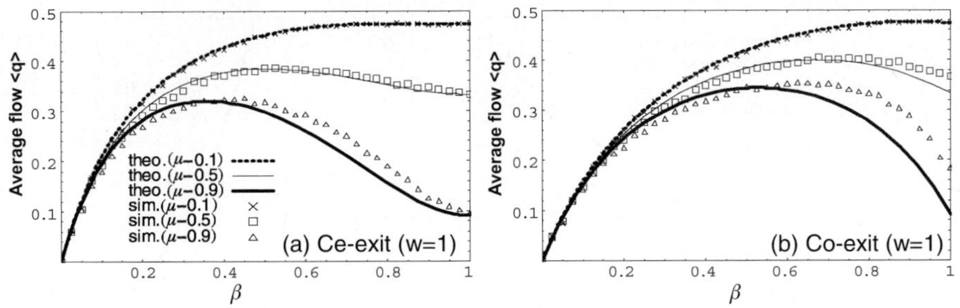

Fig. 5. Average flow $\langle q \rangle$ as a function of β for different μ, exit width w, and exit location in the case $k_s = 10$. For $\mu = 0.9$ we clearly see a maximum of flow at an optimal β.

5 The Effect of Mood and a Wall

Here we consider how the mood of pedestrians influenced the flow. The parameters are set at $\beta = 1.0$, $\mu = 0.6$ for the competitive situation and $\beta = 0.4$, $\mu = 0.0$ for the cooperative situation. Figure 6 (a) shows the average flows for variable door width w. Clearly we can observe the crossing of the two curves at a critical door width $w_c \approx 3$ in Fig. 6 (a). This means that we should cooperate with each other to increase the average pedestrian flow when the width of the exit is narrow. On the contrary, when the width of the exit is wide, we should go through the exit aggressively.

The Japanese building standards law [8] gives an average pedestrian flow 1.5[persons/(m·s)] if an exit is directly connected to the ground. We find that this value significantly changes by the pedestrians' moods, i.e., competitive and cooperative. From Fig. 6 (a) we obtain the values of the average flow through Ce-exit i.e., 1.5[persons/(m·s)] in the competitive situation and 2.0[persons/(m·s)] in the cooperative situation. The values are calculated by defining the cell size as 50[cm]×50[cm] and using a pedestrian velocity 1.3[m/s] which is according to the Japanese building standards law.

Next we compare the average flows of Ce-exit and Co-exit, and discuss how the wall has an effect on them. We obtain β_c which is the value of β that $\langle Q_{center,w} \rangle$ equals $\langle Q_{corner,w} \rangle$ as $\beta_c = 1/(1+2\mu)$ for $w = 1$ and $\beta_c = 1/(1+\mu)$ for $w \geq 2$. The curves of β_c is drawn in Fig. 6 (b). They divide the $\beta - \mu$ plane into two regions. In the lower left region, the Ce-exit flow is larger and in the upper right region the Co-exit flow is larger. We also plot the competitive and cooperative situation. The figures show that the Co-exit flow is larger in the competitive situation since the wall prevents pedestrians rushing to the exit at the same time, but the Ce-exit flow is larger in the cooperative situation. From this result, we can say that an exit should be at the center of a wall when pedestrians are in the cooperative mood, and should be at the corner of the room when people are in the competitive mood.

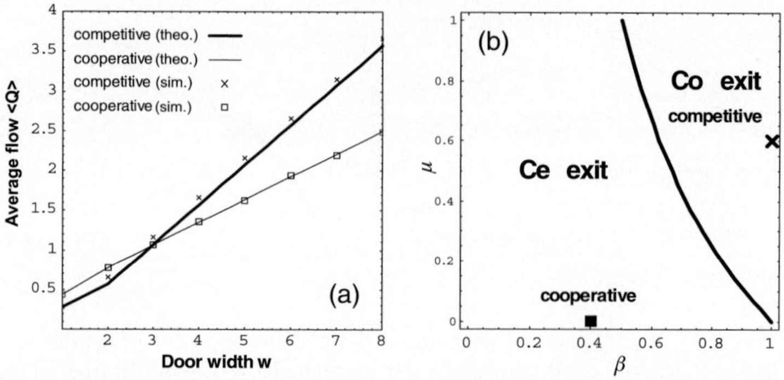

Fig. 6. (a) Average flow for various exit door width w (Ce-exit). (b) The curves of β_c on $\beta - \mu$ plane ($w \geq 2$).

6 Conclusion

We have introduced it to the FF model that the effect of slowing down of pedestrians around an exit, and obtained the analytical expression of the average flow through an exit with arbitrary w cells by employing cluster approximation. It turns out that the theoretical results agree quite well with the simulations. The effects of pedestrians' mood, a width of an exit, and wall effect are also studied. The critical exit door width, which was first obtained experimentally and was reproduced by simulations in Ref. [4], is also analytically obtained in this paper. We find that an exit should be at the center of a wall in the cooperative situation whereas it should be at the corner of the room in the competitive situation for smooth evacuation.

References

1. D. Helbing, Rev. Mod. Phys. **73**, 1067 (2001).
2. C. Burstedde, K. Klauck, A. Schadschneider, and J. Zittartz, Physica A **295**, 507 (2001).
3. A. Kirchner and A. Schadschneider, Physica A **312**, 260 (2002).
4. A. Kirchner, H. Klupfel, K. Nishinari, A. Schadschneider, and M. Schreckenberg, Physica A **324**, 689 (2003).
5. A. Kirchner, K. Nishinari, and A. Schadschneider, Phys. Rev. E **67**, 056122 (2003).
6. K. Nishinari, A. Kirchner, A. Namazi, and A. Schadschneider, Nonmembers, IEICE TRANSACTIONS on Information and Systems **E87-D** NO.3, 726 (2004).
7. Daichi Yanagisawa and Katsuhiro Nishinari, arXiv:**0708.3476** [cond-mat.stat-mech] (2007).
8. *The Explanation and the Examples of calculation of Testing the Safeness of Evacuation 2001*, edited by Ministry of Land, Infrastructure and Transport *et al* (INOUESHOIN, 2001), p. 303, 304 (in Japanese).

A Density-Dependent NaSch Model for Traffic Flow Controlled by a Traffic Light

Hui-bing Zhu[1,2], Hong-xia Ge[3], and Shi-qiang Dai[1]

[1] Shanghai Institute of Applied Mathematics and Mechanics, Shanghai University, Shanghai 200072, China sqdai@shu.edu.cn
[2] Faculty of Architectural, Civil Engineering and Environment, Ningbo University, Ningbo 315211, China zhuhuibing@nbu.edu.cn
[3] Faculty of science, Ningbo University, Ningbo 315211, China gehongxia@nbu.edu.cn

Summary. The traffic flow controlled by a traffic light on a single-lane roadway is studied by using a modified NaSch model proposed by the authors, in which the randomization is dependent on the local density of the preceding time step. The flow-density diagrams are obtained and three different traffic density regions, i.e., undersaturated traffic, saturated traffic and oversaturated traffic, are observed. While in the small cycle time four traffic density regions appear. Further it is found that mean traffic flow and mean velocity strongly depend on the cycle time of traffic lights. An application of the signal model on optimizing traffic lights in order to enhance the global throughput is studied. All the results indicate that the presented model is reasonable and has realistic meaning.

1 Introduction

Traffic light is an essential element for managing urban transportation networks. Given its importance, the research on traffic light control is by no means complete. A number of traffic light control models have been proposed in the past few years [1–4], among which some were based on the NaSch model. The NaSch model is a minimal model in the sense that any further simplification of the model leads to unrealistic behavior. For the description of more complex situations (e.g. multi-lane traffic, ramp) or for a proper modeling of the fine-structure of traffic flow, additional rules have to be added and the basic rules have to be modified. Then many cellular automaton traffic models have been proposed [5–9]. And the basic NaSch model is described as follows:

There are N vehicles moving in a one-dimension lattice of L cells with periodic boundary conditions. The number of vehicles is fixed. Each cell may either be empty or be occupied by one vehicle. Each vehicle has an integer velocity between 0 and the speed limit v_{max}. This speed limit can be different depending on the kind of vehicles under consideration. Let v_n and x_n denote the current velocity and position of the nth vehicle respectively. And let d_n be the empty sites in front of the nth vehicle, $d_n(t) = x_{n+1} - x_n - 1$. Then the state of the system at the time $t+1$ could be obtained from the state at the time t by applying the following set of rules: (1)Acceleration, $v_n \to min(v_n + 1, v_{max})$; (2) Deterministic deceleration to avoid accidents, $v_n \to min(v_n, d_n)$; (3) Randomization with probability p, $v_n \to max(v_n - 1, 0)$; (4) Update of positions, $x_n(t+1) \to x_n + v_n$. These four update rules are applied in parallel to all vehicles. Iteration over these simple rules has given realistic results such as the spontaneous occurrence of traffic jams.

2 Outline of the Model

The rules for traffic flow controlled by a traffic light using the DDR model at time $t+1$ are defined as follows:

Step 1: Determination of $p : p_n = \rho_n^r$.

Step 2: Acceleration: $v_n \to min(v_n + 1, v_{max})$.

Step 3: Braking due to other vehicles or traffic light state:

Case 1: The traffic light is red in front of the nth vehicle: $v_n \to min(v_n, d_n, s_n)$.

Case 2: The traffic light is green in front of the nth vehicle: $v_n \to min(v_n, d_n)$.

Step 4: Randomization with probability $p : v_n \to max(v_n - 1, 0)$.

Step 5: Movement: $x_n(t+1) \to x_n + v_n$.

where, v_n, x_n and d_n have been defined above; s_n is the empty sites between the nth vehicle and the traffic light ahead; p_n is randomization probability of the nth vehicle; ρ_n is the local vehicle density and expressed as $\frac{1}{d_n+1}$, here d_n is the empty sites in front of the nth vehicle of the preceding time step t; r is a parameter denoting the relationship between ρ_n and p_n, and in this paper its value is taken as 7.0 [8].

It has been shown that investigating the simpler problem of a single-lane road with one traffic light operating as a defect is sufficient to give appropriate results concerning the overall network behavior. The results can be used as a guideline to adjust the optimal traffic light periods [1]. So for simplicity, we just study a single-lane road with one traffic light using one-dimension lattice of L cells with periodic boundary conditions, considering only one kind of vehicles moving along one direction. The traffic light is set up at the one end of the road without occupying a cell. The length of a single cell is set to be $7.5m$ in accordance to the NaSch model. The maximal velocity of vehicles is set to be $v_{max} = 5$, corresponding to a typical speed limit of $67.5km/h$ in cities when 1 time step approximately corresponds to $2s$ in real time. And the value of velocity is between 0 and v_{max}. We simulate a system of length $L = 200$, which corresponds to the length of actual road around $1.5km$.

Throughout the paper, we will always assume that the duration of green light phase is equal to the duration of the red light phase and the yellow light period is ignored. The light is chosen to switch at regular time intervals $T/2$.

Initially, the vehicles are distributed randomly and their velocities are also set randomly. In the numerical simulation, the first 1×10^5 time-steps are put away in order to remove the transient effects, and then the data are recorded in successive 1×10^5 time-steps for averaging.

3 Simulation and Results

In Fig. 1, we show the simulation results of traffic flow controlled by a traffic light based on the DDR model, together with the fundamental diagrams of the DDR model and the NaSch model. The further study on the DDR model has been carried out and some meaningful results have been presented [8], among which there appears the metastable state and synchronized flow.

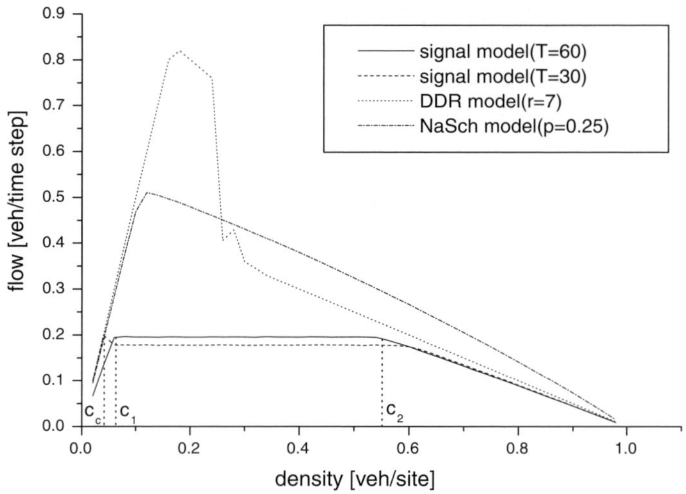

Fig. 1. The fundamental diagrams of traffic flow.

There are two fundamental diagrams of signal model with $T = 30$ and $T = 60$ (time steps) in Fig. 1. First we investigate the diagram with $T = 60$. The qualitative feature of a flat plateau (i.e., a density-independent current) has been obtained which appears in real traffic and three phases are recognized. As we know, when the state of the traffic light is red, it can be considered as the defect, just as a bottleneck on the road. Below density c_1, vehicles move forward freely and they will not aggregate much in front of the traffic light according to the cycle time presented in the diagram. The local increase of density will compensate for the reduced local velocity at the site of traffic light so that the flow around the site is identical to that below it. Then in this region traffic flow increases with the increase of density and the region is usually called undersaturated traffic [3]. In an interval of density $[c_1, c_2]$, traffic flow shows an approximate flat plateau, just as the feature of the fundamental diagram in the NaSch model with a single defect. The nontrivial phenomenon is that the mean flow J is maximum and independent of density. The traffic light has global effects whereby the traffic exhibits macroscopic phase segregation into high-density and low-density regions. This is because when the global density exceeds c_1, piling up of vehicles during the transient period leads to the phase-segregated steady state. And this region is called saturated traffic. In the high density, the traffic flow decreases linearly with the increase of vehicle density owing to lower velocity, also like that of the NaSch model. This region is called oversaturated traffic.

Then we investigate the diagram with $T = 30$, and four phases appear. Traffic flow increases linearly with the increase of density until the critical density c_c, then decreases sharply from the maximum flow and reaches the flat plateau. This phenomenon has already been observed in the study of car accidents as defects on traffic flow [5], but does not appear in the NaSch model with traffic lights. The difference should be ascribed to the fact that the probability in the DDR model applied in this simulation is dependent on local density, while it is a constant in the NaSch model.

The relationship between traffic light periods and aggregated dynamical quantities such as mean flow and mean velocity is also studied. The mean velocity v_{mean} is plotted against the cycle time T in which the density is set to be 0.06, i.e., a free-flow case (see Fig. 2(a)). The curve shows that mean velocity depends on the cycle time and reaches its maxima and minima at regular distances. The shape of the curve is similar to that obtained from the NaSch model [1].

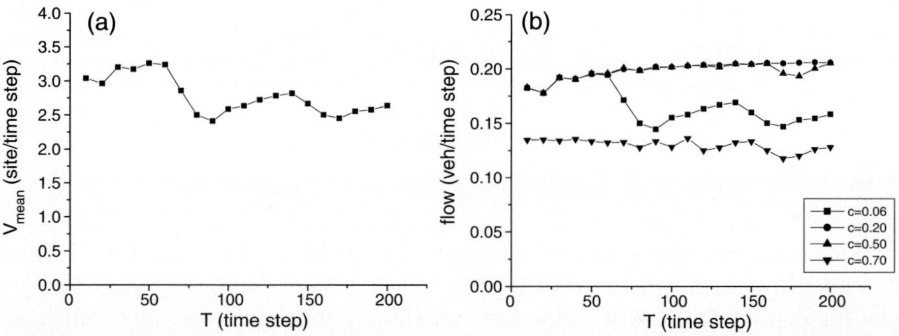

Fig. 2. The mean velocity and mean traffic flow against the cycle time T.

Fig. 2(b) shows the typical relationship between the time periods of the traffic light and the mean flow in the system. For low densities (0.06) and very high densities (0.70), we found a strongly oscillating curve with maxima and minima at regular distances. Low densities denote the free-flow densities, and then vehicles are not constricted by jamming due to the model dynamics, but rather by the red traffic light. Hence, the mean flow oscillates with switch of the traffic light and the free-flow density region shows the great potential for flow optimization. The fact that a strong dependence of the mean flow on the chosen cycle time for high densities is similar to the free-flow case was found in simulation. And this dependence is due to the movement of clusters. Since the dynamics of the system are completely determined by the movement of a jam at high densities, part of this jam is ascribed to the traffic lights, and part is ascribed to the fluctuation parameter in the model, i.e., the randomization probability p. In the DDR model, p is dependent on the local density of preceding timestep, and when local density is large p is large too. Then jams are obvious and oscillating of the mean flow against the cycle time is observed.

For densities slightly above the free-flow density ($c = 0.20$) the mean flow increases slightly with the increase of cycle time T and no characteristic maxima or minima in the mean flow is observed. The reason is that the duration of remaining jams in the system are small at this cycle time. Furthermore, when cycle time decreases large jams are divided into smaller ones by the short-cycle times and more vehicles stand. Thus the mean flow increases slightly with increasing cycle time because the number of standing vehicles decreases.

At intermediate densities (see the case of $c = 0.50$ in Fig. 2(b)) a similar phenomenon is observed as $c = 0.20$ in the smaller cycle time region. But flow breaks down at a certain higher cycle time value and grows back again to the original tendency. This is because only one jam remains in front of the light at higher cycle times. When the vehicles have to wait considerably longer than they are able to move, the breakdown finally occurs. Moreover when cycle time further increases and reaches a certain value, vehicles can all move across the traffic light in the green light phase. Then the variation of curve shows the original tendency.

Through the simulation, we want to optimize the cycle time of traffic light. The time a free-flowing vehicle requires moving from one intersection to the succeeding one (one full turn on the periodic street) is equal to:

$$T_{free} = \frac{D}{v_{free}} \tag{1}$$

where $T_{free} = T_{green} + T_{red}$, $v_{free} = v_{max} - p$ is the free-flow velocity of the underlying NaSch model [6]. That is to say, the situation that vehicles organized in a cluster can move ahead all the time in the free-flow region is possible only if Eq. (1) is satisfied. Obviously, this case corresponds to a maximum in flow whereby the green light period (or red light period) is $T_{free}/2$. Additionally, there are further maxima when $T_{free} = n(T_{green}+T_{red})$ with $n = 0, 1, 2, \ldots$.

According to the DDR model, p is dependent on the local density and is very small in the free-flow region. So we assume: $v_{free} \approx v_{max}$. As 1 time step in the simulation approximately corresponds to $2s$ in real time, then $T_{free} = L/v_{max} \approx 40$ (time step) $= 80s$ in this case. From Figs. 2(a) and 2(b), we find that the mean velocity in the free-flow case and the mean flow in different density region are stable when $T \leq 50$. In addition, taking account the fact that the bearing limit of human to the red light is $60s$, we select $T = 80s(T_{green} = T_{red} = 40s)$ as the appropriate cycle time in the presented situation.

This is an application of the signal model on optimizing the traffic light in order to enhance the global throughput. It can provide some theoretical reference for traffic engineering.

4 Conclusions

Traffic flow controlled by a traffic light on a single-lane roadway has been studied by using a modified NaSch model in which the randomization is dependent on the local density of the preceding time step. A single light situation is selected on account of the fact that investigating the simpler problem of a single road with one traffic light is sufficient to give appropriate results and to adjust the optimal traffic light periods. Simulation results show that this signal model is able to capture the characteristic features of traffic flow with a traffic light, i.e., there are three different traffic density regions in the flow-density diagrams when cycle time is large and four traffic density regions when cycle time is small. This phenomenon is different from that using the NaSch model and the underlying mechanism of these phenomena has been analyzed. Further it is found that mean traffic flow and mean velocity strongly depend on the cycle time of traffic light. On the basis of above investigations an application of the signal model on optimizing traffic lights in order to enhance the global throughput is studied. All the results indicate that the presented model is reasonable and has realistic meaning.

Acknowledgements

This work was supported by the National Basic Research Program of China (Grant No. 2006CB705500), the National Natural Science Foundation of China (Grant No. 10532060), the Special Research Fund for the Doctoral Program in Higher Education of China (Grant No. SRFDP 20040280014), the Shanghai Leading Academic Discipline Project (Grant No. Y0103) and the Natural Science Foundation of Ningbo (No. 2007A610050).

References

1. Brockfeld E, Barlovic R, Schadschneider A, Schreckenberg M (2001) Phys Rev E 64, 056132.
2. Huang DW, Huang WN (2003) Phys Rev E 67, 056124.
3. Jiang R, Wu QS (2006) Physica A 364: 493-496.
4. Yang XQ, Ma YO, Zhao YM (2004) Phys Rev E 70, 046121.
5. Barlovic R, Santen L, Schadschneider A, Schreckenberg M (1998) Eur Phys J B 5: 793-800.
6. Jiang R, Wu QS, Zhu ZJ (2001) Phys Rev E 64, 017101.
7. Ge HX, Zhu HB, Dai SQ (2005) Acta Physica Sinica, 54: 4621-4626. (in Chinese)
8. Zhu HB, Ge HX, Dai SQ, Dong LY (2007) Eur Phys J B 57: 103-108.
9. Zhu HB, Ge HX, Dai SQ (2007) Int J Mod Phys C 18(5): 773-782.

Part II

Granular Matter

Large-Scale Collective Behavior of Swimming Microorganisms at High Concentrations

Andrey Sokolov[1,2], Igor S. Aranson[1], John O. Kessler[3], and Raymond E. Goldstein[4]

[1] Argonne National Laboratory, 9700 South Cass Avenue, Argonne, IL 60439, USA
[2] Illinois Institute of Technology, 3101 South Dearborn Street, Chicago, IL 60616, USA
[3] Department of Physics, University of Arizona, Tucson, AZ 85721, USA
[4] Department of Applied Mathematics and Theoretical Physics, University of Cambridge, Wilberforce Road, Cambridge CB3 0WA, UK

Summary. Suspensions of motile bacteria such as *Bacillus subtilus* or *E. coli* form a dynamical state exhibiting extended spatio-temporal organization at concentrations near the maximum allowed by steric repulsion. The viscous liquid into which locomotive energy of individual microorganisms is transferred also carries interactions that drive the self-organization. The concentration dependence of collective swimming state correlation length is probed here with a novel technique (bacterial crowd control) that herds bacteria into condensed populations of adjustable concentration. For the free-standing thin-film geometry employed, the correlation length varies smoothly and monotonically through the transition from individual to collective behavior. Using insights from these experiments, we develop a specific model incorporating hydrodynamic interactions in thin-film geometries and show by numerical studies that it displays large scale persistently recurring vortices, as actually observed.

1 Introduction

The dynamical properties of interacting, self-propelled organisms, such as motile bacteria, sperm cells, fish, marching locusts, molecular motors [1–6] are now recognized as legitimate fundamental problem of nonequilibrium statistical physics [6] and nonlinear dynamics [7–11]. An intriguing issue is the nature of possible transitions to collective motion and its relation with highly simplified "flocking" or Vicsek-type models [8, 10], which are based on local interactions between elements and the phenomenon of collective swimming in which non-local hydrodynamic interactions are obviously important.

Concentrated suspensions of motile (swimming) bacteria such as *Escherichia coli* or *Bacillus subtilus* may serve as model systems. Experiments on soap-like free-standing thin liquid films [1] showed that small whorls and jets of cooperative swimming exceeding scale of individual bacterium produced greatly-enhanced diffusion and even superdiffusion [12] of passive tracers. More recent experiments [2–4, 13], continuum theory [14], and simulations [15] have demonstrated that the correlation length of collective swimming state can exceed the size of individual cells by more than an order of magnitude, and the collective flow speed (of the order 50-100 μm/sec) similarly exceed the speed of individual bacterium (about 15-20 μm/sec). This collective swimming is found at sufficiently high number density (or volume fraction) of the microorganisms; dilute suspensions show no collective flow and the correlation length is comparable to the size of a bacterium (of the order of 5 μm).

Two key questions involving the onset of collective swimming dynamics arise in this context. How does spatio-temporal correlation depend on the concentration of microorganisms? How can their concentration be managed as a control parameter? In our previous works [13, 14] we presented experimental and theoretical studies of collective swimming dynamics in a free-standing thin-film geometry based on a novel technique to adjust the number density of bacteria in a given experiment (bacterial crowd control). This approach allows detail investigation of the correlation length and the mean swimming speed over a range of densities *with a single bacterial colony*, greatly reducing statistical fluctuations due to the inherent physiological differences between colonies. Our theoretical studies revealed only a gradual increase of the correlation length with the increase of number density, and no sharp transition. We propose that this can be explained as a noise-induced smearing of a dynamical phase transition, with the main source of noise being that due to strong fluctuations in the orientation of individual bacteria. Here we extend our theoretical and experimental studies of collective swimming states and present additional results not included in early publications [13, 14].

2 Experiment

2.1 Description of Experimental Setup

Experiments were conducted on suspensions of strains 1085 and YB886 of *Bacillus subtilis*, a peritrichously flagellated rod-shaped bacterium ~4 μm long and with a diameter of ~0.7 μm. Spores stored on agar or sand were used to inoculate nutrient medium (T5574 from Sigma). The suspension of grown cells was then washed and centrifuged. Our experimental setup, shown in Fig. 1, is similar to that used in Ref. [1], but has a number of important modifications.

A small drop of bacterial suspension is placed between four supporting, movable fibers: two Platinum (Pt) wires (to exclude contamination of the film) and two dielectric fibers, 0.3 mm diameter. The drop was then stretched between the fibers up to the necessary thickness (about 1 micron) by simultaneously pulling all four fibers apart with a control screw. The film was placed

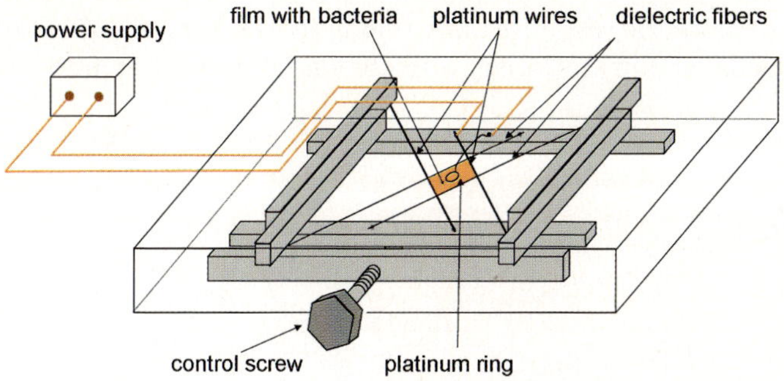

Fig. 1. Schematics of experimental setup. A thin film containing bacteria spans two adjustable Pt wires and two dielectric fibers and is stretched by a control screw. Electric current is transmitted between the Pt wires and a small Pt ring lowered onto the film.

in a controlled humidity chamber to minimize evaporation. Unlike previous experiments on *E. coli* [1], no additional surfactant was used to stabilize the liquid film. Instead, the metabolic products secreted by *B. subtilis* created the necessary surface tension and elasticity to maintain the film during the course of an experiment, typically several minutes. After that, the film either ruptured, or the secreted products solidified and motion ceased to exist. Images were obtained with a video camera operating at 100 frames/s framerate, equipped with a high-resolution, long working distance microscope objective. Image processing was performed by custom-designed software based on Matlab toolboxes. Working with a monolayer film of bacteria allows identification, counting and tracking each and every cell in the field of view.

2.2 Response of Bacteria on Electric Current: pH-Taxis

In order to concentrate bacteria, we applied a small electric voltage (≤ 2.6 V) between two platinum wires, thus creating the electrolysis, and, consequently, a change of pH level (i.e the acidity or concentration of ions) in the vicinity of electrodes. The change of pH in turns triggers chemotactic response in motile bacteria [16, 17]: these bacteria tend to swim away from the electrodes, towards the areas with a more comfortable pH level ($\sim$7.2) in the center of cell. Due to diffusion of ions, the areas of elevated/decreased pH levels expand gradually towards the middle of the experimental cell. If the applied voltage exceeds a certain threshold ($\sim$2.4V), the expansion rate of the pH level front becomes comparable with the average speed of bacteria (about 20 μm/sec), resulting in an avalanche-like accumulation and directed migration of bacteria (Fig. 2, a,b). This avalanche-like accumulation of bacteria at the pH-front is related to the fact that the bacteria begin to swim away from the front (to the center of the cell) only when the local pH gradient becomes large enough to trigger the chemotactic response. Remarkably, this technique stimulates nontrivial chemotactic response only among motile bacteria; dead or otherwise non-motile cells remain stationary and are left behind the front of congregating bacteria.

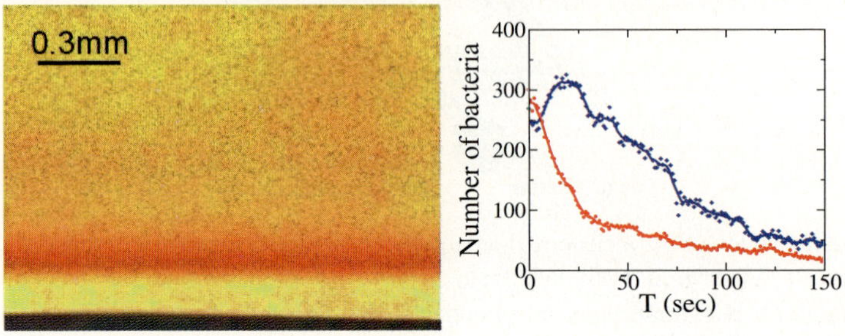

Fig. 2. (a) Formation of high bacterial density layer (seen as red) during directed migration of bacteria. (b) Total number of bacteria vs time in the region of interest ($100 \times 400\ \mu m^2$) at the distance 0μm–100μm from electrode (red) and the distance 100μm–200μm (black) after the voltage 2.4 V between platinum wires was applied.

2.3 Density Adjustment: Bacterial Crowd Control

In order to adjust the density of bacteria (the "filling fraction" ρ) and keep it almost uniform in the field of view of microscope over the entire course of one experiment, a small thin platinum ring (ring diameter ~1 mm, Pt wire diameter 0.03 mm) was gently lowered onto the stretched film containing the bacteria, and a small voltage (~2.0 V) was applied between the ring and two external Pt wires. The change of the pH level was monitored by the addition of indicator fluid Bromothymol blue to the solution. The evolution of the pH level as a function of time is illustrated in Figs. 3 a–c, in which the blue regions, corresponding to lower pH levels, expand on both sides of the ring after application of electrical current. This characteristic expansion of low pH-level region occurs due to diffusion of ions produced in the course of electrolysis. Uncomfortable with the change of pH at the ring's perimeter, bacteria inside the ring tend to swim towards the ring's center (see Fig. 2d). However, since the applied voltage was small (< 2.4 V), the migration of bacteria was relatively slow, and, therefore, only gradual concentration of bacteria at the middle of the ring was observed, while the density distribution was fairly homogeneous in the field view of the microscope (area $230 \times 230\ \mu m^2$ in the middle of the ring). Using this technique (aka "bacterial crowd control") we able to change the number density of bacteria by a factor exceeding 5.

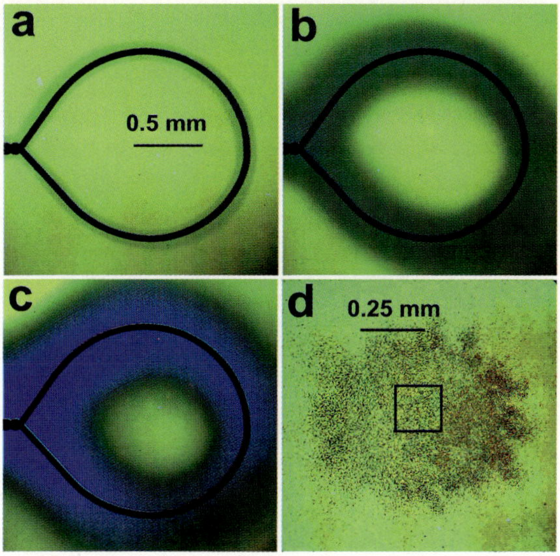

Fig. 3. Images illustrating decrease of the pH level near the electrodes as a result of transmission of electric current during 3 sec (a), 20 sec (b) and 80 sec (c). Concentrated bacteria (central part of the image), square indicates the field of view of microscope (d).

2.4 Large-Scale Collective Behavior

Representative snapshot of collective bacterial swimming pattern for filling fractions $\rho \sim 0.4$ is shown in Fig. 4 (here, the density $0 < \rho < 1$ is measured as a fraction of full surface coverage by bacteria). Large-scale collective swimming of bacteria, characterized by recurring whirls and reversing currents, occurs at the values of density ρ above 0.3–0.35. Corresponding evolution of bacterial density ρ in the course of experiment is shown in Fig. 5a.

The velocity field of the collective swimming state $\mathbf{V}$ was extracted from consecutive images by the Particle Imaging Velocimetry (PIV); bacteria themselves serving as tracers (resolution about 2 μm/sec). Since we are working with rather high bacterial filling fractions, traditional methods of flow visualization relaying on passive tracer particles are rather ineffective: the tracers tend to adhere to the bacteria. In addition to the velocity field, we also analyzed from the experimental images the (reconstructed) vector field of bacterial *orientations* $\boldsymbol{\tau}$. This was done by extracting from the images the director $\mathbf{n}$ by finding in the small subimage the angle of maximal projection of a bacterium on a certain direction. Then, assuming that the absolute value of the angle difference between the orientation vector $\boldsymbol{\tau}$ and the bacterial flow velocity field is smaller than $\pi/2$, $(\boldsymbol{\tau} \cdot \mathbf{V}) > 0$ (since bacteria swim in the direction of their orientation), one may extend the definition of the director $\mathbf{n}$ and reconstruct uniquely the vector field $\boldsymbol{\tau}$.

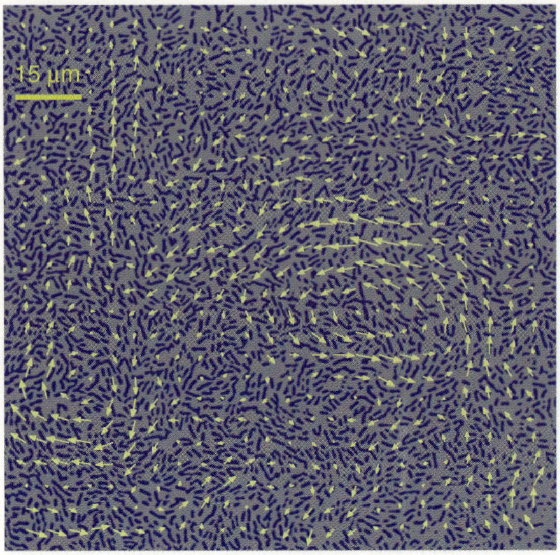

Fig. 4. Representative pattern of collective swimming in the colony of *Bacillus subtilis* confined in a thin free-suspended fluid film. Bacteria are seen as blue short stripes on grey background of the image. The instantaneous velocity field was obtained by the particle-image velocimetry, and is shown in the image by arrows (the longest arrow represents velocity of 100 μm/sec).

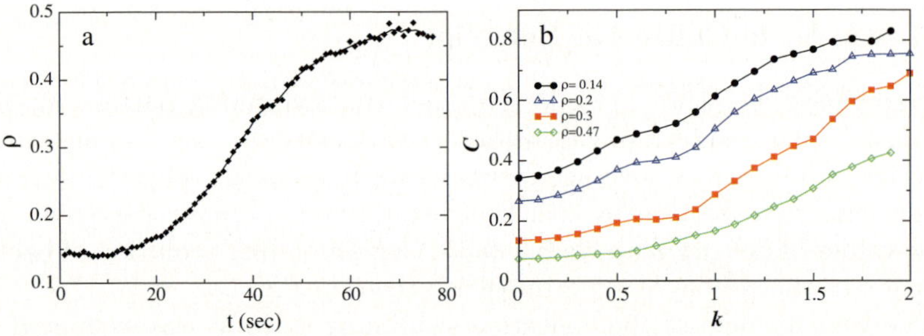

Fig. 5. (a) Evolution of bacterial density in the course of experiment; (b) Alignment coefficient C between velocity and orientation fields, see Eq. (1), as function of threshold k for different values of density ρ.

2.5 Cross-Correlation Between Velocity and Orientation

In the regime of well-developed large-scale flows (Fig. 4), the cross-correlation between the vector fields $\boldsymbol{\tau}$ and $\mathbf{V}$ over the entire span of the image yields a rather puzzling result: the correlation coefficient was smaller than 8%! However, this seeming contradiction of the fact that bacteria swim in the direction of their orientation can be resolved: for the case of well-developed chaotic flow the bacteria are advected by the fluid velocity field $\mathbf{v}_f$ created by *all bacte-*

ria. Since each bacterium swims in the direction of its orientation only in the *local frame* moving with fluid velocity $\mathbf{v}_f$, one should not expect strong correlation between $\mathbf{V}$ and $\boldsymbol{\tau}$ as long as the fluid velocity field is chaotic. In contrast, there is significant correlation between the maximal values of these fields, since more oriented patches of bacteria create faster flow. In order to demonstrate this, we characterize the alignment between these two fields by the alignment coefficient C,

$$C = \frac{\langle \cos\phi \rangle - 2/\pi}{1 - 2/\pi} \tag{1}$$

where ϕ is the angle between $\mathbf{V}$ and $\boldsymbol{\tau}$ (note that $-\pi/2 < \phi < \pi/2$ due to our choice of the direction of $\boldsymbol{\tau}$). If the directions between $\mathbf{V}, \boldsymbol{\tau}$ were random, then the value of $\langle \cos\phi \rangle = 2/\pi$ and $C = 0$, whereas perfectly aligned fields yield $C = 1$. To amplify contributions from large-amplitude regions of the fields $\boldsymbol{\tau}$ and $\mathbf{V}$ we used the following method: in the averaging procedure we excluded those points of both fields where the amplitudes were below certain variable thresholds $V_s = k\langle|\mathbf{V}|\rangle$ or $\tau_s = k\langle|\tau|\rangle$, where the threshold k measures the fraction of the mean value of each field. The alignment coefficient C vs k is shown in Fig. 5b, and one sees there that the alignment indeed decreases with the density ρ due to the effect of large-scale flow discussed above. Since at low concentrations the bacteria swim exactly in the direction of their orientation, one obtains $C \to 1$. With the increase in k the alignment coefficient C indeed increases, supporting the statement that more aligned regions are also moving faster.

2.6 Autocorrelation Functions

In order to characterize the properties of collective flow over a range of bacterial number density ρ, we independently extracted spatial autocorrelation functions of the velocity and the orientation fields. We also calculated the typical bacterial velocity $\bar{V} = \sqrt{\langle \mathbf{V}^2 \rangle - \langle \mathbf{V} \rangle^2}$, the magnitude of orientation field $\boldsymbol{\tau}$, and the radial correlation functions for velocity $K_V(r)$ and orientation $K_\tau(r)$, defined as

$$K_A(r) = \int d\mathbf{r}' \int_0^{2\pi} d\theta \left[\langle \mathbf{A}(\mathbf{r}') \cdot \mathbf{A}(\mathbf{r} + \mathbf{r}') \rangle - \langle \mathbf{A}(\mathbf{r}') \rangle^2 \right]$$

where θ is a polar angle, $\mathbf{A} = \mathbf{V}, \boldsymbol{\tau}$.

The corresponding correlation length L_A was then extracted from the functions $K_A(r)$ by a fit to an exponential, $K_A(r) \sim \exp(-x/L_A) + B, B = const.$ It appears that the correlation length for orientation field $\boldsymbol{\tau}$ was typically of the order of bacterium size (see Fig. 6) whereas the velocity correlation length was much bigger and increased with the increase of bacterial density. In the most of experiments the constant B was rather small, however near the onset of collective swimming the function $K_V(r)$ appears to exhibit oscillations,

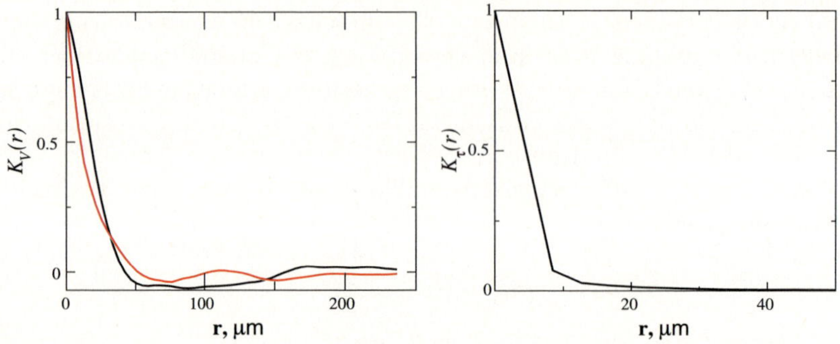

Fig. 6. Typical autocorrelation functions. Black curves depict velocity $K_V(r)$ (left) and orientation $K_\tau(r)$ (right) autocorrelations for well-developed turbulent state. Red curve shows velocity autocorrelation function near the onset of large-scale collective motion, oscillations in $K_V(r)$ tail signal the presence of large-scale vortex structure.

likely due to intermittent large-scale vortex motion. The results are shown in Fig. 6. While the form of the correlation functions $K_V(r)$ and the value of correlation length L are in agreement with previous measurements in Ref. [3], our results have the added benefit of not being contaminated by boundary effects and large-scale oxygen concentration gradients. Our measurements are performed in a film of constant thickness, whereas in previous works [2, 3] the film thickness variations were significant.

The dependence of the velocity correlation length L and the typical swimming velocity vs density ρ is shown in Fig. 7a,b. As one sees from Fig. 7, no sharp transition occurs with the increase in density; we find only a smooth (although steep) increase of the velocity $\bar{V}$ and the correlation length L. The overall changes in these quantities were about a factor of five. For even higher density ($\rho \to 1$, i.e. close 90% surface coverage) we notice complete termination of motion and the formation of a biofilm.

2.7 Fluctuations-Broadened Transition

Absence of well-defined phase transition with the increase in density ρ does not agree with the predictions of simplified Vicsek-type models of collective motion in systems of self-propelled particles [7–10]. We interpret our observation as a smearing of a phase transition by noise which can arise from a number of sources, such as spontaneous orientation fluctuations of individual bacteria due to tumbling, small-scale hydrodynamic fluctuations due to flagellum rotation, size distribution of bacteria, etc. While we did not pinpoint the primary cause for the noise, its effect appears to be very robust and qualitatively independent on the specific nature of fluctuations. To support this observation, we compared our experimental data for $\bar{V}$ vs ρ with those

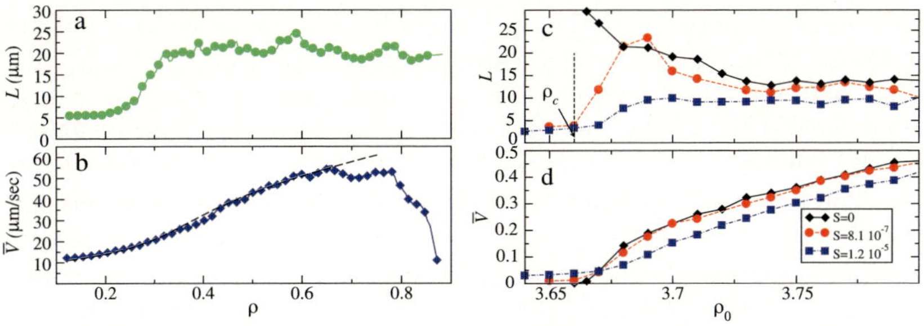

Fig. 7. Experiment: velocity correlation length L (a) and typical velocity $\bar{V}$ (b) vs bacterial number density ρ. Dashed line shows fit of $\bar{V}$ to solution Eq. (2) with $\rho_c \approx 0.28$. Model: velocity correlation length L (c) typical velocity $\bar{V}$ (d) vs average density ρ_0 for three different level of noise S and for parameters of Fig. 10.

obtained from the normal form equation for generic noisy second order phase transition

$$\partial_t V = (\rho - \rho_{cr})V - V^3 + \zeta(t) \tag{2}$$

where ζ is Gaussian white noise with the intensity S, and ρ_{cr} is critical density. In the absence of noise one obtains the mean-field result $\bar{V} \sim \sqrt{\rho - \rho_{cr}}$. With the noise one obtains a smearing of the transition point ρ_{cr}. Figure 7 shows that the fit of a solution to Eq. (2) is consistent with the experimental data shown in Fig. 7.

3 Model

The experiments [1–3, 13] have demonstrated that concentrated suspensions of swimming bacteria exhibit an intriguing collective dynamics reminiscent of high Reynolds number hydrodynamic "turbulence". Various theoretical models [11, 18, 19] have predicted that a state of true long-range order of swimming orientation, suggested by simplified models of self-propelled particles [7, 8, 10], is linearly unstable if the long-range interactions between particles are taken into account. The fate of the system beyond that instability has not been well studied within the context of those models.

On the other hand, a proposed [3] analogy with sedimenting suspensions, in which transient and recurring vortices and jets are commonly observed, suggested that the structures seen in our experiment might arise purely from hydrodynamic interactions. Direct numerical simulations [15] of ensembles of self-propelled particles, whose interactions are given solely by the hydrodynamic flow fields they generate, confirm this notion, showing large-scale swirls beyond a critical volume fraction.

Motivated in particular by these experiments on thin films of bacterial suspensions with controllable density, we proposed a continuum model describing self-propelled swimmers [14]. The model is formulated in terms of a two-dimensional master equation for the probability density $P(\mathbf{r}, \phi)$ of finding a bacterium at a particular orientational angle ϕ at position $\mathbf{r} = (x, y)$, derived from microscopic interaction rules. The master equation links the scale of individual bacterium to the macroscopic scale of collective motion, described by continuum coarse-grained equations for local bacterial density and orientation. The system is supplemented by the Navier-Stokes equations for the fluid velocity with the forcing term due to swimming of oriented bacteria. In the nonlinear regime, the model reveals a scale selection mechanism associated with deflection of bacterial swimming by the shear flow, and is in qualitative agreement with experiment.

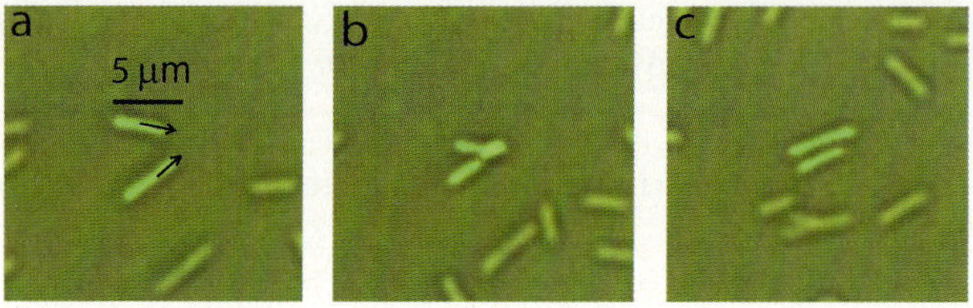

Fig. 8. Sequence of experimental images illustrating an "inelastic collision" between swimming *Bacillus subtilis* in a thin film. Colliding bacteria (highlighted) swimming from left to right begin misaligned in (a), reorient during collision (b) and swim off parallel afterwards (c).

Following previous works [14, 20], we model the bacteria as polar rods of length l, diameter d_0, subject to two rules: (i) bacteria swim at velocity v_0 with respect to the liquid in the direction of their unit orientation vector $\mathbf{n} = (\cos\phi, \sin\phi)$; (ii) in a collision of two bacteria with the angles $\phi_{1,2}$ the pair swims off in the direction of the average orientation $\bar{\phi} = (\phi_1 + \phi_2)/2$ from a location $\bar{\mathbf{r}} = (\mathbf{r}_1 + \mathbf{r}_2)/2$, the average of their immediate post-collisional locations. By analogy with the physics of granular matter, we call this a *fully inelastic* collision, and it arises from the quadrupole velocity field of swimmers [15]. It is readily seen in experiment (Fig. 8). It is well-known that inelastic collisions, resulting in build-up of spatial correlations, are precursors of collective behavior [21].

Bacteria are also subject to rotational and translational diffusion due to tumbling and small-scale hydrodynamic flows, and are advected by the fluid. The master equation is [14]

$$\begin{aligned}\partial_t P + \boldsymbol{\nabla} \cdot [(v_0 \mathbf{n} + \mathbf{v}) P] + \frac{1}{2} \Omega \partial_\phi P = D_r \partial_\phi^2 P + \partial_i D_{ij} \partial_j P \\ + \iint d\mathbf{r_1} d\mathbf{r}_2 \int_{-\pi}^{\pi} d\phi_2 W(\mathbf{r}_1, \mathbf{r}_2) P(\mathbf{r_1}, \phi_1) P(\mathbf{r_2}, \phi_2) \\ \times \left[\delta\left(\bar{\mathbf{r}} - \mathbf{r}, \bar{\phi} - \phi\right) - \delta\left(\mathbf{r}_2 - \mathbf{r}, \phi_2 - \phi\right)\right] - \gamma \left(\mathbf{E} \cdot \mathbf{n} \cdot \frac{\partial P}{\partial \mathbf{n}}\right) . \end{aligned} \tag{3}$$

The second and third terms on the l.h.s. of (3) account for the hydrodynamic advection of bacteria and their rotation by the flow, whose vorticity is $\boldsymbol{\Omega} = (\partial_y v_x - \partial_x v_y)\,\hat{\mathbf{z}}$. The first two terms in the r.h.s. of (3) describe angular and translational diffusion of rods with the diffusion tensor $D_{ij} = \left(D_{\|} n_i n_j + D_{\perp}(\delta_{ij} - n_i n_j)\right)$. D_{ij} are known in polymer physics: $D_{\|} = k_B T_e/\xi_{\|}$, $D_{\perp} = k_B T_e/\xi_{\perp}$, and $D_r = 4k_B T_e/\xi_r$, where $\xi_{\|}, \xi_{\perp}, \xi_r$ are corresponding viscous drag coefficients. For rod-like molecules, they are $\xi_{\|} = 2\pi\eta_s l/\log(l/d_0)$, $\xi_{\perp} = 2\xi_{\|}$, and $\xi_r \approx \pi\eta_s l^3/3\log(l/d_0)$, where η_s is shear viscosity and T_e is effective temperature [22]. The effective temperature T_e is understood here to arise from small-scale hydrodynamic flows and bacterial tumbling, and can exceed considerably the thermodynamic temperature. The last term in Eq. (3) describes the coupling to strain rate tensor $\mathbf{E}$ ($E_{xy} = \partial_x v_y + \partial_y v_x$) [23, 24]. Our further analysis shows that while this term has some quantitative effect, e.g. on the instability threshold, it does not change the qualitative conclusions. Thus, for simplicity we set the coupling constant $\gamma = 0$ and neglect contribution of $\mathbf{E}$.

The last term of (3) describes short-range binary interactions of rods. The two δ-functions in the collision integral describe "annihilation" of a particle with the angle ϕ_1 and "creation" of particle with the angle $\bar{\phi}$. The interaction kernel W is localized in space, and for the sake of simplicity we neglect the anisotropy of the kernel (the kernel anisotropy appears to be important, however, for self-organization of microtubules [20]). We set

$$W = (g/b^2\pi)\exp\left[-(\mathbf{r}_1 - \mathbf{r}_2)^2/b^2\right]$$

with $b \simeq l$ and g the interaction cross-section. This form of the kernel implies that only nearby bacteria interact effectively. While hydrodynamic interactions usually decay algebraically (e.g. inversely proportional to separation distance in three dimensions), exponential screening for shear perturbations occurs, for example, in the thin film geometry when the film is in contact with a frictional walls. For the free-standing thin-film geometry of our experiment, one may expect the free-slip boundary condition for the velocity on the top and the bottom surface of the film. However, as we discussed above, it appears that a layer of surfactant (produced by bacteria themselves) which quickly accumulates on both top and bottom surfaces of the film, plays the role of a frictional walls, effectively implying the *non-slip* condition for hydrodynamic velocity.

Now we introduce the coarse-grained density ρ and orientation $\boldsymbol{\tau}$:

$$\rho(\mathbf{r}) = \int_{-\pi}^{\pi} d\phi P, \quad \boldsymbol{\tau} = (1/2\pi)\int_{-\pi}^{\pi} d\phi\, \mathbf{n}(\mathbf{r}) P$$

As it was shown in [14, 20], the *spatially-homogeneous* limit of (3) exhibits onset of an oriented state above the critical density $\rho_c = (D_r/g)/(4/\pi - 1)$. Near this threshold, (3) can be simplified significantly by means of a standard bifurcation analysis, yielding a pair of coupled equations for ρ and $\boldsymbol{\tau}$. Also near threshold, P depends slowly on the variable $\mathbf{r}$, so we keep only leading terms in the expansion in spatial gradients. With proper rescalings $\mathbf{r} \to \mathbf{r}/l$, $t \to D_r t$, and $\rho \to g\rho/D_r$, we obtain

$$\partial_t \rho + \boldsymbol{\nabla} \cdot (\rho \mathbf{v}) = D_0 \nabla^2 \rho - v_0 \pi \boldsymbol{\nabla} \cdot \boldsymbol{\tau} \tag{4}$$

$$\partial_t \boldsymbol{\tau} + \mathbf{v} \cdot \boldsymbol{\nabla} \boldsymbol{\tau} + \frac{1}{2} \boldsymbol{\Omega} \times \boldsymbol{\tau} = (\epsilon \rho - 1) \boldsymbol{\tau}$$
$$- A_0 |\boldsymbol{\tau}|^2 \boldsymbol{\tau} + D_1 \nabla^2 \boldsymbol{\tau} + D_2 \boldsymbol{\nabla} \boldsymbol{\nabla} \cdot \boldsymbol{\tau} - \frac{v_0}{4\pi} \boldsymbol{\nabla} \rho \; . \tag{5}$$

Eq. (4) describes advection of the bacteria by hydrodynamic velocity $\mathbf{v}$ and diffusive spreading with the diffusion coefficient D_0. Here $D_1 = (D_\parallel + D_\perp/2)/2D_r l^2, D_2 = (D_\parallel - D_\perp)/2D_r l^2$. In the rigid rods limit $D_1 = 5/192$, $D_2 = 1/96$ [20]. For small density ρ and for the case of pure *thermal* diffusion of particles, the diffusion coefficients obey $D_0 = (D_\parallel + D_\perp)/2D_r l^2$. In the present context, this connection is not clear, especially for larger densities due to diffusive-type contributions from the collision integral in (3). In experiments, there are no significant density fluctuations observed, so we can treat $D_0 \gg D_{1,2}$ as an independent parameter in order to suppress density variations. In Eq. (5) the first term on the r.h.s. describes the orientation instability, with $\epsilon = 0.276$, $A_0 = 2.81$ for fully inelastic particles [20]. Terms proportional to v_0 arise from bacterial swimming with respect to the fluid.

The in-plane fluid velocity $\mathbf{v}$ obeys the Navier-Stokes equation with forcing due to bacterial swimming

$$\partial_t \mathbf{v} + \mathbf{v} \cdot \boldsymbol{\nabla} \mathbf{v} = \nu \nabla^2 \mathbf{v} - \nabla p - \beta \mathbf{v} + \alpha \boldsymbol{\tau} \; , \tag{6}$$

with $\boldsymbol{\nabla} \cdot \mathbf{v} = 0$ by incompressibility. In Eq. (6), $\nu = \nu_0/D_r l^2$ is renormalized viscosity, where ν_0 is the fluid kinematic viscosity, p is the pressure, and $\alpha\boldsymbol{\tau}$, with $\alpha \sim v_0$, models the forcing due to bacterial swimming. While our experiments, along with earlier ones [1], are performed in the free-standing thin film geometry, the surfactant accumulated on both surfaces of the liquid film play the role of semi-flexible walls, resulting in a nontrivial velocity profile across the film, and, consequently, viscous dissipation. The forcing term in Eq. (6) is formally different from that for the self-propelled particles proposed in Ref. [11], where the force is represented by the divergence of certain three-dimensional stress tensor σ_{ij}, but integration of that stress tensor over the film's cross-section produces a contribution $\sim \tau$ due to boundary effects. The damping term $\beta\mathbf{v}$ is generated by the thin film surface elasticity resulting in the partial slip condition for the velocity on the surface of the film

To eliminate the pressure p and satisfy continuity equation we introduce the stream function φ, with $v_x = \partial_y \varphi$, $v_y = -\partial_x \varphi$, and $\Omega = \nabla^2 \varphi$. Then Eq. (6) yields

$$\partial_t \Omega + \mathbf{v} \cdot \boldsymbol{\nabla} \Omega = \nu \nabla^2 \Omega - \beta \Omega + \alpha \left(\partial_y \tau_x - \partial_x \tau_y \right) \; . \tag{7}$$

Eqs. (4), (5), and (7) form a closed system. For flows with vanishingly small Reynolds numbers *Re* (as we deal with) the advection term $\mathbf{v} \cdot \boldsymbol{\nabla} \Omega$ can be neglected relative to the viscous dissipation $\nu\nabla^2\Omega$, but we keep it since a similar term is included in Eq. (8). While *Re* for individual swimming bacterium is exceedingly small, for the collective flows *Re* can grow significantly. To simplify the analysis we consider the constant density approximation $\rho = \rho_0$ valid for a large D_0 values. Then Eq. (5) yields

$$\partial_t \boldsymbol{\tau} + \mathbf{v} \cdot \boldsymbol{\nabla}\boldsymbol{\tau} + \frac{\boldsymbol{\Omega} \times \boldsymbol{\tau}}{2} = (\epsilon\rho - 1)\boldsymbol{\tau} - A_0|\boldsymbol{\tau}|^2\boldsymbol{\tau} + D_1\nabla^2\boldsymbol{\tau} + D_2\boldsymbol{\nabla}\boldsymbol{\nabla}\cdot\boldsymbol{\tau} \tag{8}$$

Eqs. (5), (7) have a steady-state solution corresponding to a homogeneous stream of bacteria in a certain direction (e.g. along x): $\tau_x = \tau_0 = ((\epsilon\rho - 1)/A_0)^{1/2}$, $\tau_y = v_y = 0, v_x = V = \alpha\tau_0/\beta$. The most unstable modes in the problem are longitudinal, and we examine stability of this state with perturbations of the form $(\boldsymbol{\tau}, \Omega) \sim \exp[\lambda t + ikx]$, where λ is the growthrate and k is the modulation wavenumber. Linearization of (5) and (7) shows that the equation for τ_x splits off, with a growth rate having a strictly negative real part, $\lambda = -ikV - 2\tau_0^2 - (D_1 + D_2)k^2$, while τ_y, Ω are coupled:

$$\lambda\tau_y = -ikV\tau_y - \frac{1}{2}\Omega\tau_0 - D_1k^2\tau_y \tag{9}$$

$$\lambda\Omega = -ikV\Omega - \nu k^2\Omega - \beta\Omega - ik\alpha\tau_y \tag{10}$$

They yield the two growthrates λ:

$$\lambda_{1,2} = \frac{1}{2}\Big(-(D_1 + \nu)k^2 - \beta - 2ikV \pm \sqrt{((D_1 - \nu)k^2 - \beta)^2 - 2ik\tau_0\alpha}\Big) \tag{11}$$

The instability occurs if the product $\alpha\tau_0$ exceeds a critical value, whose value can be deduced by examining the limit $k \to 0$. An expansion in powers of k of $\mathrm{Re}(\lambda)$ yields

$$\mathrm{Re}\,(\lambda) \simeq \left(\frac{\alpha^2\tau_0^2}{\beta^3} - D_1\right)k^2 + O(k^4). \tag{12}$$

Clearly, there is a long-wave instability if $(\alpha\tau_0)^2 > \beta^3 D_1$, see Fig. 9. The threshold density for this instability ρ_p always exceed critical density of the orientation transition $\rho_c = 1/\epsilon$. However, $\rho_p - \rho_c$ is small due to relative smallness of the diffusion $D_1 \approx 0.026$. Since $V = \alpha\tau_0/\beta$ is the collective steady-state swimming velocity, we can re-express the instability criterion simply as $V > V_d$, where $V_d = \sqrt{\beta D_1}$. Moreover, since $\beta \sim \nu/d^2$, where d is the film thickness, we find $V_d = \sqrt{\nu D_1}/d$. The selected wavenumber k_m can be obtained in the limit of large collective swimming speed V. Expanding (11)

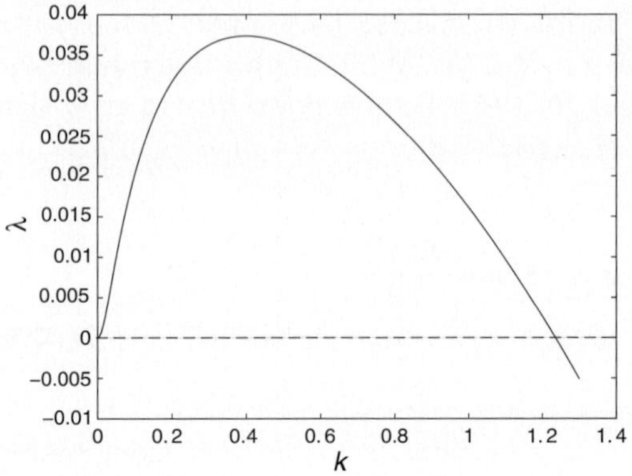

Fig. 9. Real part of growthrate, $Re(\lambda)$, vs modulation wavenumber k for representative values of parameters.

for $\alpha\tau_0 \gg 1$ we obtain

$$\mathrm{Re}\,(\lambda) \approx \frac{1}{2}\left(-(D_1+\nu)k^2 - \beta + \sqrt{|k|\tau_0\alpha}\right) + \cdots \qquad (13)$$

Then from (13) one finds (for $(D_1 + \nu)k^2 \gg \beta$) the optimal instability wavenumber:

$$k_m^{3/2} = \sqrt{\tau_0\alpha}/4(D_1+\nu)\ . \qquad (14)$$

Correspondingly, a typical scale of spatial patterns arising at the threshold of instability is given by $L_m \sim 1/k_m$. We expect that this scale will be roughly of the order of correlation length L.

We have conducted *numerical studies* of the full system (4), (5), and (7) over a range of densities ρ, with periodic boundary conditions. A typical flow pattern and distribution of orientation vectors $|\boldsymbol{\tau}|$ is shown in Fig. 10. Remarkably, over the entire computational domain the correlation between the fields $\boldsymbol{\tau}$ and $\mathbf{v}$ is close to zero, in agreement with experiment. However, there is always *local* correlation between $\boldsymbol{\tau}$ and $\mathbf{v}$ through Eq. (7).

As in experiment, the typical hydrodynamic velocity $\bar{V} = \sqrt{\langle \mathbf{v}^2\rangle - \langle \mathbf{v}\rangle^2}$, and the velocity correlation length L was calculated. The results are shown in Fig. 7c,d. The emerging picture of the transition has a strong resemblance to a second order phase transition: the typical velocity $\bar{V} \sim \sqrt{\rho - \rho_c}$, and the correlation length diverges at $\rho \to \rho_c$, consistent with the prediction of Eq. (14). In order to include effect of fluctuation, we added to the orientation equation (5) a random force $\zeta(x, y, t)$, with correlation $\langle\zeta(x, y, t)\zeta(x', y', t')\rangle = 2S\delta(x-x')\delta(y-y')\delta(t-t')$, where S is the noise amplitude. Results for various noise strengths are shown in Fig. 7c-d, where even a relatively small noise ($S = 1.2 \times 10^{-7}$) smears the transition and removes the divergence of the correlation length. For strong enough noise ($S \sim 10^{-5}$), one observes only a gradual increase of the correlation length with the density, in agreement with experiment in thin film.

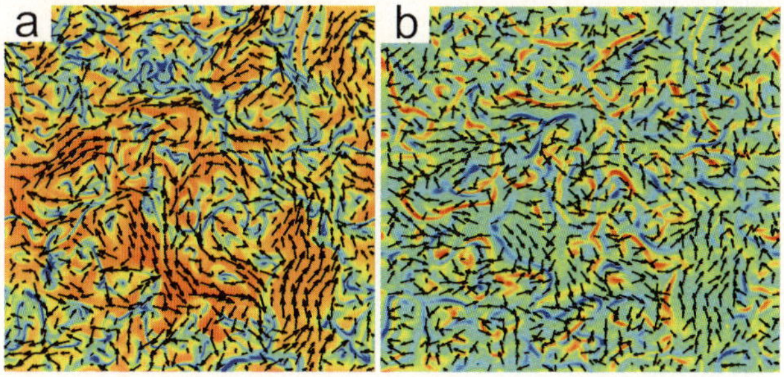

Fig. 10. Representative flow patterns obtained by solution of Eqs. (4), (5), (7) for $\rho_0 = 3.8, D_0 = 50, \nu = 3, v_0 = 0.2, \alpha = 3, \beta = 0.5$ in the periodic domain of 200×200 units. (a) magnitude of orientation field $\boldsymbol{\tau}$ superimposed with the velocity field $\boldsymbol{v}$. Red color corresponds to maximum of $|\boldsymbol{\tau}|$, and blue to $|\boldsymbol{\tau}| = 0$. Arrows depict the flow velocity $\mathbf{v}$ field. (b) Vertical vorticity field Ω shown instead of orientation $\boldsymbol{\tau}$.

4 Conclusions

We have studied collective bacterial swimming in thin free-standing liquid films where the dynamics is essentially two-dimensional and the concentration can be adjusted continuously. Our results provide a strong support for a pure hydrodynamic origin of collective swimming, rather than some chemotactic mechanisms of pattern formation (chemotactic interaction, relevant for slow "crawling" bacteria on substrate, does not play any significant role in our experiment due to very fast mixing rates in the collective flow state) [25]. For instance, the primary chemoattractant (Oxygen) would be expected to be highly uniform because of the thin film geometry with air on both side and fast stirring of fluid by the bacteria. Swimming of bacteria in the direction of their orientation is the underlying reason for the onset of large-scale chaotic flows. The technique of "bacterial crowd control" utilized for the concentration and separation of bacteria by an electric field may find interesting future applications for bioanalysis and miniaturized medical diagnostic devices. The proposed model for the large-scale flows generated by ensembles of swimming bacteria in thin films shows that the onset of coherence is attributed to the collective hydrodynamic interaction between individual objects. This work was supported by DOE grant W-31-109-ENG-38, and NSF PHY-0551742 (JOK & REG).

References

1. X.-L. Wu and A. Libchaber, Phys. Rev. Lett. **84**, 3017 (2000).
2. N. H. Mendelson et al, J. Bacteriol. **181**, 600 (1999).
3. C. Dombrowski, L. Cisneros, S. Chatkaew, R. E. Goldstein, and J. O. Kessler, Phys. Rev. Lett. **93**, 098103 (2004).

4. I.H. Riedel, K. Kruse, and J. Howard, Science **309**, 300 (2005).
5. Ch. Becco et al, Physica A **367**, 487 (2006); J. Buhl et al, Science **312**, 1402 (2006).
6. T. Feder, *Statistical Physics is for the Birds*, Physics Today, p.28, October 2007.
7. J. Toner and Y. Tu, Phys. Rev. Lett. **75**, 4326 (1995).
8. G. Grégoire and H. Chaté, Phys. Rev. Lett. **92**, 025702 (2004).
9. C. Huepe and M. Aldana, Phys. Rev. Lett. **92**, 168701 (2004).
10. T. Vicsek, A. Czirók, E. Ben-Jacob, I. Cohen, and O. Shochet, Phys. Rev. Lett. **75**, 1226 (1995); A. Czirók, H. E. Stenley, and T. Vicsek, J. Phys. A: Math. Gen. **30**, 1375 (1997).
11. R. A. Simha and S. Ramaswamy, Phys. Rev. Lett. **89**, 058101 (2002).
12. J. O. Kessler, in *International Conference on Differential Equations, vol. 2*, eds. B. Fiedler, K. Gröger, and J. Sprekels (Singapore, World Scientific), 1284 (2000).
13. A. Sokolov, I. S. Aranson, J. O. Kessler, R. E. Goldstein, Phys. Rev. Lett. **98**, 158102 (2007).
14. I. S. Aranson, A. Sokolov, J. O. Kessler, and R. E. Goldstein, Phys. Rev. E. **75**, 040901 (2007).
15. J. P. Hernandez-Ortiz, Ch. G. Stoltz, and M. D. Graham, Phys. Rev. Lett. **95**, 204501 (2005).
16. M. Kihara and R. M. Macnab, J. Bacteriol. **145**, 1209 (1981).
17. T. Minamino et al, J. Bacteriol. **185**, 1190 (2003).
18. Y. Hatwalne, S. Ramaswamy, M. Rao, and R. A. Simha, Phys. Rev. Lett. **92**, 118101 (2004).
19. R. Voituriez, J. F. Joanny, and J. Prost, Phys. Rev. Lett. **96**, 028102 (2006).
20. I. S. Aranson and L. S. Tsimring, Phys. Rev. E **71**, 050901(R) (2005); Phys. Rev. E **74**, 031915 (2006).
21. I. S. Aranson and L. S. Tsimring, Rev. Mod. Phys. **78**, 641 (2006).
22. M. Doi and S. F. Edwards, *The Theory of Polymer Dynamics*, Clarendon Press, Oxford, 1988.
23. K. Kruse et al, Phys. Rev. Lett. **92**, 078101 (2004).
24. N. A. Hill and M. A. Bees, Phys. Fluids **14**, 2598 (2002).
25. F. Keller and L. A. Segel, J. Theor. Biol. **30**, 225 (1971).

Flow and Jamming of Granular Matter Through an Orifice

Angel Garcimartín, Cristian Mankoc, Alvaro Janda, Roberto Arévalo,
J. Martín Pastor, Iker Zuriguel, and Diego Maza

Depto. de Física y Mat. Apl., Universidad de Navarra, 31080 Pamplona, Spain
http://fisica.unav.es/granular

Summary. When particles pass through an orifice, jamming can occur. The question is whether a jammed structure can be considered a new state of matter, and if the flow behaves differently when approaching jamming. An experiment, consisting of a silo filled with grains and an orifice at the base, is presented here. The jamming probability is measured, and it is shown that above a certain orifice size no jamming can occur. A power law divergence is found when that value is approached. Besides, the flow rate is different for small and large orifices. For large orifices, the Beverloo equation states that the flow depends on the diameter to the 5/2 power. But this relation breaks down for small orifices. A new functional dependence is proposed, in agreement with the experiments and the numerical simulations. Furthermore, the statistical analysis of the fluctuations for small orifices shows anomalous behavior.

1 Introduction

Jamming has been recognized as a distinctive feature of granular matter [1]. A granular material (an assembly of a large number of solid particles, or grains, among which only contact forces are relevant) is just a paradigm of a system in which a flow can be arrested due to a jamming event. Glass-forming liquids, colloids and foams, among others, share many features with granular matter when their constituent elements get fixed and jammed, forming an amorphous phase [2, 3]. Traffic and pedestrian flow, insofar as it can be considered a current of solid particles, would also belong in this category.

We will focus on a particular granular flow that can get jammed, namely, the outpouring of grains, driven by gravity, through an orifice at the base of a silo. By choosing this system, we are restricting ourselves to a particular case where some variables such as the compaction fraction –which is of utmost relevance in many situations– are left to evolve freely. But at the same time this allows us to explore methodically the features of the flow by changing just one parameter, which is the size of the outlet orifice. This opens interesting conceptual avenues to explore, such as the question of whether jamming can be defined as a phase transition. The wealth of knowledge amassed about phase transitions could then be used to predict, control, avoid, undo or just describe jamming.

Although we all have an intuitive idea of jamming, arising maybe from our everyday experience with fragmented solids, it is not easy to provide a rigorous definition of it. Despite the fact that jamming is in some aspects similar to a liquid-solid phase transition, one of the key concepts involved is the fact that the constituent particles are kept in place by external mechanical stresses, instead of internal bonds between them. The mechanically stabilized structures involve long chains of grains, some of them forming arches that are sustained by the external forces [4]. When these forces are changed or cease to act, the structures loose their stability. The formulation of this idea has led to the definition of "fragile matter" [5], which in many cases is suitable for the physical systems that can get jammed.

The experiments begin with the grains pouring freely from the aperture at the base of the silo. Eventually, if the opening is not big (we will later address the meaning of 'big' in this context) the flow will stop due to the formation of an *arch* spanning over the opening. If the arch is broken the flow will resume. Then an avalanche can be defined as the event developed between the breakage of an arch and the spontaneous formation of another one that interrupts the flow. The most readily available measurement is the size of the avalanche (the number of grains fallen) and its duration. The data obtained from a large number of avalanches allow us to describe the statistics of the jamming events. From the size of the avalanche and its duration, the mean flow rate can be obtained and analyzed depending on the size of the outlet orifice.

Three questions will be addressed in this paper. The first one is whether there are or not two different regimes, i.e. a regime in which the flow will be arrested by arches of grains, and another regime in which the outpouring of grains will never stop, and if a well defined transition separates them. We will also look into the law for the mass flow rate. Indeed, for large orifices an equation can be obtained that gives the number of beads fallen per unit time as a function of the size of the orifice. It is reasonable to ask if this relation is valid for small orifices, when jamming can occur, or if it fails, indicating that the features of the flow are different in this case. And finally, we will investigate if the grains move unusually when they are going to get jammed, by paying attention to their displacement fluctuations. This knowledge could eventually be used to devise strategies to avoid jamming. Before we proceed, we will briefly describe our experimental devices. Finally, we will gather some conclusions at the end of the paper.

2 Experimental Set-Up and Numerical Procedures

The basic assembly we use in our experiments is a scaled cylindrical silo with a circular orifice at the base. Beneath it, a scale is placed to measure the weight of the grains. The number of grains in an avalanche can then be calculated. We also measure its duration by means of a microphone that detects the sound of the falling grains.

The filling procedure of the silo is always the same, so as to begin with the same initial conditions in every run. The avalanche is started by means of an air jet aimed from below at the exit orifice. The air jet produces just a local perturbation that breaks the arch blocking the orifice, keeping the bulk of the silo undisturbed. Moreover, due to the Janssen effect [6], the pressure at the bottom of the silo does not change much -provided that the height reached by the grains is more than twice the diameter of the silo. We have also checked that the silo is wide enough so that the lateral walls do not influence the results.

The experimental set-up has previously been described, so we refer the interested reader to a previous paper [7] for additional details. A sketch is provided in Fig. 1.

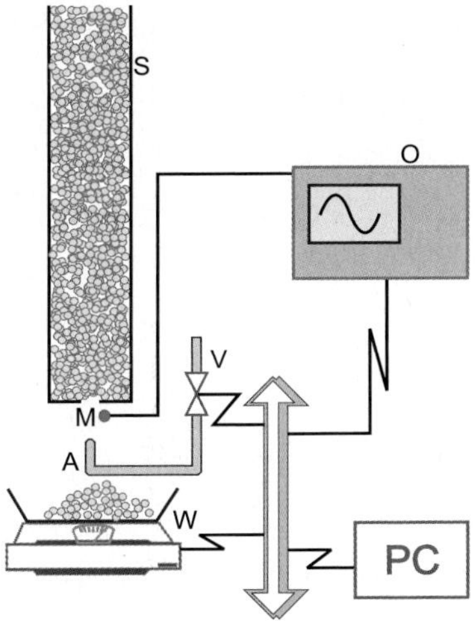

Fig. 1. Sketch of the experimental setup for the cylindrical silo. S: silo; M: microphone; A: compressed air; O: oscilloscope; V: valve; W: scales.

Experiments have also been conducted in a two dimensional silo. This container consists of two glass panes between which a gap slightly larger than the bead diameter is left. Again, a scales beneath the opening at the base is used to measure the size of the avalanche. Its duration is obtained in this case through a photodetector that registers the time that a light beam has been blocked by falling grains. The experimental procedures are therefore much similar to those in the cylindrical silo. In order to track the beads inside the silo we have used a high-speed camera (Photron model Fastcam 1024, capable of recording a window of 512x512 pixels at 3000 frames per second). In this way, we have obtained the velocity profile of the grains inside the silo, we have studied the fluctuations of grains as they move towards the outlet and we have also measured the instantaneous flow rate. This device will be described in more details in a forthcoming article [8].

In order to keep the experiment as simple as possible, we have used smooth monodisperse spherical beads most of the time (see Table 1). As the beads are big enough, we can safely ignore cohesive forces (arising from humidity, electrostatic charge, etc.). In some runs, other kinds of beads have been used to test the robustness of our results: polydisperse beads, rough spheres, and irregular granular materials, such as sand, rice, lentils and others. It will be indicated when needed. Most of the results presented in the following do not depend on the bead material or size (otherwise it will be specifically stated). In fact, the relevant parameter is $R = \frac{D}{\phi}$, the ratio between the diameter of the outlet orifice D and the diameter of the beads ϕ. We have checked this in a variety of situations [9].

Table 1. Beads used in the experiments

Material	Nominal diameter (mm)	Diameter (mm)	Density (g/cm^3)
glass	0.5	0.42 ± 0.05	2.2 ± 0.1
glass	1	1.04 ± 0.01	2.4 ± 0.1
glass	2	2.06 ± 0.02	2.2 ± 0.1
glass	3	3.04 ± 0.02	2.4 ± 0.1
lead	2	1.98 ± 0.06	11.4 ± 0.5
lead	3	3.0 ± 0.1	10.9 ± 0.5
Delrin	3	3.00 ± 0.02	1.34 ± 0.05
Stainless steel	1	1.00 ± 0.01	7.6 ± 0.3

We have also carried out computer simulations of disks in two dimensions using soft particle molecular dynamics [10]. The details of the repulsive forces and the dissipative terms used in the model can be found in [11]. The numerical experiment involves about 5000 disks (compared with more than two hundred thousands in real experiments), but it allows us to enlarge the range of parameters in which we can obtain data, and to increase the spatial resolution.

3 Jamming Probability

The first set of results from which we start is the statistics of avalanches. For each adimensional size of the orifice R we recorded typically about 3000 avalanches, although in a few cases ten times more were registered. The histogram for the number of beads s in the avalanche is shown in Fig. 2, where the number of occurrences of the avalanche of size s for an orifice of size R is called $n_R(s)$. The histograms for other orifice sizes are similar; the only conspicuous difference are the number of very small avalanches (i.e. the portion of the histogram to the left of the maximum). If we ignore these few points –which are not relevant unless the outlet orifice is of a size comparable to the bead diameter– the histogram shows an exponential decay, as evidenced in the inset of Fig. 2. This means that the phenomenon under study has a characteristic magnitude –such as, for instance, the mean avalanche size, the mean avalanche duration, or others, as long as they are related.

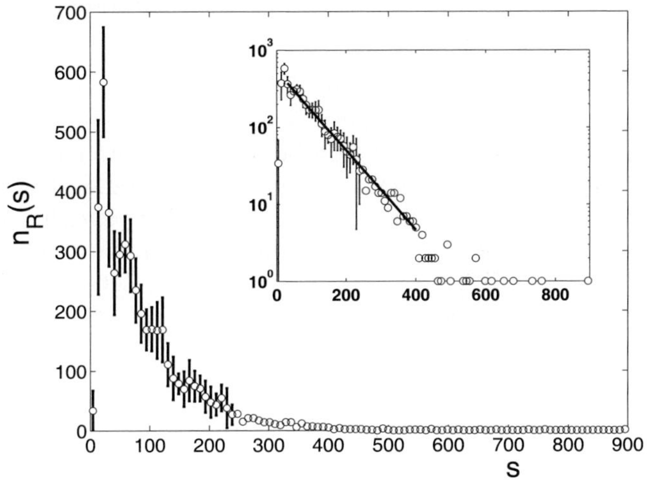

Fig. 2. Histogram of the avalanche size s (number of beads fallen) for an orifice $R = 3$. The inset shows the same data in semilogarithmic scale with a linear fit (*solid line*).

The existence of a characteristic parameter allows us to rescale the histograms, using for example the mean avalanche size $\langle s \rangle$. As expected, all the histograms collapse now in a single curve (see Fig. 3), except for very small values of $s/\langle s \rangle$, as noted before.

A simple model to describe the statistics of avalanches can be built from this ground [12], and it will lead to a simple mathematical expression of the jamming probability. The starting point is to assume that the probability

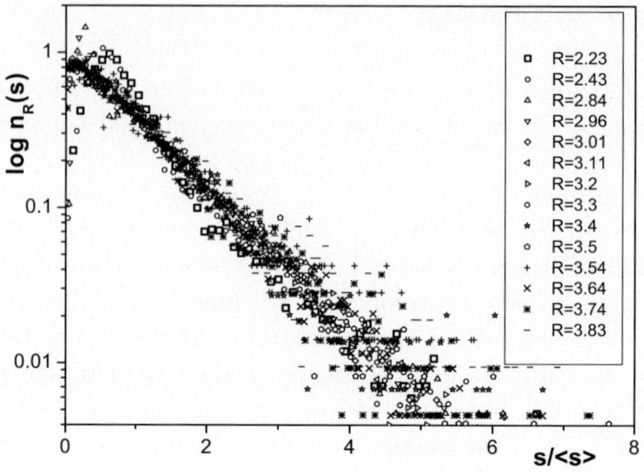

Fig. 3. Normalized histogram for a series of openings R as indicated in the legend, in semilogarithmic scale. The rescaling has been performed dividing in each histogram the avalanche size s by the mean avalanche $\langle s \rangle$ corresponding to that particular R.

of a bead to get jammed when passing through the outlet orifice does not depend on the probability that nearby beads get jammed. This amounts to consider that the events consisting of a bead passing through the orifice are independent. In this simple model, we can represent the outflow as a linear series of events, each one corresponding to a bead that either passes through the orifice without getting jammed, with probability p, or it gets jammed at the orifice with probability $1 - p$ (and the flow stops). Then, the probability of finding an avalanche of size s for a given R is equal to the probability of s beads falling through the orifice followed by a bead that gets jammed, i.e.

$$n_R(s) = p^s\,(1-p) \Rightarrow \log\left(n_R(s)\right) = s\log(p) + \log(1-p) \tag{1}$$

This expression allows us to obtain p from the linear fit of the histogram as displayed in the inset of Fig. 2: the slope of the straight line is just $\log(p)$. Note that p is also related to the mean avalanche size, because p is the total number of beads fallen in a series of avalanches divided by the number of fallen beads plus the number of jamming events, and then $p = \langle s \rangle / (\langle s \rangle + 1)$.

Let us now define the *jamming probability* J as the probability that the flow gets arrested before N beads fall. Obviously, the jamming probability depends on N and on the size of the outlet orifice R. We can write

$$J_N(R) = 1 - \sum_{s=N}^{\infty} n_R(s) \tag{2}$$

which is just the statement leading to the definition of J, i.e. the probability that the avalanche is smaller than N.

If the avalanche size is rescaled by $\langle s \rangle$, so that

$$s^* = s/\langle s \rangle \tag{3}$$

$$n_R^*(s^*) = \langle s \rangle \, n_R(s) \tag{4}$$

Considering orifices large enough, so that $\langle s \rangle \gg 1$, substituting p in terms of $\langle s \rangle$ we get

$$n_R^*(s^*) = \left(1 + \langle s \rangle^{-1}\right)^{-1} \exp\left[-s^* \langle s \rangle \, \ln\left(1 + \langle s \rangle^{-1}\right)\right] \rightarrow e^{-s^*} \tag{5}$$

and

$$J_N(R) = 1 - p^N = 1 - \exp\left[-N \ln\left(1 + \langle s \rangle^{-1}\right)\right] \rightarrow 1 - e^{-N/\langle s \rangle} \tag{6}$$

Equation (5) is just the expression of an exponential tail as presented in Fig. 3. We will now turn to (6) and see how it compares to the experimental values obtained for J. In order to do this, we have collected data for about fifty different R. If we fix N at some particular value, we can calculate from the data J_N for each R. For instance, in Fig 4(**a**) we plot J for $N = 100$ beads in a range of outlet orifices going from about $R = 1.3$ to $R = 4.3$. If we look at small R, say $R = 2$, the value of J_{100} is almost 1, meaning that for such a small orifice the probability that a jam occurs before 100 beads have fallen is very high. For large orifices (say, $R = 4$) J is almost 0, so that there is a small probability that the outpouring of beads will get arrested before 100 beads fall. At about $R \approx 3$ the jamming probability is 1/2. The solid line is the fit provided by (6). Note that there are no free parameters in the fit, as $\langle s \rangle$ is also obtained experimentally.

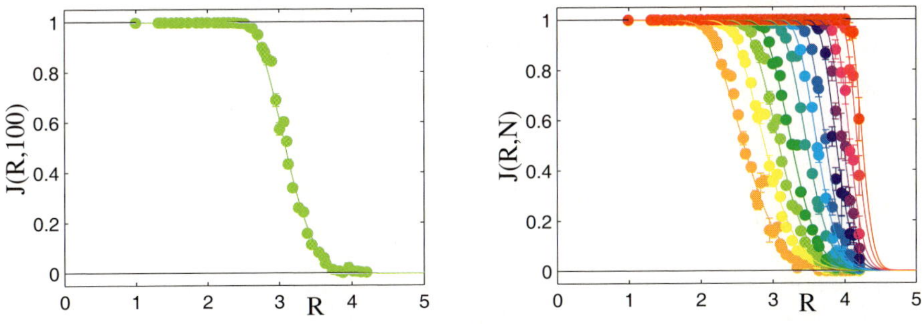

Fig. 4. The jamming probability J as a function of R. (**a**) $J(R)$ for $N = 100$. (**b**) $J(R)$ for $N = 20, 50, 100, 200, 500, 1000, 2000, 5000, 10000, 20000, 50000, 100000$ *(color changes from orange to red as N increases)*. The solid lines are the fits given by (6).

The same can be done for different values of N, which are displayed with the corresponding fits given by (6) in Fig 4(**b**). Let us remark that the agreement between the model and the experimental data does not validate the model; it merely indicates that the model can be used as a good approximation. There are indeed other more complicated models, with some theoretical foundations, that also agree with the experimental data [13].

One interesting feature that can readily be observed in Fig 4(**b**) is that J tends to a step function as N increases. Note that the series of N used in the plot is not a linear progression, so this tendency is really marked. The same idea can be staten in a different way. For a very large number of beads ($N \to \infty$), there are only two possibilities: either the orifice gets blocked, with probability very close to one, for small R, or the flow will never jam, even after waiting for very long times, if the size of the orifice is big.

The value of R separating the two situations is called the *critical radius* R_c because it is found that the mean avalanche size diverges as a power law as this value is approached [9]. Remark that although this divergency is found in some phase transitions, the inverse is not true, i.e. its existence does not warrant to conclude the presence of a phase transition. It is nevertheless enough to show the existence of two different regimes, one of them where the flow will eventually stop, and another in which it will never jam.

The value of R_c can be calculated from the mean avalanche size:

$$\langle s \rangle = \frac{A}{(R_c - R)^{\gamma}} \tag{7}$$

where A, R_c and γ are free fitting parameters. Both of them depend on the shape of the grains, but not on other characteristics such as the roughness or the density of the particles. For spherical grains, it has been obtained [7] that $R_c = 4.94 \pm 0.03$.

Even though A, R_c and γ depend on the shape of the grains, the power law is always observed (see Fig. 5), so even in these cases the existence of a critical radius can be asserted. The value of γ is usually much higher than in “typical” phase transitions; for instance, the fit gives $\gamma = 6.9 \pm 0.2$ for glass spheres.

4 The Flow Rate at the Outlet

In this section we will address the question of the mass flow rate at the outlet of the orifice. In particular, we investigate whether there are differences on the flow rate law between big and small orifices (‘small’ meaning $R \simeq R_c$, as explained in the previous section).

The most widely accepted expression for the mass flow rate of particles is commonly called *Beverloo's law* [14, 15]. This equation is valid for the outpouring of grains through an orifice due to gravity, and it can be obtained from a dimensional analysis of the problem. Earlier experiments had shown that if the dimensions of the silo, the size of the beads and the diameter of the outlet orifice fulfilled certain geometrical conditions, then the flow rate is independent on the details of the container. Let us then suppose that the mass

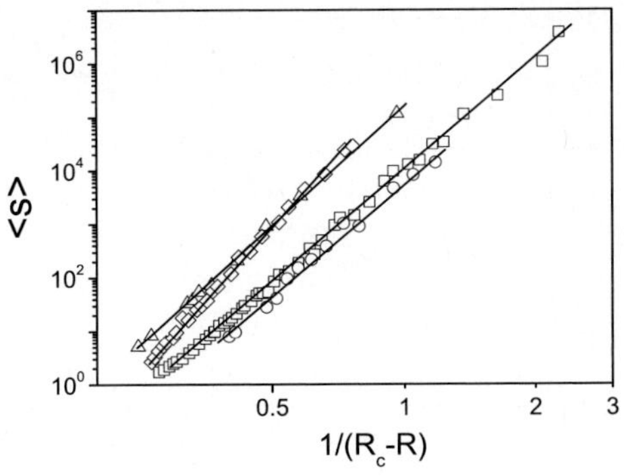

Fig. 5. The dependency of the mean avalanche size $\langle s \rangle$ on R for rice (*triangles*), lentils (*diamonds*), spherical glass beads (*squares*) and irregular pasta grains (*circles*).

flow rate W depends on the density of the granular material ϱ, the acceleration due to gravity g, the orifice diameter D and the friction coefficient μ. A simple dimensional analysis leads to the conclusion that the only permissible relationship is the following:

$$W = C(\mu)\ \varrho\ \sqrt{g}\ D^{5/2} \tag{8}$$

There is an alternative line of reasoning leading to the same formula. The flow rate must be proportional to the product of the velocity v times the area of the orifice. If it is assumed that the grains fall freely from a height $D/2$, then $W \propto v\ D^2$ and therefore $W \propto \sqrt{g}\, D^{5/2}$. Note that the relationship for a two dimensional silo with a slit at the base is $W \propto D^{3/2}$.

However, the scaling law $D^{5/2}$ is at odds with experimental results. In an attempt to rescue it, the concept of "empty annulus" [16] was taken in the formula. It means that the grains in fact do not use the whole orifice, but an effective exit aperture given by $D - k\phi$, where k is a free parameter. Using values of k between 1 and 3, many experimental results have been fitted with this equation:

$$W = C(\mu)\ \varrho\ \sqrt{g}\ (D - k\phi)^{5/2} \tag{9}$$

which is generally known as the Beverloo law.

This formula, however, has only been checked for a small range of outlet sizes, and for big values of R. We present for the first time the mass flow rate values for a large range of R. In order to combine in a single graph data from different materials, we divide the flow by the mass of one bead, so W_b is now the flow rate in number of beads per unit time. Our data are shown in Fig. 6.

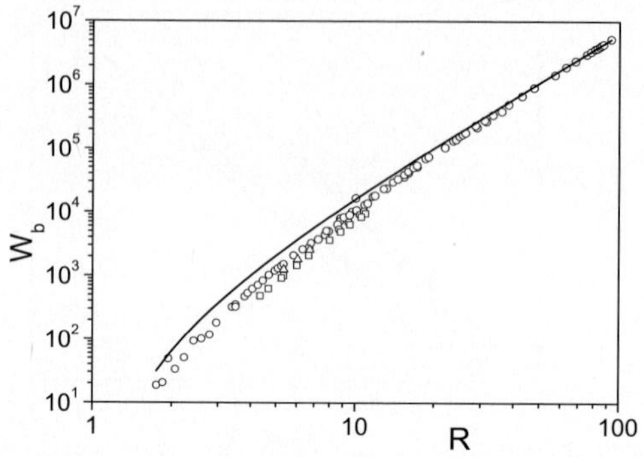

Fig. 6. The flow rate as a function of R for different spherical grains, in logarithmic scale: glass (*circles*), lead (*triangles*) and Delrin (*squares*). The solid line is a fit with (9).

As expected, there is a small dependency of W_b on the material properties, probably due to the friction coefficient, but surely not to ϱ (Delrin is lighter than glass while lead is heavier, and for both W_b is smaller than for glass).

It turns out that it is not possible to fit these data with Beverloo's equation (9). The only possible way to retain the scaling $W \propto R^{5/2}$ in agreement with the data for big R is to take $k = 1$ [8], which is the fit shown in Fig. 6. But then the flow rate predicted by (9) for small R depart noticeably from the experimental results.

The reason why Beverloo's formula (9) enjoys such a wide acceptance is because it can fit the data if only a small enough range of R is considered. This fact is more readily observed if we plot $W_b^{2/5}$ vs. R (Fig. 7). The intercept of a linear fit on the axis gives the value of k. In Fig. 7 we show such a fit for the data corresponding to $R > 50$, which is clearly unsatisfactory for small R. Any small interval of R can be reasonably fit with a straight line, and the fitting parameters provide the values for C and k. If a large interval is considered, however, a linear fit is not acceptable.

It is clear that (9) must be modified to agree with the experimental results. We have calculated the deviation of the prediction given by this formula with respect to the experimental result, and we have shown that it has the form of a negative exponential [8]. We therefore propose a new expression for the flow rate:

$$W_b = C'\left(1 - \frac{1}{2}e^{-b\cdot(R-1)}\right)(R-1)^{5/2} \tag{10}$$

The fit obtained with this equation is shown in Fig. 8.

In summary, we have proved that the parameter k, which was introduced arbitrarily in the equation for the mass flow rate, is not a valid option in the sense that it does not give an acceptable result for a large range of orifice

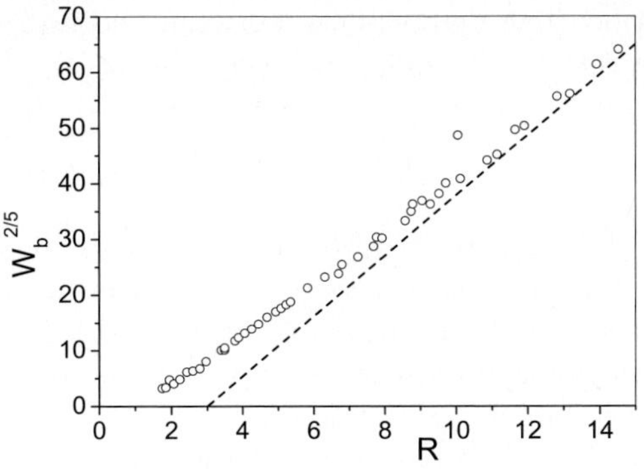

Fig. 7. $W_b^{2/5}$ as a function of R for small outlet sizes (only data for glass beads is shown). The dashed line is a linear fit taking only the data for $R > 50$.

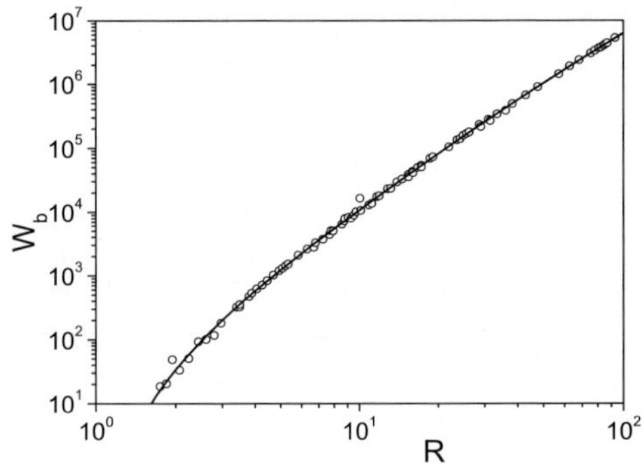

Fig. 8. The data for glass beads (same as in Fig. 6) fitted with (10). The values of the constants are $C' = 64$ and $b = 0.05$.

sizes. The only value that makes sense is $k = 1$, as W_b must tend to zero as $R \to 1$. The scaling with $R^{5/2}$ must be considered an asymptotic limit for big orifices, where a continuous picture of the granular flow becomes valid.

5 The Movement of Particles Inside the Silo

In this section the movement of individual grains inside the silo will be described. In previous works, several models have been proposed to reproduce the velocity field inside the silo . Despite the fact that these models are based on different assumptions and hypothesis, most of them end with the same expression for the mean vertical velocity. Let us follow the line of reasoning proposed by Nedderman and Tüzün [17] to derive their "kinematic model".

The simplest relation that makes sense between the horizontal and vertical components of the velocity (u and v, respectively) is

$$u = -B \frac{\partial v}{\partial x} \tag{11}$$

which means that the particles tend to be dragged to those zones where the vertical velocity is higher. Justifications to this formula have been offered in terms of the Reynolds dilatancy principle or from statistical arguments [18, 19], but its theoretical foundation remains precarious. Remark that therefore the meaning of the parameter B is unclear.

Coupling this with the assumption that the continuity equation holds for granular media, an expression for the vertical velocity is obtained:

$$v = -\frac{Q}{\sqrt{4\pi By}} \exp\left(-\frac{x^2}{4By}\right) \tag{12}$$

where Q is the flow rate through the exit orifice, and x and y are respectively the horizontal and vertical coordinates, taking the center of the orifice as the origin.

We have measured this profile by tracking the grains inside the silo. As we need visual access to the particles, we turned to the two dimensional silo described above, in which the beads can be tracked with high resolution. In order to avoid temporal or spatial short-range correlations, which are prone to appear in two dimensional silos, we have performed 24 runs, every one involving the tracking of at least 3000 particles, to obtain each velocity profile. Our results are shown in Fig. 9, which are the vertical velocities across the silo at four different heights measured for an orifice size $R = 15.8$.

It can be seen that the formula (12) reproduces quite well the experimental results. This is only true, however, for the mean velocity profile: there is a large dispersion of the velocity around its mean value. Further agreement comes from the fact that the variable Q (which was a free parameter of the fit) and the measured flow rate at the outlet differ by 10% at most.

It is interesting to study the behavior of the parameter B (sometimes called the kinematic parameter) as a function of the height above the orifice (see Fig. 10). This relationship had previously been reported, and while some authors propose an exponential growth [20] compatible with our data, others have found otherwise [21, 22].

There have been some attempts to provide a physical meaning for the parameter B. By assuming that there are characteristic time and length scales in the movement of the particles, B can be identified with a diffusive length [18, 19]. However, the predicted value is way off the experimental results. In order to correct this disagreement, an alternative model was proposed in terms of diffusion of voids inside the granular material. The idea is that a void is "shared" among several grains that move collectively, forming an ensemble called a "spot" [23]. By adjusting the size of the spot, a value of B compatible

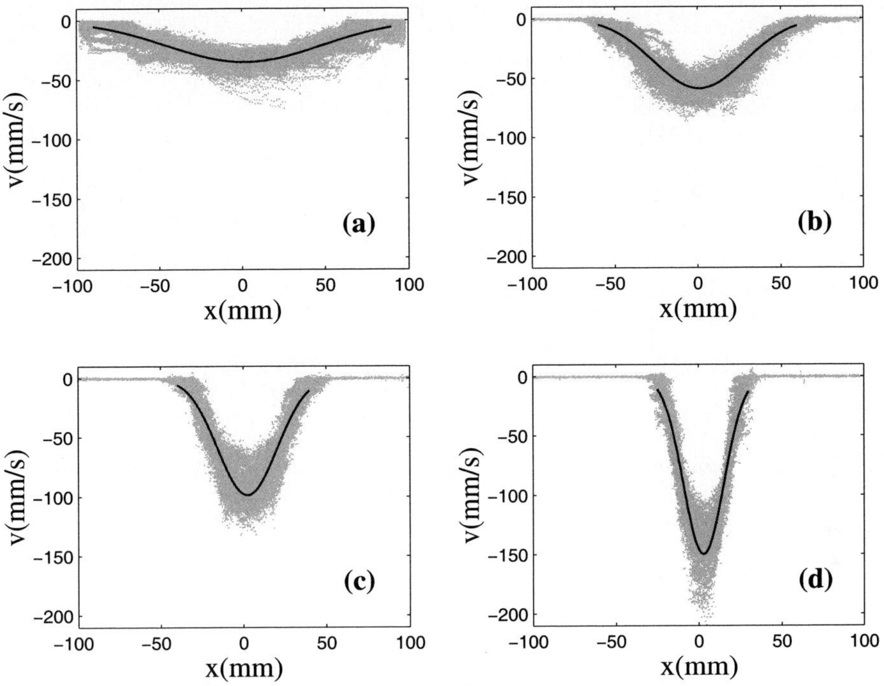

Fig. 9. Vertical velocity profiles inside the silo at different heights: **(a)** $y = 355$ mm; **(b)** $y = 155$ mm; **(c)** $y = 65$ mm; **(d)** $y = 30$ mm. All the plots are at the same scale. Solid lines are fits using (12). The diameter of the orifice is $D = 15.8$ mm and the diameter of the beads is $\phi = 1$ mm.

with experiments can be found. Nevertheless, there is no sound reason for assigning a size to the spot other than recovering the experimental result. This issue remains still unclear.

We have paid special attention to the fact that there is a large dispersion of velocities around its mean local value. All the above cited models in some way or another lead to the prediction of Gaussian fluctuations and normal diffusion for the particles as they move downwards inside the silo. Some recent results, however, do not agree with this picture [24].

We have found, both experimentally and numerically [11, 25], that the displacement fluctuations are not Gaussian. This happens for both large and small R, although we do not know whether there are differences between both situations. We show an example of these fluctuations in Fig. 11.

It is clear that anomalous behavior (in the sense of non-Gaussian fluctuations and diffusion departing from normal) is present. The extent to which this influences jamming is being actively studied.

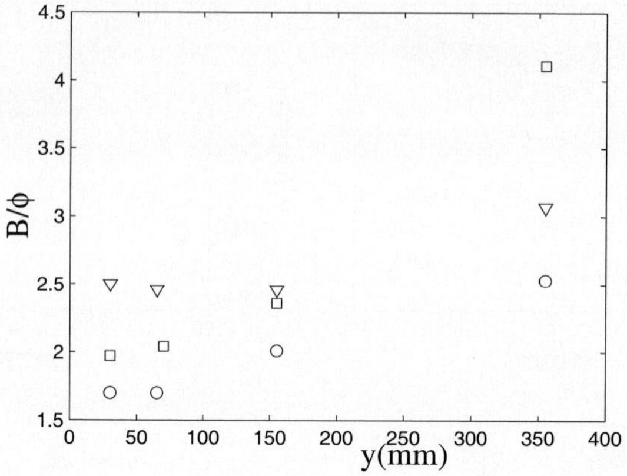

Fig. 10. The value of the parameter B (divided by the bead diameter to make it dimensionless) as a function of the height for three different orifice sizes: $R = 4.8$ (*squares*), $R = 9.5$ (*circles*) and $R = 15.8$ (*triangles*).

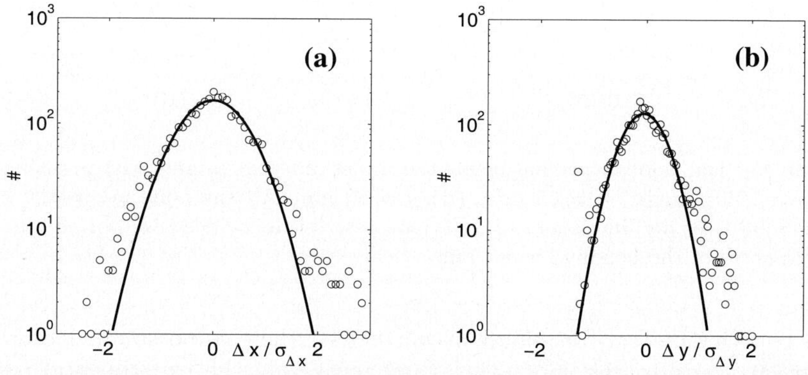

Fig. 11. The histogram for the fluctuations of the particle positions. They correspond to the fluctuations in the horizontal coordinate **(a)** and the vertical coordinate **(b)**, for $R = 15.9$. They are at the same scale and normalized by the standard deviation. The solid lines are the Gaussian best fits.

6 Conclusions

In this paper we have addressed the issue of jamming in a granular flow, driven by gravity, through an orifice. In this simple system, some conclusions can be reached. As they are quite general, it would be interesting to check their validity in other situations beyond the one presented here. Recall that in our experiments we cannot change at will the density of the granular material (i.e. the compaction fraction).

We have shown that the jamming phenomenon is governed by a characteristic parameter. As there exist a relationship among several relevant variables, one of them can be chosen without lack of generality. We have singled out the mean avalanche size $\langle s \rangle$. It has been found that $\langle s \rangle$ tends to infinity as a given value of the orifice size R_c is approached. We call this value *critical radius* because a power law divergency is found. Therefore, there are two distinct possibilities: either a jamming will appear eventually if the orifice size is smaller than R_c, or a jamming will never take place –even after waiting for a very long time– if $R > R_c$. This fact is universal and does not depend on the material properties of the grains, such as density, roughness, and so on. On the contrary, the particular value of R_c does depend on the shape of the grains [26].

The behavior of the mass flow rate W through the exit orifice has also been studied. We have shown that the scaling $W \to R^{5/2}$ is asymptotically valid for big orifices (big meaning $R \gg R_c$). Nevertheless, the correction usually introduced in Beverloo's equation (stemming from the notion of an effective aperture, and its reduction in the shape of an "empty annulus") is not valid in the sense that it cannot reproduce the experimental results spanning over a large range of R. We instead propose another formula that corrects the value of W for small orifices, by a multiplicative term involving an exponential. We cannot offer at the moment a meaning for the parameters in this correction term; this issue must be explored further.

The motion of individual particles inside the container also reveals interesting features. At long time scales, the shape of the mean velocity profile is correctly described by diffusive models. But the behavior of individual particles at short time scales shows non-Gaussian fluctuations and anomalous diffusion, which is at odds with the mentioned models. We have shown that for small R there exist particular features of the flow rate, and jamming can appear, blocking the flow. We do not know if there are specific characteristics of the motion of particles at small time scales when the exit orifice is small; we are looking into this issue, because it can give hints on how to prevent the occurrence of jamming.

Acknowledgements

We thank Eric Clément (PMMH, ESPCI, France), Luis A. Pugnaloni (CONICET, Argentine), Tom Mullin (Univ. Manchester, U.K.), and Andrés Santos (Univ. Extremadura, Spain), for discussions and suggestions. This work has been supported by Projects FIS 2005-03881 and HF 2006-0234 (Spanish Government) and PIUNA (Universidad de Navarra). A. J. thanks Fundación Ramón Areces for a grant. C. M., R. A. and J. M. P. thank Asociación de Amigos de la Universidad de Navarra for a grant.

References

1. Liu A, Nagel S (1998) Nature 396:21–22.
2. Weeks ER, Soft jammed materials, *submitted to Tohoku University Press*. Available from `http://www.physics.emory.edu/~weeks/lab/pubs.html`.
3. Liu A, Nagel S (eds.) (2001) Jamming and rheology. Taylor and Francis, London.
4. Pugnaloni LA, Barker GC, Mehta A (2001) Advances in Complex Systems 4:289–297.
5. Cates ME, Wittmer JP, Bouchaud JP, Claudin P (1999) Physica A 263:354–361.
6. Duran J (1997) Sables, poudres et grains. Eyrolles, Paris.
7. Zuriguel I, Garcimartín A, Maza D, Pugnaloni LA, Pastor JM (2005) Phys. Rev. E 71:051303.
8. Mankoc C, Janda A, Arévalo R, Pastor JM, Zuriguel I, Garcimartín A, Maza D (2007) *submitted to Granular Matter.*
9. Zuriguel I, Pugnaloni LA, Garcimartín A, Maza D (2003) Phys. Rev. E RC 68:030301.
10. Schäfer J, Dippel S, Wolf DE (1996) J. Phys I(France) 6:5–20.
11. Arévalo R, Garcimartín A, Maza D (2007) Eur. Phys. J. E 23:191–198.
12. Santos A, *private communication.*
13. To K (2005) Phys. Rev. E 71:060301R; To K, Lai PY, Pak HK (2001) Phys. Rev. Lett. 86:71–74.
14. Beverloo WA, Leniger HA, van de Velde J (1961) Chem. Eng. Sci. 15:260–269.
15. Nedderman RM (1992) Statics and Kinematics of Granular Materials. Cambridge University Press, Cambridge.
16. Brown RL, Richards JC (1960) Trans. Instn. Chem. Engrs. 38:243–256.
17. Nedderman RM, Tüzün U (1979) Powder Technol. 22:243–253.
18. Litwiniszyn J (1963) Bull. Acad. Pol. Sci., Ser. Sci. Tech. 9:61–70.
19. Mullins WW (1972) J. Appl. Phys. 43:665–678.
20. Medina A, Córdova JA, Luna E, Treviño C (1998) Phys. Lett. A 250:111–116.
21. Sielamowicz I, *private communication.*
22. Kafui KD, Thornton C (1997) Some observations on granular flow in hoppers and silos. In: Behringer RP, Jenkins JT (eds.) Powders & Grains 97. Balkema, Rotterdam.
23. Bazant MZ (2006) Mechanics of Materials 38:717–731.
24. Choi J, Kudrolli A, Rosales RR, Bazant MZ (2004) Phys. Rev. Lett. 92:174301.
25. Arévalo R, Garcimartín A, Maza D (2007) Eur. Phys. J. Special Topics 143:191–197.
26. Zuriguel I (2005) Flujo y atascos de un medio granular en la descarga de silos. PhD. Thesis, Universidad de Navarra, Pamplona.

A Fluctuating Energy Model for Dense Granular Flows

Riccardo Artoni, Andrea Santomaso, and Paolo Canu

Dipartimento di Principi e Impianti di Ingegneria Chimica "I. Sorgato"
Università di Padova. Via Marzolo 9, 35100 Padova, Italy
riccardo.artoni@unipd.it

Summary. We address the slow, dense flow of granular materials as a continuum with the incompressible Navier-Stokes equations plus the fluctuating energy balance for granular temperature. The pseudo-fluid is given an apparent viscosity, for which we choose an Arrhenius-like dependence on granular temperature; the fluctuating energy balance includes a 'mobility enhancing' term due to shear stress and a jamming, dissipative term which we assume to depend on the isotropic part of the stress tensor and on shear rate. After having proposed a 'chemical' interpretation of the phenomenology described by the model in terms of reaction rates, we report results for some 2-D standard geometries of flow, which agree semi-quantitatively with experimental and DEM observations. In particular, our model well reproduces the formation of stagnant zones of a characteristic shape (e.g. wedge-shaped static zones in a silo with flat bottom) without prescribing them a-priori with erosion techniques.

1 Introduction

Granular media exhibit a wide range of flow regimes [1], as well as a plethora of dynamical instabilities [2]. Focusing on gravity (or shear) driven flows, three regimes have been pointed out: (1) the collisional (*gas-like*) regime, where energy is dissipated by the inelasticity of the collisions, (2) the dense flowing (*liquid-like*)regime, in which particles undergo long lasting contacts, and dissipation occurs through dynamic friction, and (3) the static (*solid-like*) regime, which is capable to maintain structures due to the threshold, non-linear nature of static friction. These regimes were studied with both experiments and discrete models, the latter having experienced a great advance in the last years, starting from the work of Cundall and Strack [3].

Reliable continuum models would be of great advantage in simulating granular media, particularly when dealing with complex geometries or flows; in fact a unifying theory is still lacking. In this perspective, regimes (1) and (3) have been worked out with some success in a variety of theoretical studies, respectively with the kinetic theory of granular gases and with continuum critical state soil mechanics. For the dense regime, various theoretical approaches have been developed (and extensively reviewed in [4]); the last, more attractive one, is that proposed by the GDR MiDi based on the inertial number I (see [5–8]), the importance of which was already stated by Goddard [9].

In the following we will present our theory, which is based on conservation laws and constitutive relations for transport and dissipative terms. Then we apply the model developed to some standard configuration, and compare its results with experimental findings.

2 Conservation Laws

Following the notation of Babic [10], the macroscopic space-time weighted balance equations for mass, linear momentum and translational kinetic energy have the following form (where we have changed the convention for the stress tensor and the energy flux):

$$\partial_t(\hat{\rho}) + \nabla\cdot(\hat{\rho}\bar{\mathbf{v}}) = 0 \tag{1}$$

$$\partial_t(\hat{\rho}\bar{\mathbf{v}}) + \nabla\cdot(\hat{\rho}\bar{\mathbf{v}}\bar{\mathbf{v}}) = -\nabla\cdot\hat{\mathbf{T}} + \hat{\rho}\mathbf{g} + \mathbf{t}^F \tag{2}$$

$$\begin{aligned}\partial_t\left[\hat{\rho}\left(\epsilon^T + E^T\right)\right] + \nabla\cdot\left[\hat{\rho}\left(\epsilon^T + E^T\right)\bar{\mathbf{v}}\right] \\ = -\nabla\cdot\left(\hat{\mathbf{T}}\cdot\bar{\mathbf{v}} + \hat{\mathbf{q}}^T\right) + \hat{\rho}\mathbf{g}\cdot\bar{\mathbf{v}} - z^T + \mathbf{t}^F\cdot\bar{\mathbf{v}} + D^{TF}\end{aligned} \tag{3}$$

We neglect the equation for the angular momentum, based on the assumption that Cosserat effects are negligible, in the absence of external couples, even if particles roll on the particle scale, as demonstrated by Goddard [11].

In most situations it is possible to neglect the transport terms arising from the coupling with the interstitial fluid, *i.e.* $\mathbf{t}^F$ and D^{TF}; recalling that $E^T = (\bar{\mathbf{v}}\cdot\bar{\mathbf{v}})/2$, while $\epsilon^T = (\tilde{\mathbf{v}}\cdot\tilde{\mathbf{v}})/2$ (where $\bar{\mathbf{v}}$ is the average velocity while $\tilde{\mathbf{v}}$ is the fluctuating part), taking the product of $\bar{\mathbf{v}}$ and equation 2, and considering the tensorial relation $\nabla\cdot\left(\hat{\mathbf{T}}\cdot\bar{\mathbf{v}}\right) - \bar{\mathbf{v}}\cdot\nabla\cdot\hat{\mathbf{T}} = \hat{\mathbf{T}}^{\dagger}\colon\nabla\bar{\mathbf{v}}$ for the stress tensor $\hat{\mathbf{T}}$, where the superscript $\dagger$ stands for "transpose", we can arrive to the equation:

$$\partial_t\left(\hat{\rho}\epsilon^T\right) + \nabla\cdot\left(\hat{\rho}\epsilon^T\bar{\mathbf{v}}\right) = -\hat{\mathbf{T}}^{\dagger}\colon\nabla\bar{\mathbf{v}} - \nabla\cdot\hat{\mathbf{q}}^{\mathbf{T}} - z^T \tag{4}$$

where the energy flux vector $\mathbf{q}^{\mathbf{T}}$ and the energy dissipation rate z^T appear, that have a fundamental importance in this study.

With the help of the definition of granular temperature $\theta = < \tilde{v}^2 > /3$ [12] this turns to:

$$\frac{3}{2}\partial_t(\hat{\rho}\theta) + \frac{3}{2}\nabla\cdot(\hat{\rho}\theta\bar{\mathbf{v}}) = -\hat{\mathbf{T}}^{\dagger}\colon\nabla\bar{\mathbf{v}} - \nabla\cdot\hat{\mathbf{q}}^{\mathbf{T}} - z^T \tag{5}$$

Splitting the stress tensor as $\hat{\mathbf{T}} = p\mathbf{I} + \mathbf{\Pi}$, being p its trace, and $\mathbf{\Pi}$ its deviatoric part, Eq. 2 becomes:

$$\partial_t(\hat{\rho}\bar{\mathbf{v}}) + \nabla\cdot(\hat{\rho}\bar{\mathbf{v}}\bar{\mathbf{v}}) = -\nabla p - \nabla\cdot\mathbf{\Pi} + \hat{\rho}\mathbf{g} \tag{6}$$

Assuming the stress tensor to be symmetric, as a consequence of the absence of couple stresses, Eq. 5 becomes:

$$\frac{3}{2}\partial_t\left(\hat{\rho}\theta\right)+\frac{3}{2}\nabla\cdot\left(\hat{\rho}\theta\bar{\mathbf{v}}\right)=-p\nabla\cdot\bar{\mathbf{v}}-\mathbf{\Pi}\colon\nabla\bar{\mathbf{v}}-\nabla\cdot\hat{\mathbf{q}}^{\mathbf{T}}-z^T \tag{7}$$

As a next step, let's assume that the flow is nearly incompressible, so that $\hat{\rho}\approx$ const. The continuity equation reduces to $\nabla\cdot\bar{\mathbf{v}}=0$; the linear momentum balance changes into:

$$\hat{\rho}\partial_t\left(\bar{\mathbf{v}}\right)+\hat{\rho}\bar{\mathbf{v}}\cdot\nabla\bar{\mathbf{v}}=-\nabla p-\nabla\cdot\mathbf{\Pi}+\hat{\rho}\mathbf{g} \tag{8}$$

As we argued in a recent paper [13], taking into account the dilatancy effects assuming a compressible medium would be of great importance both for the physical coherence of the model and for final applications. As a matter of fact, in some geometries the solid fraction ν varies by more than 10% (for example, in the discharge zone of a silo); this issue is important for applications in which a gas flow is involved, in order to predict preferred paths and residence time distributions of the gas. However, for the sake of simplicity, whenever ν is known to vary less than 10%, we simplify the equations with the incompressibility hypothesis; on the other hand, when the ν field is important and its variations are not negligible in their magnitude, we shall use Eq. 7 which is more general.

Making use of this approximation, the pressure term in Eq. 7 is multiplied by $\nabla\cdot\bar{\mathbf{v}}=0$, thus Eq. 7 becomes:

$$\frac{3}{2}\hat{\rho}\partial_t\left(\theta\right)+\frac{3}{2}\hat{\rho}\bar{\mathbf{v}}\cdot\nabla\theta=-\mathbf{\Pi}\colon\nabla\bar{\mathbf{v}}-\nabla\cdot\hat{\mathbf{q}}^{\mathbf{T}}-z^T \tag{9}$$

3 Constitutive Relations

With the assumptions pointed out above, we arrived at three equations in which the stress tensor $\hat{\mathbf{T}}$, the energy flux vector $\mathbf{q}^{\mathbf{T}}$ and the energy dissipation rate z^T are unknown. In order to solve our system of equations, we need now to express some constitutive hypothesis for these variables. It is worthy to note that the equations derived here come out only from conservation principles, without taking into account the physical nature of the medium. This must be done by providing the appropriate constitutive relations.

3.1 Stress Tensor and Energy Flux

Typically we shall assume that granular temperature can propagate by a diffusion-like mechanism, thus proportional to its gradient [12],

$$\mathbf{q}^{\mathbf{T}}=\mathbf{K}\cdot\nabla\theta \tag{10}$$

where $\mathbf{K}$ is the θ diffusivity tensor; hereafter we will suppose $\mathbf{K} = k\mathbf{I}$, so that:

$$\mathbf{q}^{\mathbf{T}} = k\nabla\theta \tag{11}$$

Dense granular flows seem to exhibit a viscous-like character, whose origin is a matter of debate; Savage [12] used previous results by Hibler [14] to demonstrate that if a plasticity framework was applied to the instantaneous stress field, with the hypothesis that the fluctuations were Gaussian, the average stress tensor had a viscous-like dependence on the average strain rate tensor.

In this case, let's suppose that a general relation holds of the form:

$$\Pi_{ij} = -\eta\left(\frac{\partial v_i}{\partial x_j} + \frac{\partial v_j}{\partial x_i}\right) \tag{12}$$

such that the 'granular liquid' can be treated as a generalized Newtonian fluid. Note that only the bulk viscosity appears due to the usual approximations and the incompressibility condition.

The constitutive relations above have shifted the problem to determining the constitutive coefficients k and η, which will be, in general, functions of all the dependent and independent variables and their derivatives.

To represent the usual scaling of the flow profiles on particle diameter and on bulk density, k and η must depend on $\hat{\rho}$ and d_p in the following way (primes indicate functions of the remaining variables):

$$k = \hat{\rho} d_p^2 k' \quad \eta = \hat{\rho} d_p^2 \eta' \tag{13}$$

Granular materials are often considered to belong to the family of glassy systems, in which a transition between flowing and non-flowing behaviour can be characterized by a sharp increase in viscosity. The liquid-glass transition has been extensively studied, both experimentally and theoretically. The empirical equation proposed long ago by Doolittle [15] for the fluidity (*i.e.* the reciprocal of viscosity, η^{-1}) of a glass is $\phi = \phi_0 exp(-\gamma \frac{v_m}{v_f})$, where v_m and v_f are the volume of the molecule and the free volume respectively. This relation has been justified within a free volume approach [16].

A fully satisfactory description of granular flows in terms of a pseudocontinuum is still lacking; consequently, we attempt a semi-theoretical approximation constructing an analog of the free volume of Doolittle equation. In doing so, we could choose solid fraction (forgetting at the moment the incompressibility assumption), that is the quantity with the closest physical meaning. In fact, this analogy does not explain several aspects: solid fraction is more a static than a dynamic measure of the free volume, while we need an expression of the local 'mobility' (a *dynamic* measure of the free volume) of the medium. The role of mobility could be played by granular temperature, that, being a measure of the amplitude of velocity fluctuations, is indeed related to the local capability to move. To solve this problem for packings, Edwards [17] introduced the compactivity X, which is related to the capability of the free volume to fluctuate, but not to the fluctuations of velocity. We think that fluctuations of both the free volume and the velocity of the particles are important to describe the dynamics of granular assemblies, but at the moment we suppose that for nearly stationary, dense flows, the fluctuations in the free volume can become unimportant. With this in mind, we tentatively express the viscosity of the granular medium as:

$$\eta' = \eta_0 exp\left(\frac{E}{\theta}\right) \tag{14}$$

In the expression for η' we have neglected a direct dependence on the history of deformation (though it acts in some way through θ): while this hypothesis is useful to work with simple equations, it could be an over-simplification restricting the validity of the approach to the steady state.

For what concerns k', we do not follow Savage who suggested $k/\eta \approx$ const, extrapolating a result of Jenkins' kinetic theory, which is not valid in the dense regime under study. We consider at this moment k constant.

3.2 Energy Dissipation Rate

If we consider an over-simplified example, a pile of infinitesimal blocks sliding one above the other in the absence of gravity and subjected to constant vertical pressure, we can formulate the rate of dissipation of the specific energy:

$$\dot{\epsilon}_{diss} = \frac{d\dot{E}_{diss}}{dV} = \mu P \frac{dv}{dz} \tag{15}$$

Obviously, when applied to granular flows, this situation is quite simplistic; the original network of forces of granular materials under shear together with the bi/tri-dimensional arrangement of the particles and the threshold behaviour of microscopic friction generate a tricky problem for the analyst. We will use Eq. 15 in a phenomenological way, just to say that the specific energy dissipated per unit time is proportional to normal stresses and to the spatial gradients of the velocity, which are related to a fundamental time scale of the system. We will extend this result to higher-dimensional cases recalling that a measure of the spatial velocity gradients is clearly the shear rate $|\dot{\gamma}|$, and taking the isotropic part of the stress tensor p as a measure of normal stresses, we can write:

$$z^T = \mu p \, |\dot{\gamma}| \tag{16}$$

where μ is now an effective coefficient (different from the microscopic friction coefficient).

3.3 A 'Chemical' Interpretation

The model developed here belongs to the class of hydrodynamic models, because it makes use of the balance equations, with a particular care for the fluctuating energy equation. A significant difference between our model and previous attempts [12, 18–20] lies in the different phenomenological closures, which we have developed specifically for the intermediate regime, without extrapolating results from the rapid flow regime.

Analysing the fluctuating energy balance equation with the help of the constitutive relations, we see that it becomes:

$$\frac{3}{2}\hat{\rho}\partial_t(\theta) + \frac{3}{2}\hat{\rho}\bar{\mathbf{v}} \cdot \nabla\theta = -k\Delta\theta + \dot{Q} \tag{17}$$

where

$$\dot{Q} = -\boldsymbol{\Pi} \colon \nabla\bar{\mathbf{v}} - z^T \tag{18}$$

The viscous heating term can be represented in Cartesian notation as:

$$\eta\left(\frac{\partial v_i}{\partial x_j} + \frac{\partial v_j}{\partial x_i}\right)\frac{\partial v_i}{\partial x_j} = \eta\,|\dot{\gamma}|^2 \tag{19}$$

Thus, the entire heating and cooling term can be formulated as:

$$\dot{Q} = |\dot{\gamma}|\,(\eta\,|\dot{\gamma}| - \mu p) = |\dot{\gamma}|\,(|\tau| - \mu p) \tag{20}$$

One can see that the total net production (or consumption) of fluctuating energy is expressed in a way that involves the timescale $|\dot{\gamma}|^{-1}$ and the difference between the shear stress and a modified Coulomb criterion. Shear rate plays the role of a kinetic constant, while the distance from the modified yield criterion sounds more like an activity. If the material is not sheared, nothing changes its potential mobility, because $|\dot{\gamma}| = 0 \Rightarrow \dot{Q} = 0$. In addition, one can easily see that for the case of plane shear (where the shear stress is constant, and equal, say, to τ_0), the shear rate resembles a kinetic constant with an Arrhenius dependence on granular temperature:

$$|\dot{\gamma}| = \frac{\tau_0}{\eta} \sim exp\left(-\frac{E}{\theta}\right) \tag{21}$$

3.4 On Boundary Conditions

Given the system of equations and the closure relations, we can use our model to simulate dense granular flows once boundary conditions are provided.

Focusing on the momentum balance, we are especially interested in discussing the typical BC to apply at proximity of a wall. It is a common practice to assume a no-slip boundary condition at the walls, supported by the experimental practice of gluing particles at the walls to control friction. At first, a question comes up about the physical consistence of such a hypothesis: are we sure that the continuum, averaged interaction of mesoscopic grains is the same in the bulk and between free and glued particles at the walls, provided their totally constrained position? As a second consideration, we want to point that most of the applications of our interest (industrial, geophysical, ...) deal with walls where both slip and no-slip (and, of course, intermediate behaviors) occur. Depending on the application, one would like to favour one or the other limiting behaviour; in this perspective, a rigorous characterization of wall friction in presence of a certain wall roughness has to be done, in order to apply any model to a real situation.

A characterization of wall friction could be made with a Coulomb criterion, or within a slip-length approach. The first corresponds to impose at the wall the relation $\tau = \mu_w \sigma$ between shear and normal stress (μ_w is a wall friction coefficient), while the second (also called Navier slip condition) relates the wall tangential velocity u_t and its gradient computed on the normal to the surface with the formula $u_t = \lambda \frac{\partial u_t}{\partial n}$, where λ is a so-called *slip length*. We are convinced that investigating the influence of these (and possibly other) BCs on the prediction of models could be a useful task, as we recently demonstrated for mixing-length models in the vertical chute geometry [13].

For the fluctuating energy equation, basic research on BCs at the walls is strongly needed; in this paper we use, for simplicity, either a constant wall-temperature or an "insulation" condition (no flux of fluctuating energy).

4 Results for Some Standard Geometries

The paper by GDR MiDi [5] suggests the necessity to validate granular flow models in a number of standard geometries, for which flow patterns are known. We decided to focus our attention on 2 or 3-D geometries, following the idea that some original features (formation of static zones, hourglass effect) of granular flows are essentially 2 or 3-D and cannot be captured by 1-D reductions of the model. In this work we investigate the predictions of our model concerning flow and stress fields, in silos with flat and tilted bottom walls; the system of PDE was solved with a standard finite-element code.

4.1 Silo with Flat Bottom

In this section we present results for an axisymmetric silo with flat bottom, 800 particle diameters high and with a radius of 50 particle diameters d_p. A simple first choice for the parameters invoked by the model is resumed in table 1. Flowrate is fixed at the orifice, no slip is assumed at the walls, together with insulation for θ (*i.e.* $k\frac{\partial \theta}{\partial n} = 0$). Figure 1 summarizes the development of the velocity profiles along the silo, from Bessel-like profiles near the orifice to a plug flow with shear bands at the walls higher up the bin. In the higher part of the silo shear bands are of order 10 diameters thick. The temperature profile

Table 1. Parameters of the model

μ	.3	$adim$
E	10	s^2m^{-2}
η_0	1	s^{-1}
k_0	1	s^{-1}
ρ	10^3	kg/m^3

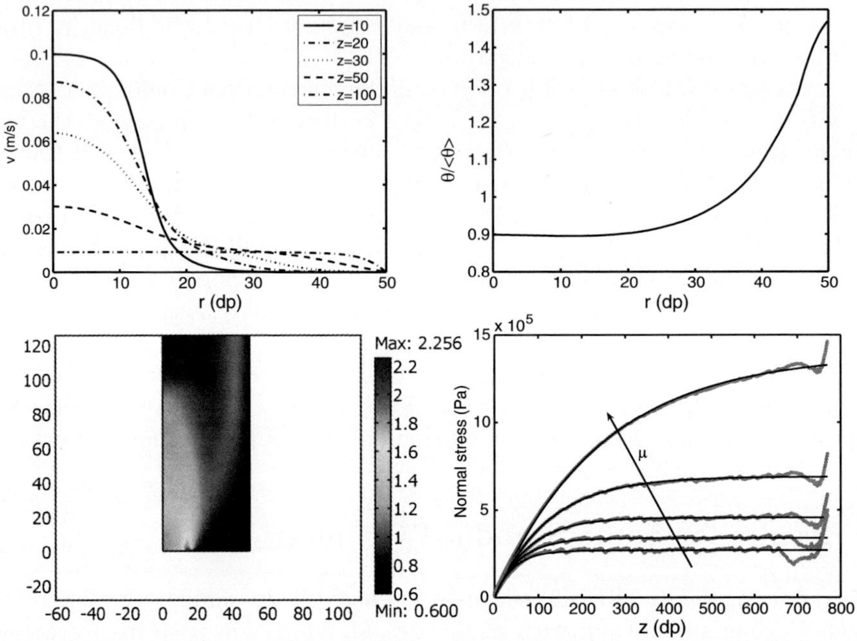

Fig. 1. Silo with flat bottom (all lengths are in d_p units): (a) Velocity profiles at different heights (b) Temperature profile far from the orifice (rescaled with the average value) (c) (Rescaled) temperature map (d) Normal stresses (grey lines) and Janssen equation (black lines) ($\sigma_w = A\,[1 - \exp(-B\ z)]$).

is qualitatively approaching data, with a peak at the wall, where material is sheared. Imposing a low temperature at the wall, we were able to reproduce the fast decaying of θ close to the wall as observed in DEM simulations [5]. Looking at velocity maps and profiles near the orifice, we can see that our model predicts well-defined static zones in the corners; this behavior can be explained by analysing temperature maps. A narrow zone at higher θ develops from the orifice to high up in the silo, finally positioning itself close to the wall. In the framework of this model, this layer has a lower viscosity that allows the two zones that it delimits to move independently one from another, thus insulating the corner, which 'cools' due to pressure rearranging action. Moreover, the map of granular temperature reveals that the flow pattern is very similar to that proposed by Brown and Hawksley [21].

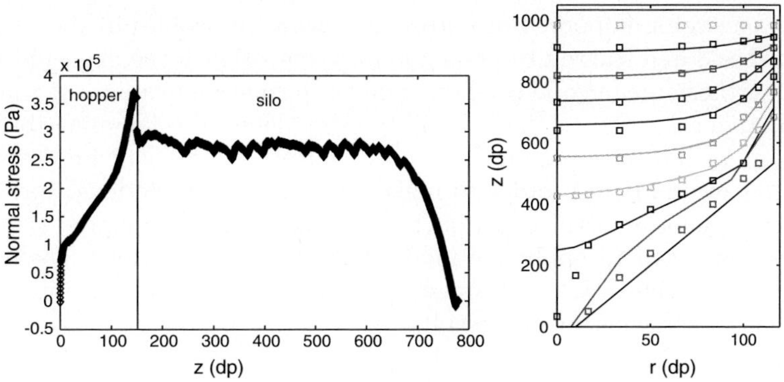

Fig. 2. Silo with converging hopper: (a) Simulation of wall normal stresses (b) Comparison between the model and a tracer.

As a next step, wall normal stresses as predicted by the model are shown in figure 1d. Perhaps surprisingly, profiles follow qualitatively Janssen's behaviour.

Finally, recent numerical simulations developed for non-stationary flow in the silo, discharging freely, show that the model predicts a constant flowrate, another original feature of dense granular flows.

5 Silo with Converging Hopper

If we tilt the flat bottom, usually for the purpose of eliminating the stagnant zones, flow and stress profiles change dramatically, as shown in Fig. 2. At first, the experimental stress field is different, with a peak close to the beginning of the converging part. Our model predicts qualitatively also this behaviour. About the flow field, stagnant zones have disappeared and a comparison between our own experimental data (for particles of size $d_p \approx 3mm$) and the prediction of the model shows that a good agreement exists between calculated and experimental results. A Navier slip condition was imposed and the slip length was fitted on the data, ending with a value of approximately ten particle diameters.

6 Conclusions

In this work we derived a hydrodynamic model for dense granular flow, based on conservation equations for mass, momentum and fluctuating energy, plus constitutive relations based on phenomenological analyses and intuitions. After having discussed some interpretation of the phenomenological picture in terms of reaction rates, we presented results for a silo with flat bottom and with a converging hopper. Both stress and velocity profiles in the two cases were analysed and showed to be in good agreement with the expected behavior [5]. Furthermore, a good prediction on the formation of stagnant zones was probed, giving insight on the possibility of modeling the jamming-to-flowing transition within a continuum approach. Future efforts will be devoted to parameter evaluation, in order to make quantitative predictions, and in this perspective some general criterion to determine the parameters is needed. Moreover, we plan to apply the model to other configurations, possibly belonging to the GDR MiDi [5] framework, to validate the model comparing the predictions with experimental results.

References

1. Jaeger, H. M., Nagel, S. R., and Behringer, R. P. (1996) *Rev. Mod. Phys.* **68(4)**, 1259–1273.
2. King, P., Lopez-Alcaraz, P., Pacheco-Martinez, H., Clement, C., Smith, A., and Swift, M. (2007) *Eur Phys J E Soft Matter* **22(3)**, 219–226.
3. Cundall, P. A. and Strack, O. D. L. (1979) *Géotechnique* **29(1)**, 47–65.
4. Pouliquen, O. and Chevoir, F. (2002) *Comptes Rendus Physique* **3**, 163–175.
5. GDR Midi (2004) *Eur. Phys. J. E* **14(4)**, 341–365.
6. Jop, P., Forterre, Y., and Pouliquen, O. (2006) *Nature* **441**, 727–730.
7. Pouliquen, O., Cassar, C., Forterre, Y., Jop, P., and Nicolas, M. (2006) In Proc. Powders & Grains 2005: A. A. Balkema, Rotterdam.
8. da Cruz, F., Emam, S., Prochnow, M., Roux, J.-J., and Chevoir, F. (2005) *Phys. Rev. E* **72**, 021309.
9. Goddard, J. D. (1986) *Acta Mechanica* **63(1-4)**, 3–13.
10. Babić, M. (1997) *Int. J. Engng. Sci.* **35(5)**, 523–548.
11. Goddard, J. D. (to appear) In Mathematical models of granular matter, Lecture Notes in Mathematics. Berlin: Springer.
12. Savage, S. B. (1998) *J. Fluid Mech.* **377**, 1–26.
13. Artoni, R., Santomaso, A., and Canu, P. (to appear) *Europhys. Lett.* e-print: cond-mat/0705.3726.
14. Hibler, W. D. (1977) *J. Geophys. Res.* **82**, 3932–3938.
15. Doolittle, A. K. (1951) *J. Appl. Phys.* **22**, 1031–1035.
16. Cohen, M. H. and Grest, G. S. Aug 1979 *Phys. Rev. B* **20(3)**, 1077–1098.
17. Edwards, S. F. and Oakeshott, R. B. S. (1989) *Physica A* **157**, 1080–1090.
18. Losert, W., Bocquet, L., Lubensky, T. C., and Gollub, J. P. *Phys. Rev. Lett.* **85(7)**, 1428–1431.
19. Bocquet, L., Losert, W., Schalk, D., Lubensky, T. C., and Gollub, J. P. *Phys. Rev. E* **65**, 011307.
20. Bocquet, L., Errami, J., and Lubensky, T. C. Oct 2002 *Phys. Rev. Lett.* **89(18)**, 184301.
21. Brown, R. L. and Hawksley, P. G. W. (1947) *Fuel* **27**, 159–173.

Influence of Particle Shape on Shear Stress in Granular Media

Emilien Azéma[1], Farhang Radjaï[1], Robert Peyroux[1], and Gilles Saussine[2]

[1] LMGC, CNRS - Université Montpellier II, Place Eugène Bataillon, 34095 Montpellier cedex 05, France azema@lmgc.univ-montp2.fr

[2] Innovation and Research Departement of SNCF, 45 rue de Londres, 75379 Paris Cedex 08, France gilles.saussine@sncf.fr

Summary. We analyze the contact and force networks in a dense confined packing of pentagonal particles simulated by means of the contact dynamics method. The particle shape effect is evidenced by comparing the data from pentagon packing and from a packing with identical characteristics except for the circular shape of the particles. A surprising observation is that the pentagon packing develops a lower structural anisotropy than the disk packing. We show in this work that this weakness is compensated by a higher force anisotropy that leads to enhanced shear strength of the pentagon packing. With the polygonal shape of the particles, the strong force chains are mostly composed of edge-to-edge contacts with a marked zig-zag aspect.

1 Introduction

The microstructure of granular materials is generically anisotropic in two respects: 1) The contact normal directions are not random; 2) The force average as a function of contact normal direction is not uniform. The corresponding fabric and force anisotropies in shear are responsible for mechanical strength at the scale of the packing [1–4]. The shear stress is fully transmitted via a "strong" contact network, materialized by force "chains" and the stability is ensured by the antagonist role of "weak" contacts which prop strong force chains [4, 5]. These features have, however, been mostly part investigated in the case of granular media composed of isometric (circular or spheric) particles.

In this paper, we consider one of the simplest possible shapes, namely regular pentagons. Among regular polygons, the pentagon has the lowest number of sides, corresponding to the least roundedness in this category, without the pathological space-filling properties of triangles and squares. We seek to isolate the effect of edge-to-edge contacts on shear stress and force transmission by comparing the data with a packing of circular particles that, apart from the particle shape, is fully identical (preparation, friction coefficients, particle size distribution) to the pentagon packing. Both packings are subjected to biaxial compression simulated by means of the contact dynamics method. The presence of edge-to-edge contacts affects both quantitatively and qualitatively the microstructure and the overall behavior during shear. These contacts do not transmit torques, but they are able to accommodate force lines that are usually unsustainable in packings of disks.

2 Numerical Procedures

The simulations were carried out by means of the contact dynamics (CD) method with pentagonal particles. The CD method is based on implicit time integration of the equations of motion and a nonsmooth formulation of mutual exclusion and dry friction between particles [6–8]. This method requires no elastic repulsive potential and no smoothing of the Coulomb friction law for the determination of forces. For this reason, the simulations can be performed with large time steps compared to molecular dynamics simulations. We used LMGC90 which is a multipurpose software developed in our laboratory, capable of modeling a collection of deformable or undeformable particles of various shapes by different algorithms [9].

We generated two numerical samples. The first sample S1, is composed of 14400 regular pentagons of three different diameters: 50% of diameter 2.5 cm, 34% of diameter 3.75 cm and 16% of diameter 5 cm. The second sample S2, is composed of 10000 discs with the same polydispersity. Both samples were prepared according to the same protocol. The gravity was set to zero in order to avoid force gradients in the samples. The coefficient of friction was set to 0.4 between grains and to 0 with the walls. At equilibrium, both numerical samples were in isotropic stress state. The solid fraction was $\phi_0 = 0.80$ for S1 and $\phi_0 = 0.82$ for S2. The aspect ratio was $h/l \approx 2$, where h and l are the height and width of the sample, respectively. Figure 1 displays snapshots of the two packings at the end of isotropic compaction.

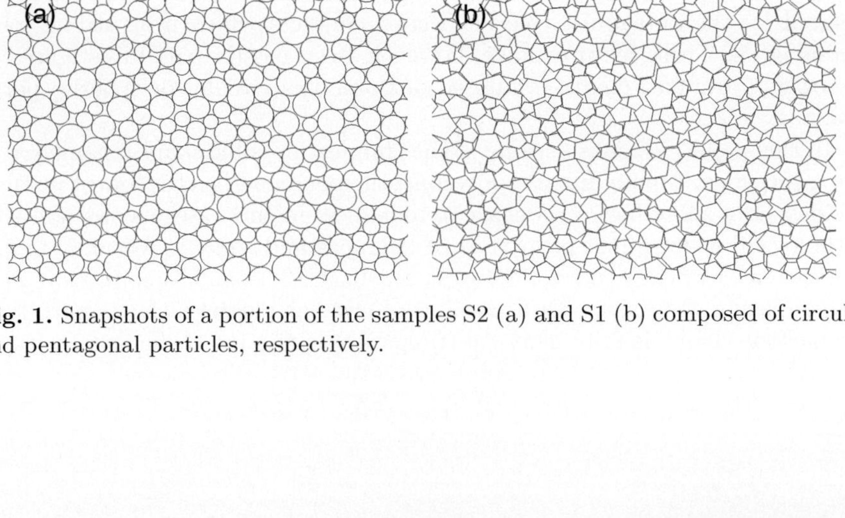

Fig. 1. Snapshots of a portion of the samples S2 (a) and S1 (b) composed of circular and pentagonal particles, respectively.

The isotropic samples were subjected to vertical compression by downward displacement of the top wall at a constant velocity of 1 cm/s for a constant confining stress σ_0 acting on the lateral walls. Since we are interested in quasistatic behavior, the shear rate should be such that the kinetic energy supplied by shearing is negligible compared to the static pressure. This can be formulated in terms of an "inertia parameter" I defined by [10]

$$I = \dot{\varepsilon}\sqrt{\frac{m}{p}}, \tag{1}$$

where $\dot{\varepsilon} = \dot{y}/y$ is the strain rate, m is the total mass, and p is the average pressure. The quasistatic limit is characterized by the condition $I \ll 1$. In our biaxial simulations, I was below 10^{-3}.

3 Shear Stress

In this section, we compare the stress-strain behavior between the packings of polygons (sample S1) and disks (sample S2). For the calculation of the stress tensor, we consider the "tensorial moment" $\boldsymbol{M}^i$ of each particle i defined by [5, 11]:

$$M^i_{\alpha\beta} = \sum_{c\in i} f^c_\alpha r^c_\beta, \tag{2}$$

where f^c_α is the α component of the force exerted on particle i at the contact c, r^c_β is the β component of the position vector of the same contact c, and the summation is run over all contacts c of neighboring particles with the particle i (noted briefly by $c \in i$). It can be shown that the tensorial moment of a collection of rigid particles is the sum of the tensorial moments of individual particles. The stress tensor $\boldsymbol{\sigma}$ for a packing of volume V is simply given by [5, 11]:

$$\boldsymbol{\sigma} = \frac{1}{V}\sum_{i\in V} \boldsymbol{M}^i = \frac{1}{V}\sum_{c\in V} f^c_\alpha \ell^c_\beta, \tag{3}$$

where $\boldsymbol{\ell}^c$ is the intercenter vector joining the centers of the two touching particles at the contact c. We extract the mean stress $p = (\sigma_1 + \sigma_2)/2$, and the stress deviator $q = (\sigma_1 - \sigma_2)/2$, where σ_1 and σ_2 are the principal stresses. The major principal direction during vertical compression is vertical, we then define the volumetric strain by:

$$\varepsilon_p = \int_{V_0}^{V} \frac{dV'}{V'} = \ln\left(1 + \frac{\Delta V}{V_0}\right), \tag{4}$$

where V_0 is the initial volume and $\Delta V = V - V_0$ is the cumulative volume change.

Figure 2 shows the normalized shear stress q/p for the samples S1 and S2 as a function of shear strain $\varepsilon_q \equiv \varepsilon_1 - \varepsilon_2$. For S2, we observe a hardening behavior followed by (slight) softening and a stress plateau corresponding to the residual state of soil mechanics [12]. For S1, we observe no marked stress peak. The residual stress is higher for polygons ($\simeq 0.35$) than for disks ($\simeq 0.28$). The higher level of q/p for the polygon packing reflects the organization of the microstructure and the features of force transmission that we analyze in more detail below.

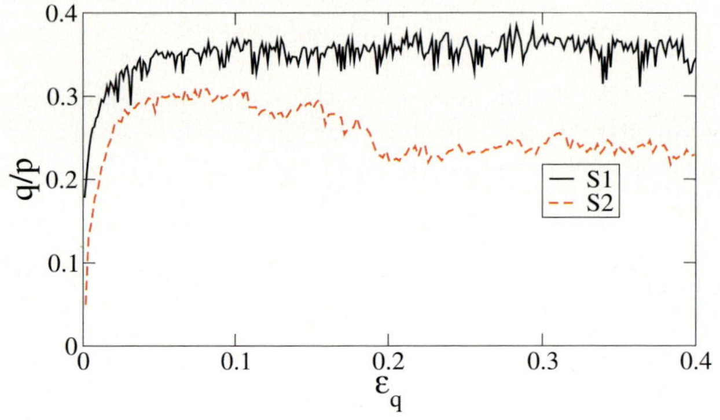

Fig. 2. Normalized shear stress q/p as a function of cumulative shear strain ε_q for the samples S1 and S2.

4 Fabric Anisotropy

The shear strength of dry granular materials is generally attributed to the buildup of an anisotropic structure during shear due to friction between the particles and as a result of steric effects depending on particle shape [13–15].

The probability density function $P_\theta(\theta)$, where θ is the orientation of the contact normal $\boldsymbol{n}$, provides the basic orientational statistical information about the granular fabric. It is π-periodic in the absence of an intrinsic polarity for $\boldsymbol{n}$. Most lowest-order information is generally given by the second moment of P_θ, called *fabric tensor* [16]:

$$F_{\alpha\beta} = \frac{1}{\pi}\int_0^\pi n_\alpha(\theta) n_\beta(\theta) P_\theta(\theta) d\theta \equiv \frac{1}{N_c}\sum_{c\in V} n_\alpha^c n_\beta^c, \quad (5)$$

where α and β design the components in a reference frame and N_c is the total number of contacts in the control volume V. By definition, $tr(\boldsymbol{F}) = 1$. The anisotropy of the contact network is given by the difference between the principal values F_1 and F_2. We define the fabric anisotropy a by

$$a = 2(F_1 - F_2). \quad (6)$$

Figure 3 displays a polar representation of $P_\theta(\theta)$ for the samples S1 and S2 at $\varepsilon_q = 0.3$. We observe a nearly isotropic distribution for the pentagon packing in spite of shearing whereas the disk packing is markedly anisotropic. This is a surprising observation in view of the higher shear strength of the pentagon packing.

The evolution of a is shown in Fig. 4 as a function of ε_q for S1 and S2. The anisotropy stays quite weak in the pentagon packing whereas the disk packing is marked by a much larger anisotropy, increasing to $\simeq 0.3$ and then relaxing to a slightly lower value in the residual state. The low anisotropy of the pentagon packing results from a particular organization of the force network in correlation with the orientations of edge-to-edge and vertex-to-edge contacts in the packing [17]. This leads us to consider the contributions of force and texture anisotropies to average shear stresses.

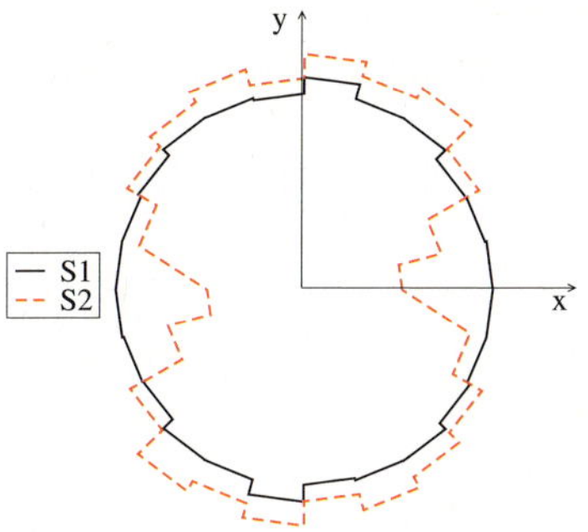

Fig. 3. Polar representation of the probability density function P_θ of the contact normal directions θ for the samples S1 and S2 in the residual state.

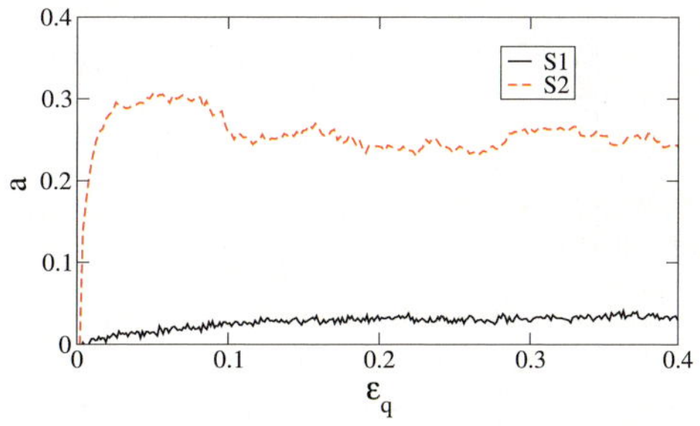

Fig. 4. Evolution of the anisotropy a with cumulative shear strain ε_q for the samples S1 and S2.

5 Force Anisotropy

The angular distribution of contact forces in a granular packing can be represented by the average force $\langle \boldsymbol{f} \rangle(\boldsymbol{n})$ as a function of the contact normal direction $\boldsymbol{n}$. We distinguish the average normal force $\langle f_n \rangle(\theta)$ from the average tangential force $\langle f_t \rangle(\theta)$. As $P(\theta)$, these two functions can be represented by their Fourier expansions truncated beyond the second term [1–4]:

$$\begin{cases} \langle f_n \rangle(\theta) = \langle f \rangle \{1 + a_n \cos 2(\theta - \theta_n)\} \\ \langle f_t \rangle(\theta) \;= \langle f \rangle a_t \sin 2(\theta - \theta_t) \end{cases} \tag{7}$$

where $\langle f \rangle$ is the average force, a_n and a_t represent the anisotropies of the normal and tangential forces, respectively, and θ_n and θ_t are their privileged directions. It is convenient to estimate the anisotropies from the following "force tensors":

$$\begin{cases} H_{\alpha\beta}^{(n)} = \int\limits_0^\pi \langle f_n \rangle(\theta) n_\alpha n_\beta d\theta, \\ H_{\alpha\beta}^{(t)} = \int\limits_0^\pi \langle f_t \rangle(\theta) n_\alpha t_\beta d\theta. \end{cases} \tag{8}$$

Then, we have $tr(\boldsymbol{H}^{(n)}) = \langle f \rangle$, and

$$\begin{cases} a_n = 2\frac{H_1^{(n)} - H_2^{(n)}}{H_1^{(n)} + H_2^{(n)}}, \\ a_t = 2\frac{H_1^{(t)} - H_2^{(t)}}{H_1^{(n)} + H_2^{(n)}}, \end{cases} \tag{9}$$

where the subscripts 1 and 2 refer to the principal values of the tensors.

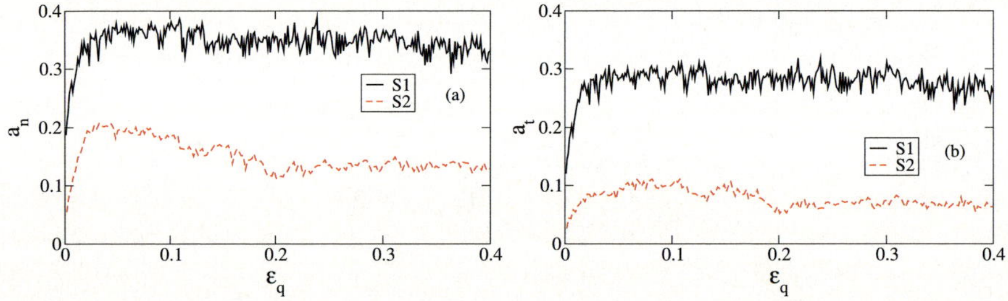

Fig. 5. Evolution of force anisotropies a_n (a) and a_t (b) as a function of cumulative shear strain ε_q in samples S1 and S2.

Figure 5 shows the evolution of a_n and a_t with ε_q in samples S1 and S2. We see that, in contrast to fabric anisotropies (Fig. 4), the force anisotropies in the pentagon packing remain always above those in the disk packing. This means that the aptitude of the pentagon packing to develop large force anisotropy and strong force chains is not solely dependent on the global fabric anisotropy of the system. Indeed, due to the geometry of the pentagons, i.e. the absence of parallel sides, the strong force chains are mostly of zig-zag shape, as observed in Fig. 6, and the stability of such structures requires strong activation of tangential forces. This explains, in turn, the large value of a_t for pentagons, very close to a_n, whereas in the disk packing a_t is nearly half of a_n.

The anisotropies a, a_n and a_t are interesting descriptors of granular microstructure and force transmission as they underlie the shear stress. Indeed, it can be shown that the general expression of the stress tensor Eq. (3) under some approximations leads to the following simple relation [3, 15]:

$$\frac{q}{p} \simeq \frac{1}{2}(a + a_n + a_t), \tag{10}$$

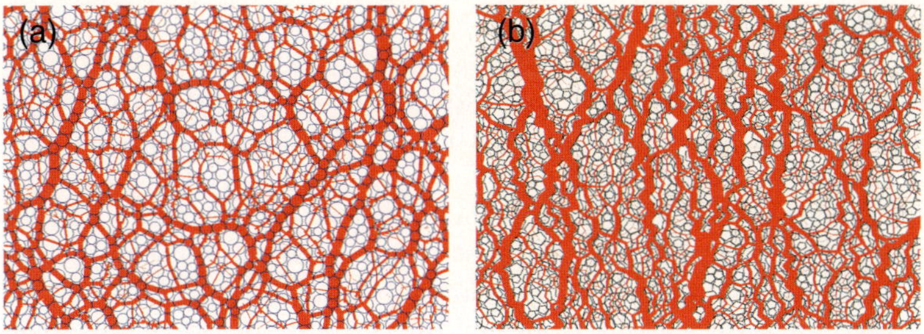

Fig. 6. Snapshots of normal forces in samples S2 (a) and S1 (b). Line thickness is proportional to the normal force.

where the cross products $a_n a$ and $a_t a$ between the anisotropies have been neglected compared to the anisotropies, and it has been assumed that the stress tensor is coaxial with the fabric tensor Eq. (5) and the force tensors Eq. (8). Fig. 7 shows that Eq. (10) holds quite well both for pentagons and disks. This equation provides an amazingly good estimate of the shear stress from the anisotropies under monotonic shearing.

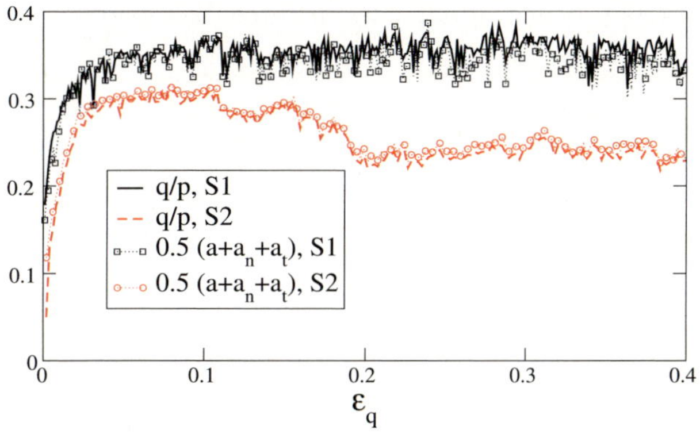

Fig. 7. Evolution of the normalized shear stress q/p for the samples S1 and S2 with ε_q together with the corresponding predictions from its expression as a function of fabric and force anisotropies Eq. (10).

A remarkable consequence of Eq. (10) is to reveal the distinct origins of shear stress in pentagon and disk packings. The fabric anisotropy provides a major contribution to shear stress in the disk packing (Fig. 4) whereas the force anisotropies are more important for shear stress in the pentagon packing (Fig. 5). In this way, in spite of the weak fabric anisotropy a, the larger force anisotropies a_n and a_t allow the pentagon packing to reach higher levels of q/p compared to the disk packing.

6 Conclusion

The objective of this paper was to isolate the effect of particle shape on shear strength in granular media by comparing two similar packings with different particle shapes: pentagons vs. disks. We observed enhanced shear strength and force inhomogeneity in the pentagon packing. But, unexpectedly, the pentagon packing was found to develop a lower structural (fabric) anisotropy compared to the disk packing under shear. This low fabric anisotropy, however, does not prevent the pentagon packing from building up a strong force anisotropy that underlies its enhanced shear strength compared to the disk packing.

This finding is interesting as it shows that the force anisotropy in a granular material depends on the particle shapes. This mechanism may be the predominant source of strength for "facetted" particles that can give rise to edge-to-edge (in 2D) contacts allowing for strong force localization along such contacts in the packing. Since the fabric anisotropy is low in a pentagon packing, the role of force anisotropy and thus the local equilibrium structures are important with respect to its overall strength properties. With pentagon packings, we were able to demonstrate the nontrivial phenomenology resulting from the specific shape of particles as compared to a disk packing. We found a similar behaviour for other regular polygons (hexagons and higher number of sides) as well as polyhedral particles in three dimensions.

References

1. N. P. Kruyt and L. Rothenburg. Micromechanical definition of strain tensor for granular materials. *ASME Journal of Applied Mechanics*, 118:706–711, 1996.
2. R. J. Bathurst and L. Rothenburg. Micromechanical aspects of isotropic granular assemblies with linear contact interactions. *J. Appl. Mech.*, 55:17, 1988.
3. L. Rothenburg and R. J. Bathurst. Analytical study of induced anisotropy in idealized granular materials. *Geotechnique*, 39:601–614, 1989.
4. F. Radjai, D. E. Wolf, M. Jean, and J. J. Moreau. Bimodal character of stress transmission in granular packings. *Phys. Rev. Letter*, 80:61–64, 1998.
5. L. Staron and F. Radjai. Friction versus texture at the approach of a granular avalanche. *Phys. Rev. E*, 72:1–5, 2005.
6. M. Jean. The non smooth contact dynamics method. *Computer Methods in Applied Mechanic and Engineering*, 177:235–257, 1999.
7. M. Jean and J. J. Moreau. Unilaterality and dry friction in the dynamics of rigid body collections. In *Proceedings of Contact Mechanics International Symposium*, pages 31–48, Lausanne, Switzerland, 1992. Presses Polytechniques et Universitaires Romandes.
8. J.J. Moreau. Some numerical methods in multibody dynamics: application to granular. *Eur. J. Mech. A/Solids*, 13:93–114, 1994.
9. F. Dubois and M. Jean. Lmgc90 une plateforme de développement dédiée à la modélisation des problèmes d'intéraction. In *Actes du sixième colloque national en calcul des structures - CSMA-AFM-LMS -*, volume 1, pages 111–118, 2003.
10. GDR-MiDi. On dense granular flows. *Eur. Phys. J. E*, 14:341–365, 2004.

11. J. J. Moreau. Numerical investigation of shear zones in granular materials. In D. E. Wolf and P. Grassberger, editors, *Friction, Arching, Contact Dynamics*, pages 233–247, Singapore, 1997. World Scientific.
12. J. K. Mitchell and K. Soga. *Fundamentals of Soil Behavior*. Wiley, NY, 2005.
13. M. Oda, J. Koshini, and S. Nemat-Nasser. Some experimentally based fundamental results on the mechanical behavior of granular materials. *Geotechnique*, 30:479–495, 1980.
14. B. Cambou. From global to local variables in granular materials. In C. Thornton, editor, *Powders and Grains 93*, pages 73–86, Amsterdam, 1993. A. A. Balkema.
15. F. Radjai, H. Troadec, and S. Roux. Key features of granular plasticity. In S. J. Antony, W. Hoyle, and Y. Ding, editors, *Granular Materials: Fundamentals and Applications*, pages 157–184, Cambridge, 2004. RS.C.
16. M. Satake. Fabric tensor in granular materials. In P. A. Vermeer and H. J. Luger, editors, *Proceedings of the IUTAM symposium on deformation and failure of granular materials, Delft*, pages 63–68, Amsterdam, 1982. A. A. Balkema.
17. E. Azéma, F. Radjai, R. Peyroux, and G. Saussine. Force transmission in a packing of pentagonal particles. *Phys. Rev. E*, 76:011301, 2007.
18. C.-h. Liu, S. R. Nagel, D. A. Schecter, S. N. Coppersmith, S. Majumdar, O. Narayan, and T. A. Witten. Force fluctuations in bead packs. *Science*, 269:513, 1995.
19. F. Radjai, M. Jean, J. J. Moreau, and S. Roux. Force distributions in dense two dimensional granular systems. *Phys. Rev. Letter*, 77:274–277, 1996.
20. S. N. Coppersmith, C.-h. Liu, S. Majumdar, O. Narayan, and T. A. Witten. Model for force fluctuations in bead packs. *Phys. Rev. E*, 53(5):4673–4685, 1996.
21. D. M. Mueth, H. M. Jaeger, and S. R. Nagel. Force distribution in a granular medium. *Phys. Rev. E.*, 57(3):3164–3169, 1998.
22. G. Lovol, K. Maloy, and E. Flekkoy. Force measurments on static granular materials. *Phys. Rev. E*, 60:5872–5878, 1999.
23. S. G. Bardenhagen, J. U. Brackbill, and D. Sulsky. Numerical study of stress distribution in sheared granular material in two dimensions. *Phys. Rev. E*, 62:3882–3890, 2000.
24. S. J. Antony. Evolution of force distribution in three-dimensional granular media. *Phys. Rev. E*, 63:011302, 2001.
25. T. S. Majmudar and R. P. Behringer. Contact force measurements and stresse-induced anisotropy in granular materials. *Nature*, 435:1079–1082, 2005.
26. L. E. Silbert, G. S. Grest, and J. W. Landry. Statistics of the contact network in frictional and frictionless granular packings. *Phys. Rev. E*, 66:1–9, 2002.

Motion of an Intruder Through a Two Dimensional Disordered Granular Medium

Pierre Cixous, Evelyne Kolb, Jean-Claude Charmet, and Chay Goldenberg

Physique et Mécanique des Milieux Hétérogenes (PMMH); CNRS UMR 7636
Université Pierre et Marie Curie (Paris VI); Université Denis Diderot (Paris VII);
Ecole Supérieure de Physique et Chimie Industrielles (ESPCI)
10 rue Vauquelin, 75231 Paris Cedex 05, France
kolb@ccr.jussieu.fr

Summary. We measured simultaneously the drag force on a cylindrical object (intruder) induced by the flow of a dense 2D bidisperse granular material around it, and the displacements of the individual grains. This type of experiment provides a way to examine the local rheological properties of the material. We found that the measured drag force increases sharply with the packing fraction near $\phi^* = 82.35\%$, indicating a transition to jamming. The dependence of the drag force on ϕ is clearly related to a decrease in the area of the empty cavity that forms behind the intruder.

1 Introduction

The flow of granular materials and the possible clogging and jamming around a solid object are of great importance in many engineering applications and natural phenomena, e.g., the flow of avalanches around obstacles, pile driving for deep foundations, and penetrometer measurements in sandy soils. From a fundamental point of view, measuring the drag force on an immersed object is a key method for characterizing the rheological properties of the surrounding medium, analogous to the Stokes experiment for viscous fluids. This type of experiment should lead to a better understanding of yielding in granular media by observing directly the structural and dynamical features of the transition between rest and flow and, more generally, provide insight into the mechanisms of plasticity in dense and disordered materials. While many studies have characterized flow and jamming in granular materials by applying a macroscopic shear or global vibration, only a few experiments (e.g., [1–4]) or numerical studies [5–7] directly tested the local rheological properties by considering slow, dense granular flow in the presence of an immersed obstacle. Albert et al. [1] studied the drag force on a rod immersed in a slowly rotating granular bed; due to the 3 dimensional (3D) geometry they could not visualize the flow of grains around the rod. They measured the fluctuations observed in the drag force and explained them using a stochastic failure model describing the formation and breaking of force chains between the grains and the immersed object [6]. These force chains were observed experimentally in a similar experiment on a 2D system by Geng and Behringer [3], in which they measured the drag force on a cylindrical object immersed in a packing of photoelastic particles, which enabled the visualization of the formation and collapse of force chains. They also analyzed the force fluctuations as a function of the packing fraction and flow velocity; however, they did not track the motion of the individual grains. Chehata et al. [2] studied the flow of dense, frictional granular matter around a cylinder in a quasi-2D vertical chute. They measured the flow field around the cylinder directly using a PIV (Particle Image Velocimetry) technique, and related it to the drag force for different flow velocities (without controlling the packing fraction). In other drag experiments, the translation (e.g., [8–11]) or rotation [12] of the obstacle was imposed, instead of that of the surrounding granular medium. Similar drag experiments and simulations were also performed in 2D [13, 14] and 3D [15] foams, as well as in colloidal suspensions [16] and viscoelastic micellar fluid [17]. The main novelty of the present work is the simultaneous measurement of the drag force on a rather small obstacle ("intruder") and the trajectories of the individual grains for different packing fractions of the surrounding medium.

2 Experimental Setup

The 2D granular material used in the experiments consists of about 6700 nickel-coated brass cylinders (see, e.g., [18]). In order to obtain a disordered packing, we use a mixture of cylinders of outer diameters $d_1 = 4$ mm and $d_2 = 5$ mm with an equal mass ratio. The cylinders are placed on a horizontal glass plate delimited by four brass walls which form a rectangular frame of constant width $W = 269.5$ mm ($54d_2$) and an adjustable length L between 457.5 mm ($91d_2$) and 470.5 mm ($94d_2$). The intruder is a brass cylinder of diameter $D = 20$ mm ($4d_2$) whose center is placed at an equal distance from the two lateral walls (see Fig. 1). The total number of cylinders is kept constant, but

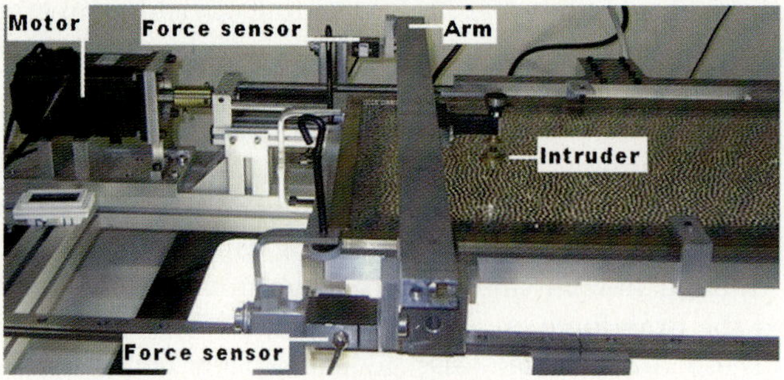

Fig. 1. The experimental setup.

the total surface $W \times L$ is varied to control the packing (area) fraction ϕ, in the range between $\phi = 80.13\%$ and $\phi = 82.41\%$ (the maximum corresponding to a nearly jammed state). The intruder is attached to a rigid transverse arm whose two extremities are connected to identical force sensors. The equivalent stiffness of the measurement device is $K_{\text{eff}} = 86000\,\text{N/m}$, large compared to the maximum measured effective stiffness of the packing. The vertical position of the intruder is fixed slightly above the horizontal glass plate, to ensure there is no friction between the intruder and the plate. The coefficient of friction between the intruder and the other grains, as well as among the grains, is $\mu_{\text{grain-intruder}} = \mu_{\text{grain-grain}} = 0.32$. The grains are in frictional contact with the bottom glass plate ($\mu_{\text{grain-bottom}} = 0.30$); the intruder is fixed in the laboratory frame, while the bottom plate is translated, inducing a motion of the grains relative to the intruder. Before each run, the grains are mixed carefully to prepare a homogeneous packing. The intruder is initially located at a distance of 110 mm from the back wall. In each run, the plate is moved using a high-torque motor at a constant speed $V_0 = 1.666\,\text{mm/s}$ until the intruder reaches a distance of 110 mm from the front wall. We performed 10 independent runs for each of the packing fractions studied.

The drag force on the grains is recorded at intervals of 1/6 mm ($d_2/30$) displacement of the plate. A CCD camera of 768×576 pixels placed above the setup records images at intervals of 1 mm ($d_2/5$), and covers a frame of $150\,\text{mm} \times 113\,mm$ ($30d_2 \times 22d_2$).

3 Results

3.1 Force Measurements

Figure 2 presents typical force signals obtained for different packing fractions ϕ. The force exhibits large fluctuations, reaching values as high as 35 N (which is larger than the maximum static friction between the grains and the plate, i.e., the force is transferred to the walls). The precision in the force measurement (corresponding to the noise level measured in the absence of grains) is 0.11 N. As shown in Fig. 2, both the mean force and the fluctuation amplitude increase with the packing fraction, exhibiting a very sharp increase for the largest packing fractions. In addition, the envelope of local minima and maxima of the force becomes wider as the intruder approaches the front wall; this effect appears similar to the increase in the force on a vertically penetrating rod observed in [8] when it approaches the floor of a container.

In order to quantify the dependence of the force on the intruder on the packing fraction, we smooth the signal by calculating a running average (with a width of $0.83D$) of the force averaged over an ensemble of 10 independent realizations. Figure 3 shows the dependence on the packing fraction of the initial drag force, $\langle F_0 \rangle_r$, calculated by further averaging the force over a

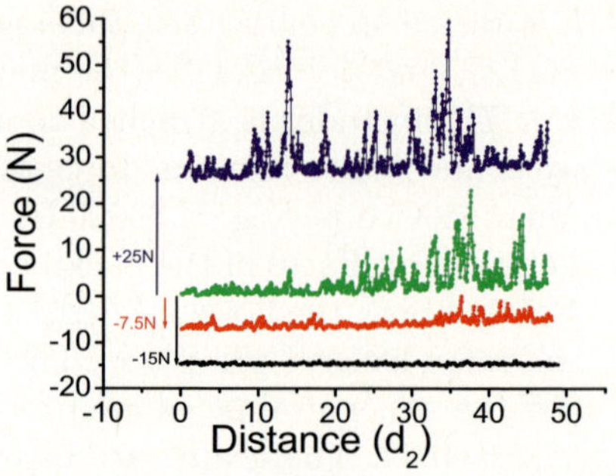

Fig. 2. Measured drag force on the intruder as a function of the plate displacement. For clarity curves for different packing fractions are shifted vertically. From bottom to top: $\phi = 81.17\%$ with a shift of the base line of $-15\,\mathrm{N}$ (black curve); $\phi = 81.88\%$ with a shift of $-7.5\,\mathrm{N}$ (red curve); $\phi = 82.05\%$, unshifted (green curve); $\phi = 82.41\%$ with a shift of $25\,\mathrm{N}$ (blue curve).

displacement interval of one intruder diameter starting with the initial displacement. This mean initial force increases with the packing fraction, with a very sharp rise near $\phi = 82.5\%$. Note that the random close packing of hard disks was estimated in simulations by various authors to be in the range $\phi = 82 - 85\%$ [19].

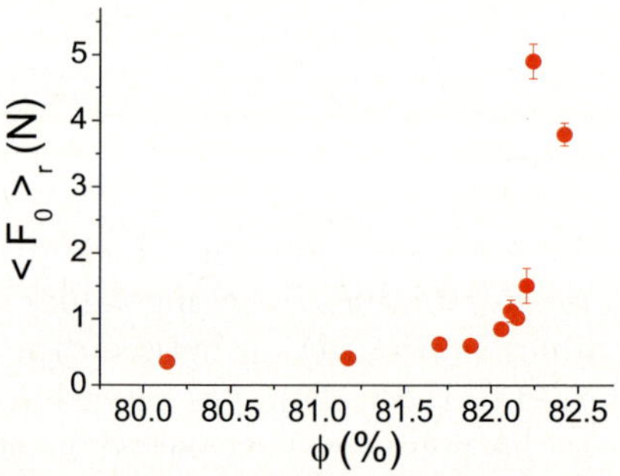

Fig. 3. Ensemble averaged and smoothed initial drag force, $\langle F_0 \rangle_r$, as a function of the packing fraction.

3.2 Displacement Measurements

As mentioned above, in parallel to the force measurements we record images of the grains. We use an image analysis technique based on the correlation of gray levels, as described in [18, 20], which determines the positions of the centers of the grains with a precision of about $5 \cdot 10^{-2}$ pixels, corresponding to about 0.01 mm, or $2 \cdot 10^{-3} d_2$. We note that the deformation of the contacts between grains for the typical measured forces are of the order of a fraction of micrometer, i.e., below our experimental resolution. A typical plot of the displacements of grains between two successive images is shown in Fig. 4a (in the laboratory frame) and in Fig. 4b (in the plate frame, subtracting the plate displacement between two successive images, $U_0 = 1\,\mathrm{mm}$). The displacement field is quite inhomogeneous. Its principal features are similar to those observed in previous experiments using a slightly different setup [18, 20]: the motion of the intruder induces displacements rather far in front of it, and a recirculating flow on both sides of the intruder. In the laboratory frame, a stagnation zone is visible ahead of the intruder, while a wake empty of particles usually forms behind it.

In order to characterize the mean flow, the displacements were averaged over the entire translation of the intruder, in square boxes of side length $1.2d_2$ and over an ensemble of 10 independent runs. The mean flow field obtained is presented in Fig. 5, for two different packing fractions. For the mean flow, the recirculation rolls are symmetrical, as expected. The displacement amplitude is largest along the vertical line $X = 0$ and decays with the distance from the intruder. Interestingly, the mean flows obtained for different packing fractions are quite similar, even though the mean initial drag force changes by a factor of almost 20 for the same range of ϕ. Figure 6 presents a more detailed comparison of the displacement fields shown in Fig. 5: the main effect of increasing the packing fraction is that the recirculation rolls are more pronounced. In particular, the Y-component of the displacements of the grains right behind the intruder are larger for the larger packing fraction. This observation emphasizes the importance of investigating the flow behind the intruder.

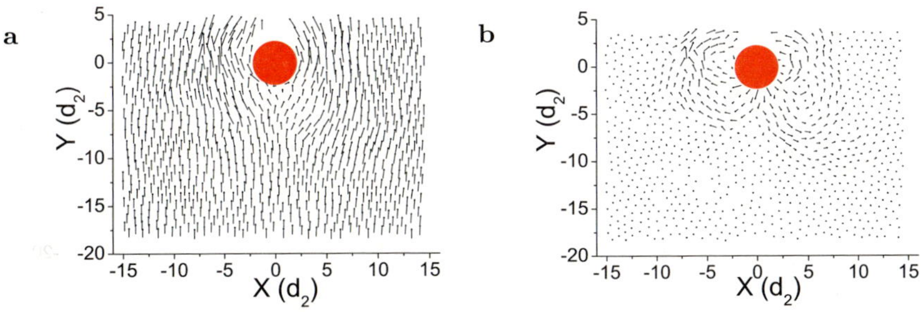

Fig. 4. Displacements of the grains between two successive images (corresponding to a plate displacement of 1 mm), multiplied by 5, for packing fraction $\phi = 82.23\%$: a. in the laboratory frame, in which the intruder is fixed and the plate moves at a constant velocity $V_0 = 1.666\,\mathrm{mm/s}$ in the positive Y direction; b. in the plate frame, in which the intruder moves at a constant velocity V_0 in the negative Y direction through a medium initially at rest.

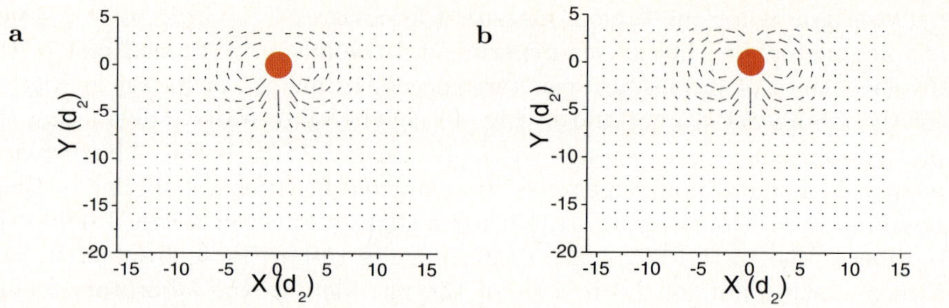

Fig. 5. Time and ensemble averaged displacement field in the plate frame, multiplied by 10, for two different packing fractions ϕ, averaged over boxes of side length $1.2d_2$; a. $\phi = 80.13\%$, b. $\phi = 82.05\%$.

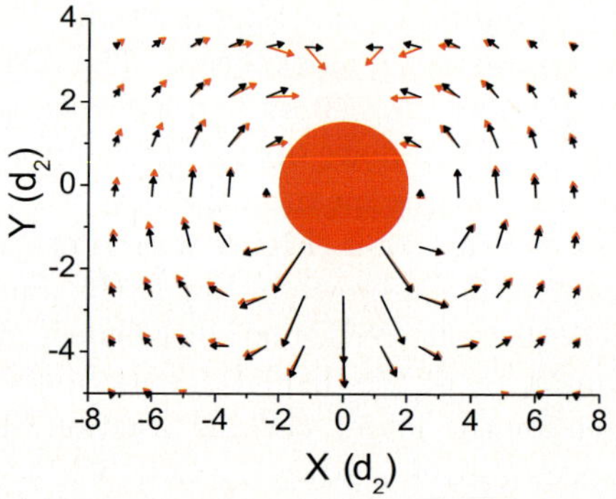

Fig. 6. Close-up view of the mean displacement field around the intruder in the plate frame, multiplied by 10; black arrows: $\phi = 80.13\%$, red arrows: $\phi = 82.05\%$.

3.3 Analysis of the Flow Behind the Intruder

Following each run, as described in Sec. 2, the particles in the packing were mixed again, and the plate was translated in the opposite direction with the same velocity, in order to image the flow behind the intruder. The images (Fig. 7) clearly show an empty wake (cavity) behind the intruder. The area of the cavity increases as the motion is started, but once established, it fluctuates around a mean value which does not vary much with the plate displacement. For the lowest packing fraction used (Fig. 7a), the cavity is long and tapered. The size of the cavity decreases with increasing packing fraction. The recirculation rolls merge behind the intruder, forming a more rounded cavity closed by a dynamic arch (Fig. 7b). At the largest packing fraction used, this arch forms right behind the intruder, and the cavity almost disappears (Fig. 7c).

The mean area of the cavity is controlled by the effective free area (which in this confined geometry is determined by the mean packing fraction). Figure 8 shows the dependence on the packing fraction of the mean area, A_w, obtained by averaging over the entire translation (except for the initial stage in which the cavity is formed) and over ten runs. The data are fit rather well by a linear dependence: $\frac{A_w}{A_{d2}} = a\,(\phi^* - \phi)$, with $a = 1140 \pm 50$ and $\phi^* = 82.35\% \pm 0.05\%$. According to this fit, the mean area of the cavity goes to zero for a packing fraction $\phi^* = 82.35\%$, the same packing fraction for which we observe a sharp increase in the force on the intruder (Fig. 3). The observation of the cavity behind the intruder therefore seems to provide a good criterion for determining the onset of jamming, which is associated with the decrease of the effective free area. We note that a similar cavity has been observed both in a 2D rotational flow of photoelastic disks [3] and in a granular chute flow (for which the length of the cavity was measured as a function of the flow velocity) [2]; however, the effect of the packing fraction on the size of the cavity was not investigated.

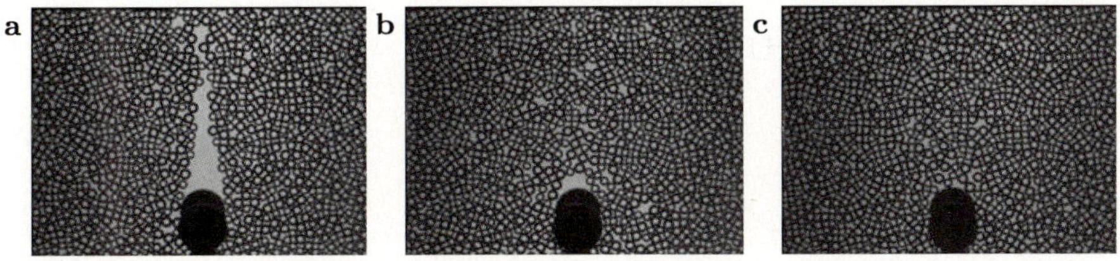

Fig. 7. Images of the packing behind the intruder for different packing fractions: a. the lowest packing fraction used, $\phi = 80.13\%$; b. $\phi = 81.70\%$; c. the highest packing fraction used, $\phi = 82.23\%$. The dark area in front of the intruder (at the bottom center of the figure) is a shadow cast by the arm holding the intruder.

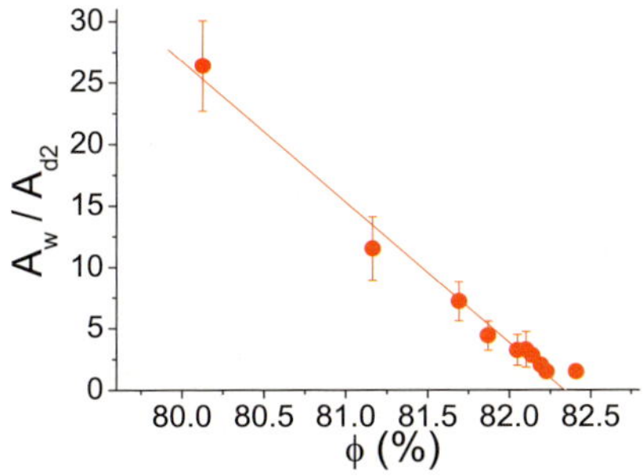

Fig. 8. The mean area of the cavity formed behind the intruder (normalized by the area of a large grain, $A_{d_2} = \pi d_2^2/4$) as a function of the packing fraction ϕ. The line corresponds to a linear fit over the first nine points, as the area for the largest packing fraction is overestimated due to a cutoff used during image processing.

3.4 Analysis of the Fluctuations

As mentioned, the area of the cavity behind the intruder fluctuates around the mean value. These fluctuations correspond to successive partial buildup and collapse of the cavity. These events are apparently related to the fluctuations observed in the force on the intruder, and to the fluctuations of individual particle displacements compared with the time and ensemble averaged flow described above. The fluctuations in the force can be decomposed into elementary events of successive jamming (force increase: Fig. 9a) and unjamming (force decrease: Fig. 9b). The large force drops (Fig. 9b) are associated with large recirculation rolls (Fig. 9d) around the intruder, i.e., large displacements of grains into the space of the cavity which (at least partially) collapses.

To analyze the fluctuations in the particle displacements, we use a coarse graining method developed in [21] (for a recent application and further details, see [22]). This method yields the following expression for the coarse grained (CG) displacement field, to leading order in the local strain:

$$\mathbf{U}(\mathbf{R}) = \frac{\sum_i m_i \mathbf{u}_i \phi[\mathbf{R} - \mathbf{r}_i^0]}{\sum_j m_j \phi[\mathbf{R} - \mathbf{r}_j^0]} \quad (1)$$

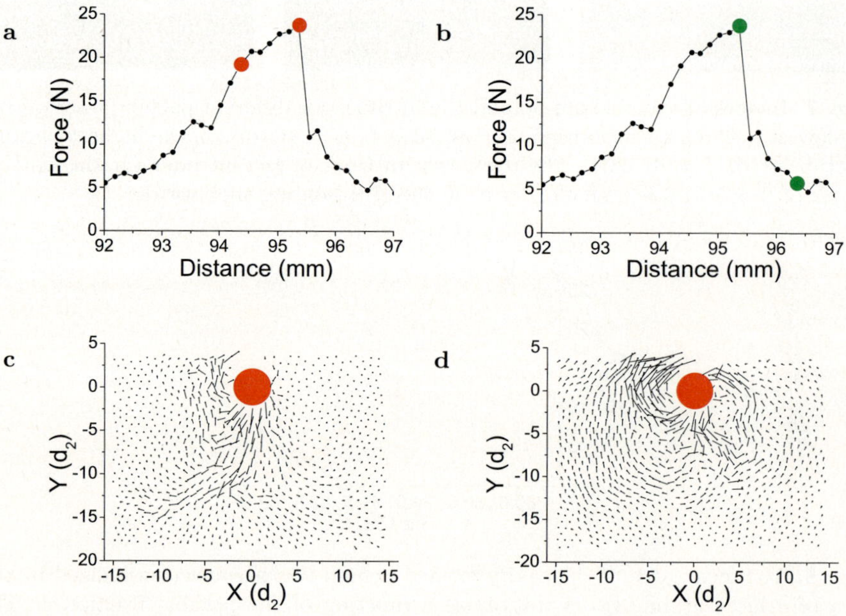

Fig. 9. The force on the intruder as a function of the plate displacement during a jamming event (a) and an unjamming event (b) and the displacements of individual grains in front of the intruder (multiplied by 10), in the plate frame (c,d). The large dots on the force curves indicate the times at which the pictures used for obtaining the displacement fields presented in the corresponding bottom panels were taken.

where $m_i, \mathbf{r}_i^0, \mathbf{u}_i$ are the mass of particle i, its position in a given frame, and its displacement between that frame and the next one, respectively, and $\phi(\mathbf{R})$ is a coarse graining function, chosen here to be a Gaussian, $\phi(\mathbf{R}) = \frac{1}{\pi w^2} e^{-(|\mathbf{R}|/w)^2}$. We use a rather small CG scale w, equal to the mean particle diameter [22]. The *displacement fluctuations* are defined as $\mathbf{u}'_i(\mathbf{r}) \equiv \mathbf{u}_i - \mathbf{U}(\mathbf{r})$. It is useful to consider, as in Fig. 10a, the displacement fluctuations at the particle positions, defined by $\mathbf{u}'_i \equiv \mathbf{u}'_i(\mathbf{r}_i^0)$. A corresponding scalar "noise" field (formally similar to the kinetic temperature) can be defined by: $\eta(\mathbf{r}) \equiv \sum_i m_i |\mathbf{u}'_i(\mathbf{r})|^2 \phi[\mathbf{r} - \mathbf{r}_i]$.

In Fig. 10 we present the CG displacement and corresponding fluctuations, as well as the noise field, obtained from two successive frames in an experiment with $\phi^* = 82.23\%$. As discussed above, the displacement field is inhomogeneous; this inhomogeneity is captured quite well by the CG field. However, the fluctuations (i.e., the deviations from a smooth displacement field) are rather localized, mostly close to the intruder. The regions of large fluctuations vary with time, and correspond to larger local rearrangements of the packing. This kind of analysis may therefore enable the identification of "plastic events", analogous to those observed in the plastic deformation of amorphous solids [23] or the T1 events observed in foams (see, e.g., [24]).

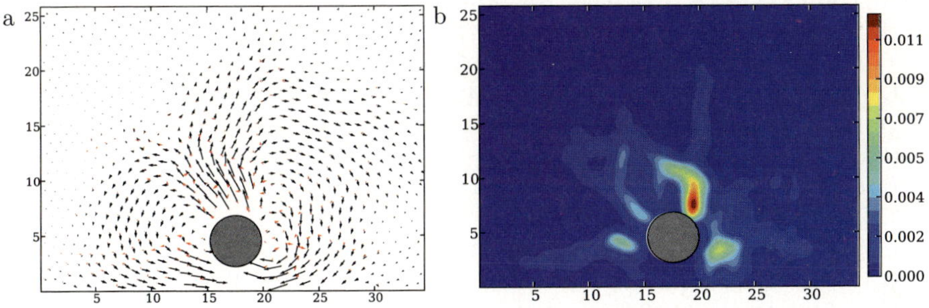

Fig. 10. a. Coarse grained displacements, $\mathbf{U}(\mathbf{r}_i)$, multiplied by 10 (black) and displacement fluctuations at the particle positions, $\mathbf{u}'_i$ (red); b. The noise field, $\eta(\mathbf{r})$, for the same two frames, in an experiment with $\phi^* = 82.23\%$. The length unit used in this figure is the mean diameter.

4 Concluding Remarks

The described experiment enables the characterization of both the mean flow behavior and the fluctuations in an inhomogeneous flow geometry. The drag force and the area of the cavity that forms behind the intruder change considerably with the packing fraction near jamming, while the mean flow in front of the intruder changes only slightly. The force, the area of the cavity and the particle displacements all exhibit considerable fluctuations with respect to their mean behavior, i.e., the average over the plate displacement, an ensemble of experiments, and space (for the displacements). This suggests that fluctuations play an important role in the flow of dense granular materials. The correlations between the fluctuations of these different quantities are currently being studied in detail.

References

1. Albert R, Pfeifer MA, Barabasi A-L, Schiffer P, Phys. Rev. Lett. **82**, 205 (1999) and references therein.
2. Chehata D, Zenit R, Wassgren CR, Phys. Fluids **15**, 1622 (2003).
3. Geng J, Behringer RP, Phys. Rev. E **71**, 011302 (2005).
4. Zhou F, Advani SG, Wetzel ED, Phys. Rev. E **69**, 061306 (2004).
5. Nouguier C, Bohatier C, Moreau JJ, Radjai F, Granular Matter **2**, 171 (2000).
6. Kahng B, Albert L, Schiffer P, Barabasi AL, Phys. Rev. E **64**, 051303 (2001).
7. Zhou F, Advani SG, Wetzel ED, Phys. Fluids **19**, 013301 (2007).
8. Stone MB, Barry R, Bernstein DP, Pelc MD, Tsui YK, Schiffer P, Phys. Rev. E **70**, 041301 (2004).
9. Hill G, Yeung S, Koehler SA, Europhys. Lett. **72**, 137 (2005).
10. Hamm E, Melo F, Europhys. Lett. **73**, 356 (2006).
11. Toiya M, Hettinga J, Losert W, Granular Matter **9**, 323 (2007).
12. D'Anna G, Europhys. Lett. **51**, 293 (2000).
13. Dollet B, Elias F, Quilliet C, Raufaste C, Aubouy M, Graner F, Phys. Rev. E **71** 031403 (2005).
14. Raufaste C, Dollet B, Cox S, Jiang Y, Graner F, Eur. Phys. J. E **23**, 217 (2007).
15. Cantat I, Pitois O, Phys. Fluids **18**, 083302 (2006).
16. Isa L, Besseling R, Poon WCK, Phys. Rev. Lett. **98**, 198305 (2007).
17. Gladden JR, Belmonte A, Phys. Rev. Lett. **98**, 224501 (2007).
18. Kolb E, Cviklinski J, Lanuza J, Claudin P, Clément E, Phys. Rev. E **69**, 031306 (2004).
19. Berryman JG, Phys. Rev. A **27**, 1053 (1983).
20. Kolb E, Goldenberg C, Inagaki S, Clément E, J. Stat. Mech.: Theory and Exp., P07017 (2006).
21. Goldhirsch I, Goldenberg C, Eur. Phys. J. E **9**, 245 (2002).
22. Goldenberg C, Tanguy A, Barrat J-L, Europhys. Lett. **80**, 16003 (2007).
23. Tanguy A, Léonforte F, Barrat J-L, Eur. Phys. J. E **20**, 355 (2006).
24. Kabla A, Debrégeas G, Phys. Rev. Lett. **90**, 258303 (2003).

Silo Collapse: An Experimental Study

Gustavo Gutiérrez[1,2], Philippe Boltenhagen[1], José Lanuza[1], and Eric Clément[1]

[1] Laboratoire de Physique et Mécanique des Milieux Hétérogènes UMR 7636/ESPCI, France
[2] Departamento de Física, Universidad Simón Bolívar, Apartado 89000, Caracas 1080-A, Venezuela gustav@usb.ve

Summary. The purpose of this experimental work is to develop some basic insight into the pre-buckling behavior and the buckling transition toward plastic collapse of a granular silo by studying different patterns of deformation generated on thin paper cylindrical shells during granular discharge. We study the collapse threshold considering the influence of the bed heights, flow rates and grain sizes. We compare the patterns that appear during the discharge of spherical beads, with those obtained in the axially compressed cylindrical shells. When the height of the granular column is close to the collapse threshold, we observe a ladder like pattern that rises around the cylinder surface in a spiral path of diamond shaped localizations, and develops into a plastic collapsing fold that grows around the collapsing silo.

1 Introduction

Cylindrical shells are used in many industrial processes to store fluids or grains; they are also used in chimneys, aircrafts and rockets. Thin shells are prone to buckle leading to their collapse. There is a vast engineering literature on empty shells [1–4], rigorous studies of cylindrical shells filled with fluids have been carried to some extent [5], and there are fewer investigations of the physics of shells filled with grains. Janssen and Vereins made a continuum model for a silo in 1895 [6]. They addressed the observation that the mean pressure at the bottom of the silo is generally significantly smaller than the hydrostatic pressure that a liquid would produce. This implies that an important part of the load is taken by the walls. This important formulation is still addressed, and has caught the attention of the physics community [7–13]. The problem of silos has gone to the physics laboratories posing new challenges to physicist. We are interested in the way thin walls of a cylindrical silo are affected by the load produced during the discharge of the grains. Thin walls become unstable and deform, sometimes globally and other times locally. These localizations and the different patterns that grow during the buckling and collapse are interesting from a fundamental point of view. We focus on the diamond shaped defects that appear on the surface of cylindrical paper shells when the silo is under discharge of the grains. This localized buckling grow into patterns similar to the ones observed in cylindrical empty shells subjected to axial compression. In our work the compression of the silo is due to the friction produced by the grains sliding down the internal wall of the shell, during the discharge by gravity through a central hole at the bottom. In section 2 we describe the experimental setup. In section 3 we indicate the parameters chosen to quantify the experimental observations. In section 4 we show our results. In section 5 we compare the patterns observed in our experiments with those due to axial compression of cylindrical shells. In section 6 we state the conclusion.

2 Experimental Set Up

In Fig. 1(a) we show a photograph of the experimental setup. We make the silos with tissue paper of thickness $t = 0.04$mm and nominal surface density $\rho_s = 20\text{g/m}^2$. The paper is wrapped around a long metal tube and glued along a narrow band to form the paper tube of diameter $D = 4$cm. The shell,

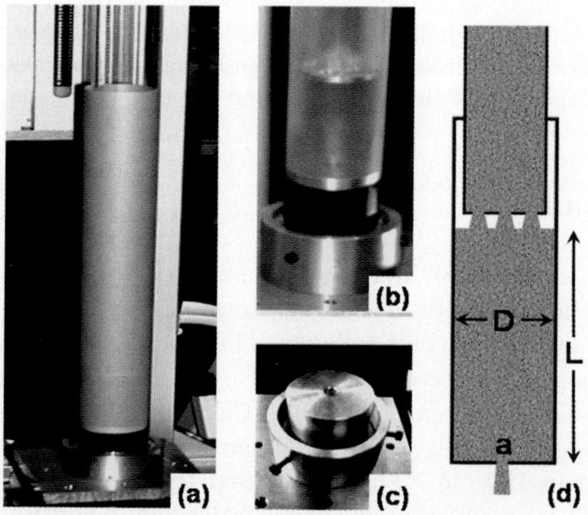

Fig. 1. (a) The photograph shows a paper silo with diameter $D = 4$cm, taped to a metal base. The silo is open at the top and a Plexiglas cylinder with an external diameter less than D is introduced from above to feed the grains. (b) A transparent plastic silo is shown; a perforated metal plate at the bottom of the Plexiglas internal tube can be seen through the walls of the silo. The level of this plate is varied to fix the effective height of the grain column. (c) The metal base shown has a center hole of depth $l = 3.8$cm. (d) Sketch of the experimental apparatus introducing the height of the granular bed acting on the silo walls, when the internal Plexiglas tube is used.

inserted into a small aluminum cylindrical base, shown in Fig. 1(c), is taped so that the thin shell is fixed at the base and open at the top. The aluminum base has a center hole of diameter a (3mm,4.5mm,6mm,8mm and 10mm). A rubber plug is inserted in the bottom hole in order to be able to fill the silo with the grains. For the initial measurements the internal tube was left inside the shell until the grains had been poured from the top and then the metal tube was carefully extracted so as to fill the silo without deforming the shell. For later measurements the long tube was extracted before pouring the grains and then a Plexiglas tube of diameter smaller than D, attached to a movable support, was inserted from above down to the desired height, as can be seen in the transparent silo shown in Fig. 1(b). The bottom of this Plexiglas tube has a metal circular plate with holes of diameter greater than a. The grains were fed from the top of the Plexiglas cylinder until it was completely filled. The plug is removed to produce the granular discharge. With the internal tube the grains can flow from it to the silo during the discharge, increasing the time the silo wall is subjected to the force produced by the grains during the discharge. The position of the internal Plexiglas tube determines the effective height of the granular bed (see fig. 1(d)). We used glass beads of different diameters d (0.2–0.3mm, 0.3–0.4mm, $\leq$ 0.63mm, 1mm, 1.5mm, 1.6mm). A CCD camera registered the transition to buckling and the collapse of the silo.

3 Control Parameters for the Column Collapse

When the plug is released and the granular flow starts, the solid friction forces on the grains are essentially polarized upward, due to the sliding on the wall. This corresponds to the condition for pressure screening as derived classically in Janssen's theory. It means that the confining pressure inside the granular bed saturates with the column height instead of growing linearly as for a fluid. The physical reason is that the walls carry the remaining load and the effective force acting on the walls has a vertical component directed downward that grows with the column height. Therefore, if we consider the silo walls as a solid cylindrical shell, we may experience buckling of the structure if the force is large enough. Along those lines, the height of the granular bed would be a simple candidate for a collapse control parameter. However the situation may not be as "simple" since we have a possible non-trivial interaction between the granular flow and the material structure. Note that the outward pressure may eventually help to stabilize the shell; the effective granular stiffness next to the wall could also change the effective rigidity of the structure. And moreover, the stress fluctuations due to the granular flow could play an important role in changing the position of the collapse onset. In this paper, which is a preliminary study, we only consider the influence of the column height, mean flow rate and granular size as possible control parameters.

From the work of Beverloo [14] and simple scaling argument, we know that when the opening radius a at the bottom of a silo is changed, the mean flow J varies as: $J = \alpha g^{1/2}(a - \kappa d)^{5/2}$ where α and κ are constants that only depend on the actual coupling between the grains and the opening. For non cohesive grains κ is about 4. Thus, we decided to change the opening size and the grain size to control the granular flux and the fluctuations.

4 Experimental Results

We performed experiments in two ways. First we filled the granular column up to a given initial height, and then we started the flow abruptly by taking the plug off at the center hole on the bottom of the silo. Second, we kept a constant height using the device described in fig. 1 (b) and (d). The existence of a threshold height L_c was recorded. The second method has the advantage that we can fix an important parameter, but in terms of the actual force on the wall the situation is less simple since we do not have a well define free surface, and we must consider an extra offset pressure to account for the junction with the inner tube.

For spherical beads with diameter $d < 0.63$mm changing the size of the bottom hole from $a = 4$mm to $a = 8$mm, keeping everything else the same, did not cause a change in the collapse threshold L_c. This occurred at $L_c \approx 20$cm. For the larger diameter a, the discharge is just faster and the collapse occurs sooner. We note that small fluctuations on the shell surface are observed when L is close to the collapse threshold. With the same conditions used above and for smaller particles (0.2–0.3mm) and a diameter of the hole $a = 3$mm, we obtained a threshold $L_c \approx 21$cm. For these smaller particles precursor fluctuations could not be observed when L was close to the height for which buckling occur. For particles of diameter $d = 1.5$mm and hole size $a = 10$mm, the threshold also occurs close to $L_c = 20$cm. For the above cases, the cylindrical shell was free at the top and no internal Plexiglas tube was used. In Fig. 2 we

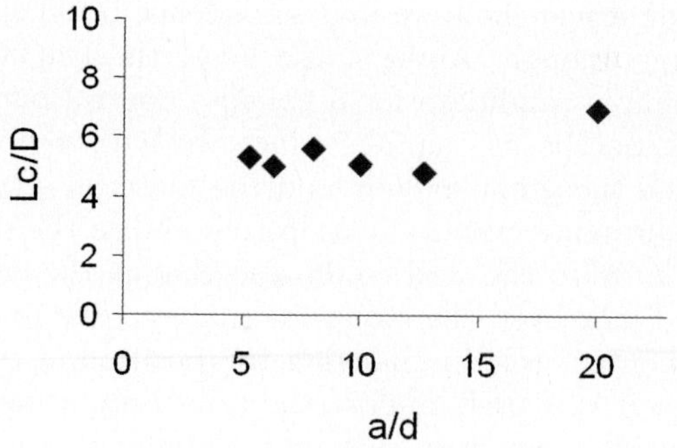

Fig. 2. Ratio of the collapse threshold to the diameter of the silo, as a function of the diameter of the bottom hole divided by the size of the glass beads.

plot the threshold value L_c divided by the diameter D of the silo as a function of the diameter a of the bottom hole divided by the grain size. For this graph we took the particle diameters d, by considering their highest nominal values, and the threshold L_c is obtained for the granular beds measured without using the internal Plexiglas tube.

For the small particles with diameters between 0.2–0.3mm, and hole size $a = 8$mm, and for big particles of size $d = 1.5$mm, we used a Plexiglas internal tube to feed more grains from above while the silo discharges (see Fig. 1(b)). With the internal tube, the threshold occurs when the tube is at a height $L \approx$ 16cm. If the threshold is taken to be around $20cm$, it is possible that the tube contributes for an effective offset of $4cm$. However, for particles of diameter $d = 1.6$mm the threshold obtained was $L_c \approx 11$cm, significantly lower than for all the other particles. This case was anomalous in the sense that huge fluctuations of the deformations on the surface of the shell appeared for L as low as $9cm$. These huge fluctuations can cause local curvatures that surpasses a local threshold value so that effectively a sufficiently large deformation is created, producing a large local defect that triggers the buckling and causes a collapse at a lower value of L. Further measurements have to be made in order to investigate the effect of these large fluctuations.

Therefore, these preliminary measurements suggest that for spherical particles and smooth walls in the ranges of particle diameters used, the threshold for collapse, for a silo which is free at the top, seems to be weakly dependent on particle diameter and the flux. The larger particles exhibit observable fluctuations of the silo surface when L is close to the collapse threshold. The precursor fluctuations in the elastic deformation of the surface of the cylindrical shell, was observed to increase with the particle size. The precursor fluctuations will be discussed in a future work. The results displayed in Fig. 2 indicate that the main factors determining the buckling and subsequent collapse of the silo is dominated should correspond essentially to a threshold in applied vertical force on the shell and should be determined at the first order by its mechanical characteristics. The presence of a granular material flowing inside could also have an influence on the value of the threshold via the effective elastic coupling or via other parameter still to be discovered. This could be explored by changing the geometrical and elastic characteristic of the silo.

5 Buckling Patterns

The focus of this section is on the different buckling patterns that were experimentally observed just prior or during collapse (see Fig. 3).

The type of buckling localizations that developed close to the threshold of collapse was common to all the cases that we measured. We observed that when the initial height L of the granular bed is sufficiently small, no change was observed at the lateral surface of the silo. For L sufficiently large some elastic fluctuations of the surface appear. Close to the collapse threshold some

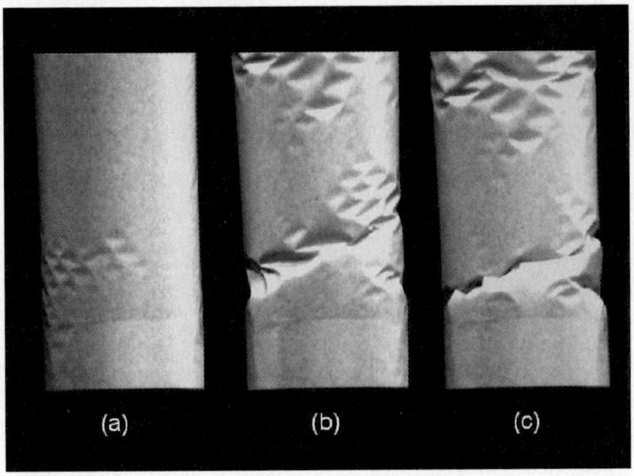

Fig. 3. The photographs show successive pictures of a collapsing silo during the discharge of small grains. (a) The buckling appears as a pattern of small diamond shaped elastic deformations occurring just before the collapse; (b) We can see the growth of bigger plastic deformations in a sequence like a spiral ladder around the cylinder; (c) During the collapse the paper folds in along a spiral that grows out of the chain of diamond shaped localizations.

persistent diamond shaped buckling occurs. This diamond shaped localized deformations are similar to the ones observed in real steel silos after damage. Some of the diamonds appear isolated and others form clusters similar to those observed in Fig. 3. During the discharge, when the Plexiglas column runs out of grains and the granular bed becomes sufficiently small these elastic deformations disappear so they seem to display some reversibility. When L is at the threshold value L_c and the collapse occur, the deformations propagate at an angle forming an "anticrack" that grows around the cylinder surface and eventually folds producing a catastrophic collapse. Very often (but not always), the first deformation appeared close to the silo bottom, in the vicinity of the boundary between the grains that remain fixed in the stagnant zone (forming a cone close to the bottom hole) and the grains that move down the wall.

6 Conclusion

We have observed different buckling patterns during the discharge of different size grains, from a central hole at the bottom of a cylindrical paper silo. These patterns are similar to the ones observed in axially symmetric compressed cylindrical empty shells. The height of the granular bed was changed to approach the threshold height L_c for different size particles and for different diameters of the discharging hole. It was found that the threshold was weakly dependent on the diameter of the discharge hole divided by the particle size. Precursor fluctuations were observed below the collapse threshold. These were found to increase with particle diameters. More experiments need to be made to fully characterize the above observations.

Acknowledgements

We would like to thank Benoit Roman, Pedro Reis and Jose-Eduardo Wesfreid for useful discussions.

References

1. Love A E H (1944) A Treatise on the Mathematical Theory of Elasticity, Dover.
2. Timoshenko S and Woinowsky-Klieger S (1959) Theory of Plates and Shells, McGraw-Hill.
3. Timoshenko S and Gere J (1961) Theory of Elastic Stability, McGraw-Hill.
4. Flugge W (1967) Stresses in Shells, Springer-Verlag.
5. Yamaki N (1984) Elastic Stability of Circular Cylindrical Shells, North-Hollan.
6. Janssen H A and Vereins Z (1895) Dtsch. Eng. 39(25): 1045.
7. de Gennes P G (1999) Reviews of Modern Physics, 71(2): S324.
8. Evesque P and de Gennes P G (1998) C. R. Acad. Sci. Paris, 326 (S. II b): 761.
9. Vanel L, Claudin Ph, Bouchaud J-Ph, Cates M E, Clément E, and Wittmer J P (2000) Phys. Rev. Lett., 84(7): 1439.
10. Vanel L, and Clément E (1999) Eur. Phys. J. B, 11(3): 525.
11. Ovarlez G, Fond C, Clément E (2003) Phys. Rev. E, 67(6): 060302(R).
12. Landry J M, Grest G S, Silbert L E, and Plimpton S J (2003) Phys. Rev. E, 67(4): 041303.
13. Bertho Y, Giorgiutti-Dauphiné F, Hulin J-P (2002) Physical Review Letters, 90(14): 144301.
14. Beverloo W A, Leniger H A and de Velde J V (1961) Chem. Eng. Sci. 15, 260.

Rough Cylindrical Object Immersed in a Granular Stream of Hard Disks

Payman Jalali and Pertti Sarkomaa

Department of Energy and Environmental Technology, Lappeenranta University of Technology, P.O.Box 20, 53851, Lappeenranta, Finland pjalali@lut.fi

Summary. We have investigated stresses on rough cylindrical objects confronted with a granular stream of inelastic hard disks. The roughness of cylindrical objects is created via covering their surfaces with hard disks of given size and material. We have employed event-driven simulations using restitution coefficients dependent of the impact velocity in a collision. We report the effect of material property (restitution coefficient) on development of granular shock wave around the object with corresponding stresses exerted on it. The role of the roughness of the object in resulting flow is also studied. Moreover, simulations are performed in two conditions with gravity and without it.

1 Introduction

Individual solid grains constituting granular materials interact through contacts or collisions. Granular materials display strange behaviors with multiple features resembling solids, liquids and gases in different conditions [1]. Certain obstacles may be put ahead of granular streams that has significant practical applications. For instance, circular pipes for heat transfer purposes are immersed in fluidized beds in contact to solid grains [2]. The formation of density shock wave around the pipe has a major role in changing the heat transfer quality to the pipe. As another example, arranging a series of solid obstacles ahead of granular streams can prevent flow jamming and arching [3]. The presence of such obstacles may also facilitate mixing of grains needed for handling and processing of powders.

Amarouchene et al. [4] performed simple experiments in two-dimensional granular flows around different obstacles. They observed shock waves forming in front of obstacles with different thicknesses depending on material properties, particles and object size and shape. Computer simulations were also performed before [5] and after [6, 7] the experiments of Amarouchene et al. in which the formation of triangular core in the upstream of the obstacle was poorly captured. The study of the dune anatomy leads to better understanding of the role of energy dissipation in the formation of the dune. The obstacle dissipates kinetic energy resulting in a steady flow. The surface roughness has also a significant effect on the dynamics of flow as it can change the overall interaction between the object and the granular stream. This paper presents some results from computer simulations obtained for dilute granular streams of finite width intercepting with circular objects having rough surfaces. The anatomy of the shock wave and the downstream flow is presented along with the analysis of the energy dissipation pattern due to the interception. The results of this paper shed some light into the physics of dilute granular flows confronting with rough objects. Moreover, the results have great impact on advanced designs of industrial facilities.

2 Geometry and Computer Simulations

In this section, we will describe the geometry and the method of simulations in two separate parts.

2.1 System Geometry

The granular stream is created as a rectangular domain of finite dimensions. The width of stream (H) has a predetermined value while its length is optional depending how long simulations are to be performed. Figure 1 demonstrates how the geometry of simulated system is built up.

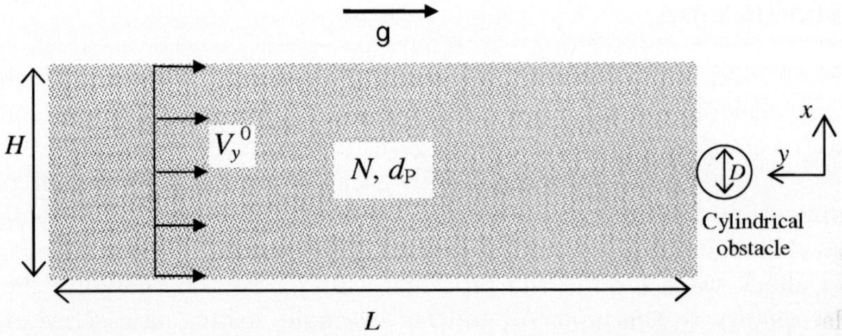

Fig. 1. Schematic of initial configuration of granular stream and a cylindrical object.

The cylindrical obstacle is initially located very close to the edge of the stream. N monodisperse hard disks are randomly distributed inside the rectangular domain ($L \times H$) resulting in an average packing fraction of 0.40. Initially, the velocity of each disk is assigned along the negative direction of y-axis. It is equal for all particles, denoted by V_y^0. The surface of obstacle is covered with hard disks as the same size and material as the particles in the stream. The diameter of hard disks is denoted by d_p. The direction of gravitational acceleration (g) is also in the negative direction of y-axis.

2.2 Simulation Method

Simulations are based on an event-driven algorithm [8] in which successive instantaneous binary collisions are determined. In each collision, the rules for inelastic collision of hard disks are implemented by noting that the restitution coefficient depends on the impact velocity [9, 10] as:

$$e(V_n) = 1 - (1 - e_0)\left(\frac{V_n}{V_0}\right)^{0.2}. \quad (1)$$

Here, e_0, V_n and V_0 represent the characteristic restitution coefficient, impact velocity magnitude and characteristic impact velocity, respectively. The value of V_0 is set to $0.5m/s$. It is worth mentioning that unlike the traditional quadratic characteristic equation for finding the collision time between any pair of particles, the characteristic equations are the fourth degree polynomial for particle-particle collisions (in the presence of gravity) and a combined polynomial with trigonometric functions for particle-wall collisions [10, 11]. These algebraic equations are solved using numerical techniques.

3 Results and Discussions

We present our results through different parts. The presented results demonstrate the effect of restitution coefficient on the basic features of the flow and the distribution of local packing density and velocity under different conditions. Moreover, the energy dissipation characteristics of the developed flow are introduced.

3.1 Flow Field Appearance

Simulations were performed using 25495 hard disks initially distributed in a region of the dimensions $L/d_p = 525$ and $H/d_p = 100$ with the average solid area fraction of $\phi = 0.40$. The radius of the object is 0.5 and 1.0 in our simulations. The surface of the object is either smooth or rough (covered with particles of the same size of free-stream particles). Figure 2 demonstrates the appearances of the flow using two different materials corresponding to the coefficients of restitutions $e_0 = 0.99$ in part (a) and $e_0 = 0.90$ in part (b). The highly-dissipative granular material ($e_0 = 0.90$) results in higher agglomeration of particles with less possibility of their scattering. Thus the shock wave becomes thinner (while denser) as the characteristic restitution coefficient decays. In addition, the highly-dissipative material leads to smaller scattering angles, which are defined as the angles between the vertical direction and the average lines through scattered branches of the flow. Note that the scattering angles might not be equal in the case of rotating obstacle.

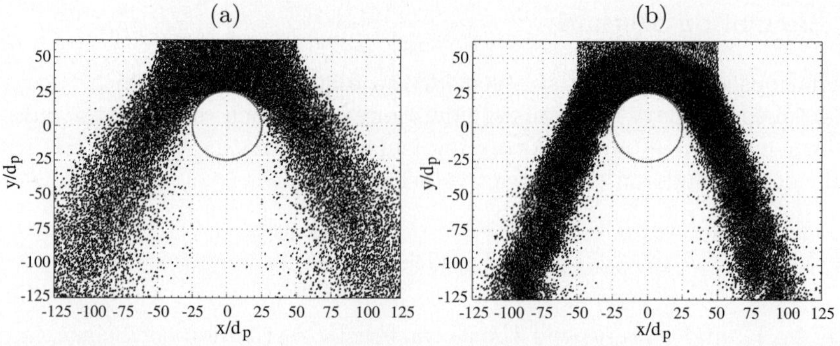

Fig. 2. View of the shock wave developed around the object with rough surface. The roughness is created by covering the surface of the obstacle by particles of the same size of the flow particles. (a) $e_0 = 0.99$ (b) $e_0 = 0.90$. Other variables are: $D/d_p = 50$, $d_p = 4cm$, $g = 10m/s^2$ and $V_y^0 = 1m/s$.

3.2 Local Distribution of Packing Density

To reveal the anatomy of the shock wave developed in front of the obstacle, we perform Voronoi tessellation for any given configuration of the system at a certain time. Figure 3 demonstrates the Voronoi tessellation in a configuration after the development of the shock wave is completed. The Voronoi cells are colored based on the packing density of each cell which is the ratio of the particle surface area to the total area of the Voronoi cell. Black cells represent those with packing densities (ϕ) larger than 0.60 and white cells for those with $\phi < 0.15$. One can see that the dense cells (identified by black color) are concentrated mostly in the front region of the obstacle. However, the dense region appears as dispersed areas within the loose background though it shows more continuity at the tip of the object.

The simulated case in Fig. 3 represents a low-dissipative material with $e_0 = 0.99$ whose packing density distribution is shown by the solid line with circles in Fig. 4. The probability density function (PDF) of ϕ has its maximum in the range of $0.10 < \phi < 0.20$ with nearly a Gaussian distribution. This is directly an indication of the scattering (expansion) of the granular branches in the downstream of the obstacle. The PDF of ϕ decays almost linearly for $\phi > 0.30$ so that the probability of finding cells with packing densities larger than 0.80 becomes extremely low. In contrast, the PDF of ϕ is completely different for a high-dissipative material such as that shown by dash line with squares in Fig. 4. In the latter, we observe two maximum values of PDF(ϕ) occurring at $\phi = 0.40$ and $\phi = 0.75$ with Gaussian distributions around each point. This result clearly indicates that lower restitution coefficient prevents scattering of particles in the downstream, which is obviously due to the coherence of particles induced by higher inelasticity [1].

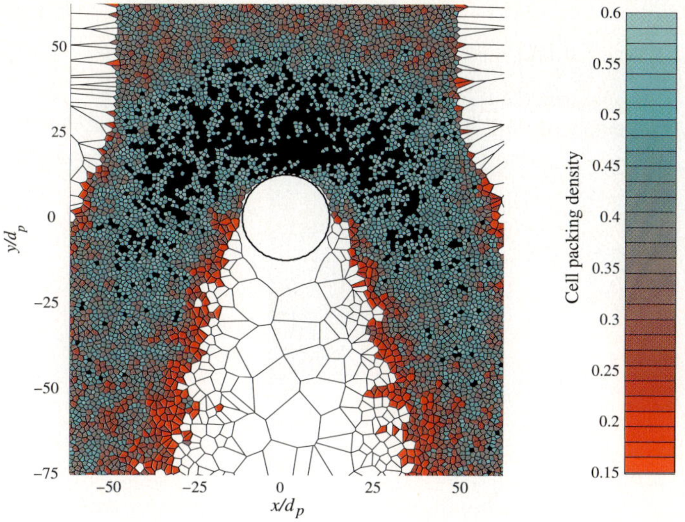

Fig. 3. Visualization of Voronoi cells colored based on the cell packing density ϕ in a configuration at steady state. As indicated in color bar, colors vary from red ($\phi = 0.15$) to cyan ($\phi = 0.60$). The cells with packing densities lower than 0.15 and higher than 0.60 are marked by white and black, respectively. Other parameters in simulations are $e_0 = 0.99$, $D/d_p = 25$, $d_p = 4cm$, $g = 10m/s^2$ and $V_y^0 = 1m/s$.

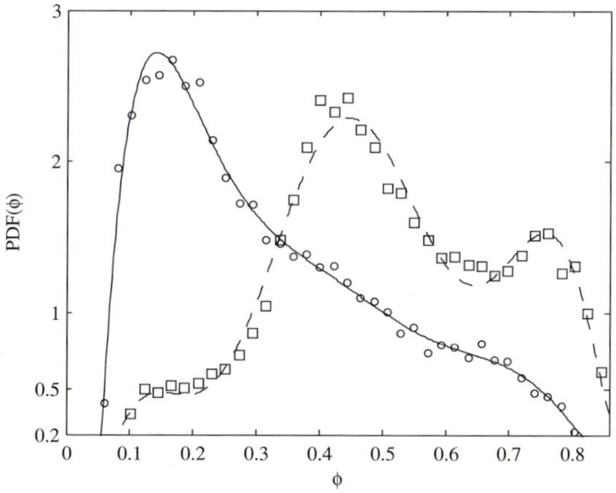

Fig. 4. Probability density function (PDF) of the packing density (ϕ) of Voronoi cells in two simulations corresponding to $e_0 = 0.99$ (fitted solid line to circles) and $e_0 = 0.90$ (fitted dash line to squares). Other parameters are as described in Fig. 3.

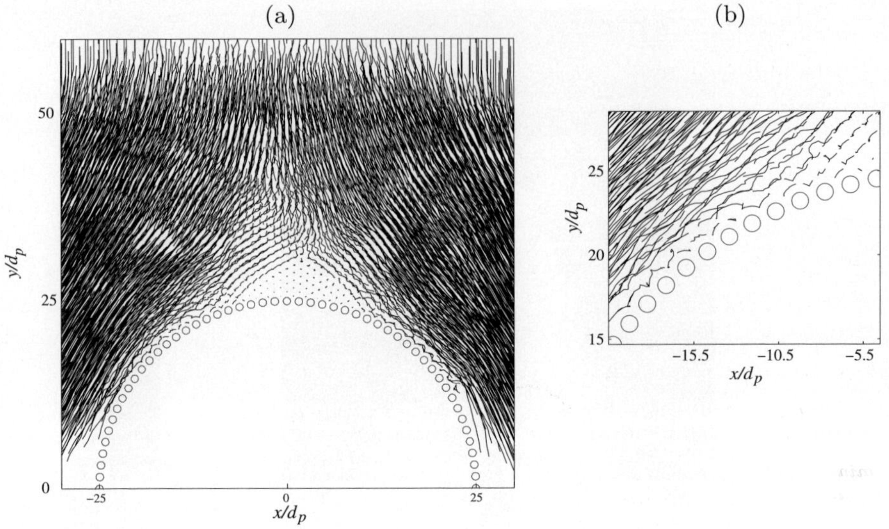

Fig. 5. (a) Particles trajectories around the object with rough surface within a dimensionless time of $\delta t^* = 2\delta t V_y^0/H = 0.5$. (b) A zoomed area of the same configuration near the rough surface. Simulation parameters are: $e_0 = 0.90$, $D/d_p = 50$, $d_p = 4cm$, $g = 10m/s^2$ and $V_y^0 = 1m/s$.

3.3 Displacement of Particles

In this section, we will demonstrate typical displacement field of particles around the rough obstacle. Figure 5 shows the lines of displacement (trajectories) for particles nearby the obstacle within a dimensionless time window of 0.5. It is clearly seen through Fig. 5(a) that there is a triangular static region at the tip of the obstacle. No considerable displacement is observed within the static region. This numerical observation is in good qualitative agreement with the experimental observations of Amarouchene et al. [4]. Specifically, Fig. 5(a) is comparable to Fig. 1(e) of the aforementioned reference [4]. Figure 5(b) represents a zoomed area of the configuration shown in Fig. 5(a), which magnifies the displacement of particles in the vicinity of the obstacle surface.

We can see that the surface roughness slows down the motion of particles next to the surface so that their displacements become very small. In other words, particles are driven by diffusion mechanism in the layer next to the surface rather than by convection. Also, one can see that there is a transitional zone between the triangular static region and the convecting zone of the flow according to Fig. 5(a).

3.4 Variation of Kinetic Energy

The shock wave developed in front of the obstacle can be characterized by the local variation of normalized kinetic energy defined as,

$$\kappa = \frac{\overline{E}_{y,\delta}}{\overline{E}_s}, \tag{2}$$

where $\overline{E}_{y,\delta}$ represents the average particle kinetic energy at position y within the bin δ, and $\overline{E}_s$ is the instantaneous kinetic energy of a particle in the free stream. In the presence of gravity, the latter will grow in time. However, κ is normalized so that its corresponding value in the free stream remains unity at any instant. Figure 6 demonstrates the variation of κ along the flow direction for a simulation with $e_0 = 0.99$. The value of κ sharply drops to a minimum from the shock front to the triangular static region that is called κ_{min}. Thereafter it increases to reach a downstream value called κ_{st}. The value of κ_{st} is 0.89 for the restitution coefficient of $e_0 = 0.99$. Meanwhile, κ_{min} is equal to 0.51 in the same case.

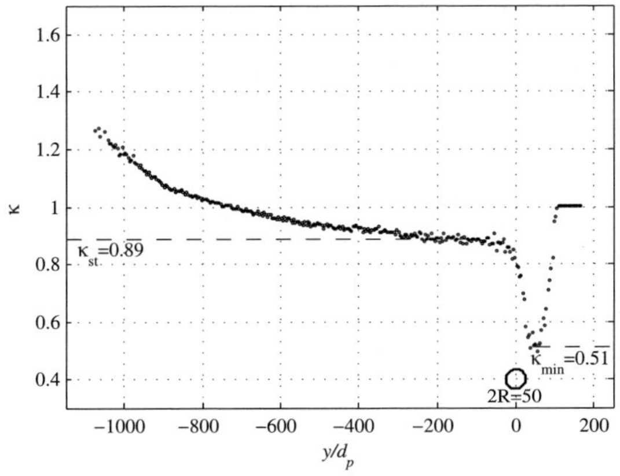

Fig. 6. Variation of normalized mean kinetic energy of particles (κ) along the flow direction, y. The circular object (with the diameter of $2R = 50$) is symbolically shown to understand its effects on the flow. Dots represent the values of κ in different averaging bins. The simulated case corresponds to $e_0 = 0.99$, $D/d_p = 50$, $d_p = 4cm$, $g = 10m/s^2$ and $V_y^0 = 1m/s$.

Note that the value of κ gradually increases in far downstream so that it finally passes unity which corresponds to the initial encounter of the stream with the obstacle. However, there is a stationary shape for the variation of κ versus y in front of the object and within a certain distance of downstream, which is not changing in time.

The values of κ_{st} and κ_{min} characterize the dissipative behavior of the flow, which are expected to be related to the mechanical properties of grains as well as the geometry and other conditions of the flow. Figures 7(a) and 7(b) display the dependence of κ_{st} and κ_{min} on e_0, respectively, for simulations with rough and smooth surfaces of obstacle in the presence and the absence of gravity (totally 4 cases).

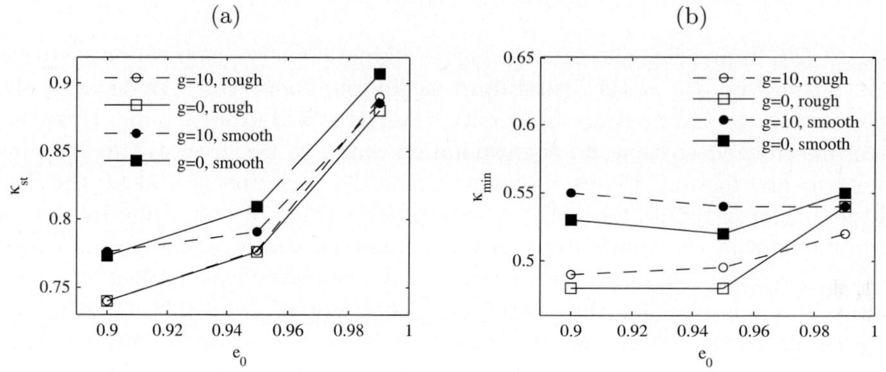

Fig. 7. Variation of (a) κ_{st} versus e_0, and (b) κ_{min} versus e_0 for four different simulations corresponding to rough and smooth objects in the presence and the absence of gravity. Symbols are introduced in the legend. Other simulation parameters are: $D/d_p = 50$, $d_p = 4cm$, and $V_y^0 = 1m/s$.

From Fig. 7, one can see that κ_{st} strongly depends on e_0 as it may vary from 0.75 to 0.90 corresponding to the variation of e_0 from 0.90 to 0.99. In contrast, κ_{min} varies at most between 0.48 and 0.55. It can be also observed that gravity has negligible effect on κ_{st} for rough objects, where κ_{st} in the presence of gravity is slightly lower than that of simulations without gravity using smooth obstacle. The effect of surface roughness is considerable through the entire range of e_0 in the absence of gravity. As generally expected, smooth objects introduce indices of dissipation lower than those of rough objects.

4 Conclusions

In the present work, some basic features of granular shock waves around a circular object (rough or smooth) are presented along with some analyzing methods applied in the simulation results using hard disk model. In this context, the coefficient of restitution for particles is found to be an effective parameter which not only alters the basic look of the developed flow, but also plays a significant role in the topological and physical characteristics of the flow. The anatomy of the shock wave from our simulations resembles the experimental observations in this regard.

The granular shock wave in front of the obstacle is characterized by a normalized index for kinetic energy. It is the ratio of the local kinetic energy to the kinetic energy of free stream at any instant. We have presented the values of this index corresponding to two key points on the stationary graph of $\kappa - y$. One point represents the minimum value of κ occurring in front of the object, namely κ_{min}, and the second point shows the downstream value of κ called as κ_{st}. Our simulations revealed that κ_{st} is far more sensitive than κ_{min} with respect to the characteristic restitution coefficient e_0. The roughness of the object decreases the indices that implies to a higher dissipation than in the case of smooth obstacle. More investigations are required to understand the effects of geometric parameters of the system on the dissipative characteristics of the system.

References

1. Jaeger HM, Nagel SR, Behringer RP (1996) Rev Mod Phys 68: 1259–1273.
2. Pearce JC, Sabersky RH (1977) Lett Heat Mass Transfer 4: 1–8.
3. Duran J (2000) Sands, Powders and Grains. Springer-Verlag, New York.
4. Amarouchene Y, Boudet JF, Kellay H (2001) Phys Rev Lett 86: 4286–4289.
5. Buchholtz V, Poschel T (1998) Granul Matter 1: 33–41.
6. Wassgren CR, Cordova JA, Zenit R, Karion A (2003) Phys Fluids 15: 3318–3330.
7. Bharadwaj R, Wassgren C, Zenit R (2006) Phys Fluids 18: 043301.
8. Allen MP, Tildesley DJ (1987) Computer Simulation of Liquids. Clarendon Press, Oxford.
9. Ramirez R, Poschel T, Brilliantov NV, Schwager T (1999) Phys Rev E 60: 4465–4472.
10. Jalali P, Li M, Ritvanen J, Sarkomaa P (2003) Chaos 13: 434–443.
11. Jalali P, Polashenski W, Tynjala T, Zamankhan P (2002) Physica D 162: 188–207.

Stability and Jamming Transition in Hard Granular Materials: Algebraic Graph Theory

Nicolas Rivier

IPCMS, Université Louis Pasteur, 67084 Strasbourg Cedex, France
nick@fresnel.u-strasbg.fr

Summary. Dry granular matter is modelled as a graph of grains linked by purely repulsive contacts. Its stability (jamming) is insured by odd circuits that prevent the grains from rolling on each other. A topological dynamical matrix is associated with the graph; it has a spectrum of low-energy excitations characteristic of dry, disordered granular matter. In the limit of large stiffness-to load ratio, dry granular matter has two possible dynamical states, dry fluid and jammed, rigid but fragile solid.

1 Introduction. Dry Granular Matter as a Graph

Dry granular matter is described naturally by a graph [1]: Two grains in repulsive contact are vertices connected by an edge. There are no attractive, cohesive forces between grains and the packing is held together by external forces (gravity or boundaries).

Hard dry granular materials (granular materials in the limit of large stiffness-to-load ratio, with infinite tangential friction between grains) can be simplified further: Two grains in contact can only roll without slip on each other, or disconnect by pulling apart. Edges represent therefore a scalar and boolean interaction between the two vertices. The grains $i = 1, \ldots, n$ are modelled as spheres of radius R_i, that must roll on each other to remain in contact. The two possible states of disordered granular matter, *dry fluid* [2–4], and *jammed*, rigid but *fragile solid* [5–7] are direct results of the topological dynamics of the graph, where the elements responsible for the jammed state are *odd circuits*, circuits with an odd number of grains in contact.

A topological dynamics can be associated with a graph, through the eigenvalues and eigenvectors of a dynamical matrix that is the adjacency matrix $\mathbf{A}$ modified by a diagonal matrix (Eq. (1) below). Odd circuits impose a non-zero lowest eigenvalue of the dynamical matrix [1], that, in disordered granular matter, scales as $1/L$ with the linear size L of the material. This role of parity explains why there can be no continuous, qualitative description of granular matter. Granular matter is the material equivalent of rational numbers.

By representing a hard granular materials as a graph, we have linearized a highly non-linear problem The nonlinearities are in the constraints, that are geometric (the edges are boolean, scalar and of constant length if they apply: they represent struts that transmit purely repulsive forces) inequalities (contacts can be broken up), with naïve number theory combinatorics (odd circuits are non trivial). The dynamics of the material is reduced to linear algebra (eigenvalues and eigenvectors) of the adjacency matrix of the graph, which contains all its generic features: contact, odd circuits and disorder.

2 Edges, Arches, Odd Circuits and Jamming

As regards statics and stability, hard granular materials differ in three ways from the classic (Maxwell) structural rigidity analysis of elastic networks as a function of the number of edges |E|, from floppy to overconstrained, through isostatic at the rigidity percolation threshold [8, 9].

a) Stresses carried by edges have a sign constraint. It follows that (repulsive) self-stresses cannot be sustained in an overconstrained network: the two grains connected by a redundant edge pull apart, |E| decreases, and the network becomes isostatic. On the other hand, a floppy network will, if it can, rearrange and add edges, to become isostatic [10, 11].
b) But hinges in a floppy network can only buckle through grains rolling on each other, a motion blocked by odd circuits.
c) Odd circuits do not occur in isolation, but surround closed loops (odd vorticity is a line defect), called R-loops for short [12]. In *disordered* granular materials, the largest R-loop extends across the entire material. By contrast, the R-loops are much smaller, of the order of the period, in a periodic granular packing.

Thus, as a function of the increasing number of edges, a disordered, hard granular material ranges from a dry fluid (underconstrained network without odd circuits) to the unrestricted, isostatic network of rigidity percolation, through the *jammed phase* [7] of underconstrained, but rigid networks with odd circuits (Fig. 1). This intermediate phase has a constant, excess density of low-energy normal modes, close to the jamming transition, as observed in (soft-sphere, but still with repulsive interactions) computer simulations [5, 6] and explained recently by a scaling analysis [7]. The essential role of disorder in the jamming transition, emphasized by Nagel *et al.*, most eloquently in the title of their paper [6]: "Jamming at zero temperature and zero applied stress: The epitome of disorder", will be elucidated here.

By plotting the number of odd circuits c as a function of the number of edges |E|, we obtain the phase diagram for hard granular matter (Fig. 1), in the limit of large stiffness to load. |E| ranges between the values DF and I. The material is unstable, both above I where it is overconstrained, and below DF where it is floppy and free to rearrange in the absence of odd circuits. At DF,

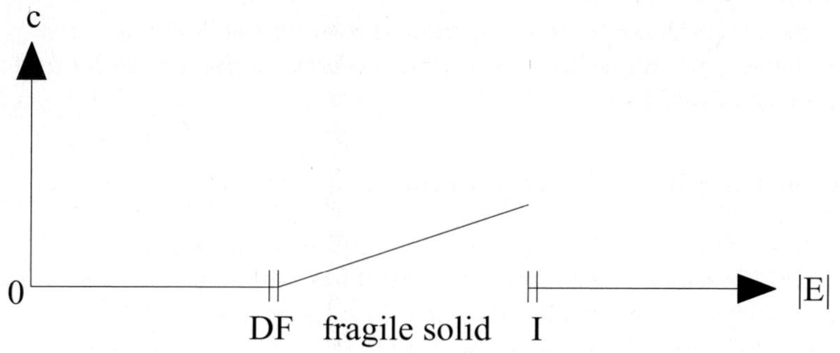

Fig. 1. The order parameter for the jammed phase, the number of odd circuits c, as a function of the number of edges $|E|$. A hard granular material is only stable in the interval between DF (dry fluid) and I (isostatic point), in the limit of large stiffness-to-load ratio. For $|E|$ above DF, the material is a jammed, fragile solid, stabilized by the odd circuits (the "arches" of the Janssen effect). At DF, the material is a dry fluid [1].

the material behaves as a dry fluid or bearing, that does not resist shear, whose constituents roll without slip on each other or on any boundary or inclusion (the point of contact between any two grains has zero velocity, even though it changes all the time). It only feels hydrostatic pressure, i.e. forces normal to any plane inside the material on any scale larger than a few grains; a solid inclusion is subject to Archimedean pull only; arches can buckle freely and make ripples [13]. Between DF and I, the material is a fragile, jammed solid, stabilized by odd circuits. The odd vorticity forms (R-) loops that are large in disordered granular solids, where the jamming transition is a true scaling phase transition, with the number of odd circuits c as an order parameter measuring geometric frustration of the jammed grains (precisely, the ratio c/n scales as $1/L$ with the linear size of the material L at the jamming transition).[1]

[1] The word *frustration* is technical [15] and *jamming* has little to do with traffic jam, although this is a conference on traffic and granular flow, with lectures on phase transition in traffic (Y. Xue) and on the drivers behaviour from a psychologists point of view (F. Saad). The jamming transition (F "blocage") is a *collective* transition, involving long ranged arches (odd circuits), that are specific of disordered, hard granular matter. The eigenvectors describing the deformation of a jammed material subjected to shear or to uniaxial stress are extended (approximated by **j** or **a**), whereas a traffic jam (F: "bouchon") has more to do with the (lack of) motion of individual cars packed bumper-to bumper. A more descriptive word would be bottleneck, which is probably the source of the French word "embouteillage" – more sophisticated than, but synonymous to "bouchon". (See also the introduction of [16]) The traffic inside biological cells, also discussed at the conference, is of the traffic jam type. On the other hand, a multi-lane freeway may exhibit a collective jamming transition similar to jamming of granular materials, as different lanes are prevented to flow independently by reckless lane change.

This is in complete contrast with ordered granular solids where the R-loops are small and c/n remains finite. Note the essential roles of *disorder* and grains (in the guise of *odd numbers*).

2.1 Even Circuits Only: Dry Fluid

In two dimensions, disks can roll on each other without slip if and only if there are no odd circuits of disks in contact. In three dimensions, spheres can roll on each other without slip if there are no odd circuits (sufficient condition) even though the axes of rotations are no longer colinear, and the centres of the spheres not coplanar [2]. This remains true when the centres of the spheres are mobile [13]. A hard granular materials without odd circuits behaves as a dry fluid, and not only as a ball bearing. The lowest eigenvalue of its dynamical matrix is zero, as we shall see.

Non-slip rotation of one sphere on another constitutes a *connection*, that can be extended from one grain to any other along a path of grains in contact. If there are no odd circuits, the connection is pure gauge, that is independent of the path [2, 13, 14].

3 Topological Dynamics of Graphs. Adjacency and Dynamical Matrices

A graph with n vertices and |E| edges is described algebraically by the *adjacency matrix* $\mathbf{A}$ (*contact matrix* in protein folding [17]), an $n \times n$ matrix $A_{ij} = 1$ if i, j in contact, 0 otherwise. $\mathbf{A}$ is related to the $n \times |\mathrm{E}|$ incidence matrix $\mathbf{D}$, through

$$\mathbf{Q} = \mathbf{D}\mathbf{D}^{\mathbf{t}} = \mathbf{\Delta} - \mathbf{A}, \tag{1}$$

where $\mathbf{\Delta}$ is a diagonal matrix $\Delta_{ij} = z_i \delta_{ij}$, with $z_i = \sum_j A_{ij}$, the valency (degree) of vertex i [18].

The matrix $\mathbf{Q}$ is the *dynamical matrix* of a physical system on the graph, where the vertices are particles of the same mass, and the edges are springs with the same stiffness. That is, the interaction between two vertices connected by an edge can have either sign.[2] The rank of $\mathbf{Q}$ is $n-1$ for a connected graph, so that its lowest eigenvalue is zero with corresponding "Woodstock" eigenvector $\mathbf{j} = (1, 1, 1, 1, \ldots, 1)^t$. (A matrix with all its elements $J_{ij} = 1$ is called Woodstock matrix, because that is how Snoopy's friend speaks).

[2] Topological dynamics of a graph where the edges are identical springs: The potential energy of the graph is $V = (1/2)k \sum_{(ij)} (x_i - x_j)^2$, where the sum is taken over all the edges (ij). The stress-free translation corresponds to $(x_i - x_j) = 0$, and the force can have either sign. The system of Euler-Lagrange equations of motion of the graph, $[-\lambda \mathbf{1} + (\mathbf{\Delta} - \mathbf{A})]\mathbf{x} = \mathbf{0}$, is the eigenvalue equation of the dynamical matrix $\mathbf{Q} = \mathbf{\Delta} - \mathbf{A}$, with the eigenvalues $\lambda = m\omega^2/k$ related to the frequencies ω of the normal modes of oscillation.

By contrast, the dynamical matrix of a hard granular system, where the vertices are grains with the same momentum of inertia, and the edges are struts, representing the non-slip rotation of the grains on each other, is $\mathbf{K} = \mathbf{\Delta} + \mathbf{A} = 2\mathbf{\Delta} - \mathbf{Q}$. Interaction has one sign only (*sign constraint*).[3]

3.1 Importance of the Sign Constraint

Because of the sign difference, the spectrum of eigenvalues of $\mathbf{Q}$ and $\mathbf{K}$, and their respective eigenvectors, are essentially different, and this difference is associated with *odd circuits*. In the absence of odd circuits, the dynamical matrix $\mathbf{K}$ is changed into $\mathbf{OKO}^{-1} = \mathbf{\Delta} - \mathbf{A} = \mathbf{Q}$ by an unitary transformation $\mathbf{O}$ that changes the sign of odd rows and columns.

Consider first a connected bichromatic graph (see [19] for an example) without odd circuits (grains in contact have different colors). Let $(-1)^i$ be the color of vertex i. The adjacency matrix has a non-zero entry $A_{ij} = 1$ only if i and j have opposite colors. The unitary transformation $O_{ij} = (-1)^i \delta_{ij}$ changes the sign of odd rows and columns, $A'_{ij} = (-1)^i A_{ij} (-1)^j$ and transforms the dynamical matrix $\mathbf{K} = \mathbf{\Delta} + \mathbf{A}$ into $\mathbf{K}' = \mathbf{OKO}^{-1} = \mathbf{\Delta} - \mathbf{A} = \mathbf{Q}$, that is the dynamical matrix for an elastic network. Now, the lowest eigenvalue of $\mathbf{Q}$ is zero with the corresponding Woodstock eigenvector $\mathbf{j}$. It is nondegenerate since $(\text{rank}\mathbf{Q}) = n - 1$ for a connected graph. Transforming back, the lowest eigenvalue of the dynamical matrix $\mathbf{K}$ of the bichromatic packing is zero, one soft mode with the alternating eigenvector

$$\mathbf{a} = (1, -1, 1, -1, \ldots)^{+} = \mathbf{Oj}. \tag{2}$$

A granular material without odd circuit is indeed a bearing of grains rotating without slip on each other [2].

[3] Topological dynamics of a graph where the edges are identical struts: The potential energy of the graph is $V = (1/2)k \sum_{(ij)} (|\theta_i + \theta_j|)^2$. The opposite rotations of two grains in contact $|\theta_i + \theta_j| = 0$ is stress-free, and the force is always repulsive, as indicated by the redundant $|\ldots|$. The system of Euler-Lagrange equations of motion of the graph, $[-\lambda \mathbf{1} + (\mathbf{\Delta} + \mathbf{A})]\mathbf{x} = \mathbf{0}$, is the eigenvalue equation of the dynamical matrix $\mathbf{K} = \mathbf{\Delta} + \mathbf{A} = 2\mathbf{\Delta} - \mathbf{A}$. The sign + is crucial. It is associated with the sign constraint in the forces, characteristic of granular matter. This simplified, topological dynamics is set up only to obtain the elementary excitations of the system: even circuits are soft modes with zero eigenvalues, each odd circuit constitutes an extended constraint (an arch) with a finite eigenvalue, that can be eliminated by breaking one contact.

(The potential energy V is not a quantitative measure of a force, but only the statements that grains i and j are in contact, and with opposite stress-free rotations. If the tangential friction is infinite, the contact force is a scalar. For an odd circuit of grains in contact, the contact force is necessarily finite, and applying shear will break the circuit by disconnecting two grains.)

3.2 Graph with Odd Circuits

Any graph containing odd circuits can be bi-colored, simply by removing one edge – (ab), say – from every odd circuit. That is, for each odd circuit, set $A_{ab} = 0$ between the two vertices a and b with the same color $(-1)^a = (-1)^b$ to be separated. The original graph is thus partitioned $\mathbf{A} = \mathbf{A^0} + \mathbf{A}^*$ into a bichromatic, vertex-spanning, edge subgraph described by the adjacency matrix $\mathbf{A^0}$, and the few, removed edges $A^*_{ab} = 1$ between same color vertices, called *critical edges* hereafter. The valency (connectivity) $z_i = \sum_j A_{ij}$ of all vertices is $z_i^0 = \sum_j A^0_{ij}$ except for the vertices that have been separated $z_a = z_a^0 + 1$.

Under the unitary transformation $\mathbf{O}$, $\mathbf{A^0}' = -\mathbf{A^0}$ changes sign, whereas $\mathbf{A}^{*\prime} = \mathbf{A}^*$ remains unchanged because it connects vertices with the same color. Thus, the dynamical matrix $\mathbf{K} = (\mathbf{\Delta} + \mathbf{A^0}) + (\mathbf{1}^* + \mathbf{A}^*)$ is transformed into

$$\mathbf{K}' = (\mathbf{\Delta} - \mathbf{A^0}) + (\mathbf{1}^* + \mathbf{A}^*) = \mathbf{Q^0} + \mathbf{J}^*, \tag{3}$$

where $(\mathbf{1}^* + \mathbf{A}^*) = \mathbf{J}^*$ is a very sparse matrix with non-zero entries 1 for the vertices a, b with the same color connected by a critical edge. $\mathbf{J}^*$ is the direct sum over all odd circuits of the matrix whose entries $J^*_{ij} = 1$ if $i, j = a, b$, and zero elsewhere.

The lowest eigenvalue of $\mathbf{Q^0}$ is zero, with eigenvector $\mathbf{j}$. Thus, the color transformation $O_{ij} = (-1)^i \delta_{ij}$ is a local, gauge transformation that sorts out even circuits by softening them. One obtains an upper bound for the lowest eigenvalue of the full dynamical matrix $\mathbf{K}$ or $\mathbf{K}'$ from the Rayleigh-Ritz variational principle,

$$\begin{aligned} 0 < \lambda_1 &\le \langle \mathbf{j}|\mathbf{K}'|\mathbf{j}\rangle / \langle \mathbf{j}|\mathbf{j}\rangle = \langle \mathbf{a}|\mathbf{K}|\mathbf{a}\rangle / \langle \mathbf{a}|\mathbf{a}\rangle \\ &= \langle \mathbf{j}|\mathbf{J}^*|\mathbf{j}\rangle / \langle \mathbf{j}|\mathbf{j}\rangle = \sum_{\text{odd circuits}} (4/n) = 4c/n, \end{aligned} \tag{4}$$

where c is the number of odd circuits (the number of disconnected contacts) and n the number of vertices in the graph.

The bichromatic network with even circuits, represented by $\mathbf{A^0}$ can itself be partitioned into c disconnected components or "blobs", by removing a few contacts $\mathbf{R}$. Each blob is connected to the next by one single critical edge, so that the set $\{\mathbf{A}\backslash\mathbf{R}\}$, represented by $\mathbf{A}$–$\mathbf{R}$, forms a chain of blobs. Each blob k has only even circuits, with a zero-eigenvalue eigenvector $(\exp[-i\mu^\square])\mathbf{j}^k$, where $j_i^k = 1$ if vertex i belongs to component $k = 0$ otherwise, with a phase $\mu^\square$. It is then easy to construct explicitly an orthonormal set of c "Bloch functions" to generalize (4) to the $\sim c/4$ lowest eigenstates [14].

4 Odd Circuits. R-Loop. Frustration

So far, so good: Odd circuits are the (extended) arches holding the material together and preventing it to flow. The lowest eigenvalue $\lambda_1 \approx 4c/n$ is a measure of the frustration generated by the odd circuits, diluted into the whole connected network. It is also the order parameter of the jammed phase (Fig. 1). The granular material is rigid, but fragile (stresses concentrate on paths – odd circuits – that form a sparse network and can be locally disconnected). To unjam the material, it is sufficient to break one contact per odd circuit. Under an applied shear stress, the disconnection is done efficiently and systematically by the motion (climb) of a topological defect line, the odd vorticity line or *R-loop*. The mechanism is similar to the glide of a dislocation that controls the plasticity of metals [20]. The location of odd circuits and their density are given by an old result [12]:

Odd circuits do not occur individually, but, like the pearls in necklaces, they are traversed by continuous lines of odd vorticity that close as loops or terminate at the surface of the material. These odd lines or *R-lines* are topological defects and sources of frustration in granular materials as well as in covalent glasses, where they were originally introduced.

Theorem ([12]). *R-lines close as loops, or terminate at the surface of the material, without passing through any irreducible even circuit. (A circuit in a graph is irreducible if there is no shorter path between any two vertices on the circuit than a path on the circuit itself. The naive picture of a reducing shortcut holds without ambiguity in granular materials. In covalent (network-) glasses, Wooten has developed an algorithm to draw the network of R-lines, based on partitioning the covalent network into cells bounded by irreducible circuits [21, 22].)*

4.1 Unjamming Under Shear

A granular material subjected to shear will, if it can, break its frustrating odd circuits, with often spectacular consequences (soil liquefaction, sliding tectonic plates, dilatancy, etc.). This is done by breaking c contacts, one per odd circuit, pulling apart the two grains. Since all odd circuits cut across once the surface(s) bounded by the set of R-loops, such a surface is the locus of all the critical contacts to be broken, and c is its area. It is therefore sufficient, and most economical, to separate those contacts located on the (minimal) film(s) attached to the R-loops of the material.

In an ordered granular material (e.g. close packed), the R-loops are small, of a size bounded by the period of the structure. Both c and the number of grains n scale as L^3, and the frustration λ_1 is finite. In disordered granular materials, the R-loops can have the size L of the material, and c scales as L^2, so that the frustration λ_1 (scaling as $1/L$) is small and vanishes at the unjamming transition.

The separation of the critical contacts requires a little space: the unfrustrated packing has less contacts than the original. In practice, this can be done by aerating gently the granular material before the start of the experiment [3], or by applying shear from a boundary [4], as happens naturally in geological faults with sliding tectonic plates or in the phenomenon of dilatancy.

This contact removal is done specifically on the $c \sim L^2$ specific edges traversing the minimal film attached to the largest R-loops of size $\sim L$. It is essential that the largest R-loop has a large size L; the best contact removal process and the partition of the graph are then (nearly) unique and realisable experimentally by shearing [3]. The contacts broken first are the ones nearest to the R-loop, which thereby shrink. The motion of the R-loop is perpendicular to the critical contact, that serves as the Burgers vector for an odd circuit. Thus, the R-loops *climb* under shear, unlike the dislocations in metals or in foams that glide under shear (their motion is parallel to the Burgers vector) [20]. Moreover, because the R-loops are large in disordered granular material, the climb surface is smooth, and unjamming is easy, certainly easier than in ordered granular materials where the R-loops are small and contact must be broken individually. This is in complete contrast to metals or foams, where disorder roughens the glide plane.

4.2 Generic Density of Eigenstates Near Jamming

The jamming transition reveals a large number $\sim c/4$ of low-energy normal modes, besides the lowest frequency one λ_1 [5–7]. These modes can be constructed explicitly, as we have seen [14]. They give rise, in disordered granular materials, to a constant density of states at the jamming transition, independent of the space dimension D and of the size L of the material. Indeed, the number of odd circuits c equals the area of the films bounded by the R-loops, scaling as $\sim L^{D-1}$. There are $c/4 \sim L^{D-1}$ low-energy normal modes (Bloch waves), of energy $\lambda \approx 4c/n \sim 1/L$ given by (4) that vanishes in the thermodynamic limit $L \sim \infty$. The density of states $D(\lambda) = \#\text{modes/vol/energy} \sim L^{D-1}/L^D/(1/L) \sim L^0$ remains finite [7, 14].

The plateau in the density of states at low energies is responsible for the high entropy of the granular packing, and for a large specific heat linear in the temperature (considerably higher at low temperatures than the Debye specific heat that increases like T^3). The elementary excitations remain extended (the eigenvectors, approximated by the Bloch functions have long wavelength) but the stress is concentrated on the frustrating odd circuits responsible for the plateau.

Moreover, the plateau provides the large specific heat (increasing linearly with the temperature) and entropy necessary to absorb the (free) energy and bring about jamming in a hard material without energy sink.

Periodic granular materials do not jam; they crystallize, with R-loops of the size of the period. Indeed, the face-centered cubic packing of spheres (filling space with tetrahedra and octahedra) is periodic, with only small trigonal circuits and small R-loops with one contact each to be broken.

5 Conclusions

Hard granular materials are described by a graph, and their physical behavior (stability, jamming) is obtained generically from the adjacency matrix. Their stability is caused by odd circuits that prevent the grains from rolling on each other. Odd circuits form linear defects (R-loops) that are large in disordered materials. The elastic behavior of isostatic, disordered granular matter is generic: the stresses (contacts to be disconnected) are localized on the odd circuits, but the displacements (soft modes) are felt on the entire network [11]. The material with only even circuits of grains in contact is a dry fluid that does not resist to shear. The jammed phase in disordered granular materials is an intermediate phase between liquid and isostatic solid, that is rigid but fragile.

Hard spheres have infinite repulsive interactions, so that both jamming and crystallisation [23] transitions are caused by the entropy carried by a large number of soft modes. But in the jamming transition, these modes are extended uniformly over the entire material. By contrast, in (fcc or bct) crystallisation, the spheres are rattling independently inside cages of small, trigonal R-loops, the packing is overconstrained and the transition is not a scaling one..

That the jamming transition is a classic, second-order phase transition dominated by scaling has been put forward by Wyart, Nagel and Witten [7], but without the odd circuits (the size of the R-loop) as the mechanism. S. Alexander [24] first pointed out the generic importance of disorder. Odd circuits are the mechanism of the jamming transition: Their density is the order parameter for the jammed phase, and the size of the R-loop, large in disordered granular materials, is the scaling agent.

Discussions with, and suggestions from J.-M. Luck, J.-F. Sadoc, M. Micoulaut, H. Crapo, S. Nagel, V. Pierron-Bohnes and C. Moukarzel are gratefully acknowledged.

References

1. N. Rivier, J. Non-cryst Solids **352**, 4505 (2006).
2. R. Mahmoodi-Baram, H.J. Herrmann, N. Rivier, Phys. Rev. Letters **93**, 044301 (2004).
3. D. Lohse, R. Rauhé, R. Bergmann, D. van der Meer, Nature **432**, 689 (2004).
4. D.A. Huerta, V. Sosa, M.C. Vargas, J.C. Ruiz-Suárez, Phys. Rev. E. **72**, 031307 (2005).
5. C.S. O'Hern, S.A. Langer, A.J. Liu, S.R. Nagel, Phys. Rev. Letters **88**, 075507 (2002).
6. C.S. O'Hern, L.E. Silbert, A.J. Liu, S.R. Nagel, Phys. Rev. E **68**, 011306 (2003).
7. M. Wyart, S.R. Nagel, T.A. Witten, Europhys. Lett. **72**, 486 (2005).
8. P. Boolchand, G. Lucovsky, J.C. Phillips, M.F. Thorpe, Phil. Mag. **85**, 3823 (2005).

9. E. Guyon, S. Roux, A. Hansen, D. Bideau, J.-P. Troadec, H. Crapo, Rep. Progr. Phys. **53**, 373 (1990).
10. C.F. Moukarzel, Phys. Rev. Letters **81**, 1634 (1998).
11. C.F. Moukarzel, in *Rigidity Theory and Applications*, M.F. Thorpe, P. Duxbury, eds., Plenum Press 1999.
12. N. Rivier, Phil. Mag. **40**, 859 (1979).
13. N. Rivier, in *Powder and Grains (2005)*, R. García-Rojo, H.J. Herrmann, S. McNamara, eds., Balkema, Leiden 2005, 29.
14. N. Rivier, Dynamics of jammed granular matter, Eur. Phys. J E (Soft Matter), to appear.
15. J.-F. Sadoc, R. Mosseri, *Geometrical Frustration*, Cambridge Univ. Press (1999).
16. F. Lechenault, *La transition de «jamming» dans un milieu granulaire bidimensionnel: statique et dynamique d'un système athermique modèle*, Thèse, Univ. Paris-Sud XI (2007).
17. J.F. Sadoc, Eur. Phys. J E, **18**, 321 (2005).
18. N. Biggs, *Algebraic Graph Theory*, Cambridge Univ. Press, 1974.
19. R. Mahmoodi-Baram, H.J. Herrmann, Fractals **12**, 293 (2004).
20. N.F. Mott, *La structure atomique et la résistance des métaux*, Dunod 1958.
21. F. Wooten, Acta Cryst. **A58**, 346 (2002).
22. N. Rivier, F. Wooten, MATCH-Comm. in Math. and Computer Chem. **48**, 145 (2003).
23. J.C. Ruiz-Suárez, Phys. Rev. Letters **95**, 018001 (2005).
24. S. Alexander, Phys. Rep. **296**, 65 (1998).

Vertical Granular Transport in a Vibrated U-Tube

Ivan Sánchez[1,2], José Ramón Darias[2], Ricardo Paredes[1], Christopher J. Lobb[3], and Gustavo Gutiérrez[2]

[1] Centro de Física, Instituto Venezolano de Investigaciones Científicas, Apartado Postal 21827, Caracas 1020-A, Venezuela ijsanche@ivic.ve
[2] Departamento de Física, Universidad Simón Bolívar, Apartado 89000, Caracas 1080-A, Venezuela
[3] Department of Physics, University of Maryland, College Park, Maryland 20742-4111, USA

Summary. We present an experimental study of the collective motion of grains inside a U shaped tube undergoing vertical oscillations, and we develop a very simple quantitative model that captures relevant features of the observed behavior. The height difference between the granular columns grows exponentially with time when the system is shaken at sufficiently low frequencies. In vacuum the exponential growth is suppressed.

1 Introduction

Vertical vibration is widely used to excite a granular system, inducing interesting phenomena as pattern formation [1], segregation [2], internal collective motion of the grains, among others [3]. The flow of granular materials through different container geometries is crucial for the improvement of several industrial processes. Tubular containers enhance the role of the walls in the interchange of energy and are extremely important from the industrial point of view. Many industrial processes involve the transport of grains through pipes and are unavoidably affected by mechanical vibrations, therefore their reliability can be greatly affected by the knowledge of the physics of vibrationally induced granular bulk flow [4]. An interesting example of vibrationally induced granular transport in a pipe can be realized in a partially filled U-tube. In this system vertical vibrations can cause the rise of one of the columns of grains until the other one is emptied, a phenomenon often referred to as the U-tube instability. The U-tube instability was first mentioned by Gutman in 1976 [5]. He pointed out that the increase of the difference in level was due to the air pressure difference acting across the bottom of the vibrated tube. Rajchenbach [6], mentioned the U-tube instability and proposed a convection based mechanism to explain it. Ohtsuki et al. [7] investigated the change in the level difference for grains in two different size vessels connected at the bottom, with experiments, simulations and a theoretical model. They found that the conditions favoring the achievement of a level difference were: low frequencies, low accelerations, large difference in chamber widths, small particles, large bed heights and large wall friction. Akiyama et al. [8] studied a similar system as [7], but with compartments of equal size in a rectangular vessel linked at the bottom. They concluded that two mechanisms were behind the level difference achieved: the dissipative structure of the granular system and the formation of a pressure gradient between the bottom of the two sections. Recently King et al. [9] discussed the formation of an instability in U-tubes and partitioned containers and proposed a similar mechanism as Gutman's. All the previously mentioned works are not conclusive, more research needs to be done to elucidate the mechanisms responsible for these instabilities and many open questions are being investigated [9–12].

Here we present an experimental study of a U-shaped tube partially filled with grains, submitted to vertical oscillations. To our knowledge this is the first systematic experimental study of the instability in a U-tube geometry, accompanied by a quantitative model. We discuss the effects of a small horizontal component in the oscillation. We present a simple model based on the idea of cyclic fluidization which captures the dynamics of the height difference between the branches of the tube.

2 Experimental Setup

In Fig. 1 we sketch the experimental setup. We used a glass U-shaped tube whose vertical arms were linked at the top and connected to a vacuum system through two outlets. The tube was fixed with plastic straps to a Plexiglas plate with a base attached to an electromagnetic shaker VTS-150. A function generator provided a sinusoidal excitation signal $a_y \sin(\omega t)$ with $\omega = 2\pi f$, where f is the frequency, a_y is the vertical amplitude and t is time. The signal was fed to an amplified connected to the shaker.

To follow the height difference Δh defined in Fig. 1(b) between the levels of the granular columns, we used a digital video camera and a stroboscopic light tuned at the same frequency as the shaker. Changing the phase difference between the frequencies of the shaker and the strobe we could choose to take the videos at the phase in which the surface of the grains was clearly defined (this phase corresponds to the portion of the oscillation cycle in which the granular system is in contact with the bottom of the tube). A critical point in this study is controlling any horizontal component in the oscillations, due to the mechanical interaction of the system, understanding horizontal as the direction normal to the vertical direction and parallel to the central axis of the bottom of the tube. For that purpose, we used a counterweight at the back of the Plexiglas base, in the form of a pendulum as shown in Fig. 1(b).

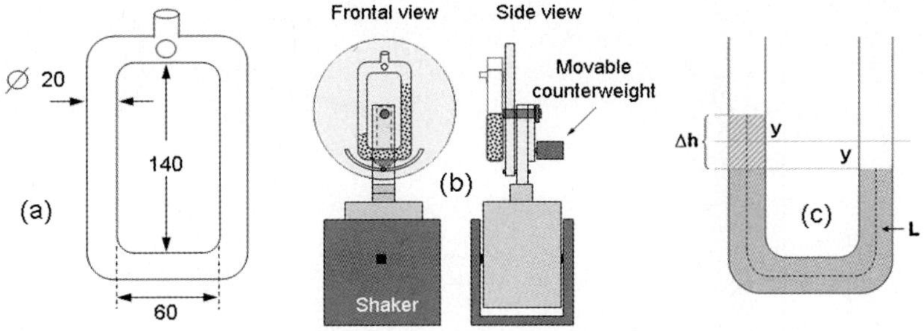

Fig. 1. (a) Dimensions of the glass U-tube (in millimeters). (b) Sketch of the experimental setup used to study the transport of grains in a vibrated U-tube, and position of the movable counterweight used to align the system and maximize the verticality of the oscillations. (c) Definition of the height difference Δh and quantities used in the model.

The angular position of the counterweight could be changed (to change the center of mass of the system) to cancel any parasite horizontal push. Once found the position that maximizes verticality, the counterweight was fixed for each experiment. To monitor the verticality of the vibrations we filmed the vibrating tube while a stroboscopic light was flashed at a frequency slightly lower than the frequency of vibration of the shaker (this was performed right after each experiment). In the majority of the cases, there was a sinusoidal horizontal component in the motion, with the same frequency of the vertical forcing (cases with a non-cyclic horizontal component were not considered since they were non-reproducible). Following a dot marked on the vessel we were able to obtain Lissajous figures for the motion of the container. In this way we could conveniently measure the horizontal amplitude a_x and the phase shift ϕ between the vertical and horizontal displacements. Besides the use of the counterweight, the weight of the base ($\approx$ 1 Kg) was 1 order of magnitude greater than the weight of the granular medium in the tube ($\approx$ 50 g), to avoid any misalignment due the unbalance of the arms. The shaker was firmly attached to a sand casted aluminum base with 3 aligning screws resting on a heavy wooden base on an optical table.

The grains used were spherical glass beads of different diameters d, between 0.18 mm and 0.36 mm, with bulk static bed density $\rho = (1.44 \pm 0.04)$ g/cm^3, and plastic balls with d = 5 mm, $\rho = (0.61 \pm 0.02)$ g/cm^3. We performed experiments at different values of Γ between 1–12, with frequencies between 6–60 Hz and amplitudes lower than 10 mm. Typically, the initial static value of L was 110 mm.

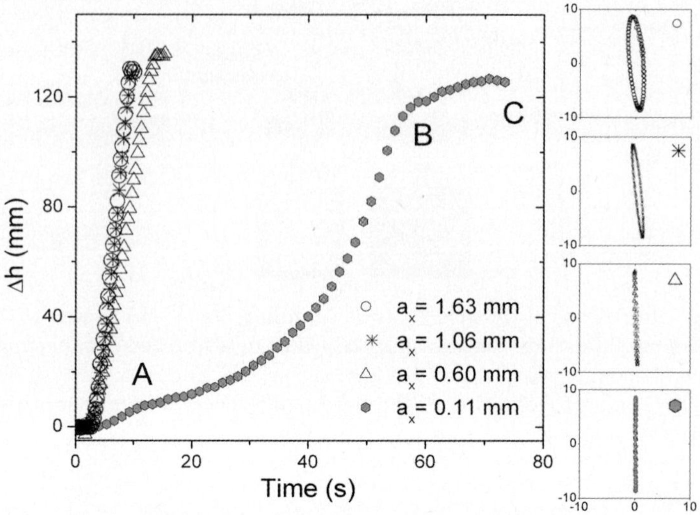

Fig. 2. Behavior of Δh with time, for glass spheres with d = (0.25–0.36) mm, vibrated at f = 8 Hz, vertical amplitude a_y = 8.5 mm and different horizontal amplitudes. Lissajous's figures of the motion of the container corresponding to each curve, are shown at right (position measured in millimeters).

3 Results

We found an exponential growth of Δh with time, for the glass beads vibrated at low frequencies (lower than 30 Hz) and specific values of amplitude (the actual limits depend on grain characteristics and the frequency of oscillation, but there were always minimum (maximum) values of amplitude under (over) which we could not observe the instability). For the cases in which we did not observe an instability we could see small fluctuations around zero of Δh when starting experiments with the columns leveled or exponential decreasing of Δh when starting the experiments with an initial height difference (this was the case for plastic balls through all the range of parameters studied). This exponential decrease is the expected behavior of a completely fluidized system (since it is the behavior of a viscous fluid).

In Fig. 2 we show the effect of the horizontal component in the behavior of Δh as a function of time, for glass beads with d = (0.25–0.36) mm, vibrated at f = 8 Hz and a_y = 8.5 mm. Different curves are produced changing the position of the movable counterweight shown in Fig. 1(b). For the most vertical case, in which a_x/a_y = 0.013, the growth of Δh is exponentially divergent in the region between A and B. As the decreasing column becomes empty, the curve deviates from exponential growth and eventually flattens (region B to C). In Fig. 3 we show the robustness of the exponential behavior. The curves corresponding to cases in which $a_x/a_y > 0.03$, showed a linear dependence of

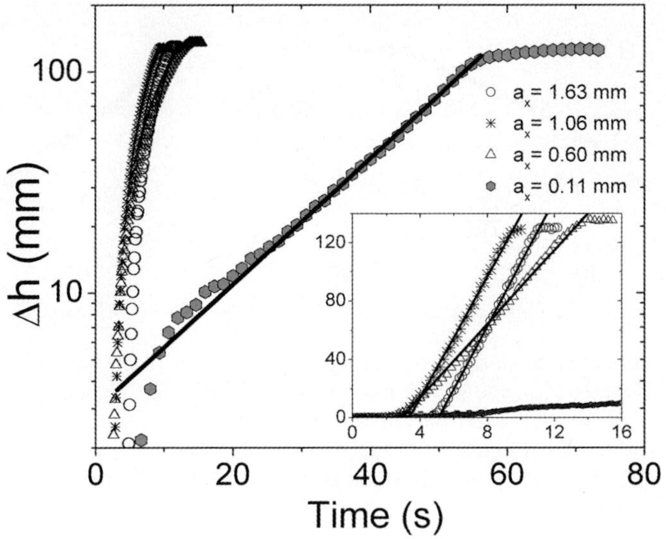

Fig. 3. The same plot as figure 2 in semi log scale, showing the robustness of the exponential behavior for the case of $a_x = 0.11$ mm. The inset shows a detail of the three cases in which there is horizontal component in the oscillation, showing that a straight line fits them well.

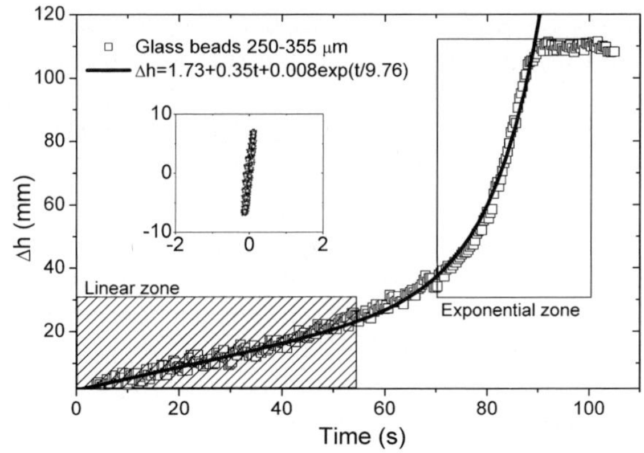

Fig. 4. When the horizontal amplitude is 2% of the vertical one, he curve of Δh with time is well fitted by the sum of an exponential plus a linear term in time.

Δh with time as can be appreciated in the inset of Fig. 3. In Fig. 4, we present an intermediate case corresponding to a rate $a_x/a_y = 0.02$. In this case the dependence of Δh with time is well fitted by an exponential plus a straight line. The curve shows a linear dominant zone at the beginning which soon

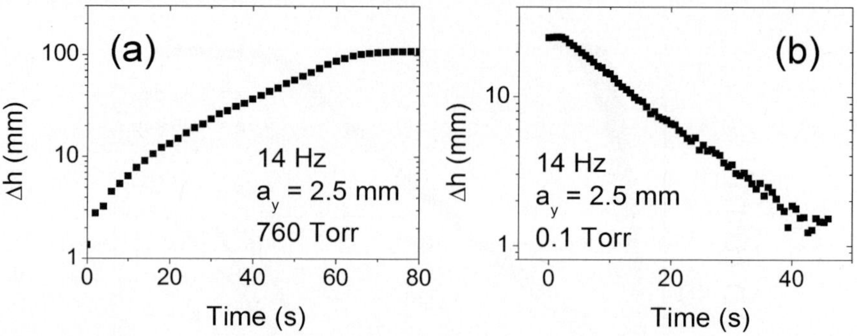

Fig. 5. The role of air is shown for glass spheres with $d = (0.255–0.350)$ mm. (a) Exponentially divergent growth at low frequencies and ambient pressure. (b) Exponential decay of Δh at the same parameters of vibration but with reduced pressure. Here the initial height difference is only of a couple of centimeters, but the decay is also exponential for higher initial height differences. If the experiment begins with arms leveled, there is no growth of Δh.

changes to an exponential dominant behavior until the decreasing column empties.

A clear way to check if the increase in Δh is due to an horizontal component is starting the shaking at a given initial height difference Δh_0 to one side, perform the experiment and repeat it with the same initial difference to the other side. If the growth occurs to the same side in both cases then there is a preferred direction clearly related to an horizontal push (as can be seen from the Lissajous's figures of the tube's displacement). If the growth happens to the higher side in both cases then the horizontal push is beaten by the instability. Even in this case, the time to reach the maximum Δh can be different, due to the fact that one side is being helped while the other is being slowed down by the horizontal component. For a given a_x there is a critical Δh_0^c below which one sees the preferred side and above which, the system increases. Except for the data in Figs. 2, 3 and 4, we report experiments for values of a_x such that Δh_0^c is very small (i.e. $a_x/a_y \leq 0.02$). In this way, starting with the arms leveled, one has to wait sometimes several minutes in order for a fluctuation to trigger the increase of Δh, which can occur to any of the branches of the tube.

The presence of air is necessary for the instability to occur. When columns of small grains are set at different levels, and the air is evacuated, Δh decays exponentially to zero instead of growing (see Fig. 5). This confirms the findings of [8]. Air, or more precisely, the interstitial fluid, plays an important role in many other vertically vibrated granular phenomena [13–20], and is a key factor in the Faraday tilting, a phenomenon that is related to instabilities as the one studied here.

4 Simple Model for Exponential Growth

The exponential growth of one of the grain columns described here, can be thought of as analogous to the draining of a viscous fluid through one of the branches of an inverted static U-tube. If the granular medium were continuously fluidized (as when shaken at sufficiently large frequencies), we would expect a typical fluid behavior: no growth when starting with leveled arms and exponential decrease of Δh when starting with unbalanced arms. This is in agreement with our findings. But at lower frequencies the granular medium does not remains continuously fluidized. It changes from an expanded to a condensed state as it slightly detaches from the container bottom. If we assume that during the expanded phase of the oscillation the system is fluidized, then we have a cyclic fluidization. With this in mind, we can describe the instability with a simple model similar to the one developed to describe the phenomenon of reverse buoyancy by Gutiérrez et al. [14].

The granular package inside the tube can only flow during the fluidized part of the oscillation, during the rest of the cycle it remains blocked. When the system is fluidized we describe it as a liquid. We assume, for simplicity, that the granular bed fluidizes during an interval of time τ that coincides with the portion of the period in which the acceleration given by the container points upward with a magnitude greater than the acceleration of gravity[4]. If we define $g_{ef} = a_y \omega^2 \cos(\omega t) - g$ as the effective acceleration of the system, then we assume that the system is fluidized while $g_{ef} \geq 0$. This effective acceleration will be approximately equal to the one felt by the granular columns, as long as the bed does not separate significantly from the bottom of the tube. The latter assumption holds for fine grains, at low frequency, in the presence of air.

The U-tube, partially filled with grains, has cross sectional area $A = \pi D^2/4$. The total mass of the granular bed is $M_g = \rho AL$ where ρ is the static bed density. The vertical position of the tube is given by $w = a_y \cos(\omega t)$. We take y in the growing side of the bed Fig. 1(c) as the only degree of freedom necessary to describe the motion of the fluidized bed. In the reference frame of the container, the 1D equation of motion for the evolution of $y(t)$ is given by:

$$M_g \ddot{y} = m(y) g_{ef} - \nu \dot{y}, \tag{1}$$

where $m(y) = 2yA\rho$ is the mass of the unbalanced section (striped zone in Fig. 1(c)) and we have included a dissipative term to account for friction losses. We integrate this equation over one period of oscillation to get the average displacement $\bar{y}$, assuming, for this simplified case, that the integral is zero if evaluated outside the interval $[-\tau/2, \tau/2]$, where $\tau = \frac{2}{\omega} \cos^{-1}(\Gamma^{-1})$. This gives us a simple differential equation for $\bar{y}$. If we consider a vertically vibrated U-tube with diameter D, partially filled with a granular medium of

[4] This assumption could be relaxed to consider the actual time that the system fluidizes.

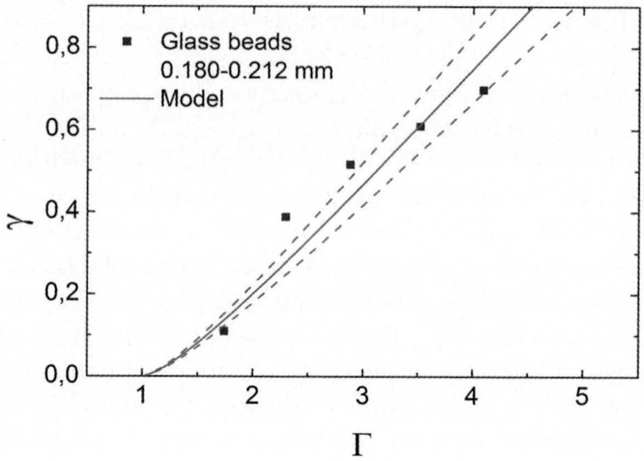

Fig. 6. Exponential growth rate of Δh, as a function of the adimensional acceleration, for glass spheres with $d = (0.18–0.21)$ mm. Three realizations at each value of Γ at a fixed frequency $f = 12$ Hz, changing the amplitude of the vibration. The solid line represents the model approximation Eq. (2), using the friction coefficient ν as a free parameter.

effective density ρ, solving equation (1), we obtain an exponential dependence of y with time, for $\Gamma \geq 1$, with a growing rate γ given by:

$$\gamma = \frac{g\rho D^2}{2\nu}\left[\Gamma \sin\left(\cos^{-1}\left(\frac{1}{\Gamma}\right)\right) - \cos^{-1}\left(\frac{1}{\Gamma}\right)\right]. \qquad (2)$$

Using the friction coefficient ν as the only free parameter we obtain fair agreement between experiment and model as can be seen in Fig. 6. The dependence on the adimensional acceleration Γ of the growing rate γ given by equation (2) appears only within the square brackets, but ρ depends on the adimensional acceleration. This dependence can be incorporated by correcting for the expansion of the granular bed as a function of Γ. Some complications may arise for some configurations because the expansion is not the same in the columns and in the link between them. The friction coefficient may also depend on Γ. The dotted lines in Fig. 6 were introduced to account for a 10% percent variation in the density that occur when the granular bed is shaken according to measurements we performed using a fast camera. In our simplified model, we are not taking into account the fact that the actual fluidization time deviates from τ with increasing Γ [5]. At higher Γ the bed tends to remain fluidized even when $g_{ef} < 0$, lowering $\bar{y}$ in each cycle, consequently lowering the value of γ.

The model proposed is crude and could be improved by considering Gutman's model to incorporate the contribution of the pressure difference across

[5] This can be seen in a rectangular vessel using a simple model like Kroll's to determine the flight time of the granular bed. See [13] for a discussion.

the bottom of the tube [5]. It seems plausible that the pressure difference across the bottom contributes to the transfer of grains between the two columns but it is not at all clear how it can act as a force to drive the larger column up in such a way that the growth of Δh is exponentially divergent. Although the pressure difference across the bottom grows with the height of the rising granular bed, the mass of the same column also grows linearly with Δh, and the resultant forces oppose each other. In our model the force due to an effective gravity points upward and increases with increasing mass, producing an exponential growth. Although the interstitial air does not appear directly in our model, it appears in the basic assumption that the fluidized part of the cycle occurs when $g_{ef} > 1$. In the absence of air, the granular package expands and flights during a longer time than in the presence of air, resulting in a drift of the fluidization time so it does not coincides with the part of the cycle in which $g_{ef} > 1$. The mechanism responsible for the exponential growth of Δh will not be as effective in vacuum as it is in the presence of air.

The fact that the range of parameters for which reverse buoyancy was reported [14] (fine grains, small frequencies and large amplitudes) is the same for which exponentially diverging growth was observed, together with the fact that the assumption of cyclic fluidization was successfully used to model both phenomena, gives support to the idea that cyclic fluidization is playing a key role in the dynamics of these interesting instabilities.

5 Conclusions

We have reported an experimental study of a U-tube partially filled with grains undergoing vertical oscillations. The evolution of the height difference between the levels of the granular material in each side of the tube shows an exponential growth in time when fine grains and low frequencies are used in the presence of air. The growth was modeled assuming cyclic fluidization, showing that this simple mechanism can be responsible for the instability regardless of the fact that wall friction and the formation of a pressure gradient at the bottom of the tube are important parameters, as shown by previous studies and by our own work. Further research is being done, both experimental and computational, to determine the effect of parameters as wall friction in the occurrence of the instability, specially for the case of grains coarser than those used here, in which air effects should not be very important. An extension of the model to include a more realistic fluidization time, and a systematic study of the effect of a small horizontal component in the oscillation are also being carried out.

Acknowledgements

We would like to thank the Decanato de Investigación of the Universidad Simón Bolívar for financial help, Leonardo Reyes for the many helpful discussions, and I. Sánchez would like to thank Leonardo Fermín for his contribution in developing the video analyzing techniques. This work was sponsored by the FONACIT under Grant S1-2000000624.

References

1. I. S. Aranson, L. S. Tsimring, Rev. Mod. Phys. 78, 641 (2006).
2. A. Kudrolli, Rep. Prog. Phys. 67, 209 (2004).
3. H. M. Jaeger, S. R. Nagel, R. P. Behringer, Rev. Mod. Phys. 68, 1259 (1996).
4. T. M. Knowlton, J. W. Carson, G. E. Klinzing, W. C. Yang, Chem. Eng. Prog. 44 (1994).
5. R. G. Gutman, Trans. Instn. Chem. Engrs. 54, 174 (1976).
6. J. Rajchenbach, Eur. Phys. Lett. 16, 2 (1991).
7. T. Ohtsuki, D. Kinoshita, Y. Nakada, A. Hayashi, Phys. Rev. E 58, 7650 (1998).
8. T. Akiyama, K. S. Shinmura, S. Murakawa, K. M. Aoki, Granular Matter 3, 177 (2001).
9. P. J. King, P. Lopez-Alcaraz, H. A. Pacheco-Martinez, C. P. Clement, A. J. Smith, M. R. Swift, Eur. Phys. J. E 22, 219 (2007).
10. T. Akiyama, T. Shimomura, Powder Technol. 66, 243 (1991).
11. Y. Maeno, Physica A 232, 27 (1996).
12. W. Chen, R. Wei, Physics Letters A 244, 389 (1998).
13. L. I. Reyes, I. Sánchez, G. Gutiérrez, Physica A 358, 466 (2005).
14. G. Gutiérrez, O. Pozo, L. I. Reyes, R. Paredes, J. F. Drake, E. Ott, Europhys. Lett. 67, 369 (2004).
15. M. A. Naylor, M. R. Swift, P. J. King, Phys. Rev. E 68, 12301 (2003).
16. P. Biswas, P. Sánchez, M. R. Swift, P. J. King, Phys. Rev. E 68, 050301(R) (2003).
17. X. Yan, Q. Shi, M. Hou, K. Lu, C. K. Chan, Phys. Rev. Lett. 91, 014302 (2003).
18. N. Burtally, P. J. King, M. R. Swift, Science 295, 1877 (2002).
19. H. K. Pak, E. Van Doorn, R. P. Behringer, Phys. Rev. Lett. 74, 4643 (1995).
20. L. Caballero, F. Melo, Phys. Rev. Lett. 93, 258001 (2004).

Dynamics of Dissipative System with Asymmetric Interaction and N-Body Problem for the Emergence of Moving Cluster

Yūki Sugiyama, Katsutoshi Masuoka, and Takahiro Ishida

Department of Complex Systems Science, Nagoya University, Nagoya 464-8601, Japan sugiyama@phys.cs.is.nagoya-u.ac.jp

Summary. Collective motion of self-driven particles is a non-equilibrium dissipative system with asymmetric interaction. Optimal Velocity Model is a minimal model formulated with Newtonian equation of particles in nonlinear asymmetric interaction with dissipative (viscous) term. Through the investigations of OV model we show the general properties in such systems: The inseparable relation between the asymmetry and dissipation. The particle-number N (or density) is a control parameter for the instability of a system. The small-N is large enough degree of freedom in such many-particle systems. They contrast sharply with the energy-momentum conserved systems.

1 Introduction

Dynamics of self-driven particles covers the wide fields in phenomena of dynamical pattern formation of physical, chemical and biological systems. Investigations of such kind of systems can be applied to traffic flow, molecular motors, urban network, internet, river-winding, pedestrians, evacuations, granular media, and collective bio-motions. From the physical viewpoint, they are non-equilibrium dissipative systems with asymmetric nonlinear interaction. A self-driven particle moves by self-driven force, which is not an external force, meaning that such a force breaks 'the action-reaction principle', which results in non-conservation of the total momentum. The inflow of energy provides a self-driven force and the energy flows out from a particle as dissipation. In the many-body system of such particles, a non-trivial macroscopic phenomenon appears in the non-equilibrium balance due to the effect of collective motions.

The general characteristic properties of such systems are, i) dynamical phase transition (bifurcation) to a non-trivial phase, ii) emergence of macroscopic scale (pattern formation), iii) emergence of macroscopic time-scale (rhythm), iv) power law behavior of fluctuations, etc. These aspects fully appear in the mathematical model, so called Optimal Velocity Model [1, 2].

OV model is a minimal 1-dimensional Newtonian system of particles in non-linear asymmetric interaction with dissipative (viscous) term, which is first introduced as a traffic flow model.

2 Asymmetric Interaction in Dissipative System

In such systems, the inseparable relation between the asymmetry of interaction and dissipation in non-equilibrium system is well understood in comparison with the energy and momentum conserved systems of interacting particles. For this purpose we investigate the "general asymmetric OV model" as the following formula.[1]

$$\frac{d^2 x_n}{dt^2} = a\left\{V(\Delta x_n) - W(\Delta x_{n-1}) - \epsilon\frac{dx_n}{dt}\right\}, \tag{1}$$

where x_n is the position of the nth particle, and $\Delta x_n = x_{n+1} - x_n$ is the headway distance. a is a control parameter, which dimension is the inverse of time, called a sensitivity constant. $V(\Delta x_n)$, so called OV-function, determines the interaction with a particle moving ahead. In addition, $W(\Delta x_{n-1})$ is introduced as the interaction with a particle moving behind. In general they are not the same function. If $W = V$, the system is ordinary nonlinear-interacting particle system, where the action-reaction principle is satisfied and the total momentum is conserved. The parameter ϵ is put on the dissipative (viscous) term for the study of the case switching off the dissipation when $\epsilon = 0$. If $W \neq V$ and $\epsilon \neq 0$, the model (1) is the expression of the simplest dissipative system with asymmetric interaction, where neither momentum nor energy is conserved. In the case that $W = 0$ and $\epsilon = 1$, the model is reduced to the original OV model (the extremely asymmetric model). In another case that $W = V$ and $\epsilon = 0$, the model is reduced to the system of both momentum and energy conservation, such as nonlinear-interacting oscillator systems (e.g. Toda Lattice).

We investigate the stability of trivial solution for the model (1), where particles are distributed uniformly with the same distance b (or the average density of particles), and they are stationary or moving with the same velocity. For this purpose we derive the dispersion relation for the linear equation of motion for small deviation $y_n = \exp(ink + zt)$ beyond the trivial solution, where $z(k) = ikz_1 + (ik)^2 z_2 + o(k)^3$ by long wave-length (small k) expansion. The result is

$$\begin{aligned} z(k) &= ik\frac{\{V'(b) - W'(b)\}}{\epsilon} \\ &\quad + k^2\left\{\frac{2\{V'(b) - W'(b)\}^2}{a\{V'(b) + W'(b)\}\epsilon^2} - 1\right\}\frac{\{V'(b) + W'(b)\}}{2\epsilon} + o(k^3), \end{aligned} \tag{2}$$

[1] The model of this type was introduced as the forward-backward looking OV model for traffic flow [3].

where V', W' are the derivatives of V, W, and supposed to be positive. The positive/negative value of the coefficient of k^2-term determines whether the solution is unstable/stable, which is controlled by a, b. In general the stability increases for $W \neq 0$ [3] in comparison with the original OV model ($W = 0$). For the symmetric interaction with dissipation ($W = V, \epsilon \neq 0$) it is absolutely stable for arbitrary b. Inversely say, the asymmetry of interaction induces the instability controlled by b. If we switch off the dissipation keeping with the asymmetry ($W \neq V, \epsilon = 0$), the behavior becomes singular, indicating that the asymmetric interaction is meaningful as long as coexisting with dissipation. The coexistence can lead to a non-trivial dynamical behavior for a dissipative system. If we tune the limit that $W \to V, \epsilon \to 0$ with the ratio fixed as Eq. (3), k^2-term vanishes and the dispersion of harmonic-oscillator chain is reproduced. In this sense, a symmetric interaction with no dissipation (both momentum and energy conserved system) is a special system defined at the singular point in the space of general systems.

$$z(k) = \pm ik\sqrt{aV'(b)} + o(k^3) \tag{3}$$

3 Dynamics of OV Model

3.1 Dynamical Phase Transition and Emergence of Moving Cluster

Here we investigate the original OV model for the purpose of studying the general properties of non-equilibrium dissipative system with asymmetric interaction. The model has the homogeneous flow solution, in which particles are moving uniformly distributed with the same velocity $V(b)$, where $b = L/N$ is the average particle-density^{-1} (N is the number of particles and L is the length of a circuit). The homogeneous flow is linearly unstable under the condition [1, 2]

$$2V'(b) > a, \tag{4}$$

which is derived from Eq. (2) with $W = 0, \epsilon = 1$. In this case, the cluster (jam) is spontaneously formed as shown in Fig. 1 (Left). After relaxation the cluster flow solution is stable. All clusters are moving backward with the same velocity opposed to the direction of the particle-motion.

3.2 Profile of Cluster Flow Solution and Induced Time-Delay

We recognize the profile of cluster flow solution by the trajectory of particles in the phase space of headway and velocity $(\Delta x_n, v_n)$ in Fig. 1 (Right). In the cluster flow solution all particles are moving along the specific closed curve, which is a kind of limit cycle [1, 2]. The limit cycle shows the non-equilibrium balance of in-and-out flow of particles through a cluster. Particles move as the same way with the time-delay τ, which is a characteristic time-scale induced

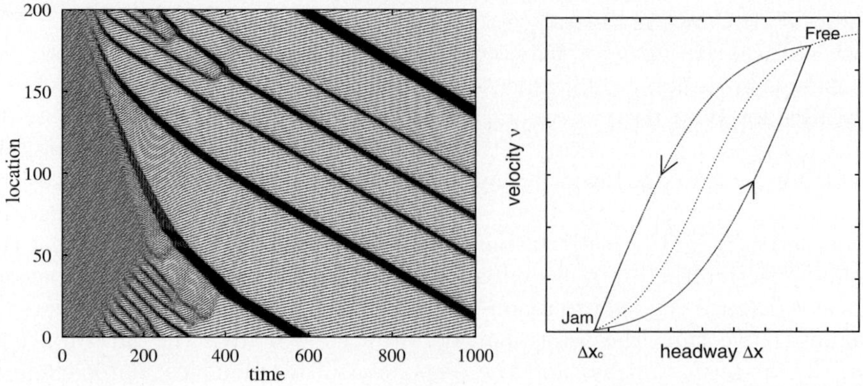

Fig. 1. (Left): The evolution of cluster formation $N = 100$ on periodic boundary condition on a lane. (Right): The profile of cluster flow solution. 'Free' and 'Jam' denote smoothly driving region and jam cluster, respectively. All vehicles move along the closed loop in the direction of arrow. A dotted curve represents OV-function.

by the effect of collective motions. It determines any characteristic behavior of moving cluster. As the result, a macroscopic object (cluster) is formed and moves with its own velocity, $v_c = -\Delta x_c/\tau$, where Δx_c is the particle-distance in a cluster as Fig. 1 (Right).

3.3 Power-Law Behavior

Clusters generated in OV model have various sizes. Figure 2 shows the probability distribution of cluster-size, which is defined by the number of particles in a cluster. $P(x)$ denotes the probability for the existence of cluster size larger than x. In the simulations of $N = 500 \sim 3000$ we observe the power-law behavior as $P(x) \propto x^{-\gamma}$, where $\gamma \sim 1$.

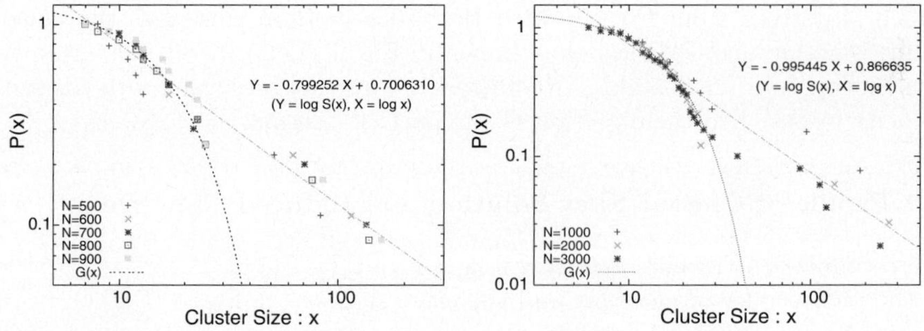

Fig. 2. The vertical axis is a log-plot of $P(x)$, and the horizontal axis is the size of cluster. A Straight line denotes a power-law, comparing with the reference curve of Gaussian distribution. (Left): $N = 500 \sim 900$. (Right): $N = 1000 \sim 3000$.

4 N-Body Problem in OV Model

The N-body problem for the emergence of moving cluster is studied both analytically and numerically. The investigation for $N = 2, 3, 4, \ldots,$ provides the general information for the dependence of the particle-number N in dynamical behaviors of non-equilibrium dissipative system.

4.1 Instability Condition

The instability of the trivial solution (the uniform flow solution) is analyzed in detail for each mode of longitudinal wave of density. The condition whether a mode makes the trivial solution unstable is

$$a < 2V'(b)\cos^2\frac{k}{2}, \; k = \frac{2\pi}{N}n, \; (n = 0, 1, 2, \ldots, N-1), \tag{5}$$

where k is a wave number (a mode of wave-length^{-1} for a frequency of density). The solution is unstable even if only one mode exists (except zero-mode) in the unstable region (shaded region) in Fig. 3 (Left).

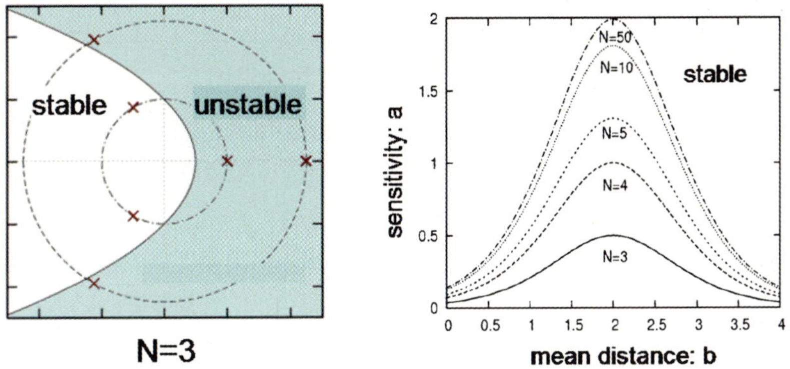

Fig. 3. (Left): Each mode is presented in the polar coordinates $(r, \theta) = (1/a, k)$, in $N = 3$ for two different values of a, for example. The critical curve is $1/r = 2V'(b)\cos^2\theta$. A mode within the shaded region makes the solution unstable. Remark that zero-mode ($k = 0$) has no relation to instability. (Right): Phase diagram of stable/unstable region in the parameter space (a, b) for the trivial solution for give N. Critical curve is derived from the condition for the existence of unstable mode.

For $N = 2$ it is impossible for any mode satisfying the condition. In contrast if $N \geq 3$ such a mode can exist for taking a small enough value of sensitivity parameter a (large radius). The instability can occur in the system that consists of only three particles, $N = 3$ as Fig. 3. The unstable region in control-parameter space (a, b) extends as N increases. We point that the difference for N shrinks rapidly. (Compare the cases for $N = 10$ and 50.) For large N even a tiny $k \sim 0$ (the long wave-length) plays a role of unstable mode, then the limit $k \to 0$ for Eq. (5) provides the instability condition as Eq. (4).

4.2 Emergence of Moving Cluster for $N = 3$

The instability of a uniform flow of particles emerges even for $N = 3$. In this case the system transits to the other state, which is a non-trivial stable state in non-equilibrium system, the state of cluster flow. Actually, we observe the "cluster" flow in Fig. 4. The profile of the solution is clearly shown as a limit cycle, as essentially the same as the cluster flow for the system of large-N particles in Fig. 1.

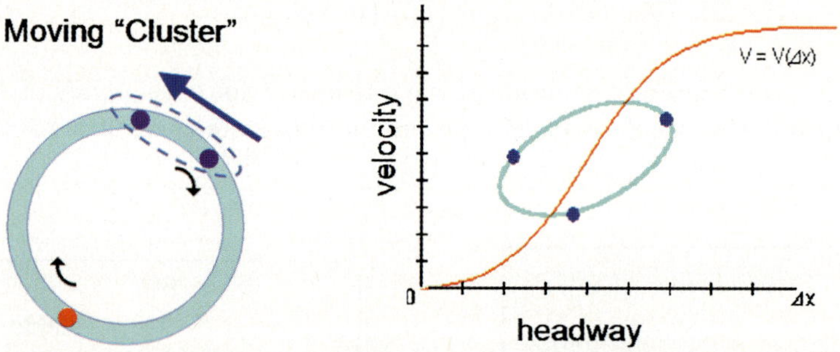

Fig. 4. (Left): A snapshot of a two-particle cluster moving backward opposed to the direction of particle movement. (Right): A limit cycle profile of a cluster flow solution in $N = 3$, together with OV-function.

4.3 Exact Solution of Moving Cluster Flow

Moreover, the N-body problem for the emergence of moving cluster is exactly solved in $N = 2, 3, 4, \ldots$, for choosing the Heavyside step-function as OV-function, such that $V(\Delta x) = 0$ for $\Delta x < d$ and $V(\Delta x) = v_{max}$ for $\Delta x \geq d$. We skip the derivation of exact cluster-flow solutions, which is presented in another paper [5]. Here, we provide the several important results as follows.

Figure 5 shows the trace of each particle in cluster flow solution in $N = 3$ for given a and b. Three particles behave just the same way with the induced time-delay τ, which is the most important value for cluster flow solution. We clearly observe a cluster propagating backward opposed to the motion of particles. The solution of trajectory of a particle has to be calculated one by one for each N. However, the τ in arbitrary N is obtained as the solution of $a\tau$ (dimensionless value) for the following general equation,

$$a\tau(1 + \mathrm{e}^{-\frac{N}{2}a\tau}) = 2(1 - \mathrm{e}^{-a\tau}). \tag{6}$$

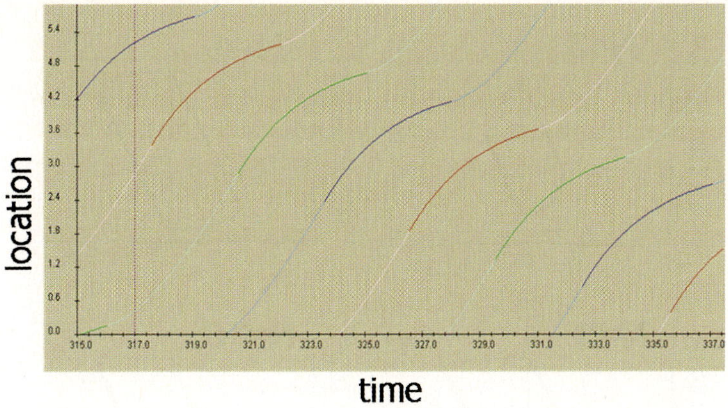

Fig. 5. The trace of each particle after relaxation in $N = 3$. The bold part of each trace represents the interval for $\Delta x < d$. The moving cluster is clearly observed.

If we put the limit $N \to \infty$, the equation (6) is reduced to the result of the previous work, which solution is $a\tau = 1.59..$ [4]. In general the solution $a\tau$ depends on N as shown in Fig. 6. For $N = 10$ the value $a\tau$ is nothing different from $N = \infty$. Thus, $N = 10$ is large enough number for quantitatively the same behavior as the system of infinite-number of particles.

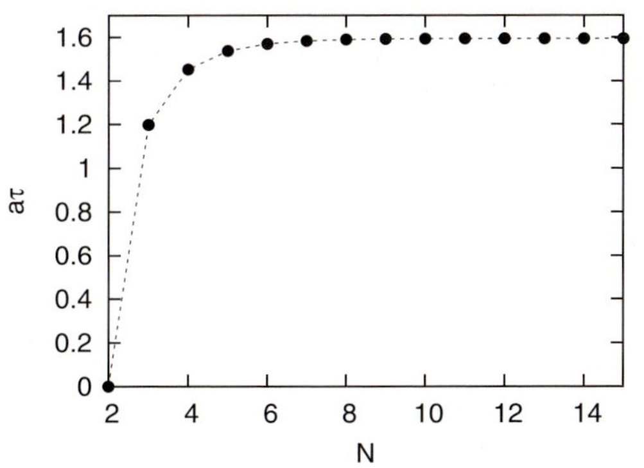

Fig. 6. N-dependence of the solution $a\tau$ of Eq. (6)

The backward-velocity of a cluster v_c is obtained by using the value of τ for given a as

$$v_c = -\frac{d - \frac{1}{2}v_{max}\tau}{\tau}. \tag{7}$$

The formula Eq. (7) is just the same as the previous result [4].

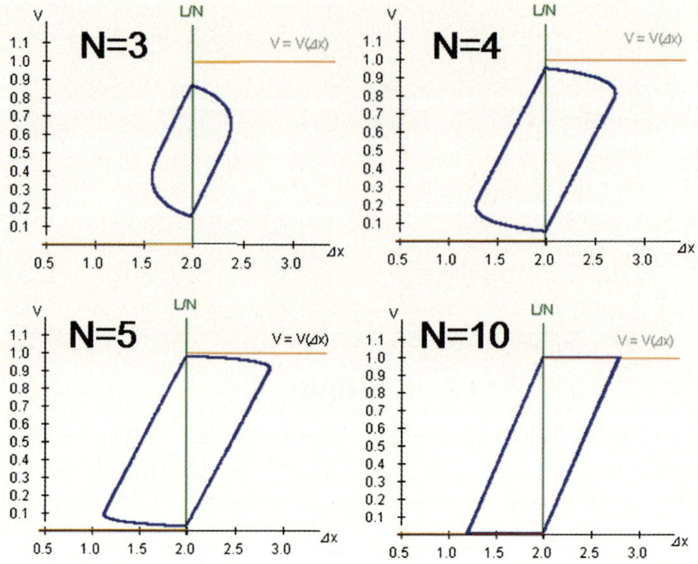

Fig. 7. The dependence of the number of particles N for limit cycle in the case for $b = d$. Heavyside step-function as OV-function is drawn together.

Figure 7 shows the limit cycles for given N. We remark that the shape of limit cycle is dependent on the number of particles N and the density b.[2] The difference among limit cycles shrinks rapidly as N increases. Actually the limit cycle for $N = 10$ is nothing different from $N = \infty$, which is already obtained in the previous work [4]. For large enough N the limit cycle is unique for given a, not depends on b.[3] We recognize that $N = 10$ is already large enough in the OV model by Fig. 7 as well as Fig. 6. The dynamical behaviors of the system of OV model for "small" number of particles are the same as many-body system.

5 Summary and Discussion

Summarizing the results of investigating OV model in previous sections, we can guess and provide the characteristic properties for a non-equilibrium dissipative system with asymmetric interaction in general, comparing with an equilibrium system with symmetric interaction (an energy-momentum conserved system). There exists the inseparable connection between the asymmetry of interaction and dissipation. If the dissipation is switched off and only the asymmetry of interaction exists, the system is singular. While, the

[2] The qualitative difference depends on $b > d$, $b = d$, $b > d$ for a given N. Any difference vanishes for infinite N.

[3] We remark that the stability of limit cycle depends on b.

dissipation with symmetric interaction leads the behavior of a system to a stationary state, which is not interesting for non-equilibrium dynamics. If and only if both the asymmetry of interaction and dissipation coexist, the system shows a non-trivial dynamics in non-equilibrium physics.

In the dissipative system with the asymmetry of interaction, a particle-density or a number of particles in the system is a control parameter, in contrast it is not in the conserved system. Instability of a trivial state and dynamical phase transition emerge even in small-N particle system (e.g. $N \geq 3$ in OV model). In contrast, in the equilibrium and conserved system, phase transition appears in large-N, strictly say, in the system with infinite degree of freedom. A small number of particles in the system investigated in this paper, is large enough number of particles in the meaning of quantitatively the same behavior for the system of infinite-number of particles. The phenomenon is characteristic for the dynamics of dissipative non-equilibrium system, which is a sharp contrast comparing with equilibrium systems. These basic properties are very important as new physical aspects appeared in non-equilibrium dissipative system with asymmetric interaction.

References

1. M. Bando, K. Hasebe, A. Nakayama, A. Shibata and Y. Sugiyama, Phys. Rev. E **51**, pp. 1035 (1995); Japan J. of Ind. and Appl. Math. **11**, pp. 203 (1994).
2. Y. Sugiyama, Proceedings of International Workshop of 'Traffic and Granular Flow' in Juelich edited by D. E. Wolf, M. Schreckenberg and A. Bachem, (World Scientific, Singapore), pp. 137-149 (1996).
3. A. Nakayama, Y. Sugiyama and K. Hasebe, Phys. Rev. E **65**, pp. 016112 (2001).
4. Y. Sugiyama and H. Yamada, Physical Review E **55**, pp. 7749 (1997).
5. Y. Sugiyama, K. Masuoka and T. Ishida, in preparation.

Leidenfrost Granular Flows

Nicolas Taberlet[1], Patrick Richard[2], and Renaud Delannay[2]

[1] Université de Lyon, Laboratoire de Physique, ENS Lyon & UFR de Physique, UCBL, 46 allée d'Italie, 69 007 Lyon, France nicolas.taberlet@ens-lyon.fr
[2] GMCM, Université Rennes 1, Campus Beaulieu, 35 000 Rennes, France

Summary. This paper presents numerical findings on rapid 2D and 3D granular flows on a bumpy base. In the supported regime studied here, a strongly sheared, dilute and agitated layer spontaneously appears at the base of the flow and supports a compact packing of grains moving as a whole. In this regime, the flow behaves like a sliding block on the bumpy base. In particular, for flows on a horizontal base, the average velocity decreases linearly in time and the average kinetic energy decreases linearly with the travelled distance, those features being characteristic of solid-like friction. This allows us to define and measure an effective friction coefficient, which is independent of the mass and velocity of the flow. This coefficient only loosely depends on the value of the micromechanical friction coefficient whereas the influence of the bumpiness of the base is strong. We give evidence that this dilute and agitated layer does not result in significantly less friction. Finally, we show that a steady regime of supported flows can exist on inclines whose angle is carefully chosen.

1 Introduction

Granular flows can display a wide variety of behaviors ranging from solid-like to gas-like and a great deal of work has been devoted to steady and fully-developed dense flows [1–4]. Here, using 2D and 3D molecular dynamics simulations we study transient granular flows on a horizontal base as well as the steady regime obtained on inclines. We focus on the supported regime, in which a strongly sheared, dilute and agitated layer spontaneously forms at the base of the flow and supports a compact packing of grains moving as a whole. This regime has been described in the geophysical literature [5–8] and in the kinetic theory framework [9], but these authors did not study either the dynamical properties of the flow or the arrest of the flow. Moreover, the influence of the bumpiness and inclination of the base remains unexplored both experimentally and numerically. The density inversion observed in supported flows is similar to the Leidenfrost effect observed when a drop of water is poured onto a hot plate. A thin layer of gas develops underneath the drop and considerably reduces the apparent friction between the drop and the plate.

Therefore, one can wonder if supported granular flows are self-lubricated. Let us mention that a density inversion has also been observed in the case of vertically shaken granular media [10, 11].

Note that an impressive number of parameters (among which grain mass, size, density, shape, polydispersity, Young modulus, restitution and friction coefficients as well as total number of grains, initial velocity direction and amplitude, initial granular temperature, initial packing fraction, bumpiness, polydispersity, order/disorder and inclination of the base) are expected to have a crucial role on the properties of the flow. Hence, it seems rather difficult to plot the simultaneous influence of all parameters. As a consequence, we decided to keep most of these parameters constant while varying those of highest interest (namely friction coefficient, total number of grains, initial velocity, bumpiness of the base and inclination).

The outline of this paper is as follow: first, the numerical method is described. Second, we present the properties of supported flows (namely the amplitude of initial velocity, packing fraction and granular temperature profiles). The temporal evolution of the supported regime is then presented and the dynamics of the arrest of the flow studied. We study the influence of the micromechanical friction and of the bumpiness of the base. Finally, we present results on inclined flows.

2 Numerical Method

This study uses 2D and 3D soft-grain molecular dynamic simulations which include friction and rotation. We model N spheres (ranging from a few up to a few tens of thousands) slightly polydisperse spheres of mean diameter, d, (size dispersion $\pm 20\%$ with constant density) and mass, m. The spring-dashpot and regularized Coulomb schemes [12] are used for the forces acting between two colliding grains. The following values of the parameters are used both in 2D and 3D: $d = 8$ mm, mass $= 0.16$ g, spring constant $k_n = 40\,000$ N.m^{-1}, viscous damping $\gamma_n = 1.2$ kg.s^{-1}, viscous regularization constant $\gamma_t = 5$ kg.s^{-1} and unless otherwise mentioned, friction coefficient $\mu = 0.3$, leading to a coefficient of restitution of $e_n = 0.9$. The equations of motion for the translational and rotational degrees of freedom are integrated with a velocity-Verlet scheme with a time step $\delta t = 1.5\,10^{-6}$ s [12].

The simulated system consists of a cell which can be inclined with an angle θ, with periodic boundary conditions in the x (flow) direction, y being perpendicular to the base. The length of the cell along x is chosen to be $L = 25\,d$ on Fig. 1 but is set to $L = 250\,d$ for more accurate measurements in the rest of the paper. The 3D simulations have periodic boundary conditions both in the flow and depth direction, and the depth of the cell, W, is also set to $25\ d$. We observed that the value of L and W does not affect the results as long as they are greater than a few particle diameters. We define the normalized number of grains, $\tilde{N} = N/(L/d)$ for 2D simulations and $\tilde{N} = N/(L\,W/d^2)$

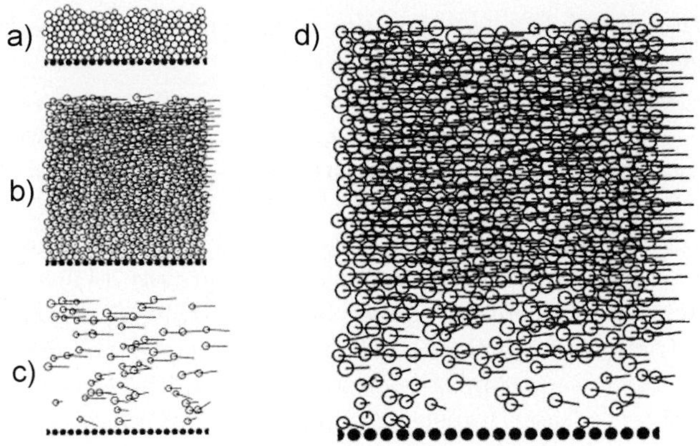

Fig. 1. Snapshots from the 2D simulation for different values of $\tilde{N}$ and V_0 on a horizontal bumpy base. a) Quasi-static regime, $\tilde{N} = 6.4$ and $V_0 = 0.2$ m/s. b) dense flow, $\tilde{N} = 24$ and $V_0 = 4$ m/s. c) gaseous regime, $\tilde{N} = 2.4$ and $V_0 = 100$ m/s. d) supported regime, $\tilde{N} = 24$ and $V_0 = 16$ m/s.

for 3D simulations, which is a measure of the initial height of the packing. The bumpy base of the cell consits of cylinders oriented perpendicularily to the flow direction. Their diameter d is the average diameter of the grains and, unless otherwise mentioned, the gap between the centers of two consecutive cylinders is set to $\delta = 1.25\ d$. The interactions between the flowing grains and the bumpy base are treated in the same fashion as the grain/grain interactions, i.e. using the same values of friction coefficient and energy restitution. The grains are initially placed on a compact square (2D) or cubic (3D) grid in the cell and given an initial velocity V_0, which is identical for all the grains and is parallel to the bottom of the cell (i.e. to the x direction).

As mentioned above, many parameters can drastically affect the properties of the flow and a comprehensive study of the inter-correlated influences of all parameters seems unfeasible. The properties of individual grains (mass, size, shape, Young modulus, energy restitution, polydispersity) are kept constant with the exception of their friction coefficient μ_{micro}. The initial velocity is always parallel to the base of the flow and uniform among the grains (i.e. no initial agitation) and the initial packing fraction always corresponds to a compact cubic (or square) grid. The bumpiness of the base is crucial in this study but once again it requires a very large number of parameters to be characterized. The base can consist of spheres or cylinders (or any other shape), can be ordered or disordered (with different degrees of disorder), can be monodisperse or polydisperse (with an infinite variety of size distribution), can be bumpy or flat (with different degrees of bumpiness) etc. Therefore, for simplicity's sake, the only parameter varied regarding the base is the distance between two consecutive cylinders δ. In summary, the only parameters that were varied in our simulations are the number of grains, $\tilde{N}$, the initial velocity,

V_0, the friction coefficient, μ_{micro}, the bumpiness of the base, δ, and the inclination of the cell, θ.

3 Flow Structure and Kinematics

In this section, the cell is kept horizontal in order to study the arrest of the flow. In this case there is no source of energy (gravity being perpendicular to the bottom) and the initial energy of the packing dissipates during the flow through friction and inelastic collisions. Therefore, the flow is bound to come to a rest after some time. Depending on the initial kinetic energy of the flow, K_0, we observed four transient regimes in our numerical simulations, as illustrated by Fig. 1 for 2D flows. For extremely low values of the initial kinetic energy, the system rapidly comes to a rest and a dense and static pile of grains is observed (Fig. 1a). For slightly higher values of the initial kinetic energy, dense flows that are similar to those studied by Prochnow [13] and Silbert *et al.* [4] are found (Fig. 1b). The packing fraction of such flows is found to be constant throughout the whole medium [4, 13]. On the contrary, for very large values of the initial kinetic energy the flows are extremely agitated and dilute (Fig. 1c). For intermediate values of K_0 (roughly 10 m/s $< V_0 <$ 50 m/s in 2D for $\mu_{micro} = 0.3$, $e_n = 0.9$ and $\tilde{N} = 24$), a remarkable regime can be observed. Since the initial packing is loose, the medium compacts when the simulation is started. This initial vertical speed creates a bouncing motion which is visible on Fig. 6. However, the bumpy base can cause a strong agitation (if the velocity of the grains is high enough) at the base of the flow. The flow then consists of a compact layer of grains supported by a very agitated gaseous one (Fig. 1d). This regime is thus extremely heterogenous, both structurally (the packing fraction is not uniform and depends on the position) and kinematically (contrary to the rest of the flow, the basal layer is extremely sheared and agitated). After the flow has slowed down, the basal layer vanishes and the flow becomes dense. Note that a high value of e_n is necessary to obtain the supported regime. This value is set to 0.9 in this paper but previous work has investigated lowers values [8].

Among the characteristics of a granular flow, the packing fraction, velocity and velocity fluctuations are the most commonly studied. Figure 2 displays instantaneous profiles of those quantities averaged over the x direction for 2D flows and figure 3 shows a snapshot and velocity and packing fraction for a 3D flow. Regarding the velocity (Fig. 2a, Fig. 3b), the first observation is that an important sliding velocity exists at the base, whereas in the case of dense flows, such a sliding velocity does not exist (except for flat bottoms). This profile also clearly demonstrates that the basal layer is extremely sheared. Above this layer, a plug flow, characterized by a uniform velocity, is observed. As expected, the packing fraction is high and almost constant in the latter region and its value is close to the packing fraction of the random close packing [14] (see Fig. 2b). With decreasing y, the packing fraction vanishes in the sheared

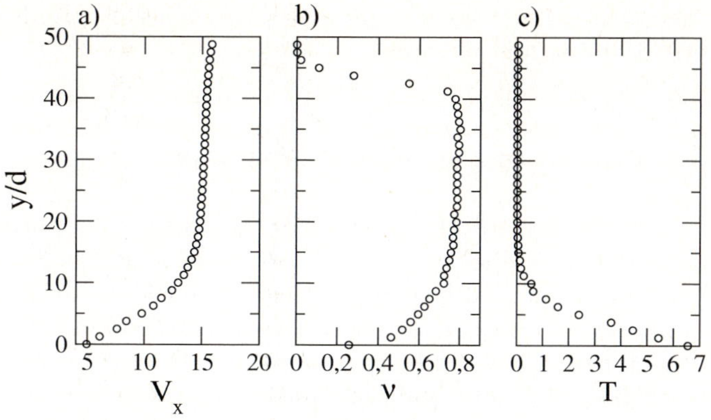

Fig. 2. Instantaneous velocity (a), packing fraction (b) and granular temperature (c) profiles for 2D flows on a horizontal base with $\tilde{N} = 40$ and $V_0 = 20$ m/s, taken after 1.5 s of running time and averaged over x.

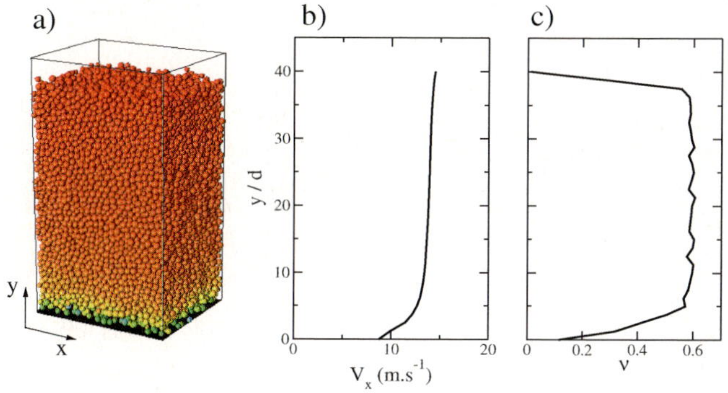

Fig. 3. Snapshot (a) and instantaneous velocity (b) and packing fraction (c) profiles for 3D flows on a horizontal base with $\tilde{N} = 40$ and $V_0 = 20$ m/s, taken after 1.5 s of running time and averaged over x. The color represents the velocity along the x direction.

layer to an almost zero value at the bottom. Another quantity of interest is the fluctuation of the velocity along x, measured by the granular temperature, $T = \langle (V_x - \langle V_x \rangle)^2 \rangle$, which is a measure of the agitation in the flow. Note that in the kinetic theory, the correct definition of the granular temperature involves all three directions and has a 1/2 prefactor but for simplicity, we will use the afore-mentioned definition. The profile reported on Fig. 2c confirms that the basal layer, contrary to the top one, is indeed agitated. The system is "heated" by the bumpy bottom and the dense part of the flow serves as a cold source. Note that the gaseous layer does not propagate through the flow: a pseudo-equilibrium can be found between the cold and the hot source (i.e. the height of the gaseous layer remains constant during a few seconds, see

Fig. 5). In the next section we will show how the velocity profile evolves as the flow slows down on a horizontal.

4 Time Evolution of the Flow Structure

Let us now study the dynamical properties of the arrest of supported flows on a horizontal base. Figure 4 shows velocity profiles obtained at different times for a 3D flow on a horizontal base with $\tilde{N} = 40$ and $V_0 = 16$ m/s. Initially, the flow is a plug flow but the basal layer appears rapidly. As the grains slow down, the velocity profile are more and more curved (a to e) and eventually the basal layer disappears and the flow becomes dense (f) until it completly stops (i).

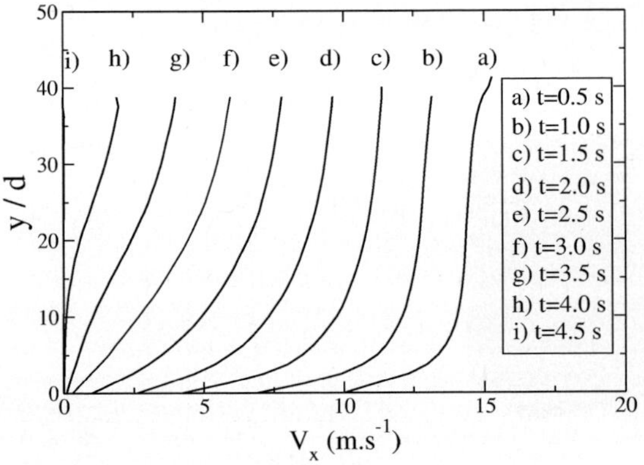

Fig. 4. Velocity profiles at different times for a 3D flow with $\tilde{N} = 40$ and $V_0 = 16$ m/s.

An interesting way to track the history of the supported regime is to define the height of the gaseous layer h_c. The packing fraction of a disordered 2D compact granular packing is always close to 0.82 [14]. The height h_c can then be defined as the height for which the packing fraction deviates from this universal value. In our simulations, h_c is defined as the height for which the packing fraction is less than 0.75. This value is arbitrary but the use of other reasonable critical values (namely 0.65 and 0.7) gave similar qualitative results. Figure 5 reports the evolution of the height of the basal layer, h_c, in time and the corresponding average velocity evolution for a 2D flow with $\tilde{N} = 48$ and $V_0 = 50$ m/s. One can observe that h_c first decreases until it reaches a plateau corresponding to about $h_c \approx 13\,d$. After some time (a few seconds), the basal layer disappears, and the flow, no longer supported, becomes dense. Note again that the transition between the supported and dense regime is rather abrupt.

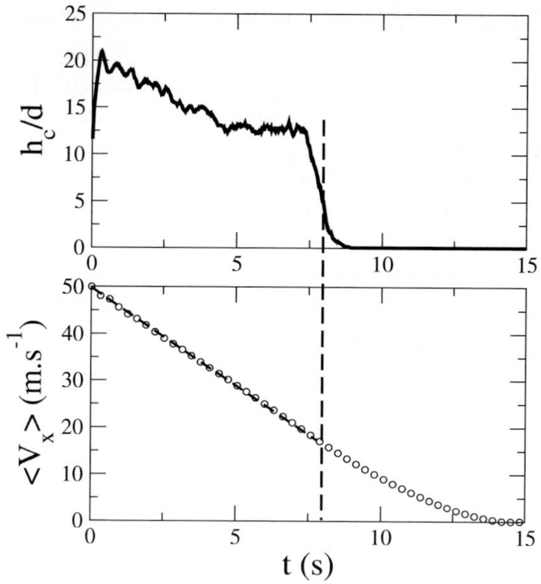

Fig. 5. Time variation of the height of the gaseous layer h_c and of the average velocity for a 2D flow with $\tilde{N} = 48$ and $V_0 = 50$ m/s. The basal layer disappears after roughly 8 s and the flow goes from supported to dense. The velocity decreases linearly in time for as long as the flow is supported (the dashed line is a linear fit to this part of the curve).

5 Energy Dissipation and Effective Friction Coefficient

5.1 Sliding-Block Analogy

One can observe on Fig. 5 that for as long as the flow is supported the velocity decreases linearly in time. The linear decrease of the velocity in time indicates that the flow behaves like a sliding block interacting with the base through solid friction. One can then define an effective friction coefficient μ_{eff} by: $dV/dt = -g\,\mu_{eff}$, where V is the average velocity per grain and g the acceleration due to gravity. When the flow becomes dense, the velocity no longer varies linearly in time, indicating that the interaction between the flow and the bumpy base is no longer sliding block-like. In conclusion, in the supported regime, and not in the dense regime, the flow behaves like a sliding block and one can define a constant effective friction coefficient. Note that this last statement is true for both 2D and 3D flows. Moreover, if supported flows behave like sliding blocks, the kinetic energy must decrease linearly with respect to the travelled distance, and the decrease rate must be independent of the initial energy: $dK/dD = -\mu_{eff}\, m\, g$, where m is the mass of the sliding block.

The average travelled distance, D, is obtained by integrating the velocity: $D(t) = \int_0^t V(t)dt$. In our simulations, most grains belong to the supported part of the flow. Indeed, the basal layer is rather thin and very dilute. The average travelled distance is therefore the distance that most grains have actually travelled. This allows us to plot the spatial evolution of the kinetic energy: $K(D)$. Figure 6 shows such plots for various values of the initial velocity and number of grains (i.e. total mass). The energy is made dimensionless by mgd and the travelled distance by d so that the slope of the curve is directly equal to the effective coefficient of friction.

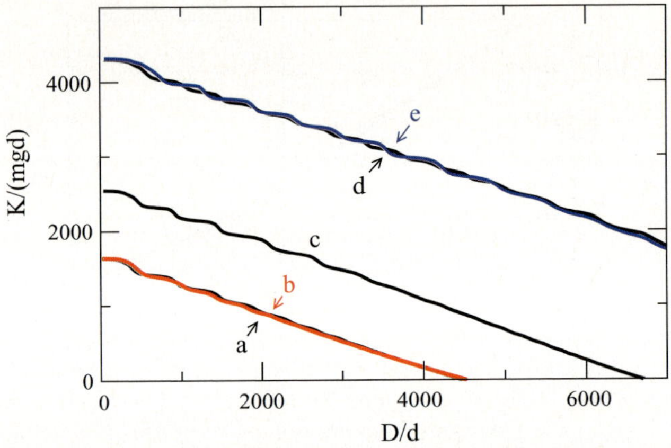

Fig. 6. (in color online) Spatial evolution of the total kinetic energy of 2D flows with various values of $\tilde{N}$ (i.e. total mass of the flow) and V_0. a) $\tilde{N} = 40$ and $V_0 = 16$ m/s, b) $\tilde{N} = 60$ and $V_0 = 16$ m/s, c) $\tilde{N} = 40$ and $V_0 = 20$ m/s, d) $\tilde{N} = 40$ and $V_0 = 26$ m/s, e) $\tilde{N} = 60$ and $V_0 = 26$ m/s.

Several conclusions can be drawn from Fig. 6. First, it confirms what was inferred from Fig. 5, that is that supported flows exhibit solid-like frictional properties since the various curves are straight lines. Second, the various lines are parallel, i.e. all slopes are identical, which means that, as in the case of a sliding block, the friction coefficient is independent of the velocity and mass. This is an important result since in the case of dense flows, the frictional properties strongly depend on the flow thickness [2]. Third, the oscillations that can be seen correspond to a bouncing motion of the dense packing on the supported layer. Note that these oscillations disappear for lower values of e_n. Finally, the value of μ_{eff} can be computed and one finds from Fig. 6, $\mu_{eff} \simeq 0.41$. This value seems rather high considering that the micromechanical friction coefficient was set to $\mu_{micro} = 0.3$.

5.2 Influence of the Micromechanical Friction Coefficient

The effective friction coefficient is a measure of the loss of kinetic energy per unit length (or velocity per unit time). Two mechanisms contribute to the dissipation of the kinetic energy in our simulations: on one hand, the sliding friction between individual grains (including those of the bumpy bottom), on the other hand, the inelasticity of the collisions. One way to identify the contributions of these two mechanisms to the overall loss of energy is to vary one while keeping the other constant. We chose to vary the value of μ_{micro} while keeping $e_n = 0.9$. The results are presented on Fig. 7 for 2D and 3D flows.

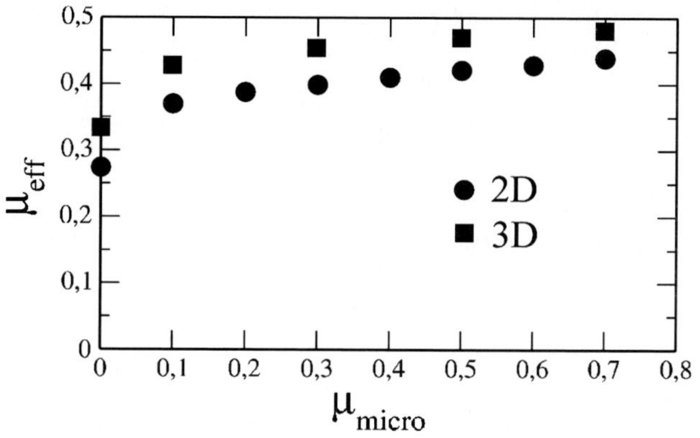

Fig. 7. Effective coefficient of friction versus the micromechanical grain-grain coefficient of friction for 2D (circles) and 3D (squares) flows.

The influence of μ_{micro} on the effective coefficient of friction is rather poor both in 2D and 3D. Even for the non-physical case of $\mu_{micro} = 0$, one obtains a non-zero effective coefficient of friction. Although it slightly increases with increasing μ_{micro}, μ_{eff} remains around the rather high value of 0.4. This shows that the energy loss is caused more by inelastic collisions than by friction, which is an indication that most of the dissipation occurs in the basal layer where the grains interact through inelastic collisions. Note that the kinetic theory applied to 3D supported flows of frictionless grains predicts an effective friction coefficient of $\sqrt{1-e_n} \approx 0.32$ [15], which is very close to that found here (see figure 7). Note also that the effective friction remains rather high (typically 0.4) which indicates that the existence of the agitated basal layer does not lead to a lubrication of the flow.

5.3 Influence of the Bumpiness of the Base

Since we have established that the collisional properties of the grains are crucial, one can expect the bumpiness of the base to be an important parameter. Let us recall here that the base of the cell consists of cylinders of diameter d (disks in 2D) regularly placed every δ perpendicularly to the main direction of the flow. The bumpiness in measured by the angle α defined on figure 8. The distance between two consecutive centers, δ, can vary from 0 (corresponding to a flat base and $\alpha = 0$) to $2d$ (maximum bumpiness, $\alpha = \pi/2$). Note that because of the small polydispersity of the flowing grains, the value $\delta = 2d$ can not be reached since the smallest grains would then fall through the base.

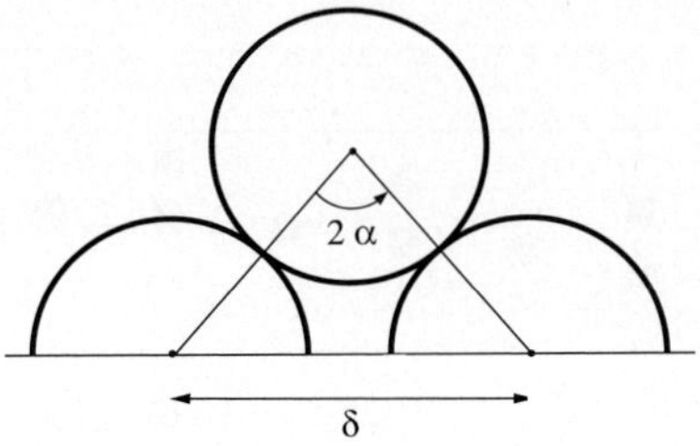

Fig. 8. Sketch of the bumpy base: the gap δ between the centers of the grains can range from 0 to $2d$. The bumpiness is measured by the angle α, where $\sin\alpha = \delta/2d$.

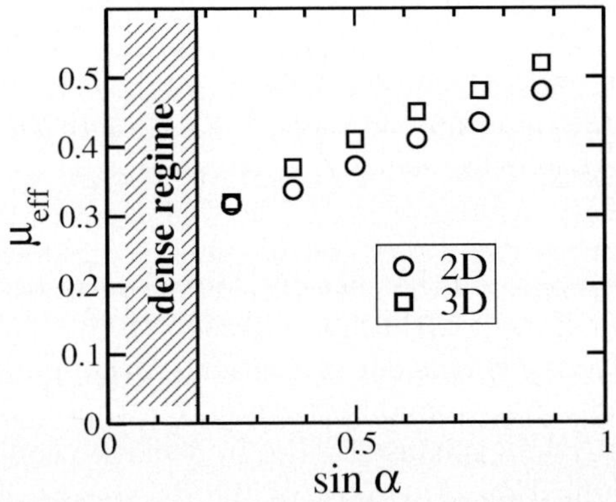

Fig. 9. Effective coefficient of friction as a function of the roughness of the base for $\tilde{N} = 40$ and $V_0 = 16$ m/s for 2D and 3D flows.

The first remark is that a minimum bumpiness is required to observe supported flows. For $\tilde{N} = 40$, $V_0 = 16$ m/s, $\theta = 0$ and $\mu = 0.3$, no supported flows can be observed for $\sin\alpha < 0.2$. In this case, the surface is rather smooth and no agitation can be triggered by the base. Note that the critical value of α below which the flow is dense is not universal and is expected to depend on θ. Figure 9 is a plot of the effective friction coefficient as a function of the bumpiness. As expected, μ_{eff} is an increasing function of α, meaning that the bumpier the base, the higher the effective friction. One can see that the 2D and 3D simulations produce very similar values of the effective friction coefficient. Note also that the range of values of μ_{eff} is wider than on Fig. 7. Yet, its value remains relatively high (above 0.3) in both 2D and 3D cases.

This again indicates that the effective friction is not reduced when the flow is in the supported regime (i.e. there is no lubrication effect).

6 Discussion and Conclusion

We have studied rapid flows of granular material over a bumpy base through 2D and 3D discrete elements simulations. We show that the regime of supported flows can exist, in 2D and 3D, as a transient on a horizontal base or as a steady state on an incline. In all cases, the flow behaves like a sliding block, which allows one to define and compute an effective friction coefficient, μ_{eff}. Our results indicate that (i) this effective friction coefficient is independent of total mass and initial velocity of the packing, (ii) it depends loosely on the micromechanical friction between individual grains and (iii) increases with increasing bumpiness of the base. We also give evidence that the steady state can be reached for one unique value of the inclination above (respectively below) which the flow accelerates (respectively decelerates). Because of the weak dependence of μ_{eff} on the micromechanical friction coefficient, for low values of the micromechanical friction coefficient one finds $\mu_{eff} > \mu_{micro}$ whereas for higher values one finds $\mu_{eff} < \mu_{micro}$. Therefore, it may seem difficult to conclude whether the supported regime is a source of lubrication. Note however that μ_{eff} remains relatively high, even in the case of frictionless grains (see figures 7 and 9).

One hypothesis to explain the existence of long-runout avalanches observed in nature (that is natural landslides that may travel many times their initial drop height) is to invoke supported flows [16]. Our work indicates that although such a regime can indeed exist, it does not yield an effective friction as low as 0.1 (instead it remains relatively high, $\mu_{eff} \simeq 0.4$). Therefore, it seems unlikely that long-runout avalanches can be accounted for by supported granular flows. Note however that the applicability of our results to geological events and direct comparison with field data are questionable. First, the range of parameters that is explored in this paper is rather small compared to that encountered in the field. It would be particularly interesting to vary the total mass of the flow over several decades. Secondly, our system is ideal and consists of an almost monodisperse assembly of spheres or disks with no interstitial fluid. These remarks leave several questions open and we hope will stimulate further work in the topic.

The authors would like to thank J.T. Jenkins and P. Eshuis for fruitful discussions.

References

1. J. Rajchenbach, Phys. Rev. Lett. **90**, 144302 (2003).
2. GdR-MiDi, Euro. Phys. J. B **14**, 341 (2004).

3. O. Pouliquen, Phys. Fluids **11**, 542 (1999).
4. L. Silbert, D. Ertas, G. S. Grest, T. Halsey, D. Levine, and S. J. Plimpton, Phys. Rev. E **64**, 051302 (2001).
5. C. Campbell, J. Geol. **97**, 653 (1989).
6. P. Cleary and C. S. Campbell, J. Geol. Res. **98**, 21911 (1993).
7. S. Straub, Geol. Rund. **85**, 85 (1996).
8. S. Straub, Geol. Rund. **86**, 421 (1997).
9. D. Hanes, J. Jenkins, and M. Richman, J. App. Mech. **55**, 969 (1988).
10. P. Eshuis, K. van der Weele, D. van der Meer, and D. Lohse, Phys. Rev. Lett. **95**, 258001 (2005).
11. B. Meerson, T. Poschel, and Y. Bromberg, Phys. Rev. Lett. **91**, 024301 (2003).
12. D. Frenkel and B. Smit (Academic Press, San Diego, 1996).
13. M. Prochnow, Ph.D. thesis, ENPC, Marne la Vallée (2002).
14. D. Bideau, A. Gervois, L. Oger, and J. Troadec, J. Phys. France **47**, 1697 (1986).
15. J. Jenkins, Private Communication (2005).
16. F. Legros, Eng. Geol. **63**, 301 (2002).

Granular Rayleigh-Taylor Instability

Jan Ludvig Vinningland[1], Øistein Johnsen[1], Eirik G. Flekkøy[1], Renaud Toussaint[2], and Knut Jørgen Måløy[1]

[1] Department of Physics, University of Oslo, P.0.Box 1048, N-0316 Oslo, Norway
[2] Institut de Physique du Globe de Strasbourg, CNRS, Université Louis Pasteur, 5 rue Descartes, 67084 Strasbourg Cedex, France

Summary. A granular instability driven by gravity is studied experimentally and numerically. The instability arises as grains fall in a closed Hele-Shaw cell where a layer of dense granular material is positioned above a layer of air. The initially flat front defined by the grains subsequently develops into a pattern of falling granular fingers separated by rising bubbles of air. A transient coarsening of the front is observed right from the start by a finger merging process. The coarsening is later stabilized by new fingers growing from the center of the rising bubbles. The structures are quantified by means of Fourier analysis and quantitative agreement between experiment and computation is shown. This analysis also reveals scale invariance of the flow structures under overall change of spatial scale.

1 Introduction

Improved understanding of granular flows would be of essential importance to a range of industrial applications, to the study of geological pattern forming processes, and, in general, to the theoretical description of disordered media.

As grains become smaller the effect of the interstitial fluid becomes more important. The result is a combination of dry granular dynamics and the hydrodynamics of the fluid. These systems give rise to a variety of exotic and most often poorly understood phenomena such as fluidization [1] and bubble instabilities [2], quicksand and jet formation [3], and sandwich structures in systems where different particle types segregate [4]. While the study of dry granular media has been extensive over the past decades, the exploration of fluid-granular systems has been of more limited scope.

A wide range of granular instabilities where various structures form along fluid-grain interfaces have been reported the last few years [2, 7]. Notably, the patterns formed by grains falling in a highly viscous liquid were investigated experimentally and theoretically by Völtz et al. [8]. The instability reported by Völtz shares its main qualitative characteristics with the classical Rayleigh-Taylor instability, i.e. a single dominating wavelength growing right from the start. In contrast, the instability discussed here arises along an air-grain interface where falling granular fingers and rising air-bubbles form [9, 10]. A coarsening of the finger-bubble pattern is observed right from the start of the instability. The coarsening subsequently stabilizes due to the continuous formation of new fingers at the centers of the rising and growing bubbles. Our experimental and numerical data compare favorably both qualitatively and quantitatively, despite the simplifications of the numerical model: Only two spatial dimensions, zero interparticle friction, no gas inertia, and a resolution of the gas-grid of 2.5 grain diameters. The finger-bubble structures further exhibit scale invariance under change of particle size, a feature which is supported both by observations and theoretical considerations.

The experiment is presented in Sec. 2, and the numerical model is outlined in Sec. 3. The analysis and the consistency check are presented in Section 4, together with the scale invariance observed when the grain diameter increases. The conclusion is given in Section 5.

2 Experiment

The experimental setup, illustrated in Fig. 1, consists of a closed Hele-Shaw cell mounted on a hinged bar which enables the cell to rotate approximately 130 degrees from a lower to an upper vertical position (from A to B in Fig. 1). The Hele-Shaw cell is made of a 1 mm thick silicone frame sandwiched between two 8 mm thick glass plates. The internal dimensions of the cell are 56 mm ×

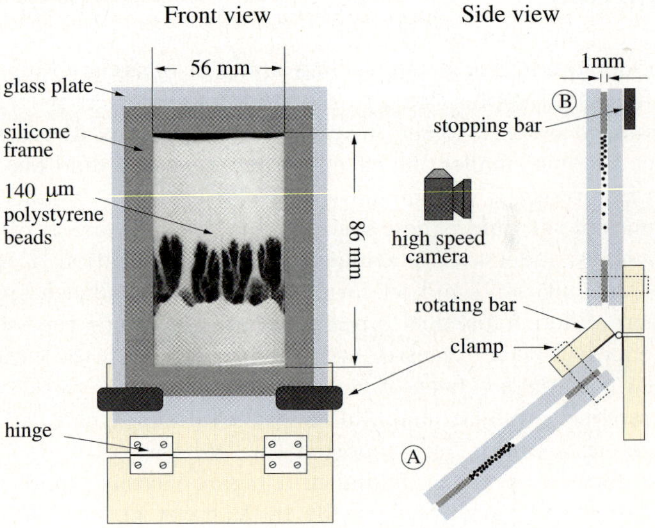

Fig. 1. Front and side view of the experimental setup. Two cell positions are superimposed in the side view to illustrate the manual rotation from position A to B.

86 mm × 1 mm, and it is filled with polystyrene beads and air at atmospheric pressure. The cell is rotated manually and it takes about 0.2 seconds to bring the cell to an upright, vertical position. The off-center pivot of the cell causes the rotation to slow down the falling motion of the grains due to centrifugal forces. A bar is mounted at position B in order to have some control of the final vertical position of the cell. The evolution of the instability is recorded by a high speed digital camera (Photron Fastcam-APX 120K) taking images with a resolution of 512 × 512 pixels at a rate of 500 frames per second.

Monodisperse polystyrene beads of 140 µm in diameter (Microbeads® Dynoseeds TS 140-51) are used in the experiments. The filling of the cell is performed with one glass plate lying down horizontally with the silicone frame adhered on top. Small portions of beads are carefully deposited on the plate and leveled with the frame before the upper plate is attached and fasted with clamps. The cell is flipped a few times after closure to allow the grains to form a random loose packed configuration before the cell is rotated.

The humidity in the lab is important in order to keep the electrostatic and cohesive properties of the grains at a suitable level to prevent the grains from clustering or sticking to the glass plates. During the filling of the cell and throughout the experiment the humidity was kept constant at about 30%.

A series of images from the experiment is shown in Fig. 2. Due to experimental imperfections the initial air-grain interface in Fig. 2(a) is not a well-defined, horizontal front. This is mainly caused by three factors: (1) Os-

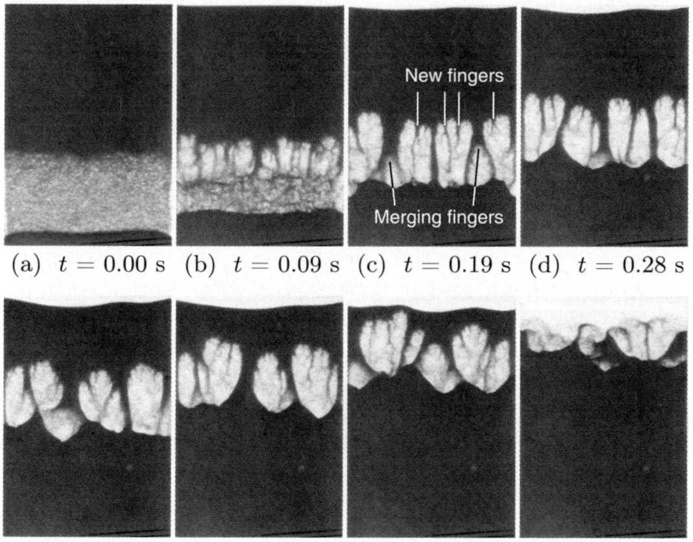

Fig. 2. Experimental images of the granular Rayleigh-Taylor instability where a layer of monodisperse polystyrene beads of 140 µm in diameter (black) displaces air (white) in a Hele-Shaw cell of dimensions 56 mm × 86 mm × 1 mm.

cillations in the cell caused by the abrupt stop as it reaches the vertical (and hits the bar), (2) the falling of grains during the rotation, and (3) the granular Boycott effect during rotation (i.e. grains and air pass each others in layers parallel to the plates) caused by the fact that up to 7 layers of beads may fit between the plates. However, all these experimental effects are non-persistent and in Fig. 2(b) the first finger-bubble structures are clearly visible. As the instability evolves a coarsening of the structures is observed. However, new fingers emerging from the center of the rising bubbles (see Fig. 2(c)) prevent the size of the structures to grow indefinitely and gives rise to a characteristic bubble-size. The coarsening is driven by fingers merging and forming an inverted Y as is indicated in Fig. 2(c).

3 Simulation

The numerical model, first introduced by McNamara *et al.* [11, 12], combines a continuum description of the air with a discrete description of the granular phase. The granular phase is a collection of rigid spheres that constitute a deformable porous media described by coarse grained solid fraction $\rho(x, y)$ and granular velocity $\mathbf{u}(x, y)$ fields on the grid where (x, y) are the two dimensional space coordinates. The continuous fields $\rho(x, y)$ and $\mathbf{u}(x, y)$ are obtained from the positions and velocities of the individual grains by means of a linear smoothing function: To calculate $\mathbf{u}(x, y)$ a fraction of $\mathbf{v}_i$ is distributed to each of the four neighbouring grid-nodes of grain i. This is illustrated in Fig. 3.

The continuum gas phase is described solely by its pressure $P(x, y)$. The inertia of the gas, and hence its velocity field, is not considered. This is justified for small particle Reynolds numbers as is the case for our system. The pressure is governed by the equation [11, 12]

$$\phi\Big(\frac{\partial P}{\partial t} + \mathbf{u} \cdot \boldsymbol{\nabla} P\Big) = \boldsymbol{\nabla} \cdot \Big(P \frac{\kappa(\phi)}{\mu} \boldsymbol{\nabla} P\Big) - P \boldsymbol{\nabla} \cdot \mathbf{u} \,, \tag{1}$$

where $\phi(x, y) = 1 - \rho(x, y)$ is the porosity, $\kappa(x, y)$ the permeability, $\mathbf{u}(x, y)$ the granular velocity field, and μ the gas viscosity. This equation is derived

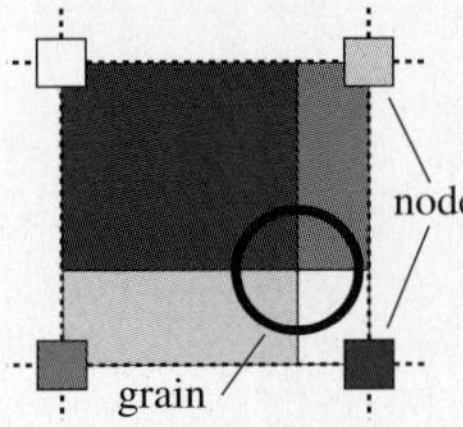

Fig. 3. Illustrating the smoothing function: A single grain i and its four neighbouring grid-nodes are shown. The four subdivisions of the cell-volume (i.e. the area between the stapled lines) correspond to the fraction of $\mathbf{v}_i$ (or m_i) that contributes to the continuous $\mathbf{u}(x, y)$ (or $\rho(x, y)$) fields at the *opposite* nodes. The gray-scale of the subdivisions indicates which node they belong to.

from the conservation of air and grain mass, using Darcy's law [16] to obtain the pressure drop over a volume with permeability κ. The Carman-Kozeny relation [17] is assumed for the permeability, and the isothermal ideal gas law is assumed for the compressible gas. See Refs. [11, 12] for a detailed derivation. The empirical Carman-Kozeny relation is given by

$$\kappa(\rho, d) = \frac{d^2}{180}\frac{(1-\rho)^3}{\rho^2}, \tag{2}$$

where ρ is the local solid fraction, and d is the diameter of the grains, and 1/180 is an empirical constant valid for a packing of beads.

The grains are governed by Newton's second law

$$m\frac{d\mathbf{v}}{dt} = m\mathbf{g} + \mathbf{F}_\mathrm{I} - \frac{V\boldsymbol{\nabla}P}{\rho}, \tag{3}$$

where m, $\mathbf{v}$, and V are respectively the mass, velocity, and volume of the grain. Contact dynamics [18] is used to calculate the interparticle force $\mathbf{F}_\mathrm{I}$ which keeps the grains from overlapping. The dynamics of the grains are simplified by neglecting particle-particle and particle-wall friction. A lower cutoff is imposed on the solid fraction because the Carman-Kozeny relation is not valid as the solid fraction drops below 0.25 [19]. This cutoff causes the permeability of the most dilute regions of the system to be slightly lower than the true permeability. The effect is a slight overestimation of the pressure forces acting on the grains in the dilute regions.

Figure 4 shows a series of snapshots from a simulation with the same grain diameter, grain density, and cell dimensions (except for the cell thickness since

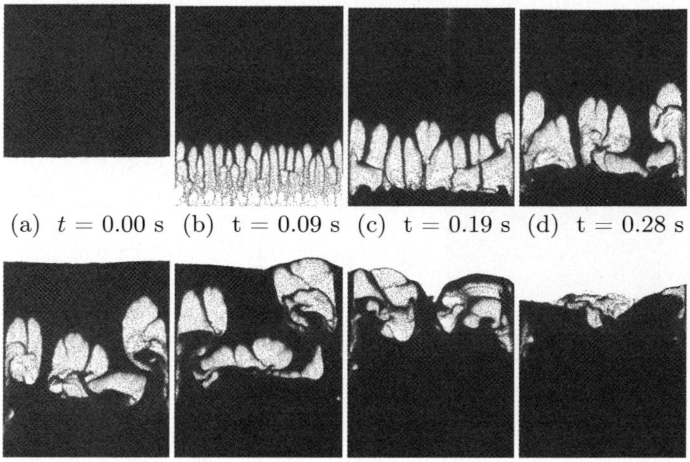

(a) $t = 0.00$ s (b) t = 0.09 s (c) t = 0.19 s (d) t = 0.28 s

(e) t = 0.38 s (f) t = 0.47 s (g) t = 0.56 s (h) t = 0.66 s

Fig. 4. Snapshots from a simulation where the input parameters are determined by the experiment in Fig. 2. The grain density is 1.05 g/cm^3, grain diameter is 140 µm, and the cell dimensions are 56 mm × 68 mm.

the simulation is 2D) as in the experiment shown in Fig. 2. A qualitative comparison of the image-series in Figs. 2 and 4 renders the simulation and experiment consistent in many respects: The sizes of the bubbles and the fingers are comparable, and the dynamical processes of finger merging and finger nucleation are observed in both cases. Nevertheless, some discrepancies are observed, particularly at the start and toward the end of the instability. The initial differences are mainly attributed to imperfections of the experimental setup: In the experiments the instability start from a homogeneous sheet of grains rather than from a sharp air-grain interface as in the simulations. Further differences are observed toward the end of the instability: In the numerical snapshots bigger bubbles reach the upper surface before the smaller bubbles (see e.g. Fig. 4(f)). In the experiment, however, the row of bubbles stays more or less horizontally aligned from start to finish. We believe this is an effect of the zero grain-wall and grain-grain friction used in the simulation.

4 Results

To study the coarsening of the observed structures quantitatively we use the Fourier spectrum of the solid fraction field to obtain a characteristic average wave number $\langle k \rangle$ by the following procedure. The discrete Fourier transform and the power spectrum of each horizontal line of $\rho(x, y)$ is calculated. The averaged power spectrum, $\bar{S}(k)$, is then obtained by averaging over all these horizontal power spectra. An average wave number is defined as $\langle k \rangle = \sum_k \bar{S}(k) \cdot k / \sum_k \bar{S}(k)$, where $1/k$ is the wavelength. Likewise, we define the squared standard deviation ${\sigma_k}^2 = \sum_k \bar{S}(k) \cdot k^2 / \sum_k \bar{S}(k) - \langle k \rangle^2$. For the experimental data the image pixel values are used to estimate the solid fraction. Figure 5(a) shows $\rho(x, y)$ for $t = 0.09$ s and three power spectra

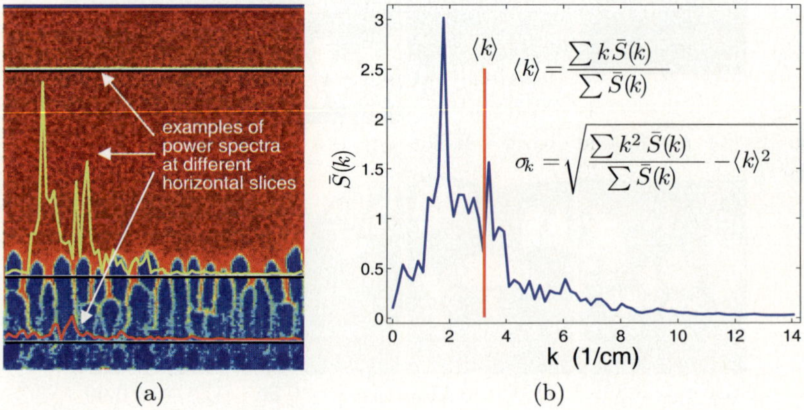

Fig. 5. (a) Power spectra superposed on the solid fraction field from which they are obtained. (b) The averaged power spectrum of the solid fraction field shown in (a) with the mean wave number $\langle k \rangle$ indicated.

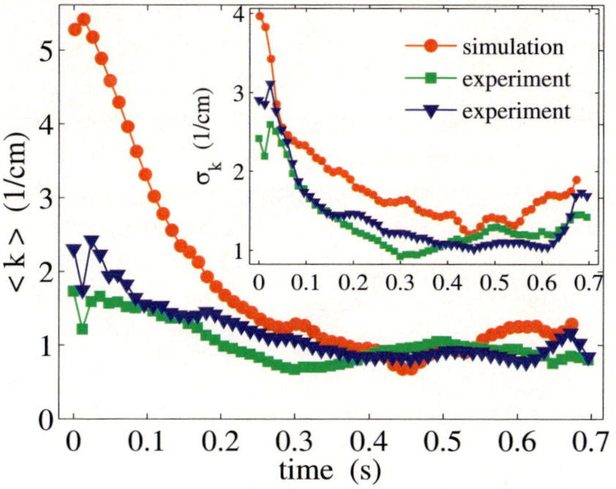

Fig. 6. Mean wave number $\langle k \rangle$ and standard deviation σ_k (inset) for two experiments and one simulation, all using polystyrene beads of 140 µm in diameter.

obtained at different vertical positions (indicated by black lines). Figure 5(b) shows the averaged power spectrum $\bar{S}(k)$ and mean wave number $\langle k \rangle$ of the solid fraction field in Fig. 5(a).

Figure 6 shows the temporal evolution of $\langle k \rangle$ and σ_k (inset) for the numerical and experimental data. An additional set of experimental data is added to the plot. The numerical curve starts out with a significantly higher wave number than the experimental curves. This behaviour is however expected given the difference in initial conditions. After about 0.2 seconds the experimental and numerical data coincide and continue to show a similar coarsening behavior. The fluctuations of $\langle k \rangle$ and σ_k are associated with the continuous nucleation of new fingers and merging of existing fingers.

We further investigate the behavior of the system as the overall spatial scale is changed: Keeping all length ratios and the particle number fixed, the size of the system will scale according to the diameter d of the grains. We measure the characteristic inverse length scale $\langle k \rangle$ as d is changed and observe a scale invariance of the evolution. A series of seven simulations are performed where d varies from 70 µm to 490 µm in steps of 70 µm. The dimension of the numerical cell confining grains of 70 µm in diameter is 28 mm × 34 mm. To compare, a series of experiments using polystyrene beads of 80, 140, 230, and 570 µm in diameter, confined in Hele-Shaw cells that scale proportionally with d in all directions, are performed.

In these simulations we have introduced the larger density of glass, rather than polystyrene in order to minimize the numerical artifacts associated with the solid fraction cutoff in the permeability. As the inertia of the grains decreases the overestimated pressure forces in the dilute regions of the system will act even stronger on the falling granular fingers.

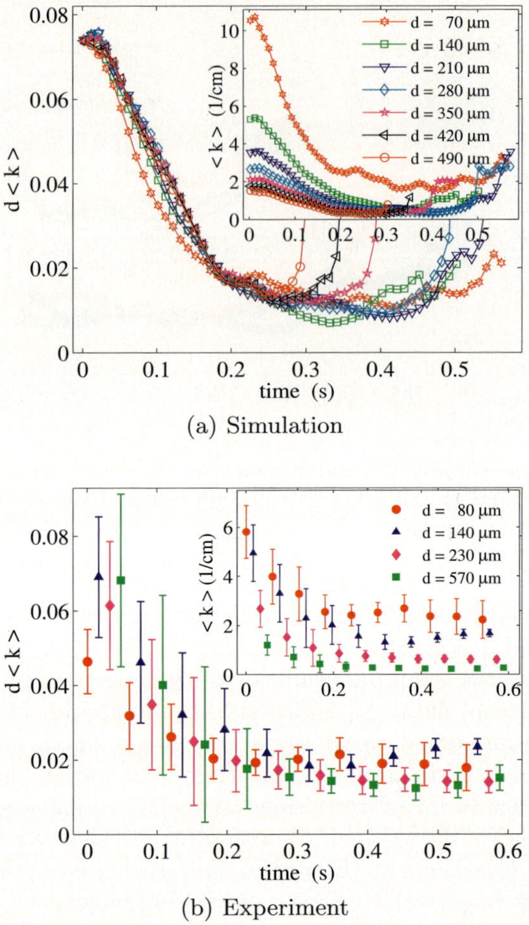

Fig. 7. Data-collapse plot of $d\langle k\rangle$ for a series of (a) simulations using glass beads, and (b) experiments using polystyrene beads. The grain diameters d are given in the legend box. The inset shows the evolution of the unscaled $\langle k\rangle$.

Data-collapse plots of the rescaled mean wave number $d\langle k\rangle$ are shown in Figs. 7(a) (simulation) and 7(b) (experiment). These plots indicate that the characteristic size of the structures is invariant when size is measured in units of d; the number of grains that spans the width of the bubbles is the same for a wide range of grain sizes.

Theoretically, the scale invariance of the product $d\langle k\rangle$ may be interpreted as follows: Compared to the other terms of Eqs. (1) and (3) the $m\,d\mathbf{v}/dt$, F_{I} and $P\boldsymbol{\nabla}\cdot\mathbf{u}$ terms may be shown to be small [20]. For that reason, these equations exhibit an approximate invariance under system size scaling. If we take δP to be the pressure deviation from the background pressure, express the velocity of grain i as $\mathbf{v}_i = \delta\mathbf{v}_i + \mathbf{u}_0$ and the locally averaged granular velocity as $\mathbf{u} = \delta\mathbf{u} + \mathbf{u}_0$, where $\mathbf{u}_0$ is the constant sedimentation velocity of a close packed system, this scaling may be expressed as $\mathbf{x} \to \lambda\mathbf{x}$, $\delta P \to \lambda\delta P$, $\mathbf{u}_0 \to \lambda^2\mathbf{u}_0$, $\delta\mathbf{u} \to \lambda\delta\mathbf{u}$ and $\kappa \to \lambda^2\kappa$, where λ is a scale factor. The structure formation of the system is governed by $\delta\mathbf{u}$ and, since this velocity scales the same way with λ as the length scales themselves, the evolution of any structure measured in units of d will be scale invariant. In particular this is true for the structures measured by the length $1/\langle k\rangle$, and so $d\langle k\rangle$ is scale invariant. However, the invariance deteriorates both when particle size is increased, and when it is decreased. In the first case, the relative effect of granular inertia is increased, in the second, the relative effect of the $P\nabla\cdot\mathbf{u}$ term is increased.

The convergence of the numerical data-collapse in Fig. 7(a) is quite good. The deviation of the 70 µm curve for small t is probably explained by the increase in the relative importance of the $P\nabla\cdot\mathbf{u}$-term. The divergences of the 350, 420, and 490 µm curves for greater t in the same plot arise because the bubbles in the coarser packings disappear before they reach the surface due to the increase of $\mathbf{u}_0$ with λ^2 [20]. The experimental data in Fig. 7(b) have a wider distribution but collapses satisfactorily given the standard deviation error bars. The experimental data are obtained by averaging over three experiments for each diameter d. The standard deviation is calculated over a time window of 0.3 seconds. The accuracy of the experiments is at its lowest during the initial coarsening of the structures, but as the mean wave number stabilizes around 0.2 seconds the accuracy improves. Nevertheless, the data points are, with a few exceptions, within a distance of one standard deviation from one another. The loss of precision for small times is most likely caused by the inaccuracy involved with the manual rotation.

5 Conclusion

In conclusion, we have presented experimental and numerical results of a gravity-driven granular flow instability which is significantly different from its classical hydrodynamic analog. The simulations reproduce the characteristic shape and size of the experimentally observed structures and provide fine patterns in the early phase of the process that are not resolved experimentally. Data-collapse plots of the mean wave number $\langle k\rangle$ indicate that the flow and the resulting structures are invariant when measured on a scale proportional to the grain diameter d for a range of diameters that spans from 70 µm to 570 µm.

References

1. H. J. Herrmann, J.-P. Hovi, and S. Luding, editors. *Physics of Dry Granular Media*, volume 350 of *NATO ASI Series E: Applied Sciences*, Dordrecht, 1998. Kluwer Academic Publishers.

2. D. Gendron, H. Troadec, K. J. Måløy, and E. G. Flekkøy. Bubble propagation in a pipe filled with sand. *Phys. Rev. E*, 64:021509, 2001.
3. D. Lohse, R. Rauhé, R. Bergmann, and D. van der Meer. Creating a dry variety of quicksand. *Nature*, 432(7018):689–690, December 2004.
4. C. Zeilstra, M. A. van der Hoef, and J. A. M. Kuipers. Simulation study of air-induced segregation of equal-sized bronze and glass particles. *Phys. Rev. E*, 74(1):010302(R), 2006.
5. G. I. Taylor. The instability of liquid surfaces when accelerated in a direction perpendicular to their planes. *Proc. Roy. Soc. A*, 201(1065):192–196, March 1950.
6. J. S. Langer. Instabilities and pattern formation in crystal growth. *Rev. Mod. Phys.*, 52(1):1–28, Jan 1980.
7. Ø. Johnsen, R. Toussaint, K. J. Måløy, and E. G. Flekkøy. Pattern formation during air injection into granular materials confined in a circular Hele-Shaw cell. *Phys. Rev. E*, 74(1):011301, 2006.
8. C. Völtz, W. Pesch, and I. Rehberg. Rayleigh-Taylor instability in a sedimenting suspension. *Phys. Rev. E*, 65:011404, 2001.
9. J. L. Vinningland, Ø. Johnsen, E. G. Flekkøy, R. Toussaint, and K. J. Måløy. Granular Rayleigh-Taylor instability: Experiments and simulations. *Phys. Rev. Lett.*, 99(4):048001, 2007.
10. J. L. Vinningland, Ø. Johnsen, E. G. Flekkøy, R. Toussaint, and K. J. Måløy. Experiments and simulations of a gravitational granular flow instability. *Phys. Rev. E*, to appear.
11. S. McNamara, E. G. Flekkøy, and K. J. Måløy. Grains and gas flow: Molecular dynamics with hydrodynamic interactions. *Phys. Rev. E*, 61(4):4054–4059, 2000.
12. D.-V. Anghel, M. Strauß, S. McNamara, E. G. Flekkøy, and K. J. Måløy. Erratum: Grains and gas flow: Molecular dynamics with hydrodynamic interactions. *Phys. Rev. E*, 74(2):029906(E), 2006.
13. E. G. Flekkøy, S. McNamara, K. J. Måløy, and D. Gendron. Structure formation and instability in a tube of sand. *Phys. Rev. Lett.*, 87(13):134302, 2001.
14. M. Strauß, S. McNamara, H. J. Herrmann, G. Niederreiter, and K. Sommer. Plug conveying in a vertical tube. *Powder Tech.*, 162:16, 2006.
15. E. G. Flekkøy, A. Malthe-Sørenssen, and B. Jamtveit. Modeling hydrofracture. *J. Geophys. Res.*, 107 (B8)(2151), 2002.
16. H. Darcy. *Les Fontaines Publiques de la Ville de Dijon.* Dalmont and Paris, 1856.
17. P. C. Carman. Fluid flow through granular beds. *Trans. Inst. Chem. Eng.*, 15:150–66, 1937.
18. F. Radjai, M. Jean, J.-J. Moreau, and S. Roux. Force distributions in dense two-dimensional granular systems. *Phys. Rev. Lett.*, 77(2):274–277, 1996.
19. A. A. Zick and G. M. Homsy. Stokes-flow through periodic arrays of spheres. *J. Fluid Mech.*, 115:13–26, 1982.
20. J. L. Vinningland, Ø. Johnsen, E. G. Flekkøy, R. Toussaint, and K. J. Måløy. Size invariance in granular flows: simulation and experiment. unpublished.

Morphology of Polydisperse Granular Media

Charles Voivret, Jean-Yves Delenne, Moulay Saïd El Youssoufi, and Farhang Radjaï

LMGC, CNRS-Université Montpellier II, Place Eugène Bataillon, 34095 Montpellier cedex, France voivret@lmgc.univ-montp2.fr

Summary. We investigate the morphology and space-filling properties of polydisperse densely-packed granular media in 2D. A numerical procedure is introduced to generate collections of circular particles with size distributions of variable size span. This particle generation procedure is used with a geometric deposition protocol in order to build large close-packed samples of prescribed polydispersity. The solid fraction is a strongly nonlinear function of the size span, and the highest levels of solid fraction occur for the uniform distribution by volume fractions. A transition occurs from a regime of topological disorder where the packing properties are governed by particle connectivity to a regime of metric disorder where pore-filling small particles prevail. The polydispersity manifests itself in the first regime through the variability of local coordination numbers. In the second regime, the material is homogeneous beyond only a few average particle diameters. We also show that the fabric anisotropy declines with size span.

1 Introduction

Granular materials in nature and industry often involve wide distributions of particles sizes. There are few studies reported in literature concerning polydisperse granular media. Numerical studies have been performed for simple size distributions (e.g. bi-disperse, power law) [1–3] but dynamic simulations are deficient in the number of particles.

Two aspects have been more extensively considered in relation to polydisperse granular media: space-filling properties, on one hand, and packing structure in terms of connectivity and structural order, on the other hand. The space-filling issue corresponds mainly to the highly polydisperse regime where numerous particles of sufficiently small size can fill the pore space between larger particles [1, 2]. The question of packing structure is often associated with the weakly polydisperse regime where structural order is drastically altered due to a weak polydispersity. This is the case in 2D where long-range order in a monodisperse packing disappears due to a narrow size distribution of particles [4, 5].

In this paper, we introduce a numerical procedure to generate collections of circular particles with a prescribed cumulate volume distribution (CVD), also called grading curve [6], and statistical representativity. Although this procedure is quite general, we will use the β-function to represent CVDs. This size generation procedure is then used within a ballistic deposition algorithm in order to build very large close-packed samples of desired polydispersity [7–9]. The close-packed samples produced by the generation procedure together with a deposition protocol allow us to investigate various descriptors of granular microstructure as a function of polydispersity parameters. We focus on how particle size distribution affects the space-filling properties and how deviation from the monodisperse limit affects structural order.

2 Size and Packing Generation Method

2.1 Size Generation

We assume that the CVD of the particles is represented by a continuous function $h(d)$ of particle diameters d varying in the range $[d_{min}, d_{max}]$. By definition, we have

$$h(d) = \frac{\int_{d_{min}}^{d} V(x)N(x)dx}{\int_{d_{min}}^{d_{max}} V(x)N(x)dx}, \tag{1}$$

where $N(d)$ is the particle size distribution and $V(x) = (\pi/4)x^2$ is the 2D volume of a particle of diameter d. We have $h(d_{min}) = 0$ and $h(d_{max}) = 1$. Since the CVD represents volume cumulate of the particles, we first discuss here how the corresponding cumulative distribution function $F(d)$ of particle diameters can be obtained.

The CVD defined over the interval $[d_{min}, d_{max}]$ is first discretized into N_c "classes" defined over subintervals $[d^i_{min}, d^i_{max}]$ of widths $\Delta d_i \equiv d^i_{max} - d^i_{min}$ and $i \in [1, N_c]$. Then, the CVD is decumulated over each interval in order to obtain the volume fraction f^i_v for each class i:

$$f^i_v = h(d^i_{max}) - h(d^i_{min}). \tag{2}$$

We require that the following two "representativity" conditions be satisfied:

1. The number of particles in each class is above a minimum ${N'}_{p/c}^{min}$.
2. The volume of each particle in a class i is small compared to the total volume f^v_i of the class.

We note that these two conditions are equivalent for a quasi-monodisperse distribution.

We further assume that the CVD is linear over each class i. This condition implies that the cumulative distribution function $F^i(d)$ of particle diameters over the class i is a normalized uniform distribution by volume fractions of particles defined over the interval $[d^i_{min}, d^i_{max}]$:

$$F^i(d) = \frac{d^i_{max}}{d^i_{max} - d^i_{min}} \frac{d - d^i_{min}}{d}. \tag{3}$$

Then, the mean diameter $d^i_m = (F^i)^{-1}(0.5)$ of the class can be estimated. This information allows us to determine the amount n^i of the particles in the class for a unit total volume:

$$n^i = \frac{4 f^v_i}{\pi (d^i_m)^2}. \tag{4}$$

Each n^i is then rescaled by a factor ${N'_{p/c}}^{min}/Min\{n^i\}$ and its integer part corresponds to the temporary number of particles N'^i in the class i:

$$N'_i = \text{int}\left\{ \frac{{N'_{p/c}}^{min}}{Min\{n^i\}}\, n^i \right\}, \tag{5}$$

where int is the integer part. This procedure ensures that each class will contain at least ${N'_{p/c}}^{min}$ particles as required by the first representativity condition.

The total number of particles $N'_p = \sum_i N'^i$ obtained at this stage is dependent on the parameters N_c and ${N'_{p/c}}^{min}$. But, N'_p may be deficient for the construction of a representative packing. Let N^{min}_p be the minimum number of particles required for the construction of the packing. We impose that the number N_p of particles should be above N^{min}_p. If $N'_p < N^{min}_p$, we rescale the N'_i by N^{min}_p / N'_p to get the number of particles N^i in each class i:

$$N^i = \text{int}\left\{ \frac{N^{min}_p}{N'_p}\, N'^i \right\}. \tag{6}$$

From the number of particles N^i in each class i, the probability distribution function $P^i(d)$ of particle diameters d for each class i is given by

$$P^i(d) = \frac{N^i}{N_p}, \tag{7}$$

and the cumulative distribution function $F(d)$ is obtained by summing up the P^i over all diameters $\leq d$. Given the population N^i in each class i, we generate the N^i diameters d in the class by inverse transform sampling of $F^i(d)$ given by Eq. (3).

2.2 Packing Generation

In order to generate polydisperse packings we need a model for the CVD. This model should be simple, i.e. involving a small number of parameters, contain particular distributions such as power laws, and be defined over a finite interval. A distribution that satisfies these requirements is the cumulative β distribution defined as [9]

$$\beta(x) = \frac{1}{B(a,b)} \int_0^x t^{a-1}(1-t)^{b-1}\, dt, \tag{8}$$

where $a > 0$ and $b > 0$ are the parameters of the distribution and

$$B(a,b) = \Gamma(a)\Gamma(b)/\Gamma(a+b), \tag{9}$$

where Γ is the Gamma function. This β distribution is defined and normalized over the interval $[0, 1]$, so that $\beta(0) = 0$ and $\beta(1) = 1$.

For using the cumulative β distribution as a model of CVD for the particle diameters d over the interval $[d_{min}, d_{max}]$, we replace x by the reduced diameter d_r defined as

$$d_r(d) = \frac{d - d_{min}}{d_{max} - d_{min}}. \tag{10}$$

Then, the CVD is defined by setting $x = d_r$ in Eq. (8):

$$h(d) = \beta(x = d_r(d); a, b). \tag{11}$$

Since only the relative particle diameters are relevant for space filling properties, we will use throughout this paper the reduced diameter d_r instead of d. In the same way, the span s of the distribution $h(d)$ in Eq. 11 will be defined as

$$s = \frac{d_{max} - d_{min}}{d_{max} + d_{min}}. \tag{12}$$

The case $s = 0$ represents a monodisperse packing whereas $s = 1$ corresponds to "infinite" polydispersity. In practice, the limit $d_{min} = 0$ never happens, and hence the span s is always strictly below unity.

In terms of reduced diameters d_r, the shape of the CVD is controlled by the parameters a and b and allows to obtain well-known distributions such as power laws or bi-disperse. Figure 1a displays several plots of CVD as defined by Eq. 11 as a function of d_r for several values of a and b. The case $a = b = 1$ corresponds to uniform distribution by volume fractions. The uniform distribution by particle diameters is obtained for $a = 3$ and $b = 1$.

For the construction of polydisperse packings we use a deposition protocol. It consists of layer-by-layer deposition of rigid particles on a substrate (a rough or smooth plane in 3D or line in 2D) according to simple geometrical rules. We determine for each particle (to be deposited) the lowest position at the free surface as a function of its diameter. In order to avoid wall effects, periodic boundary conditions were implemented in the horizontal direction. Figure 1b shows an example of small scale sample.

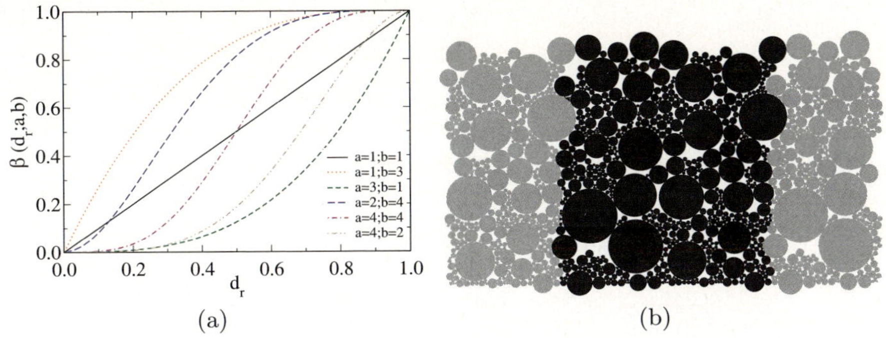

Fig. 1. (a) Cumulate volume distribution as a function of the relative diameter d_r for several values of the distribution shape parameters a and b according to Eq. 11. (b) Example of small-scale packing.

3 Solid Fraction and Radial Distribution

We consider two different space-filling properties: solid fraction ρ, and a radial volume distribution function $\rho(r)$ which describes the correlations of solid fractions. All the data presented below were obtained for samples prepared with $N_c = 10$, $N_{p/c}^{min} = 10$, $N_p^{min} = 3.10^4$ and $N_p^{max} = 10^5$. The span s will be varied from 0.02 to 0.97.

3.1 Solid Fraction

The solid fraction ρ is shown in Fig. 2a as a function of size span s for different shape parameters a and b. Except from the distribution by uniform volume fraction, we observe three regimes in variation of ρ. At very low values of s ($s < 0.1$) ρ decreases. It corresponds to the well-known order-disorder transition observed in 2D monodisperse granular media due to the perturbation of an initially triangular lattice in the presence of a small amount of polydispersity [5]. Then, it increases slowly with s in the range $0.1 < s < 0.4$. In this range, the values of ρ are sensibly the same independently of shape parameters and its value is near to that for a quasi monodisperse random close packing. Beyond $s \simeq 0.4$, ρ increases faster for distributions favoring the number of small particles ($a = 1$, $b = 3$; $a = 2$, $b = 4$) than those favoring the number of large particles ($a = 3$, $b = 1$; $a = 4$, $b = 2$). $s \simeq 0.4$ corresponds to a geometrical transition when the smallest particles can fit into the pores between largest particles. For each value of s, the solid fraction ρ for the uniform distribution by volume fractions ($a = b = 1$) is higher than those for all other values of shape parameters.

Two conditions are required to fill efficiently the pores: 1) a broad size distribution, which corresponds to higher values of s ($s > 0.4$ in our simulations), 2) a large number of smaller particles, controlled in our model by the

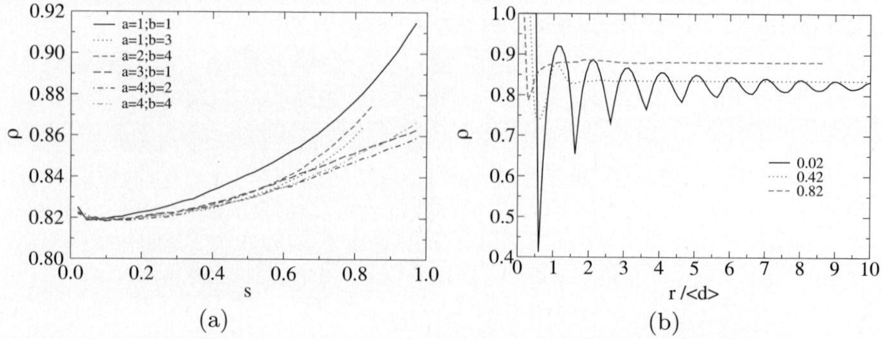

Fig. 2. (a) Solid fraction ρ as a function of size span s for several values of shape parameters a and b. (b) Average solid fraction ρ as a function of radial distance r from given particle center normalized by the average diameter $\langle d \rangle$ for uniform distribution by volume fractions and three values of size span s.

shape parameters a and b. According to Fig. 2a, these conditions appear to be optimally fulfilled for uniform distribution by volume fractions. In other words, equal volume fractions provide the best match between the volumes of particles and pores.

3.2 Spatial Order

To characterize spatial order, one uses classically the radial distribution function of the radial positions r of particle centers. Using particle centers is natural for molecular systems or monodisperse granular media where the particles have all nearly the same volume. In polydisperse granular media, we need to account for particle volumes. It seems thus more natural to consider the solid fraction $\rho(r)$ as a function of radial positions r of the particles. In practice, the solid fraction $\rho(r)$ can be calculated inside circular shells of increasing radius r centered on particle centers. Figure 2b shows $\rho(r)$ for uniform distribution by volume fractions for three values of size span s. In all cases, $\rho(r)$ is equal to 1 for $r \leq d_{min}/2$ and tends to the packing solid fraction ρ at large values of r. The long range order manifest itself through the oscillation of ρ with a period of nearly $\langle d \rangle$ at low values of s. At larger polydispersities, one mainly observes a marked valley following the initial plateau and a nearly constant increase towards ρ. This valley represents the void space between a particle and its first neighbors. In the pore-filling regime ($s > 0.4$), the material can be considered as homogeneous beyond almost two average diameters. It should be, however, borne in mind that the average diameter in this regime corresponds to many small particle diameters d_{min}.

4 Connectivity and Anisotropy

The solid fraction and radial distribution function account for the metric disorder depending on particle size distributions. However, most mechanical properties of granular media, such as force transmission and dilatancy, result from the topological disorder of the contact network. In this section, we study the influence of size distribution on the connectivity and angular ordering of the contact network.

4.1 Connectivity

Monodisperse particles can, in principal, assemble to form triangular packings with coordination number $z = 6$. However, this upper limit can never be reached without fine adjustment of particle positions due to geometrical mismatches related to steric exclusions and numerical precision. In the presence of slightest perturbation of the assembly, the coordination number collapses to $z = 4$. For example, at weak polydispersity ($s = 0.02$) we get $z = 4$, see Fig. 3a. Below a certain limit of s depending on shape parameters, z is constant and equal to 4. Beyond this limit z decreases. To explain this, we know that each deposited particle, by construction, is supported by two particles. Its implies that the decrease of z is due to particles with 2 and 3 contacts.

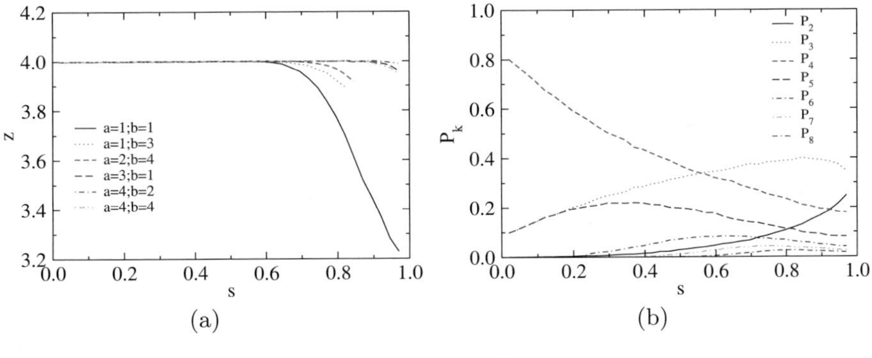

Fig. 3. (a) Coordination number z as a function of size span s for several values of distribution shape parameters a and b. (b) The connectivity functions P_k for different values of k as a function of size span s for uniform distribution by volume fractions.

In order to characterize the variations of local coordination numbers, we consider here the "connectivity function" P_k defined as the proportion of particles inside a sample with exactly k contacts. We have $\sum_k P_k = 1$ and $\sum_k k\ P_k = z$. Figure 3b shows the plots of P_k for different values of k as a function of size span s for uniform distribution by volume fractions. At very narrow spans, we have $P_4 \simeq 0.8$ and $P_3 = P_5 \simeq 0.1$. As s increases, P_3 and P_5 increase together at the expense of P_4 which declines to 0.6 at $s = 0.2$. In the range $s < 0.2$, $P_k = 0$ for all k except $k = 3$, 4, and 5. The evolution of topological disorder in this range is thus governed by the growth of two populations of particles with 3 and 5 contacts.

The populations of particles with 6 and 2 contacts appear at $s = 0.2$ and increase with s. The particles with 2 contacts are the small particles which are supported by two particles but which support none. Hence, P_2 represents the importance of geometrical arching which leads to the screening of smallest particles. This population grows rapidly with polydispersity and does not seem to saturate at highest values of s. A consequence of the growth of this population is to reduce the coordination number below 4; see Fig. 3a.

4.2 Fabric Anisotropy

The anisotropy of the contact network is known as a major mechanism of shear strength in granular media [10, 11]. This anisotropy can be expressed through the probability density function $P(\theta)$ of the orientations θ of normals $\boldsymbol{n}$ to the contact planes. Figure 4 displays $P(\theta)$ for uniform distribution by volume fractions for several values of s. This is a bimodal symmetric distribution with two peaks at $\theta = 45°$ and $\theta = 135°$ for low values of s. The peaks flatten with s and tend to the center of the distribution at $\theta = 90°$. Angular distributions of bimodal feature have also been observed in experiments and dynamic simulations of granular beds prepared by random deposition [12].

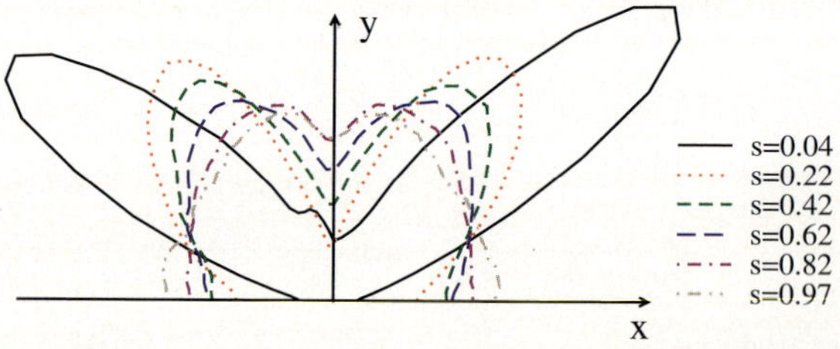

Fig. 4. Polar representation of the probability density function $P(\theta)$ of the orientations θ of contact normals for uniform distribution by volume fractions and several values of particle size span s.

The anisotropy of $P(\theta)$ can be extracted from the fabric tensor F_{ij} defined by [10]

$$F_{ij} = \int_0^{\pi} n_i(\theta) n_j(\theta) P(\theta) d\theta. \tag{13}$$

The anisotropy of the distribution $P(\theta)$ is given by $a_c = 2(F_1 - F_2)$, where F_1 and F_2 are the principal values of F_{ij}. However, this definition applied to

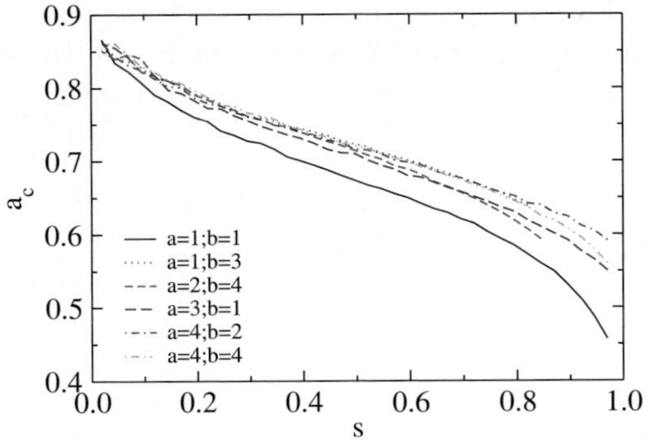

Fig. 5. Fabric anisotropy a_c over a quarter as a function of size span s for several values of shape parameters a and b.

the distributions observed in Fig. 4 yields low values of a_c as a result of the symmetry of these distributions. It is thus reasonable to restrict the definition of the fabric tensor in Eq. (13) to the sub-interval $[0, \pi/2]$. The corresponding anisotropy a_c then represents the importance of the peak in each sub-interval. Figure 5 shows this half-interval anisotropy as a function of s for several values of shape parameters. We see that a_c falls off with s from 0.85 for $s \simeq 0$ to values below 0.5 for $s \simeq 1$ in the case of uniform distribution by volume fractions. This behavior does not depend crucially on the shape parameters, the lowest anisotropies being observed for $a = b = 1$. We also remark that, the rate of variation of a_c with s involves two transitions occurring at $s \simeq 0.2$ and $s \simeq 0.8$.

5 Conclusion

In this paper, we presented a systematic investigation of microstructure in densely-packed polydisperse granular media. A model of cumulate volume distributions of the particles, based on β-distribution, was proposed. This model accounts both for size span (or the width) and the shape of size distributions. These cumulate volume distributions can be simple- or double-curved, and well-known size distributions such as monodisperse, bi-disperse, and power laws are particular instances of this function.

We then addressed in this framework two major aspects of polydisperse media: 1) space-filling properties in terms of solid fraction and radial distribution functions and 2) contact network in terms of connectivity disorder and anisotropy. The solid fraction increases in a strongly nonlinear manner with size span and a transition occurs from a basically "topological disorder" regime to a "metric disorder" regime around a particular value of size span depending on shape parameters.

In the first regime, the particle size distribution manifests itself mainly through the variability of local coordination numbers. In this regime, short-range correlations of particle positions prevail, and the solid fraction evolves slowly with size span. The metric disorder regime is governed by the aptitude of the small particles to fill the pores left by larger particles. The positions of neighboring particles are no more correlated, the only local order being the presence of a low-density region around each particle. We find that the solid fraction increases considerably with size span in this regime and the highest values of solid fraction are obtained with the uniform distribution by volume fractions. The orientational ordering of the contact network was investigated by means of the fabric anisotropy. The contact orientations define a bimodal distribution induced by the deposition protocol. The concentration of contact normals in each mode can be characterized by a fabric anisotropy. We showed that the fabric anisotropy in each mode decreases with size span. In particular, the uniform distribution by volume fractions is practically isotropic at high degrees of polydispersity.

References

1. H. J. Herrmann, R. Mahmoodi Baram, and M. Wackenhut. *Physica A: Statistical Mechanics and its Applications*, 330(1-2):77–82, December 2003.
2. T. Aste. *Phys. Rev. E*, 53(3):2571–2579, March 1996.
3. P. S. Dodds and J. S. Weitz. *Phys. Rev. E*, 65(5):056108–, May 2002.
4. M. Ammi, D. Bideau, and J. P. Troadec. *J. Phys. D: Appl. Phys.*, 20:424, 1987.
5. A. Gervois and D. Bideau. In D. Bideau, editor, *Disorder and Granular Media*, Amsterdam, 1993. Elsevier.
6. J. K. Mitchell and K. Soga. *Fundamentals of Soil Behavior, third edition*. Wiley, 2005.
7. R. Jullien and P. Meakin. *Colloids Surf. A: Physicochem. Eng. Aspects*, 165(1-3):405–422, May 2000.
8. I. Bratberg, F. Radjai, and A. Hansen. *Phys. Rev. E*, 66(3):031303–, September 2002.
9. C. Voivret, F. Radjai, J.-Y. Delenne, and M. S. El Youssoufi. *Phys. Rev. E*, 76(2):021301–12, August 2007.
10. M. Satake. In P. A. Vermeer and H. J. Luger, editors, *Proceedings of the IUTAM symposium, Delft*, 63–68, Amsterdam, 1982. A. A. Balkema.
11. F. Radjai, D. E. Wolf, M. Jean, and J.-J. Moreau. *Phys. Rev. Lett.*, 80(1):61–64, 1998.
12. F. Calvetti, G. Combe, and J. Lanier. *Mech. Coh. Frict. Materials*, 2:121–163, 1997.

Kinetic Energy Fluctuations and Diffusivity in a 2D Vibrated Granular Packing

Rim Harich[1], Geoffroy Lumay[2], and Eric Clément[1]

[1]PMMH, UMR 7636 du CNRS, ESPCI, 10 rue Vauquelin, 75005 Paris, France
[2]GRASP, Inst. de Physique B5, Univ. de Liege, B-4000, Liege, Belgique

Summary. We present an experimental realization of a 2D vibrated granular packing. The new agitation method allows a spatially non synchronized influx of energy and the study of the vibrated packing at steady state. By image analysis of fast-camera movies, we obtain the velocity fluctuation spectra at different vertical levels and then, we separate the agitation velocities from the velocity fluctuations corresponding to the "thermalized" degrees of freedom. By measuring the corresponding particle diffusivities, we show that, in spite of a large heterogeneity and anisotropy of the vibration, a relation between diffusivity and fluctuating kinetic energy can be identified.

1 Introduction

In the recent decades, there were a tremendous amount of work dealing with vibrated granular matter either from numerical, theoretical or experimental points of view (see for example all refs in bibliography and refs inside). The phenomenology is very rich as vibrated packing were found to bear the form of dissipative gases [1–4], liquids [5–8] or even a "glassy like" material [10, 12–14]. From an experimental point of view and for granular packing under gravity, the vibration procedure usually goes two ways i) an harmonically driven bottom plate [4–6, 8] or ii) successive taps separated by quiescent periods [7, 10, 12, 14]. In the case of harmonically driven packing, one has to go to very high frequencies in order to achieve desynchronization between the energy injection and the packing response [4]. Thus, in general, one gets an unsteady flux of energy (strongly modulated in time and synchronized in space). Here we present a novel mode of energy injection that breaks the spatial synchronization and allow to reach at moderate injection energy a steady state. We study a 2D model packing of cylinders and we investigate the relation between kinetic fluctuations and diffusivity of a particle.

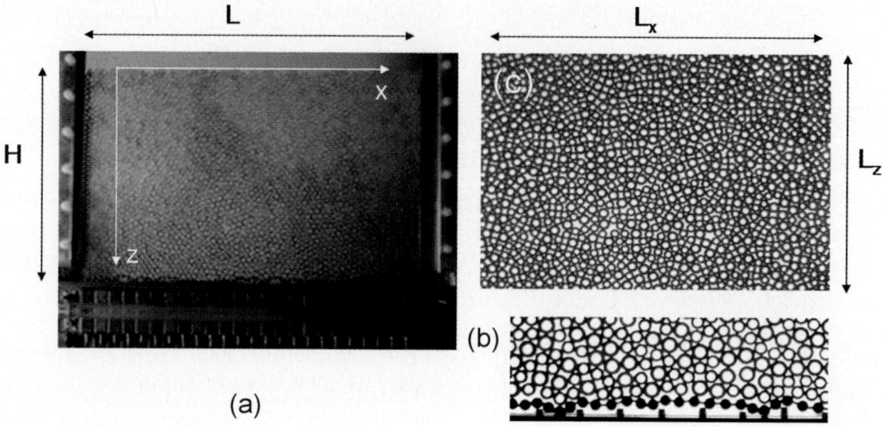

Fig. 1. Experimental set-up; (a) Picture of the cell of size $L = 275mm$, $H = 200mm$ with the 24 electromagnetic activators at the bottom; (b) Close up on the vibrating chain driven by the electromagnetic pistons; (c) 2D packing in the visualization frame of size $L_x = 118mm$, $L_z = 90mm$.

2 Experimental Setup

The packing is made of a 2D binary mixture of acrylic cylinders $d_1 = 4mm$ and $d_2 = 5mm$ which are hollow in the middle. Overall, we have 1680 small grains and 1016 large ones, mixed randomly in a cell of size $L = 279mm$ large and $4mm$ width. The packing height at rest is $H_0 = 171mm$. To input energy we inspire from what is done currently in numerical simulations [1, 3, 9], where the granular packing is in contact with a "temperature reservoir" uncoupled in space. Of course this is difficult to achieve in practice, but we tried to break as much as possible the spatial synchronization of energy transfer by the use of a vibrating chain at the bottom of the cell. The chain is activated by 24 pistons separated by $12mm$. Two neigboring pistons are vibrated at a driving frequency f_D with a phase shift of π plus a small random phase shift that can be set manually. This is not exactly a "pure thermal" wall but as we will see in the following, it allows to reach easily a steady-state where a clear separation between energy of agitation and "thermalized" degrees of freedom can be achieved. The visualization of the packing was made by a high resolution and fast CCD camera. The spatial resolution is 1632×1200 $pixel^2$ which allows a visualisation field of about 38 grains horizontally and 26 grains vertically (about 44 $pixel$ per small grain). For each picture, one uses a pattern recognition technique to locate the center of the hollow circle. This allows an individual tracking of about 900 grains at the same time. For the present report, the pistons were driven at $f_D = 20Hz$ and the camera was run at 1000 im/s during 3 seconds in order to obtain the velocity fluctuations. Thereafter, the camera was run at 1 im/s during 3000 seconds in order to obtain the long time diffusivity of the grains. To analyze the data, we choose to separate the visualization field into 6 horizontal levels and to average over boxes containing about 150 grains.

3 Velocity Agitation and Velocity Fluctuations

The fast movie acquisition allows to access the true velocity for each grain and not just the average displacements. We first compute the average quadratic velocity fluctuations at each level. The results are displayed on fig. 2(a). We observe that the driving induces a very inhomogeneous and anisotropic agitation state. The vertical degrees of freedom are much more agitated than the horizontal ones as already noticed in earlier work [15]. We also observe a strong decay of agitation from the wall to the free surface. A look at the velocity fluctuation spectrum shows that the agitation energy can be separated into two distinct parts: the input kinetic energy and the "thermalized" kinetic energy. The inset of fig. 2(b) illustrates this point. We observe that the velocity fluctuation spectrum is made of two pieces: the fundamental and the harmonics of the driving frequency (the very spiky features) and the "thermal" fluctuations with a specific wide band structure. We found that once the driving part is filtered, we could fit the velocity fluctuation spectrum $S(f)$ for each direction with a very good accuracy by the three parameters function:

$$S_{fit}(f) = S_0 \frac{\beta^2(1+(f/f_0)^2)^2}{(f/f_0)^2(1+(f/f_0)^2)^2+\beta^2} \qquad (1)$$

with S_0, f_0 and β, are the 3 fitting parameters. This relation yields a "resonance frequency": $f^* = f_0\left(2^{1/3}\beta^{2/3}-1\right)^{1/2}$, corresponding to the curve maximum. The "width" of the curve increases linearly with f^*. The velocity spectra

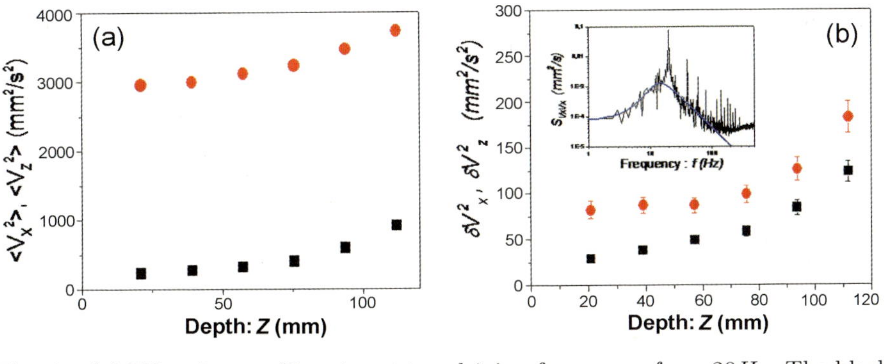

Fig. 2. (a) Vibration profiles at a piston driving frequency $f_D = 20Hz$. The black squares (red circles) are the horizontal (vertical) direction. Fig. 2(a) velocity fluctuation profiles: $< V_x^2 > (z)$ and $< V_z^2 > (z)$. (b) Velocity fluctuation profiles: $\delta V_x^2(z)$ and $\delta V_z^2(z)$. On the inset, we display the average velocity fluctuation spectrum $S(f)$ for the horizontal velocity V_x at $z = 57(\pm 9)mm$. The solid line is the best fit using equ. (1) with $S_0 = 6(\pm 0.6)10^{-5}mm^2/s$, $f_0 = 7.2(\pm 0.4)Hz$ and $\beta = 15.5(\pm 1.5)$.

are normalized in such a way that the quadratic fluctuations can be obtained via the Parseval identity: $\delta V^2 = \frac{1}{N^2}\sum S_n$, where S_n is the nth mode of the velocity spectrum identified with his fitted value. On fig. 2(b) the quadratic velocity fluctuations for the horizontal and vertical degrees of freedom, δV_x^2 and δV_z^2, are displayed. We also observe anisotropy as well as a vertical profile for the quadratic velocity fluctuations.

4 Diffusivity

The second experiment (1 image per second) allows to follow the mean square displacements for each particle over much longer time. On fig. 3(a), we present the mean square displacements for time lags τ ranging from $10^{-3}s$ to $10^2 s$. At very short time scales, we observe a ballistic dynamics then a pseudo-plateau over which the driving oscillations are visible. For longer times, we get a diffusive behavior. For each direction x or z and for each of the 6 levels, we measure the diffusivity D_i over the direction i by a linear fit of $\sigma_i^2 = A_i + 2D_i\tau$. The results for the corresponding diffusivities $D_x(z)$ and $D_z(z)$, are displayed on the inset of fig. 3. We also observe slightly anisotropic and spatially inhomogeneous variations of diffusivity.

For a simple thermalized fluid, close to equilibrium, the self-diffusion coefficient of a particle of diameter d and mass m, is related to temperature by the fluctuation-dissipation relation: $D = \frac{kT}{m}\tau_D$, where k is the Boltzman

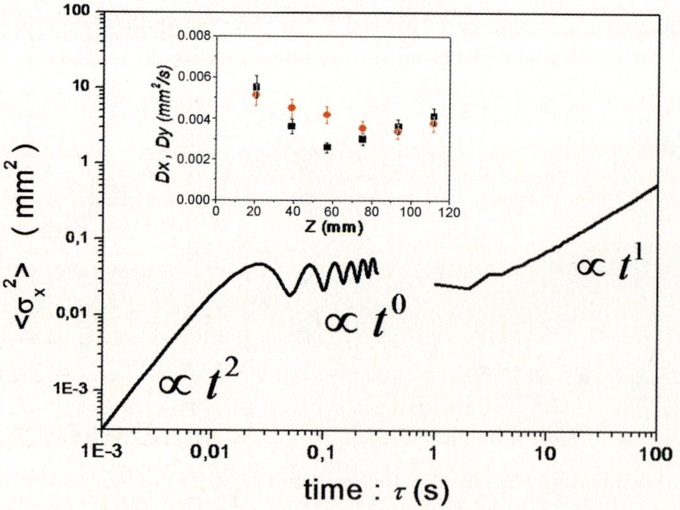

Fig. 3. Self diffusivity at a driving frequency $f_D = 20Hz$. Mean quadratic displacement $< \sigma_x^2 > (z)$ at short and long time for $z = 57(\pm 9)mm$. On the inset, we display the diffusivity profiles. The black squares (red circles) are the horizontal (vertical) direction.

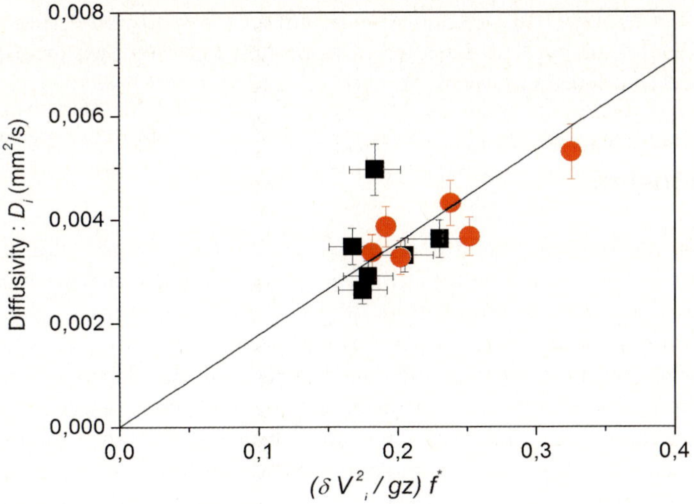

Fig. 4. Self diffusivity in horizontal (black squares) and vertical (red circles) directions as function of the rescaled frequency $\sqrt{\frac{\delta V_i^2}{P/\rho}} f^*$. The solid black line corresponds to equ. (2) with $\xi = d/30$.

constant, T the temperature and τ_D a dissipative time scale. For a Newtonian fluid this time scale is related to viscosity η by the relation $\tau_D^{-1} = 3\pi\eta d/m$. For granular gases such relations were derived rigorously [1] but for dense packing, early experimental works have tried to establish such a relation [6, 11] and essentially failed. We also tried such a rescaling for the diffusivity coefficients identifying the ratio $\frac{kT}{m}$ with the velocity fluctuations δV_i^2. Then, we obtained dissipative times $\tau_D = D_i/\delta V_i^2$ with typical values around $10^{-6}s$, much lower than any time scale a priori present in this system. However, we found a fair rescaling of our data when the diffusivities were plotted as a function of the dimensionless number $\sqrt{\frac{\delta V^2}{P/\rho}}$, where $P = \rho g z$ is the confining pressure. Furthermore, the anisotropy between the x and z directions, can significantly be reduced if one uses for the diffusivity time scale, the "resonance" frequency f^* extracted from the fit of the thermalized background. On fig. 4, we display for the directions x and z, the corresponding diffusivities as a function of the rescaled frequencies $\sqrt{\frac{\delta V^2}{P/\rho}} f^*$. We observe a fair collapse of the data onto a single linear curve that goes through zero. Therefore, this study suggests a relation between diffusivity and thermal activation, valid for every degree of freedom and of the type:

$$D_i = \sqrt{\frac{\delta V_i^2}{P/\rho}} \xi^2 f^* \tag{2}$$

where ξ is a typical length scale with a value $\xi = d/30$. Note that this scale is the order of the typical distance of approach between two grains (besides contact) in the dense compacted phase.

5 Conclusion

We presented an experimental study of a vibrated 2D granular packing of grains driven at a steady-state. In spite of the large anisotropy between the horizontal and the vertical degrees of freedom, we could extract a unique relation linking the thermal agitation and the particle diffusivity. This relation involves the confining pressure which depends on the depth below the free surface. Note that this relation also suggests in the framework of a fluctuation-dissipation theorem, an effective "viscous" drag force. Such a relation can be tested directly by observing the motion in the vibrated packing. Furthermore, it will be of great interest to study how this result holds in the limit of low agitation energy where a dense "glassy" dynamics is obtained. These are on going projects in our laboratory.

References

1. Brey J.J. et al. (2002), Phys. Rev. E **65**, 061302.
2. Wildman R.D. and Huntley J.M. (2003), Phys. of Fluids **15**, 3090.
3. Meerson B., Pöschel T., and Bromberg Y. (2003), Phys. Rev. Lett. **91**, 024301.
4. Eshuis P. et al. (2005), Phys. Rev. Lett. **95**, 258001.
5. Clement E. and Rajchenbach J. (1991), Europhys. Lett. **16**, 133; Clement E. et al. (1992), Phys. Rev. Lett. **69**, 1189; Vanel et al. (1996), Phys. Rev. E **53**, 2972.
6. Zik O., Stavans J. and Rabin Y. (1992), Europhys. Lett. **17**, 315.
7. Knight J.B. et al. (1995), Phys. Rev. E **51**, 3957.
8. Melo F. et al. (1994), Phys. Rev. Lett **72**, 172; Mujica N. and Melo F. (1998), Phys. Rev. Lett. **80**, 5121.
9. Ramirez R. et al. (2000), Phys. Rev. Lett. **85**, 1230.
10. Knight J.B. et al. (1996), Phys. Rev. E **54**, 5726.
11. d'Anna G. and Gremaud G. (2001), Phys. Rev. Lett. **87**, 254302; Nature (London) (2001) **413**, 407.
12. Philippe P. and Bideau D. (2003), Europhys. Lett. **60**, 667.
13. Caballero C. et al. (2005) in Powders and Grains 2005, p. 339 Eds S. Luding, H. Herrmann (Balkema, 2005).
14. Lumay G. and Vandewalle N. (2005), Phys. Rev. Lett. **95**, 028002.
15. van der Meer D. et al. (2006), Europhys. Lett. **74**, 384-390.

Modelling and Simulation of a Maze-Forming Process in Granular-Fluid Systems

Henning Arendt Knudsen, Bjørnar Sandnes, Eirik Grude Flekkøy, and Knut Jørgen Måløy

University of Oslo, Dept. of Physics, PB 1048 Blindern, NO-0316 Oslo, Norway
hak@fys.uio.no

Summary. Experiments on granular-fluid systems in confined geometries have been shown to produce labyrinth patterns. This is the case when the fluid from a particle/fluid dispersion in a Hele-Shaw cell is slowly withdrawn, resembling a drying process. Based on a model of capillary and friction forces, we present a simulator that reproduces the experimental behaviour.

1 Introduction

In systems driven out of equilibrium spontaneous formation of pattern and form is known to emerge in virtually every discipline of science. This can often be traced back to competing forces that govern some instability [1]. A particular pattern, governed by the competition between capillary forces and friction forces, is the labyrinthine residue of grains in confined geometries after drying out a fluid/grain mixture [2, 3]. This has been studied in a Hele-Shaw cell, where glass beads of diameter 50–100μm immersed in a water/glycerol mixture were first rapidly injected into the cell. Thereafter the fluid was pumped out again at a very low rate. Here, we emphasise the modelling of the process and its implementation into a simulator, while the experimental system and results are treated in a parallel report [4].

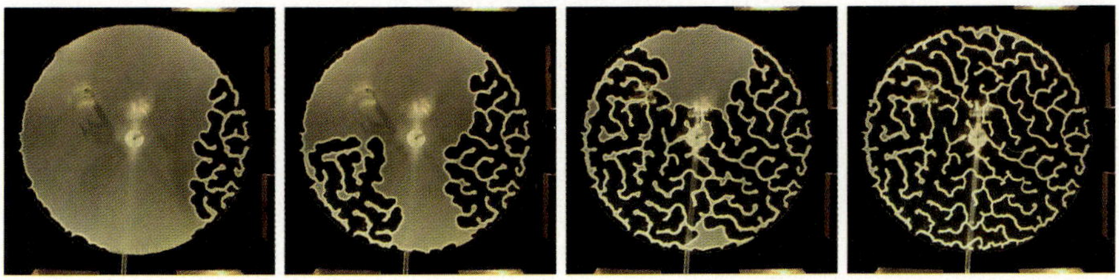

Fig. 1. A typical experiment of the draining process.

2 Experiment

As the fluid is gradually drained from the cell, the fluid/air interface at the perimeter of the circular disc starts to recede. The capillary forces between the wetting fluid and the grains gradually compile a growing layer of close-packed grains ahead of the interface as it moves. After an initial transient period where a compact layer of grains is formed all along the circular perimeter, an instability develops whereby the fluid-grain disc is slowly invaded by fingers of air. Figure 1 shows four consecutive images of this process, see [4] for more details.

3 Simulations

At every point on the interface capillary pressure and friction forces must be overcome to allow the front to move. The expression for the forces have been derived and reads [3]

$$\frac{\gamma}{R_c} + \frac{g\rho\Delta z}{2\kappa}\left[(\kappa\mu + 1)\exp\left(\frac{2\mu\kappa L}{\Delta z}\right) - 1\right], \tag{1}$$

where γ is the interfacial tension, $1/R_c$ the in-plane curvature of the front at that point, g the acceleration of gravity, ρ the relative density difference between grains and fluid, Δz the plate separation, κ the proportionality constant between vertical and horizontal stress in the front, μ the friction coefficient, and L the front thickness at that point. In the simulations the air-liquid pressure force is increased until somewhere along the interface it exceeds the threshold force given in Eq. (1). The front is then moved a tiny step in the normal direction at that point, and the procedure is repeated.

The model is based on a one-dimensional representation of the interface, where thousands of consecutive points discretize the perimeter of the fluid/grain area. In Fig. 2(a) five of these points are shown at some typical situation. Motion is from the air side towards the fluid side. Technically speaking, the fundamental information that needs to be stored at each point is the co-ordinates and the accumulated particle mass. All other quantities may be derived from these, as will be explained in the following.

The accumulated particle mass in a segment of the front is for convenience handled as a volume, the total volume of fluid and particles, which is filled by a dense packing of these particles. Numerically this corresponds to normalising the particle fraction to one in the dense region within the front. We denote the volume of the point with index i by V_i. Further, we denote the distance between neighbouring points, which is easily calculated, by l_{ij}, where i and j are neighbours. The plate separation Δz is fixed and thus the width of the front in a point can be calculated. For point no. 3 in Fig. 2(a), the front width becomes

$$L_3 = \frac{V_3}{\Delta z \left(\frac{l_{23}}{2} + \frac{l_{34}}{2}\right)} , \tag{2}$$

where the front length associated with the point is the sum of the two half-distances to its two neighbours. The resulting front width is illustrated with a vector pointing towards the fluid side in the Figure. The orientation of the vectors is always normal to the front. In the case of point 3, one sees how the vector is aligned along the dashed line. This line divides the angle $\angle 234{=}2\beta$ exactly in two equal angles β.

The curvature is also evaluated locally, only using the nearest neighbours, and it is approximated by, again for point no. 3,

$$\kappa_{c,3} = \frac{1}{R_{c,3}} = \frac{\pi - 2\beta}{\frac{l_{23}}{2} + \frac{l_{34}}{2}} . \tag{3}$$

Recall that the curvature, equal to the inverse radius of curvature, is change in tangential orientation (angle change) per arc length.

The information calculated so far suffices to evaluate all points in a given configuration and to select the point which yields first. Once selected, the point in question is moved a tiny step Δs inwards, normal to the front. This step has been chosen to be much smaller than the initial spatial resolution of the line: neighbour point distance l_{init}=1.5mm, which again is small enough to resolve the final structure of the pattern.

3.1 Moving Points

The moved point gathers volume on its way. Although the front is represented as a line, we keep in mind the fact that it is the inner side of the actual front that gathers particles. The increase in volume, again for point 3 in Fig. 2(a), becomes

$$\Delta V = \frac{\phi_{\mathrm{local}}}{1 - \phi_{\mathrm{local}}} \Delta s \left(\frac{l_{23}}{2} + \frac{l_{34}}{2}\right) , \tag{4}$$

where ϕ_{local} is the local volume fraction associated with the position of the point, to be defined below. Note that when point 3 is moved to give the situation in Fig. 2(b), only its own mass is changed. However, the front lengths of the points 2, 3, and 4 are changed, from which the front widths L of these points must be recalculated. The same is true for orientation and curvature of these three points. The simulation is implemented with an average spatial volume fraction of mass that represents the filling of grains in the Hele-Shaw cell. Disorder is included by locally allowing the volume fraction of mass to fluctuate around the average value. To estimate the local fraction, the average fraction is multiplied by a number within $I{=}(1 - \epsilon, 1 + \epsilon)$. We have placed a 120×120 virtual lattice of random numbers, which are elements in I, on top of the initial disk. The number of a given point in space is then taken as the linear interpolation between its four nearest lattice points.

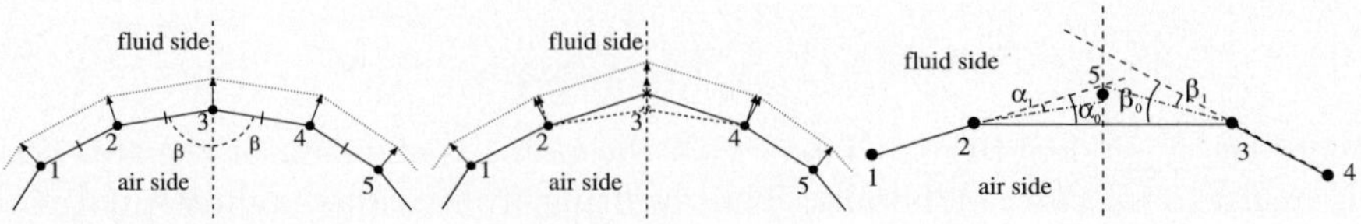

(a) before a move (b) directly after a move (c) inserting a point

Fig. 2. (a-b) Five points on the front are shown. For each point the direction normal to the front is drawn as a vector of length corresponding to front thickness L. The orientation, the curvature, and the front thickness change after a move. For clarity in the drawing the displacement in a single move is exaggerated. (c) Inserting a point: The angles α_0 and β_0 refer to the configuration before the insertion of point 5. The angles α_1 and β_1 are the ones that best preserve the curvature of the points 2 and 3, respectively.

3.2 Inserting Points

In order to maintain a constant spatial resolution, new points are added to the interface as it stretches. Numerically, a limit is set at $1.1 \times l_{\text{init}}$. Whenever two neighbours come further apart a new point is inserted in-between, equally distant from each point. See Fig. 2(c) where a new point 5 is inserted between the points 2 and 3. Along the line of possible points of equal distance (dashed vertical line), the one point is chosen which conserves the curvature. Conservation of curvature is preserved for the left and right side, respectively, when

$$\alpha_1 = \frac{l_{12} + l_{23}/2}{l_{12} + l_{23}} \alpha_0 \text{ , and } \beta_1 = \frac{l_{34} + l_{23}/2}{l_{34} + l_{23}} \beta_0 \ . \tag{5}$$

Generally, this means that the best one can do is to take a middle value, here expressed in terms of the angle χ between the lines $\mathbf{l}_{23}$ and $\mathbf{l}_{25}$,

$$\chi = \frac{1}{2} (\alpha_0 - \alpha_1 + \beta_0 - \beta_1) \ . \tag{6}$$

Volume must also be preserved upon insertion of new points. Volume from the two neighbours is transferred to the new point. Again referring to the numbering in Fig. 2(c), the volume transfer from the points 2 and 3 to 5 becomes

$$V_5 = -\Delta V_2 - \Delta V_3 = \frac{1}{2} \frac{V_2 l_{23}}{l_{12} + l_{23}} + \frac{1}{2} \frac{V_3 l_{23}}{l_{23} + l_{34}} \ . \tag{7}$$

Now, the front thickness L, the orientation and curvature is calculated anew for the points 2, 3, and 5.

3.3 Detection of Front Contact

The process of moving points and gathering mass continues as long as there are points left that are allowed to move. Points whose extended front of mass comes in mutual contact are immobilised as this represents the formation

(a) 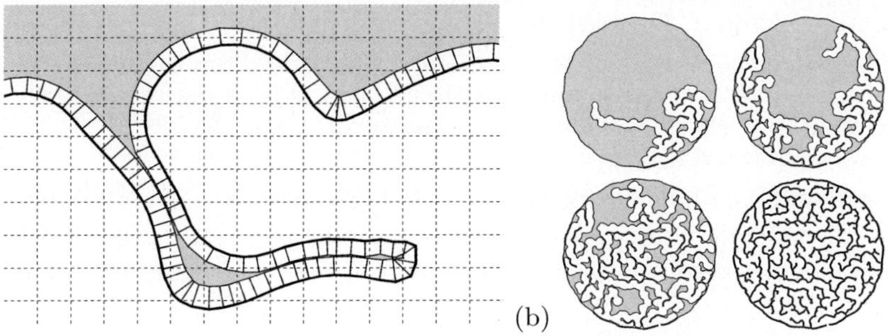 (b)

Fig. 3. (a) Snapshot of a section of a simulation. The mesh is not a part of the physical system, but merely an illustration of a numerical tool. The white area is air, the shaded area is fluid/particles. The thicker line towards the air side is the numerically stored line. A branch of grains is formed when the inside of two front segment come in contact with each-other. (b) A typical simulation of the draining process. At four different times the structure is shown, white area means air and shaded area means fluid/particles. The fourth image depicts the final structure when the whole structure is drained, numerically speaking when all points are immobilised.

of a branch of close-packed grains. Figure 3(a) shows a section of the front during the formation of a branch. The white area is air and the shaded area is fluid/particles. Note how the small pocket of isolated fluid/particles within the particle branch resemble the experimental situation. This area will be drained later when all easier parts are drained, and the pressure difference between fluid and air increases.

The numerical detection of contact is somewhat tricky in that one has to check if the line segments, representing the inside of the front of the last point moved, are crossing any other line segment along the front. In order to do this efficiently, a square grid of cells as depicted in Fig. 3(a) is used. To every cell a list of points currently inside the cell is stored. These lists are maintained upon moving and inserting points. Since collisions only need to be checked for locally, only the points in the cell belonging to last point moved and its eight neighbours must be checked. In this way the check for collisions represents a roughly constant computational cost, independent of system size and perimeter length.

4 Results and Discussion

The success of the modelling and the simulator of the process has been validated by means of visual inspection and measures of the final structure. In Fig. 3(b) for otherwise typical parameters, a simulation is shown, where the plate separation is Δz=0.4 cm and the initial volume fraction of grains is ϕ=0.15. The final structure should be compared with the final experimental structure of Fig. 1. One sees how a typical length scale appears in the structure. As a measure of this so-called wavelength we use $\Lambda = 2A/S$, where A is the initial area of the disc and S is the measured final perimeter of the structure. In Fig. 4 the final structure is shown as a function of plate separation and particle fraction, and as one can see, the typical length scale varies much. We refer to a parallel work [4] for a detailed comparison with experiments. Here we conclude that we have, based on modelling of the dominant forces in the problem, presented a simulator that captures the essential behaviour of the process.

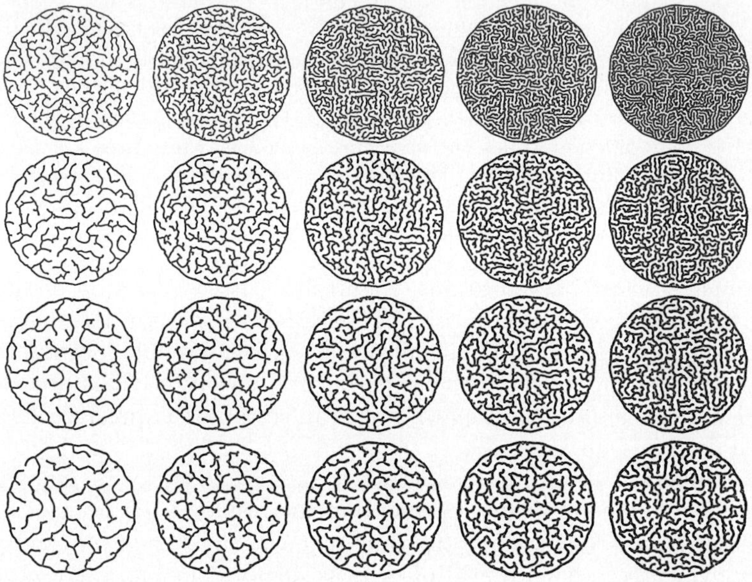

Fig. 4. Final structure as a function of: particle fraction in columns from left to right – 10%, 15%, 20%, 25%, and 30%. Plate separation in rows from top to bottom – 0.2 cm, 0.4 cm, 0.6 cm, and 0.8 cm.

References

1. M. C. Cross and P. C. Hohenberg (1993) *Pattern formation outside equilibrium*, Rev. of Mod. Phys., 65:851–1112.
2. Y. Yamazaki and T. Mizuguchi (2000) J. Phys. Soc. Jpn., 69:2387–2390.
3. B. Sandnes, H. A. Knudsen, K. J. Måløy, and E. G. Flekkøy (2007) Phys. Rev. Lett. 99:038001; and in preparation.
4. B. Sandnes, H. A. Knudsen, K. J. Måløy, and E. G. Flekkøy. Pattern formation in slowly drained granular-fluid systems. In: *Traffic and Granular Flows 2007 - Proceedings.*

Description of Wire-Reinforced Geomaterials by Numerical Discrete Experiments

Romain Laniel, Pierre Alart, and Stéphane Pagano

LMGC – UMR 5508, Université de Montpellier 2, cc 048, 34095 Montpellier cedex 5, France laniel@lmgc.univ-montp2.fr

1 Motivations

Civil engineers are often expected to build larger and larger constructions; the ground needs assistance in supporting such works. This paper focuses on a particular process: *TexSol*™. It is a soil reinforcement process created in 1984 by Leflaive, Khay and Blivet from the LCPC (Laboratoire Central des Ponts et Chaussées) [1, 2]. Its originality lies in combining the soil (sand) with wires. Although the wire volume is negligible compared to that of the sand, the wire becomes a strong reinforcement when it tangles up inside the geomaterial. They lead this mixed material friction angle to be larger than sand by 0° to 10° [2]. In the literature, we find two different continuous modellings. The model suggested in [3] is non local and includes remote interactions, but requires identification of their parameters using macroscopic experiments. Villard proposes a simpler local model in [4]. It couples a standard model of sand and an equivalent unilateral elastic stiffness contribution corresponding to the wire network. This last contribution is activated only on the tension directions because of the unilateral behavior of wire. This study interest lies in multi-scale theoretical contributions of unilateral structures with long internal length. Thus, we propose to clearly define thermo-dynamical potentials of the Villard local model. Such a stage is useful before carrying out a homogenization procedure applied to an untypical material. In the absence of physical experiments, the identification of macroscopic model will be performed using discrete numerical experiments. Those allow to study in detail the microstructure and reinforcement interactions on a microscopic scale.

2 Thermodynamical Modelling in a Local Formalism

The local model proposed by Villard couples an elastic-plastic model with an isotropic and kinematic hardening combination for the sand phase, with an equivalent unilateral elastic stiffness for the wire network phase. The unilateral characteristic of this feature means that stiffness is only activated in tensile directions. To "superpose" the elasto-plastic model of the sand and the unilateral elastic model of the wire network, some mechanical assumptions have to be considered: the strain rates equalities $\dot{\boldsymbol{\varepsilon}}_s = \dot{\boldsymbol{\varepsilon}}_w = \dot{\boldsymbol{\varepsilon}}$ and the stress additivity $\boldsymbol{\sigma} = \boldsymbol{\sigma}_s + \boldsymbol{\sigma}_w$. Those two assumptions on the two phases are quite restrictives and they would have to be verified by discrete simulations before carrying out an identification procedure. We choose $\boldsymbol{\varepsilon}$ as the state variable combined with the internal variables $\boldsymbol{\varepsilon}^{\mathrm{p}}$, $\boldsymbol{\alpha}$ and p describing the plasticity, kinematic and isotropic hardening respectively. The model elastic parameters are the sand elasticity tensor $\mathbb{K}_s$, depending on the sand elasticity modulus E_s and ν_s, and the wire network coefficients of Lamé λ_w and μ_w. The hardening ones are H_i and H_k the isotropic and kinematic hardening respectively. The *TexSol*™ free energy is written as,

$$\begin{aligned}
&\psi(\boldsymbol{\varepsilon}, \boldsymbol{\varepsilon}^{\mathrm{p}}, \boldsymbol{\alpha}, p) = \psi_s(\boldsymbol{\varepsilon}, \boldsymbol{\varepsilon}^{\mathrm{p}}, \boldsymbol{\alpha}, p) + \psi_w(\boldsymbol{\varepsilon}) \\
&\psi_s(\boldsymbol{\varepsilon}, \boldsymbol{\varepsilon}^{\mathrm{p}}, \boldsymbol{\alpha}, p) = \tfrac{1}{2}\left(\boldsymbol{\varepsilon} - \boldsymbol{\varepsilon}^{\mathrm{p}}\right) : \mathbb{K}_s \left(\boldsymbol{\varepsilon} - \boldsymbol{\varepsilon}^{\mathrm{p}}\right) + \tfrac{H_{\mathrm{k}}}{2} \boldsymbol{\alpha} : \boldsymbol{\alpha} + \tfrac{H_{\mathrm{i}}}{2} p^2 \\
&\psi_w(\boldsymbol{\varepsilon}) = \tfrac{\lambda_w}{2} \langle \mathrm{tr}(\boldsymbol{\varepsilon}) \rangle^2 + \mu_w \boldsymbol{\varepsilon}^{\geq} : \boldsymbol{\varepsilon}^{\geq} \ ,
\end{aligned} \tag{1}$$

where $\langle . \rangle = \max(., 0)$. The operator $(.^{\geq})$, described more precisely in [5], is define by $\boldsymbol{\varepsilon}^{\geq} = \boldsymbol{P} \langle \mathrm{diag}(\varepsilon_1, \varepsilon_2, \varepsilon_3) \rangle \boldsymbol{P}^T$, where ε_1, ε_2, ε_3 and $\boldsymbol{P}$ are the principle values and the passage matrix of $\boldsymbol{\varepsilon}$ respectively. The wire network model is non dissipative, consequently the *TexSol*™ and the sand dissipation potentials are the same. We thus write the Legendre – Fenchel transform of the dissipation potential as,

$$\varphi^*(\boldsymbol{\sigma}^{ir}, \boldsymbol{A}, \boldsymbol{\chi}, R) = \mathrm{I}_{\{\mathbf{0}\}}(\boldsymbol{\sigma}_s^{ir}) + \mathrm{I}_{\Omega(\boldsymbol{\chi}, R)}(\boldsymbol{A}) \ , \tag{2}$$

where $\boldsymbol{\sigma}^{\mathrm{ir}}$, $\boldsymbol{A}$, $\boldsymbol{\chi}$ and R are thermodynamical forces associated to $\boldsymbol{\varepsilon}$, $\boldsymbol{\varepsilon}^{p}$, $\boldsymbol{\alpha}$ and p respectively. The indicative function of a $\mathcal{D}$ domain is noted $\mathrm{I}_{\mathcal{D}}(.)$. In the principle stresses space, Ω represents the elastic domain which is bound by a Drucker – Prager criterion depending on the internal friction angle θ_f and the cohesion C_0, as mentioned previously. The state laws and complementary laws associated to potentials (1) and (2) have been implemented and tested in the finite element software *Cast3M*™. Three different models are tested: a sand alone called "sand", a sand bilaterally reinforced called "reinforced sand" and a sand unilaterally reinforced called "texsol". A test is carried out to highlight the repartition of the wire network stress. We thus simulate a crushing test of 0.1% on a cylinder meshed by 400 linear elements. The equivalent Von-Mises stress σ^{eq} map on a axial/radial section of the sample ($\sigma^{\mathrm{eq}} = \sqrt{J_2(\boldsymbol{S})}$ is considered as the pseudo norm of the deviatoric stress tensor $\boldsymbol{S}$). The wire network equivalent stress level of the "texsol" is bounded by the two other models but the repartition is close to that of "reinforced sand". The wire network pressure map is shown in Figure 1 and we remark that the "texsol" generates only negative pressure in the wire network phase, due to its unilateral definition.

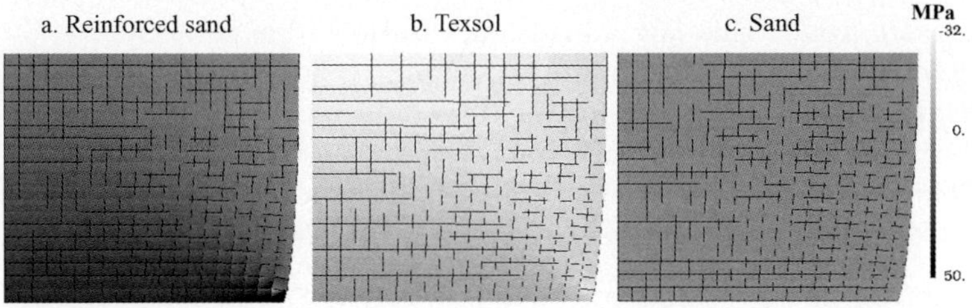

Fig. 1. Wire network pressure.

3 Numerical Discrete Investigations

In our approach, the measured fields, which are used to update continuous parameters, are replaced by discrete numerical experiments. Thus, all the particle displacements and contact forces can be extracted from the sample. By locally averaging the previous fields, the equivalent stress and strain tensors of the granular material are built up [6]. These provide more accurate information than the experimental method.

3.1 A Discrete Modelling of the *TexSol*™

The NSCD is a discrete element method used in the *LMGC90* code which simulates multibody vs. multicontact problems, privileging velocity fields [7]. For a single contact α problem, the NSCD evaluates the external forces and dynamic effects on the contactor point. To make such a transformation, $\boldsymbol{H}_\alpha$ and $\boldsymbol{H}_\alpha^T$ are used to move variables from the local contact frame to the global body and vice-versa. For a time step $i+1$, a linear relation between the relative velocity u_{i+1}^α and the averaged impulse p_{i+1}^α over $[t_i, t_{i+1}]$ is found; this is associated to a contact condition. In this way, for a frictionless problem with a Signorini contact condition, this system reveals to be a standard Linear Complementarity Problem (LCP). For a frictional contact problem, tangential reactions and tangential velocities have to verify a similar non smooth relation. A Gauss – Seidel loop computes all contact reactions until convergence.

In the *NSCD* framework, the wire network must be discretized. So, it is broken up into a collection of equidistant material points, with the wire mass equal to the sum of all point masses. All these points must be connected by a behavior law which accounts for a small segment of wire. The wire must keep its free flexion and unilaterality properties. Consequently, a wire contact law concerns only the normal direction and there is no constraint on the tangential directions. Thus, four laws, supported by contactor *points*, can be implemented in *LMGC90*, corresponding to four different wire behaviors:

- ***"Rigid rod"***: this couples the normal velocity of both candidate and antagonist particles.

- ***"Elastic rod"***: this law adds regularisation (due to elasticity k) to the contact problem in both compression and tensile directions.
- ***"Rigid wire"***: this is a **unilateral** law which couples the normal velocity of both particles only if the element strain tend to be positive.
- ***"Elastic wire"***: this includes unilaterality and the wire stiffness parameter k; *"elastic rod"* and *"rigid wire"* are coupled into this law.

An advantage of the "unilateral" laws is that the returned reaction on the contact element is only tensile, better accounting for the wire behavior.

3.2 Numerical Experiments

Once the *TexSol*™ sample clearly generates as a granular media, mechanical tests can be carried out. Large vs. small strain tests are distinguished to emphasize the differences in reinforcement mechanisms. A qualitative test on a *TexSol*™ slope consists in depositing a geometrically densified sample on a rubber plane. The initial sample includes a wire network, quasi-equiprobably distributed, discretized by beads which are connected with an *"elastic wire"* contact law. The simulation is carried out by *LMGC90* until sample kinetic energy is close to zero. Indeed the reinforcement structure was mobile and subsided following the sand particles. But this transformation leads the wire to form horizontal "stoppings" around the divergent particle flow which prevent sand circulation under gravity. The wire network becomes orthotropic. The slope friction angle of the *TexSol*™ become higher than the sand one; numerical and experimental values coincided [2]. This test dealt with reinforcement mechanisms in large transformations. But with a small strain background the wire network is not as mobile and may not generate long distance interactions as previously. Therefore, a triaxial test was carried out on a box-shaped sample. We also define a tool which is able to emphasize the interaction length inside a material (i.e. the *characteristic length* discriminating the local or non local behavior).

Definition. *We call* $\overline{\lambda}^m$ *the length of a wire segment with a tension higher than* mp_{avg}*, where* p_{avg} *is the average tension of the wire network and* $m \in \mathbb{R}^+$*. Moreover we use the maximum length* $\overline{\lambda}^m_{\max}$ *such as,*

$$\overline{\lambda}^m_{\max} = \sup_{s_1,s_2} \left\{ \int_{s_1}^{s_2} ds \,\middle|\, \forall s \in [s_1, s_2] \text{ with } s_1, s_2 \in [0, L], |p(s)| \geq mp_{\text{avg}} \right\} ,$$

where s*,* s_1 *and* s_2 *are curvilinear coordinates.* $m = 3$ *gives a good compromise to vanish weak segments and to highlight strong segments.*

We compare in Figure 2 the *characteristic length*, assimilated to $\overline{\lambda}^3_{\max}$, of the two tests; we also give in Table 1 the number of mid-ray particles of the granular media which is equal to this *characteristic length.* The test implying large transformations, leads to changes of the reinforcement structure which

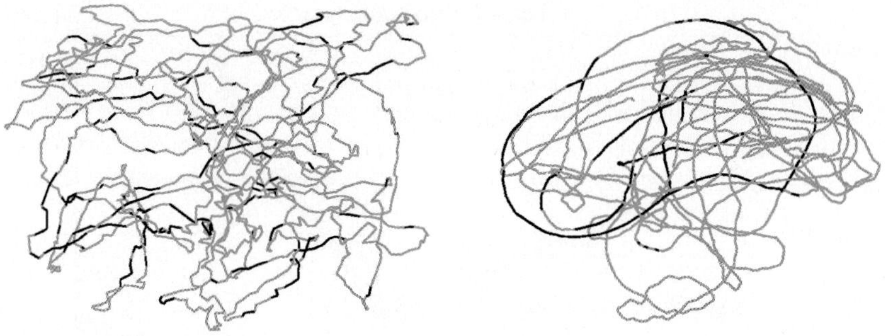

(a) Active wire segments after a triaxial test

(b) Active wire segments in a deposit slope

Fig. 2. Active wire segments distributions for both small and large strain assumptions.

Table 1. *Characteristic length* comparison for different tests

Tests	$\overline{\lambda}^3_{\max}$	Equivalence number of mid-ray particles
Slope deposit	21.56 mm	54
Triaxial test	4.29 mm	4

tend to linearize itself. This transformation is not alleviating, since it supports the propagation of the tension and then increases the remote effects. Quite contrary, the loading applied on triaxial test induces small strains which involve the reinforcement quasi-staticity and generate several small wire active elements. Consequently, this test is considered as a local one and dimensions of a representative elementary volume of *TexSol*™ must be, at least, higher than the *characteristic length*. The parameters of the wire discrete model may also have an influence on reinforcement mechanisms, especially the diameter of the wire beads. We simulate triaxial tests where the wire beads diameter are decreasing and we observe that the average strains of both phases diverge when wire beads are too small. The related physical phenomenon is the sudden large sliding of the wire with respect to the sand. This sliding occurs in the same direction leading to a non symmetry of the two horizontal wire strains. The thinner the wire, the more it slips. These relative slidings conflict with the assumptions of the continuous model. Thus, if an identification approach is performed using discrete element investigations, the validity limits can be so defined. During the triaxial process, the wire network stress level increases linearly as shown in Figure 3. The reinforcement is considered as a linear elastic structure for both "elastic" and "rigid" contact laws, which is surprising for the rigid network. The first has a stress shift due to the sample preparation above. This increases with wire stiffness but a strain shift appears, especially

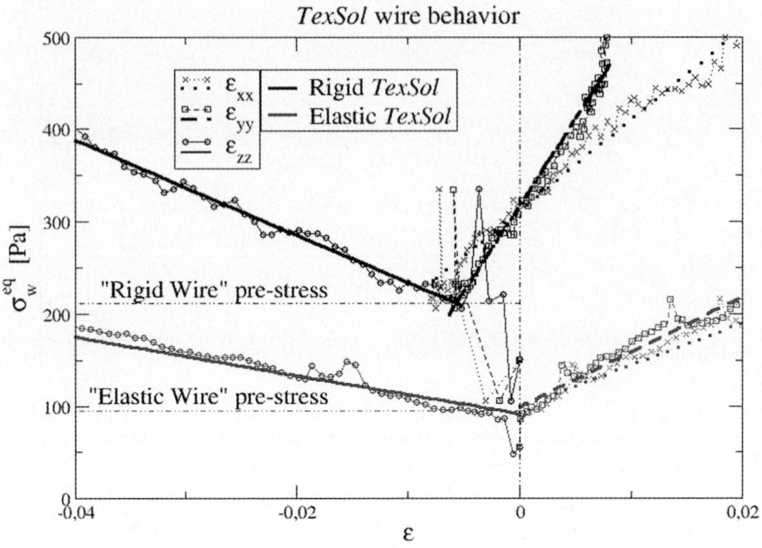

Fig. 3. Wire equivalent stress vs. strains plots for *TexSol*™ samples.

for the "rigid wire". From the very start of the triaxial test, wire behavior is disturbed by a brutal contracting reorganization amplified for "rigid wire". The slope of the $(\varepsilon_{zz}, \sigma_w^{eq})$ plot may represent a macro-stiffness. This one increases with the micro-stiffness k but not linearly. Indeed, it tends to stabilize around a limit macro-stiffness corresponding to that of "rigid wire".

References

1. E. Leflaive, M. Khay, and J-C. Blivet. Un nouveau matériaux: le texsol. *Travaux*, (602):1–3, 1985.
2. M. Khay and J-P. Gigan. Texsol - ouvrage de soutènement. Technical report, LCPC, 1990.
3. M. Fremond. *Non-Smooth Thermo-mechanics*. Springer-Verlag, Berlin Heidelberg New York, 2002.
4. P. Villard and P. Jouve. Behavior of granular materials reinforced by continuous threads. *Computers and Geothechnics*, 7:83–98, 1989.
5. R. Laniel, P. Alart, and S. Pagano. Consistent thermodynamic modelling of wire-reinforced materials. *European Journal of Mechanics - A/Solids*, 26:854–871, 2007.
6. B. Cambou and M. Jean. *Micromécanique des matériaux granulaires*. Hermès, Science-Paris, 2001.
7. M. Jean. The non smooth contact dynamics method. *Computer Methods in Applied Mechanic and Engineering*, (177 (Special issue)):235–257, 1999.

Granular Flow Under the Action of Centrifugal Force: A Critical Dynamic Friction Coefficient

Aurélie Le Quiniou[1], François Rioual[1], and Yuri Lapusta[2]

[1] Cemagref, TSCF, Domaine des Palaquins, 03150 Montoldre, France
aurelie.le-quiniou@cemagref.fr
[2] IFMA, LaMI, CNRS TIMS, Campus les Cézeaux, 63175 Aubière, France

Summary. We performed numerical simulations of the impact of a bead on a rotating vane. Considering different realistic contact laws at the collision chosen from experiments, we define a typical trajectory as a bouncing regime followed by a sticking regime. We show that there exists a critical friction coefficient between the particle and the wall for which the particle rolls without any significative sliding in a second phase. We deduce that the friction of the grains with the vane may have a minor effect because the bead rolls without any sliding after a transient stage. This may have important outcomes in spreading applications.

1 Introduction

The rheology of dilute granular flows is mostly unknown because of the lack of knowledge of the laws of interaction between the constitutive particles. This is however through collisions that the energy is transmitted and a prerequisite is necessary to explore the different modes of transmission of the energy through the shocks. We would like here to analyze the motion of a spherical bead that enters in collision with a rotating plate. This is the usual way in particular in agricultural applications to spread a granular material in the field (Fig. 1). One wishes to understand the physical mechanisms of the granular flow on a spreader. Actually, the dynamic friction coefficient is regarded as the parameter influencing more the trajectory. We deduce from our numerical study that the dynamic friction coefficient influences the dynamics of the sphere (rolling with sliding (R+S) regime or rolling without sliding (R-S) regime) i.e. it is possible that friction induced segregation in a flow of various particles. We address different issues following this numerical study which may have a strong relevance in the context of granular spreading technologies.

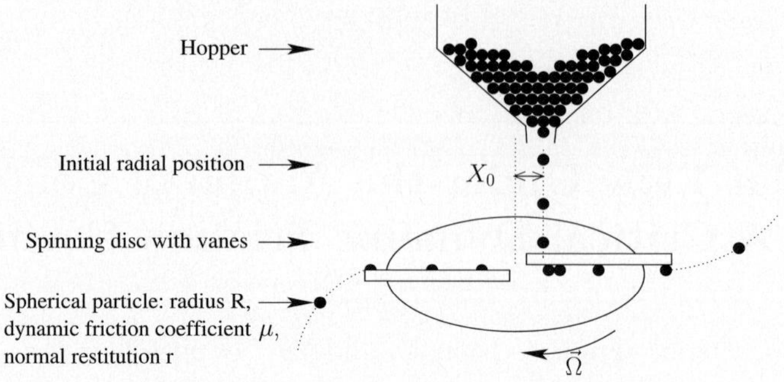

Fig. 1. Schematic view of the spreader.

2 D.E.M. Modelling

The model is developed on the discrete element method under the quasi-static assumption (Cundall and Strack (1979) [1]). We suppose that the material goes through a succession of steady states following Hertzian Contact theories. Within the quasi-static assumption, Kuwabara and Kono 1987 [2] and Brillantov and Pöschel 1999 [3] generalized the Hertzian theories (in the elastic case) to viscoelastic materials. This assumption has to be checked because the particles can reach high speeds and undergo fast deformations, dependent on the speed of request ($\sim 10m/s$). The particles are PVC beads with as characteristics a mass m of 0.2g, a radius R of 3mm, a Young's modulus E of 2.8GPa, a coefficient of normal restitution r of 0.5 and a dynamic friction coefficient μ of 0.523 (obtained with rheometric measurements). For low flow rates, the particles are distant one from another after the collision stage so it is possible to use the single particle approximation along the vane.

For a sake of simplicity in this problem, we will model the interaction by the widely used model introduced by P.A. Cundall and O.D.L. Strack [1]: it represents the linear viscoelastic interaction between the particle and the vane by a linear spring in parallel with a dashpot for the normal component (N). The tangential component (T) is modelled with a spring and a dashpot with Coulomb friction limit.

$$N = K_n \delta_n - b_n \dot{\delta}_n \tag{1}$$

$$T_{+1} = \min(T + K_t \Delta\delta_t - b_t \dot{\delta}_t, \mu N) \tag{2}$$

K and b are respectively the contact stiffness and the damping parameter: $K_n = 10^9 N/m$, $K_t = \frac{2(1-\nu)}{2-\nu} K_n$ with $\nu = 0.3$ the Poisson ratio (Vu-Quoc et al 2000 [4]), $b_n = \frac{2 \ln r \sqrt{m K_n}}{\sqrt{\pi^2 + \ln r^2}}$ and $b_t = \frac{2 \ln r \sqrt{m K_t}}{\sqrt{\pi^2 + \ln r^2}}$ (Ting et al 1989 [5]).

δ_n is the normal contact displacement also termed the overlap of the contacting bodies. The increment of tangential contact displacement $\Delta\delta_t$ is found by integrating the projection on the contact plane of the relative contact velocity.

Dimensionless numbers of the problem:
The important parameters are:

- the dynamic friction coefficient μ,
- the coefficient of normal restitution r between the bead and a static vane,
- and the centrifugal number N_i.

This centrifugal number is defined by the ratio between the velocity of the vane at the position of impact equal to $X_i \Omega$ and the velocity of the particle after the impact which can be written as R/τ:

$$N_i = \frac{X_i \Omega}{R/\tau} \tag{3}$$

The collision time τ is expressed as a function of the previous mechanical parameters for the Cundall model and a static vane: $\tau^2 = m(\frac{\pi^2 + (\ln r)^2}{K_n})$ (Ting et al 1989 [5]). With our PVC beads, the centrifugal number is about 0.01. This number almost does not influence the phase of the particle. In this case, the coefficient of normal restitution controls the possibility for the bead to bounce on the plane. With particle with smaller contact stiffness (Ammonium Nitrate $K = 3 * 10^5 N/m^{-3/2}$) the centrifugal number is close to 1. The possibility for the bead to bounce on the plane is thus controlled by the centrifugal number. Hence if $N_i \gg 1$, the particle will not be able to bounce on the rotating wall and remains in contact with the wall. On the contrary, for $N_i < 1$, the effect of the rotating plate on the characteristics of the impact is negligible. In this study we consider particle with very small centrifugal number.

3 Dynamics of a Particle Along the Vane

We suppose a spherical bead impacting a rotating vane without any significant initial spin. We neglect the component of initial vertical velocity. We explore

A typical trajectory (Fig. 2) consists firstly in the succession of jumps along the vane. At each impact, the particle gains some spin through the shocks. The particle then reaches a sticking regime where it remains in contact with the vane.

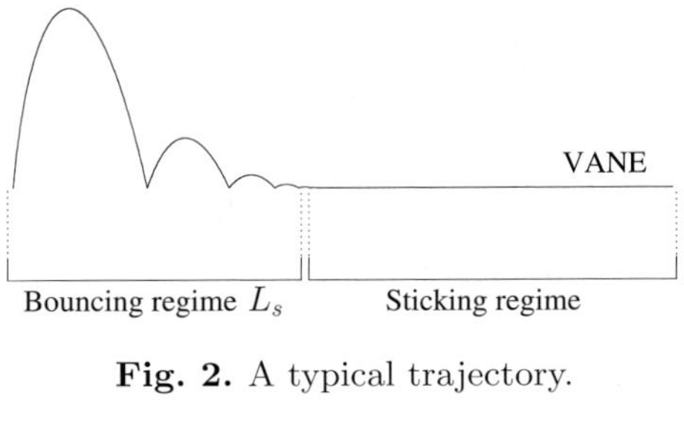

Fig. 2. A typical trajectory.

the influence of the friction coefficient in the bouncing regime and then in the sticking regime.

3.1 The Bouncing Regime

The bouncing distance:
We represent the bouncing distance for different angular velocities of the vane Ω, initial radial positions of the particle X_0 and friction coefficients μ as a function of the restitution coefficient r (See Fig. 3): The bouncing distance

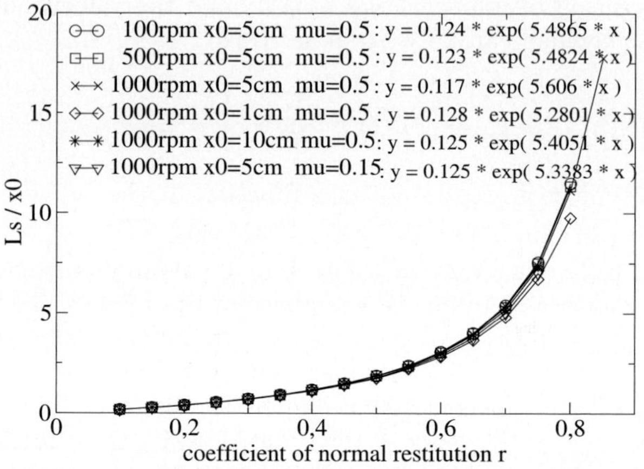

Fig. 3. The bouncing distance as a function of the control parameters.

increases linearly with X_0 and exponentially with the coefficient of normal restitution r. The phase of rebound L_s is very high for a coefficient of restitution of 1. For given mechanical parameters, the bouncing distance can be expressed directly as:

$$L_s = X_0 * \alpha * \exp(\beta r) \tag{4}$$

with α and β some constants of value 0.12 and 5.4. This shows that the bouncing distance is independent of the friction coefficient. L_s is controlled mainly by the coefficient of restitution.

The spin gained during the impacts:
The spin gained $\Delta\dot{\theta}$ is defined by:

$$\Delta\dot{\theta} = \frac{R}{I}\Delta T \tag{5}$$

with $I = \frac{2}{5}mR^2$ the moment of inertia for a sphere about its central axis and T the tangential contact force between the particle and the vane. One measures the spin gained during impacts and the tangential velocity V_i of the particle as a function of the radial position for different friction coefficients (See Fig. 4): The friction increases the amount of spin gained by the particle through an impact and decreases the amount of the tangential velocity of the center of mass V_i during an impact.

3.2 The Sticking Regime

One studies the sliding velocity along the vane defined by:

$$V_g = V_i - R|\dot{\theta}| \tag{6}$$

with V_i the tangential velocity of the center of mass, as a function of μ with the Cundall model (Fig. 5). Two distinct phases appear. If the dynamic friction coefficient μ is higher than a critical value μ^*, the sliding velocity is null after the bouncing regime: a rolling without sliding (R-S) regime is predicted. But if the dynamic friction coefficient μ is lower than this critical value μ^*, the sliding velocity increases after the bouncing regime: it is a rolling with sliding (R+S) regime (Fig. 5). One can explain this: at the beginning of the vane, the spin is zero: $V_i > R|\dot{\theta}|$, it is the (R+S) regime. Then the dynamic friction increases the amount of spin gained by the particle and decreases the amount of radial velocity through an impact. At $\mu = \mu^*$, $V_i = R|\dot{\theta}|$, it is the (R-S) regime. The radial position of the particle increases so the sliding velocity is always positive or null. At $\mu > \mu^*$, the spin and the radial velocity remain constant, it is always a (R-S) regime.

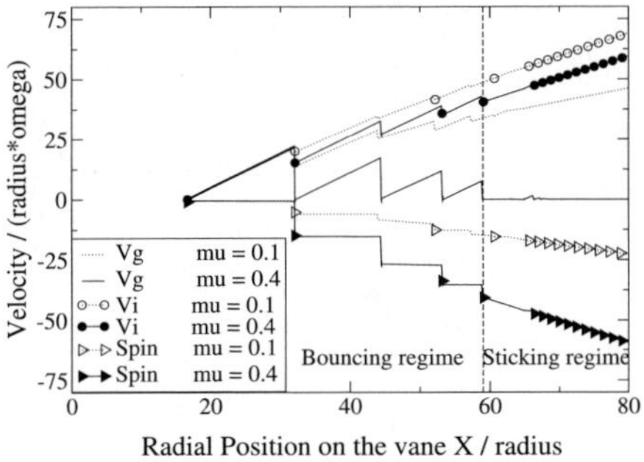

Fig. 4. Velocities as a function of the radial distance of the particle.

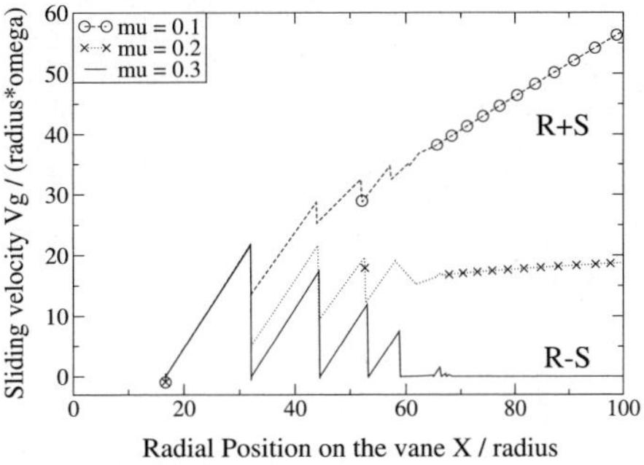

Fig. 5. Sliding velocity as a function of the radial distance of the particle.

We also show below in a phase diagram, the dependence on this critical friction parameter as a function of Ω, r and the initial position X_0 (see Fig. 6). The two possible modes for the particle appear clearly: *Rolling and sliding* (R+S), *Rolling without sliding* (R-S). This critical friction parameter increases linearly with the initial radial position. For a ratio between X_0 and the length of the vane which is higher to a critical value L^*, the particle always rebounds. L^* is also dependent on the coefficient of restitution. We also show that Ω and r may have a minor effect on this critical friction parameter.

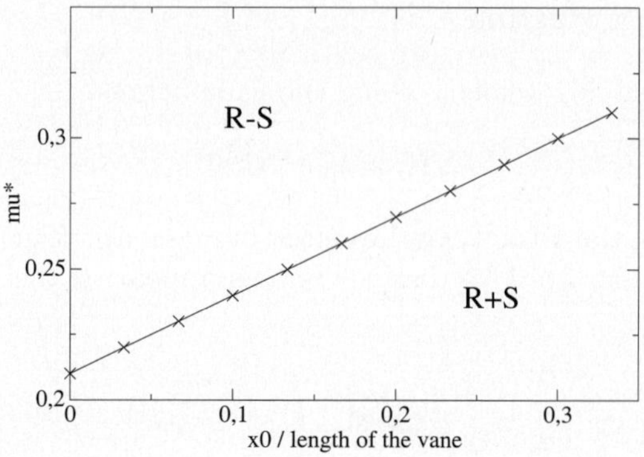

Fig. 6. Dynamic friction coefficient as a function of the initial radial distance.

4 Conclusion

The study carried out on the influence of the dynamic friction coefficient shows that there is a critical value μ^* for which the acquired spin is sufficient to have a (R-S) regime. With a flow of different particles, the friction may induce segregation in the flow between the particles which roll with sliding and the particles which roll without sliding. The dynamic friction coefficients, obtained with rheometric measurements for materials used in spreading, are $\mu = 0.552$ for Ammonium Nitrate: NH_4NO_3, $\mu = 0.466$ for Potassium Chloride: KCl and $\mu = 0.523$ for PVC used in pilot experiments. One notices that these particles have a coefficient higher than the critical value μ^*. For these values, the model would predict that our particles roll without sliding. This is interesting for spreading applications: the friction is not the parameter which provides the ejection characteristics of the particle because the friction coefficient may have a minor effect. A great care has to been taken to the influence of the other parameters of the problem such as the coefficient of restitution.

Acknowledgements

This work has benefitted from interesting discussions with E. Tijskens, T. Biben, S. Bonelli, M. Naiim.

References

1. Cundall PA, Strack ODL (1979) Geotechnique 29:47–69.
2. Kuwabara G, Kono K (1987) Jpn. J. Appl. Phys. Part 1 26(8):1230–1233.
3. Brillantov NV, Pöschel T (1999) Eur. Phys. J. B 12:299–301.
4. Vu-Quoc L, Xhang X, Lesburg L (2000) J. Appl. Mech. 67(2):363-371.
5. Ting JM, Corkum BT, Kauffman CR, Greco C (1989) J. Geotech. Eng. 115(3):379–398.

Interfacial Instability of a Confined Suspension Under Oscillating Shear

Maniya Maleki[1,2], Héctor Pacheco[3], Carlos Ruiz Suárez[3], and Eric Clément[1]

[1] PMMH, UMR 7636 du CNRS, ESPCI, 10 rue Vauquelin, 75005 Paris, France
maniamaleki@yahoo.com
[2] Institute for Advanced Studies in Basic Sciences (IASBS), Zanjan 45195, P.O. Box 45195-1159, Iran
[3] Departamento de Fisica, CINVESTAV-IPN, Merida, AP 73, Cordemex, Yucatan 97310, Mexico

Summary. We study experimentally the interfacial stability of a polystyrene particle suspension sheared harmonically by an upper flow free of particles. The suspension is confined in a rectangular Hele-Shaw cell and initially sedimented. The suspension can be made almost no buoyant by addition of salt in water. When an oscillating flow is applied to the upper flow, we monitor as a function of the amplitude and frequency, the interfacial dynamics. We observe at increasing shear, the interface going from compaction to expansion, and at higher shear, up to ripple formation. We show that the amplitude threshold for ripple formation decreases with frequency. We also study the effect of the buoyancy when tilting the cell or by adding salt.

1 Introduction

When two simple fluids form an interface and move with respect to each other, a Kelvin-Helmholtz instability may occur. The theory developed independently by Kelvin and Helmholtz allows to predict the onset of the instability and the transition to a turbulent flow. If one of the fluids is a gravitationally settled suspension and the other the clear fluid on the top, a similar phenomenon of wavy instability can be observed when the sediment suspension is sheared by a continuous or an oscillatory flow [1, 2]. An important feature for the interface is a resuspension process due to shear, first described by Leighton and Acrivos [3]. In the case of an erodible bed of dense and buoyant granular material, like a sand-bed, the resuspension process is almost invisible but a ripple instability may occur as one observes currently along the beaches and on the river beds. This ripple instability is also present for either steadily [4] or oscillatory shearing flows [5]. Different experimental geometries have been used to study these instabilities: a Couette geometry [6], the plane Couette [7] and an annular channel [8, 9]. Here we try to bridge these two approaches by studying the interfacial motion of an initially sediment suspension confined in an Hele-Shaw cell. The buoyancy contrast is much lower that for usual beds of granular materials and can be set to almost zero by the progressive addition of salt in the fluid.

2 Experimental Setup

The Hele-Shaw cell is made of two parallel plexiglass plates separated by a distance $s = 400\mu m$. The suspension is confined in a rectangular area with length $L = 30cm$ and width $W = 5cm$. At the upper ends of the rectangular cell, two inlet/outlet $8mm$ diameter holes are connected to rigid tubes. The suspension is made of water or a water/salt solution and of polystyrene spheres with a diameter $d = 80\mu m$, a density of $\rho_0 = 1.06 g/cm^3$ and an initial volume fraction of $\nu_{susp} = 9\%$. The suspension is injected in the cell via one of the tubes. Then, the cell is put vertically to let the polystyrene beads sediment at the bottom and form a dense packing around $\nu_0 \approx 55\%$ over an initial height $H_0 = 8mm$. Then the tubes are half filled with water or the water/salt solution. One of the tubes connects the cell to a syringe, and the other one is open at the atmospheric pressure. The syringe is attached to a loud-speaker that can be vibrated at different frequencies and amplitudes. Therefore, the syringe pumps the air into the tube and creates an oscillating flow inside the Hele-Shaw cell. The available frequency f range is 1.5–$10Hz$. Movies of the oscillating suspension are taken with a CCD camera and by image analysis based on the contrast between the suspension and the clear flow, the average sediment height H is extracted. The flow amplitude is obtained by putting a narrow pipe on the free tube and filming it. Therefore, the volume of the liquid displaced during each oscillation cycle is then obtained and we divide this volume by the thickness of the cell and by the height of the liquid above the sediment layer to obtain a control parameter: A, describing the mean flow displacement amplitude in the upper free fluid layer. The advantage of the Hele-shaw geometry if that in the upper layer the flow if homogeneous vertically and the active shear of the suspension by the pure fluid only occurs in the vicinity of the interface at distances lower or equal to the plate separation s.

3 Interfacial Dynamics

Figure 1 shows the relative changes in sediment height $(H - H_0)/H_0$ as a function of the rescaled time (tf) for different driving amplitudes A and at a frequency of oscillation $f = 2Hz$. The liquid is pure water (buoyancy contrast 6%). As it can be seen from this figure, the amplitude of the upper flow drives two different dynamics for the interface motion: compaction ($A < 0.6mm$) and expansion ($A > 0.6mm$). We noticed that when the sediment layer start

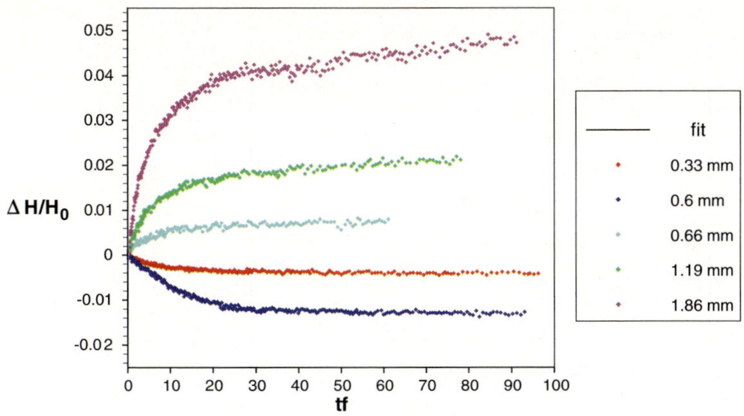

Fig. 1. Relative changes in height $(H - H_0)/H_0$ of the sediment layer as a function of the rescaled time tf for different driving amplitudes A. The liquid is water and the particles are polystyrene beads of diameter $80\mu m$. The frequency of oscillation is $f = 2Hz$. The solid lines are the fits according to equation 1.

to be sheared, the grains at the interface start to move and thereafter, the motion penetrates inside the granular layer. For very low driving amplitudes A, the small oscillations of the grains result in a compaction process but for larger amplitudes, the grains need more space in order to move relatively to each others and then, the layer expands possibly due to dilatancy effects and hydrodynamic resuspension forces. The interface average motion over time can be simply described by an equation of the type:

$$\Delta H/H_0 = B(1 - exp(-(t/\tau)^{\beta})) \tag{1}$$

with τ being the time scale for the compaction/expansion process, B represents the relative average expansion for long times ($t \gg \tau$) and the exponent β characterizes a deviation from a typical first order process when $\beta < 1$ (stretch exponential). These parameters extracted from a fit are displayed on Fig. 2. Note that we see on this figure that the expansion process is in competition with a compaction process which seems to overcome at low shearing rate. This defines a clear threshold for the onset of expansion. The compaction process is very well characterized by a simple exponential but for the expansion, it seems that we have a radical change of dynamics. We see a slow stretched exponential dynamics with a typical time scale $\tau \approx 10(\pm 2)s$ and an exponent $\beta \approx 0.67 \pm 0.05$.

Ripple formation – Actually, there is another threshold of shear above which the interface becomes longitudinally unstable and ripples start to form. The ripple formation threshold is about $A_c = 1.5mm$. It can be seen that above A_c, the average expansion B increases more rapidly with the driving

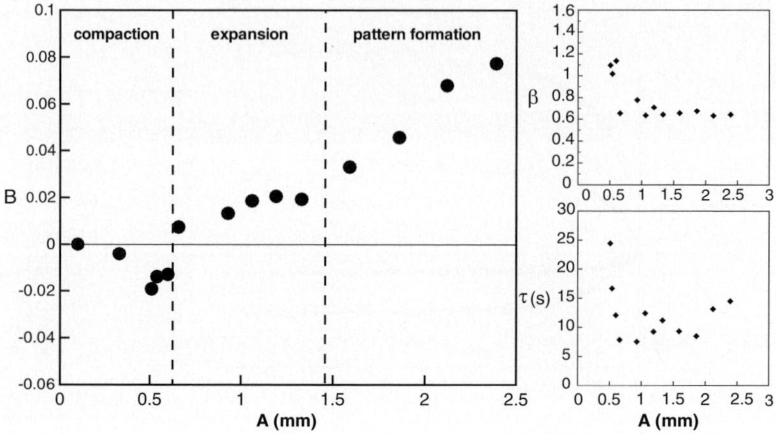

Fig. 2. Fit parameters of equation (1) as a function of the flow driving amplitude A. Left: expansion amplitude B. The height is defined the for ripple formation regime as the mean height of the layer. Up-right: stretched exponential exponent β. Down-right: expansion time scale τ.

amplitude A but interestingly, the typical time scale τ and the stretch exponential exponents β stay of about of the same values.

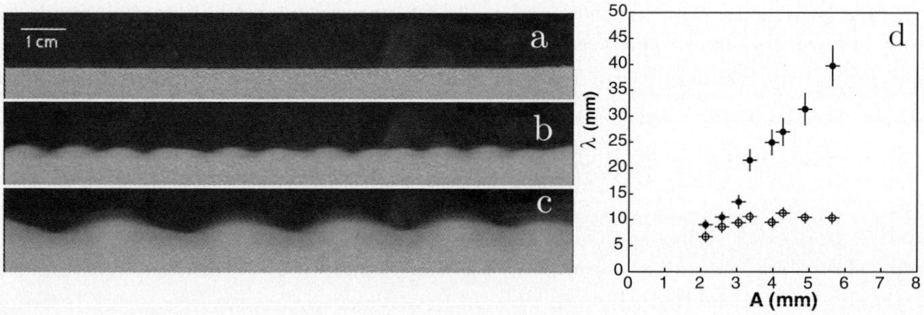

Fig. 3. a) The sediment layer before shearing; b) ripple formation; c) coarsening. The frequency is $f = 3Hz$ and the amplitude is $A = 2.6mm$. Fig. 3(d), initial wavelength (empty circles) and wavelength after $10min$ (full circles) in the same driving conditions.

The ripples evolve in time and a coarsening process happens in which the wavelength of the pattern increases and then saturates to a final value (Figure 3(a,b,c)). The initial and final wavelength of the ripples λ increase with the shearing amplitude A. Figure 3(d) shows the initial and final wavelengths as a function of the driving amplitude A. Here, the final wavelength is taken after $10min$. The initial wavelength first increases with the amplitude and then remains almost constant, but the final value is always an increasing function of the amplitude.

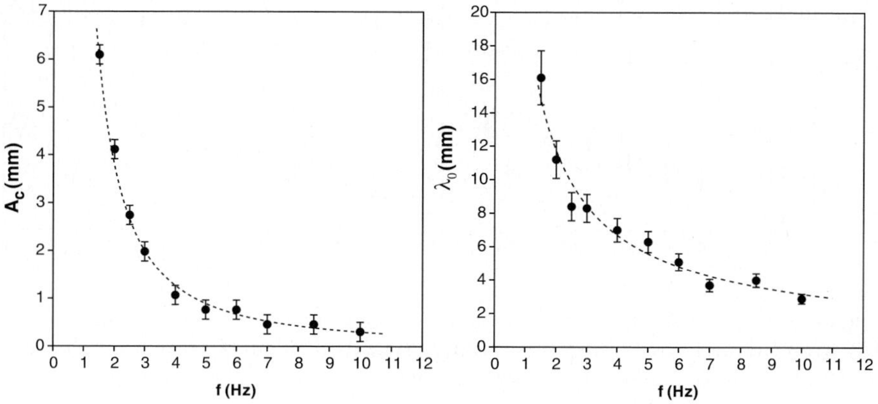

Fig. 4. Left- The threshold amplitude for ripple formation as a function of frequency. **Right**- The initial wavelength of the ripples at the threshold as a function of frequency.

We also measured the threshold amplitude A_c of ripple formation at different frequencies. The result is shown in Fig. 4(left). The threshold amplitude is a decreasing function of the frequency of oscillation and can be fitted with a power-law curve $A_c = af^b$ with $b = -1.6 \pm 0.2$ (dashed line in Figure 4). We measured the initial wavelength of the ripples at the threshold and we observed that the wavelength is also a decreasing function of the frequency. The result is shown in Figure 4(right). The data can be fitted with a power-law $\lambda_0 = af^b$ with $b = -0.8 \pm 0.1$.

4 Effect of Buoyancy

Two forces act on the grains: buoyancy and the hydrodynamic forces. The hydrodynamic force includes the drag force of the liquid, and the hydrodynamic interaction between the grains. To study the effect of the buoyancy force, we proceed in two ways. The first procedure consists in dissolving different amounts of salt in the water in order to obtain a fluid with different densities and approach very closely to the polystyrene value. The effect of the density on the ripple formation threshold can be seen in Figure 5(left). As we expect, the threshold increases with density difference $\Delta\rho$ between the grains and the solution.

The other way to change the buoyancy force on the grains is by tilting the cell and by using an effective value of the gravity ($g \sin\theta$) as the actual restoring mechanism for sedimentation. As we change the angle θ between the

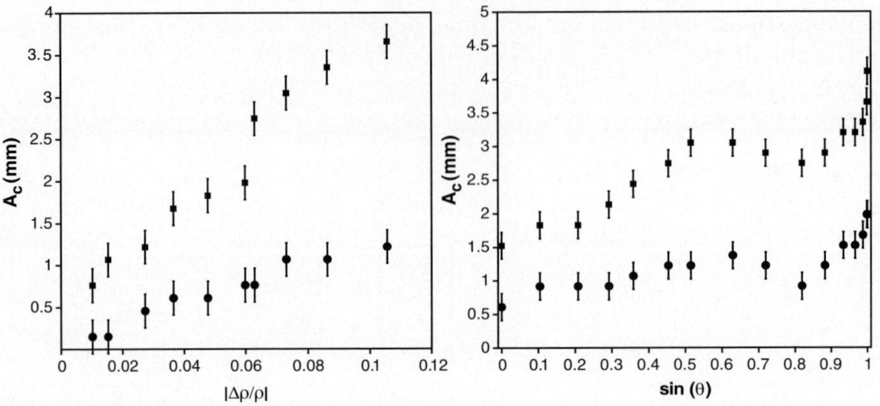

Fig. 5. Left- The ripple formation threshold as a function of density difference for $f = 3Hz$ (squares) and $f = 5Hz$ (circles). **Right**- The ripple formation threshold as a function of the angle for $f = 2Hz$ (squares) and $f = 3Hz$ (circles).

cell and the horizontal surface before shearing, the ripple formation threshold is measured. The result is shown in Figure 5(right). The behavior in this case is complex, with a sudden decrease in the threshold when deviating a little from the vertical position. The other important feature is that there is a non-zero threshold even for the horizontal case ($\sin\theta = 0$).

5 Conclusion

We present a preliminary study of the interface dynamics of a sediment suspension confined in a Hele-Shaw cell. The suspension interface is destabilized under oscillating shear and we evidence a competition between a compaction process and a resuspension. In both cases (compaction and expansion), a steady state is reached for low shearing amplitude. However, if the shearing amplitude is too large a ripple instability occurs which shows coarsening dynamics. We have studied the initial wavelength selection which was found to decrease with the driving frequency. Also, we have shown that at low buoyancy contrast, the expansion and the instability processes were still present. This opens the way to a systematic study of this interfacial instability as well as a observation of the granular motion in the bulk of the sediment.

Acknowledgements

We thank Bruno Andreotti, Eduardo Wesfreid and Joachim Kruithof for many stimulating discussions.

References

1. Gadala-Maria, F.A. and Acrivos, A. (1979) J. Rheol. 24:789–814.
2. Schaflinger, U., Acrivos, A. and Zhang, K. (1990) Int. J. Multiphase Flow 16:567–578.
3. Leighton, D.T. and Acrivos, A. (1987) J. Fluid Mech. 177:109–131.
4. Charru, F. and Mouilleron-Arnould, H. (2002) J. Fluid Mech. 452:303–323.
5. Blondeaux, P. (2001) Mechanics of coastal forms. Ann. Rev. Fluid Mech. 33:339–370.
6. Wallner, J. and Schaflinger, U. (1998) Acta Mechanica 127:147–153.
7. Lenoble, M., Snabre, P. and Pouligny, B. (2005) Phys. Fluids, 17, 073303:1–14.
8. Rousseaux, G., Yoshikawa, H., Stegner, A. and Wesfreid, J.E. (2004) Phys. of Fluids 16:1049–1058.
9. Rousseaux, G., Stegner, A. and Wesfreid, J.E. (2004) Phys. Rev. E 69:031307.

Pattern Formation in Slowly Drained Granular-Fluid Systems

Bjørnar Sandnes[1,2], Henning Arendt Knudsen[1], Knut Jørgen Måløy[1], and Eirik Grude Flekkøy[1]

[1] Department of Physics, University of Oslo, P.O.Box 1048 Blindern, 0316 Oslo, Norway bsand@fys.uio.no

[2] School of Chemical and Biomolecular Engineering, University of Sydney, NSW 2006, Australia

Summary. We present a new pattern formation process where labyrinthine structures emerge during slow drainage of a confined granular-fluid mixture. Capillary and frictional forces govern the process, and the resulting pattern has a characteristic wavelength that is a function of both the initial volume fraction of granular material in the mixture, and also the system thickness.

1 Introduction

There are many examples of pattern formation in granular systems driven out of equilibrium: ripples in wind-blown sand, segregation of grains in granular flows [1], the Rayleigh-Taylor instability in falling grains [2], and various patterns in vertically and horizontally vibrated granular layers [3].

Here we present a new pattern formation process where the central ingredients are a granular medium submerged in a Newtonian fluid [4–6]. The granular-fluid system is confined between the parallel glass plates of a Hele-Shaw cell, and as the cell is gradually drained, the receding fluid-air interface gathers and pushes the grains ahead of it as fingers of air invade the system. When the cell is fully drained, all the granular material that was originally uniformly distributed in the cell has been reorganized into a branching labyrinth structure of compacted grains.

From a combination of experiments, simulations and theory we characterize the governing forces in the system, and show that a well defined wavelength develops as a compromise between capillary forces and friction. Emphasis is here placed on the experimental system and results, while the simulations are given a similar treatment in a parallel report [7].

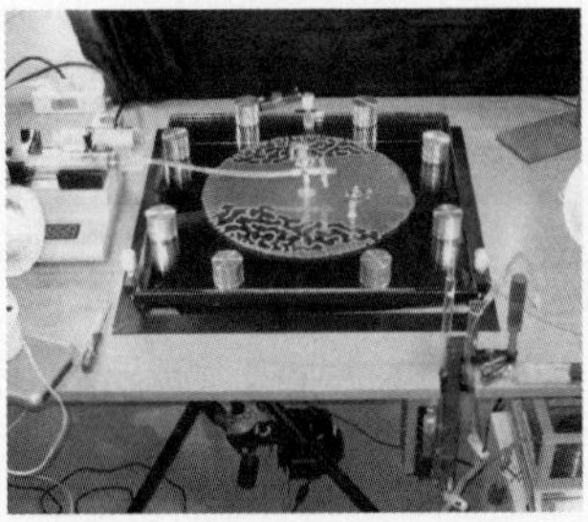

a)

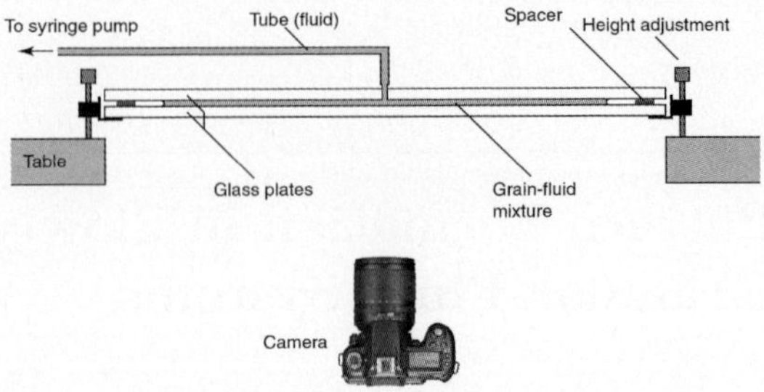

b)

Fig. 1. a) Picture of the experimental setup. b) Illustration of a cross-section of the Hele-Shaw cell including outlet and height adjustment mechanism.

2 Experimental Setup

Figure 1a) shows a picture of the experimental setup including the Hele-Shaw cell, a syringe pump (to the left in the picture), and a Nikon D100 camera placed underneath the cell acquiring images in time lapse mode. A schematic drawing of the cell is shown in Fig. 1b). The dimensions of each of the two glass plates are $50 \times 50 \times 1$ cm, and they are kept separated by metal spacers of well defined thickness. The cell is mounted in a frame with three adjustment screws resting on a table such that the cell can be leveled accurately. A square "window" is cut in the table allowing imaging from underneath where no tubes or other equipment blocks the view.

A small hole, approximately 5 mm diameter, is drilled through the top glass plate, and a brass tube connector is glued on. This serves as both inlet during loading of the granular-fluid mixture into the cell, and as outlet when the fluid is subsequently withdrawn. A plastic tube connects the cell to the syringe pump.

In the experiments reported here we use a granular medium consisting of poly-disperse glass beads with diameters in the range of 50–100 μm. The beads are weighed up and added to a 50% glycerol/water mixture in a large syringe such that the total volume of the mixture is measured. The syringe is shaken vigorously and attached to the inlet hole of the cell, and the mixture is then injected into the cell at a high rate while the beads remain suspended in the viscous fluid. As a result of the fast injection, the granular-fluid mixture spreads out in a circular disc in the gap between the plates, where the radius depends on the volume of the injected mixture. Example: 10 g of glass beads are dispersed in 50% glycerol/water such that the total volume is 40 ml and the volume fraction of beads is 20%. After injection, the circular disc of granular-fluid mixture measures 35 cm in diameter in a cell where the glass plates are separated by 0.4 mm thick spacers.

Shortly after the granular-fluid medium is loaded into cell, the grains sediment out, and fall to rest on the surface of the lower glass plate. An additional small amount of beads are deposited around the outlet hole in order to prevent any fingers of air sweeping past the outlet channel. We are now ready to proceed with the drainage experiment. The syringe pump is set to withdraw fluid from the interior of the cell at an extremely low rate such that viscous forces are negligible and also so that we are able to drain fluid through the developing granular branches. At a typical withdrawal rate of 0.01 ml/min, the experiment in the above described example takes close to 2.5 days to complete.

3 Pattern Formation

As the fluid is gradually drained from the cell the under-pressure with respect to the ambient air increases, and the fluid-air interface starts to recede. The granular material at the perimeter is pushed ahead of the slowly moving interface by capillary forces, and soon the beads start to stack up, building up a layer of compacted grains that span the gap between the two plates. Due to the low flow rate, the advancement of the front occurs in a creeping fashion, and only a small section of the front is in motion at any one time. Initially the interface contracts all along the circular perimeter, but after a while, when a layer of compacted grains has been compiled all along the perimeter, the interface becomes unstable and air proceeds to invade the granular-fluid system in a slow fingering process (Fig. 2).

The invading air fingers move the grains ahead of them, and reorganize the granular material locally into thin branches of close-packed beads (Fig. 3a) and b)). The initially circular interface deforms and stretches, but no pinch-off of the interface occurs during the experiment, and consequently, the final labyrinth structure is simply-connected. The invading fingers have a well defined width, which results in a uniform characteristic wavelength in the labyrinth pattern that is perceivable to the eye as an average distance between the grain branches.

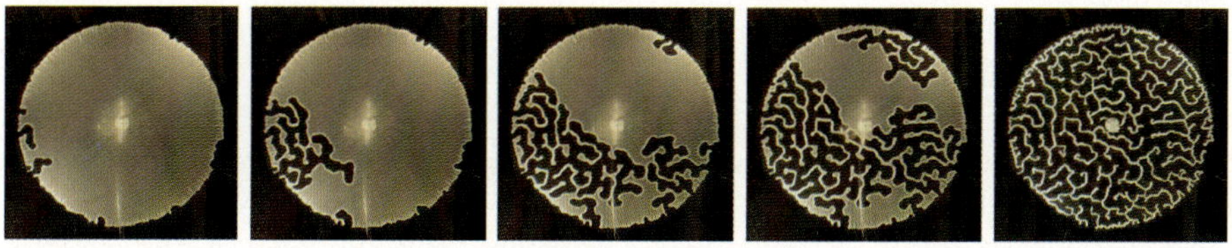

Fig. 2. Time series of pictures taken from underneath the cell during the pattern formation process. The elapsed time is 3, 11, 28 and 42 hours for the first four pictures (left to right) and the last picture is taken some time after the cell was fully drained. The residue of granular material has been compiled into a branching labyrinthine structure seen as the bright pattern in the final picture. The initial granular-fluid disc is approximately 35 cm in diameter.

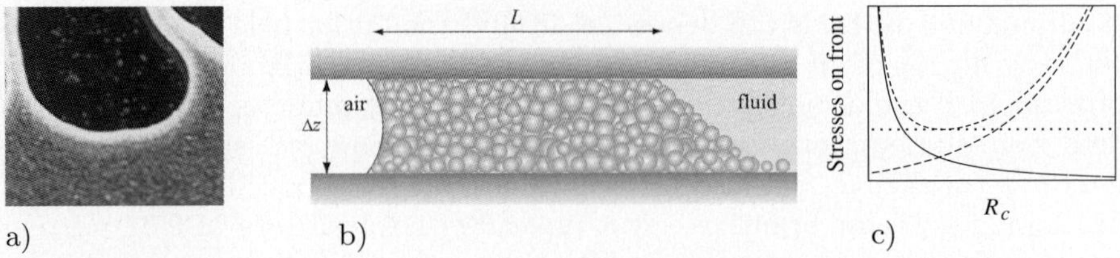

Fig. 3. a) Top view of an air finger invading the granular-fluid system. The layer of compacted grains is visible as a bright zone ahead of the interface. b) Sketch of a cross-section of the air-fluid-grain front. The length of the compact layer is L, and the plate spacing is Δz. c) Schematic representation of the dominant forces in the system plotted against finger curvature radius: surface tension (solid), friction (long-dashed) and the total force required to move the interface at that specific location along the front (short-dashed). The dotted line shows the pressure required to move the front, the yield pressure.

4 Physical Mechanisms

The pattern formation develops as a result of a competition between two forces: surface tension and friction. The surface tension is responsible for moving the collected mass by imposing a force on the wetting beads that perturb the interface. But capillary forces also have a more profound influence on the pattern formation since surface tension acts to minimize surface area and thereby also minimize curvature during the fingering process. Surface tension alone therefore favors wide fingers of invading air as this minimizes the total creation of fluid/air surface.

Friction, on the other hand, opposes the inwards displacement of granular material as the interface recedes. A wide finger would gather mass from a large area and build up a thick layer of compacted granular material ahead of it, resulting in a high frictional dissipation. Friction therefore favors narrow fingers that gather less mass.

The pattern formation is driven by the pressure difference across the meniscus as fluid is continuously drained from the system. Movement of the front requires that this pressure overcomes the local yield pressure set up by interfacial and frictional forces, and since the process is slow, we can assume that movement occurs only at small segments of the front at any one time. The balance of forces for a front segment that is set in motion can to a first approximation be written $\Delta P = a\gamma/R + \sigma$, where $a\gamma$ represents an effective surface tension for the air/fluid/grains interface, R is the local in-plane curvature of the interface, and σ is the frictional stress from the granular packing ahead of the moving interface [4, 5].

Figure 3c) shows the forces plotted against radius of curvature for an invading finger. Capillary forces dominate for small radius of curvature (narrow fingers) and friction dominates for large (wide fingers). Since the dynamics of the fingering process is governed by a selection of lowest yield pressure any-

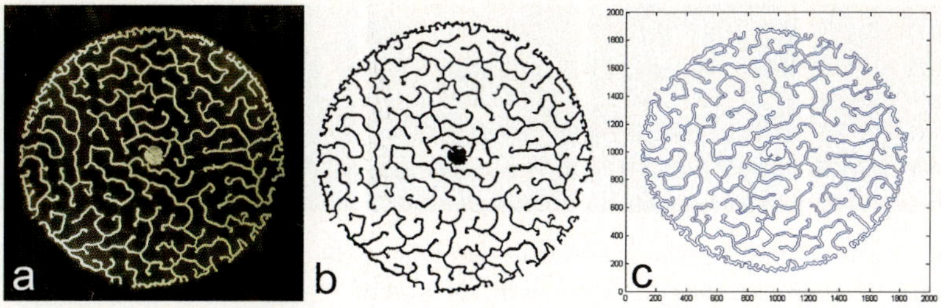

Fig. 4. a) Raw picture of granular labyrinth. b) Black/white bitmap of isolated simply-connected grain cluster. c) The circumference of the cluster is measured.

where along the front (minimizing the pressure plotted in Fig. 3c)), this also results in a selection of a characteristic length. The theoretical results quoted below are based on this mimimization of yield pressure.

5 Analysis of Length Scale

The labyrinth pattern is random, simply-connected, and has a well defined characteristic length scale. In this first study, we focus on the perhaps most visually pronounced feature of the pattern, the wavelength, loosely defined as the average separation between neighboring branches.

Figure 4 illustrates the method we have used to measure characteristic lengths in the patterns. An image of a fully developed labyrinth is calibrated and, using the Matlab image analysis toolbox, thresholded to a bitmap array. The simply-connected cluster of granular material is identified and isolated, and using built in Matlab functions we measure a series of parameters including the circumference of the branching cluster and the area of the cluster. The average branch thickness, W, is simply the cluster area, A_{cluster}, divided by half the circumference S: $W = 2A_{\text{cluster}}/S$. The wavelength, or what the eye perceives as the "density" of the pattern, is given by $\Lambda = 2A_{\text{disc}}/S$, where A_{disc} is the area of the circular granular-fluid disc that initially filled the cell.

6 Characteristic Lengths

Figure 5 shows pictures of the final granular structures obtained in experiments where the initial volume fraction of granular material, φ, increases from left to right as labeled. (φ is defined as 1 for the close packed granular branches.) Figure 6 shows the characteristic wavelengths measured for these patterns plotted as a function of the volume fraction of grains for experimental data points (filled squares), data obtained from simulations (circles) and also predictions from an analytical model (solid line) [4, 5]. As the eye perceives and the measurements show: the characteristic wavelength of the pattern decreases with increasing amounts of granular material in the initial

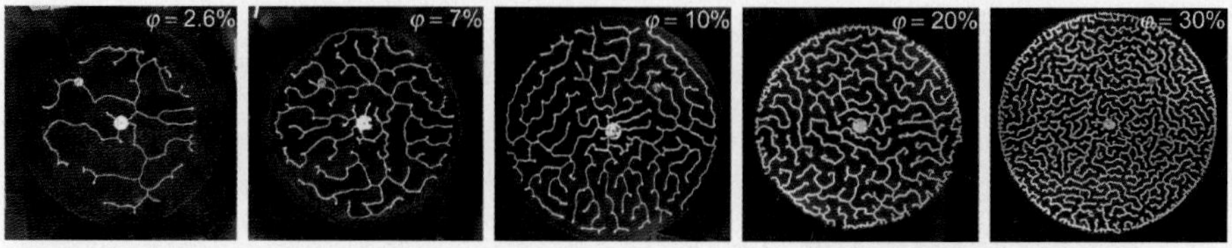

Fig. 5. Granular labyrinths in experiments where the initial filling fraction of granular material increases from left to right as labeled.

mixture. Likewise, the inset in Fig. 6 shows that the wavelength increases with increasing plate separation ($\varphi = 20\%$). The close match between experiments, simulations and theory shows that the physical mechanisms outlined above capture the essence of the problem, and that the pattern formation process is well described incorporating only capillary and frictional forces.

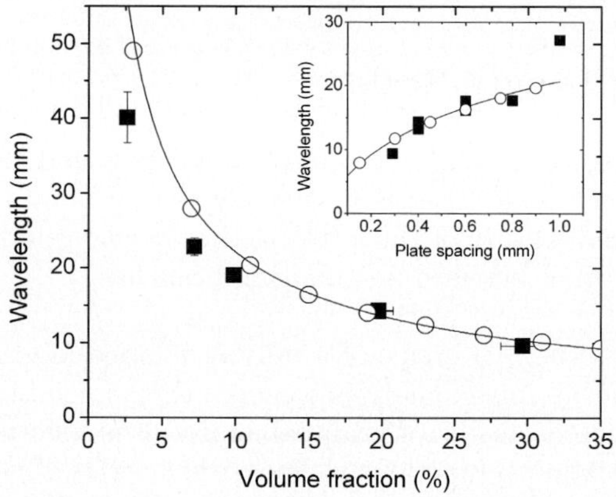

Fig. 6. Measured characteristic wavelengths as a function of volume fraction grains for experiments (filled squares), simulations (circles) and theoretical predictions (solid line). The inset shows wavelengths for different plate separations (constant volume fraction of 20%).

References

1. Aranson I S, Tsimring L S (2006) Rev Mod Phys 78:641.
2. Vinningland J L, Johnsen Ø, Flekkøy E G, Toussaint R, Måløy K J (2007) Phys Rev Lett 99:048001.
3. Bizon C, Shattuck M D, Swift J B, McCormick W D, Swinney H L (1998) Phys Rev Lett 80:57–60.
4. Sandnes B, Knudsen H A, Måløy K J, Flekkøy E G (2007) Phys Rev Lett 99:038001.
5. Knudsen H A, Sandnes B, Flekkøy E G, Måløy K J (2007) In preparation.
6. Yamazaki Y, Mizuguchi T (2000) J. Phys. Soc. Jpn. 69:2387.
7. Knudsen H A, Sandnes B, Flekkøy E G, Måløy K J (2007) In: *Traffic and Granular Flow 2007 - Proceedings.*

Emerging Stripe Patterns in Drying Suspension Droplets

Bjørnar Sandnes[1,2] and David Molenaar[1]

[1] Department of Physics, University of Oslo, P.O. Box 1048 Blindern, 0316 Oslo, Norway bsand@fys.uio.no

[2] School of Chemical and Biomolecular Engineering, University of Sydney, NSW 2006, Australia

Summary. When a droplet of silica particles suspended in water dries out on a horizontal surface, one can observe an emerging pattern of interconnected stripes of sedimented mass. From experimental observations and numerical simulations we find that the mechanism for the pattern formation is a combination of meniscus shape perturbations caused by pinned particles and a reorganization of the evaporation driven flow.

1 Introduction

Many types of patterns occur in thin film flows of complex fluids [1–4]. One fascinating class of such patterns are drying patterns in fluid-particle mixtures. Driven by evaporation of the fluid component, the pattern is usually a result of the accumulation of the solid particles at temporarily pinned fluid contact lines. For instance, when a spilled drop of coffee dries out on a horizontal surface, most of the dissolved coffee particles accumulate at the edge of the droplet, leaving a ring-shaped stain after the remaining fluid has evaporated [1]. The ring deposit forms as the contact line of the drop becomes pinned to the substrate while ongoing evaporation generates a radially outward-directed flow, carrying with it particles that are subsequently deposited at the pinned droplet edge.

In this paper we describe the formation of stripe patterns during the evaporation of a thin film suspension of silica particles deposited on a substrate. The silica particles, carried by the evaporation driven flow, aggregate at so-called pinning points, which are formed when the largest particles sediment out under the influence of gravity, at random locations within the flow. Consequently, pattern formation occurs throughout the entire area of the flow, in contrast to coffee-ring patterns and related processes where the deposition of particles occurs at the pinned contact line. Indeed, the mechanism of pattern formation described here is, in its essential features, similar to the physics of pattern formation in thin film particulate suspensions draining down an inclined surface, as described by Buchanan et al. [4]. By means of experimental observations and numerical simulations we show here that the mechanism for evaporation driven pattern formation is a combination of meniscus shape perturbations and reorganization of the flow caused by the compiled mass of arrested particles, where a self-organized focusing of the flow through gradually smaller channels induces the growth of stripes perpendicular to the overall direction of the flow.

2 Evaporating Thin Film Suspensions

2.1 Experiments

A dispersion of silicon dioxide powder in water was prepared by adding 0.05 g silicon dioxide particles (Aldrich) sieved with mesh size -325 (maximum particle diameter 44 μm) to 20 ml purified water (capillary length $L_c = 0.0027$ m). The sample was sonicated for 45 minutes prior to the experiments in order to break up particle aggregates within the mixture. 1 ml of the dispersion was spread out on a horizontally placed (or slightly inclined) hydrophilic microscope glass slide (8×8 cm), and left undisturbed under ambient temperature and humidity conditions. The subsequent evaporation process was imaged using a digital Nikon D200 camera mounted on a Zeiss Stemi 2000-C microscope. Images were acquired at regular time intervals (5 or 10 s). For each magnification setting, the spatial dimensions were calibrated by imaging an object micrometer.

2.2 Observed Process

The non-Brownian particles sediment out and fall to rest on the substrate surface. Initially, evaporation sets in from the fluid surface with no observable effects on the particles. After a while, the thickness of the fluid film becomes comparable to the diameter of the largest particles sedimented out onto the substrate. The fluid wets the particles, and the meniscus is therefore perturbed; pulled up the sides of the particle like water in a capillary tube. The meniscus then acts with a force on the particle, pinning it to the substrate where it is immobilized by friction. Once the interface becomes curved locally, a pressure difference builds up across the meniscus; the Laplace pressure given by $\Delta P = \gamma(1/r_1 + 1/r_2)$, where γ is the surface tension and r_1 and r_2 are the radii of curvature in orthogonal planes. If the film thickness, and hence local curvatures, varies over the area due to differences in evaporation rates, pressure gradients are set up within the fluid film generating internal fluid flows. This flow carries with it those particles that are not pinned to the substrate by the meniscus or are otherwise too heavy to be moved by the flow. Evaporation rates are higher at the perimeter of the fluid film, and therefore

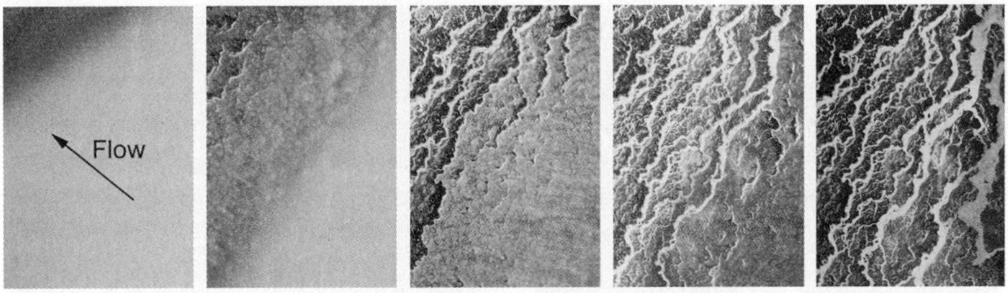

Fig. 1. Time sequence (left to right) of the pattern formation. The evaporation driven flow has an overall direction as indicated by the arrow in the first image. The height of each image is 15 mm, and the time period from first to last image is 18 minutes.

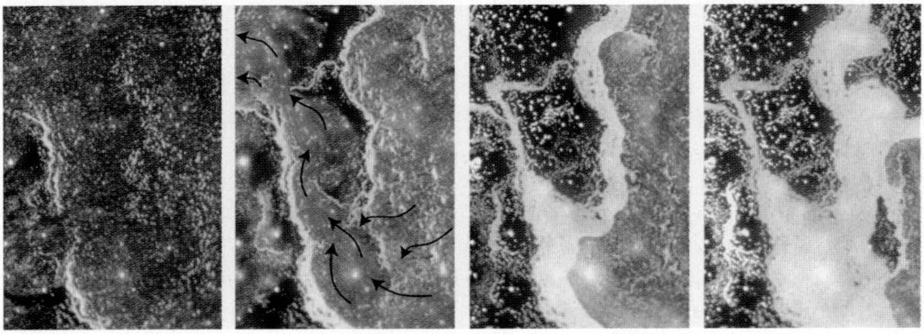

Fig. 2. A close-up of the nucleation and growth of stripes. The overall flow direction is from right to left in these images. The aggregated bands of particles form obstacles for the fluid flow which in turn transports new particles to their vicinity. The arrows drawn into image 2 illustrates the reorganization of flow into gradually smaller channels as more mass becomes deposited onto already formed stripes. Naturally, the stripes grow in thickness on the upstream side as more material becomes deposited here. Height of images: 1.9 mm, time period: 6 minutes.

one typically has an overall flow direction, which is radially outwards from the center towards the edges of the droplet.

As the evaporation progresses, one can after a while observe emerging stripes, first towards the edges, then gradually closer to the center of the droplet, following the thinning of the fluid film (Fig. 1). The pinned large particles act like nucleation centers where the growth of particle aggregates starts. Gradually smaller particles, brought to the vicinity of the pinned particles by the internal flow in the fluid film, start to aggregate on to already immobilized particles. These clusters form obstacles for the flow which becomes deflected to their sides, and the shape of the meniscus becomes perturbed by the hydrophilic particles and the flow within the film. A pressure bulge develops on the upstream side of an immobile particle cluster, and, likewise, a through develops on the downstream side, significantly altering the film surface. The

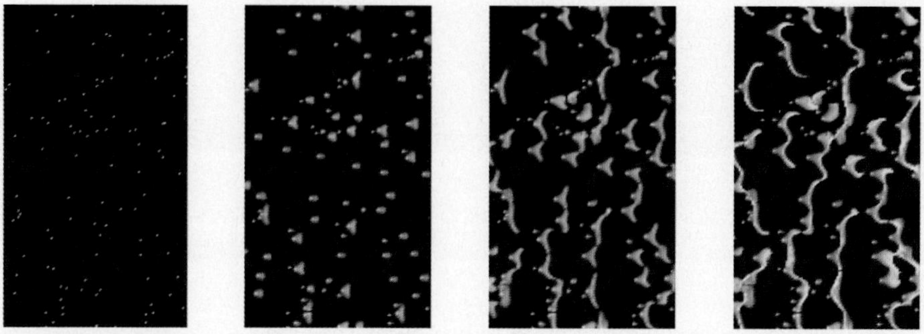

Fig. 3. Time sequence (left to right) of pattern formation in the numerical model. Snapshots of the particle concentration ranging from low (black) to high (white) are shown. The mean flow is from right to left in each image. Height of images: one capillary length.

change in the film surface allows additional, smaller particles to get trapped, leading to a self organized process where particles carried within the flow aggregate onto trapped clusters, and where the clusters reorganize the flow to their sides into gradually narrower channels (Fig. 2). Thus, immobile particle clusters nucleate, grow and coalesce together until the final stripe-pattern emerges. Naturally, the stripes show considerable local variation though they have an overall orientation orthogonal to the mean flow direction.

3 Numerical Simulations

Preliminary simulations of the particle concentration field in a film segment are shown in Fig. 3. We compute the evolution of both the free film surface and a scalar particle concentration, a strongly nonlinear system equations, as detailed in a forthcoming publication [5]. Pinning of particles is mimicked by a locally diverging mixture viscosity, which, in turn, alters local flow rates (the reorganization process). Starting from an initial field with a critical film thickness and a few, randomly distributed pinning points (left-most image), clear structures develop, oriented roughly perpendicular to the mean flow direction. The system is driven by an imposed uniform velocity field (from right to left in the images), mimicking the evaporation driven flow. However, once the largest band structures have formed, the mean flow is blocked completely. The associated capillary number, defined as the squared ratio between film thickness and the capillary length, equals $Ca = 0.001$; the number of gridpoints is 200 by 100.

4 Conclusion

Evaporation generates flow in a thin fluid film held up by the capillary action of pinned silica particles. A pressure bulge develops on the upstream side of an immobile particle cluster, and, likewise, a through develops on the downstream side, significantly altering the film surface. The change in the film surface allows additional, smaller particles to get trapped, leading to a self organized process where particles carried within the flow aggregate onto trapped clusters, and where the clusters reorganize the flow to their sides into gradually narrower channels. Thus, immobile particle clusters nucleate, grow and coalesce together until the final stripe-pattern emerges. The stripes show considerable local variation, but have an overall orientation orthogonal to the flow direction.

References

1. Deegan R D, Bakajin O, Dupont T F, Huber G, Nagel S R, Witten T A (1997) Nature 389:827.
2. Short M B, Baygents J C, Beck J W, Stone D A, Tooney III R S, Goldstein R E (2005) Phys Rev Lett 94:018501.
3. Zhou J, Dupuy B, Bertozzi A L, Hosoi A E (2005) Phys Rev Lett 94:117803.
4. Buchanan M, Molenaar D, DeVilliers S, Evans M (2007) Langmuir 23:3732.
5. Molenaar D, Buchanan M, DeVilliers S, Evans M (2007) *draft.*

Behaviour of Railway Ballast Under Cycling Loading

Gilles Saussine and Pierre Etienne Gautier

Innovation and Research department of SNCF, 45 rue de Londres, Paris Cedex 08, France gilles.saussine@sncf.fr, pierre-etienne.gautier@sncf.fr

1 Introduction

Ballast is widely employed for the construction of railway track because of its mechanical properties and its flexibility from the point of view of construction and maintenance. This granular material, produced by crushing of hard rocks, is confined between concrete sleepers supporting the rails and the platform.

The ballast layer needs to be raised from time to time because of its cumulative settlement under repeated loading [1, 2]. The increasing of cost of such operations with the advent of high speed trains has motivated a number of studies on physical origins of settlements.

The ballast layer is a thin granular interface with a thickness that is below the correlation length of contact forces. These correlations are expected to be long ranged because ballast grains are faceted and angular, and there is strong wall effect due to the confinement of ballast between the sleepers and the sublayer.

In this paper we investigate the mechanical behaviour of a model portion of railway track submitted to 60000 loading cycles. The aim of this study is to characterize the stabilisation phase and to propose a settlement model.

2 Numerical Procedure

In order to take into account the angular shape of ballast grains, we use polygonal grains in a two-dimensional space with a number of sides between 5 and 7. The grains are confined in an open box of length L with deformable base composed of polygons with a square shape supported by visco-elastic springs. In order to take into account the specific configuration of railway track, we have modeled the sleeper with a polygon. Numerical samples are composed of approximately 1000 grains with specific distribution size: 50% of diameter equal to 2.5 cm, 34 % of diameter equal to 3.75 cm, 16% of diameter equal to 5 cm. The friction coefficient between grains is $\mu_g = 0.6$ and equal to

0.6 or 0 between the top and bottom side walls. The simulations were carried out by means of the Contact Dynamics method which is based on implicit time integration of the equation of motion and a nonsmooth formulation of mutual exclusion and dry friction between particles [3–6].

Fig. 1. Sample after deposit (color represent the pressure).

Grains first fall freely under gravity onto the sublayer until they reach an equilibrium state (Fig. 1). After this deposit phase, 10 vertical loading cycles are applied on the sleeper in order to obtain a better connectivity with the ballast grains.

$$F(Q,V,t) = \frac{Q}{2}(X^{(\frac{(Vt-a)}{d})^2} + X^{(\frac{(Vt-a-L)}{d})^2}) \tag{1}$$

When the sample has reached an equilibrium state, loading cycles are applied in order to simulate the running of trains. The applied signal has the form presented in the equation 1 [7], with is axle load, V the velocity, X a real number which characterize the soil properties.

3 A Settlement Model

In general it is considered that the main source of settlement which occurs on railway track is due to ballast. In our case the settlement represents the sum of each vertical increment after one loading cycle of the sleeper. In experimental case, it is possible to distinguish two evolution phases [8]:

From the numerical computations presented in this article, we can identify two average behaviours:

- one between 0 and 10000 cycles, where the size increment and the total settlement represent more than 60% of the final settlement,
- one between 10000 and 60000 cycles, where the cumulative settlement evolves slightly.

The evolution of settlement samples are extremely sensible to the initial configuration of grains under the sleeper. However we obtain for the different samples after 60000 loading cycles a cumulative settlement about 80 mm. This value is more important than in reality because we never impose compaction under the sleeper to restore its initial vertical position like in maintenance operations.

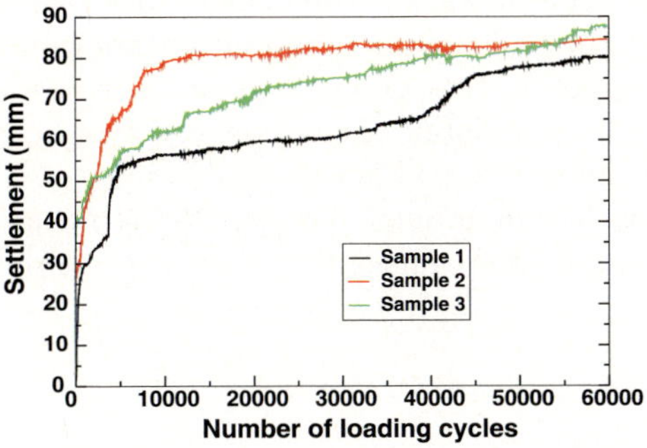

Fig. 2. Evolution of the cumulated settlement of the sleeper.

From these numerical computations, it is possible to propose a curve fit in order to predict settlement evolution. We can propose as a first approximation:

$$\tau_f = (\tau_0)^{\frac{Q}{Q+Q_0}} \tag{2}$$

with Q, the cumulative total load apply on the system, and Q_0 a reference value. The total load is more representative because it depends on the kind of train. τ_0 and Q_0 are evaluated from the linear regression of the curve $(Q, \frac{Q}{Ln(\tau)})$ with value obtained from numerical simulations. It is important to notice that for all linear regressions, we obtain a correlation coefficient of about 99%.

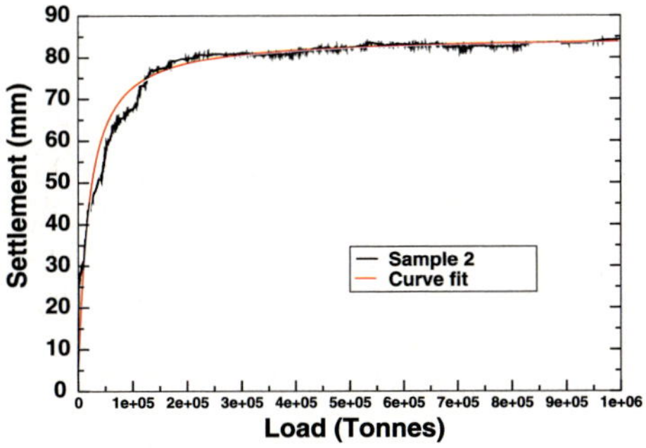

Fig. 3. Comparison between result of numerical simulations and proposed fit.

The proposed model describe very well the evolution of settlement, and allows us to predict the settlement by considering the evolution of the sample only for 10000 loading cycles. This model is proposed only for samples which have not been submitted to a high number of loading cycles.

4 Characterisation of the Stabilisation Phase

The characterisation of the stabilised phase is very important in order to decrease computation effort by preparing sample with properties of this phase. In real case the evolution of track settlement is governed by the properties of the subballast layer, the ballast state and maintenance operation employed to restore the initial geometry of the track. In general, the ballast has been always submitted to a very large number of circulations, and we have to identify the origins of track defects in this conditions.

From our numerical simulations, it is possible to characterise the ballast properties in the stabilisation phase in order to prepare sample with these properties.

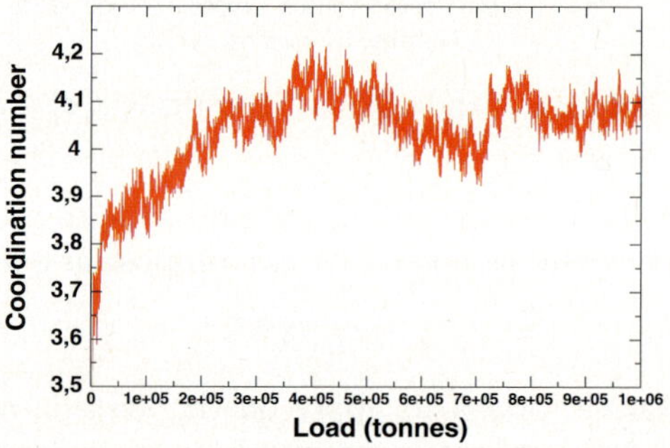

Fig. 4. Evolution of the coordination number.

The connectivity of the particles is described at lowest order by the average number z of contact neighbours per grains. The large value of z presented in figure 4 characterizes a sample of angular particles which reflects the enhanced arching effect [9]. In our case, the coordination number under the sleeper increase slightly to an average value 4.1 with large fluctuations. These fluctuations characterize the lost of stability of the system due to the creation and destruction of rigid paths identified in previous studies (localised rigid structures) [2].

It is important to notice that the average motion of grains under the sleeper is very small. At the time scale of one loading cycle we can propose that the grains follows a quasi-static solid motion depending on the sublayer deformations and the contact network follows a dynamic.

The organisation of simple and double contacts enhances the arching effects and a high heterogeneity of the sample. In the stabilisation phase it is important to notice that the strong double contact represent 30% of the contact set (fig. 5). The double contact are strongly involved in the stability of the system submitted to a vertical solicitation but these rigid paths are strongly influenced by shear solicitations which can be induced by the deformations of the sublayer.

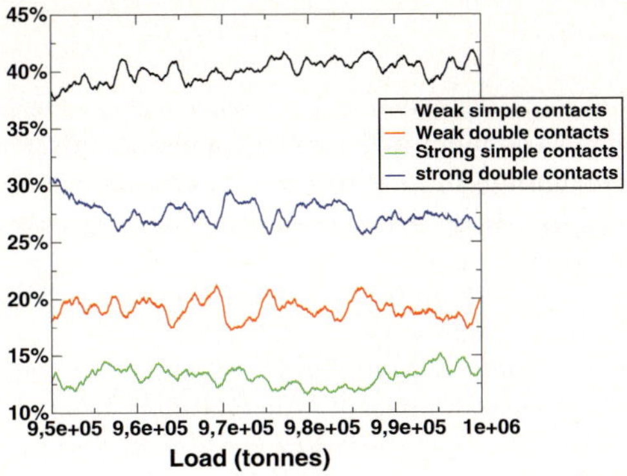

Fig. 5. Evolution of the fraction of double and simple contacts from strong and weak network.

The figure 5 presents a repartition of double contact in weak and strong force network which is different from classical VER analysis [9]. This state characterize a thin granular layer with angular particles.

5 Conclusion

The objective of this paper was to characterise the stabilisation of ballast layer by means of numerical simulations in order to propose a settlement model and identify the properties of stabilised sample. The identification of these properties allows to decrease the computational effort for further study on ballast which is mostly in stabilised state in real case.

From the numerical simulations we observed that it is possible to find a settlement model which allows to predict final settlement through the simulation of only 10000 loading cycles. It will be possible to evaluate the total settlement with an error less then 10%, which is an acceptable result.

The stabilised state of ballasted track model is characterized by a high value of the coordination number which reflects the angular shape of the grains and the history of loading cycles applied on the system. From our results, the average value of the coordination number in this phase is 4.1.

In this phase we observed a particular organisation of the force network: the double contact, side/side particle contact represent 30% of the strong force network. It is the second most important force network after the weak simple contact. This important role implies a strong heterogeneity of the thin layer with a high probability to create localised rigid structures.

From this work, it is possible to decrease our computational effort to create sample with stabilised properties and investigates the influence of increase of speed for example.

We thank Frédéric Dubois for assistance with the LMGC90 [10] platform used for simulations and Réseau Ferré de France which funded this project.

References

1. X. Oviedo. *Etude du comportement du ballast par un modéle micromécanique.* PhD thesis, Ecole Nationale des Ponts et Chaussées, 2001.
2. G. Saussine. *Contribution la modélisation de granulats tridimensionnels: application au ballast.* PhD thesis, Université des Sciences et Techniques du Languedoc, Montpellier, 2004.
3. M. Jean. *The non smooth contact dynamics method.* Computer methods in applied mechanics and engineering, 177:235–257, 1999.
4. A. Curnier, editor. *Unilaterality and dry friction in the dynamics of rigid bodies collection.* Presses Polytechniques et Universitaires Romanes, Lausanne, 1992.
5. J.J. Moreau. *Unilateral contact and dry friction in finite freedom dynamics.* In J.J. Moreau and P.D. Panagiotopoulos, editors, Nonsmooth Mechanics and Applications, chapter CISM Courses and Lectures, vol. 302, pages 1–82. Springer-Verlag, 1988.
6. G. Saussine, F. Dubois, C. Bohatier, C. Cholet, P.E. Gautier, and J.J. Moreau. *Modelling ballast behaviour under dynamic loading.* Part 1: a 2d polygonal discrete element method approach. Computer methods in applied mechanics and engineering, 2004.
7. R. Sauvage. *Railway track static behaviour.* 1993. INRETS-LTN.
8. V. Bodin. *Comportement du ballast des voies ferrées soumises à un chargement vertical et latéral.* PhD thesis, Ecole Nationale des Ponts et Chaussées, 2001.
9. E. Azema, R. Radjai, R. Peyroux, and G. Saussine. *Force transmission in a packing of pentagonal particles.* 2007.
10. F. Dubois and M. Jean. *Lmgc90 une plateforme de développement dédiée à la modélisation de problèmes d'interaction.* In M. Potier-Ferry, M. Bonnet, and A. Bignonnet, editors, 6ème colloque national en calcul des structures, pages 111–118. Imprimerie de l'école polytechnique, 2003.

Penetration of a Projectile by Impact into a Granular Medium

Antoine Seguin, Yann Bertho, and Philippe Gondret

Univ Paris-Sud, Université Pierre et Marie Curie-Paris6, CNRS
Lab FAST, Bât 502, Campus Univ, Orsay, F-91405, France

Summary. When a solid sphere drops in a granular medium, an impact crater is created. The penetration of the projectile in the granular bed as a function of the impact energy is evaluated theoretically using a simple model including a friction law between the projectile and the grains, a viscous dissipation in the bed and a force from the collisions between the projectile and the granular material. This model is observed to be in agreement with our quasi-2D experimental results and suggests that the penetration depth is a power law of the total drop distance.

For millions of years, meteorites have crashed on Earth, ejecting rocks at huge velocity and creating enormous craters. Although these collisions happened during a short time, they may have biological and geophysical aftermaths. That is why the knowledge of the projectile and the impacted medium is significant. The creation of impact craters is an intricate dynamics and the penetration of a projectile needs to be modelled with conservation law to reproduce the behavior of a granular medium. When a solid projectile impacts a granular layer, a crater is created and the projectile stops at a penetration depth δ. The morphology and the scaling of this crater depend on the granular medium and many features of the impacting projectile (diameter, density, impact velocity...) [1].

Recent experiments have brought out the dependance of the penetration depth δ with experimental parameters such as the projectile diameter d and density ρ, and the drop height [2, 3]. For a spherical projectile impacting a 3D granular layer, the evolution of the penetration δ with these parameters is usually described by power laws. Introducing the free fall height h and the total drop distance $H = h + \delta$, the depth is generally adjusted by a power law of the form $\delta \propto H^{1/3}$. In a recent work, Katsuragi and Durian [4] propose that the total force on the projectile results from the sum of three contributions: the gravity, a Coulomb solid friction and an inertial drag, such as

$$m\frac{dv}{dt} = mg - F_f - \alpha_c v^2. \tag{1}$$

From their experimental measurements, they suggest that the friction force F_f can be reduced to a simple linear behavior, $F_f = \alpha_f |z|$, where α_f mainly depends on the projectile diameter d and density ρ, and the friction coefficient in the granular medium.

Otherwise, experiments and numerical simulations have been undertaken in a two-dimensional case, to study more specifically the dynamics of the projectile during a crater formation [5–7]. In particular, they show that the time taken for a projectile to slow to a stop in the granular medium is independent of its impact velocity. Moreover, the mean drag force on the projectile dropped into the granular medium is observed to be constant during most of the projectile's trajectory and proportional to the impact velocity v suggesting that fluid-like properties might be important.

In this paper, we chose to focus on the penetration dynamics of a projectile in a granular medium in a quasi-2D experiment. Here we are going to expose an other model which describes our results.

1 Experimental Setup and Procedure

A packing of 1 mm glass beads (density $\rho_g = 2.5 \times 10^3$ kg m^{-3}) is confined into a vertical Hele-Shaw cell consisting of two parallel glass plates separated by a gap of 15 mm. In order to have reproducible measurements, the initial grain piling of height 56 mm is gently stirred with a thin rod before each experiment, allowing us to consider that a random close packing has been reached (we checked that measurements are well reproducible in our experiments).

A cylindrical steel projectile (density $\rho = 7.7 \times 10^3$ kg m^{-3}, mass $m \simeq 8.9$ g) of diameter $d = 10$ mm is hold initially by a magnet at a distance h above the granular surface. This apparatus allows us to drop the projectile without any initial velocity and spinning motion. The length of the projectile is a little smaller than the gap width to reduce the friction with the plates, and to prevent beads to be wedged between the glass plates and the projectile during the collision process. The impact speed is tuned by varying the drop height h from 75 to 305 mm corresponding to impact velocities ranging from 1.2 to 2.5 m s^{-1}.

The dynamical properties of the impact are analyzed by means of a high speed video camera at a frame rate of 2000 Hz. Using image analysis, the position of the projectile is then extracted during the free fall stage and as it penetrates inside the granular medium. In the following, the presented results correspond to an average over ten experiments to reduce the fluctuations of the measurements.

2 Experimental Results

Figure 1(a) displays the position of the projectile as a function of time for three different drop heights h (note that $t = 0$ corresponds to the beginning

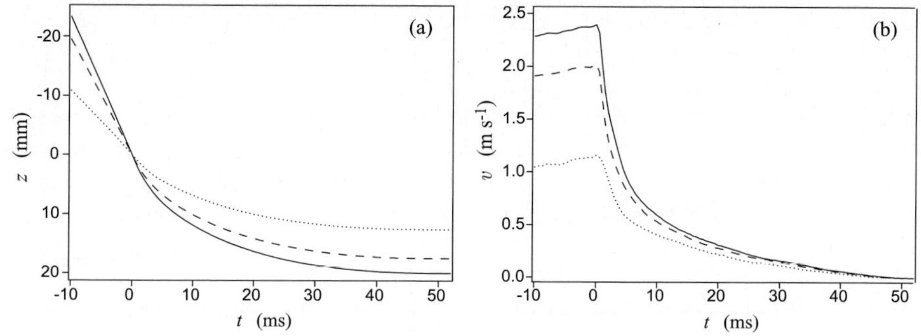

Fig. 1. (a) Position z and (b) velocity v of the projectile versus time for three different drop heights: ($\cdot\ \cdot$) $h = 75\,\mathrm{mm}$, (– –) $h = 215\,\mathrm{mm}$, (—) $h = 305\,\mathrm{mm}$.

of the penetration phase). Despite the fact that the projectile undergoes two very different phases – the free fall motion in the air and the penetration inside the granular layer – the transition does not appear clearly regarding the position z. As observed in Fig. 1(a), z tends towards a finite value δ (the penetration depth), all the more higher than the drop height is important. The curve is then derived numerically to get the variation of the velocity v as a function of time [Fig. 1(b)]. After a first stage where the velocity increases linearly during the free fall phase with a slope equals to gravity acceleration g, the velocity decreases abruptly with a discontinuity in acceleration as soon as the projectile hits the granular surface. These behaviors are in agreement with those observed both in two and three dimensional experiments reported elsewhere [4, 5]. Does the model proposed by Katsuragi and Durian [see Eq. (1)] to describe the forces at work during the impact reproduce the behaviors observed in our 2D experiment?

3 Modelling

We did not succeed to fit our results with the model of Katsuragi and Durian [4] as if something missed. So we add another friction force of viscous type so that the equation of the projectile during the penetration phase is

$$m\frac{d^2z}{dt^2} = mg - \alpha_f|z| - \alpha_v v - \alpha_c v^2. \tag{2}$$

The term $\alpha_f|z|$ represents a solid friction, without which the projectile would not stop at a finite penetration depth. The term $\alpha_c v^2$ can be representative of a dynamic frictional force which includes collision of the grains, as already presented in Katsuragi and Durian's work [4]. The term $\alpha_v v$ has been used in the pioneering work of Allen *et al.* for the penetration of a bullet in sand [8] and is now justified by the recent dense granular rheology where the friction coefficient increases linearly with the velocity [9].

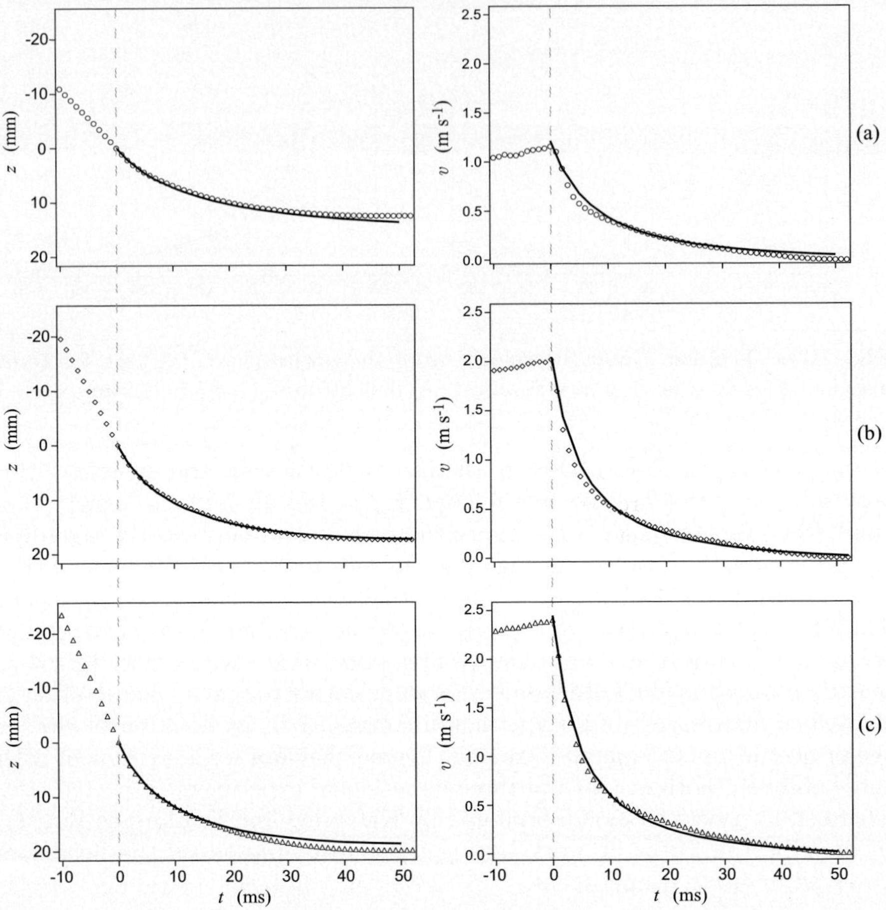

Fig. 2. Position z and velocity v of the projectile as a function of time for (a) $h = 75\,\text{mm}$, (b) $h = 215\,\text{mm}$, (c) $h = 305\,\text{mm}$. (—) Best fit of experimental data with numerical values $\alpha_f = 6\,\text{kg s}^{-2}$, $\alpha_c = 0.7\,\text{kg s}^{-1}$ and $\alpha_v = 0.58\,\text{kg m}^{-1}$. (- -) Time at which the impact occurs.

The solution of this model obtained with a numerical solver is displayed in Fig. 2 and compared with our experimental measurements for three different drop heights. The numerical coefficients corresponding to the best fit of the experimental curves for the position and the velocity are $\alpha_f = 6\,\text{kg s}^{-2}$, $\alpha_c = 0.7\,\text{kg m}^{-1}$ and $\alpha_v = 0.58\,\text{kg s}^{-1}$. We can directly compare these values with the ones of Katsuragi and Durian [4]; in their 3D experiment, they estimate $\alpha_f = 7.2\ 10^{-3}\,\text{kg s}^{-2}$ and $\alpha_c = 0.79\,\text{kg m}^{-1}$ using a spherical projectile of mass 62.9 g and a granular layer composed of spherical glass beads (diameter ranges from 250 to 350 μm). We obtain the same order of magnitude for α_c, whereas the value of α_f is completely different. The differences might be

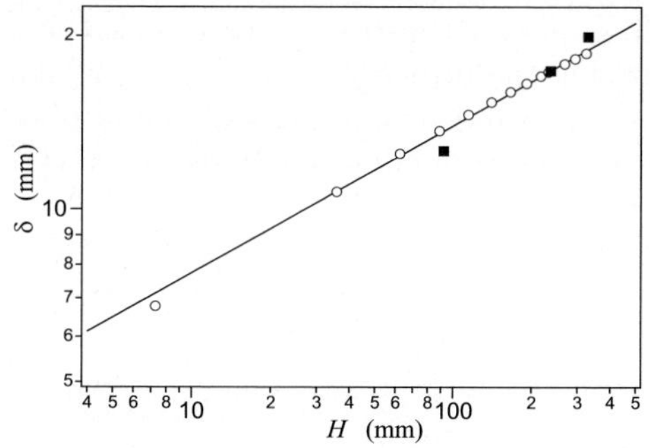

Fig. 3. Penetration depth δ versus the total drop distance H. (■) Experimental measurements; (○) numerical values; (—) the adjustment is a power law of the form $\delta = AH^{\lambda}$ with $\lambda \simeq 1/4$ and $A \simeq 4.3\,$S.I..

explained by the differences in the experimental conditions – such as the bead and projectile diameters or the 2D/3D nature of the experiment.

Note that each parameter has its own action field. At the beginning of the penetration, where the projectile velocity is important, the term $\alpha_c v^2$ dominates the other terms. At the end of the penetration, the projectile slows down and both terms $\alpha_c v^2$ and $\alpha_v v$ become negligible compared with $\alpha_f |z|$. Our model fits the experimental results reasonably even if small differences are still existing. Note also that α_f, α_v and α_c depend on many physical parameters (such as density, geometry...) but the study of such dependencies is far from the scope of this paper.

Moreover, our model allows one to extract the penetration depth δ for a given drop height h. The plot of the penetration depth versus the total drop distance H (where $H = h + \delta$), show that the numerical results can be adjusted by a power law $\delta = AH^{\lambda}$ with $\lambda \simeq 1/4$ and $A \simeq 4.3\,$S.I. (Fig. 3). These results display the same behavior as the 3D experimental results of Ambroso, De Bruyn and coworkers [2, 3], but with a smaller exponent, as they found $\lambda \simeq 1/3$ in 3D. This may be attributed to confinement effects in our Hele-Shaw cell.

4 Conclusion and Outlook

In summary, we have presented the penetration of a projectile impacting in a granular layer in a quasi-2D experiment. First, we succeed to find that the penetration depth δ follows a scaling law similar to the 3D experimental results of the form $\delta = AH^{\lambda}$ but with a smaller exponent λ probably due to confinement effects. Then, the dynamics of penetration is well adjusted by a model where many forces act: the gravity, a solid friction force $\alpha_f |z|$, a viscous force $\alpha_v v$ and a dynamic frictional force $\alpha_c v^2$.

Ultimately, additional experiments will be realized by modifying the granular layer and modifying the projectile (with different diameters d and densities ρ) in order to test the validity of the model and to observe the dependance of the three coefficients α_f, α_v and α_c on these parameters.

References

1. Walsh A. M., Holloway K. E., Habdas P. and De Bruyn J. R. (2005) Phys. Rev. Lett. 91:104301.
2. Ambroso M. A., Santore C. R., Abate A. R. and Durian D. J. (2005) Phys. Rev. E 71:051305.
3. De Bruyn J. R. and A. M. Walsh (2004) Can. J. Phys. 82:439–446.
4. Katsuragi H. and Durian D. J. (2007) Nature Phys. 3:420–423.
5. Ciamarra M. P., Lara A. H., Lee A. T., Goldman D. I., Vishik I. and Swinney H. L. (2004) Phys. Rev. Lett. 92:194301.
6. Tsimring L.S. and Volfson D. (2005) Modeling of impact cratering in granular media. In Powders & Grains 2005, R. García-Rojo, H. J. Herrmann, and S. McNamara (Eds.), A.A. Balkema, Rotterdam, 1215–1223.
7. Hou M., Peng Z., Liu R. and Lu K. (2005) Phys. Rev. E 72:062301.
8. Allen W. A., Mayfield E. B. and Morrison H. L. (1956) J. Appl. Phys. 28(3):370–376.
9. GDR MiDi (2004) Eur. Phys. J. E 14:341–365.

Rheological Transition in Granular Media

Zahra Shojaaee*, Lothar Brendel, and Dietrich E. Wolf

Department of Physics, University of Duisburg-Essen - 47048 Duisburg, Germany
*zahra.shojaaee@uni-duisburg-essen.de

Summary. The planar shearing of a two-dimensional system of bidisperse, non-cohesive discs at constant shear velocity is simulated using Contact Dynamics. The granular medium is confined between two parallel walls, which are pushed inwards by constant normal forces. The flow as a function of the shear velocity shows characteristics of a first order phase transition. The order parameter, the center of mass velocity in units of the shear velocity, undergoes a spontaneous symmetry breaking below a certain shear velocity, which does not depend on the distance between the walls.

1 Introduction

The simplest geometry to study the rheology of a granular material is planar shear. In simulations one can use Lees-Edwards boundary conditions [1] to shear the system without any walls [2]. However the walls are important in an experimental setup and can lead to shear localization [3]. In this work we always use walls to shear the system and investigate the dependence of flow properties on shear velocity and system size.

Using Contact Dynamics [4] we simulate a two dimensional bidisperse system, in which the particles are hard discs interacting by Coulomb friction forces parallel to, and volume exclusion forces perpendicular to the contact surfaces. The microscopic friction coefficient is $\mu = 0.5$, and the collisions are totally inelastic.

The particles are confined between two walls perpendicular to the y-axis, which are smooth but have friction. They are pushed inwards with normal forces of magnitude F_y and move in x-direction with the same constant shear velocity V in opposite directions. In order to satisfy these boundary conditions an anisotropic mass is assigned to the walls,

$$M = \begin{pmatrix} M_{xx} & M_{xy} \\ M_{yx} & M_{yy} \end{pmatrix} = \begin{pmatrix} \infty & 0 \\ 0 & M_{yy} \end{pmatrix} .$$

In order to preserve the symmetry of the system, care had to be taken during the preparation of the initial configuration. The distance L_y between the walls was fixed, and the system was closed by an additional wall perpendicular to the x-axis. Then a given number of frictionless particles settled under gravity parallel to the x-axis. The right surface of the filling was smoothed by means of a piston, determining L_x. Then the left wall and the piston were replaced by periodic boundary conditions, gravity was switched off, friction was switched on, L_x was kept fixed, and L_y was allowed to evolve, resulting in fluctuations with a small amplitude depending on the force F_y and the shear velocity V. Particles were given a linear velocity profile at the beginning, which extrapolated to the wall velocities on both sides.

The diameter of the large particles is taken as length unit ($d_{\max} = 1[L]$). Similarly the mass density of the particles is set to unity ($\rho = 1[M]/[L]^2$). The time unit is chosen such that the normal forces applied to the walls have a value $F_y[T]^2/[M][L] = 5$. A time step in the simulation is equal to $0.008[T]$.

2 High and Low Velocity Regimes

Figure 1(a) shows the time evolution of the velocity profile of a system of initial height $L_y = 40$ as it is sheared with a velocity $V = 0.5$. One sees, that after a short transient the velocity profile adopts a symmetric s-shape with only weak fluctuations in time. There is symmetric shear localization at both walls including slip. In contrast, if the system is sheared with low velocities (e.g. $V = 0.03$, see Fig. 1(b)), the velocity profile fluctuates strongly and is generally asymmetric. Shear is still localized at the walls, but no longer symmetric. In particular it happens that most of the filling moves together like one block with one wall. In the following we address the question, how to characterize this symmetry breaking transition, when one goes from high to low velocities.

3 Behavior of Center of Mass Velocity

3.1 Order Parameter

We propose to take the center of mass velocity V_S divided by V, $m = V_S/V$, as an order parameter to describe the symmetry breaking. For a symmetric profile m is zero, whereas for the block-like motion described above it is close to ± 1.

Note that for a finite system a spontaneously broken symmetry is restored dynamically, i.e. the order parameter switches back and forth between $m \approx 1$ and $m \approx -1$ with a characteristic switching time τ that should diverge with system size. As a consequence, $\langle m \rangle \to 0$ also in the symmetry broken phase, if one averages over long enough times $t \gg \tau$, and even more so, if one averages over the ensemble of equivalently prepared random samples as well ($\langle . \rangle$ denoting both averages).

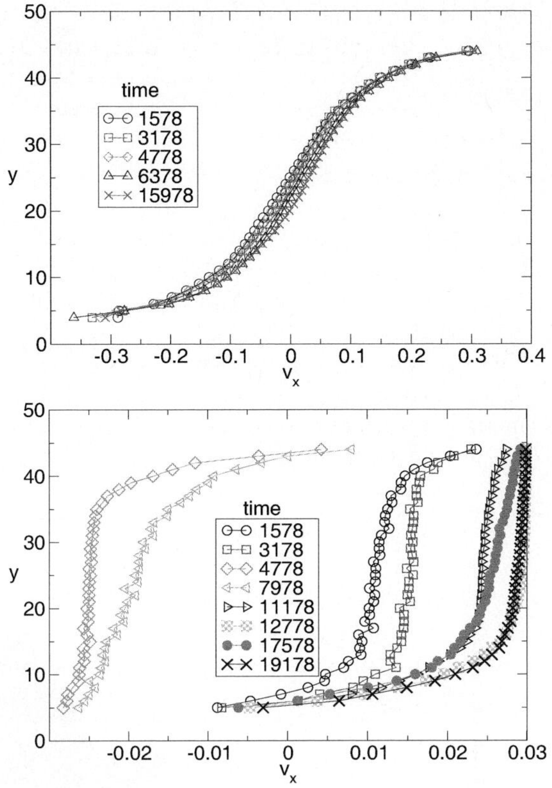

Fig. 1. Velocity profiles at different times for shear velocities (a) $V = 0.5$ and (b) $V = 0.03$.

3.2 Order Parameter Fluctuations

Figure 2 shows the order parameter fluctuation $\chi = \langle m^2 \rangle - \langle m \rangle^2$ as a function of shear velocity V for different system sizes L_y. The averaging was done over a time $t = 20000$ and 10 different runs so that $\langle m \rangle \approx 0$ in all cases. This explains why in the low velocity regime χ approaches a constant value which should be approximately $\langle |m| \rangle^2$.

Interestingly, for increasing system size the fluctuations decrease for velocities $V > V_c \approx 0.1$, while they increase below V_c. This behavior indicates a phase transition at $V_c \approx 0.1$. Furthermore, right at V_c the order parameter fluctuation does not depend on system size noticeably.

3.3 Order Parameter Histograms

Figure 3 shows four histograms of the order parameter $m(t)$, accumulated over a long time. For large velocities (Fig. 3 a)) the order parameter values

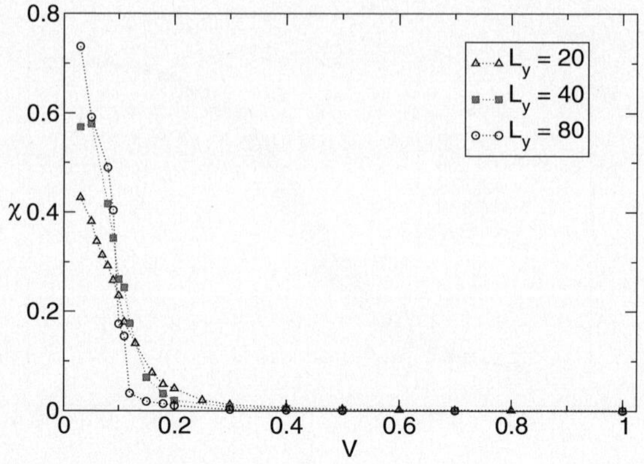

Fig. 2. Order parameter fluctuation vs. shear velocity for systems of different initial heights $L_y = 20$, $L_y = 40$ and $L_y = 80$.

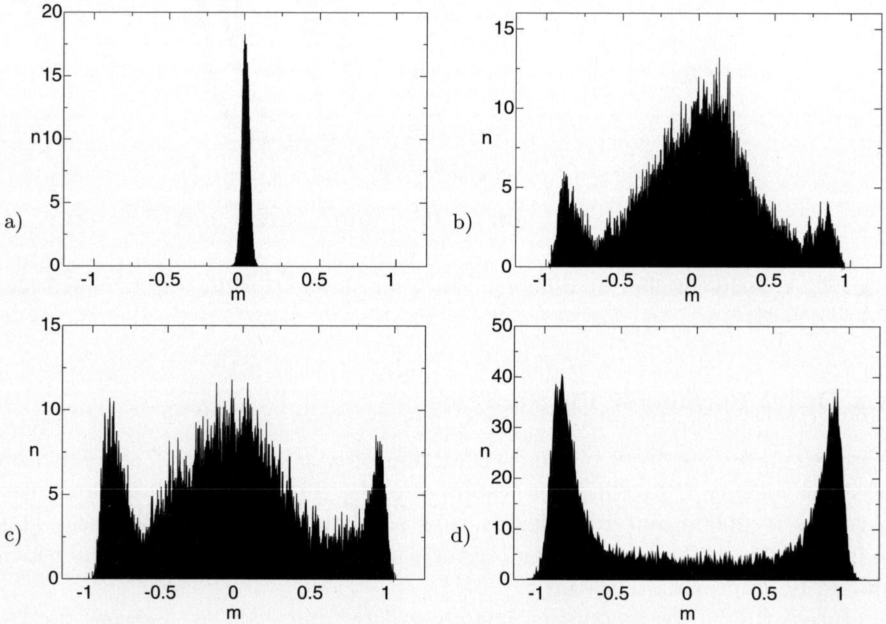

Fig. 3. Histograms of the order parameter of a system of height $L_y = 40$ with shear velocities a) $V = 0.5$, b) $V = 0.12$, c) $V = 0.1$ and d) $V = 0.05$.

are concentrated at $m = 0$. For small velocities (Fig. 3 d)) one has two peaks close to $m = \pm 1$. They are symmetric as a result of ensemble averaging. For large systems the histograms for individual runs remained asymmetric, because the reversal of the order parameter became too rare to restore the symmetry within the recording time. Close to the transition the central peak widens and decreases, while the outer peaks grow in place, i.e. without moving continuously outwards from the center. This is the phenomenology of a first order phase transition. With increasing system size the interval of shear velocities, for which the histogram shows three peaks, becomes narrower.

4 Ergodic Time τ

The ergodic time τ is the average time until a jump between the two outer histogram peaks (from positive to negative side and vice versa) happens. This quantity is plotted against shear velocity for different system sizes in fig. 4. τ is

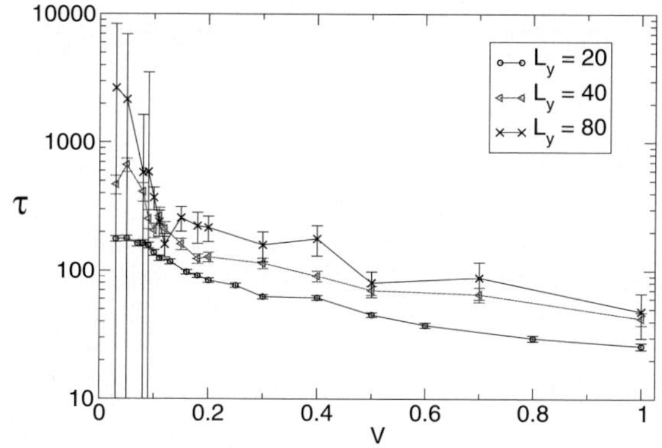

Fig. 4. Ergodic time against shear velocity for systems of different heights $L_y = 20$, $L_y = 40$ and $L_y = 80$.

likely to grow to infinity for an infinite system below the critical shear velocity, which corresponds to symmetry breaking in the thermodynamic limit.

5 Shear Localization

According to Fig. 2 the shear velocity V_c at the transition seems to be independent of L_y. This means that the transition is not governed by the global shear rate $\dot{\gamma}_{\text{global}} = 2V/L_y$. Another length scale must be relevant, which we identify with the width of the shear zone, as shown in figure 5. There we selected velocity profiles in the symmetry broken phase, where the whole system moves together with one of the two walls, apart from a shear zone at the other wall. Obviously the width of this shear zone, W, does not depend on L_y. Hence, when describing the transition in terms of a dimensionless parameter, the inertial number I [5], it is the shear rate $\dot{\gamma} = 2V/W$ which must be employed.

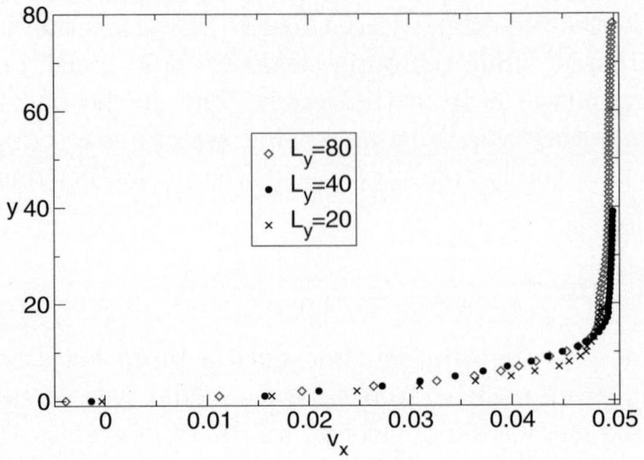

Fig. 5. Velocity profiles of systems with heights $L_y = 20$, $L_y = 40$ and $L_y = 80$ sheared with the shear velocity of $V = 0.05$ proofing the L_y-independence of the shear band height.

Acknowledgements

We acknowledge partial support by the DFG-Collaborative Research Center SFB 445 "Nanoparticles from the Gas Phase – Formation, Structure, Properties". DW thanks F. Chevoir, E. Clement, and J.-N. Roux for their hospitality during a research stay.

References

1. A. W. Lees and S. F. Edwards. The computer study of transport processes under extreme conditions. *Journal of Physics C*, 5(15):1921–1928, 1972.
2. P. G. Rognon, J.-N. Roux, D. E. Wolf, M. Naaïm, and F. Chevoir. Rheophysics of cohesive granular media. *Europhysics Letters*, 74(4):644–650, 2006.
3. E. Aharonov and D. Sparks. Shear profiles and localization in simulations of granular materials. *Physical Review E*, 65:051302, 2002.
4. L. Brendel, T. Unger, and D. E. Wolf. Contact dynamics for beginners. In H. Hinrichsen and D. E. Wolf, editors, *The Physics of Granular Media*, pages 325–343, Berlin, 2004. Wiley-VCH.
5. F. da Cruz, S. Emam, M. Prochnow, J.-N. Roux, and F. Chevoir. Rheophysics of dense granular materials: Discrete simulation of plane shear flows. *Physical Review E*, 72:021309, 2005.

Stress Transmission in a Multi-Phase Granular Packing

Vincent Topin[1,2], Jean-Yves Delenne[1], Farhang Radjaï[1], and Frédéric Mabille[2]

[1] LMGC, CNRS-Université Montpellier 2, Place Eugène Bataillon, 34095 Montpellier cedex, France topin@lmgc.univ-montp2.fr
[2] IATE, CNRS-INRA, 2 place Pierre Viala, 34060 Montpellier cedex 1, France

Summary. We analyze stress transmission in granular media involving an interstitial cementing matrix of variable volume fraction. We rely on a lattice-type discretization of both the particles and cemented matrix. We show that the stress chains are essentially guided by the cementing matrix in tension and by the particulate backbone in compression. The signature of granular structure appears clearly on the probability density functions of node stresses. We can discern large, intermediate and weak stresses. The stress distributions are increasingly wider for a decreasing matrix volume fraction in tension. Finally, we compare the contact force network computed from stresses localized at the matrix bridges between particles with that computed by means of the discrete element method with cohesive interactions and for the same configuration of the particles. We show that the two methods yield similar force patterns at low matrix volume fraction.

1 Introduction

Cemented granular materials (CGMs) form a wide class of materials encountered in various fields of science and engineering. Some examples are mortars, concrete and asphalt (aggregates of various sizes glued to each other by a cement paste) [1], solid propellants and high explosives (large volume of energetic particles in a polymeric binding matrix) [2], sedimentary rocks (sandstones, conglomerates and breccia) [3], and some biomaterials such as the wheat endosperm (starch granules forming a compact structure bound together by a protein matrix) [4, 5]. These materials have a similar microstructure consisting of densely packed particles and a solid matrix filling the interstitial space and sticking to the particles.

In this paper we analyze stress transmission in CGMs by means of a lattice-type discretization of both the particles and cementing matrix . We are interested in stress transmission under different loadings and for different matrix volume fractions. A detailed comparison of the contact force network between particles observed from the lattice model and from a discrete element method with cohesive interactions will be presented.

2 Numerical Method

Lattice-type discretization has been extensively used for the statistical mechanics of fracture in disordered media [6], and applied to study the fracture properties of concrete [7], ceramics [8] and biomaterials like wheat endosperm [5]. The space is discretized as a grid of points (nodes) interconnected by one-dimensional elements (bonds). Each bond can transfer normal force, shear force and bending moment up to a threshold in force or energy. When several phases are present as in a cemented granular medium, each phase and its boundaries are materialized by lattice elements sharing the same properties and belonging to the same portion of space. For a cemented granular material, there are three *bulk* phases; the particles, the matrix and the voids, denoted respectively here 'p', 'm'; and 'v', and two kinds of interface phases; particle-particle interface, denoted 'pp', and particle-matrix interface, denoted 'pm'.

We use linear elastic-brittle elements, each element characterized by a Hooke constant and a breaking force threshold. The bonds transmit only normal forces between the lattice nodes and thus the strength of the lattice in shear and distortion is ensured only by the high connectivity of the nodes. A sample is defined by its contour and the configuration of the phases in space. The samples are deformed by imposing displacements or forces to nodes belonging to the contour. The initial state is the reference (unstressed) configuration. The total elastic energy of the system is a convex function of node displacements and thus finding the unique equilibrium configuration of the nodes amounts to a minimization problem. Performing this minimization for stepwise loading corresponds to subjecting the system to a quasistatic deformation process. The method used here can be found in more detail in reference see [9].

3 Stress Transmission

The Cauchy stress tensor σ makes sense only for a sufficiently large number of material points inside a control volume such that the surface density of forces is well defined. Following a general framework, first introduced by Moreau, we can attribute a stress tensor σ^i to each node i of the lattice network [9, 10]. The physical content of this tensor remains the same whether applied to a node i or to a portion of space including several nodes, and it tends to the Cauchy stress tensor at large scales. In order to analyze the stress distributions in our numerical samples, we use these "node stresses" whose components are represented by a proportional color intensity or grey level over the elementary hexagonal cell centered on each node.

We first focus on the vertical stresses for a simple uniaxial compression test. We use a large dense packing of rigid disk-like particles compressed isotropically by a discrete element method. A rectangular portion of this two-dimensional packing containing about 500 particles is then discretized on a triangular lattice. The matrix volume fraction $\rho^m \simeq 0$ here, corresponding to dry cohesive granular media.

Figure 1 shows the vertical stress fields σ_{yy} in the undamaged state. The strong stress paths that we observe in this map are reminiscent of force chains observed both in cohesive and cohesionless granular media [11, 12].

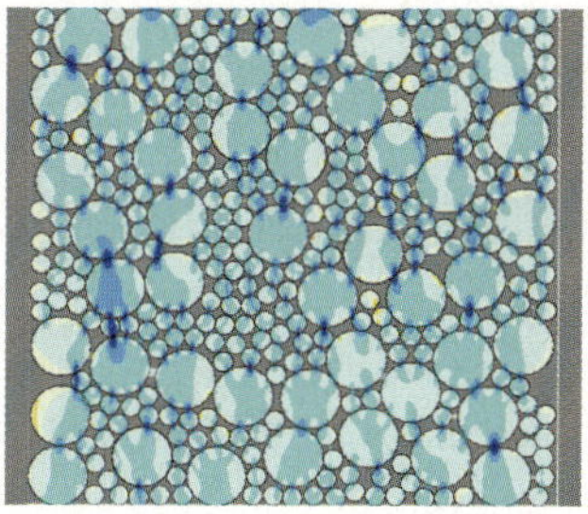

Fig. 1. Vertical stresses represented in levels of color intensity. Blue and red colors represents compressive and tensile stresses, respectively.

The probability density function (pdf) of the vertical stresses is shown in Fig. 2 a). The stresses are normalized by the total stress σ_{yy} of the lattice. From the shape of the pdf, we distinguish three parts: 1) Large stresses fall off exponentially as observed for large contact forces in dry granular media [11, 12]; 2) Weak stresses have nonzero probability reflecting the arching effect; 3) Intermediate stresses are centered on the mean and define a Gaussian distribution. Figure 2 b) shows a typical map of vertical stresses where strong, weak and intermediate stresses are represented by different colors. We see that the large stresses do mostly concentrate at the contact zones and they form well-defined chains that cross the particles. The intermediate stresses are almost fully localized inside the particles.

We now consider the influence of loading for the same configuration of particles as but for $\rho^m = 0.12$. Our studies show that, due to the presence of the granular backbone, the stress chains are essentially guided by the cementing matrix in tension and by the particulate backbone in compression. The probability density functions (pdf) are shown in Fig. 3 a). It is remarkable that, for the same matrix volume fraction, the distribution of large stresses is broader in compression than in tension. In the same way, the weak stresses are far more numerous in compression than in tension. This means that stress distribution is far more inhomogeneous in compression than in tension. The

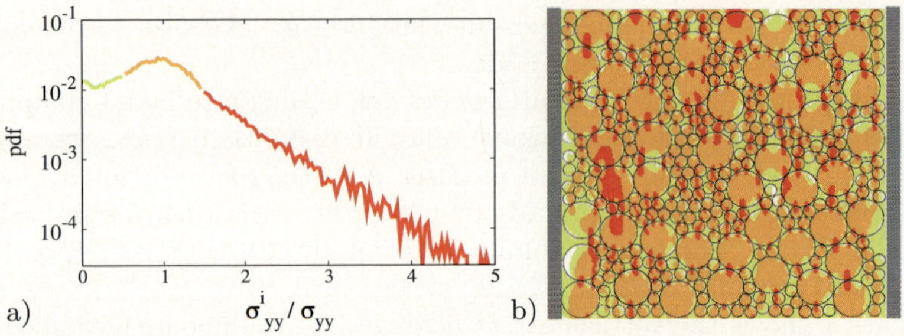

Fig. 2. a)Probability distribution function of vertical bond stresses normalized by the average stress; b)Tricolor map of vertical stresses in compression with weak, intermediate and strong stresses represented in yellow, orange and red, respectively.

range of large stresses corresponds to stress chains or, more precisely, to a "strong network" as in dry granular media where a well-defined subset of contact forces (the strong force network) transmits nearly the whole stress deviator sustained by the system.

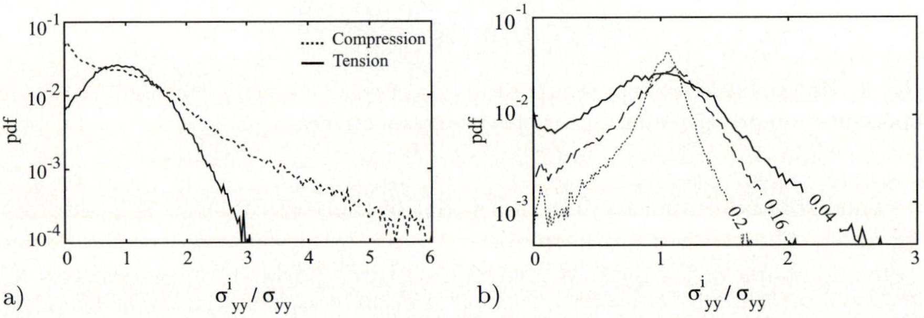

Fig. 3. a) Probability distribution function of vertical bond stresses normalized by the average stress in compression and tension. b) Probability distribution functions of vertical node stresses in tension for three values of the matrix volume fraction.

The effect of the matrix content is to redistribute more homogeneously the node stresses. Figure 3 b) shows the pdf's of vertical stresses for different values of ρ^m in tension and for identical stiffness between the particles and the matrix. The pdf is increasingly wider for decreasing matrix content so that the stresses are more and more concentrated in the binding bridges between the particles. On the other hand, at the limit where the pores are filled with the matrix, the distribution is peaked on the mean stress. Since there is no contrast in stiffness and the pores are filled, the only source of inhomogeneity in this limit is the presence of non cohesive zones between particles.

4 Force Distribution: Comparison with DEM

We compare here the contact force network between particles computed from the lattice method with that computed by means of the discrete element method (DEM) with cohesive interactions. A sample of 5000 particles is simulated both by lattice method at very low matrix volume fraction $\rho^m \simeq 0$ and DEM. In the lattice-based model, the particles are bonded together by the matrix bridges which provide the mechanical integrity of the structure. Forces between particles are calculated by evaluating the total stress localized in the matrix bridges projected on the normal direction, see Fig. 4 a) and b). Concerning the DEM simulations, cohesive interaction between two particles implies resistance to relative motion (normal displacement d_n, tangential displacement d_t, and angular displacement γ) of two edge points belonging, respectively, to the two particles and coinciding initially with the contact point; see Fig. 4 c). The corresponding contact actions are the normal force f_n, the tangential force f_t, and the contact torque M. Details of the contact cohesion model used for the present studies can be found in [13].

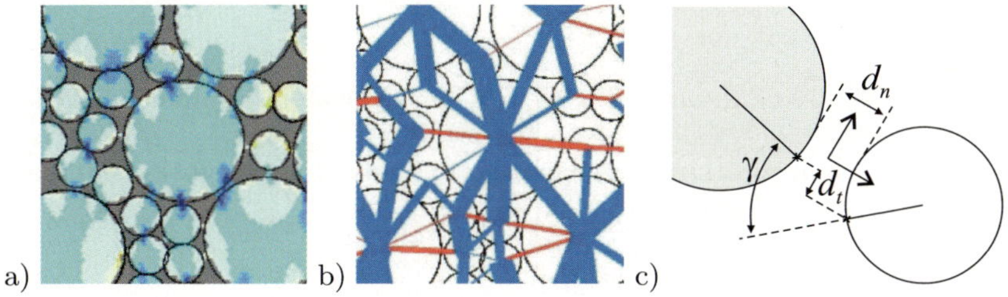

Fig. 4. a) Vertical stresses represented in levels of color intensity, b) Forces between particles calculated from stresses localized in the matrix bridges, c) Relative displacements between two edge points belonging to two particles and coinciding initially with their contact point.

The pdf's of normal forces from lattice method and DEM are shown in Fig. 5. We observe that the two pdf's coincide over nearly the whole range of forces. This agreement between the two methods is all the more interesting that in DEM the particles are assumed to be rigid and the stresses inside the particles are not involved in the calculation of contact forces. The pdf has well-known features of force distributions in dry granular media. Indeed, we note that the data for forces lower than the mean seem to have a nearly uniform distribution whereas the data for forces larger than the mean represent a nearly exponential decay. Note that the pdf's of normal forces contain no Gaussian peak as that observed in Fig. 2 for intermediate forces. This indicates that the Gaussian peak corresponds to the stresses inside the particles.

Our results suggest that lattice discretization can be used for the simulations of dense granular materials with or without a solid matrix. In particular,

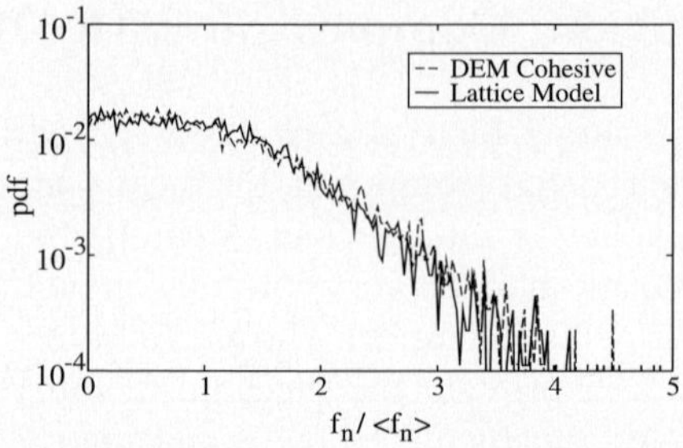

Fig. 5. Probability distribution function of positive normal forces normalized by the average normal force with lattice method and discrete element method with cohesion.

the forces at the contact zones compare well with discrete element simulations. The contact force distribution from the lattice model can be considered as a validation of that obtained from discrete element simulations.

References

1. E. Schlangen, J.G.M. van Mier, Cement and Concrete Composites **14**(2), 105 (1992).
2. H. Tan, Y. Huang, C. Liu, G. Ravichandran, H. Inglis, P. Geubelle, International Journal of Solids and Structures **44**(6), 1809 (2007).
3. E.J. Tarbuck, F.K. Lutgens, *Earth – An introduction to Physical Geology* (Pearson Education, New Jersey, 2005).
4. *Wheat: Chemistry and technology* (AACC, St Paul USA, 1988).
5. V. Topin, F. Radjai, J.Y. Delenne, A. Sadoudi, F. Mabille, Journal of Cereal Science **In Press, Corrected Proof** (2007).
6. H.J. Herrmann, S. Roux, eds., *Statistical Models for Fracture in Disordered Media* (North Holland, Amsterdam, 1990).
7. E. Schlangen, E.J. Garboczi, Engineering Fracture Mechanics **57**(2-3), 319 (1997).
8. A. Delaplace, G. Pijaudier-Cabot, S. Roux, Journal of the Mechanics and Physics of Solids **44**(1), 99 (1996).
9. V. Topin, J.Y. Delenne, F. Radja, L. Brendel, F. Mabille, The European Physical Journal E **In Press, Corrected Proof** (2007).
10. J.J. Moreau, *Numerical Investigation of Shear Zones in Granular Materials*, in *Friction, Arching, Contact Dynamics*, edited by D.E. Wolf, P. Grassberger (World Scientific, Singapore, 1997), pp. 233–247.
11. F. Radjaï, M. Jean, J.J. Moreau, S. Roux, Phys. Rev. Lett. **77**(2), 274 (1996).
12. D.M. Mueth, H.M. Jaeger, S.R. Nagel, Phys. Rev. E **57**, 3164 (1998).
13. J.Y. Delenne, M.S.E. Youssoufi, F. Cherblanc, J.C. Bénet, International Journal for Numerical and Analytical Methods in Geomechanics **28**(15), 1577 (2004).

Part III

Molecular Motors and Motion in Biology

Non-Equilibrium Collective Transport on Molecular Highways

Andrea Parmeggiani

Cellular Biophysics and Biomimetics Team, DIMNP, UMR 5235, cc. 107, Pl. E. Bataillon, 34095 Montpellier Cedex 5, France
andrea.parmeggiani@univ-montp2.fr

Summary. Intracellular motion driven by motor proteins on cytoskeletal filaments provides an interesting example for studying non-equilibrium transport phenomena. We will discuss a 3-dimensional lattice gas model, based on the generalization of a totally asymmetric exclusion process with non conserved dynamics (TASEP coupled with Langmuir Kinetics). In particular, we will focus on emergent physical properties, specific to the three-dimensional geometry, concerning traffic jam and phase separation phenomena, with interesting biological implications for real intracellular transport processes.

1 Physical Properties of Intracellular Transport

In recent years, progress in monitoring and measuring *in vitro* and *in vivo* properties of biological systems has stimulated as never before the interest of the physics community toward biology. Characterized by a multitude of molecular machines and constituents that continuously assemble, operate and communicate in a controlled and coordinated manner, the complexity of living matter is a result attained on evolutionary time scales. This is a genuine challenge for disciplines like condensed matter and statistical mechanics, for such organization originates in a strongly fluctuating environment like the world at "room temperature, but at nanometer scales". The physico-chemical description of small systems, their non-equilibrium thermodynamics and how they organize collectively are some fundamental questions in statistical mechanics which find in biology a field to be systematically explored.

Intracellular transport (IT) is one of the "amazing" phenomena we are used to discuss and study as physicists involved in biology, taking advantage of an unprecedented possibility of analysis by using, for example, fluorescent microscopy or single molecule techniques. IT provides the main mechanism in order to transport matter, energy and information through and outside cells. Among the different processes involving translocation of cellular material, we discuss here IT of cargos driven by *motor proteins* in *eukaryotic* cells [1].

More particularly, for the sake of brevity, we shall focus on transport driven by *conventional kinesins* over *microtubule* filaments [1, 2].

In the first place, we shall investigate how many of these motor proteins dynamically organize their collective transport on a single filament, through their stochastic and off-equilibrium stepping walk. We shall see how predictions on traffic jams, based on a simple one-dimensional (1D) model, can be useful to learn some physics about this phenomenon.

We shall then study the role of viscosity for motor *jam* formation on the filament by considering the cytoplasm around the filament and the diffusion of motors in it. We shall show how stochastic modeling, in a three-dimensional (3D) version of the 1D model, describes counterintuitive traffic jam phenomena, possibly relevant to cell physiology.

Recently, efforts in this direction have revealed interesting physics, bridging biological transport description to other transport phenomena like in vehicular/pedestrian traffic or in complex fluids-materials [3]. The origin of jam formation is, indeed, one key topic for research fields such as intracellular traffic, traffic sciences and granular materials. In biology, in particular, jams represent phases in which transport of cellular components is no longer efficient, thus representing a danger for cell integrity and life. Hypotheses in this sense have been invoked to explain neuro-degenerative pathologies like Alzheimer's Disease, for example, [4].

Some Notion About IT. Transport of matter inside a cell takes place essentially via two kinds of mechanisms. On a short scale in space, from nanometers up to one micrometer and when the cytoplasm is locally homogeneous and isotropic, *Brownian diffusion* [2, 5, 6] allows an efficient process for distributing and targeting cytoplasm components. On larger scales or when the internal organization of the cytoplasm becomes more complex, *directed transport* driven by specialized motor proteins along a network of biopolymers, the *cytoskeleton*, becomes necessary to assure efficient displacement and addressing of cargos in the cell [2].

The cytoskeleton is essentially a cellular scaffold embedded in the cytoplasm, connecting nodal regions for cell life, development and division [1]. Together with *actin* and *intermediate* filaments, microtubules are the major "quasi-linear" protein assemblies (biopolymers) forming the cytoskeleton. Microtubules are periodic and helical hollow cylinders made by parallel arrangements of linear *protofilaments* of $\alpha\beta$-tubulin dimers. They have an intrisic *polarity*, a consequence of the $\alpha\beta$-tubulin dimer structural asymmetry.

Kinesins are motor proteins, nanomachines which convert the free energy of the ATP-hydrolysis reaction[1] into directed mechanical motion and force. Due to their high bending rigidity, microtubules are molecular "highways" crossing the cytoplasm. Their structural polarity makes kinesin motors moving along them from the center to the periphery of the cell (*anterograde motion*), while *dynein* motors execute the transport in the opposite direction (*retrograde*

[1] Adenosin-Tri-Phosphate $\rightarrow$ Adenosin-Di-Phosphate + inorganic Phosphate.

motion). Indeed, filament polarity imposes a preferential orientation in the attachment of the motor to the filament, and fixes the direction of the motor stochastic stepping (approximately, a *1D biased-like Brownian walk* [2, 7]).

Kinesin Moving on a Microtubule: Generic Properties. From enzymology, structural biochemistry and *in vitro* single molecule experiments [2] we know that conventional kinesin is a double-headed motor of typical size (hydrodynamic radius) $r_h \sim 10$–$20\ nm$. It alternates a 1D *directed walk* phase of hundred steps along microtubule protofilaments with unbinding and binding events between microtubules and the cytoplasm. During the detached phase, the motor performs essentially a 3D Brownian walk in the cytoplasm.

Each motor step is the result of very complex mechanochemical processes such as *ATP hydrolysis-induced conformational changes.* That way, the motor hydrolyses in average one hundred molecules per second, i.e. at a frequency $\omega_s = 100\ s^{-1}$, at saturating (high) ATP concentration. Experiments have shown that for each hydrolysis event and different values of applied external forces, the protein steps by $l_s \simeq 8\ nm$ in length [8], i.e. the same microtubule periodicity imposed by the $\alpha\beta$-tubulin dimer size. A kinesin, *in vitro* and at room temperature, then moves with a typical average speed $v \sim 1\ \mu m/s$.

The 3D diffusional search is in principle a complex process, considering that it takes place in a complex environment like the cytoplasm [5]. In first approximation, we can consider the cytoplasm as a medium with an average *effective* viscosity η_m. A change of η_m implies a change of the translational diffusion constant D of the motor, estimated by using the Stokes-Einstein relation $D = kT/6\pi\eta_m r_h$ (where k is the Boltzmann constant, T the temperature of the medium, η_m the medium viscosity, and r_h the protein hydrodynamic radius). Typical diffusion constants in cytoplasm can vary depending on the effective viscosity η_m and on the protein size r_h (plus eventually the size of the cargo attached). For a protein size of 10 nm, D ranges approximately between 0.1 and 10 $\mu m^2/s$, the higher value for aqueous environment, the lower one for more viscous media such as membrane-like or crowded cell compartments.

The stepping rate ω_s is associated with a (rather) *slow* event once compared with the typical rate ω of the motor to diffuse in water over a typical distances like the step size l_s. The rate ω can be 3–4 orders of magnitude larger than ω_s. This can be estimated by using the classical formula of the mean square displacement of a 3D Brownian walk: $\omega \simeq 6D/l_s^2 \simeq 10^6\ s^{-1}$.

However, from the Stokes-Einstein relation, ω depends also on the medium viscosity η_m. For example[2], for a medium viscosity η_m hundred times higher than water viscosity, $\eta_w \simeq 1\ mPa.s$, ω will be only 1–2 times larger than ω_s.

Depending on the kinesin double-head structure and its specific biochemistry, *processivity* is another (complex) property of this motor protein to con-

[2] A similar argument applies in case of micrometer-sized organelles attached to the motor. In this case, the hydrodynamic radius of the complex "motor-cargo" is ten/hundred times bigger than the motor: the diffusion rate ω decreases accordingly to the formula used above.

sider. A processive kinesin at saturating ATP concentration is able to hydrolyse in average hundred ATP molecules per second without detaching from the microtubule [2]. This implies an average "residence" time, during the kinesin walk on the filament, of about 1 s, i.e. a detachment frequency of $\omega_d \simeq 1\ s^{-1}$.

These rates are average values that depend not only on the continuous action of the thermal kicks of the surrounding fluid, but also on the external force applied to the motor [9, 10]. In general, the motor mechanochemical coupling implies that an external mechanical force can affect directly the motor biochemistry, thus also speed and processivity, too. For our purposes, it is useful to know that the maximal force produced by the motor before detachment, the so-called "stall force", is of the order of $f_s = 6\text{–}7\ pN$ [2, 8, 9].

Motor Ability to Move in Viscous Environments. Motor proteins are powerful nanoscopic devices for transport in viscous media. To make the point, a simple argument can be inferred by the Stokes relation, which defines the friction of a bead of hydrodynamic radius r_h in a medium of viscosity η_m as $\xi = 6\pi\eta_m r_h$. For typical values of water viscosity, the viscous load on a bead of 1 μm of diameter and moving in water at a speed $v = 1\ \mu m/s$ is negligible: $f_v \propto \xi v \sim 0.01\ pN$. A bead of 100 μm in diameter would produce a viscous drag on the motor which is about only 20% of its stall force f_s!

A rigorous analysis was performed by Hunt and collaborators [11]. By using motility assay experiments on microtubules driven by kinesins (fixed on a glass surface), they provided experimental evidence that the speed of a 10–20 μm microtubule long, driven by one or few kinesins, decreases only by a factor 2 after a 100-fold increase in medium viscosity. In this respect, speed and processivity of motor proteins like kinesins are practically insensitive to changes in medium viscosity, up to 3 order of magnitude greater than η_w.

This fact is biologically meaningful. Cytoplasm is a highly dynamic and heterogeneous medium, made up of active elements moving and mutually interacting (specifically or non-specifically). Its effective viscosity η_m ranges spatially and temporally from water value η_w up to a hundred or a thousand times η_w (depending also on the zone of the cell considered). Reasonably, motor activity should not be particularly sensitive to most of the viscosity changes. On the contrary, motor activity and speed should be more strongly affected by biochemical regulation or sources of stalling like the action of other motors, mechanical constraints, crowding of materials, *et cetera*.

2 TASEP Coupled to LK: A New Driven Lattice Gas

Generic properties of motor protein collective transport can be mathematically described in terms of driven lattice gases at low dimensionality. These models have a rich phenomenology, which can be analyzed through computer simulations and different analytical methods [12–14].

Recently, we have studied the non-equilibrium properties of the flow of a collection of motors (particles) on a single filament (*N-sites* lattice), in contact with the cytoplasm (particle reservoir) [15, 16]. Over the lattice, particles move with microscopic dynamical rules inspired by the essential features of motor protein motion, see Figs. 1(a),(b):

- unidirectional and stochastic stepping of fixed length at a given rate ω_s;
- binding/unbinding dynamics between lattice and reservoir with rates ω_a, ω_d, respectively;
- "hard sphere interaction": a site can be occupied by one particle only (*exclusion*);
- at the left (right) boundaries particles can enter (exit) with rate α (β).

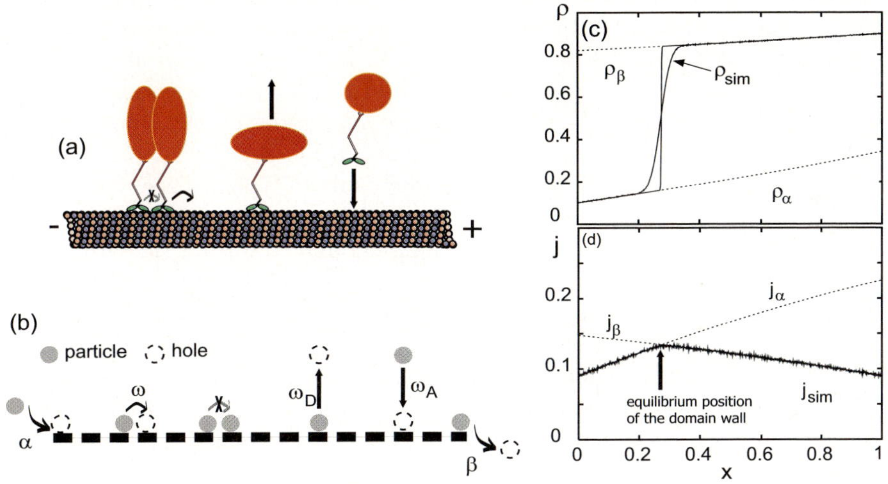

Fig. 1. (a) Sketch of kinesin transport over a microtubule; (b) Related TASEP+LK 1D model. (c) ρ_{sim} is the average density profile along the filament coordinate $x = i/N$ obtained from MC simulations. ρ_α, ρ_β are the analytic mean field solutions, respecting boundary conditions, in the thermodynamic limit $N \to \infty$. (d) j_{sim} is the average current related to ρ_{sim}. The DW equilibrium position is located at the maximum of the current profile, i.e. when left and right currents match, $j_\alpha = j_\beta$.

The model generalizes the well-known *Totally Asymmetric Simple Exclusion Process* (TASEP) [13, 14] to a *non-conserved* on-lattice dynamics. Indeed, binding/unbinding events break particle current conservation on the lattice.

Originally proposed to investigate ribosome traffic on messenger-RNA [17], TASEP has been extensively studied by physicists in the last decade because of its richness in non equilibrium phenomena like phase transitions and phase coexistence [13, 14]. For example, an interface between regions of low and high density, i.e. a *Domain Wall* (DW), can form because of the on site-exclusion of particles (hard-spheres) during their directed motion on the lattice. Such "mesoscopic" structure, in general, performs a (biased) random walk on the lattice with specific boundary conditions and effective rates, which depend on the model microscopic dynamics [18–20].

Mesoscopic Limit, Phase Separation and Coexistence. The simple but fundamental difference of our model compared with TASEP is the particle binding/unbinding dynamics in the lattice bulk, i.e. a Langmuir Kinetics (LK). Non-conservation of particle current on the lattice implies a completely new dynamics in the configuration space of the system and, therefore, new properties in the model phase diagram [16]. When TASEP and LK are *competing* processes, one can observe DW formation and, in particular, its *spatial localization* (phase separation and coexistence of low and high density domains) despite a completely homogeneous microscopic dynamics on the lattice [15, 16]. "Competing", here, means that the particle fluxes, at the filament boundaries and in the bulk, become comparable: since currents at the boundaries are not *extensive* quantities, attachment/detachment currents in the bulk should scale with the lattice size N as $1/N$. In this condition and for large system sizes (*mesoscopic limit*), conservation of particles is assured for long times only locally, defining a *quasi-conserved dynamics* on the whole lattice.

Interestingly, this regime naturally occurs in IT: processive kinesins walk over distances which are comparable to microtubule typical sizes, 10 μm, in a cell. In this regime, processive motion implies that motors on the filament are in mutual interaction for a lag of time long enough to form a (localized) DW before detachment. This feature can be understood by looking at the average density and current profiles along the lattice, Figs. 1(c),(d). The average current j is given by a left and a right components, j_α and j_β, each depending on their relative boundary condition. Local particle conservation forces the DW to be localized on the current maximum. Any other position, indeed, would in average imply a predominance of j_α or j_β. Excess of one of these two currents would displace the DW to the maximum where j_α and j_β match again.

Questions Arising from the 1D Model. The 1D model does not consider the role of the motor Brownian diffusion in the cytoplasm surrounding the filament: it neglects fluctuations in the particle density around the filament, thus implying homogeneous binding and unbinding rates along the lattice.

Intuitively, 3D diffusion in volume should destroy phase separation on the filament as observed in TASEP+LK. For example, whenever a particle detaches from a high density region, it will have a much higher probability to reattach in the back of the lattice, at the left of the DW, where filament density is low and binding is more probable. This is even more true when the motion occurs in a *closed* volume. In this case the total current of motor along the filament longitudinal direction should be zero. If motors walk on the filament in one direction at some speed, in volume an opposite backward current should arise, thus increasing the probability of attachment in front of the DW. In general, as our first intuition would tell, diffusion works against concentration gradients like the DW generated by particle interactions and directed transport over the lattice. The faster is the diffusion in volume compared to the motor dynamics on the filament (stepping), the more difficult it should be to build density gradients on the filament. In conclusion, one might suppose that the appearance/localization of the DW is unique to the 1D model TASEP + LK, while it disappears once diffusion dynamics in volume is introduced.

Different models have considered the transport of motors along filament and their mutual exchange between filament and cytoplasm [21–24]. To answer the previous "paradox", we extend the 1D model to 3D by introducing the diffusion of motor in volume and performing numerical simulations in the context of the model recently introduced by Klumpp *et al.* [23, 24].

Coupling Diffusion with Directed Transport. The microtubule is placed in the middle of the experimental chamber, Fig. 2. Motors can diffuse freely in the volume, binding or unbinding, and move unidirectionally along the filament once attached. For simplicity, we consider that the motor-microtubule binding rate has the same value of Brownian diffusion rate ω with no preferential binding sites on the filament (no specific entry or exit sites). Moreover, the motor leaves the filament end with step rate ω_s ("open end"). Space is dis-

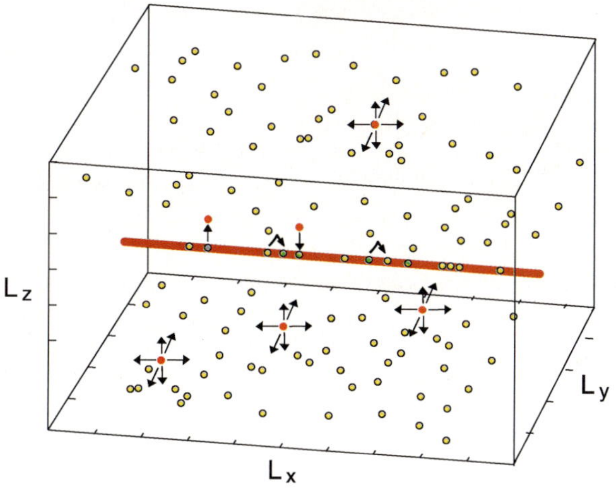

Fig. 2. Scheme of the 3D model implementing TASEP on filament with Langmuir binding kinetics of particles on the filament and 3D Brownian diffusion in the volume surrounding the filament (cytoplasm).

cretized with lattice size corresponding to the motor step $l_s \sim 10\ nm$. We consider a closed box with size comparable to typical scales in cells: $L_x = 10\ \mu m$ along the axis (1000 sites) for a filament of 8 μm in length (800 sites), while $L_y = L_z = 1\ \mu m$ (100 sites). Motor concentration c_m varies from 1 nM (nanomolar) up to 40 nM, typical values for *in vitro* experiments (6 up to 250 motors in the volume chosen). The diffusion rate ω is estimated by the mean square displacement for a 3D Brownian motion and the Stokes-Einstein relation $\omega \sim kT/\pi\eta_m r_h l_s^2$, while all other parameters are those typical for kinesins as defined above: stepping rate $\omega_s \sim 100\ s^{-1}$ and detachment rate $\omega_d \sim 1\ s^{-1}$ (at room temperature). In cytoplasm, viscosity can increase up to

10–100 times water viscosity, while motor speed and processivity essentially remain constant, both being practically viscous drag independent.

Some Result from MC Simulations. We performed kinetic Monte-Carlo simulations implementing a random sequential updating technique [25]. Figure 3(a) shows the *kymograph* of the distribution of motors along the filament in time, for $c_m = 40\ nM$ and medium viscosity equal to water viscosity, $\eta_m = \eta_w$.

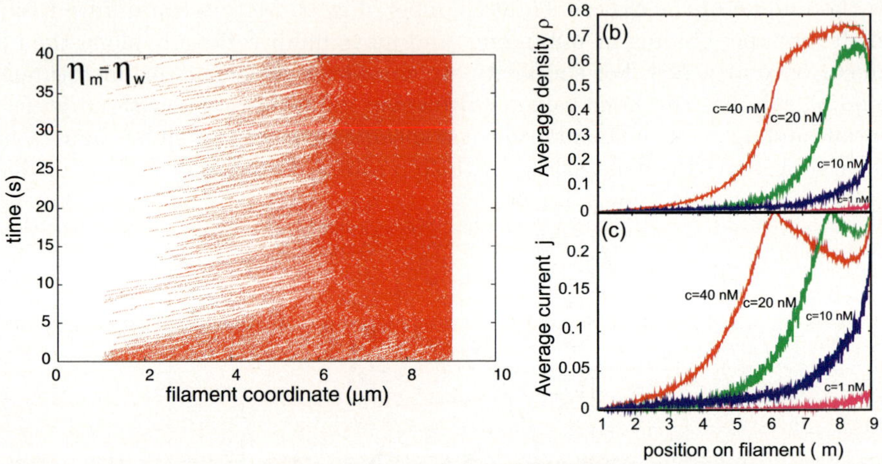

Fig. 3. (a) Kymograph of the particle distribution for $\eta_m = \eta_w$ and $c_m = 40\ nM$. A jam forms on the filament bulk, $\sim$ 6.1 μm, after a transitory time. Line slopes on left of the jam correspond to 1 $\mu m/s$, the typical average speed for conventional kinesins. (b) Average density profiles of particles on the filament and (c) average currents, for different concentrations of motors in volume and $\eta_m = \eta_w$. Phase separation occurs at typical concentrations of *in vitro* experiments. The jam is localized around the maximum of the average current as predicted by TASEP+LK. What changes considerably from the 1D model are the density profile and the shape of the region between the low and high density domains [25].

After a transitory time, a jam occurs at around 6.1 μm of the filament coordinate. Interestingly, the kymograph displays a jammed region localized in space with strong fluctuations of the front profile due to the reattachment of particles at the left of the jam. The analysis of the average density profile along the filament shows the presence of a transition region (with exponential profile [24]) connecting low and high density phases and localized around the maximum of the average current. The principle of localization of the DW found in TASEP+LK is valid also for this 3D model, although the front profile is different and fluctuations due to the particle dynamics in volume have to be taken specifically into account [24, 25].

Even more surprising is the jam behavior when medium viscosity increases by 10 or 100 times, see Figs. 4(a),(b). The jam moves toward the right end of the filament, respectively at $\sim 7.1\ \mu m$ for $\eta_m = 10\eta_w$ and $\sim 8.8\ \mu m$ for $\eta_m = 100\eta_w$. Contrary to what one would expect, diffusion influences jam formation, but fast diffusion clearly promotes jams in the filament bulk.

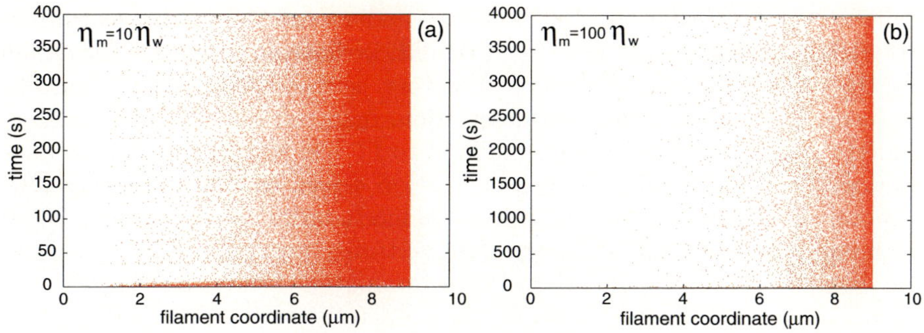

Fig. 4. Kymographs for $c_m = 40\ nM$ and medium viscosity (a) $\eta_m = 10\eta_w$ and (b) $\eta_m = 100\eta_w$. Contrary to our intuition, localized jams shift to the filament right end (fixed by the motor directed of motion) when medium viscosity increases.

This is a new outcome specific to this 3D model. Brownian diffusion here couples with stepping and detachment processes. When diffusion is fast, also the attachment of particles to the filament will be fast. Due to the small detachment rate (processivity), the filament then behaves as a "trap" for the particles during a time lag, which is long when compared to the time necessary for finding the filament via 3D diffusion in volume. Once many of the particles are bound to the filament and move, they can mutually interact and form jams. On the contrary, at higher viscosity, particles attach to the filament more slowly (at lower frequency). On average they populate less the filament, which never reaches transport saturation and jams are avoided.

Biological Consequences. This result can be relevant also from a physiological perspective: typical cell viscosities seem to prevent jam formation, allowing motors to transport cargos at maximal speed on the filament.

Moreover, from the analysis of the kymographs in Fig. 4, another interesting feature appears: as medium viscosity increases, microtubule sites are in general not crowded by motors. Since it is known that microtubules play multiple roles in cell life and organization, binding a multitude of specific and non-specific agents and proteins, a microtubule "congested" by transport would be unavailable for these kinds of interactions.

Specific experiments in controlled conditions as well as more theoretical modeling are needed to test these results. It is evident that jams can form as soon as the volume is excessively crowded with motors, but our simulations show that this should not be the usual situation for transport on microtubules in a "healthy" cell for typical values of cytoplasmic viscosity. Only anomalies in overexpressing transporters or cargos, in some specific pathological condition, should be a possible reason for intracellular jams on cytoskeletal filaments.

3 Conclusions

We have presented some principles describing the physical properties of intracellular transport (IT). We have shown how simple driven lattice gas models in one and three dimensions can be used to understand and predict features of intracellular traffic phenomena. In particular, we have discussed the formation and the stability of jammed phases of motors, as they might occur in eukaryotic cells. Beyond these kinds of description, many efforts have already been made to consider: complex events related to IT like inter-filament force production [26] or microtubule depolymerization driven by motor proteins [27]; or motor specific features like their internal mechano-chemical coupling [28] or the number of motor units forming a single transporter (not only monomers, but dimers) [29] and configurations like competing *k-mers* moving a cargo on the filament [30]; not to mention the need for studying and modeling the number of "cofactors" regulating the motor activity.

The physics of intracellular transport is a fast-developing field due to progress made in single molecule experiments *in vitro* and now also *in cellulo* [31, 32]. It is a challenging research field for it aims at describing the collective nature of non equilibrium processes that involve complexity at two different scales: one concerning complex molecular objects like motor proteins, the other concerning the collective behavior of such molecular machines.

Acknowledgements

A.P. acknowledges Norbert Kern, Paolo Pierobon and Giovanni Cappello for stimulating discussions, Francesca and Alberto for the careful reading of the manuscript.

References

1. Alberts A., Bray D., et al. (1994), The Molecular Biology of the Cell. Garland, New York.
2. Howard J. (2001), Mechanics of Motor Proteins and the Cytoskeleton. Sinauer Associates, Sunderland, MA, and references therein.
3. Chowdhury D., Schadschneider A. and Nishinari K. (2005) Phys. of Life Rev. 2:318.
4. Stokin G.B. et al. (2005) Science 307:1282.
5. Berg H.C. (1993), Random Walks in Biology. Princeton University Press. Princeton.
6. Frey E. and Kroy F. (2005) Ann. Phys. (Leipzig) 14:20.
7. Frey E., Parmeggiani A. and Franosch T. (2004) Genome Informatics 15:46.
8. Schnitzer M.J. and Block S.M. (1997) Nature 388:386.
9. Schnitzer M.J., Visscher K. and Block S.M. (2000) Nat. Cell Biol. 2:718.
10. Parmeggiani A., Jülicher F., Peliti L. and Prost J. (2001) Europhys. Lett. 56:603.

11. Hunt A.J., Gittes F. and Howard J. (1994) Biophys. J. 67:766.
12. Schmittmann B. and Zia R. (1995) In: Phase Transitions and Critical Phenomena, vol. 17, C. Domb and J. Lebowitz (eds), Academic Press, London.
13. Nonequilibrium Statistical Mechanics in One Dimension (1997), V. Privman (ed), Cambridge University Press, Cambridge, and Derrida B. and Evans M., chap. 14, 277 therein.
14. Schütz G.M. (2001), In: Phase Transitions and Critical Phenomena, vol. 19, C. Domb and J. Lebowitz (eds), Academic Press, San Diego.
15. Parmeggiani A., Franosch T. and Frey E. (2003) Phys. Rev. Lett. 90:086601-1.
16. Parmeggiani A., Franosch T. and Frey E. (2004) Phys. Rev. E 70:046101-1.
17. MacDonald C., Gibbs J. and Pipkin A. (1968) Biopolymers 6:1.
18. Kolomeisky A., Schütz G., Kolomeisky E. and Straley J. (1998) J. Phys. A 31:6911.
19. Santen L. and Appert C. (2002) J. Stat. Phys. 106:187.
20. Pierobon P., Parmeggiani A., von Oppen F. and Frey E. (2005) Phys. Rev. E 72:036123.
21. Smith D.A. and Simmons R.M. (2001) Biophys. J. 80:45.
22. Nedelec F., Surrey E. and Maggs A.C. (2001) Phys. Rev. Lett. 86:3192.
23. Lipowsky R., Klumpp S. and Nieuwenhuizen T. (2001) Phys. Rev. Lett. 87:108101-1.
24. Klumpp S. and Lipowsky R. (2003) J. Stat. Phys. 113:233.
25. Parmeggiani A., *to be submitted.*
26. Kruse K. and Sekimoto K. (2002) Phys Rev E 66:031904.
27. Klein G.A., Kruse K., Cuniberti G. and Jülicher F. (2005) Phys Rev Lett 94:108102.
28. Nishinari K., Okada Y., Schadschneider A. and Chowdhury D. (2005) Phys. Rev. Lett. 95:118101.
29. Pierobon P., Franosch T., and Frey E. (2006) Phys. Rev. E 74:031920.
30. Klumpp S. and Lipowsky R. (2005) PNAS 17284:102 and Klumpp S., Müller M.J.I. and Lipowsky R. (2006) Biophys. Rev. Lett. 1:353.
31. Courty S., Luccardini C., Bellaiche Y., Cappello G. and Dahan M. (2006) NanoLetters 6:1491.
32. Watanabe T.M. and Higuchi H. (2007) Biophys. J. 92:4109.

Traffic of Molecular Motors: From Theory to Experiments

Paolo Pierobon

Institut Curie, CNRS UMR 168, 11 Rue P. et M. Curie, F-75231 Cedex 05, France
pierobon@curie.fr

Summary. Intracellular transport along microtubules or actin filaments, powered by molecular motors such as kinesins, dyneins or myosins, has been recently modeled using one-dimensional driven lattice gases. We discuss some generalizations of these models, that include extended particles and defects. We investigate the feasibility of single molecule experiments aiming to measure the average motor density and to locate the position of traffic jams by mean of a tracer particle. Finally, we comment on preliminary single molecule experiments performed in living cells.

1 Introduction

Living cell is a highly organised structure that constantly needs to move its constituent parts from one place to another. It is therefore provided with a complex and accurate distribution system: the *cytoskeleton*, the network of biopolymers (actin, microtubules and intermediate filaments) that gives the cell its structural and mechanical features, functions as road system for the transport of organelles and vesicles; the motion of these objects is entrusted to *motor proteins* moving along these filaments. These motors are enzymes that convert the energy obtained by hydrolysis of an ATP (adenosin-triphosphate) molecule into a work (displacement of their cargo). Myosin V (on actin filament) and kinesin or dynein (on microtubules) are well known examples of these so called *processive* motors [1, 2].

It has been observed that these motors can act cooperatively or interact with each other giving rise to collective phenomena. In particular, in some phase of the cell cycle, the motors can be expressed in high concentrations: it seems therefore natural to investigate analogies and differences with the traffic observed in a city. Experimental techniques as single molecule and fluorescence imaging have just started giving some hints on the complex behaviour of this system, but are still far from giving quantitative description of traffic situations. While waiting for experimental data some theoretical models have been developed to physically describe intracellular transport.

2 Driven Lattice Gases: Models for Intracellular Transport

A simple model to capture the behavior of many motors on a filaments needs to include three fundamental features: (i) the motors move in a step-like fashion on a one dimensional track and bind specifically to the monomer constituting the cytoskeletal filaments; (ii) cytoskeleton filaments present a specific polarization: the chemical properties of the track guarantee that the motion is always directed towards one of the two ends of the filament; (iii) the particles move according to a stochastic rule, i.e. they move upon a chemical reaction, the hydrolysis of ATP, which occurs randomly with a typical rate.

The Total Asymmetric Simple Exclusion Process (TASEP) is a stochastic process first introduced to describe the motion of ribosomes on mRNA substrate [3] and encodes all these features. It rapidly became a paradigm of non-equilibrium statistical mechanics and one of the few example of exactly solvable systems [4, 5]. In this model each particle occupies a site on a one dimensional lattice and advances stochastically and in one direction. The most obvious observable is the average density profile of particles along the lattice.

The system with open boundaries where particles enter the lattice with rate α at one end and leave with rate β at the other, shows a non trivial phase diagram where three distinct non-equilibrium steady states appear: a *low density* phase controlled by the left boundary, a *high density* phase controlled by the right boundary and a *maximal current* phase, independent of the boundaries.

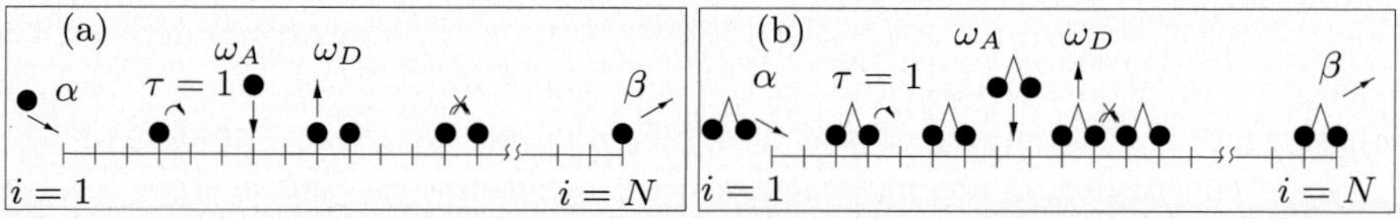

Fig. 1. Schematic representation of the TASEP with Langmuir kinetics in the case of (a) monomers and (b) dimers. The allowed moves are: forward jump (with rate $\tau = 1$), entrance at the left boundary (with rate α), exit at the right boundary (with rate β), attachment (with rate ω_A), and detachment (with rate ω_D) in the bulk.

In a first attempt to construct a minimal model for molecular intracellular transport, one needs to add to the TASEP the fact that the tracks are embedded in the cytosol with a reservoir of motors in solution. This allows the motors to attach to or detach from the track. This led to the construction of the TASEP with Langmuir (i.e. attachment/detachment) kinetics or TASEP/LK model [6, 7] depicted in Fig. 1a. According to this model, in addition to the TASEP properties, particles enter (leave) the system with rate ω_A (ω_D) also in the bulk. All over the lattice they obey exclusion: two particles cannot occupy the same site.

According to the rules described above, the rate equation in the average density at site $i = 2 \dots N$, $\langle n_i \rangle$ can be written as:

$$\partial_t \langle n_i \rangle = \langle n_{i-1} (1 - n_i) \rangle - \langle n_i (1 - n_{i+1}) \rangle + [\omega_A (1 - \langle n_i \rangle) - \omega_D \langle n_i \rangle] \, . \quad (1)$$

The first two brackets describe the average current and the second one the on-off kinetics (source and sink terms). At the boundaries this equation reads:

$$\partial_x \langle n_1 \rangle = \alpha (1 - \langle n_i \rangle) - \langle n_1 (1 - n_2) \rangle \, , \quad (2)$$

$$\partial_x \langle n_N \rangle = \langle n_{N-1} (1 - n_N) \rangle - \beta \langle n_N \rangle \, . \quad (3)$$

Equation (1) shows a non-closed hierarchy in the correlation functions (i.e. $\langle n_i \rangle$ depends on $\langle n_i n_j \rangle$ and so on): this suggests the use of the mean field approximation $\langle n_i n_{i+1} \rangle \approx \langle n_i \rangle \langle n_{i+1} \rangle$. In the stationary state, the leading term of the continuum limit ($N \to \infty$, $\langle n_i \rangle \to \rho(x)$) of Eq. (1) reads:

$$- \partial_x [\rho(1 - \rho)] + [\Omega_A (1 - \rho) - \Omega_D \rho] = 0 \, , \quad (4)$$

supplemented by two boundary conditions: $\rho(0) = \alpha$ and $\rho(1) = 1 - \beta$ [6]. When the solution of Eq. (4) cannot be matched continuously with the left and right boundaries, the density profile displays a localized discontinuity (or *shock*) in the bulk (Figs. 2b–c). This translates into the emergence of mixed phases that were not present in the simple TASEP. For some sets of parameters the phase diagram can exhibit up to 7 kinds of coexistence (see Fig. 2a).

A key point in the study of TASEP/LK is the introduction of a *mesoscopic limit* where local adsorption-desorption rates $\omega_{A,D}$ have been rescaled in the limit of large but finite systems $\omega_{A,D} = \Omega_{A,D}/N$, such that the macroscopic rates are comparable to the injection-extraction rates at the boundaries [6]. Far from being only a simple mathematical trick that allows the competition of the directed motion with the on-off kinetics, this limit captures the fact that the motors explores a significant fraction of the track before detaching. This is precisely the limit of highly processive motors that biologically motivated our studies. Surprisingly only in the mesoscopic limit the density profiles show the shock.

3 Extending the Model: Dimers and Defects

3.1 Dimers

Many processive molecular motors (kinesins, dyneins and myosin V) are composed of two heads that bind specifically each to a subunit of the molecular track. A natural extension of the previous model towards a more realistic one consists in introducing non-pointlike particles in the system such as *dimers* (see Ref. [8] and Fig. 1b). There are several challenging aspects in this problem: (i) the TASEP of particles of size ℓ (or ℓ-TASEP) is known to have a non-

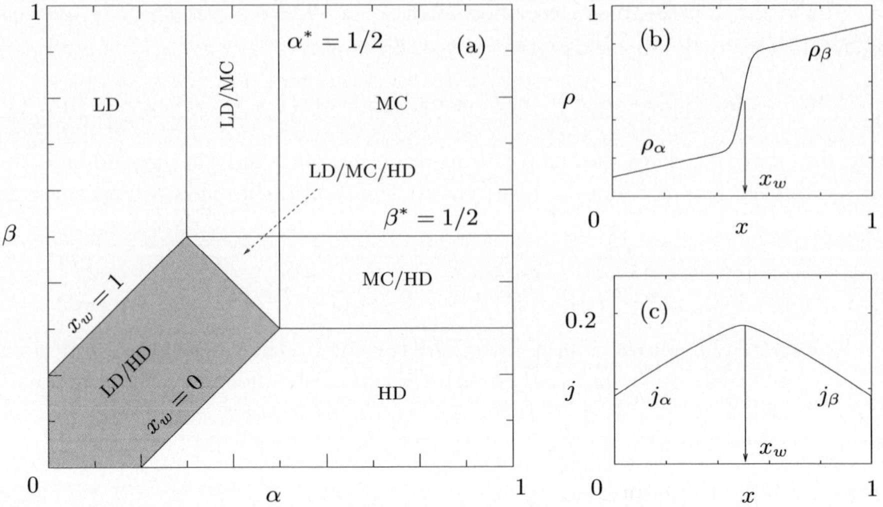

Fig. 2. (a) Phase diagram of TASEP/LK for $\omega_A = \omega_D$. One recognizes seven phases: in addition to the TASEP low density (LD), high density (HD) and maximal current (MC) phases; there are four more coexistence phases, namely the LD/HD, LD/MC, MC/HD and LD/MC/HD phases. The shaded region highlights the LD/HD coexistence where a localized domain wall appears. (b) Typical density profile in the LD/HD phase and (c) the corresponding current profile. At the matching point x_w between the left (j_α) and right (j_β) currents a domain wall develops and connects the left (ρ_α) and right (ρ_β) density profile.

trivial current-density relation $j(\rho)$, different from the model with monomers [3, 9, 10]; (ii) even the simple on-off kinetics of dimers exhibits non-trivial dynamics, for example the stationary state is reached from an empty system through a double step relaxation process [11, 12]. The coupling of an equilibrium process with two intrinsic relaxation regimes (on-off kinetics) to a genuine driven process (the ℓ-TASEP) suggests interesting dynamical phenomena likely to result in new phases and regimes.

In absence of exact solution, the main challenge has been to construct a refined mean field theory, based on probability theory, and to prove the consistency of the approximation for the two competing process (TASEP and the on-off kinetics). We have solved the mean field equation within the mesoscopic limit in the stationary state:

$$-\partial_x \left[\frac{\rho(1-2\rho)}{1-\rho}\right] + \left[\Omega_A \frac{(1-2\rho)^2}{(1-\rho)} - \Omega_D \rho\right] = 0\,. \qquad (5)$$

The complete equation is formally similar to the one for pointlike particles (Eq. 4): the current term (first bit in square brackets) must be balanced by the on-off term (second part).

Exploiting the analytical properties of the solution of the mean field equation and the local continuity of the current, we constructed the global density profile. We have performed extensive stochastic simulations and the agreement with the mean field solution is excellent. As in the case of TASEP/LK of monomers a new phase coexistence region appears for some parameters. The main effect of extended nature of dimers on the phase behavior of the system is related to the breaking of the (*particle-hole*) symmetry of the model. This does have quantitative but not qualitative consequences on the density profile and on the phase diagram, which remains topologically unchanged. The origin of the *robustness* of the picture found for monomers can be traced back to the form of the stationary density profile which depends exclusively on the form of the current-density relation and of the isotherm of the on-off kinetics. In both the monomer and the dimer case the current-density relation is concave and presents a single maximum, while the isotherm is unique and constant: these two features are enough to determine the topology of the phase diagram. This robustness suggests that the TASEP dynamics washes out the interesting two-step relaxation dynamics that characterizes the on-off kinetics of dimers: the non-trivial outcome is that, in these systems, the diffusion (yet asymmetric) always dominates the large time-scale relaxation.

3.2 Defect

Another question that often arises in the study of these models concerns the role of some kind of randomness: the motion of the particles can be altered by structural defects of the track or by microtubule associated proteins (MAP). A good modeling requires to introduce some sort of disorder in the system: site-related or particle related, quenched or annealed. While disorder in TASEP has been treated with exact and approximated methods [13–15],the role of bottlenecks was never coupled to the TASEP/LK. As a preliminary study of the role of quenched disorder, it becomes particularly interesting to investigate the influence of an isolated defect (i.e. point-wise disorder) on the stationary properties of the TASEP/LK. In Ref. [16] we extensively studied this model where the defect has been characterized by a reduced hopping rate $q < 1$ (see Fig. 3).

As a consequence of the competition between the TASEP and LK dynamics, the effects of a single bottleneck in the TASEP/LK model are much more dramatic than in the simple TASEP [17] (or the TASEP for extended objects [18]), where a localized defect was shown to merely shift some transition lines in the phase-diagram, but do not affect its topology. Here, new and mixed phases induced by the bottleneck have been obtained.

As a key concept of our analysis, we have introduced the *carrying capacity*, which is defined as the maximal current that can flow through the bulk of the system. In contrast to the simple TASEP the spatial dependence of the current, caused by the Langmuir kinetics, makes the carrying capacity nontrivial: the defect depletes the current profile within a distance that we called *screening length.* This quantity increases with the strength of the defect and decreases with the attachment-detachment rates. The competition between the current imposed at the boundaries and the one limited by the defect determines the density profiles and the ensuing phase-diagram. When the boundary currents are dominant, the phase behavior of the defect-free system is recovered. Also, above some critical entrance and exit rates, the system transports the maximal current, independently of the boundaries. Between these two extreme situations, we have found several coexistence phases, where the density profile exhibits stable shocks and kinks. Indeed, above some specific parameter values the phase-diagram is characterized by *bottleneck phases.* Depending on the screening length imposed by the defect, which can cover the entire system or part of it, different phase-diagrams arise. The latter are characterized by four, six or even nine bottleneck phases, which have been quantitatively studied within the mean-field theory introduced in Sec. 2: in fact for entrance and exit rates that exceed the critical value $q/(1+q)$ (for which the bottleneck becomes relevant) the system can be split in two sublattices that can be treated as independent TASEP/LK with effective exit/entrance rates at the junction (Fig. 3a).

Our results were checked against numerical simulations, which brings further arguments in favor of the validity of mean-field approaches for studying the TASEP/LK-like models (Figs. 3b–c). The somewhat surprising quantitative validity of this approximate scheme can be traced back to the current-density relationship, which is correctly predicted by the mean-field theory.

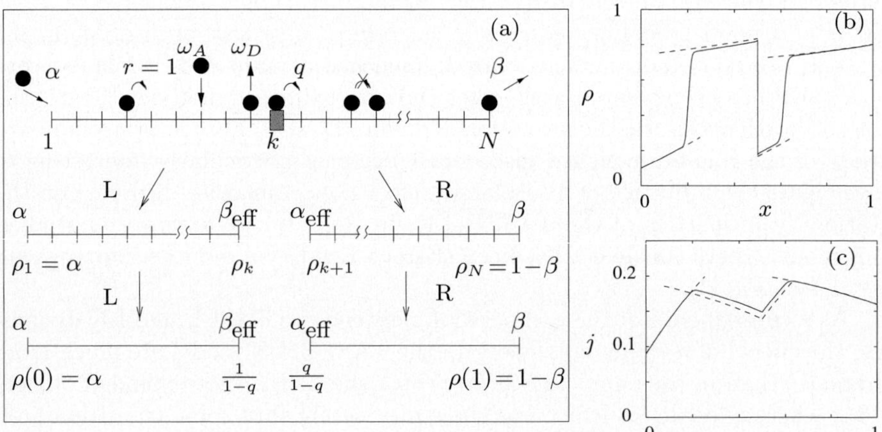

Fig. 3. (a) Cartoon representing the TASEP/LK with a bottleneck: above a threshold density the system can be divided into two lattices where the results of TASEP/LK are valid. (b) Density and (c) current in a case where the bottleneck is relevant and competes with the boundaries leading to the formation of a double domain wall, theoretical results (dashed lines) are superimposed to numerical simulations (bold line).

Eventually, we think that this study showed clearly that the presence of disorder in the TASEP/LK model, even in its simplest form, generally gives rise to quite rich and intriguing features and should motivate further studies of more 'realistic' and biophysically relevant situations, as in the presence of clusters of competing defects or quenched site-wise randomness (where the motors are slowed down at several points in the system).

4 Towards the Experiments

4.1 Tracer Dynamics

All the models described so far would be rather useless without the possibility to measure quantitatively the density and to localize the shock. To this aim, simple bulk fluorescence imaging would be difficult to apply to quantitative experiments: single molecule analysis are far more promising. We proposed hence a method, based on single particle tracking, to measure the density of the system [19]. The idea is to use the simple TASEP (exact) results to reconstruct the density from the velocity or the diffusion constant of the tracer. It is known that the average velocity $v \equiv \frac{d\langle x(t)\rangle}{dt}$ is related to the density ρ through the relation $\rho = 1 - v$. Exact results on a ring show that the same relation holds for the diffusion constant $D \equiv \frac{\langle x^2\rangle - \langle x\rangle^2}{t} = 1 - \rho$ [20].

Since the rates scales with the size of the system, according to the mesoscopic limit, we suppose that the influence of attachment detachment on this relation to be negligible and the density to be locally continuous. We simulated the system TASEP/LK and measured the position x as a function of the time for several particles, in order to construct the probability density function $P(x, t, |0, 0)$ to find a tracer particle at site x after a time t from its entrance in the system at site 0. From this function we can measure the velocity and the diffusion constant. As shown in Fig. 4, the shock is localized within a 10% precision through both the indicators. While the density is well reconstructed from the information on the velocity (Fig. 4a), the hypothesis $\rho = 1 - D$ works well only in the low density phase before the particle arrives at the shock. Once the particle has passed through the shock the hypothesis on the continuity of the density breaks down and the relation is not valid anymore while the loss of particles in the high density phase makes the system subdiffusive. Yet, the quantity $1 - D$ can be used to localized the shock (Fig. 4b).

Numerical simulations show that, on system of realistic size (i.e. $1\mu m$, roughly 100 sites), analysing the trajectories of a hundred particles is enough to reconstruct the density profile and localize the shock (after a time moving average resulting on a smoothing of the data). This suggests that single molecule experiments observing local features could give information on the global scale.

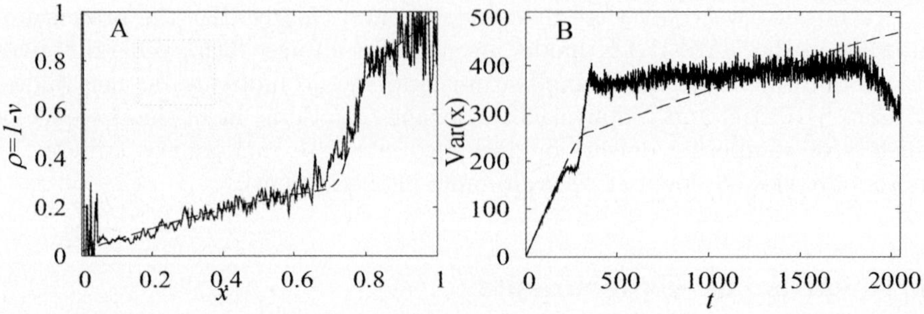

Fig. 4. (a) Density profile of TASEP/LK derived from the analysis of 100 tracer particles: wiggly line match rather well the theoretical prediction (dashed line). (b) Superposition of the first two moments of $P(x,t)$, $\langle x \rangle (t)$ (dashed) and $Var[x(t)]$ (wiggly) in time: according to the simple TASEP results the two quantities should be the same; this is true only until the particles have reached the shock, afterwards the variance deviates.

4.2 Single Molecule in the Cell

So far much information concerning processive molecular motors was provided by single molecule experiments. Typically a latex bead is attached to the motor to make it visible and to manipulate it: this allowed us to measure not only velocity and processivity of the motors, but even the forces they exert, the steps and (in some lucky cases) substeps [21–24]. In-vitro observations can be quite controversial: some experiments did not show any difference upon increasing the concentrations of motors, as if traffic effects were not relevant [25]. However the situation in the overcrowded environment of the cell is hardly reproducible in vitro and could reserve more surprises.

Since single molecule experiments have often been criticized because of their distance from the real systems, many groups are cautiously moving the single molecule experiments directly into the cell [26–29]. In this situation the main problem is to have a good signal-noise ratio which is hardly achievable with the usual fluorescent probes. Recently a method to mark and observe single molecule in vivo by using quantum dots (QDs) has been proposed [28, 30]. In contrast with the usual fluorescent probes, the QDs do not bleach, are excitable on all the visible spectrum and show a very narrow emission band. The only drawback is that they blink without a typical timescale: this inconvenient can be avoid by taking longer series of images. In Ref. [30] kinesins are biotinilated to bind to a streptavidinated QD. The conjugated particles so obtained are introduced in Hela cell by *pinocytosis* followed by osmotic shock and the motion of the QDs is observed with a customized fluorescence microscopy setup and a fast CCD camera. The QDs signal on 25 pixels is fit with a Gaussian to obtain sub-pixel resolution (FIONA [31]). The blinking of moving QDs is taken as an evidence that we are working in single molecule. The results on kinesin speed ($570 \pm 20 nm/s$) and processivity ($1.73 \pm 0.06 s$,

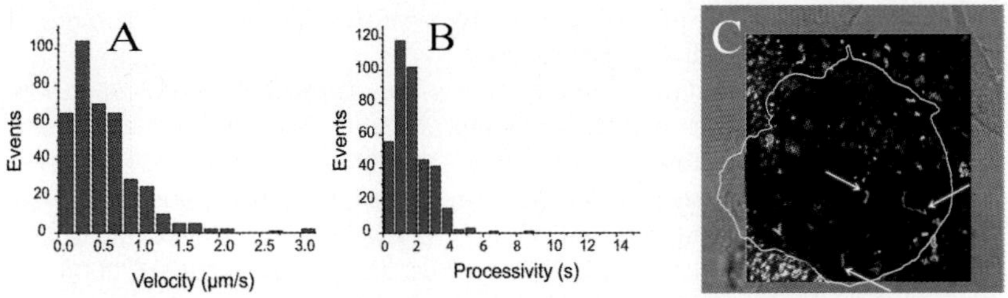

Fig. 5. Histograms of the speed (A) and processivity (B) of kinesin in Hela cells. The results confirmed in vitro measurements. (C) Picture of the trajectories (arrows) of a QD labeled Kinesins: image obtained choosing the maximal intensity over many frames of a movie with long integration time (100ms).

i.e.$\sim 1\mu m$ in space) are compatible to the ones known from in-vitro experiments (Fig. 5a–b)[1].

5 Conclusion

In this proceeding we have reviewed a model introduced to describe intracellular transport [6] and its possible extensions towards a more realistic picture: we investigated how the introduction of dimers [8] and the presence of disorder on the track [16] would affect the known results. These theoretical approaches while aiming to a realistic description of the traffic phenomena, are inspiring statistical models interesting in its own right (see e.g. [32, 33]). The latest experimental successes in tracking individual motors in living cells combined with an appropriate analysis, presented in the last section, could not only confirm the theories but also provide ispiration for a complete picture of the cell logistic.

Acknowledgements

I would like to thank the coauthors of Refs. [8, 16, 24]. I benefit from useful discussions with A. Parmeggiani, S. Achouri, A. Dunn and L. Sengmanivong.

References

1. B. Alberts, A. Johnson, J. Lewis, M. Raff, K. Roberts, and P. Walter. *Molecular Biology of the Cell.* Garland Science, New York, NY, USA, 4th edition, 2002.

[1] Additional experiments, using drugs, were carried out to ensure that the motion was actually due to motors.

2. J. Howard. *Mechanics of motor proteins and the cytoskeleton.* Sinauer Associates, Inc., Sunderland, MA, USA, 2001.
3. C.T. MacDonald, J.H. Gibbs, and A.C. Pipkin. *Biopolymers*, 6:1–25, 1968.
4. B. Derrida, E. Domany, and D. Mukamel. *J. Stat. Phys.*, 69:667–687, 1992.
5. G. Schütz and E. Domany. *J. Stat. Phys.*, 72:277–296, 1993.
6. A. Parmeggiani, T. Franosch, and E. Frey. *Phys. Rev. Lett.*, 90:086601, 2003.
7. A. Parmeggiani, T. Franosch, and E. Frey. *Phys. Rev. E*, 70:046101, 2004.
8. P. Pierobon, T. Franosch, and E. Frey. *Phys. Rev. E*, 74(031920), 2006.
9. T. Chou and G. Lakatos. *Phys. Rev. Lett.*, 93:19, 2004.
10. L.B. Shaw, R.K.P. Zia, and K.H. Lee. *Phys. Rev. E*, 68:021910, 2003.
11. J.D. McGhee and P.H. von Hippel. *J. Mol. Biol.*, 86:469–489, 1974.
12. E. Frey and A. Vilfan. *Chemical Physics*, 284:287–310, 2002.
13. R.J. Harris and R.B. Stinchcombe. *Phys. Rev. E*, 70:016108, 2004.
14. R. Juhász, L. Santen, and F. Iglói. *Phys. Rev. Lett.*, 94:010601, 2005.
15. K.M. Kolwankar and A. Punnoose. *Phys. Rev. E*, 61(3/4):2453–2456, 2000.
16. P. Pierobon, M. Mobilia, R. Kouyos, and E. Frey. *Phys. Rev. E*, 74(031906), 2006.
17. A.B. Kolomeisky. *J. Phys. A*, 31:1152–1164, 1998.
18. L.B. Shaw, A.B. Kolomeisky, and K.H. Lee. *J. Phys. A*, 37:2105–2113, 2004.
19. P. Pierobon. PhD thesis, Ludwig Maximilian Universität München, 2006.
20. B. Derrida, M.R. Evans, and D. Mukamel. *J. Phys. A*, 26:4911–4918, 1993.
21. C. Veigel, F. Wang, M.L. Bartoo, J.R. Sellers, and J.E. Molloy. *Nat. Cell. Biol.*, 4:59, 2002.
22. S. Uemura, H. Higuchi, A.O. Olivares, E.M. De La Cruz, and S. Ishiwata. *Nat Struct. and Mol. Biol.*, 11:877, 2004.
23. A.R. Dunn and J.A. Spudich. *Nat. Struct. Mol. Biol.*, 14:246–8, 2007.
24. G. Cappello, P. Pierobon, C. Symonds, L. Busoni, C. Gebhardt, M. Rief, and J. Prost. *Proc. Natl. Acad. Sci. USA*, 104(39):15328–15333, 2007.
25. A. Seitz and T. Surrey. *The EMBO Journal*, 25(2):267–77, 2006.
26. M. Dahan, S. Lévi, C. Luccardini, P. Rostaing, B. Riveau, and A. Triller. *Science*, 302:302, 2003.
27. C. Kural, H. Kim, S. Syed, G. Goshima, V.I. Gelfand, and P.R. Selvin. *Science*, 308:1469–1472, 2005.
28. X.S. Xie, J. Yu, and W.Y. Yang. *Science*, 312:228, 2006.
29. T.M. Watanabe and H. Higuchi. *Biophys. J.*, 92(11):4109–4120, 2007.
30. S. Courty, C. Luccardini, Y. Bellaiche, G. Cappello, and M. Dahan. *Nanoletters*, 6:1491–1945, 2006.
31. A. Yildiz, J.N. Forkey, S.A. McKinney, T. Ha, Y.E. Goldman, and P.R. Selvin. *Science*, 300:2061–5, 2003.
32. D. Chowdhury, A. Schadschneider, and K. Nishinari. *Physics of Life reviews*, 2:318–352, 2005.
33. M. Mobilia, T. Reichenbach, H. Hinsch, T. Franosch, and E. Frey. `arxiv:cond-mat/0612516`, 2006.

Numerical Investigations on Coupling of Asymmetric Exclusion Process with Zero Range Process

Rui Jiang[1], Bin Jia[2], Mao-Bin Hu[1], Ruili Wang[3], and Qing-Song Wu[1]

[1] School of Engineering Science, University of Science and Technology of China, Hefei 230026, China
[2] School of Traffic and Transportation, Beijing Jiaotong University, Beijing 100044, China
[3] Institute of Information Sciences and Technology, Massey University, New Zealand

Summary. Recently, the coupling of driven diffusive process with other processes has attracted the interests of physicists. In this paper, we investigate the coupling of asymmetric exclusion process (ASEP) with zero range process (ZRP) in a closed ring system consisting of two equally sized compartments. In one compartment, the dynamics is governed by an ASEP and in the other compartment, the dynamics is governed by a ZRP. The particle exchange is allowed both at the two ends and in the bulk. It is found with the increase of exchange rates, the moving direction of the domain wall changes several times.

1 Introduction

In recent years, the driven diffusive system has attracted the interests of physicists because it shows a variety of nonequilibrium effects [1, 2]. A very prominent example is the asymmetric simple exclusion processes (ASEPs), which are discrete non-equilibrium models that describe the stochastic dynamics of multi-particle transport along one-dimensional lattices. Each lattice site can be either empty or occupied by a single particle. Particles interact only through hard core exclusion potential. ASEPs were introduced in 1968 as theoretical models for describing the kinetics of biopolymerization [3] and have been applied successfully to understand polymer dynamics in dense media [4], diffusion through membrane channels [5], gel electrophoresis [6], dynamics of motor proteins moving along rigid filaments [7], the kinetics of synthesis of proteins [8], and traffic flow analysis [9, 10]. The simplest limit of ASEP is that particles can only move in one direction. This is called the totally asymmetric simple exclusion process (TASEP).

The zero-range process (ZRP) is another particular driven diffusive system [11]. In the ZRP, particles hop from site to site with a hop rate which depends, most generally, on the site from which it hops and the number of particles at the departure site. It has been widely applied as a model for the dynamics of avalanches, granular systems, interface growth, polymer dynamics, and various transport processes.

The coupling of TASEPs with other processes has led to many unusual and unexpected phenomena. For example, Parmeggiani et al. investigated the interplay of TASEP with the creation and annihilation of particles (Langmuir kinetics, LK) [12]. The phenomenon of localized density shocks was produced and was explained by applying a phenomenological domain wall theory [13–16]. Pierobon et al. studied the bottleneck-induced transitions in the TASEP coupled with LK, where a localized point defect acts as a bottleneck [17]. Evans et al. considered the disordered effect in the TASEP coupled with LK, in which creation and annihilation of particles occur on randomly selected sites [18].

The coupling of TASEPs with symmetric diffusive process has also been studied by Hinsch and Frey [19] in periodic boundary conditions and Klumpp and Lipowsky [20] in open boundary conditions. It is shown in open boundary conditions, a diffusive bottleneck at the boundary of the system leads to the absence of a maximal current phase, while a diffusive bottleneck in the interior of the system leads to a new phase characterized by different densities in the two active compartments adjacent to the diffusive one and to a maximal current defined by the bottleneck [20]. In periodic boundary conditions, an interesting correlation effect is observed and phenomenologically explained [19].

Inspired by the previous works, we study the coupling of TASEPs with ZRP in this paper. Our model is defined in a closed ring system consisting of two equally sized compartments. Each compartment is a lattice of N sites (Fig. 1). For the lower compartment, the dynamics is governed by TASEP; for the upper compartment, the dynamics is governed by ZRP. Moreover, parti-

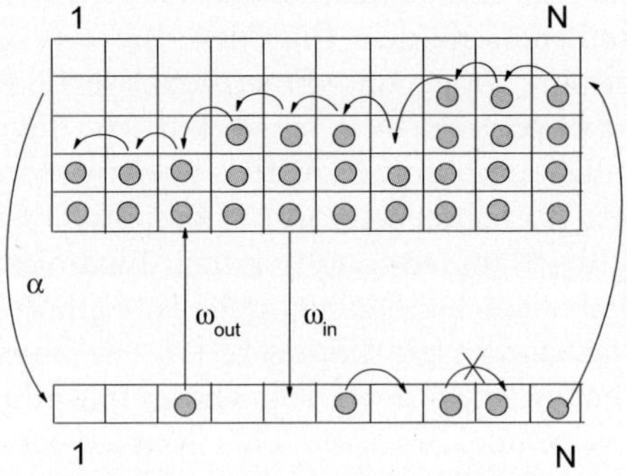

Fig. 1. The sketch of the coupling of ASEP with ZRP.

cle exchange happens between the two compartments in both ends and in the bulk. We employ Monte Carlo (MC) simulations to characterize the emerging nonequilibrium steady states, and various interesting nonlinear effects are revealed.

2 Model

We introduce an occupation variable τ_i where $\tau_i = 1$ (or $\tau_i = 0$) indicates that the state of the ith site in the lower compartment is occupied (or vacant). A second variable y_i is introduced to denote the number of particles in the ith site in the upper compartment and $y_{\max}$ the maximum value of y_i.

We apply the following dynamical rules. For each time step, a site i in the lower compartment is chosen at random.

- If $i = N$. If $\tau_N = 1$, the particle enters the Nth site of the upper compartment with probability $\beta = 1 - y_N/y_{\max}$.
- If $1 \leq i < N$. If $\tau_i = 1$, the particle firstly jumps to site i in the upper compartment with probability ω_{out} if $y_i < y_{\max}$. If it does not jump to the upper compartment, it moves into site $i+1$ if $\tau_{i+1} = 0$. If $\tau_i = 0$, a particle enters the site from site i in the upper compartment with probability ω_{in} if $y_i > 0$.

Then a site i in the upper compartment is chosen at random.

- If $i = 1$. If $y_1 > 0$ and $\tau_1 = 0$, a particle enters the 1st site of the lower compartment with probability $\alpha = 1/y_1$.
- If $1 < i \leq N$. If $y_i > 0$ and $y_{i-1} < y_{\max}$, then with probability $1/y_i$, $y_i \to y_i - 1$, $y_{i-1} \to y_{i-1} + 1$.

3 Simulation Results

In our simulations, the parameters are set $N = 1000, y_{\max} = 100$. The density profiles in both compartments are investigated in the space of ω_{in} and ω_{out}, and various kinds of density profiles are identified. Firstly, we consider the case that the total number of particles is $N_p = 30000$ and focus on the density in the bulk (i.e., far from the boundaries). Four thresholds for ω_{out} are identified. In the upper compartment, the density in the left bulk is high and the density in the right bulk is very small when $\omega_{out} = 0$. The different densities are separated by a domain wall. With the increase of ω_{out}, the density increases in the right bulk and the domain wall moves left (Fig. 2(a)). At the first threshold $\omega_{out} = \omega_{out,1}$, the domain wall reaches the left boundary. For $\omega_{out} > \omega_{out,1}$, a very small density appears in the left bulk. With the increase of ω_{out}, the domain wall moves right and the density in the right bulk continues to increase (Fig. 2(b)). When ω_{out} is larger than a second threshold $\omega_{out,2}$, a zero density will appear in the left bulk and the density in the right bulk reaches one (for example, $\omega_{out} = 0.1$ in Fig. 2(b)). The state remains if ω_{out} is smaller than a third threshold $\omega_{out,3}$. When $\omega_{out} > \omega_{out,3}$, the density begins to increase with ω_{out} in the left bulk and the domain wall moves right. The density in the right bulk is still one (Fig. 2(c)). At the fourth threshold $\omega_{out,4}$, the domain wall reaches the right boundary. When $\omega_{out} > \omega_{out,4}$, a zero density appears in the right bulk. The domain wall moves left and the density in the left bulk still increases with the increase of ω_{out} until $\omega_{out} = 1$ (Fig. 2(d)).

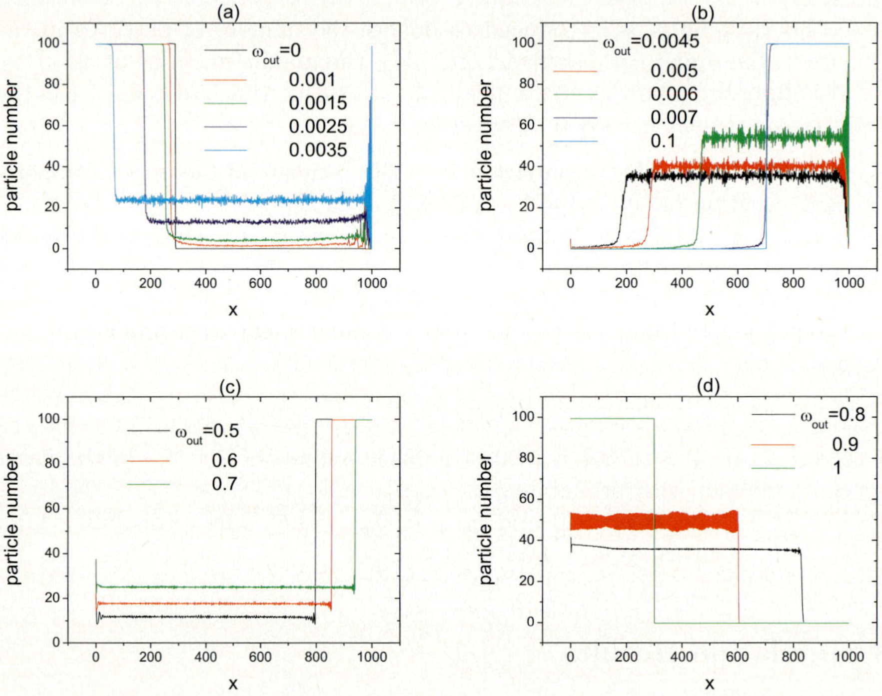

Fig. 2. Density profile in the upper compartment. Here $\omega_{in} = 0.5$.

In the lower compartment, the density is very high when $\omega_{out} < \omega_{out,1}$, despite a small density jump exists in the bulk (Fig. 3(a)). When $\omega_{out,1} < \omega_{out} < \omega_{out,2}$, the density is monotonically increasing with x in the left bulk and it is high in the right bulk (Fig. 3(b)). When $\omega_{out,2} < \omega_{out} < \omega_{out,3}$, a zero density is reached in the left bulk and density one is reached in the right bulk as in the upper compartment (see $\omega_{out} = 0.1$ in Fig. 3(b)). Then similar results as in upper compartment is observed with the increase of ω_{out} (Figs. 3(c) and (d)).

Our simulations show the above results remain qualitatively the same provided the total number of particles is in the range $10000 < N_p < 80000$. The

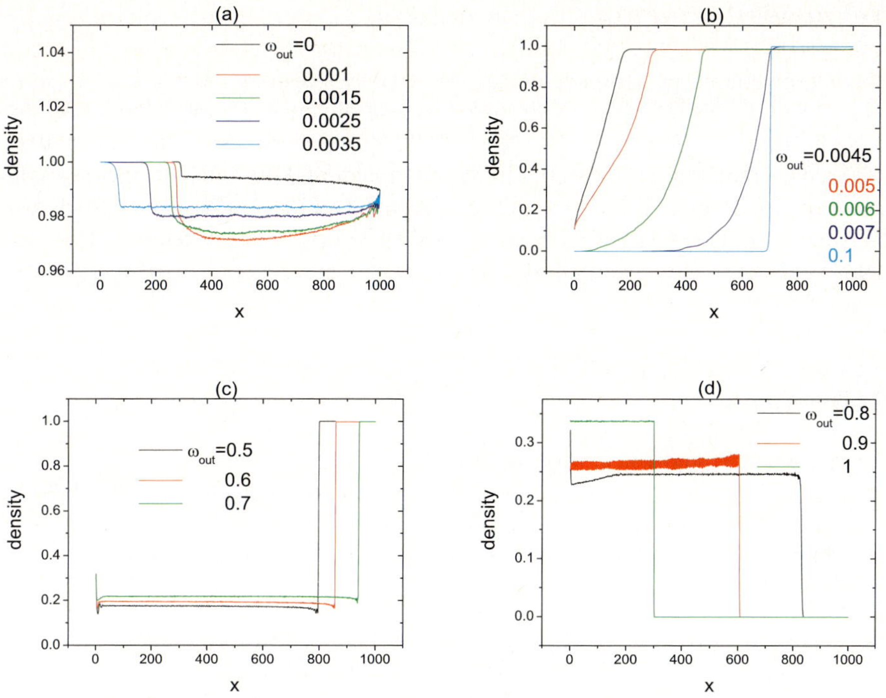

Fig. 3. Density profile in the lower compartment. Here $\omega_{in} = 0.5$.

results from $N_p > 80000$ and $N_p < 10000$ are different and will be reported in future work. Similarly, the results remain qualitatively the same provided $y_{\max}$ is not very small or very large. We will also investigate the situations arising from very small and very large $y_{\max}$ in future work. We also observe the fluctuation of densities on both compartments when $\omega_{out} = 0.9$. This will also be investigated in future work.

4 Conclusion

ASEP and ZRP are two typical nonequilibrium driven diffusive systems. In this paper, we have investigated the coupling of ASEP with ZRP. We study a closed ring system consisting of two equally sized compartments. In one compartment, the dynamics is governed by a TASEP and in the other compartment, the dynamics is governed by a ZRP. The particle exchange is allowed both at the two ends and in the bulk. It is shown that the moving direction of domain wall changes several times, depending on the exchange rates Ω_{in} and Ω_{out}. A more detailed investigation will be reported elsewhere [21].

Acknowledgements

We acknowledge the support of National Basic Research Program of China (No. 2006CB705500), the National Natural Science Foundation of China (NNSFC) under Key Project No. 10532060 and Project Nos. 10404025, 70501004, 70601026, 10672160, the CAS President Foundation, and the Chinese Postdoc Research Foundation (No. 20060390179). R. Wang acknowledges the support of the ASIA:NZ Foundation Higher Education Exchange Program (2005), Massey Research Fund (2005), and International Visitor Research Fund (2007).

References

1. Derrida B (1998) Phys. Rep. **301**, 65.
2. Schütz GM, in *Phase Transitions and Critical Phenomena*, Vol. 19, edited by C. Domb and J.L. Lebowitz (Academic Press, San Diego, 2001).
3. MacDonald JT, Gibbs JH and Pipkin AC (1968) Biopolymers **6**, 1.
4. Schütz GM (1999) Europhys. Lett. **48**, 623.
5. Chou T (1998) Phys. Rev. Lett. **80**, 85.
6. Widom B, Viovy JL and Defontaines AD (1991) J. Phys. I **1**, 1759.
7. Klumpp S and Lipowsky R (2003) J. Stat. Phys. **113**, 233.
8. Shaw LB, Zia RKP and Lee KH (2003) Phys. Rev. E **68**, 021910.
9. Chowdhury D, Santen L and Schadschneider A (2000) Phys. Rep. **329**, 199.
10. Helbing D (2001) Rev. Mod. Phys. **73**, 1067.
11. Evans MR and Hanney T (2005) J. Phys. A **38**, R195 and references therein.
12. Parmegianni A, Franosh T and Frey E (2003) Phys. Rev. Lett. **90**, 086601; Phys. Rev. E **70**, 046101 (2004).
13. Popkov V et al. (2003) Phys. Rev. E **67**, 066117.
14. Evans MR, Juhász R and Santen L (2003) Phys. Rev. E **68**, 026117.
15. Levine E and Willmann RD (2004) J. Phys. A **37**, 3333.
16. Juhász R and Santen L (2004) J. Phys. A **37**, 3933.
17. Pierobon P, Mobilia M, Kouyos R and Frey E (2006) Phys. Rev. E **74**, 031906.
18. Evans MR, Hanney T, and Kafri Y (2004) Phys. Rev. E **70**, 066124.
19. Hinsch H and Frey E (2006) Phys. Rev. Lett. **97**, 095701.
20. Klumpp S and Lipowsky R (2004) Phys. Rev. E **70**, 066104.
21. Jiang R et al., (to be submitted).

Traffic by Small Teams of Molecular Motors

Melanie J.I. Müller[1], Janina Beeg[1], Rumiana Dimova[1], Stefan Klumpp[2], and Reinhard Lipowsky[1]

[1] Max Planck Institute of Colloids and Interfaces, Science Park Golm, 14424 Potsdam, Germany mmueller@mpikg.mpg.de

[2] Center for Theoretical Biological Physics, University of California San Diego, La Jolla, CA 92093-0374, USA

Summary. Molecular motors transport various cargos along cytoskeletal filaments, analogous to trucks on roads. In contrast to vehicles, however, molecular motors do not work alone but in small teams. We describe a simple model for the transport of a cargo by one team of motors and by two teams of motors, which walk into opposite directions. The cooperation of one team of motors generates long-range transport, which we observed experimentally *in vitro*. Transport by two teams of motors leads to a variety of bidirectional motility behaviour and to dynamic instabilities reminiscent of spontaneous symmetry breaking. We also discuss how cargo transport by teams of motors allows the cell to generate robust long-range bidirectional transport.

1 Introduction: Traffic of Molecular Motors and the Need for Motor Teams

Molecular motors are protein molecules which power various transport processes in cells [1]. Their traffic is in many ways similar to car traffic. While cars drive on roads, molecular motors walk along tracks provided by cytoskeletal filaments. While cars consume petrol, molecular motors use energy from the hydrolysis of adenosine triphosphate (ATP) in order to perform mechanical work. Prominent examples are the kinesin and dynein motors traveling along filaments called microtubules which form a 'highway network' inside the cell. These are 'one-way highways': dynein motors walk preferentially to one end of the microtubules (called 'minus' end) while most kinesins walk to the opposite microtubule 'plus' end. Just like cars, motors *can* walk backwards, but are rather bad at it: they usually do so only slowly and if forced. Of course, there are also important differences between cars and motors. An obvious difference is the length scale: While cars are several m in size and travel km distances, molecular motors are only about 100 nm in size and travel μm distances. As a consequence, molecular motors work in an environment dominated by thermal noise, which for a car would be comparable to permanently driving in a hurricane. This leads to another interesting feature of molecular motors: they can 'fly', i.e. unbind from their track. However, upon unbinding they lose their ability to perform directed motion and randomly diffuse in the surrounding solution until they finally rebind to a filament. Due to these unique features, the traffic of molecular motors has become an attractive problem for traffic modeling and studies of non-equilibrium transport [2–4].

In this paper we address another aspect of cargo transport by molecular motors. While in road traffic a single truck usually suffices to carry its cargo, in cellular traffic this is not so. As mentioned, due to thermal noise molecular motors unbind from their track from time to time. For the molecular motor kinesin this happens on average after a 'run length' of about $1\,\mu$m. A cellular cargo, however, must accomplish distances of tens of μm, and in some extremely large cells like neurons even up to a meter [5]. Furthermore, the cellular surrounding is very viscous, leading to high frictional forces which can become too large for a single motor. A third problem for the cell is bidirectional transport. In a cell, the 'one-way' microtubule tracks are usually arranged in an isotropic way, pointing from the cell center to the cell periphery [6]. A single motor walks in only one direction along these tracks. However, many cellular cargos travel bidirectionally [7, 8], as has to be the case in order not to accumulate cargo at either the cell periphery or the cell centre [9, 10].

The cell solves all three problems by using several molecular motors rather than a single motor to transport a cargo. In order to obtain large run lengths and forces, several motors of the same species can work together as 'one team'. In order to accomplish bidirectional transport, motors with different directionalities transport a single cargo as 'two teams'. The number of motors in a team is small, typically between 1 and 10 [11, 12]. In this paper we review recent theoretical analyses [13, 14] and *in vitro* experiments [15] performed in our group that studied the cooperation of small teams of molecular motors pulling a single cargo along a unidirectional filament network. After defining our model, we will first examine the transport by one cooperating team of molecular motors of the same species and then the transport by two antagonistic teams of molecular motors that walk into opposite directions.

2 Theoretical Modelling: From One to Many Motors

We consider a cargo which is transported by fixed numbers of N_+ plus and N_- minus motors. Because of thermal fluctuations, a motor stochastically unbinds after some time. Therefore the cargo is pulled by a fluctuating number of motors, see Fig. 1(b). The state of the cargo is determined by the numbers n_+ and n_- of pulling plus and minus motors. In the simplest case, the motors work independently of each other, and one can deduce the cargo behaviour from the behaviour of a single motor. A single motor can bind to the filament with the binding rate π_0, walk along it with the forward velocity v_F, and unbind from it with rate ϵ_0. If the motor has to work under a force F, which can be caused by

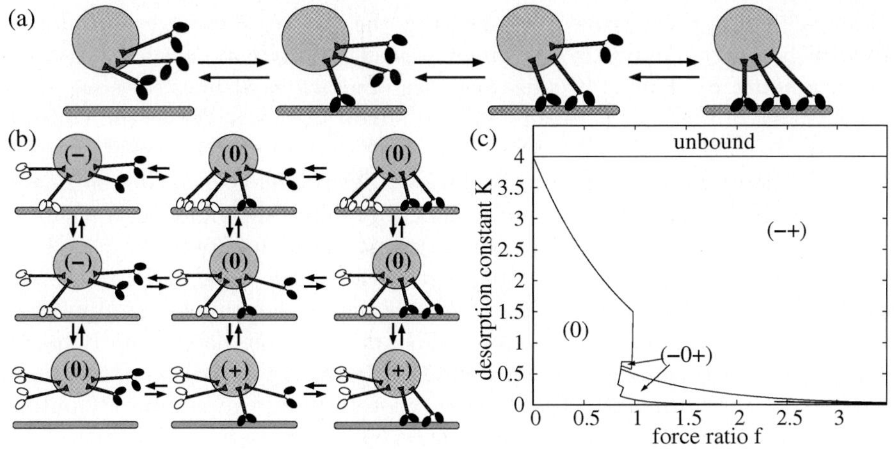

Fig. 1. Transport by small teams of motors. (a) A cargo with $N = 3$ motors is pulled by a fluctuating number of motors. (b) A cargo with $N_+ = 2$ (black) plus motors and $N_- = 2$ (white) minus motors is pulled by a fluctuating number of plus and minus motors. States with only plus motors bound (+), only minus motors bound (−) and both types of motors bound (0) correspond to fast plus, fast minus and slow motion, respectively. (c) Motility diagram for the symmetric tug-of-war of 4 against 4 motors with the same single-motor parameters (except their preferred direction). Depending on the single-motor force ratio $f = F_s/F_d$ of stall and detachment force and desorption constant $K = \epsilon_0/\pi_0$, the cargo is in one of three motility states (0), (−+) or (−0+) as explained in the text. For high desorption constants, the cargo is unbound.

opposing motors, Stokes friction, or an optical trap, these rates become force-dependent. The unbinding rate increases exponentially with the forces [16] as $\epsilon(F) = \epsilon_0 \exp(F/F_d)$, where the force scale is set by the detachment force F_d. The velocity decreases linearly [17], $v(F) = v_F(1 - F/F_s)$ until it reaches zero at the stall force F_s. For higher loads, the motor walks backwards [17] with the velocity $v(F) = -v_B(1 - F/F_s)$, with a very small backward velocity v_B.

3 Transport by One Team of Motors

We first consider a cargo transported by $N = N_+$ plus motors and no minus motors, $N_- = 0$. The number $n = n_+$ of motors which are bound to the filament fluctuates between 0 and N, see Fig. 1(a). As the motors work independently, the unbinding and binding rates for one motor in the cargo state with n pulling motors are simply $n\epsilon_0$ and $(N - n)\pi_0$ with the single motor unbinding rate ϵ_0 and binding rate π_0. This leads to a Markov process on the states $n = 1, \ldots, N$, for which we have obtained a number of analytical results [13].

In particular, the average run length, i.e. the distance a cargo moves along a filament before it unbinds from it, increases essentially exponentially with the motor number N. This is because the cargo particle continues to move along the filament unless all N motors unbind simultaneously. When the cargo is transported by the molecular motor kinesin, 3 kinesins suffice to cross a cell of 50 μm diameter, and 7–8 kinesins lead to average run lengths in the centimeter range [13]. The corresponding run length probability distribution is a sum of exponentials, which develops fat tails for large motor numbers N.

The increase of run length with increasing motor number has been observed *in vitro* [18–20], but it has been difficult to determine the number of motors pulling the cargo. To overcome this limitation, we have recently used a combination of dynamic light scattering (DLS) and a comparison of measured and theoretical run length distributions to determine the number of pulling motors [15]. In our experiments, latex beads were incubated in solutions with different concentrations of kinesin motors. The kinesins bound stochastically to the beads and pull them along an array of immobilized isopolar microtubules within a glass channel [21]. In such an assay, the maximal number of motors which are available for binding to the microtubule and pulling the cargo is not constant, but varies from bead to bead. The theoretical run lengths were therefore weighted with a truncated Poissonian distribution [15], and then fitted to the measured run length distributions for 9 different kinesin concentrations using only 2 fit parameters. The agreement of theory and experiment allowed to calculate the maximal number N of motors which were available for bead transport. This result was found to correspond well to the motor number independently estimated from the DLS measurement.

4 Transport by Two Teams of Motors

Next we consider a cargo with N_+ plus and N_- minus motors attached. The numbers n_+ and n_- of bound plus and minus motors, change stochastically as shown in Fig. 1(b). As the motors are assumed to act independently, the rates for unbinding and binding of a single motor when the cargo is in the state (n_+, n_-) can be deduced from the corresponding single motor rates. The opposing motors exert force on each other, so that each plus motor feels the force F_+ and each minus motor the force F_-. Newton's third law requires $n_+F_+ = n_-F_-$. Furthermore, as both motor types are bound to the same cargo, the velocity of each plus motor under force F_+ must equal the velocity of each minus motor under force F_-. The force and velocity balance determine the motor forces F_+ and F_- and therefore the motor binding and unbinding rates and the cargo velocity [14].

It is instructive to consider the symmetric tug-of-war of $N_+ = N_-$ plus and minus motors which have the same single motor parameters and differ only in their preferred direction. In this case the direction of motion in each of the cargo states shown in Fig. 1(b) is simply given by the majority motor type and is indicated by (+), (-) or (0) for plus motion, minus motion and slow motion, respectively. The probability $p(n_+, n_-)$ that the cargo is in the state with n_+ bound plus and n_- bound minus motors can have either 1, 2 or 3 maxima, depending on the single motor parameters, see Fig. 1(c). As the cargo spends most of its time in configurations with high probability, these maxima characterize the large-scale cargo motion. For 'weak' motors with a low ratio $f = F_s/F_d$ of stall force to detachment force, the probability distribution $p(n_+, n_-)$ has only one maximum at a configuration with $n_+ = n_-$, which corresponds to no motion [no motion motility state (0)]. When the motors have a high force ratio f, on the contrary, the motor number probabilities exhibit two maxima at $(n, 0)$ and $(0, n)$. In this parameter range, the cargo switches stochastically between fast plus and fast minus motion [$(-+)$ motility state]. In an intermediate range of f, all three types of maxima appear, and the cargo switches between fast plus motion, fast minus motion and pauses [$(-0+)$ motility state].

For large force ratios f, the appearance of two maxima at $(n, 0)$ and $(0, n)$ in a situation symmetric with respect to plus and minus motors is reminiscent of spontaneous symmetry breaking during continuous phase transitions. The reason for its appearance is a dynamic instability caused by the nonlinearity in the force-dependence of the single motor unbinding rate. The time for switching between the two non-symmetric maxima increases exponentially with the motor number $N_+ = N_-$, indicating a non-equilibrium phase transition in the infinite system.

If the tug-of-war is non-symmetric, the dynamic instability persists, and the cargo switches stochastically between fast plus motion, minus motion and / or pauses. However, now the plus-minus motor symmetry is lost, biased plus or minus motion is possible. Thus, even though the motors are engaged in a tug-of-war, fast motion into plus or minus direction can be generated.

5 Discussion: Robustness and Regulation

Why should cells use a team of motors instead of one strong motor which rarely unbinds from the filament? And why should cells use two teams engaged in a tug-of-war instead of one team only which is substituted by a team of opposite-directional motors when appropriate? The reasons may be robustness and sensitivity to regulation. A team of motors is more robust against failure of a single motor. And a team can be easily regulated by simply regulating the number of motors involved in the team. Two teams of motors can carry a cargo into two directions instead of only one. A bidirectionally moving cargo can search for its target, bypass obstacles and correct targeting errors. Furthermore, a cargo with two teams of motors engaged in a tug-of-war is very sensitive to regulatory mechanisms: because of the dynamic instability, small changes in the molecular properties (or of the number) of one or both motor types can qualitatively change the characteristics of cargo motion. The cargo can move into one direction faster or for a longer time and it may show net plus or net minus motion. In this way the cell can easily target its cargos as appropriate.

In summary, we have described a simple model for cargo transport by one or two teams of molecular motors. Despite its simplicity, the model exhibits a rich variety of motility behaviours and explains how the cell might satisfy its need for long-range bidirectional transport.

Acknowledgements

The authors thank R. Serral Gracià and E. Unger for enjoyable collaborations. SK was supported by Deutsche Forschungsgemeinschaft (Grants KL818/1-1 and 1-2) and by the National Science Foundation through the Physics Frontiers Center-sponsored Center for Theoretical Biological Physics (Grants PHY-0216576 and PHY-0225630).

References

1. Howard J (2001) Mechanics of Motor Proteins and the Cytoskeleton. Sinauer Associates, Sunderland (Mass.).
2. Lipowsky R, Klumpp S (2005) Physica A 352:53–112.
3. Lipowsky R, Klumpp S, Nieuwenhuizen TM (2001) Phys. Rev. Lett. 87:108101.
4. Parmeggiani A, Franosch T, Frey E (2003) Phys. Rev. Lett. 90:086601.
5. Goldstein LSB, Yang Z (2000) Annu. Rev. Neurosci. 23:39–71.
6. Lane J, Allan V (1998) Biochim. Biophys. Acta 1376:27–55.
7. Gross SP (2004) Phys. Biol. 1:R1–R11.
8. Welte MA (2004) Curr. Biol. 14:R525–R537.
9. Klumpp S, Nieuwenhuizen TM, Lipowsky R (2005) Biophys. J. 88:3118–3132.
10. Müller MJI, Klumpp S, Lipowsky R (2005) J. Phys.: Condens. Matter 17:S3839–S3850.
11. Habermann A, Schroer TA, Griffiths G, Burkhardt JK (2001) J. Cell. Sci. 114:229–240.
12. Gross SP, Vershinin M, Shubeita GT (2007) Curr. Biol. 17:R478–R486.
13. Klumpp S, Lipowsky R (2005) Proc. Natl. Acad. Sci. USA 102:17284–17289.
14. Müller MJI, Klumpp S, Lipowsky R (2008) Proc. Natl. Acad. Sci. USA 105:4609–4614.
15. Beeg J, Klumpp S, Dimova R, Serral Gracià R, Unger E, Lipowsky R (2008) Biophys. J. 94:532–541.
16. Schnitzer MJ, Visscher K, Block SM (2000) Nature Cell Biol. 2:718–723.
17. Carter NJ, Cross RA (2005) Nature 435:308–312.
18. Block SM, Goldstein LSB, Schnapp BJ (1990) Nature 348:345–352.
19. Coy DL, Wagenbach M, Howard J (1999) J. Biol. Chem. 274:3667–3671.
20. Seitz A, Surrey T (2006) EMBO J. 25:267–277.
21. Böhm KJ, Stracke R, Mühlig P, Unger E (2001) Nanotechnology 12:238–244.

Traffic Flow on Ant Trails: Empirical Results vs. Theoretical Predictions

Alexander John[1], Andreas Schadschneider[1,2], Katsuhiro Nishinari[3], and Debashish Chowdhury[4]

[1] Institut für Theoretische Physik, Universität zu Köln, 50937 Köln, Germany aj@thp.uni-koeln.de, as@thp.uni-koeln.de
[2] Interdisziplinäres Zentrum für komplexe Systeme, 53117 Bonn, Germany
[3] Department of Aeronautics and Astronautics, Faculty of Engineering, University of Tokyo, Tokyo, 113-8656, Japan tknishi@mail.ecc.u-tokyo.ac.jp
[4] Department of Physics, Indian Institute of Technology, Kanpur 208016, India debch@iitk.ac.in

Summary. We present empirical results for traffic flow on ant trails. The observed spatio-temporal organization of the ants as well as quantitative results for the fundamental diagram and headway distributions are compared with predictions of a cellular automaton model. In sharp contrast to highway traffic, no jammed phase is observed and the average velocity is almost independent of the density of ants.

1 Introduction

Ants form large trail systems [1] that share many features with human transportation networks. Belonging to the class of eusocial insects ants exhibit cooperation to a very high degree [1, 2] and one can assume that evolution has in some sense lead to an optimization of the transport processes in ant colonies. Nevertheless, *collective* properties of traffic on ant trails have not been studied empirically with the exception of [3, 4]. We therefore apply tools from traffic engineering (e.g. [5, 6]) to collect quantitative empirical data for traffic on unidirectional trails which have striking similarities with highway traffic.

2 Empirical Ant-Traffic Data

Unlike in vehicular traffic all 'vehicles' in our field study are of the same size and have the same properties. This is ensured by choosing the monomorphic species *Leptogenys processionalis* for investigations [1, 7]. Generally we avoid manipulations of the natural situation in order to focus on the generic effects.

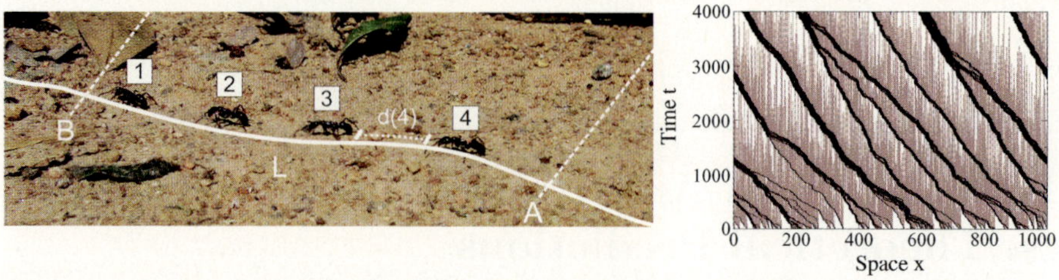

Fig. 1. Left: Photography illustrating the observed trail section. The nth ant entering is also the nth one leaving. Right: Space-time plot obtained from simulating our unidirectional ant-trail model is shown. Platoons of ants (black) are followed by a trace of pheromone marks (grey). Due to evaporation this trace has finite length.

2.1 Material and Methods

The observed trail section exhibits a constant shape for the period of observation (several hours) and has no intersections or branchings that could strongly influence the dynamics. To reduce the effect of boundaries and external disturbances, we focus on a data set recorded at a bulk section of one specific trail over a continuous time interval of about 13 minutes. Based on comparisons with ten other trails of the same type we believe that the observations described here are generic [8] for trails of the species.

The length L of the observed section (Fig. 1 left) has been measured by tracing one single ant passing through the section. Putting marks to a transparency on the video screen we obtained $L = 17$ bl in units of the body-length (1 bl $\approx$ 18 mm) of a single ant.

The most surprising qualitative observation is the absence of overtaking. Although ants temporarily left the trail and were passed by succeeding ones we never observed an ant speeding up in order to overtake. This observation can be used to extract 'single vehicle' data by cumulative counting (by hand [8]) based on video recordings of the observed section.

2.2 Extraction of Traffic Data

Since no overtaking takes place, ants can uniquely be specified by the order n in which they enter the observation section at A at time $t_+(n)$ (Fig. 1 left). They will leave the section at B in the same order at times $t_-(n)$. An efficient tool for analyzing such data is the *cumulative plot* (Fig. 2 left) showing the numbers $n_+(t)$ and $n_-(t)$ of ants that have passed the point A and B, respectively, up to time t. The two resulting curves (*arrival function* and *departure function*) can be obtained by inverting $t_\pm(n)$.

Based on cumulative counting the travel time $\langle \Delta T(n) \rangle$ and the average velocity $v(n)$ of the nth ant can be extracted (Fig. 2, lower inset):

$$\langle \Delta T(n) \rangle = t_+(n) - t_-(n) \quad ; \quad v(n) = \frac{L}{\langle \Delta T(n) \rangle} \qquad (1)$$

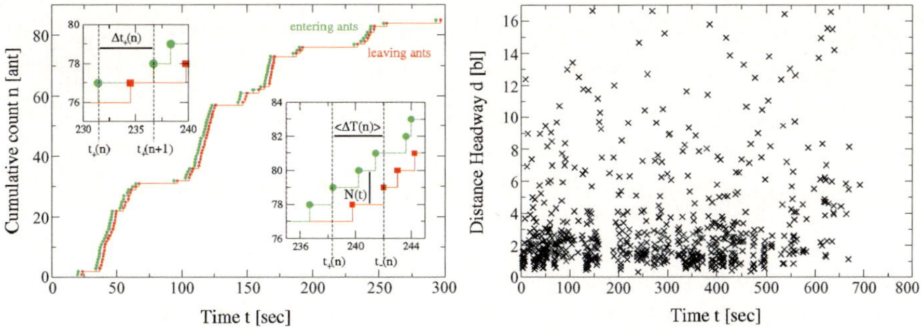

Fig. 2. Left: Schematic figure illustrating the technique employed for collecting empirical data. Generally one obtains a pairwise structure of datapoints for entering (•) and leaving (□) ants. The lower inset shows the travel time $\langle \Delta T \rangle$ for the 79th ant. On the upper inset the time headway of the 77th ant is shown. Right: Also distance headways were calculated, indicating a characteristic headway of $\bar{d} \approx 1.8$ bl.

The density of ants is determined by the number $N(t)$ of ants within the observed section at time t. $N(t)$ only changes if ants enter or leave the observed section (Fig. 2, lower inset), i.e. at times $t_i \in \{t_\pm(n)\}$. Then

$$\langle N_{av}(n) \rangle = \frac{1}{\langle \Delta T(n) \rangle} \sum_{i=i_+}^{i_- -1} N(t_i)(t_{i+1} - t_i) \quad \text{with} \quad \rho(n) = \frac{\langle N_{av}(n) \rangle}{L}. \quad (2)$$

Here i_+ and i_- are defined by $t_{i_+} = t_+(n)$ and $t_{i_-} = t_-(n)$. So the sum is over all 'events' during the time the nth ant is in the observed trail section.

The flow is then calculated using the hydrodynamic relation $F(n) = \rho(n)v(n)$ which gives the fundamental diagram (Fig. 3).

The time headway of two succeeding ants is obtained easily for example at the entrance point (Fig. 2, upper inset) as $\Delta t_+(n) = t_+(n) - t_+(n-1)$. Assuming a constant velocity throughout the observed section, the distance headway (Fig. 2 right) for entering ants is $\Delta d(n) = \Delta t_+(n)v(n-1)$.

2.3 Empirical Results

As seen in the cumulative count, the arrival- and departure function show clustering of datapoints at small time headways (Fig. 2 left). Similarly, the time-series of distance headways shows clustering for small distances (Fig. 2 right). This is in accordance with the observation that ants tend to move in platoons. Small distance headways are equivalent to intra-platoon distances whereas larger ones correspond to inter-platoon distances.

The density-dependence of the average velocity (Fig. 3 left) observed here is quite different from vehicular traffic. The jammed branch, characterized by a monotonic decrease of flow with increasing density, is completely missing and flow $F = \rho v$ is linearly increasing over the whole observed density regime.

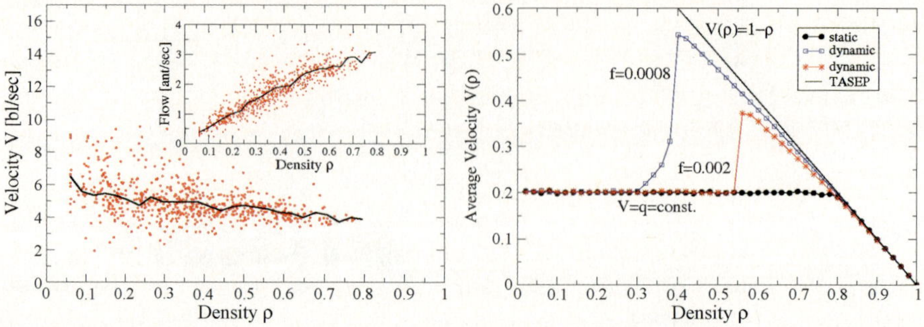

Fig. 3. Fundamental diagrams from empirical observations (left) and from computer simulations (right). Both exhibit a nearly constant average velocity for small to intermediate densities. The empirical fundamental diagram neither exhibits a sharp increase nor a monotonic decrease at high densities. Nevertheless depending on the evaporation rate f the sharp increase can be suppressed completely.

Effects like mutual blocking which usually are expected to become dominant for larger densities are strongly suppressed. Also additional effects known from highway traffic like hysteresis or synchronized flow [6] are not observed.

3 CA-Modelling Approach

The empirical data presented allow a direct comparison to a simple cellular automaton model. This model is based on the well known TASEP. The quasi-one-dimensional movement of ants is mapped to the hopping of particles along a one-dimensional lattice.

At each update-step a site i is chosen at random. If site i is occupied with a particle, ants in our case, and site $i+1$ is not blocked by another ant hopping is possible. If site $i+1$ is additionally marked with a pheromone hopping takes place at rate Q and at rate q otherwise (with $q < Q$).

An ant hopping from site i to $i+1$ leaves a pheromone mark at site i. If a marked site is chosen for updating the pheromone is removed at rate f. For more details of the model definition, see [9–11].

3.1 Theoretical Results

As a consequence of employing two different hopping-rates depending on the presence or absence of pheromone marks, so-called particle-wise disorder is induced dynamically. Each ant or particle is followed by a trace of pheromones which is of finite length (see Fig. 1 right) depending on the evaporation rate f. A succeeding ant perceiving that trace catches up to the preceding one since $Q > q$. Like in case of static particle-wise disorder [12, 13] clusters or platoons of ants are formed (see Fig. 1 right).

Biologically this mechanism is quite speculative as the complex pheromone system is very difficult to access experimentally. Like already applied in case of pedestrians dynamics also the interpretation as some kind of distance-dependent interaction mediated by virtual pheromones is possible.

The fundamental diagram exhibits features very similar to those known from the case of static disorder (see Fig. 3 right). The average velocity stays constant at $V = q$ for small to intermediate densities. But unlike in the static case a sharp increase is observed beyond a critical density. Obviously a TASEP-like regime is attained. This is a direct consequence of the dynamic disorder caused by the pheromones. At sufficiently large densities the average distance between the ants is of the order of magnitude of the pheromone-trace. Therefore nearly all ants are affected by the pheromones and hopping predominantly takes place at rate Q. Obviously the corresponding TASEP-case with $p = Q$ is recovered. The evaporation rate determines the critical density for the transition to the TASEP-case.

3.2 Empirical Results vs. Theory

On a qualitative level platoons of ants are found on the real as well as for the simulated trail (see Fig. 1). In accordance to that pattern one finds the same qualitative features in the fundamental diagrams (see Fig. 3). For an appropriate density regime the average velocity stays constant.

Nevertheless also differences are found. The empirical fundamental diagram exhibits no transition to a jammed phase (see Fig. 3 left). Typically one would expect this to occur near bottlenecks. The single-ant velocities show a density dependent scattering which can be interpreted as a consequence of two intrinsic velocities. The average velocity corresponds to q whereas a second and higher one would be equivalent to Q. At low densities the initial distance headway of two succeeding ants is quite large. So an up catching ant could pass the observed section nearly without being slows down by a slower preceding one. Depending on the initial distance headway one finds a continuum of measured average velocities $v(n) \in [q, Q]$. Nevertheless the mechanism of assigning the two intrinsic velocities and hence the particular kind of disorder is unclear.

4 Summary and Conclusions

We presented empirical and simulation results for ant-traffic flow on a unidirectional single lane trail. Drawing analogies between vehicular- and ant-traffic we apply tools from traffic engineering to an ant trail. The extracted fundamental diagram basically shows a free-flow state. Despite the striking absence of overtaking, no jamming transition even at large densities $\rho \approx 0.8$ is observed. This is a result of the spatial distribution of ants along the trail. In accordance with direct observations one finds that ants predominantly tend to move in platoons.

The experimental data presented give no conclusive picture about the particular mechanism of platoon formation although the data are in accordance with the existence of two different intrinsic velocities.

One possible explanation is based on the fact that the pheromone trace following each ant or platoon basically introduces a characteristic length scale. This interaction-range should in principle be reflected in the scattering: at very low densities (large distance headways) movement would predominantly take place at $v = q$. In contrast, very high velocities are mainly found for this density regime. As pointed out this could be a consequence of the finite length of the observed section so that most catching-up events occur outside of it. Ants will predominantly move at $v = Q$ but without being slowed down to $v = q$ within the observed section. Nevertheless an alternative explanation based on a static assignment of intrinsic velocities due to some behavioral pattern is also conceivable.

Acknowledgements

The empirical data were collected in a field study supported by the German Academic Exchange Service (DAAD) through a joint Indo-German research project. We would also like to thank R. Gadagkar, T. Varghese, M. Kolatkar, P. Chakroborty and M. Burd for enlightning discussions.

References

1. B. Hölldobler and E. O. Wilson. *The Ants*. Cambridge, Belknap, 1990.
2. E. Bonabeau, M. Dorigo, and G. Theraulaz. *Nature*, 400:39–42, 2000.
3. M. Burd, D. Archer, N. Aranwela, and D. J. Stradling. *American Natur.*, 159:283–293, 2002.
4. K. Johnson and L. F. Rossi. *J. Theor. Biol.*, 2006.
5. A. D. May. *Traffic Flow Fundamentals*. Prentice Hall, 1990.
6. B. Kerner. *The Physics of Traffic*. Springer, 2004.
7. K. N. Ganeshaiah and T. Veena. *Behav. Ecol. Sociobiol.*, 29:263–270, 1991.
8. A. John. *Physics of Traffic on Ant Trails and Related Systems*. PhD thesis, Universität zu Köln, 2006.
9. D. Chowdhury, V. Guttal, K. Nishinari, and A. Schadschneider. *J. Phys.A: Math. Gen.*, 35:L573–L577, 2002.
10. K. Nishinari, D. Chowdhury, and A. Schadschneider. *Phys. Rev. E*, 67:036120, 2003.
11. A. John, A. Schadschneider, D. Chowdhury, and K. Nishinari. *J. Theor. Biol.*, 231:279, 2004.
12. J. Krug and P. A. Ferrari. *J. Phys. A*, 29:L465, 1996.
13. M. Barma. *Physica*, A372:22, 2006.

Part IV

Networks

Modeling Metropolis Public Transport

Christian von Ferber[1,2], Taras Holovatch[3], Yurij Holovatch[4,5], and Vasyl Palchykov[4]

[1] Applied Mathematics Research Centre, Coventry University, Coventry CV1 5FB, UK C.vonFerber@coventry.ac.uk
[2] Physikalisches Institut, Universität Freiburg, 79104 Freiburg, Germany
[3] Ivan Franko National University of Lviv, 79005 Lviv, Ukraine
[4] Institute for Condensed Matter Physics of the National Academy of Sciences of Ukraine, 79011 Lviv, Ukraine
[5] Institut für Theoretische Physik, Johannes Kepler Universität Linz, 4040 Linz, Austria hol@icmp.lviv.ua

Summary. We present results of a survey of public transport networks (PTNs) of selected 14 major cities of the world with PTN sizes ranging between 2000 and 46000 stations and develop an evolutionary model of these networks. The structure of these PTNs is revealed in terms of a set of neighbourhood relations both for the routes and the stations. The networks defined in this way display distinguishing properties due to the constraints of the embedding 2D geographical space and the structure of the cities. In addition to the standard characteristics of complex networks like the number of nearest neighbours, mean path length, and clustering we observe features specific to PTNs. While other networks with real-world links like cables or neurons embedded in two or three dimensions often show similar behavior, these can be studied in detail in our present case. Geographical data for the routes reveal surprising self-avoiding walk properties that we relate to the optimization of surface coverage. We propose and simulate an evolutionary growth model based on effectively interacting self-avoiding walks that reproduces the key features of PTN.

1 Introduction

Urban public transport networks (PTNs) share general features of other transportation networks like airport, railroad networks, power grids, etc. [1]. The most obvious common features are their embedding into a 2D space, evolutionary growth, and optimization. The evolution of specific local transportation networks is closely related to that of the town or region in which they are embedded. However, statistically, some general overall features such as their fractal dimensions have been observed [2]. Modeling different aspects of transportation network structure and functioning helps to understand and optimize various processes that occur on these networks as well as to improve their planning. In this paper, we will consider a PTN from the point of view of complex network theory [1, 3–5] quantifying statistical features of their structure and proposing a model that reproduces their key features. To exemplify these, we survey the empirical analysis of PTNs of 14 major cities of the world (see Refs. [13, 16] for more details). While the empirical analysis of PTNs of different cities has been subject of several studies [6–13, 15] (see Sec. 2), we are not aware of previous attempts to specifically model the network evolution of a PTN in the way we propose here. The set-up of the paper is as follows: in the next section we discuss common features of PTNs and list some numbers that quantify these features; section 3 we describe our model for PTN evolution and give supporting arguments; in section 4 we compare some characteristics of real and simulated PTNs; conclusions and an outlook are given in section 5.

2 Empirical Analysis of PTNs: Statistical Properties

A distinct feature of our study is that we interpret the PTN as a network of *all* means of public transport (buses, trams, subway, etc.) offered in a given city. Several previous studies have analyzed specific sub-networks of PTNs. The Boston [6–9] and Vienna [9] subway networks may serve as examples. However, each particular sub-network (e.g. the network of buses, trams, or subways) is not a closed system: it is a subgraph of a wider transportation system of a city, or as we call it here, of a PTN. Therefore to understand and describe the properties of transport in a city it is important to deal with the complete PTN, not restricting the analysis to specific parts. Indeed, extending the restricted subway network to the "subway + bus" network drastically changes the network properties, as it was shown for Boston [7, 8].

Another important quantity that certainly restricts the reliability of generalizations drawn from trends observed for specific PTNs of different cities is the statistics, i.e. the size of the observed PTNs. The numbers of stations N in the PTNs analyzed so far ranged from several decades (subway networks of Boston, $N = 124$ [8], and Vienna, $N = 76$ [9]) to several thousands as in the PTN analysis of 22 Polish cities with up to 2811 stations [11] or bus-transport networks of three Chinese cities with up to $N = 3938$ stations. In our sampling [10, 13] we have chosen PTNs of 14 major cities of the world of various size with N ranging from 1544 (Düsseldorf) to 46244 (Los Angeles), see Table 1.

Here, we define different graph representations ('spaces') for a given PTN in terms of nodes (vertices) and links (edges) as illustrated in Fig. 1. The primary network topology is defined by a set of routes each servicing an ordered series of given stations, see the sketch labeled $\mathbb{L}'$ in Fig. 1 Based on this we define a graph ($\mathbb{L}$-space) representing each station by a node and linking any two that are served consecutively by at least one route. More generally linking any two stations serviced by a common route we define the $\mathbb{P}$-space [11, 17] representation. The interrelation of the routes in turn is described by a com-

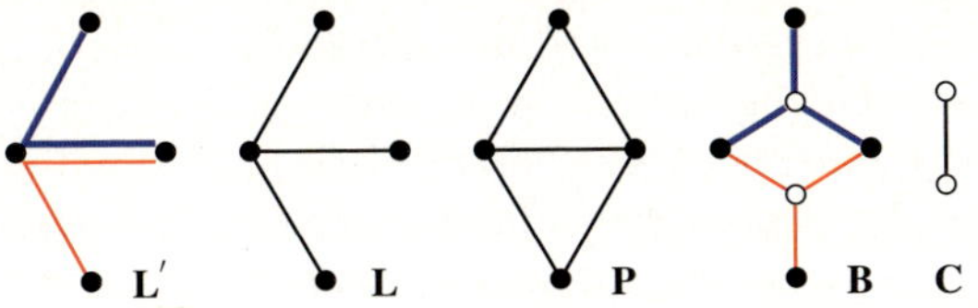

Fig. 1. Different representations of a PTN. $\mathbb{L}'$-space: nodes correspond to stations, links show different routes. $\mathbb{L}$-space: the same as $\mathbb{L}'$ but multiple links are reduced to single ones. $\mathbb{P}$-space: any two stations (nodes) are connected when they are serviced by a common route. $\mathbb{B}$-space: routes (open discs) are linked to those stations (filled discs) that they service. $\mathbb{C}$-space: nodes correspond to routes, links show that two routes share a common station. Note that by keeping multiple links in $\mathbb{P}$ and $\mathbb{C}$ one may also define $\mathbb{P}'$ and $\mathbb{C}'$-spaces.

plementary $\mathbb{C}$-space representation where now the nodes represent routes and linking any two that service a common station. From Fig. 1 it is easy to verify, that the last two representations correspond to one-mode projections of a bipartite $\mathbb{B}$-space represenation with nodes of two types, representing either a station or a route linking each route with all stations it services.

Table 1. PTN characteristics in $\mathbb{L}$- and $\mathbb{P}$- representations, see text for definitions. N: number of stations, M: number of routes, the scale $\hat{k}$ or the exponent γ of fits to the node degree distribution $p(k)$ to the laws (1) or (2) respectively, $\mathcal{C}$: ratio of the mean clustering coefficient to its random graph value, maximal $\hat{\ell}$, and mean $\overline{\ell}$ shortest path length. Characteristics in $\mathbb{P}$-space representations are indicated by a subscript 'p'. See Refs. [13, 16] for more data and sources.

City	N	M			$\mathcal{C}$	$\mathcal{C}_p$	$\hat{\ell}$	$\overline{\ell}$	$\hat{\ell}_p$	$\overline{\ell_p}$
Berlin	2996	218	$\hat{k} = 1.24$	$\hat{k}_p = 38.5$	52.85	42.03	68	18.61	5	2.93
Dallas	6571	131	$\gamma = 4.99$	$\hat{k}_p = 76.9$	17.26	63.00	269	85.84	10	3.78
Düsseldorf	1544	124	$\hat{k} = 1.12$	$\hat{k}_p = 58.8$	22.45	20.97	56	13.18	5	2.58
Hamburg	8158	708	$\hat{k} = 1.47$	$\hat{k}_p = 55.6$	262.92	133.99	158	39.74	11	4.79
Hong Kong	2117	321	$\hat{k} = 2.60$	$\hat{k}_p = 125.0$	58.98	12.51	60	11.11	4	2.26
Istanbul	4043	414	$\gamma = 4.04$	$\hat{k}_p = 71.4$	41.40	41.54	131	29.69	6	3.09
London	11012	2005	$\gamma = 4.58$	$\gamma_p = 4.39$	326.17	90.00	107	26.68	6	3.26
Los Angeles	46244	1893	$\gamma = 4.88$	$\gamma_p = 3.92$	588.44	427.06	247	43.55	14	4.60
Moscow	3755	679	$\hat{k} = 2.12$	$\hat{k}_p = 50.0$	128.23	41.93	28	7.08	5	2.52
Paris	4003	232	$\gamma = 2.61$	$\gamma_p = 3.70$	85.90	71.75	47	7.22	5	2.79
Rome	6315	681	$\gamma = 4.39$	$\hat{k}_p = 45.5$	68.61	76.93	93	29.64	8	3.58
São Paolo	7223	998	$\gamma = 2.72$	$\hat{k}_p = 200.0$	268.83	38.32	33	10.34	5	2.66
Sydney	2034	596	$\gamma = 3.99$	$\hat{k}_p = 38.5$	81.62	34.92	35	12.76	7	3.03
Taipei	5311	389	$\hat{k} = 1.75$	$\hat{k}_p = 200.0$	186.23	15.38	74	20.86	6	2.35

In our analysis, we are interested in different features of the PTN as measured when represented in the above defined spaces. It is worth to mention here, that these standard network characteristics measured in different spaces turn out to be specific of practical value in judging about the quality of public transport in a given city. The particular quantities we analyze here are the maximal and mean shortest path length $\hat{\ell}$ and $\overline{\ell}$, the clustering coefficient C, and the betweenness C_B; for definitions see Ref. [14] in this volume. Table 1 gives some of these quantities for the cities analyzed in $\mathbb{L}$- and $\mathbb{P}$- representations. One can see that the above networks are highly clustered small worlds characterized by small shortest path lengths and large ratios $\mathcal{C}$ of the mean clustering coefficient relative to its value $C_{ER} = 2M/N^2$ on a random graph with the same numbers of nodes N and links M.

To classify the node degree distributions $p(k)$ we performed fits to both an exponential function

$$p(k) = Ae^{-k/\hat{k}}, \tag{1}$$

as well as to a power law:

$$p(k) = Bk^{-\gamma}. \tag{2}$$

The result of the better fit together with the value of the fit parameters $\hat{k}$ or γ are shown in table 1. One can see that the $\mathbb{L}$-space node degree distribution of a part of these PTNs (8 out of 14) is governed by power laws, indicating scale-free properties. For the PTNs of several cities, this fact has also been described in Refs. [10, 11]. The remarkable feature of the data shown in table 1 is that some PTNs (3 out of 14) show scale-free behavior even in $\mathbb{P}$-space. As examples compare the degree distributions $p(k)$ of the PTNs of Paris and Sydney in Fig. 2. Whereas the Paris PTN is scale-free both in $\mathbb{L}$- and $\mathbb{P}$-space, the PTN of Sydney is scale-free in $\mathbb{L}$-space only. Note that to reduce the noise in the data we plot the $\mathbb{P}$-space cumulative degree distribution

$$P(k) = \sum_{q=k}^{\infty} p(q). \tag{3}$$

In Fig. 3 plot the mean betweenness value $C_B(k)$ (see [14] in this volume) as function of the node degree k in $\mathbb{L}$, $\mathbb{P}$, $\mathbb{C}$, and $\mathbb{B}$-spaces (similar behaviour is found for the networks of other cities). One definitely sees a pronounced correlation and a tendency for power-law behaviour: a phenomenon observed previously for several other networks [11, 18, 19].

In addition to the standard characteristics of complex networks discussed above, one can introduce some characteristics which are specific to PTNs. One of them is due to the fact that very often several routes go in parallel and pass L consecutive stations. In particular, the notion of *harness* was introduced in Ref. [13] to quantify this behaviour. In Fig. 4 we show the harness distribution $P(L, R)$: the number of sequences of L consecutive stations that are serviced by R parallel routes for the PTNs of Paris (a) and of Sydney (b). The log-log

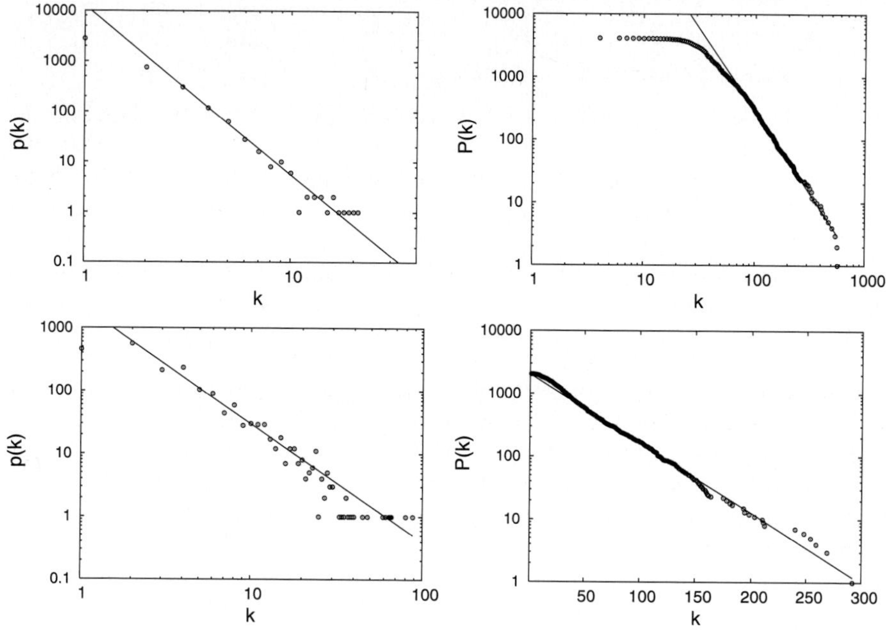

Fig. 2. Node degree distributions for Paris (first row) and Sydney (second row) PTNs. Left column: degree distributions $p(k)$ in $\mathbb{L}$-space. Right column: cumulative degree distributions (3) in $\mathbb{P}$-space.

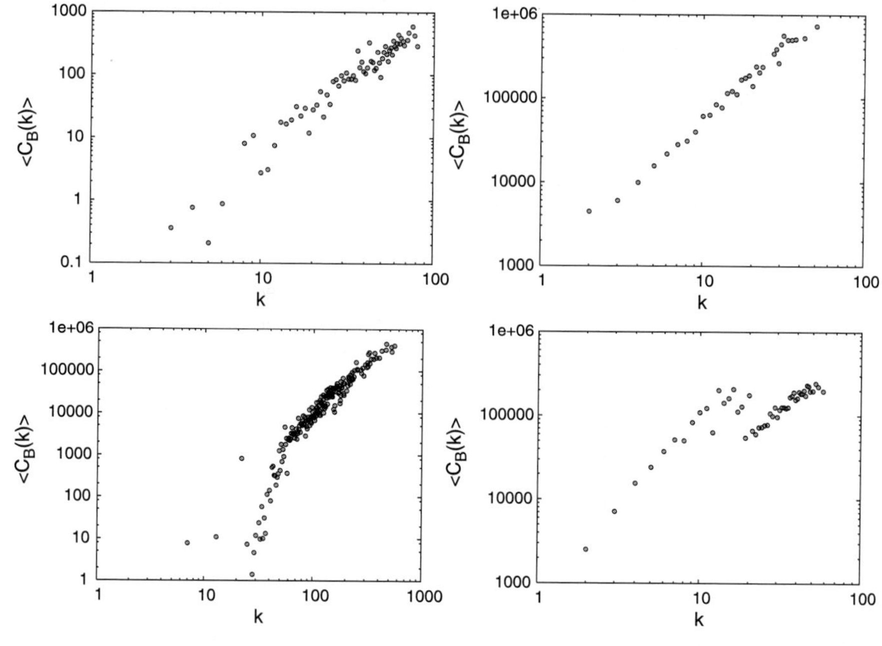

Fig. 3. Betweenness-degree correlations for Paris PTN in different representations.

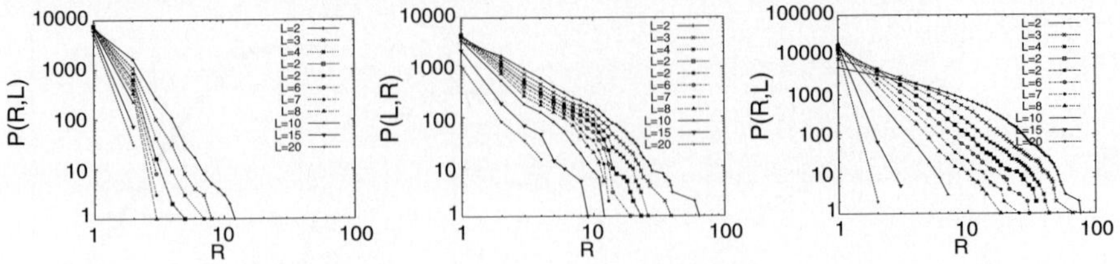

a) b) c)

Fig. 4. Harness distribution for Paris (**a**), for Sydney (**b**), and for a simulated city (**c**).

plot indicates scale free properties of this distribution. We have found similar behaviour for the majority of the cities under consideration.

Following this analysis of PTNs of different cities of the world and having at hand the numerical data on different features of these networks, let us propose a model, that may reproduce most of these features. In particular, the model should be capable to discriminate between different types of behaviour observed so far in PTNs (an example may be the exponential and power-law node degree distributions for $p(k)$ observed for different cities).

3 Evolutionary Model of PTNs

To model the properties of PTNs we have proposed [13] an evolutionary growth model for these networks along the following lines: We model the grid of streets by a square lattice and allow every lattice site $\boldsymbol{r}$ (street corner) to be a potential station visited by say $k(\boldsymbol{r})$ routes. The routes are modeled as self-avoiding walks (SAWs) on this lattice. The rules of our model are the following:

1. First route: construct a SAW of length n starting at an arbitrary site.
2. Subsequent routes:
   - (i) choose a terminal station on lattice site $\boldsymbol{r}$ with probability $q \propto k(\boldsymbol{r}) + a$;
   - (ii) choose a subsequent station of this route at a neighboring site $\boldsymbol{r}'$ with probability $q \propto k(\boldsymbol{r}') + b$;
   - (iii) repeat step (ii) until the walk has reached n stations, in case of self-intersection discard the walk and restart with step (i).
3. Repeat step 2 until M routes are created.

The above rules resemble the preferential attachment growth rule [20] favoring high-degree nodes when linking new nodes to a network. The principal difference of our algorithm is that at each step (2ii) we link an *existing* station to a neighboring site which does not need to be empty. New stations are then only added at the frontier of the PTN cluster while high degree nodes (hubs)

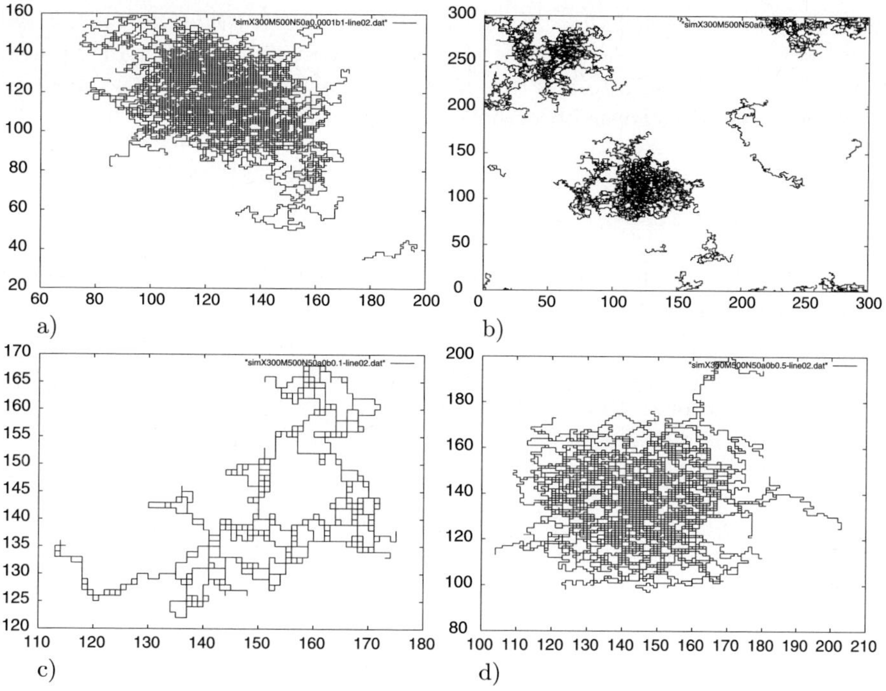

Fig. 5. Different simulated PTN maps of 'cities' of size 300×300 with $M = 500$ routes of $n = 50$ stations each. **a)**: $a = 0.0001, b = 1$, **b)**: $a = 0.001, b = 1$, **c)**: $a = 0$, $b = 0.1$, **d)**: $a = 0, b = 0.5$.

accumulate at its center. The choice of a SAWs to model the routes may seem odd at first sight. however, the fractal dimensions measured in PTNs [2, 13] are compatible with 2D SAW behaviour for the routes and it is obvious that a single PTN route very seldom intersects itself. Moreover, the scaling properties of SAWs on disordered lattices do not change, provided the disorder is short-range correlated [21]. In our application this means, that even the presence of certain geographical constraints and deviations from the square lattice still allow for a SAW description of a PTN route.

The above described generating procedure results in a simulated PTN of M routes, each consisting of n stations. In Fig. 5 we show several typical PTN configurations of simulated cities on a 300×300 square lattice with $M = 500$, $n = 50$ and different values of a and b. The parameters a and b allow to discriminate between different regimes of the network evolution. Setting $a = 0$ limits every new route to start from an already existing station. On the contrary, for $a \neq 0$, the terminal station of a new route may be situated at any lattice site. Therefore, increasing a allows for PTNs that consist of more than one component (c.f. Figs. 5a and 5b). The parameter b on the other hand tunes the evolution of the routes: a small value of b forces the routes to propagate in parallel (*harnessed)* while increasing b results in routes that cover more sites of the lattice (see Figs. 5c, 5d correspondingly). In the next section we will investigate how the numerical characteristics of the modeled PTNs correspond to those observed for real cities.

4 Simulation Results

Postponing a more detailed analysis of the numerical simulations based on the model of section 3 to a separate publication [16], we focus here on several principal features of PTNs and demonstrate how they are reproduced by our model. One of them is the behaviour of the node degree distribution $p(k)$. As it was shown in section 2, the behaviour of this function varies for PTNs of different cities. In $\mathbb{L}$-space one often observes $p(k)$ to be of a power-law type (2) [10–13], however sometimes it is governed by an exponential decay (1). Moreover, recently these two types of behaviour were also observed in $\mathbb{P}$-space [13]. A distinguished feature of our model is that depending on the values of the evolution parameters a, b it discriminates between a power-law and an exponential $p(k)$ in $\mathbb{P}$-space. Note that on the square lattice the $\mathbb{L}$-space degrees are limited to $k \leq 4$ ruling out such an analysis. As an example, in Fig. 6 we show the cumulative node degree distribution $P(k)$, see Eq. 3, for PTNs of two simulated cities in the $\mathbb{P}$-space. Changing the value of parameter b for fixed a one passes from an exponential (a straight line in the log-linear plot in Fig. 6a) to a power-law regime (a straight line in the log-log plot in Fig. 6b).

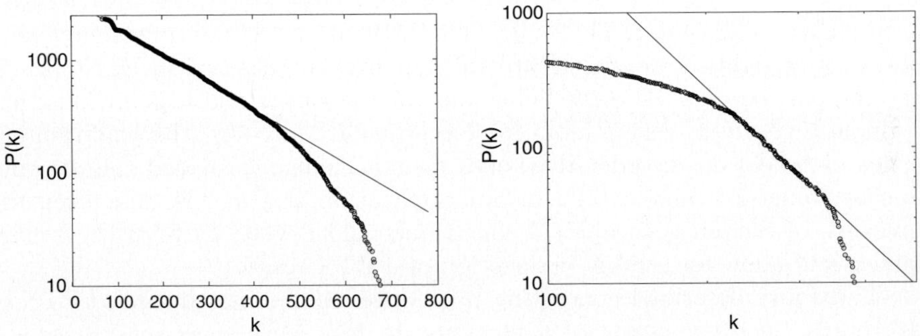

a) b)

Fig. 6. $\mathbb{P}$-space cumulative degree distribution $P(k)$ for different simulated cities of 300×300 sites with $M = 500$ and $n = 50$. **a)** $a = 0$, $b = 0.5$: exponential in $\mathbb{P}$-space, **b)** $a = 0$, $b = 0.1$: power law in $\mathbb{P}$-space.

Another specific feature of real world PTNs that is nicely reproduced by our model is the harnessing effect. In Fig. 3c we show the harness distribution $P(L, R)$ for the PTN of a simulated city on 300×300 sites with $M = 500$

and $n = 50$ at $a = 0$, $b = 0.5$ in comparison with the same quantity for the PTNs of Paris (Fig. 3a) and of Sydney (Fig. 3b). Again, as in the case of real-world PTNs one may speculate about power-law behaviour. Note however, that neither for the node degree nor for the harness distribution we so far find a simple relation between the exponents that may govern the scaling and the model evolution parameters a, b. Finally, let us compare the betweenness-degree correlation. In the same way as for the PTN of Paris (Fig. 3), we plot this function in Fig. 7 for different representations for a simulated city of 300×300 sites with $M = 500$ and $n = 50$ for the evolution parameters $a = 0, b = 0.5$. One can see an overall qualitative agreement between the behaviour observed for the real-world network and the simulated one (for $\mathbb{L}$ see discussion above). In this context it is worth to mention that the simultaneous use of different representations (spaces) serves as a useful tool to quantify the correspondence and differences between real word networks as well as simulated ones.

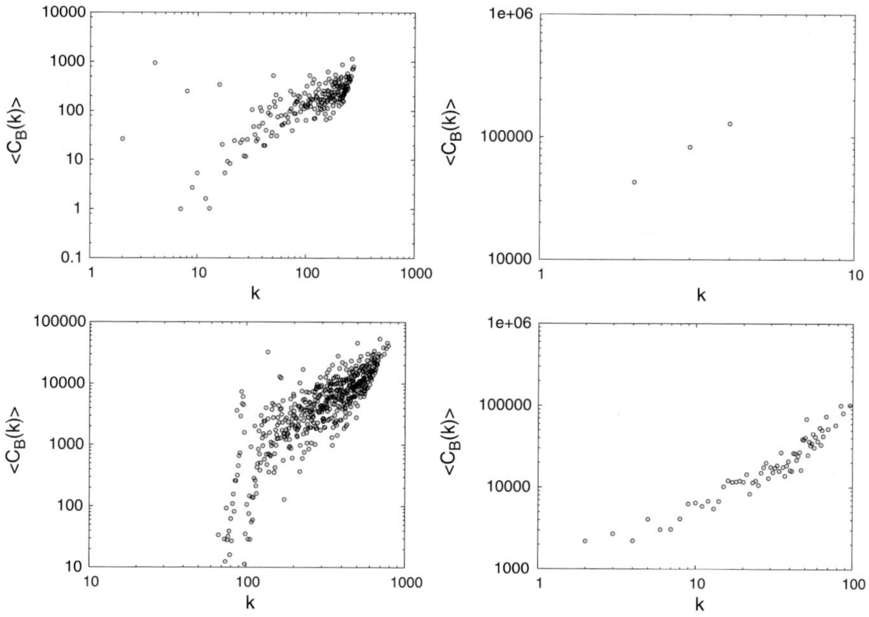

Fig. 7. Betweenness-degree correlation for the simulated city of 300×300 sites with $M = 500$ and $n = 50$ in $\mathbb{C}$-, $\mathbb{L}$-, $\mathbb{P}$-, and $\mathbb{B}$-spaces. $a = 0, b = 0.5$.

5 Conclusions and Outlook

The small-world properties of public transport networks are an everyday experience: it is easy to reach almost any given place in a city with only a small number of changes of transport (from table 1 it follows e.g. that the connection between any two stations in Paris implies on average $\overline{\ell_{\mathrm{P}}}-1 = 1.8$ changes), the scale-free properties of these networks are not that evident. Even more, scale free properties for these networks have sometimes been doubted. The results of our empirical analysis of PTNs of 14 major cities of the world [13, 15, 16] together with the empirical data for several other cities [6–12] however give a strong evidence of the fact that scale-free behaviour emerges in many PTNs. Apart from our model we currently cannot give more general arguments for this behaviour. Furthermore, not all PTNs seem to exhibit power-law node degree distributions, as some display rather an exponential decay of $p(k)$, see Table 1. Inspired by this observation we developed an evolutionary model of self-avoiding walks on 2D lattice which discriminates between the above types of behaviour and recovers a number of other basic features of the PTNs. The model applies the idea of the preferential attachment scenario [20], however with specific differences to that standard scenario: as far as the PTNs constitute an example of ever evolving networks, such a mechanism is not unlikely to play a role in their growth. In more general terms, scale-free networks have been shown in certain situation to minimize both the effort for communication and the cost for maintaining connections [22, 23]. Similar optimization was shown to lead to the small world properties [24] and used to explain the appearance of power laws [25]. Therefore one may expect the observed scale-free behaviour of PTNs to naturally emerge from obvious optimization objectives followed in their design.

One of the specific features of the PTNs we have analyzed is the harnessing effect: very often several routes go in parallel and pass together several consecutive stations. While other networks with real-world links like cables or neurons embedded in two or three dimensions often show similar behavior, these can be studied in detail in our present case. Our empirical analysis of the *harness* distribution that quantifies this behavior indicates power-law behaviour. The same behaviour is inherently recovered by our model. We found that a useful tool to classify PTNs as well as to find correspondence between real-world and simulated PTNs is a comparison of the observables in different representations ('spaces'). Furthermore, the standard network characteristics as represented in different spaces turn out to be natural measures for the quality of public transport in a city.

Of the many interesting further questions that are related to this study let us only mention the vulnerability of PTNs to random failures and targeted attacks (see our contribution [14] on this subject in this volume) as well as the correlation between the topological properties of a PTN and its geographical embedding.

Acknowledgements

Support by the Austrian Fonds zur Förderung der wissenschaftlichen Forschung, Project P19583 (Yu.H.) and by the EC, Project MTKD-CT-2004-517186 (C.v.F) is gratefully acknowledged. We thank Dietrich Stauffer for making us aware of Ref. [2].

References

1. M. E. J. Newman: SIAM Review **45**, 167 (2003).
2. L. Benguigui: J. Phys. I France **2**, 385 (1992); L. Benguigui, M. Daoud: Geogr. Analysis **23**, 362 (1991).
3. R. Albert, A.-L. Barabási: Rev. Mod. Phys. **74**, 47 (2002).
4. S. N. Dorogovtsev, S. N. Mendes: *Evolution of Networks*, (Oxford University Press, Oxford, 2003).
5. Yu. Holovatch, A. Olemskoi, C. von Ferber et al.: J. Phys. Stud. **10**, 247 (2006).
6. M. Marchiori, V. Latora: Physica A **285**, 539 (2000).
7. V. Latora, M. Marchiori: Phys. Rev. Lett. **87**, 198701 (2001).
8. V. Latora, M. Marchiori: Physica A **314**, 109 (2002).
9. K. A. Seaton, L. M. Hackett: Physica A **339**, 635 (2004).
10. C. von Ferber, Yu. Holovatch, V. Palchykov: Condens. Matter Phys. **8**, 225 (2005).
11. J. Sienkiewicz, J. A. Holyst: Phys. Rev. E **72**, 046127 (2005).
12. X. Xu, J. Hu, F. Liu, L. Liu: Physica A **374**, 441 (2007).
13. C. von Ferber, T. Holovatch, Yu. Holovatch, V. Palchykov: Physica A **380**, 585 (2007).
14. C. von Ferber, T. Holovatch, Yu. Holovatch: Attack Vulnerability of Public Transport Networks. In *Traffic and Granular Flow '07*. Springer (2007).
15. T. Holovatch: Modeling statistical properties of public transport networks. Diploma Thesis, Ivan Franko National University of Lviv, Lviv, Ukraine (2007).
16. C. von Ferber, T. Holovatch, Yu. Holovatch, V. Palchykov: in preparation.
17. P. Sen, S. Dasgupta, A. Chatterjee et al.: Phys. Rev. E **67**, 036106 (2003).
18. A. Vázquez, R. Pastor-Satorras, A. Vespignani: Phys. Rev. E **65**, 066130 (2002).
19. K.-I. Goh, E. Oh, B. Kahng et al.: Phys. Rev. E **67**, 017101 (2003).
20. A.-L. Barabási, R. Albert: Science **286**, 509 (1999); A.-L. Barabási, R. Albert, H. Jeong: Physica A **272**, 173 (1999).
21. A.B. Harris: Z. Phys. B **49**, 347 (1983); Y. Kim: J. Phys. C **16**, 1345 (1983).
22. R. Ferrer i Cancho, R. V. Solé: e-print cond-mat/0111222; S. Valverde, R. Ferre Cancho, R. V. Solé: Europhys. Lett. **60**, 512 (2002); R. Ferrer i Cancho, R. V. Solé. In: *Statistical mechanics of Complex Networks*, ed by R. Pastor-Satorras, M. Rubi, A. Diaz-Guilera (Lecture Notes in Physics Vol 625, Springer, Berlin, 2003), p. 114.
23. M. T. Gastner, M. E. J. Newman: Eur. Phys. J. B **49**, 247 (2006).
24. N. Mathias: Phys. Rev. E **63**, 021117 (2001).
25. R. Ferrer i Cancho, R. V. Solé: Proc. Natl. Acad. Sci. USA. **100**, 788 (2003); R. Ferrer i Cancho: Physica A **345**, 275 (2005).

Attack Vulnerability of Public Transport Networks

Christian von Ferber[1,2], Taras Holovatch[3], and Yurij Holovatch[4,5]

[1] Applied Mathematics Research Centre, Coventry University, Coventry CV1 5FB, UK C.vonFerber@coventry.ac.uk
[2] Physikalisches Institut, Universität Freiburg, 79104 Freiburg, Germany
[3] Ivan Franko National University of Lviv, 79005 Lviv, Ukraine
[4] Institute for Condensed Matter Physics of the National Academy of Sciences of Ukraine, 79011 Lviv, Ukraine
[5] Institut für Theoretische Physik, Johannes Kepler Universität Linz, 4040 Linz, Austria hol@icmp.lviv.ua

Summary. The behavior of complex networks under attack depends strongly on the specific attack scenario. Of special interest are scale-free networks, which are usually seen as robust under random failure or attack but appear to be especially vulnerable to targeted attacks. In a recent study of public transport networks of 14 major cities of the world we have shown that these networks may exhibit scale-free behaviour [Physica A **380**, 585 (2007)]. Our further analysis, subject of this report, focuses on the effects that defunct or removed nodes have on the properties of public transport networks. Simulating different attack strategies we elaborate vulnerability criteria that allow to find minimal strategies with high impact on these systems.

1 Introduction

A number of different phenomena related to complex networks [1] may be described in terms of percolation theory [2]. Take for example a network built following given construction rules. Then, how should the rules be tuned such that an infinite connected component is constructed with finite probability and what are the properties of this class of networks when the parameters reach the corresponding percolation threshold? Taken that percolation is in general seen as a critical phenomenon one may expect to find power laws in the vicinity of this point. The network (class) being described by more than one parameter, there are also many scenarios to cross the threshold exhibiting different behavior of the observables. Related questions are: how do infections spread on a network and are there optimal immunization strategies? These and similar questions are best formulated within percolation theory [2] generalized from its original formulation for regular grids to general network graphs.

In this paper we intend to apply concepts of complex network theory [1] to analyze the behaviour of urban public transport networks (PTNs) under successive removal of their constituents. In particular, continuing our recent study of PTNs of 14 major cities of the world [3, 4], we analyse their resilience against targeted attacks following different scenarios.

It has been observed before that the behaviour of a complex network under an attack that removes nodes or links may drastically differ from that of regular lattices (i.e. from the classical percolation problem). Early evidence of this fact was found analysing real world scale-free networks: the www and the internet [5, 6], as well as metabolic [7], food web [8], and protein [9] networks. In these studies, the interest was in the robustness of these networks subject to the removal of their nodes. It appeared that these networks display an unexpectedly high degree of robustness under random failure. However, if the scenario is changed towards "targeted" attacks, the same networks may appear to be especially vulnerable [10, 11].

To check the attack resilience of a network, different scenarios of attacks have been proposed: e.g. a list of vertices ordered by decreasing degree may prepared for the unperturbed network and the attack successively removes vertices according to this original list [12, 13]. In a slightly different scenario the vertex degrees are recalculated and the list is reordered after each removal step [5]. In initial studies only little difference between these two scenarios were observed [11], however further analysis showed [14, 15] that attacks according to recalculated lists often turn out to be more harmful than the attack strategies based on the initial list, suggesting that the network structure changes as important vertices or edges are removed. Other scenarios consider attacks following an order imposed by different 'centralities' of the nodes, e.g. the so-called betweenness centrality [15]. In particular for the world-wide airport network, it has been shown recently [16, 17] that nodes with higher betweenness play a more important role in keeping the network connected than those with high degree.

As it turns out, the behavior under attack of different real-world networks, even if they are scale-free differ considerably; e.g. computer networks behave differently than collaboration networks, see [15]. Therefore, it is important to investigate in how far the behaviour under attack of different real-world networks is consistent or shows strong variations. Below we present some results of our analysis for the PTNs of 14 major cities of the world (see Ref. [3] and chapter [4] of this volume for a detailed description of the included PTNs). A more complete survey will be a subject of a separate publication [18].

2 Observables and Attack Strategies

In the analysis presented below we consider the PTNs of the following cities: Berlin (number of stations $N = 2996$, number of routes $M = 218$), Dallas ($N = 6571$, $M = 131$), Düsseldorf ($N = 1544$, $M = 124$), Hamburg ($N =$

8158, $M = 708$), Hong Kong ($N = 2117$, $M = 321$), Istanbul ($N = 4043$, $M = 414$), London ($N = 11012$, $M = 2005$), Los Angeles ($N = 46244$, $M = 1893$), Moscow ($N = 3755$, $M = 679$), Paris ($N = 4003$, $M = 232$), Rome ($N = 6315$, $M = 681$), São Paolo ($N = 7223$, $M = 998$), Sydney ($N = 2034$, $M = 596$), Taipei ($N = 5311$, $M = 389$). This sampling includes cities from different continents, with different concepts of planning and different history of the evolution and growth of the city and its PTN. For the purpose of this paper let the PTN of a given city be given by the routes offered in this network. Each route services a given ordered list of stations. Representing the PTN in terms of a graph, we apply the following mapping: each station is represented by a node; any two nodes that are successively serviced by at least one route are connected by a single link. We note that there are several other ways to represent a PTN as a graph [3, 4, 19, 20]. The particular representation that we use here is referred to as a $\mathbb{L}$-space in Refs. [3, 4, 20].

The importance of a node i of a given network $\mathcal{N}$ may be measured by calculating a number of graph theoretical indicators. Besides the node degree k_i, which in our representation equals the number of nearest neighbours $z_1(i)$ of a given node i, different centralities of the node may be defined as follows (see e.g. [21]):

$$\text{closeness centrality} \qquad C_C(i) = \frac{1}{\sum_{t\in\mathcal{N}} \ell(i,t)}, \tag{1}$$

$$\text{graph centrality} \qquad C_G(i) = \frac{1}{\max_{t\in\mathcal{N}} \ell(i,t)}, \tag{2}$$

$$\text{stress centrality} \qquad C_S(i) = \sum_{s\neq i\neq t\in\mathcal{N}} \sigma_{st}(i), \tag{3}$$

$$\text{betweenness centrality} \qquad C_B(i) = \sum_{s\neq i\neq t\in\mathcal{N}} \frac{\sigma_{st}(n)}{\sigma_{st}}. \tag{4}$$

In Eqs. (1)–(4), $\ell(i,t)$ is the shortest-path length between a pair of nodes i, t that belong to a network $\mathcal{N}$, σ_{st} is the number of shortest paths between two nodes $s, t \in \mathcal{N}$, and $\sigma_{st}(i)$ is the number of shortest paths between nodes s and t that go through the node i. When observing a network under attack we will also record the next nearest neighbours $z_2(i)$ and the clustering coefficient $C(i)$ of all remaining nodes n. The latter is the ratio of the number of links E_i between the k_i nearest neighbours of i and the maximal possible number of mutual links between them:

$$C(i) = \frac{2E_n}{k_i(k_i - 1)}. \tag{5}$$

Note that the mean values of all the above introduced quantities are well-defined for a connected network $\mathcal{N}$. However, some of the analysed PTNs consist of several disconnected components even before any perturbation is applied. Moreover, the number of components naturally increases when nodes are removed. Therefore, we restrict averages of the observables to the largest network component $GCC \subset \mathcal{N}$. We will indicate these averages by an overline. Nevertheless, some of quantities are also well defined for the whole network, the corresponding average will be denoted by angular brackets. An example we note the inverse shortest path length:

$$\langle \ell^{-1} \rangle = \frac{2}{N(N-1)} \sum_{i>j} \ell^{-1}(i,j) \tag{6}$$

where the summation spans over all N sites of the (possibly disconnected) network and defining $\ell^{-1}(i,j) = 0$ if nodes i,j are disconnected. Note that in this case $\langle \ell \rangle$ is obviously ill-defined.

In what follows, we will pursue a number of different attack strategies or selection rules and criteria to remove the nodes (vertices). In particular, the scenarios are the following. "Random vertex" (RV): vertices (nodes) are removed in random order. "Random neigbour" (RN): one by one, a randomly chosen neighbour of a randomly chosen node is removed. This scenario appears to be effective for immunization problems [22] and it is based on the fact, that this way nodes with a high number of neighbors will be selected with higher probability. In further scenarios nodes are removed according to the lists prepared in the order of decreasing node degrees (k), centralities ($C(C)$, $C(G)$, $C(S)$, $C(B)$), the number of their second nearest neighbours (z_2), and increasing clustering coefficient (C). The latter seven scenarios can be either implemented according to lists prepared for the initial PTN before the attacks (we indicate the corresponding scenario by a subscript i, e.g. $C_i(C)$) or the list is built by recalculating the order of the remaining nodes after each step. This way we follow sixteen different strategies in attacking the networks. The observed changes of the properties of the PTN under these attacks are described in the next section.

3 Numerical Results

The theory of complex networks is concerned with the properties of ensembles of networks (graphs) that are characterized e.g. by common construction rules. Such an ensemble is said to be in the percolation regime if even the infinite graphs in this ensemble have a connected component that contains a finite fraction of their nodes. This component is referred to as the giant connected component GCC. If the ensemble properties are controlled by some parameter, e.g. the concentration of active nodes, then the percolation threshold in terms of this parameter is defined as the value at which the network ensemble enters the percolation regime. In the present case of finite networks we denote by GCC the largest connected component of a given network. For the finite networks defined by the PTN we analyze the behaviour of the their largest component that contains N_{GCC} nodes. We introduce the normalized largest component size S by:

$$S = \frac{N_{\mathrm{GCC}}}{N} \times 100\%. \tag{7}$$

In Fig. 1 we show the behavior of S for the attack strategies described above for the PTNs of Dallas and Paris. At each step of the attack 1% of the nodes

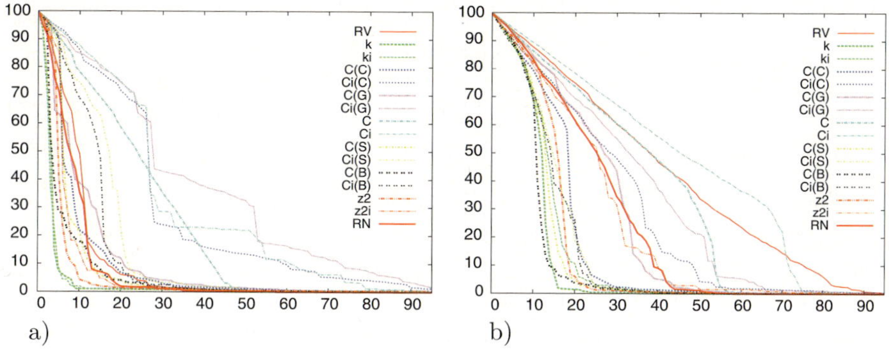

Fig. 1. Attacks on PTNs of (a) Dallas and (b) Paris. Each curve corresponds to a different attack scenario as indicated in the legend, see text. Horizontal axis: percents of removed nodes, Vertical axis: normalized size S of the largest component.

is successively removed following the selection criteria of the given scenarios. The effectiveness of the attack scenarios may be judged by their impact on the value of S. As it is clearly seen from Fig. 1, the least effective is the scenario of removing random nodes (RV): it is characterized by the slowest decrease of S. Another obvious conclusion is that scenarios based on lists calculated for the initial network (marked by a subscript i) appear to be less harmful than those, that are based on recalculated lists. Note however that the difference between 'initial' and 'recalculated' scenarios is less evident in the strategies based on the local characteristics, as e.g. the node degree or the number of second nearest neighbours (c.f. curves for k, k_i and z_2, z_{2i}, respectively). The above difference is even more pronounced for the centrality-based scenarios. A principal difference between attacks on the highest degree nodes on the one hand, and on the highest betweenness nodes on the other hand is that the first quantity is a local, i.e. is calculated from properties of the immediate environment of each node, whereas the second one is global. Moreover, the first strategy aims to remove a maximal number of edges whereas the second strategy aims to cut as many shortest paths as possible. Our analysis shows that the most effective are those scenarios that are either targeted at nodes with the highest values of the node degree k, the betweenness centrality C_B, the next nearest neighbour number z_2, or the stress centrality C_S recalculated after each step of the attack. Figures 2, 3 show that the order of destructiveness of these scenarios differ for PTNs of different cities. However, among the scenarios analyzed so far these four appear to be the most effective ones.

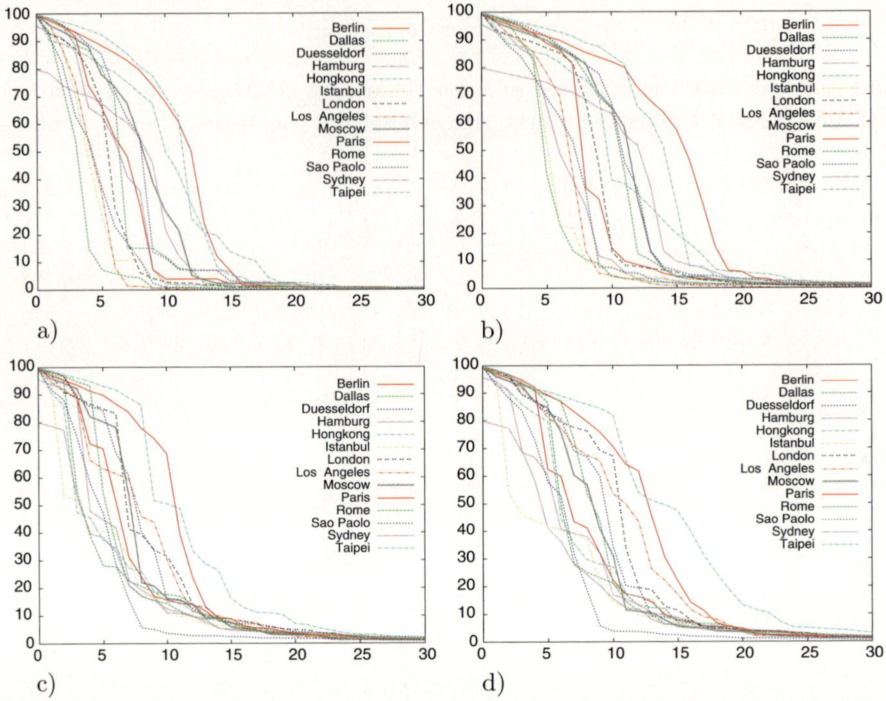

Fig. 2. Four attack scenarios for different PTNs (with recalculation): attacks targeted at nodes of the highest (a) degree k, (b) number of second neighbours z_2, (c) betweenness centrality C_B, or (d) stress centrality C_S. Vertical and horizontal axis as in Fig. 1.

Another interesting quantity that we may deduce from Fig. 2 is the vulnerability of the network in terms of the level of destruction at which the largest network component breaks down. We observe that this is strongly correlated to the initial value of the so called Molloy-Reed parameter $\kappa = \overline{z}_2/\overline{z}_1$ of the unperturbed network. Considering model networks that are randomly built from sets of nodes with given degree distributions it has been shown that the value of $\kappa_c = 1$ represents the percolation threshold in such networks [22, 23]. A value much larger than κ_c then indicates a significant distance from the threshold. The values of this parameter for the PTN studied here are: Dallas ($\kappa = 1.28$), Istanbul (1.54), Los Angeles (1.59), Hamburg (1.85), London (1.87), Berlin (1.96), Düsseldorf (1.96), Rome (2.02), Sydney (2.54), Hongkong (3.24), São Paolo (4.17), Paris (5.32), Moscow (6.24). Comparing in particular with Fig. 2 a) we find indeed that the higher the initial κ value the less vulnerable the network appears to be.

To more precisely define the threshold region for the concentration of removed nodes we observe the behaviour of the maximal $\ell_{\max}$ and mean $\overline{\ell}$ shortest path lengths under attack, as shown in Fig. 3. We focus on the recalculated degree scenario (k). Both maximal and average path lengths display similar behaviour: initial growth and then an abrupt decrease when a certain threshold is reached. Obviously, removing the nodes initially increases the path lengths as deviations from the original shortest paths need to be taken into account. Further removing nodes then at some point leads to the breakup of the network into smaller components on which the paths are naturally limited by the boundaries which explains the sudden decrease of their lengths. For the PTN of Paris we observe that this threshold is reached for both $l_{\max}$ and $\overline{\ell}$ at the same value of $c_{\text{segm}} \simeq 13\%$. The average shortest path on all components of the network, $\langle \ell \rangle$, also possesses a maximum in the same region (for the PTN of Paris it occurs at $c \simeq 13\%$). However, the values of c_{segm} differ for different cities (see Fig. 3b) and obviously strongly depend on the attack scenario.

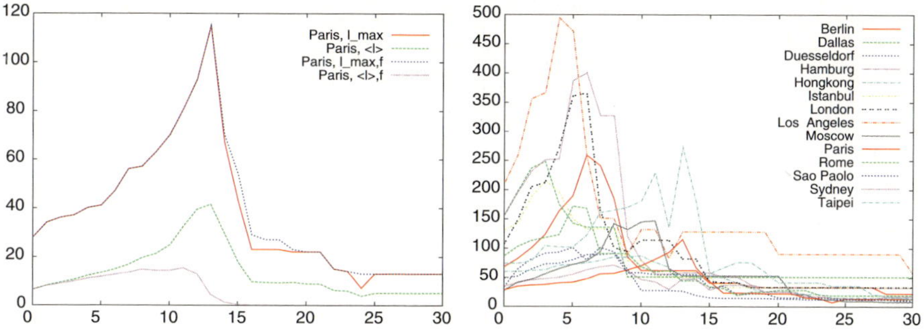

a) b)

Fig. 3. Highest degree scenario. Horizontal axis as in Fig. 1. (a) Behavior of the maximal and mean shortest path lengths for the Paris PTN calculated for the largest component ($\ell_{\max}$, $\overline{\ell}$) and for the whole network ($\ell_{\max,\text{f}}$, $\langle \ell \rangle_{\text{f}}$). Note a sharp maximum occurs at 13 % of removed nodes (stations) for $\ell_{\max}$, $\overline{\ell}$, $\ell_{\max,\text{f}}$. (b) Behavior of the maximal shortest path length $\ell_{\max}$ for the PTNs of different cities.

As discussed the observed maximum in $\ell_{\max}$ (or in $\overline{\ell}$) appears to be a suitable criterion to identify the values of c (or at least the region in c), where the segmentation of a network occurs. Other observables which resemble an 'order parameter', are the above described largest connected component size S, Eq. (7), or the average value of the inverse shortest path $\langle \ell^{-1} \rangle$ (6) are less suitable for this purpose because of their rather smooth behaviour. In Fig. 4 we show for PTNs of fourteen cities the behavior of $\langle \ell^{-l} \rangle$ under attacks following the four most harmful scenarios, i.e. the recalculated highest k, C_B z_2 and C_S scenarios. Comparing the impact of different attack scenarios (as seen in particular in Fig. 3, 4) one notices that the apparent relative impact strongly depends on the choice of the observable (e.g. S or $\langle \ell^{-l} \rangle$).

It is worth to note the statistical origin of the data exposed so far. Different instances of the same scenario may differ to some extent. This is obvious for the random RV or RN scenarios, where the nodes are removed according to a random procedure. However, it remains true even for the attacks following pre-ordered lists of nodes. Obviously, several nodes may have the same value with respect to a given characteristic (e.g. k, z_2, or one of the centrality indices). Then, the choice between these nodes is random. To check the dispersion of the results, Figs. 5, 6 show the results of 10 complete attack sequences for the same scenario. Figure 5 shows the change in the largest connected component S of the PTNs of Dallas (a), Hongkong (b), and Paris (c) for the random vertex (RV) scenario. The scatter of the curves in each figure provides an

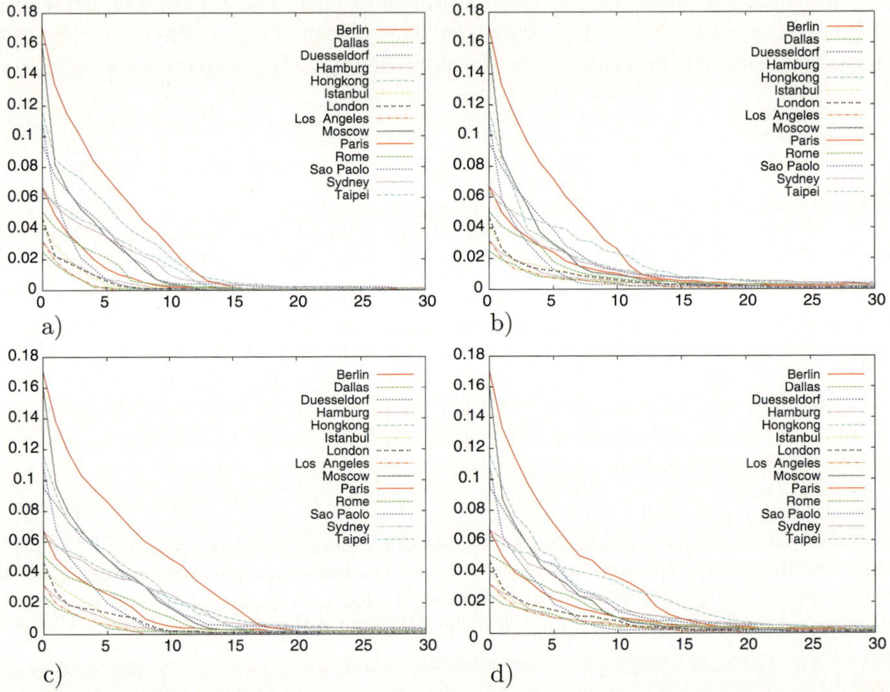

Fig. 4. Behaviour of $\langle \ell^{-l} \rangle$ for PTNs of different cities under attack following four different scenarios, see text: a) highest k, b) highest C_B, c) highest z_2, d) highest C_S. Horizontal axis as in Fig. 1.

idea about the deviations between individual samples. The figures also clearly show that even attacked randomly, PTNs of different cities may display a range of different behaviour: from the comparatively fast decay of the largest connected component (as in the case of Dallas, Fig. 5a) to very slow, nearly linear decay (as in the case of Paris, Fig. 5c).

The dispersion in the largest connected component size S is much less for sequences of targeted attacks. In Fig. 6 we show the behavior of the largest cluster size and the maximal and mean shortest path lengths for the Paris PTN for ten complete attack sequences following the recalculated degree (k) scenario. Besides a rather narrow scattering of the data for S one notes, that

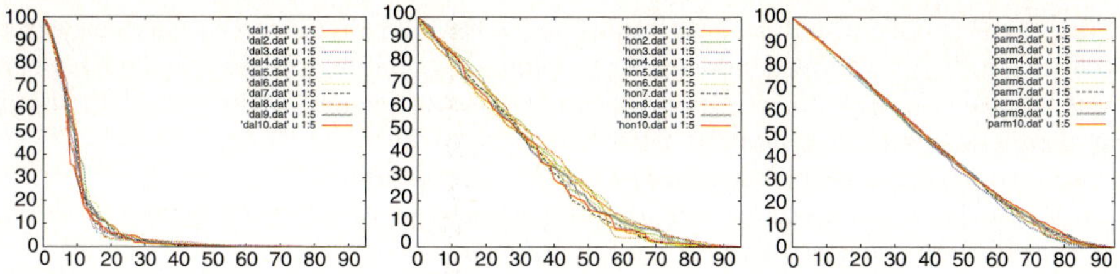

a) b) c)

Fig. 5. Impact and variance of the random vertex (RV) scenario on the normalized size S of the largest component for the PTNs of (a) Dallas, (b) Hongkong, and (c) Paris. Ten curves of different colour indicate different instances of the same scenario for each city. Vertical and horizontal axis as in Fig. 1.

within the current resolution the locations of the maxima in $\ell_{\max}$ and $\overline{\ell}$ are very robust.

To give an idea for the numerical values of different characteristics of the PTN as monitored during our analysis we display in Table 1 some data for the PTN of Paris for the recalculated degree scenario for some points of the sequence between the unperturbed network and the vicinity of the threshold (maximum of the shortest path lengths).

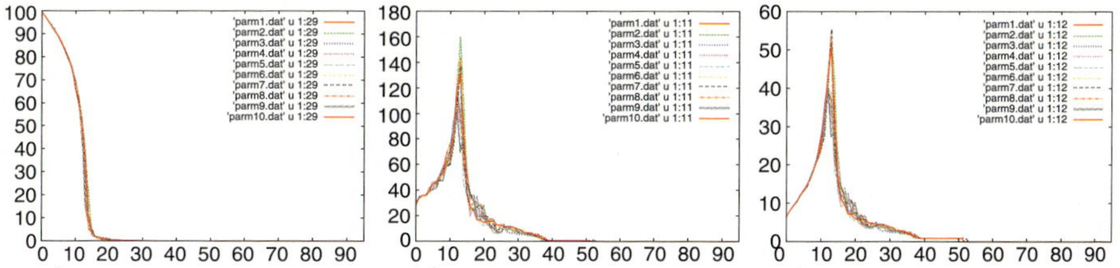

a) b) c)

Fig. 6. Ten instances of the recalculated highest degree scenario for the PTN of Paris, observing: a) the largest connected component size S, b) the maximal shortest path length $\ell_{\max}$, c) the mean shortest path length $\overline{\ell}$. Horizontal axis as in Fig. 1.

4 Conclusions

In this paper we reported on some results concerning the behavior of PTNs under attacks. Similar to other real-world and model complex networks [5–9, 15], the PTNs manifest very different behaviour under attacks of different scenarios. With some notable exceptions they appear to be robust to random attacks but more vulnerable to attacks targeted at nodes with particular importance as measured by the values of certain characteristics (the most significant being the first and second neighbour numbers, as well as the betweenness and

Table 1. PTN of Paris during an attack sequence following the recalculated degree scenario. c: % of removed nodes; N: number of remaining nodes; $\overline{k} = \overline{z}_1$: mean node degree; $\overline{z}_2/\overline{z}_1$: ratio of the mean second to the mean first nearest neighbour number; $\ell_{\max}$: maximal shortest path length; $\overline{\ell}$: mean shortest path length; $\langle \ell^{-1} \rangle$: mean inverse shortest path length (for all of the remaining network); $\overline{C}_C$, $\overline{C}_G$, $\overline{C}_S$, $\overline{C}_B$: mean closeness, graph, stress, and betweenness centralities; $\overline{C}$: mean clustering coefficient; S: normalized largest component size.

c	N	$\overline{k} = \overline{z}_1$	$\overline{z}_2/\overline{z}_1$	$\ell_{\max}$	$\overline{\ell}$	$\langle \ell^{-1} \rangle$	$\overline{C}_C$	$\overline{C}_G$	$\overline{C}_S$	$\overline{C}_B$	$\overline{C}$	S
0	3728	3.73	5.32	28	6.41	0.17	0.004	5.47	38167	10062	0.079	99.87
1	3691	3.25	3.40	34	8.08	0.13	0.003	4.64	40419	12912	0.073	97.85
5	3543	2.52	2.05	41	13.35	0.07	0.002	3.60	50496	20439	0.062	88.81
10	3358	2.00	1.43	70	24.84	0.03	0.002	2.02	53654	30406	0.044	68.45
12	3284	1.84	1.25	93	39.44	0.01	0.001	1.42	56218	36097	0.036	50.40
13	3247	1.77	1.19	115	41.49	0.01	0.003	1.21	31803	18404	0.039	24.41
14	3210	1.70	1.13	67	29.69	0.00	0.008	1.90	11915	6598	0.022	12.37

stress centralities). The observed difference between attack scenarios based on the initial and the recalculated distributions shows that the network structure changes essentially during the attack sequence. This is necessarily to be taken into account in the construction of efficient strategies for the protection of these network.

As a suitable criterion to identify the level of resilience, i.e. the number of removed nodes that leads to segmentation it is useful to observe the behaviour of the maximal shortest path length $\ell_{\max}$. For the majority of PTNs networks we have analyzed here this observable displays a sharp maximum as function of the removed node concentration which indicates the breakup of the network. Other 'order-parameter-like' variables like the largest connected component size S or the average value of the inverse shortest path $\langle \ell^{-1} \rangle$ are less suitable for this purpose because of their smooth behaviour. Another observation is that in the recalculated highest-degree attack scenario for the segmentation often occurs at a value of $\kappa = \overline{z}_2/\overline{z}_1 \sim 1$ (see e.g. Table 1 for Paris). Although the PTNs are correlated structures, the above estimate resembles the Molloy-Reed [23] criterion for randomly built uncorrelated networks. Further investigation is needed to understand the mechanisms that lead to higher resilience against random failure as observed e.g. for the Paris network and how this behavior is related to the network architecture.

As mentioned in the introduction, there are different graph representations, also called 'spaces', for a given PTN [3, 4, 19, 20]. These will also lead to different connectivity relations and path lengths between nodes. The resilience of PTNs in these more general 'spaces' will be discussed elsewhere [18].

Acknowledgements

Yu.H. acknowledges financial support of the Austrian Fonds zur Förderung der wissenschaftlichen Forschung under Project P19583. C.v.F. was supported in part by the EC under the Marie Curie Host Fellowships for the Transfer of Knowledge MTKD-CT-2004-517186.

References

1. M. E. J. Newman: SIAM Review **45**, 167 (2003); R. Albert, A.-L. Barabási: Rev. Mod. Phys. **74**, 47 (2002); S. N. Dorogovtsev, S. N. Mendes: *Evolution of Networks*, (Oxford University Press, Oxford, 2003); Yu. Holovatch, A. Olemskoi, C. von Ferber et al.: J.Phys. Stud. **10**, 247 (2006).
2. J. W. Essam: Rep. Prog. Phys. **43**, 833 (1980); D. Stauffer, A. Aharony: *Introduction to Percolation Theory*, (Taylor & Francis, London, 1991).
3. C. von Ferber, T. Holovatch, Yu. Holovatch, V. Palchykov: Physica A **380**, 585 (2007).
4. C. von Ferber, T. Holovatch, Yu. Holovatch, V. Palchykov: Modeling Metropolis Public Transport. In *Traffic and Granular Flow '07*. Springer (2007).
5. R. Albert, H. Jeong, A.-L. Barabási: Nature (London) **406**, 378 (2000).
6. Y. Tu: Nature (London) **406**, 353 (2000).
7. H. Jeong, B. Tombor, R. Albert, Z. N. Oltvai, A.-L. Barabási: Nature (London) **407**, 651 (2000).
8. R. V. Solé, J. M. Montoya: Proc. R. Soc. Lond. B **268**, 2039 (2001).
9. H. Jeong, S. P. Mason, A.-L. Barabási, Z. N. Oltvai, Nature (London) **411**, 41 (2001).
10. R. Cohen, K. Erez, D. ben-Avraham, S. Havlin: Phys. Rev. Lett. **85**, 4626 (2000).
11. D. S. Callaway, M. E. J. Newman, S. H. Strogatz, D. J. Watts: Phys. Rev. Lett. **85**, 5468 (2000).
12. A.-L. Barabási, R. Albert: Science **286**, 509 (1999); A.-L. Barabási, R. Albert, H. Jeong: Physica A **272**, 173 (1999).
13. A. Broder, R. Kumar, F. Maghoul, et al.: Comput. Netw. **33**, 309 (2000).
14. M. Girvan, M. E. J. Newman: Proc. Natl. Acad. Sci. USA **99**, 7821-7826 (2002).
15. P. Holme, B. J. Kim, C. N. Yoon, S. K. Han: Phys. Rev. E **65**, 056109 (2002).
16. R. Guimera, L. A. N. Amaral, Eur. Phys. J. B **38**, 381 (2004).
17. R. Guimera, S. Mossa, A. Turtschi, L. A. N. Amaral: Proc. Nat. Acad. Sci. USA **102**, 7794 (2005).
18. C. von Ferber, T. Holovatch, Yu. Holovatch: in preparation.
19. P. Sen, S. Dasgupta, A. Chatterjee et al.: Phys. Rev. E **67**, 036106 (2003).
20. J. Sienkiewicz, J. A. Holyst: Phys. Rev. E **72**, 046127 (2005).
21. U. Brandes: J. Math. Sociology, **25**, 163 (2001).
22. R. Cohen, S. Havlin, D. ben-Avraham: Phys. Rev. Lett. **91**, 247901 (2003).
23. M. Molloy, B. A. Reed: Random Struct. Algorithms **6(2/3)**, 161 (1995); M. Molloy, B. Reed: Combinatorics, Probability and Computing **7**, 295 (1998).

Traffic Dynamics Based on Local Routing Strategy in a Weighted Scale-Free Network

Mao-Bin Hu[1,2], Yong-Hong Wu[1], Rui Jiang[2], Qing-Song Wu[2], and Wen-Xu Wang[3]

[1] Department of Mathematics and Statistics, Curtin University of Technology, Perth WA6845, Australia humaobin@ustc.edu.cn
[2] School of Engineering Science, University of Science and Technology of China, Hefei 230026, P.R. China
[3] Department of Electronic Engineering, City University of Hong Kong, Hong Kong SAR, P.R. China

Summary. In this paper, the traffic flow on weighted scale-free networks is investigated based on local routing strategy using link weights: $P_{l\to i} = \frac{w_{li}^{\alpha}}{\sum_j w_{lj}^{\alpha}}$. The capacity of links is controlled by $max(\beta w_{lj}, 1)$. It is shown by simulations that two critical threshold β_{c1} and β_{c2} exist. When $\beta > \beta_{c1}$, both the network capacity and the corresponding α_c value remain unchanged. When $\beta_{c1} > \beta > \beta_{c2}$, the network capacity decreases and the critical value of α_c increases with the decrease of β. When $\beta < \beta_{c2}$, α_c decreases with the decrease of β. The behaviour can be explained by investigating the average number of packets on nodes and delivered through links.

1 Introduction

Traffic flow problem is of great importance for modern society and for the safety and convenience of metropolitans. And due to the importance of large communication networks such as the Internet and WWW, many investigations have been focused on ensuring free traffic flow and avoiding traffic congestion on networks [1–7].

Recently, empirical evidences show that many traffic systems can be described by complex networks with the small world [8] and/or scale-free properties [9]. A wide range of systems in nature and society can be related to complex networks. The prototypes include urban road [10], bus route [11, 12], railway [13] and airline transport systems [14]. The traffic flow can be modelled by introducing packets generating rate R and random origin and destination (OD) for each packet [5–7, 15–17]. And the capacity of networks are measured by a critical generating rate R_c. At this critical rate, a continuous phase transition from free flow state to congested state occurs. The free flow state corresponds to the case that the numbers of created and delivered packets are balanced. While the jammed state means that the number of accumulated packets increases with time due to the limited delivering capacity or finite queue length of each node. And packets are forwarded following the random walking [15, 18], the shortest path [16], the efficient path [17], the nearest-neighbor search strategy [19–21], the next-nearest-neighbor search strategy [22], the local information [19–21] or the integration of local static and dynamic information [23]. These routing methods are more or less similar to the real cases in engineering.

In most of the real cases, the networks are often associated with a large heterogeneity in the capacity and intensity of the connections, i.e., most networks are weighted networks. The traffic load is often represented by the weight of links also [24]. However, the traffic dynamics have been investigated only in unweighted networks up to now. And the previous traffic routing strategies have been focused on how to navigate the packets by using local or global node information (mainly by using node degree). In this paper, a study of traffic dynamics in a weighted scale-free network is carried out with a new traffic model in which packets are guided based on local link information with a single tunable parameter α. To maximize the capacity of the networks which can be measured by the critial packet generating rate R_c, the optimal α_c is sought out. And the traffic load distribution among nodes and links are also studied to give an explanation for the optimal α value.

2 Network and Traffic Model

To generate the underlying network infrastructure, this paper uses the weighted scale-free network model proposed by Wang et al [25], in which the power-law distributions of degree, weight, and strength are all in good accordance with real observation of weighted technological networks. In this model, starting from m_0 nodes fully connected by links with assigned weight w_0, the system is driven by two mechanics: (1) the strength dynamics: the weight of each link connecting i and j is updated as $w_{ij} \to w_{ij} + 1$, with probability $P_{ij} = W \times p_{ij} = W \times \frac{s_i s_j}{\sum_{a<b} s_a s_b}$; (2) the topological growth: a new node n is added with m links that are randomly attached to a node i according to the strength preferential probability $\Pi_{n \to i} = \frac{s_i}{\sum_j s_j}$, where j runs over all existing nodes.

The traffic dynamics is modelled as follows. At each time step, R packets are generated homogeneously on the nodes in the system. We treat all the nodes as both hosts and routers and assume that node i can deliver at most s_i packets per time step towards their destinations, where s_i denotes strength of node i, i.e., the capacity of the nodes are set to $C_i = s_i$. To navigate the packets, all the nodes perform a parallel local search among their immediate neighbours. If a packet's destination is found within the searched area of node l, i.e. the immediate neighbours of l, the packets will be delivered from l directly to its target and then removed from the system. Otherwise, the packet will be delivered to a neighboring node i with link weight preferential probability:

$$P_{l\to i} = \frac{w_{li}^{\alpha}}{\sum_j w_{lj}^{\alpha}}, \tag{1}$$

where w_{li} is the weight of the link connecting nodes l and i, the sum runs over the immediate neighbours of the node l, and α is an introduced tunable parameter characterizing the preferential probability in chosing links to forward packets. The capacity (or bandwidth) of the link connecting nodes l and i is set to $B_{li} = \max(\beta w_{li}, 1)$, i.e., the link can handle at most B_{li} packets from each end per time step. When the link capacity is reached, the delivery of packets will be delayed and wait for next time step. During the evolution of the system, the FIFO (first-in-first-out) rule is applied on the nodes.

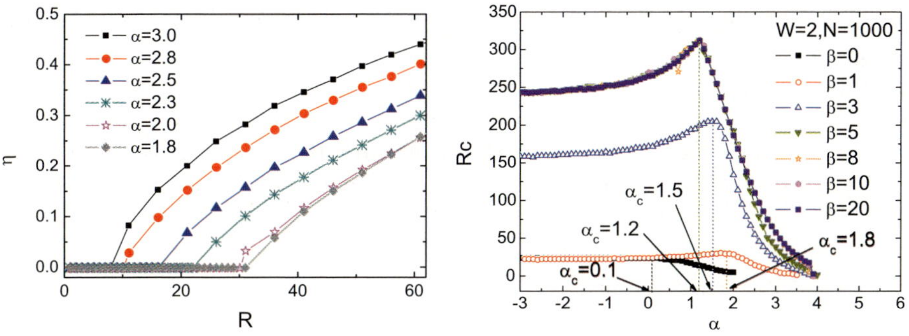

Fig. 1. (a) The order parameter η versus R for weighted scale-free networks with different routing parameter α. Other parameters are networks size $N = 1000$, $m0 = m = 5$, $W = 2$ and $\beta = 1$. (b) Overall network traffic capacity characterized by the critical value of R_c with fixed $W = 2$ and different β values. The optimal values of α_c corresponding to the maximum R_c are sought out.

3 Simulation and Results

To characterize the network capacity, we investigate the order parameter:

$$\eta(R) = \lim_{t\to\infty} \frac{1}{R}\frac{\langle \Delta N_p\rangle}{\Delta t}. \tag{2}$$

Here $\Delta N_p = N_p(t + \Delta t) - N_p(t)$, $\langle \ldots \rangle$ denotes taking the average over a time window of width Δt, and $N_p(t)$ is the number of packets in the system at time t. As shown in Fig. 1(a), when $R < R_c$, $\langle \Delta N\rangle = 0$ and $\eta(R) = 0$, corresponding to the case of free-flow state, in which the numbers of added and removed packets are balanced; while $\eta(R)$ increases suddenly from zero at $R = R_c$. Therefore a phase transition occurs at $R = R_c$ when congestion emerges and spreads in the system, and packets will continuously accumulate in the system. Hence, the system overall handling and delivering capacity can be measured by the critical value of R_c.

Our simulations show that when $\beta \to \infty$, i.e., when the capacity of every link is very large, the maximum network capacity is reached when α is slightly larger than 1.0. Figure 1(b) shows the overall traffic capacity measured by R_c vs α with different β values. One can see that both the network capacity and the value of α_c remain unchanged provided β is larger than a threshold $\beta_{c1} \approx 5$. The maximum capacity always appears at $\alpha_c = 1.2$ with $R_c \approx 313$. This behaviour implies that when $\beta > \beta_{c1}$, the links are operating efficiently under their maximum capacity, and the network capacity is mainly controlled by the node capacity. When $\beta_{c1} > \beta > \beta_{c2} \approx 1$, the system's overall capacity decreases rapidly, and the α_c value increases with the decrease of β. This is because the capacity of some links will be reached from time to time, and thus the delivery of some packets are delayed, and accordingly the network capacity decreases. Nevertheless, when β becomes smaller than the second threshold $\beta_{c2} \approx 1$, α_c begins to decrease with the decrease of β. As we can see in Fig. 1(b), when $\beta = 0$, i.e., all links have the same capacity that is equal to one, the system's maximum capacity appears at $\alpha_c = 0.1$ with $R_c = 25$.

In addition, simulations also show that R_c tends to be a constant with the decrease of α. This constant decreases with the decrease of β for $\beta_{c2} < \beta < \beta_{c1}$. When $\alpha < 0$ and $|\alpha|$ is very large, the probability that links with large weight are chosen to deliver packets is very small.

Recently, it was found that in a typical Barabasi-Albert network, the maximum traffic capacity appears at $\alpha_c = -1.0$ when the node capacity is set to $C = const$ [19]. This means to repel the packets from the central nodes and to make them move along the periphery of the network. When considering the heterogeneity of node capacity, they found that $\alpha_c = 0.0$ when $C = k$ [19]. This means that random walk is the best strategy for the packets. Different from previous results, it is shown here that using links with large value of weight is better for the navigation of packets on weighted networks.

In the following, we give a heuristic explanation for the optimal α_c value by investigating the traffic load distribution on the network. In Fig. 2(a), we investigate the average visits per node divided by the node strength (denoted as ω) which can be useful to analyze the traffic load distribution among the nodes. When $\alpha < 1.2$, ω self-organizes to a power law, which implies that the traffic burden of high-degree nodes is alleviated, while when $\alpha > 1.2$, ω is an increasing function with respect to S, which may lead to the collapse of hub-nodes. When $\alpha = 1.2$, ω essentially remains constant with respect to S. At this point, the balance of node capacity with traffic load is achieved for all nodes. This balance is useful as in this state no node will be more easy to jam than others and thus the optimal traffic capacity will be achieved. This

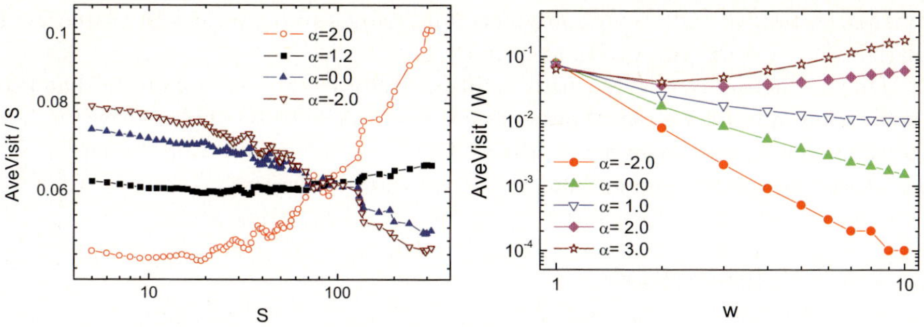

Fig. 2. (a) Average visits per node divided by the node strength, (b) Average visits per link divided by the link weight. The other parameters are $N = 1000$, $m0 = m = 5$, $W = 2$, $R = 20$ and $\beta = 1$.

conclusion is confirmed by our previous simulation result that the optimal α_c value is always located at 1.2 when $\beta \geq 5$.

To explain why $\alpha_c > 1.2$ for low β value, we investigate the average visits per link divided by the link weight (denoted as ζ) which reflects the traffic load distribution along the links. Figure 2(b) shows ζ vs w (link weight) for different α values. One can see that when $\alpha \approx 2.0$, ζ is roughly constant for all the link weights. Thus we can see that the capacity of the links will be more fully used when α is set to the value close to 2.0 for the case of $\beta = 1$, and therefore the overall capacity of the system will be maximized. We also note that ζ remains almost at the same value for links with weight $w = 1$. As in our network model, the weights of most of the links in the system ($\approx 95\%$ links in the simulation) are equal to one, this behaviour explains why the system's capacity will remain constant for negative α value.

Finally, we briefly introduce the effect of link weight growth rate W on the packet traffic capacity. In general, the system's overall capacity will increase with the increase of W. Another interesting phenomenon is the optimal value of routing parameter remains at the same value for different W values.

4 Conclusion and Discussions

In conclusion, the traffic on weighted scale-free network is studied with a routing strategy based on local weight information. The study reveals some new characteristics on weighted scale-free networks, which are different from the traffic on well-organized lattice, on regular or random networks, and on unweight scale-free networks. In general, the overall capacity decreases when the bandwidth parameter of links (β) is below a critical value β_{c1}. In most cases, the optimal value of local routing parameter α appears at $\alpha_c > 1.0$. And the α_c value also depends on the bandwidth parameter. We give explanations for the variation of α_c by investigating the average visits per node and average visits per link, which are also important for studying the traffic load distribution among the nodes and links.

Due to the many new features mentioned above, we can conclude that the traffic dynamics on weighted networks is worthwhile to be explored by further efforts of physics community. And the results reported here may be useful for the design of traffic systems and the navigation of drivers.

Acknowledgements

This work is funded by National Basic Research Program of China (No. 2006 CB705500), the NNSFC under Key Project No. 10532060 and No. 10635040, Project Nos. 70601026, 10672160, 10404025, the CAS President Foundation, and the China Postdoctoral Science Foundation (No. 20060390179). Y.-H. Wu acknowledges the support of Australian Research Council through a Discovery Project Grant.

References

1. Arenas A, Díaz-Guilerà A, Guimerà R (2001) Phys Rev Lett 86:3196.
2. Guimerà R, Arenas A, Díaz-Guilera A, Giralt F (2002) Phys Rev E 66:026704.
3. Guimerà R, Díaz-Guilera A, Vega-Redondo F, Cabrales A, Arenas A (2002) Phys Rev Lett 89:248701.
4. Guimerà R, Danon L, Díaz-Guilerà A, Giralt F, Arenas A (2003) Phys Rev E 68:065103(R).
5. Tadić B, Thurner S, Rodgers GJ (2004) Phys Rev E 69:036102.
6. Tadić B, Thurner S (2004) Physica A 332:566.
7. Tadić B, Thurner S (2005) Physica A 346:183.
8. Watts DJ, Strogatz SH (1998) Nature 393:440.
9. Barabási AL, Albert R (1999) Science 286:509.
10. Rosvall M, Trusina A, Minnhagen P (2005) Phys Rev Lett 94(2):028701.
11. Sienkiewicz J, Holyst JA (2005) Phys Rev E 72(4):046127.
12. Li P, Xiong X, Qiao ZL et al (2006) Chin Phys Lett 23(12):3384.
13. Li KP, Gao ZY, Mao BH (2006) Inter J Mod Phys C 17(9):1339.
14. Li W, Cai X (2004) Phys Rev E 69:046106.
15. Noh JD, Rieger H (2004) Phys Rev Lett 92:118701.
16. Zhao L, Park K, Lai YC (2004) Phys Rev E 70:035101(R).
17. Yan G, Zhou T, Hu B, Fu ZQ, Wang BH (2006) Phys Rev E 73:046108.
18. de Moura APS (2005) Phys Rev E 71:066114.
19. Wang WX, Wang BH, Yin CY, Xie YB, Zhou T (2006) Phys Rev E 73:026111.
20. Hu MB, Wang WX, Jiang R, Wu QS, Wu YH (2007) Phys Rev E 75:036102.
21. Hu MB, Wang WX, Jiang R, Wu QS, Wu YH (2007) Europhys Lett 79:14003.
22. Yin CY, Wang BH, Wang WX, Yan G, Yang HJ (2006) Eur Phys J B 49:205.
23. Wang WX, Yin CY, Yan G, Wang BH (2006) Phys Rev E 74:016101.
24. Barrat A et al (2004) Proc Natl Acad Sci USA 101:3747; Macdonald PJ, Almaas E, Barabaási AL (2007) Europhys Lett 79:14003.
25. Wang WX, Wang BH, Hu B, Yan G, Ou Q (2005) Phys Rev Lett 94:188702.

Performance Evaluation of VANET Under Realistic Vehicular Traffic Assumption

Mabiala Muriel, Anthony Busson, and Véronique Véque

Institut d'Electronique Fondamentale, Bât 220, Université Paris-Sud, F91405 Orsay cedex, France nom@u-psud.fr

Summary. Vehicle-to-vehicle communication has become a very challenging topic in recent years. Vehicles equipped with devices capable of short-range wireless connectivity can form a mobile ad hoc network, called a Vehicular Ad-hoc NETwork (VANET). In such networks, two vehicles which are not in the radio range of each other can communicate by using other vehicles as relay.

Our work focuses on the efficiency of routing algorithm in VANET. A routing algorithm is a distributed algorithm used by the nodes to learn the route to potential destinations. There exists standardized routing protocols designed for ad-hoc networks. VANET is a particular ad hoc network where the nodes (vehicles) move very fast. This introduces a high dynamism in the network topology which is difficult to manage. So, we compare the performances of different existing routing protocols in this context. We use two different simulators. The first is a traffic simulator which emulates the vehicular traffic on a highway. It uses microscopic simulations. The second is a network simulator (NS2) which simulates all the protocols (wifi, IP, UDP and the routing protocols) used by the nodes to communicate. The simulations show that reactive protocols are most efficient compared with geographical routing in a VANET.

The remainder of the paper is organized as follows. In Section 1, we present the microscopic traffic simulator. We present the network simulator and simulation parameters in Section 2. Different routing protocols are presented in Section 3. The simulation results are discussed in Section 4.

1 Traffic Simulator

The objective of our study is to analyze and evaluate the performance of ad hoc routing protocols with a realistic simulation of traffic. The simulator we use to model the traffic is a microscopic simulator. It emulates the individual behaviour of vehicles taking into account the interaction between them.

We use a microscopic model in our simulations because we are interested about the impact of individual characteristics (speed, geographical location, ...) of vehicles on the performance of routing protocols. The location of vehicle is defined by the co-ordinates x, y, where the parameter y represents the lane of vehicle.

The model we use is presented in [1]. The trajectory of a vehicle is determined by two models: lane changing model and a car-following model. The car-following model includes three modes: emergency model, car-following and free flow regime. If the time of contact with the vehicle ahead is smaller them a certain threshold, the vehicle is in the emergency regime. The acceleration a_n is given by:

$$a_n = \begin{cases} min(a_n^-, a_{n-1} - 0.5\frac{(v_n - v_{n-1})^2}{g_n}), & v_n > v_{n-1}; \\ min(a_n^-, a_{n-1} + 0.25a_n^-), & v_n \leq v_{n-1}. \end{cases}$$

If the time of contact with the vehicle ahead is larger than another threshold, the car is in free flow regime. In this regime, the vehicle is not obstructed by other vehicle. So, it tries to reach a certain desired speed. The acceleration a_n is then given by:

$$a_n = \begin{cases} a_n^+, \text{ if } v_n < v_n^{target}; \\ 0, \quad \text{if } v_n = v_n^{target}; \\ a_n^-, \text{ if } v_n > v_n^{target}. \end{cases}$$

When the time of contact is between the two thresholds, the vehicle is in car-following regime. The acceleration is calculed by: $a_n = \alpha^{+-}.\frac{v_n^{\beta+-}}{1}(v_{n-1} - v_n)$.

The other component of microscopic traffic simulator is the lane changing model [1]. The location of vehicle in the lane and the changing of the lane influences the traffic. The drivers take into account the vehicles on their lane and on adjacent lane to improve their displacement.

The lane changing of a vehicle is modeled in the following way:

- The vehicle checks, whether it is satisfied with the condition of traffic on its current lane,
- If it is satisfied, it stays in its lane,
- If not, it checks, if the conditions are more satisfactory, on the adjacent lanes,
- If the conditions in the left lane are satisfactory, it examines the various positions, which separate it from the vehicles which are behind and in front on the target lane,
- If these distances do not allow displacement, then it continues its movement on current lane.

2 Network Simulator

We use NS-2 (Network Simulator version 2) [2] for simulations. The result of simulation is a trace which lists the various events. For the mobility of nodes, we use a microscopic simulator of traffic to generate the movement of nodes. The movements are inserted in NS2.

For the data traffic, we assume that each vehicle is equipped with a wifi card in ad hoc mode. The number of packets per second are defined, which makes it possible to determine the arrival interval of packets. In the MAC layer, each vehicle transmits using 802.11 characteristics with data rate of 2 Mbit/s and radio range of 250 meters. The mechanism RTS/CTS is supported. The RTS/CTS (Request to Send/Clear to Send) is broadcasted between a source and destination before sending data. RTS is a request to use of the medium and CTS is the favorable answer. We use Two-Ray ground propagation reflection model, which takes into account, the direct way and reflection of the ground. The connections are etablished between vehicles randomly chosen in the highway. For each connection a constant bit rate traffic is generated.

The NS-2 simulations generate a trace file, containing the list of events (sending, receptions of the packets). We use these files to analyze the performance of routing protocols in terms of:

- Packet delivery Ratio (PDR) is the ratio of the amount of data received by the destination and sent by the source.
- Delay is a time between the emission of data by the source and its reception at the destination.
- Overhead routing load is the total number of control packets transmitted by the nodes.

The overhead allows us to observe the scalability of a protocol. A protocol which has a large number of routing packets has a higher probability of collisions which decreases the capacity of the network.

Packet Delivery Ratio allows us to observe the ability of the protocols to find a route to the destination.

3 Presentation of Protocols

We evaluate the performance of DSR, GPSR and AODV. GPSR (Greedy Perimeter Stateless Routing protocol) [3] is a geographical routing protocol. In GPSR, nodes are supposed to know their geographical location. The nodes know their geographical location by using a geolocalization system like GPS. The data packets are marked by the source to geographical location of destinations obtained by a location service protocol [4]. A source or an intermediate node forwards the packet in the greedy mode. The choice of next hop is the neighbour geographically closest to the destination. The process is repeated until the destination is reached. DSR (Dynamic Source Routing protocol) and AODV (Ad Hoc On demand Distance Vector) are two reactive routing protocol. In DSR, when a node has data to send to another node, it checks in its cache if it has a route to reach the destination. If it has, it uses this path to forward the data packet, otherwise, it broadcasts a request packet. The detailed description of DSR can be found in [5]. In AODV (Ad hoc On demand

Distance Vector), when a source node wants to transmit a packet, it checks if it has a route to the destination in the routing table. If it doesn't have route, it broadcasts a route request (RREQ) packet. Nodes receiving this request check if they have a route to reach the destination. If any node has a route, it unicasts a RREP back to the source. Otherwise, it rebroadcasts the RREQ.

4 Results

We simulate a highway of 14 km with 3 lanes (only one direction is taken into account). The duration of each simulation is 150 second. Each point of the figure represents values obtained from 50 samples. On all the graphs, the vertical bars correspond to 95% confidence interval.

In Figure 1, we compare PDR of DSR, AODV and GPSR. The three protocols DSR, AODV and GPSR show similar results. The losses in the case of these three protocols are due primarily to the fact that the network is not dense enough. The traces show that in DSR, AODV and GPSR the losses are due mainly to the fact that there is no path between source and destination. When densities of vehicle are high, the connectivity in network is significantly better.

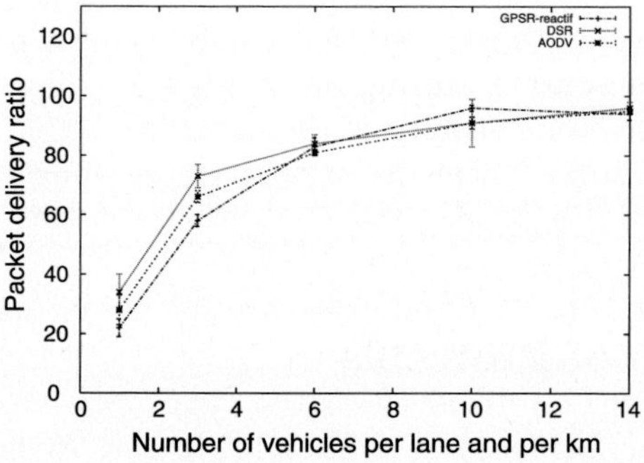

Fig. 1. PDR as a function of number of vehicle per lane and per km.

In Figure 2, we examine the average delay of data packets. DSR has the highest delay with nearly all simulated data traffic. Its is due to the fact that its route discovery process takes a long time compared to AODV and GPSR, when we compare it with protocol GPSR without service of localization. When a service of localization is added, we notice that the delay is more significant in GPSR.

In Figure 3, we vary the number of connections. We notice that Packet delivery ratio of all protocols decreases with number of connections. AODV, DSR and GPSR generates a lot of control packets which cause congestion in the network.

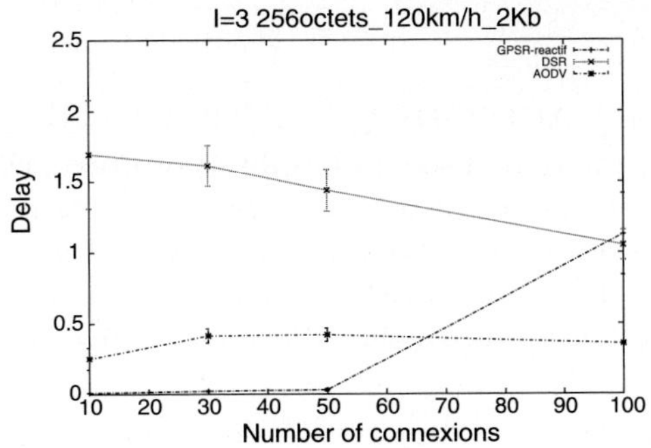

Fig. 2. Delay as a function of the number of connections.

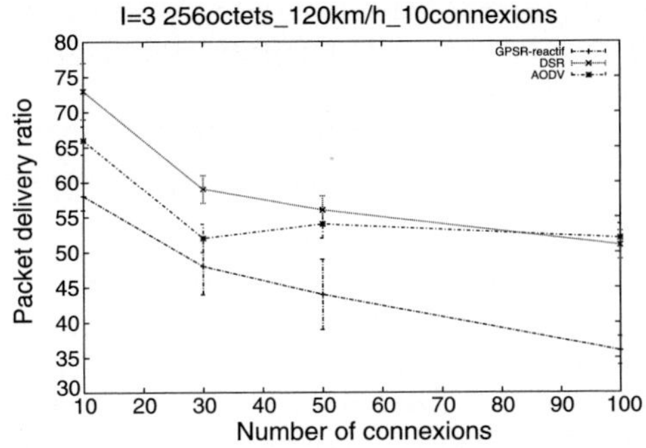

Fig. 3. PDR as a function of the number of connections.

In Figure 4, we plot the routing overhead with number of connections. DSR demonstrates significantly lower routing overhead than AODV and GPSR. AODV and DSR routing is stable with the number of connections which is a favorable scalability property. In case of GPSR however, the increase in routing packets is linear with the number of connections.

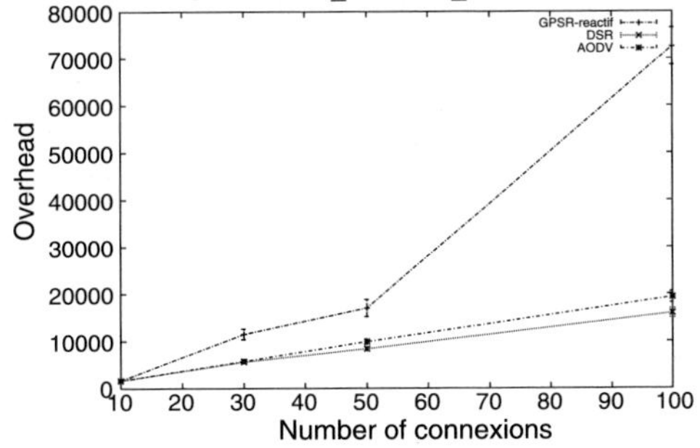

Fig. 4. Overhead in function of number of connections.

5 Conclusion

A comparison between AODV, DSR and GPSR shows that reactive protocols offer better performances than the geographical protocol. DSR and AODV not only a very low overhead but also show a better rate of reception of packets. The delay is always better with GPSR. The explanation of our results comes owing to the fact that we use GPSR without service of localization but when a service of localization is added, the end to end delay increases.

References

1. Kazi Iftekhar Ahmed (1999). Modeling Drivers Acceleration and Lane Changing Behavior. MA Thesis, Massachusetts Institute of Technology, Massachusetts.
2. http://www.isi.edu/nsnam/ns/.
3. B. Karp and H. T. Kung (2000). GPSR: Greedy Perimeter Stateless Routing for wireless networks. In: Proc. of 6th ACM/IEEE Int. Conf. on Mobile Computing and Networking, pp 243–254.
4. T. Camp, L. Wilcox, and J. Boleng (2002). Location information services in mobile ad hoc networks – Available at toilers.mines.edu.
5. David B. Johnson and David A. Maltz (1996). Dynamic Source Routing in Ad Hoc Wireless Networks. In: Mobile Computing, Ed. T. Imielinski and H. Korth, Kluwer Academic Publishers, Dordrecht, The Netherlands.
6. T. Camp, J. Boleng, B. Williams, L. Wilcox, and W. Navidi (2002). Performance comparison of two location based routing protocols for ad hoc networks. In: Proceedings of Infocom 2002.
7. http://www.cn.uni-duesseldorf.de/staff/kiess/software/hls-ns2-patch.

Packet Flow and its Temporal Properties in the Internet

Shin-ichi Tadaki

Computer and Network Center, Saga University, Saga 840-8502, Japan
tadaki@cc.saga-u.ac.jp

Summary. The Internet has been reported to have scale-free structures in some features. Its dynamical properties also have been reported to obey a power law. The power-law fluctuations can be observed in longer time-scale than some days. A scale-free structure of a network can not affect power-law behavior with simple processes.

1 Introduction

The Internet is one of the most important media for modern communication. On the contrary to the importance of the Internet, it has no global centers for controlling the global structure and the data transmission. It is an autonomous evolving network supported by modern technologies. The structure and transport properties of the Internet have been attracting scientific interests from the viewpoints of statistical physics [1].

One of the interesting features in the Internet is its scale-free structure in physical and logical features. The degree distribution in IR (Internet Router) or AS (Autonomous Systems) levels has been reported to obey a power law [2]. The power-law degree distribution of web links also has been reported [3].

Internet traffic had been thought to be modeled by a Poisson process. The validity of this assumption has clearly lost on the basis of various experimental measurements [4]. Power-law properties of Internet traffic have been investigated instead [5, 6].

The Internet is a man-made communication system. So the traffic on the Internet is affected by human social activities. It contains daily periodicity reflecting ones of social activities. To avoid the effect of the daily periodicity, previous works have focused their attention to the time scale shorter than a day. Can we find the power-law fluctuation in the longer time scale than a day?

The power law fluctuations can be found in various observations. The origins of such fluctuations have not be discussed clearly. We do not know that the scale-free structure of the Internet is an origin of the power-law fluctuations in packet flow. Some simulation results will be shown in this report.

2 Detrended Fluctuation Analysis

Power-law fluctuations have been discussed also in expressway traffic flow since the pioneering work by Musha and Higuch [7]. The power-law fluctuations have been analyzed in the longer range than a day by using the Detrended Fluctuation Analysis (DFA) [8].

DFA is one of methods for analyzing non-stationary time series. It was first developed for analyzing the long-range correlations in deoxyribonucleic acid (DNA) sequences [9, 10].

The simplest form of the method applied to flow data is as follows. First the profile $y(t)$ of the flow $q(t)$ at time t is defined:

$$y(t) = \sum_{\tau=0}^{T-1} \left[q(\tau) - \langle q \rangle \right], \tag{1}$$

where $\langle q \rangle = (1/T) \sum_{t=0}^{T-1} q(t)$ is the time average of the flow and T is the length of the time sequence.

The entire sequence of length T is divided into T/l non-overlapping segments of length l. In the n-th segment the *local trend* $\tilde{y}_n(t)$ is defined by fitting the profile $y(t)$ in the segment. We employ the linear least-squares method to fit the profile. This is called *first-order* DFA. For the entire sequence, the *detrended profile* $y_l(t)$ is defined as the deviation of the profile $y(t)$ from the local trend $\tilde{y}_n(t)$:

$$y_l(t) = y(t) - \tilde{y}_n(t), \text{ if } nl \leq t < (n+1)l. \tag{2}$$

The standard deviation of the detrended sequence is defined as the mean square of the detrended profile:

$$F^2(l) = \frac{1}{T} \sum_{t=0}^{T-1} y_l^2(t). \tag{3}$$

We find the long-range correlations in the sequence by analyzing the dependence of the deviation $F(l)$ on the segment length l. If the deviation $F(l)$ behaves obeying a power law of l,

$$F(l) \sim l^{\alpha}, \tag{4}$$

the power spectrum $P(k)$ of the sequence $q(t)$ also obeys the power law

$$P(k) \sim k^{-\beta}, \ \beta = 2\alpha - 1. \tag{5}$$

3 Internet Packet Flow

We observe Internet packet flow at a gateway of Saga University to Kyushu University [11]. There we connect out university local area network (LAN) to the science information network (SINET), the Japanese backbone network for academic organizations. The bandwidth of the line is 100 MBps. The data set was collected by a multi router traffic grapher (MRTG) system from September 2005 to May 2006. The beginning part of the data is shown in the left of Fig. 1.

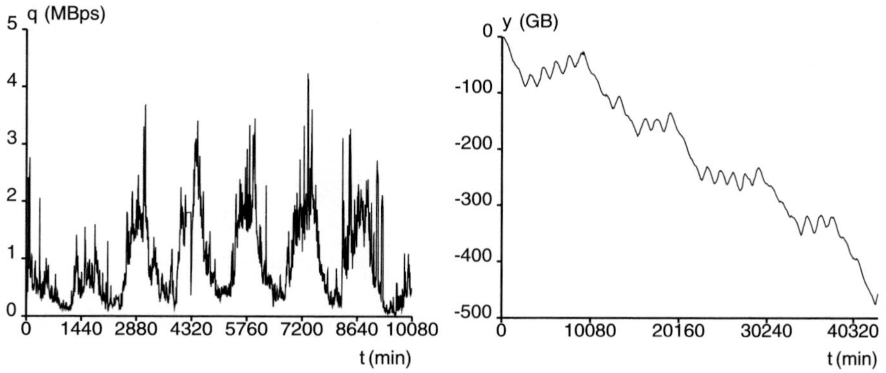

Fig. 1. The packet flow observed by MRTG (left) and its profile (right).

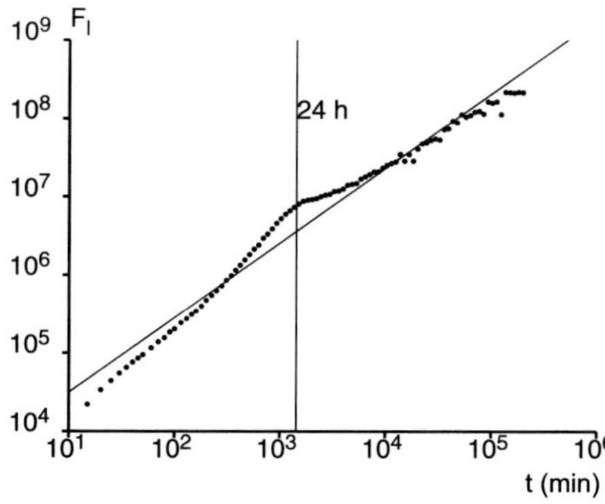

Fig. 2. The standard deviation $F(l)$ of MRTG data.

As discussed in Sect. 2, first we calculate the profile $y(t)$ as shown in the right of Fig. 1. Steep decreasing parts correspond to weekend inactiveness. The standard deviation $F(l)$ is shown in Fig. 2.

The deviation $F(l)$ shows a typical dependence on the segment length l. This is the typical feature of periodic time sequences with power-law fluctuations [12]. The bending point at one day long corresponds the dominant periodicity.

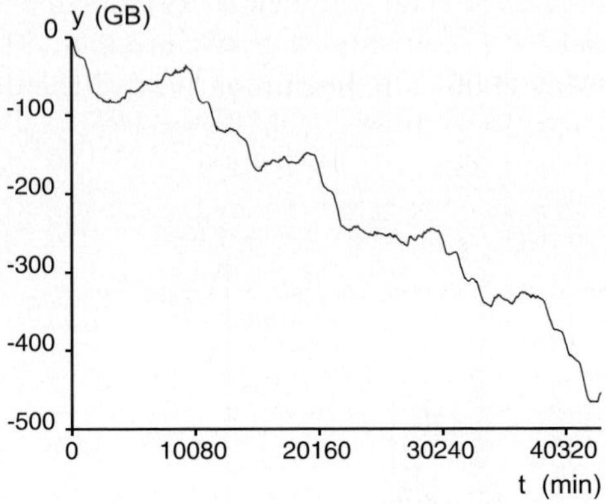

Fig. 3. The profile of the modified packet flow.

The modified flow data $q'(t)$ is defined by extracting the daily periodicity from the raw flow data $q(t)$. The profile for the modified data is shown in Fig. 3. The daily periodicity almost vanishes. By applying DFA to the modified data, the standard deviation is obtained as shown in Fig. 4. The exponent is obtained as $\alpha \sim 0.96$. Namely the fluctuations around the daily periodicity obey the power-law with α.

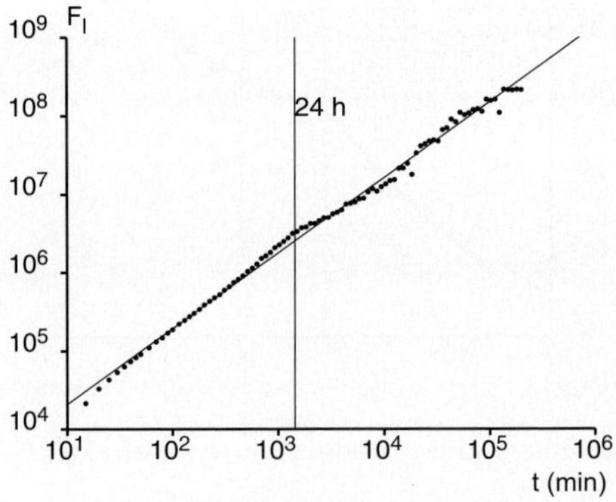

Fig. 4. The standard deviation $F(l)$ of modified MRTG data.

4 Random Walk in Networks

As shown in the previous section, the packet flow in the Internet shows the power-law fluctuations. Power-law fluctuations can be observed in various systems. We do not know any systematic explanation of the origins of power-law fluctuations.

For Internet packet flow, there are some exclusion and queuing mechanisms such as queuing in routers and exclusion in Ethernets. These mechanisms are pointed to be a key feature of power-law behavior [6].

On the contrary, the relation between the scale-free structure and power-law behavior in the Internet has not been discussed well. In this section, very simple simulation results are shown.

A Barabási-Albert scale-free network is prepared [3]. As the Initial state of a network growth process, the network has two nodes connected with two links. Every time step of the growth a new node is created. The new node selects one of existing nodes with probability proportional to its degree. The growing process continues till the number of nodes reaches to the fixed size N. For comparison, a regular two-dimensional lattice network is also prepared.

At the Initial state of the dynamics, every node has ten packets in it. Every time step, each packet selects one of neighboring nodes as the next hop. All packets hop to their next node simultaneously. Namely the simulation neglects effects of finite queue size of routers, routing and source-destination data of packets, which are fundamental features of the Internet. So the simulation results reflect the network structure.

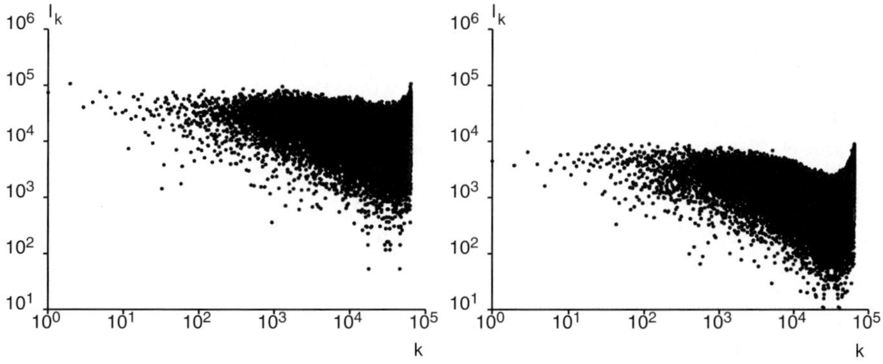

Fig. 5. Power spectrum of the load of a central node for Barabási-Albert (left) and regular lattice (right) networks.

The number of packets stored at a central node is observed as the load of the node. Its power spectrum is shown in Fig. 5. We can not observe power-law features in both networks. Namely scale-free structure can not simply affects its dynamical properties.

5 Summary

Long-term power-law fluctuations are observed at a gateway to the Internet. This report shows the DFA result for the in-going packet flow. The out-going packet flow also shows the power-law fluctuations. So the power-law behavior seems to be universal in the Internet packet flow.

There are some possible candidates as the origin of power-law fluctuations in the Internet. This report provides limited results with simulations. The simulations show that a scale-free structure of a network can not simply induce power-law dynamical properties.

Acknowledgements

A part of this work is financially supported by a Grant-in-Aid 18500215 from the Ministry of Education, Culture, Sports, Science and Technology, Japan.

References

1. R. Pastor-Satoras and A. Vespignani (2004) Evolution and Structure of the Internet, Cambridge University Press, Cambridge.
2. M. Faloutsos, P. Faloutsos, and C. Faloutsos (1999) ACM SIGCOMM Comp. Commun. Rev. 29:251–263.
3. A.-L. Barabási and R. Albert (1999) Science 286:509–512.
4. V. Paxson and S. Floyd (1995) IEEE/ACM Trans. Networking 3:226–244.
5. I. Csabai (1994) J. Phys. A: Math. Gen. 27:L417–L421.
6. M. Takayasu, H. Takayasu, and T. Sato (1996) Physica A 233:824–834.
7. T. Musha and H. Higuchi (1976) Jpn. J. Appl. Phys. 15:1271–1275.
8. S. Tadaki, M. Kikuchi, A. Nakayama, K. Nishinari, A. Shibata, Y. Sugiyama, and S. Yukawa (2006) J. Phys. Soc. Jpn. 75:034002.
9. C.-K. Peng, S. V. Buldyrev, S. Havlin, M. Simons, H. E. Stanley, and A. L. Goldberger (1994) Phys. Rev. E49:1685–1689.
10. C.-K. Peng, S. Havlin, H. E. Stanley, and A. L. Goldberger (1995) Chaos 5:82–87.
11. S. Tadaki (2007) J. Phys. Soc. Jpn. 76:044001.
12. K. Hu, P. Ch. Ivanov, Z. Chen, P. Carpena, and H. E. Stanley (2001) Phys. Rev. E64:011114.

List of Contributors

Tomas Apeltauer
Faculty of Civil Engineering
Brno University of Technology
Brno, CZ-602 00
Czech Republic
apeltauer.t@fce.vutbr.cz

Cécile Appert-Rolland
Laboratoire de Physique Théorique
CNRS (UMR 8627)
Université Paris-Sud 11
Bâtiment 210
F-91405 Orsay Cedex
France
Cecile.Appert-Rolland@
th.u-psud.fr

Igor S. Aranson
Argonne National Laboratory
9700 South Cass Avenue
Argonne, IL 60439
USA

Riccardo Artoni
Dipartimento di Principi e Impianti
di Ingegneria Chimica "I. Sorgato"
Università di Padova
Via Marzolo 9
35100 Padova
Italy
riccardo.artoni@unipd.it

Emilien Azéma
LMGC, CNRS
Université Montpellier II
Place Eugène Bataillon
34095 Montpellier cedex 05
France
azema@lmgc.univ-montp2.fr

Yann Bertho
Univ Paris-Sud
Université Pierre et Marie Curie-
Paris 6
CNRS, Lab FAST
Bât 502
Campus Univ
F-91405 Orsay
France

Alexander P. Buslaev
Moscow State Automobile
and Road Technical University
64, Leningradsky pr.
Moscow
Russia
apal2006@yandex.ru

Mario C. Campanella
Delft University of Technology
Stevinweg 1
2628 CN Delft
The Netherlands
m.c.campanella@tudelft.nl

Partha Chakroborty
Department of Civil Engineering
Indian Institute of Technology
Kanpur
Kanpur 208016
India
partha@iitk.ac.in

Nicolas Chiabaut
Laboratoire Ingénierie Circulation
Transport
INRETS / ENTPE
Université de Lyon
Rue Maurice Audin
69518 Vaulx En Velin CEDEX
France
chiabaut@entpe.fr

Pierre Cixous
Physique et Mécanique des Milieux
Hétérogenes (PMMH)
CNRS UMR 7636
Université Pierre et Marie Curie
(Paris VI)
Université Denis Diderot (Paris VII)
Ecole Supérieure de Physique et
Chimie Industrielles (ESPCI)
10 rue Vauquelin
75231 Paris Cedex 05
France

Winnie Daamen
Delft University of Technology
Stevinweg 1
2628 CN Delft
The Netherlands
w.daamen@tudelft.nl

Nadir Farhi
INRIA
Rocquencourt
Domaine de Voluceau
78153 Le Chesnay, Cedex
France
Nadir.Farhi@inria.fr

M. Ebrahim Fouladvand
Department of Physics
Zanjan University
P.O. Box 45196-313
Zanjan
Iran
foolad@iasbs.ac.ir

Angel Garcimartín
Depto. de Física y Mat. Apl.
Universidad de Navarra
31080 Pamplona
Spain
http://fisica.unav.es/granular

Hong-xia Ge
Faculty of science
Ningbo University
Ningbo 315211
China
gehongxia@nbu.edu.cn

Diego Luis González
Departamento de Física
Universidad de Los Andes
A. A. 4976 Bogotá
Colombia
die-gon1@uniandes.edu.co

Philip Greulich
Institut für Theoretische Physik
Universität zu Köln
50937 Köln
Germany
pg@thp.uni-koeln.de

Gustavo Gutiérrez
Laboratoire de Physique et
Mécanique des Milieux Hétérogènes
UMR 7636/ESPCI
France
Departamento de Física
Universidad Simón Bolívar
Apartado 89000
Caracas 1080-A
Venezuela
gustav@usb.ve

Habib Haj-Salem
INRETS/GRETIA
2 av du General Malleret-Joinville
F-94114 Arcueil
France
haj-salem@inrets.fr

Samer H. Hamdar
Northwestern University
The Transportation Center
Chambers Hall
600 Foster Street
Evanston, Illinois 60208
USA
hamdar@northwestern.edu

Xianglin Han
Shanghai Institute of Applied Mathematics and Mechanics
Shanghai University
Shanghai 200072
China
xlhan653@163.com@shu.edu.cn
School of Science
Huzhou University
Huzhou 313000
China

Rim Harich
PMMH
UMR 7636 du CNRS
ESPCI
10 rue Vauquelin
75005 Paris
France

Serge P. Hoogendoorn
Faculty of Civil Engineering and Geosciences
Delft University of Technology
Stevinweg 1
2628 CN Delft
The Netherlands
s.p.hoogendoorn@tudelft.nl

Mao-Bin Hu
Department of Mathematics and Statistics
Curtin University of Technology
Perth WA6845
Australia
School of Engineering Science
University of Science and Technology of China
Hefei 230026
P.R.China
humaobin@ustc.edu.cn

Ding-wei Huang
Department of Physics
Chung Yuan Christian University
Chung-li
Taiwan
dwhuang@phys.cycu.edu.tw

Wei-neng Huang
Department of Physics
Chung Yuan Christian University
Chung-li
Taiwan
wnhuang@phys.cycu.edu.tw

Payman Jalali
Department of Energy and Environmental Technology
Lappeenranta University of Technology
P.O.Box 20
53851, Lappeenranta
Finland
pjalali@lut.fi

Bin Jia
School of Traffic and Transportation
Beijing Jiaotong University
Beijing 100044
P.R.China
bjia@bjtu.edu.cn

Rui Jiang
School of Engineering Science
University of Science and Technology of China
Hefei 230026
China

Arne Kesting
Technische Universität Dresden
Institute for Transport & Economics
Andreas-Schubert-Straße 23
D-01062 Dresden
Germany

Victor Knoop
Delft University of Technology
Stevinweg 1
2628 CN Delft
The Netherlands
v.l.knoop@tudelft.nl

Henning Arendt Knudsen
University of Oslo
Dept. of Physics
PB 1048 Blindern
NO-0316 Oslo
Norway
hak@fys.uio.no

Romain Laniel
LMGC – UMR 5508
Université de Montpellier 2
cc 048, 34095 Montpellier cedex 5
France
laniel@lmgc.univ-montp2.fr

Jean-Marc Lasgouttes
INRIA Paris-Rocquencourt
Project Team IMARA
Domaine de Voluceau – BP. 105
78153 Le Chesnay Cedex
France
jean-marc.lasgouttes@inria.fr

Aurélie Le Quiniou
Cemagref
TSCF
Domaine des Palaquins
03150 Montoldre
France
aurelie.le-quiniou@cemagref.fr

Jean-Patrick Lebacque
INRETS-GRETIA
2 avenue du Général Malleret Joinville
F 94114 Arcueil
France
lebacque@inrets.fr

Ludovic Leclercq
Laboratoire Ingénierie Circulation Transports
ENTPE / INRETS
Université de Lyon
Vaulx-en-Velin
France
leclercq@entpe.fr

Xingli Li
Shanghai Institute of Applied Mathematics and Mechanics
Shanghai University
Shanghai 200072
China
lxl326@163.com

Ihor Lubashevsky
A.M. Prokhorov General Physics Institute of Russian Academy of Sciences
Vavilov str., 38
Moscow, 119311
Russia
ialub@fpl.gpi.ru

Tai-Yu Ma
INRETS/GRETIA - 2
Avenue du General-Malleret Joinville
F-94114 Arcueil France
tai-yu.ma@inrets.fr

Reinhard Mahnke
Rostock University
Institute of Physics
D–18051 Rostock
Germany
reinhard.mahnke@uni-rostock.de

Maniya Maleki
PMMH
UMR 7636 du CNRS
ESPCI
10 rue Vauquelin
75005 Paris
France
maniamaleki@yahoo.com
Institute for Advanced Studies in Basic Sciences (IASBS)
Zanjan 45195
P.O. Box 45195-1159
Iran

Bertrand Maury
Laboratoire de Mathématiques
Université Paris-Sud
91405 Orsay Cedex
France
Bertrand.Maury@math.u-psud.fr

Yeliu Mo
Institute of Physical Science and Engineering
Guangxi University
Nanning, 530004
China
yuxuegxu@gxu.edu.cn

Melanie J.I. Müller
Max Planck Institute of Colloids and Interfaces
Science Park Golm
14424 Potsdam
Germany
mmueller@mpikg.mpg.de

Mabiala Muriel
Institut d'Electronique Fondamentale
Bât 220
Université Paris-Sud
F91405 Orsay cedex
France
nom@u-psud.fr

Akihiro Nakayama
Department of Physics
Meijo University
Nagoya 468-8502
Japan

Andrea Parmeggiani
Cellular Biophysics and Biomimetics Team
DIMNP, UMR 5235
cc. 107, Pl. E. Bataillon
34095 Montpellier Cedex 5
France
andrea.parmeggiani@univ-montp2.fr

Paolo Pierobon
Institut Curie
CNRS UMR 168
11 Rue P. et M. Curie
F-75231 Cedex 05
France
pierobon@curie.fr

Nicolas Rivier
IPCMS
Université Louis Pasteur
67084 Strasbourg Cedex
France
nick@fresnel.u-strasbg.fr

Christian Rogsch
University of Wuppertal
Institute for Building Material Technology and Fire Safety Science
Pauluskirchstr. 7
D-42285 Wuppertal
Germany
christian@rogsch.de

Ivan Sánchez
Centro de Física
Instituto Venezolano de
Investigaciones Científicas
Apartado Postal 21827
Caracas 1020-A
Venezuela
ijsanche@ivic.ve

Bjørnar Sandnes
Department of Physics
University of Oslo
P.O.Box 1048 Blindern
0316 Oslo
Norway
bsand@fys.uio.no

Ludger Santen
Fachrichtung Theoretische Physik
Universität des Saarlandes
Postfach 151150
66041 Saarbrücken
Germany
santen@lusi.uni-sb.de

Gilles Saussine
Innovation and Research
Department of SNCF
45 rue de Londres
Paris, Cedex 08
France
gilles.saussine@sncf.fr

Andreas Schadschneider
Institut für Theoretische Physik
Universität zu Köln
50937 Köln
Germany
as@thp.uni-koeln.de
Interdisziplinäres Zentrum für
komplexe Systeme
53117 Bonn
Germany

Michael Schreckenberg
Physik von Transport und Verkehr
Universität Duisburg-Essen
47048 Duisburg
Germany
michael.schreckenberg@uni-due.de

Armin Seyfried
Jülich Supercomputing Centre
Research Centre Jülich
52425 Jülich
Germany
a.seyfried@fz-juelich.de

Zahra Shojaaee
Department of Physics
University of Duisburg-Essen
47048 Duisburg
Germany
zahra.shojaaee@uni-duisburg
-essen.de

Florian Siebel
Sebastian-Andre-Weg 8
D-82362 Weilheim
Germany
florian.siebel@gmx.de

Yūki Sugiyama
Department of Complex Systems
Science
Nagoya University
Nagoya 464-8601
Japan
sugiyama@phys.cs.is.nagoya
-u.ac.jp

Anton Šurda
Institute of Physics
SAS
Dubravska 9
Bratislava
Slovakia
anton.surda@savba.sk

Nicolas Taberlet
Université de Lyon
Laboratoire de Physique
ENS Lyon & UFR de Physique
UCBL
46 allée d'Italie
69 007 Lyon
France
nicolas.taberlet@ens-lyon.fr

Shin-ichi Tadaki
Computer and Network Center
Saga University
Saga 840-8502
Japan
tadaki@cc.saga-u.ac.jp

Akiyasu Tomoeda
Department of Aeronautics and Astronautics
University of Tokyo
Bunkyo 113-8656
Japan
tt67055@mail.ecc.u-tokyo.ac.jp

Vincent Topin
LMGC, CNRS
Université Montpellier 2
Place Eugène Bataillon
34095 Montpellier cedex
France
topin@lmgc.univ-montp2.fr
IATE
CNRS-INRA
2 place Pierre Viala
34060 Montpellier cedex 1
France

Martin Treiber
Technische Universität Dresden
Institute for Transport & Economics
Andreas-Schubert-Straße 23
D-01062 Dresden
Germany

Hans van Lint
Traffic and Transport Department
Faculty of Civil Engineering and Geosciences
Delft University of Technology
Stevinweg 1
2628 CN Delft
The Netherlands
j.w.c.vanlint@tudelft.nl

Rosa M. Velasco
Department of Physics
Universidad Autónoma Metropolitana
09340 México
rmvb@xanum.uam.mx

Jan Ludvig Vinningland
Department of Physics
University of Oslo
P.O.Box 1048
N-0316 Oslo
Norway

Charles Voivret
LMGC, CNRS
Université Montpellier II
Place Eugène Bataillon
34095 Montpellier cedex
France
voivret@lmgc.univ-montp2.fr

Christian von Ferber
Applied Mathematics Research Centre
Coventry University
Coventry CV1 5FB
UK
C.vonFerber@coventry.ac.uk

Hans Weber
Luleå University of Technology
Department of Physics
SE–97187 Luleå
Sweden
Hans.Weber@ltu.se

Marko Woelki
Universität Duisburg-Essen
Lotharstraße 1
D-47057 Duisburg
Germany
woelki@ptt.uni-due.de

Daichi Yanagisawa
Department of Aeronautics and Astronautics
School of Engineering
The University of Tokyo
Hongo
Bunkyo-ku, Tokyo 113-8656
Japan
tt66421@mail.ecc.u-tokyo.ac.jp

Yasushi Yokoya
Japan Automobile Research Institute
2530 Karima
Tsukuba
Ibaraki 305-0822
Japan
yyokoya@jari.or.jp

Xue Yu
Institute of Physical Science and Engineering
Guangxi University
Nanning, 530004
China
yuxuegxu@gxu.edu.cn
Shanghai Institute of Applied Mathematics and Mechanics
Shanghai University
Shanghai, 200072
China

Hui-bing Zhu
Shanghai Institute of Applied Mathematics and Mechanics
Shanghai University
Shanghai 200072
China
Faculty of Architectural
Civil Engineering and Environment
Ningbo University
Ningbo 315211
China
zhuhuibing@nbu.edu.cn

Printing: Krips bv, Meppel, The Netherlands
Binding: Stürtz, Würzburg, Germany